CALCULUS

SECOND EDITION

EDWIN E. MOISE
Harvard University

ADDISON-WESLEY PUBLISHING COMPANY

Reading, Massachusetts · Menlo Park, California · London · Don Mills, Ontario

This book is in the
ADDISON-WESLEY SERIES IN MATHEMATICS

Consulting Editor: LYNN H. LOOMIS

Author's Note on the Second Edition

The preface to the first edition was an explanation of the author's intentions, and since these intentions have not changed, the original preface is reprinted after this one. But the present edition is a thorough revision of the first, with many major changes and even more minor ones. Some of these are as follows.

1. The use of language has been simplified throughout. Excessive colloquialisms have been eliminated.

2. Many problems have been added, most of them being easy. In cases where several problems form a sequence, they have been combined into a single problem, with parts (a), (b), (c), ... Thus it is now safe to assign all odd-numbered problems, without checking to make sure that problem-sequences are not being broken up.

3. Various long sections have been divided into two parts.

4. The classical definition of a limit has been restored. Exploratory problems, dealing with limits and continuity in terms of "boxes," have been eliminated, and this material has been inserted in later portions of the text.

5. Section 5.8, on the derivative of one function with respect to another, has been completely recast, following suggestions of Professor Hugh Thurston, of the University of British Columbia. The new version is mathematically straightforward, and it bridges the gap between the modern concepts of function and derivative and the "fractional" notation du/dv commonly used in physics.

6. Chapter 8, on the conic sections, has been shortened and simplified, by omitting various topics not ordinarily covered in a first course in calculus. In particular, the section on the geometry of the ellipse has been omitted. In a way it is a pity to leave this out, because it is very good mathematics, but in a first course in calculus we barely have time for essentials.

7. In the chapters on vector spaces, the standard use of the terms "vector space" and "inner product space" has been restored.

8. The old Chapter 10, on number theory and partial fractions, has been omitted. The above remarks about the geometry of the ellipse also apply here.

9. The chapter on infinite series has been completely recast. In the first edition, the idea of uniform convergence was built into the presentation, almost from the outset; it was used, in various special cases, to justify term-wise integration, long before the general definition of uniform convergence was stated. This treatment had advantages, for some students, but it had a serious disadvantage: it meant that the hardest part of the study of infinite series could not be skipped, or even postponed. The chapter has now been arranged in such a way that the hardest parts of it come

last. Term-wise integration and differentiation of power series are introduced early, and play a central part throughout the chapter; but the justification of these processes is saved for the end.

The construction of the complex numbers (using congruence classes of real polynomials modulo $1 + x^2$) has been moved to an appendix.

10. The chapter on linear transformations, matrices, and determinants has been recast and simplified in various ways. For example, the idea of isometries between subspaces has been omitted (from the text and therefore from the problems.)

11. In the chapter on functions of several variables, the Leibniz notation for partial derivatives has been introduced, in parallel with the subscript notation $f_x, f_y, \ldots$. The former notation is of course universal in physics, and it cannot be denied that it makes the chain rule easier to remember.

These examples should make it plain that this is not a perfunctory revision. The intent of the revision is to make the book more teachable and more flexible, without weakening its mathematical content. Some sections (as indicated above) have been omitted outright. Some chapters have been recast in such a way that more topics can be omitted at the teacher's discretion. But the main substance of the book, and the conception of calculus that it attempts to teach, have not been changed. All the hard problems in the first edition have been retained (except for one very embarrassing case, in which I asked the student to prove a false theorem).

Most of the calculus books now in print are of one of the following three types:

1) Some are written on a high plateau of austerity and rigor, and the Devil take the hindmost.

2) Some are "quick calculus" books. A typical device, in this sort of book, is to use the Fundamental Theorem of Integral Calculus as a *definition* of the definite integral. This enables the student to imitate the behavior of mathematicians, in calculating definite integrals, without sharing the mathematicians' conception of what the problem meant in the first place.

3) Some are a combination of types 1 and 2, with exact theoretical material included in the text but not in the problems. A course based on a book like this is evidently intended to play a double game, in which some of the students learn the text, but most of them study, in effect, a "quick calculus" course based on the same sort of problems used in books of type 2.

The present book is of none of these three types. It is addressed to all students who ordinarily take calculus courses and pass them, and it is designed to teach the ideas of calculus, at least in some form, to these students. In a sense we are playing a double game, because the teacher has a great deal of choice. If the entire text is taught, then the course is logically complete, and nothing that it includes needs to be taught again later. But many of the theoretical sections, especially at the ends of chapters, giving proofs of "foundational" theorems, can easily be omitted. For example, if we omit Sections 5.6, 5.7, and 5.8, then something is subtracted from the course, but nothing is disrupted or confused. And in Chapter 10 we can stop at almost any point. But even if such omissions are made on a large scale, the remaining hard core of the book conveys the ideas of calculus in conceptual forms.

The rock-bottom minimum, in any good mathematics course, is for the student to attach conceptual meanings to the problems that he solves and to the "answers" that he computes. If we settle for less than this, we are making a bad bargain. The hard fact is that "practical" calculus courses are not practical. In real life it seldom, if ever, happens that a mathematical problem takes the form of a homework exercise which can be solved by copying the pattern of the "solved problems" that immediately precede it. In physics it is the *conceptual* definite integral that is crucial, and numerical valuations are often done by computers. Thus the art of setting up integrals is often more useful than the art of calculating them by elementary methods. The same principle applies very widely. When people put their mathematical training to practical use, they seldom need the logical refinements that appear in a thorough treatise, but they nearly always use their conceptual grasp—at some level—of mathematical ideas. Obviously, many techniques are needed, and in this book we have worked hard to teach them. But as the same time we have tried to produce the sort of conceptual grasp of mathematics that can be put to work in real life.

New York City, N.Y.   E. E. M.
October 1971

Preface

The mathematical content of the first ten chapters of this book is familiar and easy to describe. These chapters present, more thoroughly than is customary, the material normally covered in one-year introductions to college calculus, and end with a chapter on infinite series. (This portion of the book is being published separately, under the title *Elements of Calculus*.) In the last four chapters of the complete edition, the choice of material is nowhere nearly so traditional. In particular, we have laid heavy stress on the methods of linear algebra.

In the latter portion of this preface, we explain the considerations on which the selection of topics in the last few chapters is based. Most of the novelties in the *Elements* are in the style of treatment; and the ideas underlying them may best be explained by means of numerous examples.

1. THE SPIRAL PROCESS

The central concepts of the calculus are deep. It is not to be expected that they can be learned all at once, in the forms in which a modern mathematician thinks of them. Therefore, in this book, the more difficult ideas are presented in a series of different forms, in ascending order of difficulty, generality, and exactitude. Thus the idea of the definite integral makes its first and simplest appearance in Section 2.10; it is generalized in Section 3.7; and it is not presented in final form (using Riemann sums) until Section 7.1, where Riemann sums are needed, in the calculation of arc length.

Similarly, the chain rule for derivatives appears first in Section 3.6, for powers and square roots of functions; it is proposed, in more general forms, in Problem Sets 3.6, 3.8, 4.3, and 4.5; and it appears in final form only in Section 4.6.

The mean-value theorem is first stated, in geometric terms, in Section 3.2, before any formal definition of the derivative. It is used freely thereafter. Finally, in Section 5.7, it is proved, after the ideas needed in the proof have been used and motivated in other ways.

The idea of the limit of a function appears first in Section 2.7. The formal definition is in Section 3.3. Earlier sections include a lengthy preparation for the formal definition, designed to eliminate in advance as many of its difficulties as possible. This purpose is served by the text of Sections 1.4 and 2.5. Thus the style of treatment is such that an inspection of isolated sections of the book is likely to lead to an overestimate of the difficulty of the course. The point is that the sections are not isolated: difficult discussions have been provided with elaborate foundations, in the text and especially in the problems.

The spiral treatment, in which concepts appear in various forms as the theory develops, is intended to make the concepts easier to learn. But this is not its only purpose. The processes by which special ideas are generalized, and heuristic ideas are made concrete and exact, are part of the substance of what we ought to be teaching. Thus the heuristic treatment of exponentials and logarithms, in Section 4.9, is not given merely in order to make the student's life easier. The transition from Section 4.9 to Sections 4.10 and 4.11 (in which the theory is based on the definition $\ln x = \int_1^x (dt/t)$) is valuable in itself, as an illustration of a recasting process which is essential both in the growth of mathematics and in the growth of the people who use it.

2. MOTIVATION

The desire to solve interesting puzzles is very strong; there is no maturity level at which it disappears; and we should appeal to it continually. Most of the time, however, when new ideas are introduced, they ought to be motivated by a sense of power, and by the light that they throw on ideas already regarded as significant. For example, if we present Riemann sums, in full generality, long before we deal with problems in which they are needed, it is not reasonable to expect the student to master their complications. Similarly, the completeness of the real number system, in the sense of Dedekind, is not needed at all in the theory of pointwise limits: this theory takes exactly the same form in the rational domain as in the real domain. If we postpone the idea of completeness until the point where it is needed, in the study of functions continuous on an interval, it is more likely to be understood, partly because it is more likely to get the student's attention.

The problem of motivating the idea of the limit of a function involves a peculiar difficulty. The only cases in which $\lim_{x \to a} f(x)$ is easy to calculate are those in which f is a continuous function, described by a simple formula. In these cases, the formula works just as well for $x = a$ as for other values of x; in practice, it turns out that the limit is $f(a)$; and the student is likely to get the idea that the expression $\lim_{x \to a} f(x)$ is merely a devious and pretentious description of $f(a)$. If we avoid this trouble by starting with significant cases, such as

$$\lim_{x \to 0} \frac{\sin x}{x},$$

then the technical difficulties are formidable, and workable problem material is hard to come by. If we choose, instead, to discuss limits of sequences, then we have evaded the issue by changing the subject: in the differential calculus, limits of functions are what we need.

But there is a fourth alternative: we can introduce the idea of a limit not as a subject in its own right, but as a device for solving a problem. In Section 2.7 we mention limits for the first time, in finding the limit of a *linear* function. To pass to the limit, in this case, we merely plug the hole in a punctured line. This process has no intrinsic significance. But in the context of Section 2.7, it has an extrinsic significance, because it is used to solve a nontrivial problem, namely, the problem of finding the slope of the tangent to a parabola. Similarly, in Section 2.10, we use the idea of the limit of a sequence, in a technically simple case, in order to find the area of a parabolic segment. (A formal definition of the limit of a sequence finally appears in Section 10.1). There are many other points at which ideas are introduced, in simple forms, in connection with a discussion of something else.

3. BLACK BOXES

It is generally agreed that in a physics laboratory the student should build as much as possible of his own equipment. Nobody learns very much by watching the performance of the proverbial "black box." In mathematics the situation is similar: we do not learn mathematical principles by hearing them mentioned once, no matter how elegantly; we need to live with them and use them. Therefore, in this book certain extremely powerful theorems have been proved long before being stated. That is, the proof has been presented, in the form of a method of solving a certain class of problems; and after the student has learned the idea by using it on many problems, we have summed up the situation by stating the general theorem that the proof proves. This scheme costs very little time, even in the short run; and in the long run it is likely to save a great deal of time. The point is that if we allow recipes to take the place of ideas, in a first course, then the ideas need to be taught all over again later; and the second attempt may be harder, because the problem-solving motivation for these particular ideas has already been used up.

There are good reasons for not giving examples of this technique.

It should be understood that the avoidance of black boxes has no particular connection with the pursuit of logical rigor. Indeed, if we have to choose, it is better to master an idea in an heuristic form, by using it repeatedly, than to listen once to a rigorous exposition, and then forget it.

4. PROBLEMS

In a quick examination of a textbook, it is not a good idea to read the text and skip the problems; it is better to read the problems and skip the text. The problems represent the life that the student leads when he studies the course; and any ideas that do not appear in them are unlikely to be learned, no matter how much preachment may be devoted to them.

In this book, a variety of problems are used for a variety of purposes. There are:

1) Technical problems, as, for example, in the chapter on the technique of integration. These are carefully graded, and often they form sequences, in which the answer to one problem can be used in the solution of the next.

2) Theoretical problems, some easy, some hard. Vigorous attempts have been made to find easy ones, so as to avoid a dichotomy between techniques (which the student really uses) and "theory" (of which he is intermittently a spectator).

3) Puzzle problems.

4) Sketching exercises, in which the student is asked to translate back and forth between analytic ideas and visual images.

5) Discovery problems, which anticipate, in special cases, ideas which will later be explained in the text.

There is wide general agreement on the content of the first year course in college calculus; and in writing the *Elements,* the author was in the happy position of working on the basis of a consensus with which he was fully in sympathy. But there is no such general agreement on the content of a course in intermediate calculus. In the past decade, calculus courses have tended to grow, by including various topics from advanced calculus and linear algebra. But it is not easy to decide which of these topics should be included, and what relative stress should be placed on them; and in fact there is no reason to suppose that such questions have unique answers.

On the other hand, every book and every course must make *some* choices, and then stick to them long enough to permit a valid learning process. If the pursuit of flexibility turns an intermediate calculus book into an anthology, then its little pieces are unlikely to have any lasting effect. For example, if the treatment of infinite series is sketchy, then its residuum in the mind of the student may include hardly more than the ratio test. And the dangers presented by brief treatments of linear algebra are worse.

Modern algebra is modern because its motivations and its applications came late. Today, there are very good reasons for studying groups, rings, fields, vector spaces, normed vector spaces, inner product spaces, linear transformations, matrices, and so on. But the logical simplicity of the rudiments of these theories is misleading. For example, the manipulative process of multiplying matrices can be taught to almost anybody, at almost any level; but the significant applications of this process are another matter entirely. In a short treatment of axiomatic and linear algebra, at the freshman or sophomore level, we cannot presuppose knowledge of the significant applications, and we have no time in which to present them. Thus we may fall into a peculiar form of use-mention confusion: the reader hopes that the ideas of modern algebra are going to be used, but in the end he sees that they have merely been mentioned.

For these reasons we have tried, throughout, never to state an algebraic definition until the reader already knows at least one important instance of the idea that the definition describes; and once an algebraic idea has been introduced, we have tried, throughout, to put it to work for the purposes that it is good for. Thus, for example, matrices are introduced as a shorthand for handling linear transformations; and thereafter the treatment of the two is closely tied together. The Schwarz inequality is first introduced (on page 521) as a theorem in Cartesian three-space, and for this case it is proved by the trivial observation that $\cos^2 \theta \leq 1$ for every θ. Later, on page 536, it is proved in the general case, and thereafter it is used in a great variety of ways, to trivialize problems which would not otherwise be trivial. It appears in disguised forms in many problems (which should not be listed here). These examples are typical of the style of Chapters 11 through 13. It appears to the author that the nature, the purposes, and the power of algebraic methods are not likely to be understood unless they are conveyed to the student by some such extended experience.

The most impressive, but also the most difficult, of these applications occurs in Chapter 12, on Fourier series. This topic is not ordinarily included in intermediate courses; and if something must be omitted, in teaching a course from this book, Chapter 12 is an excellent candidate for omission. (None of the material in it is used later.)

Chapters 1 through 13 amount to more than 600 pages; something had to be shortened; and so the treatment of functions of many variables is shorter than might have been expected, and there is no separate chapter on differential equations. It should be noted, however, that there is a substantial treatment of linear differential equations at the end of Chapter 13, and that the viewpoint of differential equations has been stressed throughout. (Recall, for example, the treatment of the fundamental theorem of integral calculus, and of the elementary functions, in the *Elements*.) In Chapter 10, the standard method of showing that a given series converges to a given function is first to show that the series and the function satisfy the same differential equation, and then to show that the differential equation (with initial condition) has only one solution. Usually, the series is derived from the differential equation, and so the student is not likely to be surprised when the same process is applied later to equations whose solutions were not previously known. For this sort of reason, the book conveys much more of the spirit and methodology of differential equations than the table of contents would suggest.

Moreover, it appeared to the author that the natural sequels of the material in Chapter 14 would grow exponentially more difficult, and that they rightly belong in an advanced calculus course. The hard fact is that multivariate calculus, once we get past its beginnings, is not an elementary subject; and if we try to make it seem elementary, we are likely to give up both intuition and logic in favor of a bewildering formalism. Thus it appeared, at the end of Chapter 14, that we should say either much more, or no more at all; and since every book—even a calculus book—has got to end somewhere, the choice was clear.

The above discussion is an attempt to indicate some of the author's objectives, and some of the methods used in pursuing them. Obviously no such discussion can prove anything about the extent of the contribution that the text makes to the achievement of these objectives. A great deal has happened, in the teaching of calculus, in the past decade, and it remains to be seen how much more can be accomplished, and how.

New York City, N.Y.   E. E. M.
October 1971

Contents

Chapter 1 Inequalities

  1.1 Introduction 1
  1.2 Products which are equal to zero 2
  1.3 Order 3
  1.4 Absolute values. Intervals on the number line 9

Chapter 2 Analytic Geometry

  2.1 Introduction 16
  2.2 Coordinate systems. The distance formula 16
  2.3 The graph of a condition. Equations for circles 21
  2.4 Equations of lines. Slopes, parallelism, and perpendicularity . . . 26
  2.5 Graphs of inequalities. And, or, and if . . . then 33
  2.6 Parabolas 38
  2.7 Tangents 43
  2.8 A shorthand for sums 49
  2.9 The induction principle and the well-ordering principle 51
 2.10 Solution of the area problem for parabolas 57

Chapter 3 Functions, Derivatives, and Integrals

  3.1 The idea of a function 63
  3.2 The derivative of a function, intuitively considered 69
  3.3 Continuity and limits 75
  3.4 Theorems on limits 82
  3.5 The process of differentiation 89
  3.6 The process of differentiation: roots and powers of functions . . . 97
  3.7 The integral of a nonnegative function 102
  3.8 The derivative of the integral 109
  3.9 Uniformly accelerated motion 119
*3.10 Proof of the formula for the derivative of the integral 124

Chapter 4 Trigonometric and Exponential Functions

  4.1 Directed angles. Trigonometric functions of angles and numbers . . 128
  4.2 The law of cosines and the addition formulas 135
  4.3 The derivatives of the trigonometric functions; the differences Δx and Δf; the squeeze principle 139

4.4 The approximation of differences by differentials 148
4.5 Composition of functions 154
4.6 The chain rule 159
4.7 Invertible functions. The inverse trigonometric functions 165
4.8 Simpson's rule. The computation of π 176
4.9 Exponentials and logarithms 185
4.10 The functions ln and exp 191
4.11 Exponentials and logarithms. The existence of e 197

Chapter 5 The Variation of Continuous Functions

5.1 Intervals on which a function increases, or decreases 206
5.2 Local maxima and minima, direction of concavity, inflection points . 211
5.3 The behavior of functions at infinity 216
5.4 The introduction of functions into geometric problems; the use of
     existence theorems as shortcuts 223
5.5 The use of functional equations as shortcuts 232
5.6 The completeness of **R** and the existence of maxima 238
5.7 The mean-value theorem and the no-jump theorem 246
5.8 The derivative of one function with respect to another 250

Chapter 6 The Technique of Integration

6.1 Introduction 254
6.2 Independent variables and indefinite integrals 255
6.3 Integrals leading to the logarithm and the inverse secant. Algebraic
     devices 265
6.4 Integration by parts 273
6.5 Integration of powers of trigonometric functions 278
6.6 Integration by substitution 284
6.7 Algebraic substitutions 291
6.8 Algebraic devices: completing the square and partial fractions . . 297

Chapter 7 The Definite Integral

7.1 The problem of arc length 303
7.2 The definite integral, defined as a limit of sample sums 308
7.3 The calculation of volumes, by the method of disks 315
7.4 The general method of cross sections, and the method of shells . . 321
7.5 The area of a surface of revolution 327
7.6 Moments and centroids. The theorems of Pappus 335
7.7 Improper integrals 344
*7.8 The integrability of continuous functions 350

Chapter 8 The Conic Sections

8.1 Translation of axes 356
8.2 The ellipse 360
8.3 The hyperbola 366
8.4 The general equation of the second degree. Rotation of axes . . . 372

Chapter 9   Paths and Vectors in a Plane

9.1   Motion of a particle in a plane 381
9.2   The parametric mean-value theorem; l'Hôpital's rule 385
9.3   Other forms of l'Hôpital's rule 393
9.4   Polar coordinates 397
9.5   Areas in polar coordinates 402
9.6   The length of a path 405
9.7   Vectors in a plane 409
9.8   Free vectors 415
9.9   Velocity, acceleration, and curvature 422
9.10   Concluding remarks on vector spaces and inner product spaces . . 430

Chapter 10   Infinite Series

10.1   Limits of sequences 431
10.2   Infinite series. Convergence. Comparison tests 437
10.3   Absolute convergence. Alternating series 445
10.4   Estimates of remainders 448
10.5   Termwise integration of series. Power series for Tan^{-1} and ln . . 453
10.6   The ratio test for absolute convergence. Applications to power series 457
10.7   Power series for exp, sin, and cos 463
10.8   The binomial series 468
10.9   Taylor series 473
10.10 Taylor's theorem. Estimates of remainders 477
10.11 The complex number system 479
10.12 Sequences and series of complex numbers. The complex exponential
       function 484
10.13 De Moivre's theorem 489
*10.14 The radius of convergence. Differentiation of complex power series 493
*10.15 Integration and differentiation of real power series 499

Chapter 11   Vector Spaces and Inner Products

11.1   Cartesian coordinate systems in three-dimensional space 508
11.2   Direction cosines. The directed normal form 512
11.3   Three-dimensional space, regarded as an inner-product space . . . 518
11.4   The dimension of a vector space. Various ways to form a basis . . 526
11.5   Orthonormal bases 530
11.6   The Schwarz inequality. More general concepts of norm and distance 533

Chapter 12   Fourier Series

12.1   Projections into a subspace, trigonometric polynomials and Fourier
       series 541
12.2   Uniform approximations by trigonometric polynomials 549
12.3   Integration of Fourier series. The uniform convergence theorem . . 556

Chapter 13 Linear Transformations, Matrices, and Determinants

13.1 Linear transformations . . . . ' . . . 563
13.2 Composition of linear transformations and multiplication of matrices 570
13.3 Formal properties of the algebra of matrices. Groups and rings . . 577
13.4 The determinant function 582
13.5 Expansions by minors. Cramer's rule and inversion of matrices . . 590
13.6 Row and column operations. Linear independence of sets of functions 596
13.7 Linear differential equations 601
13.8 The dimension theorem for the space of solutions. The nonhomo-
       geneous case 607

Chapter 14 Functions of Several Variables

14.1 Surfaces and solids in R^3 614
14.2 The quadric surfaces 620
14.3 Functions of two variables. Slice functions and partial derivatives . 626
14.4 Directional derivatives and differentiable functions 634
14.5 The chain rule for paths 641
14.6 Differentiable functions of many variables. The chain rule. . . . 644
14.7 Directional derivatives and gradients 648
14.8 Interior local maxima and minima, for functions of two variables
       Level curves 651
14.9 Double integrals, intuitively considered 660
14.10 Cylindrical coordinates in space. The definition of the integral . . 666
14.11 Moments and centroids of nonhomogeneous bodies 674
14.12 Line integrals 680

Appendix A   The Shorthand of Logic and Set Theory 687
Appendix B   Algebraic Operations with Limits of Functions 690
Appendix C   Algebraic Operations with Limits of Sequences 695
Appendix D   The Error in the Approximation $\Delta f \approx df$ 697
Appendix E   The Continuity of Composite Functions 700
Appendix F   The Error in Simpson's Rule 702
Appendix G   The Idea of a Measurable Set 705
Appendix H   Proof of the Northeast Theorem 707
Appendix I   Proof of the Formula for Path Length 711
Appendix J   A Method for Constructing the Complex Numbers 713
Appendix K   Iterated Limits. Mixed Partial Derivatives 717
Appendix L   Possible Peculiarities of Functions of Two Variables 721
Appendix M   Maxima and Minima for Functions of Two Variables 725
Appendix N   An Exact Definition of the Idea of a Function 727
              Selected Answers 733
              Index 759

1   Inequalities

In this book it is assumed that you know elementary geometry and the algebra of the real number system. Theorems of plane geometry will be used only occasionally, and there is no need to reexamine the subject as a whole.

Inequalities, however, are another matter. We shall be using them constantly, and they are tricky. We shall therefore handle them with care. To derive the laws that govern them we first need to recall the elementary laws of the number system. These are as follows.

We have given the set **R** of real numbers, with the operations of addition and multiplication. Thus the number system is a triplet

$$[\mathbf{R}, +, \cdot].$$

Addition and multiplication are subject to the following laws:

Closure. For every a and b in **R**, $a + b$ and ab are in **R**.

Associativity. For every a and b,

$$a + (b + c) = (a + b) + c,$$

and

$$a(bc) = (ab)c.$$

Commutativity. For every a and b,

$$a + b = b + a \qquad \text{and} \qquad ab = ba.$$

Distributive Law. For every a, b, and c,

$$a(b + c) = ab + ac.$$

Existence of 0 and 1. There are two different numbers 0 and 1 such that

$$a + 0 = a \qquad \text{and} \qquad a \cdot 1 = a$$

for every a.

Existence of Negatives. For every a there is a number $-a$ such that $a + (-a) = 0$.

Existence of Reciprocals. For every $a \neq 0$ there is a number $1/a$ such that $a \cdot 1/a = 1$.

These laws are called the *field postulates*; and any number system which satisfies them is called a *field*. There are many such number systems: the real numbers form a field, and so do the complex numbers. For a long time to come, however, we shall be working only with the real numbers. Therefore, when we speak of *numbers*, we mean *real* numbers, unless the contrary is stated.

We shall assume not only the field postulates but also the familiar laws based on them. For example, we know that $(a - b)(a + b) = a^2 - b^2$, and that $a \cdot 0 = 0$ for every a.

1.2 PRODUCTS WHICH ARE EQUAL TO ZERO

When we perform calculations, we shall not stop to justify them on the basis of the field postulates. But the following principle is worth special mention, because it is used in reasoning processes which don't involve calculations:

Theorem 1. If $ab = 0$, then either $a = 0$ or $b = 0$.

Proof.

1) If $a = 0$, there is nothing to prove.
2) If $a \neq 0$, then a has a reciprocal. Therefore

$$\frac{1}{a}(ab) = \frac{1}{a} \cdot 0, \qquad \left(\frac{1}{a} \cdot a\right)b = 0, \qquad 1 \cdot b = 0,$$

and
$$b = 0.$$

Thus either $a = 0$ or $b = 0$.

Obviously it is possible that a and b are both $= 0$. In Theorem 1 (and everywhere else in mathematics) when we say *either . . . or . . .* , we allow the possibility of *both*.

PROBLEM SET 1.2

1. Show that if $x^2 = 0$, then $x = 0$.

2. a) Obviously the numbers 1 and -1 are roots of the equation

$$(x - 1)(x + 1) = 0.$$

   How do you know that no other number is a root of the equation?
   b) Show that 2 and 3 are the only roots of the equation

$$x^2 - 5x + 6 = 0.$$

3. If 0 had a reciprocal, then its reciprocal would be a root of the equation

$$0 \cdot x = 1.$$

   Show that this equation has no root.

4. a) If $ab = ac$, does it follow that $b = c$? Why or why not?
   b) If $ab = ac$, and $a \neq 0$, does it follow that $b = c$? Why or why not?

5. a) Show that if $abc = 0$, then $a = 0$ or $b = 0$ or $c = 0$.

    b) Show that 1, 2, and 3 are the only roots of the equation

$$x^3 - 6x^2 + 11x - 6 = 0.$$

6. a) If $a^2 = b^2$, does it follow that $a = b$? Why or why not?

    b) If $a^2 = b^2$, what can you conclude about the relation between a and b? Why?

7. Under what conditions (if any) is it true that

$$\frac{1}{x} + \frac{1}{a} = \frac{1}{x + a}?$$

8. a) Under what conditions (if any) is it true that

$$(a + b)^2 = a^2 + b^2?$$

    b) Under what conditions (if any) is it true that

$$(a + b)^3 = a^3 + b^3?$$

*9. Consider the "number system" which has only two elements 0 and 1, with addition and multiplication defined by the following tables:

+	0	1
0	0	1
1	1	0

·	0	1
0	0	0
1	0	1

Which of the field postulates hold true, in this system? Which, if any, fail to hold? (The answer to this question suggests that the field postulates are not, in themselves, a very adequate description of the real number system.)

*10. Consider the number system in which the "numbers" are 0, 1, 2, and 3, with addition and multiplication defined by the following tables:

+	0	1	2	3
0	0	1	2	3
1	1	2	3	0
2	2	3	0	1
3	3	0	1	2

·	0	1	2	3
0	0	0	0	0
1	0	1	2	3
2	0	2	0	2
3	0	3	2	1

Exactly one of the field postulates fails to hold in this number system. Find out which one. [*Hint:* Don't bother to test the Associative and Distributive Laws; in fact, they hold true in this system, although the verifications are extremely tedious.]

    Does Theorem 1 hold true in this system? Why or why not?

1.3 ORDER

We think of the real numbers as being arranged on a line, like this:

When we write $a < b$, this means (roughly speaking) that a lies to the left of b on the number line. Thus what we have in mind is a system

$$[\mathbf{R}, +, \cdot, <],$$

where $<$ is a relation having the following properties:

O.1. (*Trichotomy*) For every a and b in $\mathbf{R}$, one and only one of the following conditions holds:

$$a < b, \quad\text{or}\quad a = b, \quad\text{or}\quad b < a.$$

O.2. (*Transitivity*) If $a < b$ and $b < c$, then $a < c$.

A relation satisfying O.1 and O.2 is called an *order relation*, and an expression of the form $a < b$ is called an *inequality*. We write $b > a$ to mean $a < b$; $a \leq b$ means that either $a < b$ or $a = b$; and $a \geq b$ means that either $a > b$ or $a = b$. A number a is *positive* if $a > 0$; a is *negative* if $a < 0$. Zero is neither positive nor negative.

But O.1 and O.2 do not, by themselves, enable us to handle inequalities. We need to know how $<$ is related to $+$ and $\cdot$. The laws are the following:

MO. If $a > 0$ and $b > 0$, then $ab > 0$.

AO. If $a < b$, then $a + c < b + c$ for every c.

These four laws, in combination, tell the whole story: all of the elementary laws of inequalities can be derived from them. You will carry out this process, in the following problem set. Meanwhile we state the theorems without proof.

Theorem 1. If $a > 0$, then $-a < 0$.

Theorem 2. If $a < 0$, then $-a > 0$.

Theorem 3. If $a < b$, and $c < d$, then

$$a + c < b + d.$$

Theorem 4. An inequality is *preserved* if both sides are multiplied by the same positive number.

That is, if $a < b$ and $c > 0$, then $ac < bc$.
Similarly,

Theorem 5. An inequality is *preserved* if both sides are divided by the same positive number.

That is, if $ac < bc$ and $c > 0$, then $a < b$.

Theorem 6. An inequality is *reversed* if both sides are multiplied by the same negative number.

That is, if $a < b$ and $c < 0$, then $ac > bc$.

Theorem 7. An inequality is *reversed* if both sides are divided by the same negative number.

That is, if $bc < ac$, and $c < 0$, then $b > a$.

Consider now an inequality involving an unknown number x, for example,

$$3x + 4 < 5x + 7.$$

An expression like this, involving a variable, is called an *open sentence;* in an open sentence, x marks the spot where numbers are to be inserted. Some numbers, when substituted for x, may give true statements, and other numbers may give false statements. For example,

$$3 \cdot 2 + 4 < 5 \cdot 2 + 7$$

is true, because $10 < 17$; but

$$3(-5) + 4 < 5(-5) + 7$$

is false, because $-11 > -18$.

In simple cases like this, it is easy to find out what numbers satisfy the inequality. If

$$3x + 4 < 5x + 7, \tag{1}$$

then

$$4 < 2x + 7, \tag{2}$$

by AO. (We have added $-3x$ to each side of the inequality.) Therefore

$$-3 < 2x, \tag{3}$$

by AO; and so

$$x > -\tfrac{3}{2}, \tag{4}$$

by Theorem 4. (We have multiplied, on each side, by $\tfrac{1}{2}$, and then written the inequality backwards, to put x on the left.)

Thus every number which satisfies (1) also satisfies (4). And all of our steps can be reversed. If

$$x > -\tfrac{3}{2}, \tag{4}$$

then

$$-3 < 2x, \tag{3}$$

by Theorem 4; therefore

$$4 < 2x + 7, \tag{2}$$

by AO; and so

$$3x + 4 < 5x + 7, \tag{1}$$

by AO. Therefore every number which satisfies (4) also satisfies (1). We can sum all this up briefly by writing

$$3x + 4 < 5x + 7 \quad \Leftrightarrow \quad x > -\tfrac{3}{2}.$$

Here the symbol $\Leftrightarrow$ is pronounced "is equivalent to." When we write $\Leftrightarrow$ between two inequalities (or any two open sentences of any kind) we mean that whenever one of them is satisfied, so is the other.

We use a single-headed arrow to indicate that one condition *implies* another. For example,

$$x > 0 \quad \Rightarrow \quad x^2 > 0.$$

This is true. (Why?) But

$$(?) \, x > 0 \quad \Leftrightarrow \quad x^2 > 0 \, (?)$$

is false, because $x = -1$ satisfies the second inequality but not the first. Similarly, $a = b \Rightarrow a^2 = b^2$ is true, but

$$(?) \, a = b \quad \Leftrightarrow \quad a^2 = b^2 \, (?)$$

is false, because if $a \neq 0$ and $b = -a$, then the second inequality holds, but the first does not.

The shorthand symbols $\Leftrightarrow$ and $\Rightarrow$ are worth learning and using. The reason is that when we write down strings of formulas, in solving a problem, we ought to indicate what the connection between them is supposed to be. We are more likely to do this if we have a way of doing it briefly.

Using the symbols $\Rightarrow$ and $\Leftrightarrow$, we can restate some of the theorems of this section in a more efficient way. For example, AO says that

$$a < b \quad \Rightarrow \quad a + c < b + c. \tag{5}$$

And given $a + c < b + c$, we can add $-c$ to both sides, preserving the inequality. Therefore

$$a + c < b + c \quad \Rightarrow \quad a < b. \tag{6}$$

These fit together to give:

The Addition Law of Order. $a < b \quad \Leftrightarrow \quad a + c < b + c.$

We shall refer to this, for short, as ALO. Similarly, Theorem 4 says that

$$\text{for } c > 0, \quad a < b \quad \Rightarrow \quad ac < bc.$$

Theorem 5 says that

$$\text{for } c > 0, \quad ac < bc \quad \Rightarrow \quad a < b.$$

These fit together to give:

The Multiplication Law of Order. For $c > 0$, $a < b \Leftrightarrow ac < bc$.

This will be referred to as MLO. Theorems 6 and 7 say that

$$\text{for } c < 0, \quad a < b \quad \Rightarrow \quad ac > bc, \tag{7}$$

and

$$\text{for } c < 0, \quad bc < ac \quad \Rightarrow \quad b > a. \tag{8}$$

And (8) can be rewritten in the form

$$\text{for } c < 0, \quad ac > bc \quad \Rightarrow \quad a < b. \tag{8'}$$

Thus Theorems 6 and 7 fit together to give:

Reversal of Order. For $c < 0$, $a < b \quad \Leftrightarrow \quad ac > bc$.

We sum all this up in the short form on the next page. The meanings of the abbreviations should be plain.

> **Trich.**   For every a and b in **R**, one and only one of the following
> conditions holds:
>
> $$a < b, \quad \text{or} \quad a = b, \quad \text{or} \quad b < a.$$
>
> **Trans.** $a < b$ and $b < c \;\Rightarrow\; a < c.$
>
> **MO.** $a > 0$ and $b > 0 \;\Rightarrow\; ab > 0.$
>
> **AO.** $a < b \;\Rightarrow\; a + c < b + c.$
>
> **Theorem 1.** $a > 0 \;\Rightarrow\; -a < 0.$
>
> **Theorem 2.** $a < 0 \;\Rightarrow\; -a > 0.$
>
> **Theorem 3.** $a < b$ and $c < d \;\Rightarrow\; a + c < b + d.$
>
> **ALO.** $a < b \;\Leftrightarrow\; a + c < b + c.$
>
> **MLO.**   For $c > 0$, $a < b \;\Leftrightarrow\; ac < bc.$
>
> **RO.**   For $c < 0$, $a < b \;\Leftrightarrow\; ac > bc.$

The last three of these are convenient in solving inequalities; they enable us to
write $\Leftrightarrow$ at each stage, instead of working first forward and then backward. For
example, the solution of the illustrative problem above can now be written like this:

$$3x + 4 < 5x + 7$$

$$\Leftrightarrow \qquad 4 < 2x + 7 \qquad \text{by ALO}$$

$$\Leftrightarrow \qquad -3 < 2x \qquad \text{by ALO}$$

$$\Leftrightarrow \qquad -\tfrac{3}{2} < x \qquad \text{by MLO}$$

$$\Leftrightarrow \qquad x > -\tfrac{3}{2} \qquad \text{by definition of } >.$$

A linear inequality is said to be *solved* when we find an equivalent inequality of
the form $x < a$ or $x > a$.

PROBLEM SET 1.3

Solve the following inequalities, by writing a chain of equivalent inequalities, and giving
on the right the reason for each step, as in the text.

1. $5 - 3x > 17 + x$ 2. $5x - 3 < 17x + 1$

3. $5x + 3 > 17x + 1$ 4. $5 + 3x < 17 + x$

5. $-3x - 7 < x + 5$ 6. $-4x - 8 < 2x + 6$

7. $6x - 10 > 5x + 3$ 8. $3 - 2x < 4 - 3x$

9. $2x - 6 < 2 - 2x$ 10. $6x - 2 < 3 + x$

11. $2x + 6 < 3 + x$ 12. $6(x - 2) > x - 3$

In the following problems, we develop the theory in which all of the results of this
section are derived from Trich., Trans., MO, and AO. Therefore, at the start, these are

the only statements that can be given as reasons in proofs. In each problem, however, you may assume that the results given in the preceding problems are known and you may cite them as reasons.

13. Following are the steps in the proof of Theorem 1. Complete the proof by giving a reason for each step.

a) $a > 0 \Rightarrow 0 < a$ b) $0 < a \Rightarrow -a + 0 < -a + a$

c) $-a + 0 < -a + a \Rightarrow -a < 0$ d) $a > 0 \Rightarrow -a < 0$

14. Following is an outline of the proof of Theorem 2. Complete the proof by giving a reason for each $\Rightarrow$.

$$a < 0 \Rightarrow -a + a < -a + 0 \Rightarrow 0 < -a$$
$$\Rightarrow -a > 0.$$

15. a) Give a reason for the statement

$$a < b \Rightarrow a + c < b + c.$$

b) Similarly, for

$$c < d \Rightarrow b + c < b + d.$$

c) Prove Theorem 3.

16. a) Show that

$$a < b \Leftrightarrow b - a > 0.$$

(More than one step is needed here.)

b) Give a reason for the statement

$$c > 0 \quad \text{and} \quad b - a > 0 \Rightarrow (b - a)c > 0.$$

c) Prove Theorem 4.

17. Show that

$$x^2 \geq 0, \qquad \text{for every } x.$$

[*Hint:* By Trich., there are three cases to be considered: $x > 0$, or $x = 0$, or $x < 0$. Show that in each of these cases we have either $x^2 > 0$ or $x^2 = 0$.]

18. Show that

$$y^2 - 2y + 1 \geq 0, \qquad \text{for every } y.$$

19. a) Everybody knows that $1 > 0$. Prove it, on the basis of the theory that we have developed so far. (You may assume, of course, that $1 \neq 0$.)

b) Show that

$$a > 0 \Rightarrow \frac{1}{a} > 0.$$

That is, the reciprocal of every positive number is positive. [*Hint:* By Trich., it will be sufficient to show that the conditions $1/a = 0$ and $1/a < 0$ are impossible. Remember that $a \cdot 1/a = 1$.]

20. Show that

$$c > 0 \quad \text{and} \quad ac < bc \Rightarrow a < b.$$

(This is Theorem 5.)

21. Give the reason for each step in the following proof of Theorem 6.

$$a < b \qquad \text{and} \qquad c < 0$$
$$\Rightarrow \quad b - a > 0 \qquad \text{and} \qquad c < 0$$
$$\Rightarrow \quad b - a > 0 \qquad \text{and} \qquad -c > 0$$
$$\Rightarrow \quad (b - a)(-c) > 0$$
$$\Rightarrow \quad ac - bc > 0$$
$$\Rightarrow \quad ac > bc.$$

22. Give the reason for each step in the following proof of Theorem 7.

$$bc < ac \qquad \text{and} \qquad c < 0$$
$$\Rightarrow \quad bc < ac \qquad \text{and} \qquad -c > 0$$
$$\Rightarrow \quad bc < ac \qquad \text{and} \qquad \frac{1}{-c} > 0$$
$$\Rightarrow \quad ac - bc > 0 \qquad \text{and} \qquad \frac{1}{-c} > 0$$
$$\Rightarrow \quad \frac{1}{-c}(ac - bc) > 0$$
$$\Rightarrow \quad b - a > 0$$
$$\Rightarrow \quad b > a.$$

23. Is there a positive number which is smaller than all other positive numbers? Why or why not?

24. Is there a negative number which is larger than all other negative numbers? Why or why not?

*25. Is it possible to define, for the complex numbers, a relation $<$ which obeys the laws O.1 and O.2? (That is, can an order relation be defined for the complex numbers?) Why or why not?

*26. Is it possible to define, for the complex numbers, a relation $<$ which satisfies not only O.1 and O.2 but also MO and AO? [*Hint:* Since $i \neq 0$, we must have $i > 0$ or $-i > 0$.]

The language in which these problems are stated ought to suggest what the answers are. The answer to Problem 26 indicates why it is that arranging the complex numbers in an order is not a useful proceeding. In the complex number system, no theory of inequalities can be made to work.

1.4 ABSOLUTE VALUES. INTERVALS ON THE NUMBER LINE

The *absolute value* $|x|$ of a number x is defined by the following two conditions:

1) If $x \geq 0$, then $|x| = x$.

2) If $x < 0$, then $|x| = -x$.

Thus under Condition (1) we have

$$|2| = 2, \qquad |\pi| = \pi;$$

and under Condition (2) we have

$$|-2| = -(-2) = 2, \qquad |-\pi| = -(-\pi) = \pi.$$

Thus the operation | | leaves positive numbers unchanged, and replaces each negative number by the corresponding positive number. On this basis it is easy to see that the following theorem holds.

Theorem 1. For every x,

$$|x| \geqq 0.$$

Proof. There are two cases to consider.

Case 1. $x \geqq 0$. Here $|x| = x$, by definition of $|x|$. Therefore $|x| \geqq 0$ in Case 1.

Case 2. $x < 0$. Here $|x| = -x$, by definition of $|x|$; and $-x > 0$, by Theorem 2 of the preceding section. Therefore $|x| > 0$ in Case 2.

Thus in each case we have $|x| \geqq 0$.

Theorem 2. For every x,

$$|x|^2 = x^2.$$

This is true because $|x|$ is either x or $-x$, and $(-x)^2 = x^2$.

A number x is a square root of a number a if $x^2 = a$. For each $a > 0$, $\sqrt{a}$ is the positive square root of a. Thus, for example, 9 has two square roots, 3 and -3; and $\sqrt{9}$ is 3, which is the positive square root. We define $\sqrt{0} = 0$. Here and hereafter, we are assuming that positive numbers have roots of all orders—square roots, cube roots, and so on.

Theorem 3. For every x,

$$|x| = \sqrt{x^2}.$$

Proof. By Theorem 2, $|x|^2 = x^2$, and so $|x|$ is a square root of x^2. By Theorem 1, $|x| \geqq 0$. Therefore $|x| = \sqrt{x^2}$, by definition of $\sqrt{x^2}$.

Theorem 4. For every x,

$$|-x| = |x|.$$

This is true because $|-x| = \sqrt{(-x)^2} = \sqrt{x^2} = |x|$.

Theorem 5. For every x and y,

$$|xy| = |x| \cdot |y|.$$

This is true because $|xy| = \sqrt{(xy)^2} = \sqrt{x^2 y^2} = \sqrt{x^2} \cdot \sqrt{y^2} = |x| \cdot |y|$.

Theorem 6. For every x,

$$x \leqq |x|.$$

Here, as in the proof of Theorem 1, we need to consider two cases.

Case 1. $x \geqq 0$. Here $x \leqq |x|$, because $|x| = x$.

Case 2. $x < 0$. Here $|x| = -x$, and $-x > 0$. (Why?) Thus $x < 0 < -x = |x|$, and so $x < |x|$.

Theorem 7. (*The triangular inequality*) For every x and y,

$$|x + y| \leqq |x| + |y|.$$

The trouble with this theorem, if we try to prove it by brute force, is that there are too many cases to consider: each of the numbers x and y may or may not be negative; and if x and y have different signs, $x + y$ may or may not be negative. It turns out, however, that we can get a proof by examining only two cases:

Case 1. $x + y \geqq 0$. In this case

$$|x + y| = x + y.$$

Since

$$x \leqq |x|, \quad \text{and} \quad y \leqq |y|,$$

we have

$$x + y \leqq |x| + |y|,$$

and so

$$|x + y| \leqq |x| + |y|.$$

Case 2. Suppose that $x + y < 0$. Then $-x - y > 0$. Therefore

$$|x + y| = |-x - y| = |(-x) + (-y)| \leqq |-x| + |-y|,$$

by the result of Case 1. Since $|-x| = |x|$ and $|-y| = |y|$, we have

$$|x + y| \leqq |x| + |y|,$$

which was to be proved.

Theorem 8. Given $d > 0$. Then

$$|x| < d \quad \Leftrightarrow \quad -d < x < d.$$

This is geometrically obvious: $|x|$ is "the distance between 0 and x, on the number line"; and the points that lie within a distance d of the origin are the numbers between $-d$ and d. We get a more general result by using any given point a instead of the origin.

Theorem 9. Given $d > 0$, and any number a. Then

$$|x - a| < d \quad \Leftrightarrow \quad a - d < x < a + d.$$

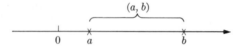

Proof. In Theorem 8, substitute $x - a$ for x. This gives

$$|x - a| < d \quad \Leftrightarrow \quad -d < x - a < d.$$

And

$$-d < x - a < d \quad \Leftrightarrow \quad a - d < x < a + d.$$

(Reason?)

If $a < b$, then the set of all numbers between a and b is called an *open interval*, and is denoted by (a, b).

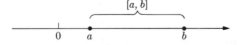

There is a shorthand for this sort of statement:

$$(a, b) = \{x \mid a < x < b\}.$$

The expression on the right denotes the set of all objects that satisfy the condition following the vertical bar. This is called the *solution set* of the open sentence $a < x < b$. Similarly, the set of all positive numbers is the solution set of the open sentence $x > 0$; this is denoted by $\{x \mid x > 0\}$. Thus two open sentences are equivalent if they have the same solution set.

Sometimes it turns out than an open sentence never gives a true statement, no matter what we substitute for x. In such cases, the solution set is empty. The empty set is denoted by $\{\ \}$. For example,

$$\{x \mid \sqrt{x^2 + 1} = x - 1\} = \{\ \}.$$

The notation $\{\ \}$ is designed to suggest its meaning: we describe sets in the brace notation; and when there is nothing written between the braces, this means that the set has nothing in it.

If we add to the open interval (a, b) the endpoints a and b, we get a *closed interval*, denoted by $[a, b]$.

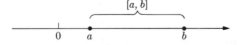

Thus

$$[a, b] = \{x \mid a \leqq x \leqq b\}.$$

We shall also be dealing with "infinite intervals." In the first figure below, the "infinite interval" is

$$(a, \infty) = \{x \mid a < x\}.$$

Similarly,

$$(-\infty, a) = \{x \mid x < a\},$$

as shown in the second figure below.

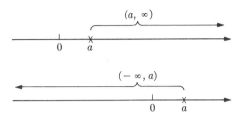

This notation, in which "∞" is used as if it denoted a number, is not very logical, but it is convenient. To keep track of the notation, you should think of fictitious "numbers" $-\infty$ and ∞ as the "ends" of the number line, as shown below.

We also use "half-open" intervals:

$$[a, b) = \{x \mid a \leqq x < b\} \quad \text{and} \quad (a, b] = \{x \mid a < x \leqq b\},$$

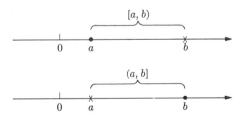

and "closed infinite" intervals:

$$[a, \infty) = \{x \mid x \geqq a\}, \quad \text{and} \quad (-\infty, a] = \{x \mid x \leqq a\}.$$

Finally, we may refer to the whole real number system **R** as the interval $(-\infty, \infty)$. Thus we have a total of nine kinds of interval:

$$(a, b), \quad [a, b], \quad [a, b), \quad (a, b],$$
$$(a, \infty), \quad [a, \infty), \quad (-\infty, a), \quad (-\infty, a],$$
$$(-\infty, \infty).$$

In some of the problems below, you may find it convenient to use the following:

Theorem 10. If $|x| = |y|$, then $x = y$ or $x = -y$.

Proof.

$$|x| = |y| \quad \Rightarrow \quad |x|^2 = |y|^2$$
$$\Rightarrow \quad x^2 = y^2$$
$$\Rightarrow \quad x^2 - y^2 = 0$$
$$\Rightarrow \quad (x - y)(x + y) = 0$$
$$\Rightarrow \quad x = y \quad \text{or} \quad x = -y.$$

(The converse is obvious.)

PROBLEM SET 1.4

Describe each of the following sets in the interval notation. Your answers should be in a form like the following:

$$\{x \mid 3x + 4 < 5x + 7\} = (-\tfrac{3}{2}, \infty).$$

(This is the example discussed in Section 1.3 of the text.)

1. $\{x \mid 3x + 4 > 4x + 5\}$
2. $\{x \mid 3 - x > x - 3\}$
3. $\{x \mid |x| < 1\}$
4. $\{x \mid |x - 3| \leq 2\}$
5. $\{x \mid |x - 5| < 5\}$
6. $\{x \mid 2x + 3 \geq 6x - 4\}$
7. $\{x \mid |1 - x| \geq 2\}$
8. $\{x \mid |2 + x| < 0\}$

9. a) Is it true that $\sqrt{x^2} = x$ for every x? Why or why not? Describe the set

$$\{x \mid \sqrt{x^2} = x\},$$

   in the interval notation.
   b) Describe the set

$$\{x \mid \sqrt{(x + 1)^2} = x + 1\},$$

   in the interval notation.

Find out for what numbers x (if any) each of the following conditions holds. In each case in which the solution set is an interval, the answer should be given in the interval notation.

10. $\sqrt{x^2 - 2x + 1} = x - 1$
11. $|x^2 - 5x + 6| = |x - 3| \cdot |x - 2|$
12. $|x^2 - 5x + 6| = x^2 - 5x + 6$
13. $|x - 5| = |2x - 3|$
14. $|x + 1| = |1 - x|$
15. $\sqrt{x^2 + 1} = x$
16. $\sqrt{x^2 - 1} = x$
17. $|2x - 1| + |x + 3| \geq |3x + 2|$

18. $|7x + 3| + |3 - x| \geq 6\,|x + 1|$ 19. $\sqrt{2x - x^2} = 1$

20. $|2x - x^2| = x + 2x^2$ 21. $|x + 1| + |2x + 3| > 2$

22. $|x - 1|\,|x^2 + x| = |x^3 - x|$

Indicate graphically, on a number scale, the places where the following conditions hold; describe the graphs in the interval notation if possible.

23. $|x| < 2$ 24. $|x - 2| < \frac{1}{2}$

25. $|2x - 3| \leq \frac{1}{2}$ 26. $|x - 1| < \frac{1}{2}$ and (also) $|x - 2| \leq 1$

27. $|3 - 2x| \leq \frac{1}{2}$ 28. $|x - 2| < \frac{1}{4}$ and $x \geq 2$

29. $|1 - x| \leq 2$ 30. $|2x - 4| < 1$

31. $|x - 1| < 2$ and $|x| \geq \frac{1}{2}$ 32. $|2x - 1| \geq 1$ and $|2x - 1| \leq 1$

33. a) Show that if $b \neq 0$, then

$$\left|\frac{1}{b}\right| = \frac{1}{|b|}\,.$$

(By definition, the reciprocal of $|b|$ is the number y such that $|b| \cdot y = 1$. Therefore it is sufficient to show that $|b| \cdot |1/b| = 1$.)

b) Show that if $b \neq 0$, then

$$\left|\frac{a}{b}\right| = \frac{|a|}{|b|}\,.$$

34. a) Show that for every a and b,

$$|a - b| \geq |a| - |b|.$$

(There is a short proof.)

b) Show that for every a and b,

$$|a + b| \geq |a| - |b|.$$

(The proof is short.)

35. For what numbers a is the fraction $a/|a|$ defined? What is this fraction equal to, for various values of a?

36. Sketch

$$\{x \mid |x - 2| + |7 - x| = 5\}$$

on the number line, and describe this set in the interval notation.

2   Analytic Geometry

2.1   INTRODUCTION

This chapter includes various topics which serve as a preparation for calculus. Some of these topics are familiar to you, at least in some form. In such cases you should still read the text carefully, in order to learn the terminology that will be used hereafter.

2.2   COORDINATE SYSTEMS. THE DISTANCE FORMULA

We shall now apply algebra to the study of geometry. We start with a plane, in the usual sense of Euclidean geometry; and we suppose that a unit of distance has been chosen, once for all, so that the distance between two points P and Q is a well-defined nonnegative number. The distance between the points P and Q is denoted by PQ. (We say merely that PQ is nonnegative, rather than $PQ > 0$, because we are allowing the case $P = Q$, and in this case $PQ = 0$.)

To set up a coordinate system in a plane, we first need to assign number-labels to the points of a line. We choose a point O as the origin; it is given the label 0.

$$x_2 = -OP_2 < 0. \qquad x_1 = OP_1 > 0.$$

Each point P_1 to the right of O is labeled with the distance $x_1 = OP_1$, which is positive. And each point P_2 to the left of O is labeled with the number $x_2 = -OP_2$, which is negative. Thus we have a matching scheme, under which each point of the line is matched with exactly one real number.

For the points marked in the figure, the matching pairs are

$$P \leftrightarrow -2, \qquad Q \leftrightarrow -1, \qquad R \leftrightarrow 1,$$
$$S \leftrightarrow \sqrt{2}, \qquad T \leftrightarrow 2, \qquad U \leftrightarrow \pi.$$

Here the double arrow $\leftrightarrow$ is pronounced "is matched with." Every such pair has the form $P \leftrightarrow x$, where P is a point and x is a number. A one-to-one matching scheme, between the elements of one set and the elements of another, is called a *one-to-one correspondence* between the two sets.

If the correspondence is set up in the way that we have just described, then we can compute the distance between any two points by means of the formula

$$P_1 P_2 = |x_2 - x_1|.$$

Here $P_1 \leftrightarrow x_1$ and $P_2 \leftrightarrow x_2$. This distance formula holds no matter how the points P_1 and P_2 are situated on the line:

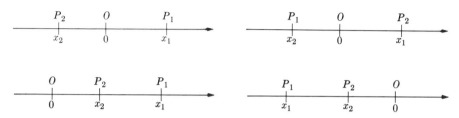

and so on; in every case, $P_1 P_2 = |x_2 - x_1|$. Thus we have a one-to-one correspondence $P \leftrightarrow x$, between the points of the line and the real numbers, such that the distance formula holds for every pair of points. Such a correspondence is called a *coordinate system* for the line. If $P \leftrightarrow x$, then x is called the *coordinate* of P.

These ideas are summed up in the following postulate.

The Ruler Postulate. Every line has a coordinate system. And given any two points O and P of the line, there is a coordinate system in which the coordinate of O is 0 and the coordinate of P is positive.

On the basis of the ruler postulate, it is easy to set up a coordinate system in the plane. We take two perpendicular lines X and Y, intersecting in a point O. On each of the two lines we set up a coordinate system, in such a way that $O \leftrightarrow 0$; that is, the coordinate of O is zero on each of the lines X and Y. X is called the *x-axis*, Y is called the *y-axis*, and the point O is called the *origin*.

Given any point P of the plane, we drop a perpendicular from P to the x-axis, ending at a point M. The point M has a coordinate x, on the line X. If $M \leftrightarrow x$, then x is called the *x-coordinate* of P.

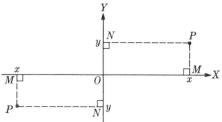

Similarly, we drop a perpendicular from P to the y-axis, ending at a point N. If $N \leftrightarrow y$, then y is called the *y-coordinate* of P. Thus we have a matching scheme

$$P \leftrightarrow (x, y)$$

between the points P of the plane and the ordered pairs (x, y) of real numbers. The order in which we write the numbers makes a difference. In the left-hand figure below,

$$P \leftrightarrow (1, 2) \quad \text{and} \quad Q \leftrightarrow (2, 1).$$

We may speak of "the point $(1, 2)$" or "the point (x, y)," meaning "the point which is matched with $(1, 2)$" or "the point which is matched with (x, y)." Thus we may write $P = (x, y)$, meaning $P \leftrightarrow (x, y)$.

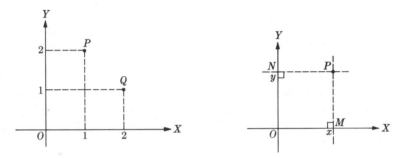

Obviously x and y are determined when P is known. And P is determined when x and y are known, because the vertical line through M and the horizontal line through N intersect in exactly one point. Thus we have a one-to-one correspondence

$$P \leftrightarrow (x, y)$$

between the points of the plane and the ordered pairs of real numbers. Such a correspondence is called a *coordinate system* for the plane. We need to see how the algebra in this situation is related to the geometry.

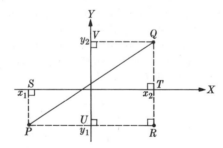

Consider first the question of distance. If we know the coordinates (x_1, y_1) and (x_2, y_2) of two points P and Q, then the points are determined, and so the distance between them is determined. The following theorem gives a formula for the distance.

Theorem 1. If

$$P \leftrightarrow (x_1, y_1) \qquad \text{and} \qquad Q \leftrightarrow (x_2, y_2),$$

then

$$PQ = \sqrt{(x_2 - x_1)^2 + (y_2 - y_1)^2}.$$

Proof. Draw the vertical line through Q and the horizontal line through P, meeting at the point R. Let S and T be the feet of the perpendiculars to X, from P and R respectively. Then

$$PR = ST,$$

because opposite sides of a rectangle have the same length. And

$$ST = |x_2 - x_1|,$$

by definition of a coordinate system on a line. Therefore

$$PR = |x_2 - x_1|.$$

For the same sort of reason,

$$RQ = UV = |y_2 - y_1|.$$

But $\triangle PQR$ is a right triangle, with its right angle at R. Therefore, by the Pythagorean theorem,

$$PQ^2 = PR^2 + RQ^2$$

$$= |x_2 - x_1|^2 + |y_2 - y_1|^2.$$

Therefore

$$PQ = \sqrt{|x_2 - x_1|^2 + |y_2 - y_1|^2}.$$

This is not quite the formula given in Theorem 1, because it uses absolute-value signs instead of parentheses. But this makes no difference, because

$$|x_2 - x_1|^2 = (x_2 - x_1)^2,$$

and

$$|y_2 - y_1|^2 = (y_2 - y_1)^2.$$

(Why? We need a theorem from Section 1.4.)

In the previous figures, we have shown the x-axis going positively from left to right, and the y-axis going positively from bottom to top. Logically speaking, we could equally well have put the axes in any of a number of other positions:

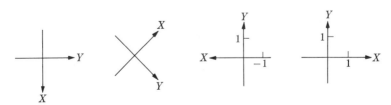

But the axes are usually drawn as shown on the right above. This figure shows the minimum that must be indicated when graph paper is used for drawing pictures of coordinate systems. That is, the axes must be labeled, and the number scale must be shown on each axis, by indicating the coordinate of at least one point.

The two axes separate the plane into four parts, called *quadrants*. The quadrants are numbered I, II, III, IV. That is, the *first quadrant* is the set of all points (x, y) of the plane for which $x > 0$ and $y > 0$; the second quadrant is the set of all points (x, y) for which $x < 0$ and $y > 0$; and so on.

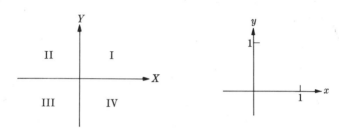

We have used the letters X and Y in order to have convenient names for the x- and y-axes. The axes are more commonly labeled as on the right above.

PROBLEM SET 2.2

Calculate the distances between the following pairs of points. Then plot the points and check the plausibility of your answers.

1. a) $(1, 2)$ and $(3, 4)$ b) $(-2, -4)$ and $(4, 2)$
   c) $(7, -5)$ and $(5, -7)$ d) $(1, 0)$ and $(0, 1)$

2. a) $(3, 7)$ and $(-3, -7)$ b) $(1, 3)$ and $(-2, 7)$

3. Obviously, $PQ = QP$ for every pair of points P, Q; the distance between two points does not depend on the order in which the points are named. Therefore any correct distance formula has the property that when we interchange the two points, the formula gives the same answer. Check algebraically that our distance formula has this property.

4. Find out whether or not the points $(-10, 10)$, $(14, 3)$, and $(38, -4)$ are vertices of an isosceles triangle. Is the triangle equilateral?

5. Find all points (x, y) such that $(0, 0)$, $(2, 2)$, and (x, y) are the vertices of an equilateral triangle.

6. Find out whether the points $(-2, 3)$, $(0, 1)$, and $(3, 4)$ are the vertices of a right triangle. Then plot the points and check for plausibility. (This problem can and should be worked by the use of distances alone. The use of slopes is *not* necessary.)

7. Find the coordinates of the point which is equidistant from $(0, 0)$, $(1, 2)$, and $(3, -1)$. Find the radius of the circle which passes through the three given points.

8. What point on the y-axis, if any, is equidistant from $(-1, -2)$ and $(2, 3)$?

9. a) Give a formula for the perpendicular distance between (x, y) and the x-axis.
   b) Give a formula for the perpendicular distance between (x, y) and the y-axis.

10. Find out whether the points $(-1, -1)$, $(0, 1)$, and $(2, 5)$ are collinear. Then plot and check for plausibility. (The remarks following Problem 6 also apply here.)

11. Find a point on the x-axis which is collinear with the points $(1, 2)$ and $(0, 3)$. (The remarks following Problem 6 also apply here.)

The following problems are a review of the main theorems of elementary geometry that we have been using so far.

12. Show that an exterior angle of a triangle is greater than either of its remote interior angles.

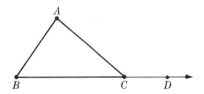 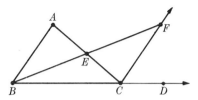

That is, show that in the left-hand figure we have $\angle ACD > \angle A$. The proof is based on the figure on the right. [*Query:* If you know that $\angle ACD > \angle A$, how do you infer that $\angle ACD > \angle B$?]

13. Show that there is only one perpendicular to a given line, from a given external point. That is, show that the left-hand figure below is impossible for $A \neq B$. (We needed this in order to explain what was meant by the x-coordinate of a point; A must be determined when P is known.)

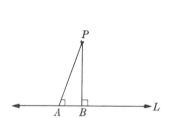 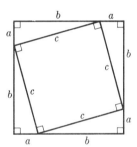

14. Write the proof of the Pythagorean theorem suggested by the figure on the right above.

15. The proof of Theorem 1 of this section was incomplete: it discussed only the most significant case and neglected to mention two other cases. The point is that if P and Q lie on the same horizontal line, or the same vertical line, then there is no such thing as $\triangle PQR$, and so the Pythagorean theorem cannot be used.

Show that the distance formula holds in the case $x_1 = x_2$, and also in the case $y_1 = y_2$.

2.3 THE GRAPH OF A CONDITION. EQUATIONS FOR CIRCLES

Given a point P and a positive number r, the circle with center P and radius r is the set of all points of the plane whose distance from P is equal to r. That is, a point Q is on the circle if $PQ = r$.

This is the first and simplest example of the idea of the *graph of a condition*. If we state a condition which every point of the plane either satisfies or doesn't satisfy, then the *graph* of the condition is the set of all points of the plane that satisfy it. (Thus the graph is simply the solution set of an open sentence; we use the word *graph* when the solution set is a set of points.) In this language, we say that the graph of the condition $OQ = r$ is the circle with center at the origin and radius r.

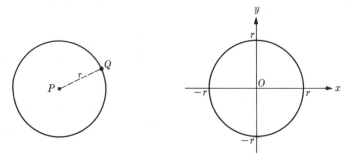

The *interior* of the circle with center P and radius r ($r > 0$) is the set of all points Q such that $PQ < r$. Thus the interior is the graph of the inequality $PQ < r$. We indicate such graphs in figures by means of shading or cross-hatching.

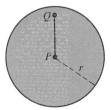

Sometimes the condition takes the form of an algebraic equation. For example, if $Q \leftrightarrow (x, y)$, then the distance formula tells us that

$$OQ = \sqrt{x^2 + y^2}.$$

Therefore the condition

$$OQ = r \tag{1}$$

can be written in the equivalent form

$$\sqrt{x^2 + y^2} = r, \tag{2}$$

or

$$x^2 + y^2 = r^2. \tag{3}$$

The point (x, y) is on the circle if and only if x and y satisfy (2). And

$$\sqrt{x^2 + y^2} = r \iff x^2 + y^2 = r^2 \quad (r > 0).$$

Thus the circle with center at the origin and radius 2 is the graph of

$$\sqrt{x^2 + y^2} = 2 \iff x^2 + y^2 = 4;$$

and the interior of this circle is the graph of

$$\sqrt{x^2 + y^2} < 2 \iff x^2 + y^2 < 4.$$

Similarly, the first quadrant is the graph of the condition $x > 0$ and $y > 0$.

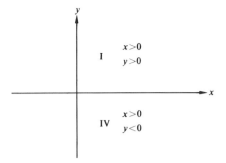

The fourth quadrant is the graph of the condition $x > 0$ and $y < 0$.

We found that the circle with center at the origin and radius r is the graph of the equation

$$x^2 + y^2 = r^2.$$

Consider, more generally, the circle with center at $Q = (a, b)$ and radius r. By definition, the circle is

$$\{P \mid QP = r\}.$$

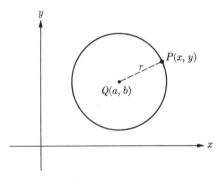

Using the distance formula to express QP algebraically, we get

$$QP = r$$
$$\Leftrightarrow \quad \sqrt{(x - a)^2 + (y - b)^2} = r$$
$$\Leftrightarrow \quad (x - a)^2 + (y - b)^2 = r^2.$$

Thus:

Theorem 1. The circle with center at (a, b) and radius r is the graph of the equation

$$(x - a)^2 + (y - b)^2 = r^2.$$

An equation written in the above form is easy to interpret. For example, given

$$(x + 2)^2 + (y - 5)^2 = 4,$$

we see by Theorem 1 what the graph is. On the other hand, if such an equation is "simplified" algebraically, it may look like this:

$$x^2 + y^2 + 4x - 10y + 25 = 0.$$

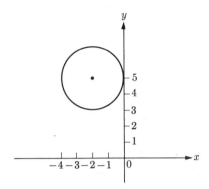

To find out what the graph is, we first "unsimplify" by completing the square:

$$x^2 + 4x + y^2 - 10y = -25$$

$$\Leftrightarrow \quad x^2 + 4x + 4 + y^2 - 10y + 25 = -25 + 4 + 25$$

$$\Leftrightarrow \quad (x + 2)^2 + (y - 5)^2 = 4.$$

In the general case, for equations of the form

$$x^2 + y^2 + Dx + Ey + F = 0,$$

there are three possibilities for the graph. In some cases, the graph is a circle. But

$$x^2 + y^2 = 0$$

is also an equation of this form, and its graph is not a circle, but a single point, namely the origin. And the equation

$$x^2 + y^2 + 1 = 0$$

is never satisfied, for any x and y. Its graph is therefore the empty set { }.

By completing the square, starting with the general form, we shall show that these three possibilities—a circle, a point, and the empty set—are in fact the only ones:

$$x^2 + y^2 + Dx + Ey + F = 0$$

$$\Leftrightarrow \quad x^2 + Dx + \left(\frac{D}{2}\right)^2 + y^2 + Ey + \left(\frac{E}{2}\right)^2 = -F + \left(\frac{D}{2}\right)^2 + \left(\frac{E}{2}\right)^2$$

$$\Leftrightarrow \quad \left(x + \frac{D}{2}\right)^2 + \left(y + \frac{E}{2}\right)^2 = \frac{D^2 + E^2 - 4F}{4}.$$

If the fraction on the right, in the last equation, is positive, then it is $= r^2$ for some positive number r, and so the graph is the circle with center at $(-D/2, -E/2)$ and radius r. If the fraction on the right is $= 0$, then the equation takes the form

$$\left(x + \frac{D}{2}\right)^2 + \left(y + \frac{E}{2}\right)^2 = 0,$$

and the graph contains only the point $(-D/2, -E/2)$. Finally, if the fraction on the right is negative, then the equation is never satisfied, for any x and y, and so the graph is the empty set $\{\ \}$.

To sum up:

Theorem 2. The graph of an equation of the form

$$x^2 + y^2 + Dx + Ey + F = 0$$

is a circle, a point, or the empty set.

PROBLEM SET 2.3

Problems 1 through 6.

In the illustration below six figures are drawn. For each of these figures, state a condition which has the given figure as its graph. In the figure, the arrowheads merely indicate that the line is supposed to go infinitely far in the indicated direction. Thus (1) and (2) are entire lines; (3) is a *ray*, going infinitely far on the right, but stopping at the point (0, 4) on the left; and (6) is a *segment*, with endpoints (1, −3) and (4, −3).

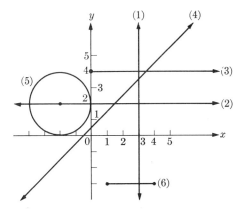

Problems 7 through 10.

Follow the same directions as in the previous problems for the illustration below.

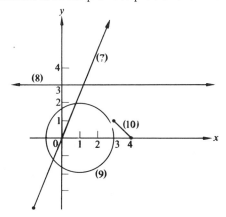

Sketch the graphs of the following conditions, using cross-hatching to indicate regions.

11. $x^2 + y^2 = 1$ 12. $x^2 + y^2 < 1$ 13. $x^2 + y^2 > 1$

14. $x = 2$ and $0 \leqq y \leqq 2$ 15. $x = -3$ 16. $y = 2$ 17. $y = x/|x|, x \neq 0$

18. $x^2 + y^2 \leqq 1$ 19. $xy \geqq 0$ 20. $x > 0, \; y = 3$

21. a) Sketch the graph of the condition "(x, y) is equidistant from the points $(0, 1)$ and $(1, 0)$."
    b) Write this condition in the simplest possible algebraic form.

22. Write the simplest equation that you can get, for the set of all points that are equidistant from $(1, 2)$ and $(0, 3)$. What sort of a figure is this graph? How is it related to the segment from $(1, 2)$ to $(0, 3)$?

23. Same problem, for the set of all points that are equidistant from $(1, 2)$ and $(2, 2)$.

24. Same problem, for the set of all points that are equidistant from $P_1 = (x_1, y_1)$ and $P_2 = (x_2, y_2)$.

*25. Describe and sketch the graph of the equation

$$\sqrt{(x - 1)^2 + (y - 2)^2} + \sqrt{(x - 4)^2 + (y - 7)^2} = \sqrt{34}.$$

[*Hint:* If you do a lot of algebra, you will probably get the wrong answer; the graph is *not* an ellipse.]

*26. Describe the graph of the equation

$$\sqrt{x^2 + (y - 1)^2} + \sqrt{(x - 2)^2 + y^2} = 1.$$

[The same hint as for Problem 25 applies here.]

27. Draw the graph of the equation

$$x^3y + y^3x - xy = 0.$$

28. Draw the graph of the equation

$$x^2y + xy^2 - xy = 0.$$

29. Consider the set of all points that are twice as far from the origin as from the point $(3, 0)$. Find an equation for this graph, and sketch.

2.4 EQUATIONS OF LINES. SLOPES, PARALLELISM, AND PERPENDICULARITY

Every line is the graph of an equation of the form

$$Ax + By + C = 0,$$

where A and B are not both $= 0$. The proof is as follows.

Every line is the perpendicular bisector of some segment. If L is the perpendicular bisector of the segment from $Q \leftrightarrow (a_1, b_1)$ to $R \leftrightarrow (a_2, b_2)$, then

$$L = \{P \,|\, PQ = PR\}.$$

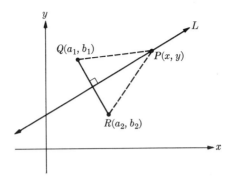

(Remember your geometry.) Therefore L is the graph of the equation

$$\sqrt{(x - a_1)^2 + (y - b_1)^2} = \sqrt{(x - a_2)^2 + (y - b_2)^2}$$

$$\Leftrightarrow \quad x^2 - 2a_1 x + a_1^2 + y^2 - 2b_1 y + b_1^2 = x^2 - 2a_2 x + a_2^2 + y^2 - 2b_2 y + b_2^2$$

$$\Leftrightarrow \quad 2(a_2 - a_1)x + 2(b_2 - b_1)y + a_1^2 + b_1^2 - a_2^2 - b_2^2 = 0.$$

This has the desired form

$$Ax + By + C = 0,$$

with

$$A = 2(a_2 - a_1), \qquad B = 2(b_2 - b_1),$$

and

$$C = a_1^2 + b_1^2 - a_2^2 - b_2^2.$$

The numbers A and B cannot both be $= 0$, because $a_2 - a_1$ and $b_2 - b_1$ cannot both be $= 0$; the number pairs (a_1, b_1) and (a_2, b_2) are the coordinates of Q and R, and $Q \neq R$, because Q and R are the endpoints of a segment.

An equation of this type is called a *linear equation in x and y*. Thus we have

Theorem 1. Every line is the graph of a linear equation in x and y.

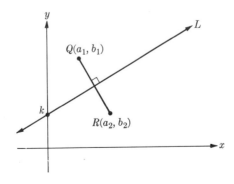

If the line is not vertical, we can say more. In this case, the perpendicular segment from Q to R is not horizontal, and this means that $b_2 - b_1 \neq 0$. Therefore $B \neq 0$, and we can divide by B and solve for y. This gives

$$y = -\frac{A}{B}x - \frac{C}{B},$$

which has the form

$$y = mx + k,$$

where

$$m = -\frac{A}{B} = -\frac{a_2 - a_1}{b_2 - b_1}.$$

In the figure, the label k on the y-axis is correct, because k is the y-coordinate of the point where L crosses the y-axis $(m \cdot 0 + k = k)$. The number k is called the *y-intercept* of the line.

The number m also has a geometric meaning, as we shall soon see. If $P_1 \leftrightarrow (x_1, y_1)$ and $P_2 \leftrightarrow (x_2, y_2)$ are any two points of a nonvertical line, then the *slope* of the segment from P_1 to P_2 is defined to be the fraction

$$\frac{y_2 - y_1}{x_2 - x_1}.$$

The denominator $x_2 - x_1$ is marked Δx in the figures below; it is pronounced "delta x," and stands for the *difference* in x. Similarly, $y_2 - y_1$ is marked Δy, which stands for the *difference* in y. Here Δy and Δx are not necessarily distances in the sense of elementary geometry, because they may be negative.

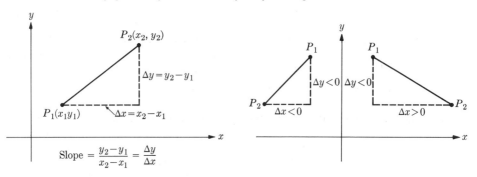

We shall show that all segments of the same line have the same slope, and that this slope is the number m which appears in the equation $y = mx + b$.

Given two points $P_1 \leftrightarrow (x_1, y_1)$ and $P_2 \leftrightarrow (x_2, y_2)$, on the line

$$y = mx + k,$$

then

$$y_2 = mx_2 + k \qquad \text{and} \qquad y_1 = mx_1 + k.$$

Therefore

$$y_2 - y_1 = m(x_2 - x_1),$$

and

$$\frac{y_2 - y_1}{x_2 - x_1} = m.$$

(In this calculation we do not care whether $y_2 - y_1$ and $x_2 - x_1$ are positive or negative. The algebra takes care of all cases at once.)

The number m is called the *slope* of the line. And we have proved the following theorem:

Theorem 2. The graph of the equation

$$y = mx + k$$

is the nonvertical line with slope m and y-intercept k. All segments of this line have slope $= m$.

The equation given in this theorem is called the *slope-intercept* form of the equation of the line.

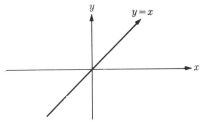

A line can be described by many different equations. For example, the bisector of the first and third quadrants above is the graph of each of the following equations:

$$y = x$$
$$\Leftrightarrow \quad x - y = 0$$
$$\Leftrightarrow \quad 3x - 3y = 0$$
$$\Leftrightarrow \quad (x - y)^2 = 0$$
$$\Leftrightarrow \quad (x - y)^{177} = 0,$$

and so on. But there is only one equation, in the slope-intercept form, for every nonvertical line, because when the line is named, its slope and its y-intercept are determined.

Often a line will be described by its slope m and the coordinates x_1, y_1, of one of its points. We can then find an equation for it in the following way. If (x, y) is any other point of the line, then

$$\frac{y - y_1}{x - x_1} = m,$$

because all segments of the line have the same slope m. Therefore

$$y - y_1 = m(x - x_1).$$

The graph of this equation contains (x_1, y_1), because $0 = m \cdot 0$. And the graph is a line with slope $= m$, because the equation has the form

$$y = mx + (y_1 - mx_1) = mx + k.$$

Thus:

Theorem 3. The graph of the equation $y - y_1 = m(x - x_1)$ is the line which has slope $= m$ and contains the point (x_1, y_1).

For example, the graph of the equation $y - 3 = -2(x + 1)$ is the line which has slope $= -2$ and passes through the point $(-1, 3)$. Solving for y, we get the slope-intercept form $y = -2x + 1$.

Theorem 4. Two nonvertical lines are parallel if and only if they have the same slope.

Given:

$$L_1: y = m_1x + k_1, \qquad L_2: y = m_2x + k_2,$$

we need to prove two things: (1) If the slopes are the same, and the lines are different, then the lines are parallel. (2) If the slopes are different, then the lines are not parallel.

1) If $m_1 = m_2$, then $k_1 \neq k_2$, because the lines are different. Therefore the lines are parallel, because the two equations are inconsistent: they take the form

$$y = m_1x + k_1, \qquad y = m_1x + k_2.$$

Since $k_1 \neq k_2$, these equations have no common solution.

2) If $m_1 \neq m_2$, the lines cannot be parallel, because the equations always have a common solution. By subtraction we get

$$0 = (m_1 - m_2)x + (k_1 - k_2), \qquad x = -\frac{k_1 - k_2}{m_1 - m_2},$$

and we now find the y-coordinate of the point of intersection by substituting in either of the original equations.

Theorem 5. If two nonvertical lines are perpendicular, then their slopes are negative reciprocals of each other.

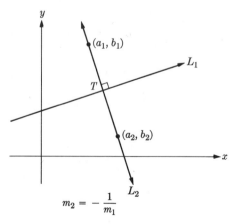

Proof. Given L_1 with slope m_1 and L_2 with slope m_2, intersecting at right angles at T. Let (a_1, b_1) and (a_2, b_2) be points of L_2 which are equidistant from T. Then L_1 is the perpendicular bisector of the segment between these points. As we found earlier, the slope of L_1 is

$$m_1 = -\frac{a_2 - a_1}{b_2 - b_1}.$$

But we can calculate the slope m_2 of L_2 by the slope formula, using the points (a_1, b_1) and (a_2, b_2). This gives

$$m_2 = \frac{b_2 - b_1}{a_2 - a_1}.$$

Obviously $m_2 = -1/m_1$.

This also works the other way around:

Theorem 6. Given two lines L_1, L_2, with slopes m_1, m_2. If

$$m_2 = \frac{-1}{m_1},$$

then L_1 and L_2 are perpendicular.

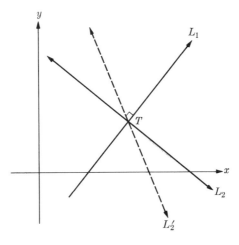

Proof. First we observe that the lines cannot be parallel, because m_2 cannot be $= m_1$. (Why?) Let T be the point where they intersect. Let L_2' be the line through T, perpendicular to L_1. Then L_2' has slope $m_2 = -1/m_1$. But through a given point there is only one line with a given slope. (Why?) Therefore L_2' is L_2, and L_2 is perpendicular to L_1.

Probably you have seen these theorems proved before, in different ways. The treatment given above is intended to avoid repetitions and also to furnish some practice in drawing geometric conclusions by algebraic methods.

PROBLEM SET 2.4

Find point-slope equations, and slope-intercept equations, for the lines containing the following pairs of points.

1. $(-3, 2), (2, 1)$ 2. $(3, -4), (1, 2)$ 3. $(1, 0), (3, 3)$ 4. $(-1, 1), (2, -2)$

5. Find an equation for the tangent to the graph of

$$x^2 + y^2 = 25,$$

at the point $(3, 4)$.

6. Given that $P_1 \leftrightarrow (x_1, y_1)$ lies on the circle

$$x^2 + y^2 = a^2,$$

with

$$x_1 \neq 0.$$

Let P_2 be the point where the tangent at P_1 crosses the x-axis. Find the distance $P_1 P_2$. [*Warning:* Geometric distances are never negative.]

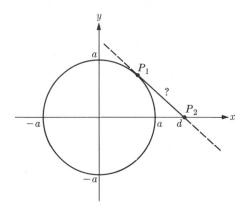

7. Find the points P on the circle $x^2 + y^2 = 2$ so that the tangent line to the circle at P passes through the point $(2, 0)$. (You may use the fact that, at any point P on a circle, the tangent and the radius are perpendicular.)

8. Sketch the graph of the equation

$$x^2 + y^2 + 1 + 2xy + 2x + 2y = 0.$$

9. Sketch the graph of the equation

$$x^2 + 4y^2 + 1 - 4xy + 2x - 4y = 0.$$

10. Sketch the graphs of the following equations.

a) $y = |x|$ b) $y = -|2x|$ c) $y = 1 - |x - 1|$

For this problem we offer a hint which applies equally well to a very large number of other problems. If you didn't know the meaning of the symbol $|x|$, you would have no hope of sketching the graph. This suggests that you should recall the definition of $|x|$, and use it.

11. Sketch the graph of

$$|x| + |y| = 1.$$

[*Hint:* As a first step, sketch the portion of the graph that lies in the first quadrant.]

12. Sketch the graph of the equation

$$y = x + |x| + 1.$$

13. Sketch the graph of the equation

$$|x| - |y| = 1.$$

14. Sketch the graph of the equation

$$\sqrt{(x - 1)^2 + (y - 3)^2} + \sqrt{(x - 4)^2 + (y - 2)^2} = \sqrt{10}.$$

15. Let C be the set of all points P such that the segment from $(-1, 0)$ to P is perpendicular to the segment from P to $(2, 1)$. What sort of figure is C? Sketch. (In answering this one you should bear in mind that the endpoints of a segment are always different. That is, there is no such thing as the segment from P to P.)

*16. Let $A = (-2, 0)$, let $B = (2, 0)$, and let G be the set of all points P such that $\angle APB$ is an angle of $60°$. What sort of figure is G? Sketch. (You will have to remember and use some plane geometry, to do this one. If you have suitable drawing instruments, you ought to be able to do a good sketch.)

2.5 GRAPHS OF INEQUALITIES. AND, OR, AND IF . . . THEN

We have found that the graph of the equation

$$(x - 1)^2 + (y - 1)^2 = 1$$

is the circle with center at $A = (1, 1)$ and radius 1. The interior of the circle is the graph of the condition $AP < 1$. This is the region marked R_1 in the figure. It is the graph of the inequality

$$(x - 1)^2 + (y - 1)^2 < 1.$$

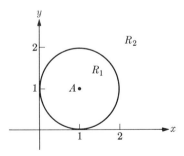

Similarly, the exterior R_2 is the graph of the condition $AP > 1$, so that

$$R_2 = \{(x, y) \,|\, (x - 1)^2 + (y - 1)^2 > 1\}.$$

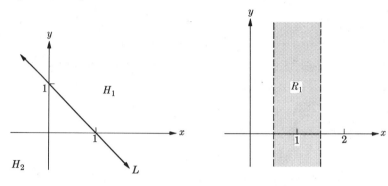

The graph of the equation

$$y = 1 - x$$

is a line L. The points lying above L form a set H_1, called a *half-plane*. Evidently H_1 is the graph of the inequality

$$y > 1 - x.$$

The points lying below L form a half-plane H_2; and H_2 is the graph of the inequality

$$y < 1 - x.$$

Consider now the double inequality

$$\tfrac{1}{2} < x < \tfrac{3}{2}.$$

The graph is an infinite vertical strip R_1, lying between the lines

$$x = \tfrac{1}{2} \quad \text{and} \quad x = \tfrac{3}{2}.$$

Similarly, the graph of

$$\tfrac{1}{2} < y < 1$$

is an infinite horizontal strip, as shown on the left below.

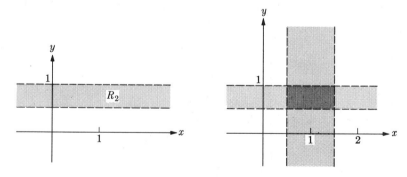

Consider next the condition

$$\tfrac{1}{2} < x < \tfrac{3}{2} \quad \text{or} \quad \tfrac{1}{2} < y < 1.$$

The graph of the condition using *or* is an infinite cross-shaped region. This region R' is the *union* of an infinite vertical strip R_1 and an infinite horizontal strip R_2; it contains all points of the plane that belong to R_1 or to R_2.

(In mathematics, when we say that one condition holds *or* another condition holds, we allow the possibility that both conditions hold. If we mean ". . . but not both," we have to say so.)

Similarly, the graphs of the conditions

$$y > x, \qquad y > -x$$

are two half-planes H_1 and H_2. They are respectively to the left of the line $y = x$ and to the right of the line $y = -x$, as shown in the figure on the left below. The graph of the condition

$$y > x \qquad \text{and} \qquad y > -x$$

is the intersection of these two half-planes. This is the interior R_1 of $\angle AOB$.

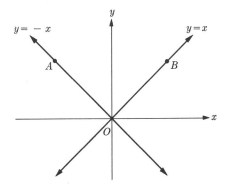 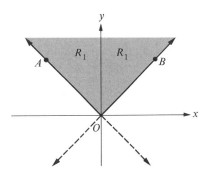

The graph of the condition

$$y > x \qquad \text{or} \qquad y > -x$$

is the *union* of the two half-planes.

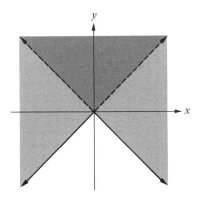 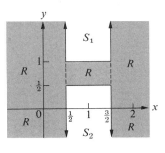

Let us now see what sort of graph we get when we combine two inequalities by "if . . . then." Consider the condition

$$\tfrac{1}{2} < x < \tfrac{3}{2} \;\Rightarrow\; \tfrac{1}{2} < y < 1.$$

This says that

$$if \quad \tfrac{1}{2} < x < \tfrac{3}{2}, \quad then \quad \tfrac{1}{2} < y < 1.$$

Let R be the graph. We assert that R looks like the drawing on the right above. That is, R contains all points that do *not* lie in either of the two vertical strips marked S_1 and S_2. The reason is as follows:

1) If (x, y) is a point of R, and $\tfrac{1}{2} < x < \tfrac{3}{2}$, then we must have $\tfrac{1}{2} < y < 1$. Therefore the part of R that lies between the lines $x = \tfrac{1}{2}$ and $x = \tfrac{3}{2}$ must be the interior of a rectangle, as indicated by the dashed lines in the figure.

2) On the other hand, if x is *not* between $\tfrac{1}{2}$ and $\tfrac{3}{2}$, then the condition for the graph imposes no restriction on y at all. Therefore R contains all points to the left of the line $x = \tfrac{1}{2}$ and all points to the right of the line $x = \tfrac{3}{2}$. R also contains these two vertical lines, for the same reason.

The reasoning in (2) may seem a little tricky, but may be clarified by an analogy from everyday life. The law in most places requires that *if* a person has seriously defective vision, *then* he must wear corrective glasses when driving a car. A person with normal vision automatically obeys this law; its restrictive clause does not apply to him. In the same way, the "law"

$$\tfrac{1}{2} < x < \tfrac{3}{2} \;\Rightarrow\; \tfrac{1}{2} < y < 1$$

imposes a restriction only on points (x, y) for which $\tfrac{1}{2} < x < \tfrac{3}{2}$; all other points automatically obey the "law," because its restrictive clause does not apply to them.

Thus the "law" holds under each of the following three conditions:

1) $\tfrac{1}{2} < x < \tfrac{3}{2}$ and $\tfrac{1}{2} < y < 1,$

2) $x \leqq \tfrac{1}{2},$

3) $x \geqq \tfrac{3}{2}.$

The graph of (1) is the rectangular region in the middle of the figure; the graph of (2) is the infinite region to the left of the line $x = \tfrac{1}{2}$; and the graph of (3) is the infinite region to the right of the line $x = \tfrac{3}{2}$.

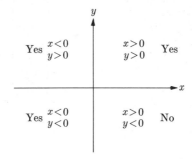

Similarly, the graph of the condition

$$x > 0 \quad \Rightarrow \quad y \geqq 0$$

contains all of the plane except for the fourth quadrant. It is only in the fourth quadrant that $x > 0$ holds and $y \geqq 0$ does not hold; and the possibility $x > 0$, $y < 0$ is the only possibility that is ruled out by the condition $x > 0 \Rightarrow y \geqq 0$.

In each of the following cases, the shaded region is the graph of the condition appearing below it.

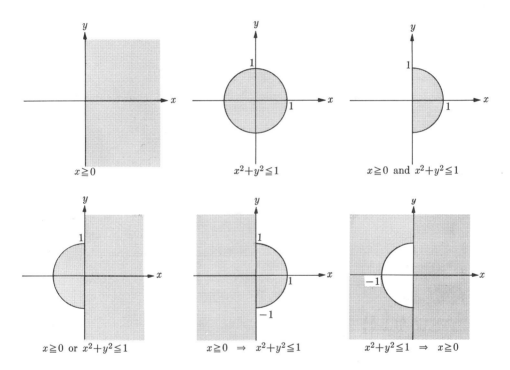

There is no need to use graph paper in the following problem set. Reasonably neat freehand sketches, with cross-hatching used to indicate regions, are sufficient.

PROBLEM SET 2.5

Sketch the graphs of the following conditions:

1. $\frac{3}{4} < x < \frac{5}{4}$ 2. $|x - 1| < \frac{1}{4}$

3. $\frac{19}{10} < y < \frac{21}{10}$ 4. $|y - 2| < \frac{1}{10}$

5. $|x - 1| < \frac{1}{4}$ and $|y - 2| < \frac{1}{10}$ 6. $|x - 1| < \frac{1}{4} \Rightarrow |y - 2| < \frac{1}{10}$

7. $|y - 2| < \frac{1}{10} \Rightarrow |x - 1| < \frac{1}{4}$ 8. $y \geqq |x|$

9. $|y| \leqq |x|$ 10. $x^2 + y^2 \leqq 1$ and $(x - 1)^2 + y^2 \leqq 1$

11. $x^2 + y^2 \leqq 1$ or $(x - 1)^2 + y^2 \leqq 1$

12. $x^2 + y^2 \leq 1 \Rightarrow (x - 1)^2 + y^2 \leq 1$

13. $(x - 1)^2 + y^2 \leq 1 \Rightarrow x^2 + y^2 \leq 1$

14. $x^2 + y^2 \leq 4$ and $(x + 1)^2 + y^2 \leq 1$

15. $x^2 + y^2 \leq 4$ or $(x + 1)^2 + y^2 \leq 1$

16. $x^2 + y^2 \leq 4 \Rightarrow x^2 + y^2 \leq 1$ 17. $x^2 + y^2 \leq 1 \Rightarrow x^2 + y^2 \leq 4$

18. $x^2 + y^2 \leq 1$ and $y \geq x$ 19. $x^2 + y^2 \leq 1$ and $x \leq |y|$

20. $x + y \leq 1$ 21. $x - y \leq 1$

22. $-x + y \leq 1$ 23. $-x - y \leq 1$

24. $|x| + |y| \leq 1$ 25. $|x| - |y| \leq 1$

26. $|x + y| \leq 1$ 27. $|x - y| \leq 1$

28. $x = \frac{1}{2} \Rightarrow y = 1$ 29. $|x - 2| < \frac{1}{10} \Rightarrow |y - 1| < \frac{1}{2}$

30. $|x - 3| < \frac{1}{4} \Rightarrow |y - 2| < \frac{1}{2}$

31. Suppose you know that (a) $P \Rightarrow Q$ and (b) P is false. What, if anything, can you infer about Q?

32. Suppose you know that (a) $P \Rightarrow Q$ and (b) Q is true. What, if anything, can you infer about P?

33. Suppose you know that (a) $P \Rightarrow Q$ and (b) Q is false. What, if anything, can you infer about P?

34. Suppose you know that $P \Rightarrow Q$. Which of the following are possible?

   a) P is true and Q is true.   b) P is true and Q is false.

   c) P is false and Q is true.   d) P is false and Q is false.

35. Suppose you know that $P \Leftrightarrow Q$. Which of the combinations (a), (b), (c), and (d) in Problem 34 are possible?

2.6 PARABOLAS

The *distance* from a point to a line is the length of the perpendicular from the point to the line. Given a point F and a line D not containing F, *the parabola with focus F and directrix D* is the set of all points of the plane that are equidistant from F and D.

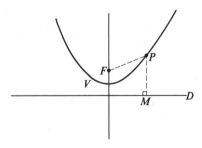

The parabola is the graph of the condition

$$FP = MP,$$

where M is the foot of the perpendicular from P to D. The perpendicular line to D

through F is called the *axis* of the parabola. The point where the axis crosses the parabola is called the *vertex*. (There is only one such point, because any such point is midway between the focus and the directrix.)

The first step in the study of parabolas is to get equations for them.

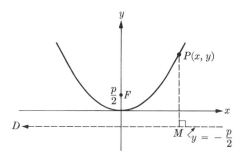

In setting up our axes, we take the vertex as the origin, and the x-axis parallel to the directrix, in such a way that D is below the x-axis and the focus is above it. The number p is the distance from the focus to the directrix. Now let $P \leftrightarrow (x, y)$ be a point of the parabola. Then

$$FP = \sqrt{x^2 + \left(y - \frac{p}{2}\right)^2},$$

and

$$MP = \sqrt{\left(y + \frac{p}{2}\right)^2}.$$

Therefore

$$FP = MP$$

$$\Leftrightarrow \quad \sqrt{x^2 + \left(y - \frac{p}{2}\right)^2} = \sqrt{\left(y + \frac{p}{2}\right)^2}$$

$$\Leftrightarrow \quad x^2 + \left(y - \frac{p}{2}\right)^2 = \left(y + \frac{p}{2}\right)^2$$

$$\Leftrightarrow \quad x^2 + y^2 - py + \frac{p^2}{4} = y^2 + py + \frac{p^2}{4}$$

$$\Leftrightarrow \quad x^2 = 2py$$

$$\Leftrightarrow \quad y = \frac{1}{2p} x^2.$$

This has the form

$$y = ax^2,$$

where $a = 1/2p$, and

$$p = \frac{1}{2a}.$$

Thus we have proved the following theorem:

Theorem 1. The graph of the equation

$$y = ax^2$$

is a parabola, with focus at $(0, 1/4a)$ and directrix

$$y = -1/4a.$$

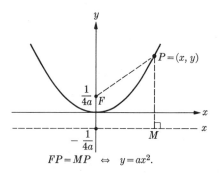

$$FP = MP \quad \Leftrightarrow \quad y = ax^2.$$

If a parabola is situated like this, relative to the axes, then the parabola is said to be in *standard* position. The use of standard position simplifies the equation considerably. For example, if F is the point $(2, -1)$ and D is the line $y = 3$, then the parabola is the graph of the equations

$$FP = MP$$
$$\Leftrightarrow \quad \sqrt{(x - 2)^2 + (y + 1)^2} = \sqrt{(y - 3)^2}$$
$$\Leftrightarrow \quad x^2 - 4x + 4 + y^2 + 2y + 1 = y^2 - 6y + 9$$
$$\Leftrightarrow \quad x^2 - 4x - 4 = -8y$$
$$\Leftrightarrow \quad y = -\tfrac{1}{8}x^2 + \tfrac{1}{2}x + \tfrac{1}{2}.$$

It is not hard to check, in general, that if the directrix is horizontal, then the equation always takes the form

$$y = Ax^2 + Bx + C, \qquad A \neq 0.$$

And if the directrix is vertical, we get

$$x = Ay^2 + By + C, \qquad A \neq 0.$$

If the directrix is neither horizontal nor vertical, then the equation involves, in general, terms in x^2, y^2, and xy, as well as linear terms and a constant. In this case it is hard to derive the equation when the focus and directrix are given; and it is even harder, when the equation is given, to see that the graph is a parabola. This case will be discussed in Chapter 8.

For a long time to come, however, we shall deal only with the simplest case, in which the directrix is horizontal.

Parabolas arise in a variety of contexts which appear at first to be unrelated. Following are a few.

1) If a right circular cone is cut by a plane parallel to an element of the cone, the resulting curve is a parabola. This was the viewpoint from which the Greeks studied parabolas; and it is for this reason that a parabola is one of the *conic sections*. There are other kinds of conic sections, obtained by slicing cones by planes in various positions.

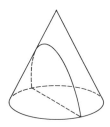 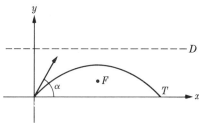

2) If a theoretical projectile is fired from the surface of the earth, in any direction other than straight upward, the path that it moves along is a portion of a parabola. In the figure on the right above, the *x*-axis lies along the surface of the earth, the *y*-axis is vertical, $\angle \alpha$ represents the angle at which the gun is aimed, and *T* is the point where the projectile hits the ground. We say, "a theoretical projectile," because to get this result you must assume both that the weight of the projectile is independent of its altitude and that the air makes no resistance. These assumptions are false, but they are good approximations to the truth, if the projectile is not going very fast or very high. For high-speed, long-range projectiles, both assumptions are quite unrealistic, and the situation is more complicated.

3) If you rotate a parabola around its axis, you get a surface which is called a *paraboloid of revolution*. The mirror in a reflecting telescope is a paraboloid of revolution, as is the reflector in an automobile headlight. The reason is that if a ray of light travels along a line parallel to the axis, and is reflected in the usual way, it always hits the focus. And conversely, if a ray of light starts at the focus, hits the surface and is reflected, it always continues along a line parallel to the axis. The first of these principles is used in telescopes, and the second in headlights.

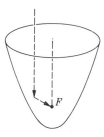

4) Suppose that you fire a "theoretical projectile" vertically upward. It moves up a vertical line, for a certain distance *h*, and then comes down again along the same line. Thus the *path* of motion is simply a segment. Suppose now that we label our

horizontal axis as the *t*-axis; we measure time starting at the moment of firing; and we plot, for each time *t*, the height of the projectile at time *t*.

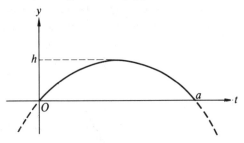

The resulting graph is a portion of a parabola. In the figure above, *a* is the time at which the projectile hits the ground. Note that the graph that we have been discussing is not all of the parabola: one minute before firing, the projectile was in the gun; it was not underground. And at *t* = *a* the projectile hits the ground, and the motion stops. Therefore, in the figure there is a solid arc, which indicates the portion of the parabola that is related to the physical problem; the irrelevant part of the curve is indicated by dashed arcs to the left and right.

This example indicates that geometric ideas come up in physics in unexpected ways; the uses of geometry are not limited to the study of figures in space.

PROBLEM SET 2.6

1. Take a full-size sheet of graph paper; draw the *y*-axis in the center; and draw the *x*-axis near the bottom of the paper. Then choose the largest uniform scale that you can, on the axes, in such a way that *x* ranges from -2 to 2 and *y* ranges from $-\frac{1}{4}$ to 4. Now sketch the graph of $y = x^2$. First plot the points corresponding to the following values of *x*:

$$x = 0, \quad x = 0.1, \quad x = 0.2, \quad \ldots, \quad x = 0.9, \quad x = 1,$$
$$x = 1.2, \quad x = 1.4, \quad \ldots, \quad x = 1.8, \quad x = 2.$$

Then draw the curve, freehand, as smoothly as you can. If this is done carefully, it will really look as if *FP* = *MP* at every point of the curve.

One of the reasons for doing this is that it will give you an accurate idea of what a parabola really looks like.

2. Show that

$$0 < x_1 < x_2 \Rightarrow x_1^2 < x_2^2.$$

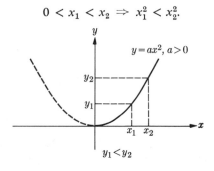

This means that the right-hand half of a parabola in standard position rises as we go from left to right along the curve.

3. Show that
$$x_1 < x_2 < 0 \;\Rightarrow\; x_1^2 > x_2^2.$$

   What does this tell us about parabolas in standard position?

4. Find the focus and the directrix of the graph of the equation $y = x^2$.

5. Same problem, for the equation $y = 3x^2$.

6. Same problem, for the equation $y = \frac{1}{2}x^2$.

7. Show that the graph of the equation $y = x^2 + 1$ is a parabola. To show this, you must find its focus F and directrix D. You can then check by deriving the equation of the parabola with focus F and directrix D.

8. Show that the graph of the equation $y = (x - 2)^2$ is a parabola.

9. Same problem, for the equation $y = (x - 2)^2 + 1$.

10. Show that the graph of $y = x^2 - 2x$ is a parabola.

11. Show that the graph of the equation $y = (x + 1)^2$ is a parabola. (Find the focus F and directrix D.)

12. Same problem, for $y = (x + 1)^2 - 1$.

13. Same problem, for $y = (2x + 1)^2$.

2.7 TANGENTS

In geometry, tangent lines to circles are defined as follows.

Definition. A *tangent* to a circle is a line (in the same plane) which intersects the circle in one and only one point. This point is called the *point of contact.*

It is then shown that a line is tangent to the circle if and only if the line is perpendicular to the radius drawn to the point of contact. (In fact, the latter condition is probably the one that you used to find the slopes of tangent lines to circles, in Problem Set 2.4.)

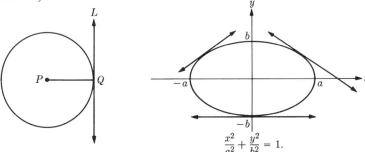

$$\frac{x^2}{a^2} + \frac{y^2}{b^2} = 1.$$

Tangency can be defined in the same way for an ellipse. Ellipses will be studied in Chapter 8. Meanwhile we observe that an ellipse is an oval curve, of the sort shown in the right-hand figure above, and the tangents to it are the lines that intersect it in one and only one point.

But for some curves, tangents cannot be described by the definition that we use for circles. Consider, for example, a parabola, as shown in the figure below. The tangent to the parabola, at the point (x_1, y_1), intersects the curve only at (x_1, y_1). But the vertical line through (x_1, y_1) has the same property; and the vertical line is not a tangent.

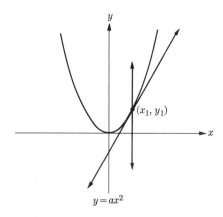

We may try to get around this trouble by providing that the tangent line must only touch the curve, without crossing it. But for many curves, this won't work either. The graph of $y = x^3$ is shown below. The tangent to this curve at the origin turns out to be the x-axis; and the x-axis crosses the curve, at the point of tangency. In other cases a tangent line may cross a curve in many points.

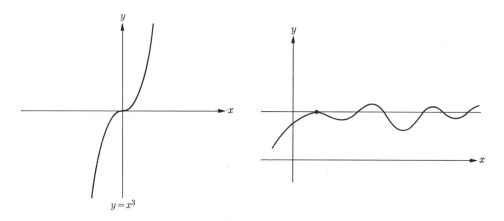

The geometric idea of tangency is obvious in all these cases. But the above examples indicate that the mathematical definition that works for circles does not work in general. To find the tangents to other curves, we need a better definition.

Consider first the graph of $y = x^2$, and the fixed point $(1, 1)$ at which we want to find the slope of the tangent. For every other point (x, x^2) of the curve, let L_x be the secant line through $(1, 1)$ and (x, x^2). Then the slope of L_x is

$$m_x = \frac{x^2 - 1}{x - 1} \qquad (x \neq 1).$$

Here the restriction $x \neq 1$ reflects the geometric fact that it takes two different points to determine a line. It also refers to the algebraic fact that fractions with denominator 0 have no meaning.

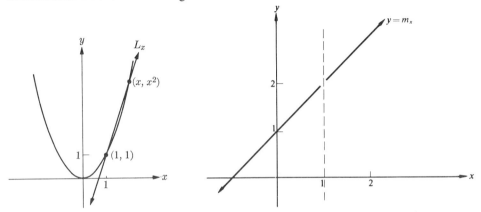

We shall now draw the graph of $y = m_x$ $(x \neq 1)$. We have

$$y = m_x = x + 1 \qquad (x \neq 1).$$

The graph is a line from which one point has been deleted. For $x = 1$, there is no such thing as "the secant line through $(1, 1)$ and $(1, 1^2)$"; and for $x = 1$, there is no such thing as the "fraction" $m_1 = 0/0$. But this causes no trouble, because it is easy to see that m_x is very close to 2 when x is very close to 1. We express this by writing

$$\lim_{x \to 1} m_x = 2.$$

This is read: "The limit of m_x, as x approaches 1, is equal to 2." Later we shall give a general definition of the idea of a limit. But in the present case, the meaning of the limit is clear, and so we use it in the definition of the tangent to the parabola.

Definition. The tangent to the graph of

$$y = ax^2 + bx + c,$$

at a point (x_0, y_0) of the graph, is the line through (x_0, y_0) with slope

$$S_{x_0} = \lim_{x \to x_0} m_x,$$

where m_x is the slope of the secant line passing through the points (x_0, y_0) and $(x, ax^2 + bx + c)$ $(x \neq x_0)$.

Even in the general case, the slope is easy to calculate on the basis of this definition. We have

$$m_x = \frac{ax^2 + bx + c - ax_0^2 - bx_0 - c}{x - x_0} \qquad (x \neq x_0)$$

$$= \frac{a(x^2 - x_0^2) + b(x - x_0)}{x - x_0} \qquad (x \neq x_0)$$

$$= a(x + x_0) + b \qquad (x \neq x_0)$$

$$= ax + (ax_0 + b). \qquad (x \neq x_0)$$

The graph of $y = m_x$ is a line with one point missing. The line from which the point is missing is shown on the left below. The graph of $y = m_x$ is on the right.

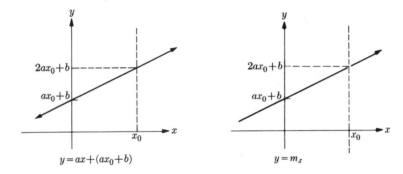

$$y = ax + (ax_0 + b) \qquad\qquad y = m_x$$

Here again, the limit of m_x is simply the y-coordinate of the point that is missing from the graph. Thus we have:

Theorem 1. Let (x, y) be a point of the graph of

$$y = ax^2 + bx + c.$$

Then the slope of the tangent to the graph, at (x, y), is

$$S_x = 2ax + b.$$

For some curves, there is no tangent. Consider, for example, the graph of $y = |x|$, at the point $(0, 0)$. For each $x \neq 0$,

$$m_x = \frac{|x| - |0|}{x - 0} = \frac{|x|}{x}.$$

Thus:

$$m_x = 1 \quad \text{for} \quad x > 0, \qquad m_x = -1 \quad \text{for} \quad x < 0.$$

(Remember the definition of $|x|$.) Therefore the graph of $y = m_x$ looks like the

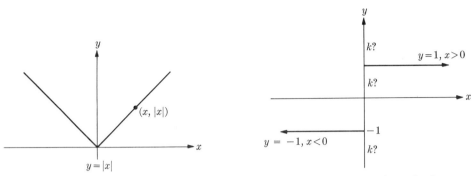

$y = |x|$

drawing on the right above. For this graph there is no *one* number that y is close to, whenever x is close to 0. Therefore, there is no such thing as

$$(?) \lim_{x \to 0} m_x (?);$$

for every number k, the statement

$$(?) \lim_{x \to 0} m_x = k (?)$$

is false.

Geometrically it is obvious that the origin is the only point of the curve at which things go wrong; at every other point $(x, |x|)$, the curve has a tangent; the slope of the tangent is 1 for $x > 0$ and -1 for $x < 0$.

PROBLEM SET 2.7

1. You already have a carefully drawn graph of the equation $y = x^2$. At each point (x, y) of the graph, the slope of the tangent ought to be $2x$. Check this graphically by drawing lines of the proper slope at the points where $x = 0.2, 0.4, 0.6, 0.8$, and 1.

2. Given $y = x^2 - 4x + 4$. Find the slopes of the tangents at the points where $x = -2$, $x = 0$, and $x = 2$, and sketch, showing all three of these tangents.

3. Same problem, for $y = x^2 + x + 1$, using the points where $x = 0$, $x = \frac{1}{2}$, and $x = 1$.

4. By completing the square, show that

$$y = ax^2 + bx + c$$

can be expressed in the form

$$y = a(x - A)^2 + B.$$

For $a > 0$, this means that the point where $x = A$ is the lowest point on the curve. Find the slope of the tangent at this point.

5. a) Given the graph of

$$y = ax^2$$

and a point (x_0, y_0) of the curve. Show that the tangent at (x_0, y_0) is the only non-vertical line which passes through (x_0, y_0) and has no other point in common with the parabola. That is, show that, if the graph of

$$(y - y_0) = m(x - x_0)$$

intersects the parabola only at (x_0, y_0), then

$$m = 2ax_0.$$

b) Prove the corresponding theorem for the graph of

$$y = ax^2 + bx + c.$$

6. a) Get a plausible answer for the slope of the tangent to the graph of $y = x^3$, at the point $(1, 1)$. Sketch the graph of $y = m_x$, explain what sort of graph it is, and explain as well as you can why your value for the slope is plausible.
   b) Do the same for $y = x^3$, at an arbitrary point (x_0, x_0^3).

7. a) Show that, if $m < 0$, then the line through the origin with slope m meets the graph of $y = x^3$ at precisely one point.
   b) Show that, if $m > 0$, then the line through the origin with slope m meets the graph of $y = x^3$ at precisely three points.

8. Sketch the graph of

$$y = x\,|x|,$$

and describe this curve in terms of types of curve that we already know about. At which points does this graph have a tangent? What is S_2? What is S_{-2}? Give, if possible, a general formula for S_x. Is there such a thing as S_0?

9. Consider the graph of

$$y = x^3 - 4x.$$

Where does this cross the x-axis? At which points is the tangent horizontal? What is the slope of the tangent at $(0, 0)$? For what values of x is $y > 0$? For what values of x is $y < 0$? Use this information to draw a reasonable sketch of the graph, plotting only *five* points.

10. Carry out the steps of Problem 9 for the equation

$$y = 2x^3 - 6x.$$

11. Show that every parabola has the reflecting property. In the figure, T is the tangent at P, and you need to show that $\alpha = \beta$. The key to the proof is that the quadrilateral $FPRQ$ is a rhombus. (That is, all four sides have the same length.)

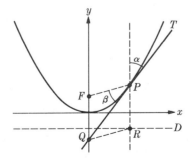

2.8 A SHORTHAND FOR SUMS

An *arithmetic series* is a sum of the form

$$S_n = a + (a + d) + (a + 2d) + \cdots + (a + [n - 1]d).$$

A geometric series is a sum of the form

$$T_n = a + ar + ar^2 + ar^3 + \cdots + ar^{n-1}.$$

There is a shorthand for sums, which makes them easier to handle. Given a sum

$$S_n = a_1 + a_2 + \cdots + a_n,$$

we write

$$S_n = \sum_{i=1}^{n} a_i.$$

(This is pronounced: "The summation from 1 to n of a_i.") That is, when we write $\sum_{i=1}^{n}$ and follow it with an expression involving i, this means that we are to substitute all (integral) values of i, from 1 to n, and add the results.

1) The geometric series

$$S_n = a + ar + ar^2 + \cdots + ar^{n-1}$$

can be written as

$$\sum_{i=1}^{n} ar^{i-1}.$$

The shorthand can be checked by substituting the values of i from 1 to n.

$i =$	1	2	3	4	$\cdots$	n
$ar^{i-1} =$	ar^0	ar^{2-1}	ar^{3-1}	ar^{4-1}	$\cdots$	ar^{n-1}
$=$	a	ar	ar^2	ar^3	$\cdots$	ar^{n-1}

When we add these, we get the geometric series.

2) Consider the sum

$$R_n = 1^2 + 2^2 + 3^2 + \cdots + n^2.$$

In the short form,

$$R_n = \sum_{i=1}^{n} i^2.$$

3) An arithmetic series can be written in the form

$$S_n = \sum_{i=1}^{n} [a + (i - 1)d].$$

This can be checked by means of a table of the sort that we gave above for the case of the geometric series.

In each case, the formula after $\sum$ gives the ith term; $i = 1$ gives the first term, $i = 2$ gives the second term, and so on. This will always be true so long as we are taking the sum from $i = 1$ to n. However, we also write such sums as

$$\sum_{i=2}^{5} i^3.$$

Here we take all values of i from $i = 2$ to $i = 5$ inclusive and add the results. Therefore

$$\sum_{i=2}^{5} i^3 = 2^3 + 3^3 + 4^3 + 5^3 = 8 + 27 + 64 + 125 = 224.$$

In general, for $m \leqq n$,

$$\sum_{i=m}^{n} a_i = a_m + a_{m+1} + \cdots + a_n.$$

Thus

$$\sum_{i=2}^{4} (a_i^3 + 1) = (a_2^3 + 1) + (a_3^3 + 1) + (a_4^3 + 1)$$

$$= a_2^3 + a_3^3 + a_4^3 + 3 = \sum_{i=2}^{4} a_i^3 + 3.$$

Note that $\sum$ applies only to the expression immediately after it; in the last line, we are told to add the numbers a_i^3 (from $i = 2$ to $i = 4$) and add 3 to the result. The parentheses in the formula $\sum_{i=2}^{4} (a_i^3 + 1)$ indicate that 1 is part of every term of the sum.

PROBLEM SET 2.8

Find each of the following sums numerically:

1. $\displaystyle\sum_{i=1}^{3} i^2$ 2. $\displaystyle\sum_{i=1}^{3} (i - 1)^2$ 3. $\displaystyle\sum_{i=2}^{5} (i^2 - 1)$ 4. $\displaystyle\sum_{i=1}^{4} i^2$ 5. $\displaystyle\sum_{i=2}^{3} (i^3 - 1)$ 6. $\displaystyle\sum_{i=1}^{3} \frac{1}{i}$

Each of the sums below is of the form $\sum_{i=m}^{n} a_i$. Write each of them in the long form $a_m + a_{m+1} + \cdots + a_n$.

7. $\displaystyle\sum_{i=3}^{5} (b_i + 2c_i)$ 8. $\displaystyle\sum_{i=2}^{3} (3b_i^2 + d)$ 9. $\displaystyle\sum_{i=m}^{n} i^7$

Convert each of the indicated sums to the short form:

10. $1^2 + 2^2 + 3^2 + \cdots + n^2 + (n + 1)^2$ 11. $3^2 + 4^2 + \cdots + k^2$

12. $k^2 + (k + 1)^2 + (k + 2)^2 + \cdots + (n - 1)^2$

13. $\frac{1}{2} + \frac{1}{3} + \cdots + \dfrac{1}{k - 1} + \dfrac{1}{k}$

14. Is is true that

$$\sum_{i=1}^{n} (a_i + b_i) = \sum_{i=1}^{n} a_i + \sum_{i=1}^{n} b_i?$$

Why or why not?

15. Is it true that

$$\sum_{i=1}^{n} k a_i = k \sum_{i=1}^{n} a_i?$$

Why or why not?

16. Is it true that

$$\sum_{i=1}^{n} \left(\frac{hi}{n}\right)^2 \cdot \frac{h}{n} = \frac{h^3}{n^3} \sum_{i=1}^{n} i^2?$$

Why or why not?

17. For $0 \le k \le b$, $\binom{n}{k}$ is the number of subsets with exactly k elements, in a given set with n elements. For example, $\binom{52}{13}$ is the number of possible 13-card bridge hands; $\binom{52}{5}$ is the number of possible 5-card draw poker hands. Show that $\binom{5}{4} = \binom{5}{1}$.

*18. Show that $\binom{52}{13} = \binom{52}{39}$.

*19. Show that

$$\sum_{k=0}^{n} \binom{n}{k} = 2^n.$$

2.9 THE INDUCTION PRINCIPLE AND THE WELL-ORDERING PRINCIPLE

Consider the following game. We have three spindles, of the sort used as targets in quoits. On the first spindle is a stack of wooden disks, diminishing in size from bottom to top. (See the figure.) The disks are numbered $1, 2, 3, \ldots, n$, from top to bottom; in the figure, $n = 5$.

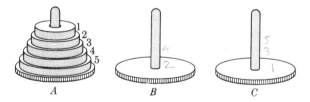

A B C

A *legal move* consists in taking the topmost disk from one spindle and placing it on one of the other spindles, providing that we must not, at any stage, place a disk above a smaller disk.

At the start, all the disks are on spindle A. The object of the game is to get all the disks onto spindle B, by a series of legal moves.

For example, we might begin by taking disk 1 off spindle A and putting it on spindle B. There would then be three possibilities for the second move: (1) Put disk 1 back on spindle A, (2) put disk 1 on spindle C, and (3) put disk 2 on spindle C. It would not be legal to put disk 2 on spindle B, because disk 2 would then be above disk 1, which is smaller.

We shall see that the game can always be completed, no matter how large the positive integer n may be. For each positive integer n, Let P_n be the proposition that the game can be completed, starting with n disks. What we need to show is that *all* of the propositions P_n are true.

Lemma 1. P_1 is true.

(A lemma is a sort of subtheorem, used as a step in the proof of a harder theorem.)

Proof of Lemma 1.   Move the one and only disk from spindle A to spindle B.   Then the game is over.

Lemma 2. P_2 is true.

Proof of Lemma 2.   (1) Move disk 1 to spindle C.   (2) Move disk 2 to spindle B. (3) Move disk 1 to spindle B.   Then the game is over.

Lemma 3. P_3 is true.

Proof of Lemma 3.   By Lemma 2, disks 1 and 2 can be moved to spindle C.   (Lemma 2 really means that any two disks at the top of a stack can be moved to any other spindle.)   Do this.   Then move disk 3 to spindle B.   By Lemma 2, disks 1 and 2 can then be moved to spindle B, whereupon the game is over.

A pattern is now appearing, suggesting the following lemma. This lemma states that *if* the game with n disks can be completed, *then* the game with $n + 1$ disks can also be completed.

Lemma 4.   For each n, $P_n \Rightarrow P_{n+1}$.

Proof of Lemma 4.   We are given $n + 1$ disks on spindle A, and we are given by hypothesis that P_n is true.   Therefore the stack consisting of disks $1, 2, \ldots, n$ can be moved to spindle C by legal moves.   Do this.   (Disk $n + 1$ causes no trouble; it can be regarded as the base of the spindle on which it lies, because it is larger than any of the disks being moved.)   Then move disk $n + 1$ from spindle A to spindle B. By P_n, we know that disks $1, 2, \ldots, n$ can be moved from spindle C to spindle B. Then the game is over.

Lemma 4 gives us an infinite chain of implications:

$$P_1 \Rightarrow P_2 \Rightarrow P_3 \Rightarrow P_4 \Rightarrow \cdots \Rightarrow P_n \Rightarrow P_{n-1} \Rightarrow \cdots$$

And Lemma 1 tells us that the *first* statement in the chain is true.   Therefore all of the statements $P_1, P_2, \ldots$ are true.   This idea is conveyed mathematically as follows:

> **The Induction Principle.**   Let $P_1, P_2, \ldots$ be a sequence of propositions (one for every positive integer).   If
>
> a) P_1 is true, and
> b) $P_n \Rightarrow P_{n+1}$ for every n,
>
> then all of the propositions $P_1, P_2, \ldots$ are true.

The problem of the disks is probably the clearest illustration of what the induction principle means.   The principle is used continually, in all branches of mathematics. In this section, we shall use it to get short formulas for certain sums.

Theorem 1. For every n,

$$\sum_{i=1}^{n} i = \frac{n}{2}(n + 1).$$

Proof. For each n, let P_n be the proposition that

$$\sum_{i=1}^{n} i = \frac{n}{2}(n + 1).$$

a) P_1 is true, because

$$\sum_{i=1}^{1} i = 1 = \tfrac{1}{2}(1 + 1).$$

b) $P_n \Rightarrow P_{n+1}$ for every n, because

$$\sum_{i=1}^{n} i = \frac{n}{2}(n + 1)$$

$$\Rightarrow \sum_{i=1}^{n} i + (n + 1) = \frac{n}{2}(n + 1) + (n + 1)$$

$$\Rightarrow \sum_{i=1}^{n+1} i = \left(\frac{n}{2} + 1\right)(n + 1)$$

$$\Rightarrow \sum_{i=1}^{n+1} i = \frac{n + 1}{2}(n + 2).$$

In this chain of implications, the first equation is P_n and the last is P_{n+1}. Therefore $P_n \Rightarrow P_{n+1}$. By the induction principle, P_n is true for every n, which was to be proved.
In fact, there is a simpler way of getting this result. If

$$S_n = 1 + 2 + 3 + \cdots + (n - 1) + n,$$

then

$$S_n = n + (n - 1) + (n - 2) + \cdots + 2 + 1;$$

and adding terms in pairs, we get

$$2S_n = (1 + n) + (1 + n) + \cdots + (1 + n) + (1 + n),$$

to n terms. Therefore

$$2S_n = n(n + 1) \quad \text{and} \quad S_n = \frac{n}{2}(n + 1),$$

as before. This device is neat but very special. Consider now the problem of calculating

$$S_n = \sum_{i=1}^{n} i^2 = 1^2 + 2^2 + 3^2 + \cdots + n^2.$$

We have just found that the sum of the first n positive integers is a polynomial in n, of degree 2. This suggests that S_n is a polynomial of degree 3. That is, we conjecture that

$$S_n = An^3 + Bn^2 + Cn + D,$$

for some numbers A, B, C, D. The problem is to find A, B, C, and D, and prove by induction that they work. Let P_n be the proposition that

$$P_n: \sum_{i=1}^{n} i^2 = An^3 + Bn^2 + Cn + D.$$

Then P_{n+1} asserts that

$$P_{n+1}: \sum_{i=1}^{n} i^2 + (n+1)^2 = A(n+1)^3 + B(n+1)^2 + C(n+1) + D.$$

We want $P_n \Rightarrow P_{n+1}$, to make the induction proof work. This means that

$$An^3 + Bn^2 + Cn + D + (n+1)^2 = A(n+1)^3 + B(n+1)^2 + C(n+1) + D.$$

If this equation holds, then $P_n \Rightarrow P_{n+1}$. (Check the algebra.) Collecting coefficients, we get the equivalent equation

$$An^3 + (B+1)n^2 + (C+2)n + D + 1$$
$$= An^3 + (3A+B)n^2 + (3A+2B+C)n + A + B + C + D,$$

or

$$(1 - 3A)n^2 + (2 - 3A - 2B)n + 1 - A - B - C = 0.$$

This holds if

$$A = \tfrac{1}{3}, \qquad B = \tfrac{1}{2}(2 - 3A) = \tfrac{1}{2}, \qquad C = 1 - A - B = \tfrac{1}{6}.$$

This gives

$$P_n: \sum_{i=1}^{n} i^2 = \tfrac{1}{3}n^3 + \tfrac{1}{2}n^2 + \tfrac{1}{6}n + D.$$

Thus, for any D, $P_n \Rightarrow P_{n+1}$. For $D = 0$, P_1 is true. We take $D = 0$; and we know by the induction principle that

$$\sum_{i=1}^{n} i^2 = \tfrac{1}{3}n^3 + \tfrac{1}{2}n^2 + \tfrac{1}{6}n,$$

for every n. Taking a common denominator on the right and factoring, we get:

Theorem 2. For every n,

$$\sum_{i=1}^{n} i^2 = \frac{n}{6}(n+1)(2n+1).$$

For some purposes, the following idea is easier to use than the Induction Principle.

The Well-Ordering Principle. Every nonempty set of positive integers has a least element.

(See, for example, Problems 10 and 12 below.) The Well-Ordering Principle and the Induction Principle are equivalent. (See Problems 14 and 15 below.)

PROBLEM SET 2.9

1. Prove by any method that for every n, the sum of the first n odd numbers is n^2. That is,

$$\sum_{i=1}^{n} (2i - 1) = n^2.$$

This can be shown by induction, but there are at least two other ways.

2. Prove by induction that

$$1 + r + r^2 + \cdots + r^n = \frac{r^{n+1} - 1}{r - 1} \qquad (r \neq 1).$$

3. Prove by induction that

$$\sum_{i=1}^{n} i^3 = \left(\frac{n(n + 1)}{2}\right)^2.$$

4. Find by any method a formula for

$$\sum_{i=1}^{n} (3i - 1).$$

5. Find by any method a formula for

$$\sum_{i=1}^{n} (4i - 2).$$

6. Find a formula for

$$\sum_{i=1}^{n} (i^2 + i + 1).$$

7. Find a formula for

$$\sum_{i=1}^{n} (i^2 - i).$$

8. Assume that if A_1, A_2, A_3 are points, then

$$A_1A_2 + A_2A_3 \geqq A_1A_3.$$

Prove that for every $n \geqq 3$ we have

$$A_1A_2 + A_2A_3 + \cdots + A_{n-1}A_n \geqq A_1A_n.$$

This is known as the *polygonal inequality*.

9. a) Let p_n be the number of moves required to complete the game with n disks. Show that for every n,

$$p_{n+1} = 2p_n + 1.$$

b) Let p_n be as in (a). Show that for each n,

$$p_n = 2^n - 1.$$

(Since $2^{10} = 1024$, this means that the game with 20 disks requires over a million moves. Thus, if you want to verify that P_{20} is true, the easiest way to do it is to show by induction that P_n is true for every n, and then set $n = 20$.)

*10. Throughout this problem, the numbers under discussion are positive integers. If $a = bc$ for some c, then b is called a *factor* of a (or a *divisor* of a). If $p > 1$, and the only positive factors of p are p and 1, then p is a *prime*. Obviously every prime has a prime factor, namely, itself. Prove that every number greater than 1 has a prime factor. [Beginning of the proof: "Let K be the set of all numbers which are greater than 1 and have no prime factors. We need to show that K is empty. If K is not empty, then . . ."]

*11. Following is the beginning of Euclid's proof that there are infinitely many primes. Suppose that there are only a finite number of primes, say

$$p_1, p_2, p_3, \cdots, p_n.$$

Consider the number

$$N = p_1 p_2 p_3 \cdots p_n + 1.$$

Complete Euclid's proof, by showing that this situation is impossible.

*12. Show that every rational number can be expressed as a fraction in lowest terms. [*Hint:* Try the Well-Ordering Principle.]

13. In the song "The Twelve Days of Christmas," gifts are sent on successive days according to the following scheme:

   First day: a partridge in a pear tree.
   Second day: another partridge, and two turtledoves.
   For each i, let G_i be the number of gifts sent on the ith day. Then

$$G_i = G_{i-1} + i.$$

(Which we have just observed for $i = 2$.)

   Let T_n be the total number of gifts sent on the first n days of Christmas. Get a formula for T_n, in the form

$$T_n = \frac{?(? + ?)(? + ?)}{?}.$$

As a check, the final value is $T_{12} = 364$. (I am indebted, for this problem, to Professor Thomas F. Banchoff.)

*14. Show that, if the Well-Ordering Principle is taken as a postulate, then the Induction Principle can be proved as a theorem. [Start of the proof: Suppose that not all of the propositions P_n are true, and let

$$K = \{n \mid P_n \text{ is false.}\}.$$

Then $K \neq \{\ \}$. Therefore . . .]

*15. Show conversely that, if the Induction Principle is taken as a postulate, then the Well-Ordering Principle can be proved as a theorem. [Start of the proof: For each n, let P_n be the proposition that *none* of the integers $1, 2, \ldots, n$ belongs to K . . . .]

The diagram below is related to one of the problems in this section.

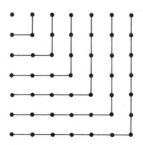

2.10 SOLUTION OF THE AREA PROBLEM FOR PARABOLAS

If a line intersects a parabola in two points, then it cuts off a region called a *parabolic sector*. In the left-hand figure below, the sector is the region lying above the parabola and below the line. In the third century B.C., Archimedes discovered a method for finding the area of a parabolic sector. In this section we shall give an easier solution of the problem.

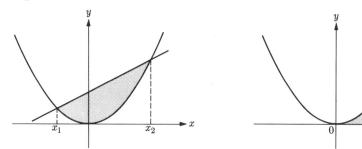

The problem will be solved if we can find the area of a "curvilinear triangle" of the type shown on the right above. If we can do this, then we can find the area of the trapezoid in the other figure, and subtract the areas of the two curvilinear triangles. The result will be the area of the sector.

We shall attack the area problem, for the graph of $y = x^2$, by approximating the region with rectangles, like this:

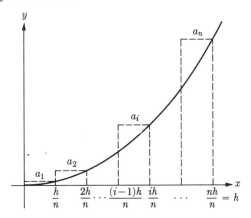

We cut the closed interval $[0, h]$ into n little intervals of equal length, using the division points

$$0, \frac{h}{n}, \frac{2h}{n}, \ldots, \frac{(i-1)h}{n}, \frac{ih}{n}, \ldots, \frac{(n-1)h}{n}, h.$$

This gives a sequence of closed intervals

$$\left[0, \frac{h}{n}\right], \left[\frac{h}{n}, \frac{2h}{n}\right], \ldots, \left[\frac{(i-1)h}{n}, \frac{ih}{n}\right], \ldots, \left[\frac{(n-1)h}{n}, h\right].$$

With each of these intervals as base, we construct a rectangle, using as altitude the height of the parabola at the right-hand endpoint. The right-hand endpoint of the ith interval is ih/n. Therefore the altitude of the ith rectangle is $(ih/n)^2$. Therefore the area of the ith rectangular region is

$$a_i = \left(\frac{ih}{n}\right)^2 \frac{h}{n} = \frac{h^3 i^2}{n^3}.$$

Let R_n be the union of all these rectangular regions. Then the area of R_n is

$$A_n = \sum_{i=1}^{n} a_i = \sum_{i=1}^{n} \frac{h^3 i^2}{n^3} = \frac{h^3}{n^3} \sum_{i=1}^{n} i^2.$$

We want to find out what limit A_n approaches as n becomes very large. If we find this limit, then our problem is solved, because the limit is the area of the region R that we started with.

We found, in Theorem 2 of Section 2.9, that

$$\sum_{i=1}^{n} i^2 = \frac{n}{6}(n + 1)(2n + 1).$$

Therefore

$$A_n = \frac{h^3}{n^3} \cdot \frac{n}{6}(n + 1)(2n + 1) = \frac{h^3}{3}\left(1 + \frac{1}{n}\right)\left(1 + \frac{1}{2n}\right).$$

As n becomes large without limit, it is easy to see that

$$\frac{1}{n} \to 0, \quad 1 + \frac{1}{n} \to 1, \quad \frac{1}{2n} \to 0, \quad \text{and} \quad 1 + \frac{1}{2n} \to 1,$$

so that

$$\left(1 + \frac{1}{n}\right)\left(1 + \frac{1}{2n}\right) \to 1,$$

and

$$A_n = \frac{h^3}{3}\left(1 + \frac{1}{n}\right)\left(1 + \frac{1}{2n}\right) \to \frac{h^3}{3}.$$

Therefore the area under the parabola, from 0 to h, is

$$A = \frac{h^3}{3}.$$

It would have been equally natural to approximate the area from the inside. We shall see that this procedure leads to the same answer as before. Here we have cut up the interval $[0, h]$ into the same little intervals as before; but on each little interval we have set up a rectangle whose altitude is the height of the parabola at the

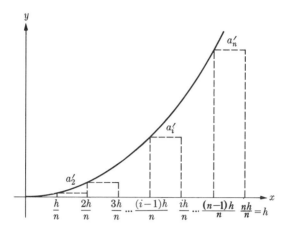

left-hand endpoint. Therefore, on $[0, h/n]$ our "rectangle" is merely the base interval, with area 0; and thereafter the area of the ith rectangle is

$$a'_i = \left[\frac{(i-1)h}{n}\right]^2 \cdot \frac{h}{n} = \frac{h^3}{n^3}(i-1)^2.$$

Let R'_n be the union of these rectangular regions. Then the area of R'_n is

$$A'_n = \sum_{i=1}^{n} a'_i = \sum_{i=1}^{n} \frac{h^3}{n^3}(i-1)^2 = \frac{h^3}{n^3}\sum_{i=1}^{n}(i-1)^2 = \frac{h^3}{n^3}\sum_{i=1}^{n-1} i^2.$$

To see why the last equation holds, observe that each of the indicated sums is the sum of the squares of the integers from 1 to $n-1$. Therefore

$$A'_n = \frac{h^3}{n^3}\left[\frac{n}{6}(n+1)(2n+1) - n^2\right] = A_n - \frac{h^3}{n}.$$

As n increases, $A_n \rightarrow h^3/3$ and $h^3/n \rightarrow 0$. Therefore $A'_n \rightarrow h^3/3$, and we get the same limit as before. To sum up:

Theorem 1. Let

$$R = \{(x, y) \mid 0 \leq x \leq h \quad \text{and} \quad 0 \leq y \leq x^2\}.$$

Then the area of R is $h^3/3$.

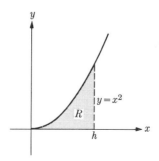

It is easy to extend this result to the case in which the parabola is the graph of $y = kx^2$, $k > 0$.

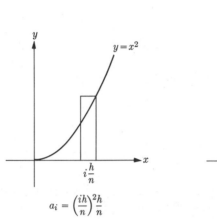

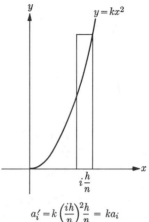

$$a_i = \left(\frac{ih}{n}\right)^2 \frac{h}{n}$$

$$a_i' = k\left(\frac{ih}{n}\right)^2 \frac{h}{n} = ka_i$$

When we multiply y by k, this multiplies the area of each approximating rectangle by k. Thus, if $A_n = \sum_{i=1}^{n} a_i$, as before, and

$$B_n = \sum_{i=1}^{n} a_i',$$

we have

$$B_n = \sum_{i=1}^{n} ka_i = k \sum_{i=1}^{n} a_i = kA_n.$$

Since $A_n \to h^3/3$, we have

$$B_n \to \frac{kh^3}{3}.$$

Therefore we have the following theorem:

Theorem 2. Let

$$R = \{(x, y) \mid 0 \leq x \leq h \quad \text{and} \quad 0 \leq y \leq kx^2\},$$

with $k > 0$. Then the area of R is $kh^3/3$.

In general, for $a < b$ let

$$A_a^b kx^2$$

be the area of the region under the graph of $y = kx^2$, from $x = a$ to $x = b$. Then we have the following:

Theorem 3.

$$A_a^b kx^2 = \frac{k}{3}(b^3 - a^3).$$

(Proof? There are three cases to consider: $a < b \leq 0$, $a < 0 \leq b$, $0 \leq a < b$.)

PROBLEM SET 2.10

Find the area under the graph of $y = 5x^2$, between the following limits.

1. From 0 to 4   2. From 0 to 2   3. From -2 to 0
4. From 2 to 4   5. From -2 to 2

Find the area under the graph of $y = 2x^2 + 1$, between the following limits.

6. From 0 to 4   7. From -1 to 0   8. From -1 to 3
9. Find the area of the parabolic sector between the graphs of $y = 2x^2$ and $y = x + 1$.
10. Same problem, for $y = x^2$ and $y = x$.
11. Find the area of the sector between the graphs of $y = x^2 - 1$ and $y = -x^2 + 1$.
12. Same problem, for $y = x^2$ and $y = 2x^2 - 1$.
13. Solve, for the general case, the problem of Archimedes, stated at the beginning of this section.
14. a) For each n, let

$$A_n = 1 + \frac{1}{\sqrt{n}}.$$

Obviously $A_n > 1$ for every n. Under what condition for n can you be sure that

$$A_n < \tfrac{1}{10}?$$

b) Under what condition for n can you be sure that

$$A_n - 1 < \frac{1}{10,000,000}?$$

c) Let ϵ be any positive number. Under what condition for n can you be sure that $A_n - 1 < \epsilon$?

15. a) For each n, let

$$B_n = \frac{2n - 2}{n^3 - 1}.$$

Under what condition for n can you be sure that $B_n < \tfrac{1}{10}$?
b) Under what condition for n can you be sure that

$$B_n < \frac{1}{10^{10}}?$$

c) Given any positive number ϵ, under what condition for n can you be sure that $B_n < \epsilon$?

16. For each n, let

$$C_n = \left(1 + \frac{1}{n}\right)\left(4 + \frac{1}{2n^2} + \frac{1}{3n^3}\right).$$

Obviously $C_n > 4$ for every n. Given a positive number ϵ, show that $C_n - 4 < \epsilon$ whenever n is sufficiently large.

17. a) For each n, let $D_n = \dfrac{n + 1}{n^2 + 3n + 2}$. Under what condition for n can you be sure

   that $D_n < \dfrac{1}{10^2}$?

   b) Given any positive number ϵ, under what condition for n can you be sure that $D_n < \epsilon$?

18. Given an ellipse, find its area.

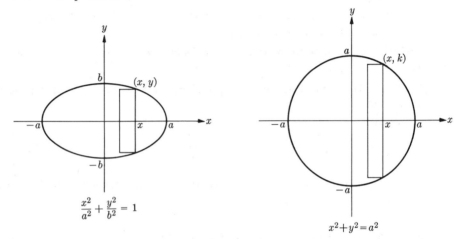

$$\frac{x^2}{a^2} + \frac{y^2}{b^2} = 1$$

$$x^2 + y^2 = a^2$$

This can be done by a method somewhat similar to one used in the preceding section of the text. [*Hint:* In the figures, what is the relation between y and k?]

19. In the discussion preceding Theorem 1, we found that $A'_n = A_n - h^3/n$. Verify this statement geometrically, without using a formula for either A_n or A'_n.
   Hint: Draw a figure showing both the inner and outer rectangles, and explain why

$$A_n - A'_n = \frac{h}{n} \cdot h^2.$$

*20. a) Find a formula for

$$\sum_{i=1}^{n} i^3.$$

   b) Find the area of the region under the graph of $y = x^3$, from 0 to 1.

*21. a) Let

$$A_n = \sum_{i=1}^{n} a_i = \frac{h^3}{n^3} \sum_{i=1}^{n} i^2,$$

   as in the text, and let

$$E_n = A_n - \frac{h^3}{3} .$$

   Thus E_n is the error in the approximation $A_n \approx h^3/3$, and $E_n > 0$ for each n. Calculate E_n and show that $E_n < h^3/n$ for each n.

   b) Show that for every $\epsilon > 0$, $E_n < \epsilon$ when n is sufficiently large. That is, find a number N such that $E_n < \epsilon$ whenever $n > N$.

3 Functions, Derivatives, and Integrals

3.1 THE IDEA OF A FUNCTION

Roughly speaking, a *function* is a law of correspondence under which to each element of one set there corresponds one and only one element of another set. Consider some examples.

1) Suppose that we have set up a coordinate system in a plane E. Then to each point P of E there corresponds a number x which is the x-coordinate of P. Thus we have a function

$$E \to \mathbf{R}$$

which matches points P of E with elements x of $\mathbf{R}$.

2) Similarly, every point P has a unique y-coordinate y. Thus we have *another* function

$$E \to \mathbf{R}.$$

To distinguish these two functions, we give them different names, say, X and Y. Thus

$$X: E \to \mathbf{R},$$
$$: P \mapsto x$$

is the "x-coordinate function," and

$$Y: E \to \mathbf{R},$$
$$: P \mapsto y$$

is the "y-coordinate function." When we write $P \mapsto x$ (with the vertical bar on the left-hand end of the arrow), this means that each point P is matched with its x-coordinate x. Thus we write $\to$ between *sets* and $\mapsto$ between *elements* of the sets.

3) If the real number x is known, then x^2 is determined. Thus we have a function

$$f: \mathbf{R} \to \mathbf{R},$$
$$: x \mapsto x^2.$$

4) Every nonnegative real number has one and only one nonnegative square root. Thus we have a function

$$g: \mathbf{R}^+ \to \mathbf{R},$$
$$: x \mapsto \sqrt{x},$$

where $\mathbf{R}^+$ denotes, as usual, the set of all nonnegative real numbers.

5) If $x \geq 2$, then $x - 2$ is nonnegative, and so has one and only one nonnegative square root. Thus we have a function

$$h: [2, \infty) \to \mathbf{R},$$
$$: x \mapsto \sqrt{x - 2}.$$

6) The absolute value $|x|$ of x is defined by the conditions

$$|x| = x \quad \text{for} \quad x \geq 0 \qquad \text{and} \qquad |x| = -x \quad \text{for} \quad x < 0.$$

In either case, if x is known, then $|x|$ is determined. Thus we have a function

$$i: \mathbf{R} \to \mathbf{R},$$
$$: x \mapsto |x|.$$

In each of these six cases we have a function

$$f: A \to B,$$

where A and B are sets of some kind. The elements of A are the objects *to which* things are going to correspond. The set A is called the *domain* of the function f. In each case, B is a set which contains all of the objects which correspond to elements of A. The set B is called the *range* of the function f. Finally, to have a function f, we must have a rule under which to each element of A there corresponds a unique element of B. Under these conditions, we have a *function of A into B*.

We can sum up the preceding examples in the following table.

Example	Function	Domain	Range	Rule
1	X	E	$\mathbf{R}$	$P \mapsto x$
2	Y	E	$\mathbf{R}$	$P \mapsto y$
3	f	$\mathbf{R}$	$\mathbf{R}$	$x \mapsto x^2$
4	g	$\mathbf{R}^+$	$\mathbf{R}$	$x \mapsto \sqrt{x}$
5	h	$[2, \infty)$	$\mathbf{R}$	$x \mapsto \sqrt{x - 2}$
6	i	$\mathbf{R}$	$\mathbf{R}$	$x \mapsto |x|$

It is not required that all the elements of the range actually get used. Thus, in Example 3, $x^2 \geq 0$ for every x, and so we could equally well write

$$f: \mathbf{R} \to \mathbf{R}^+,$$
$$: x \mapsto x^2,$$

using $\mathbf{R}^+$ as the range instead of $\mathbf{R}$.

Often functions are defined by algebraic formulas, but some of the most important functions are defined in other ways. Consider the following example.

7) Given the parabola, shown below, which is the graph of the equation $y = x^2$. For each point P of the parabola, the arc of the curve from the origin O to P has a certain length. If to each x we let correspond the length of the arc from $O = (0, 0)$ to $P = (x, x^2)$, then we have a function

$$j: \mathbf{R} \to \mathbf{R}^+.$$

(Here we are talking about simple geometric length, independent of direction, and so the length of the arc is never negative.) Later we shall find that this function can be described by a formula. But we don't need to know this, let alone find the formula, to know that we are dealing with a function.

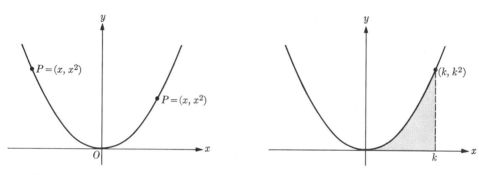

8) Given the same parabola. To each number $k \geq 0$ there corresponds a number A which measures the area of the shaded region in the right-hand figure. To be exact, the region is

$$R_2 = \{(x, y) \mid 0 \leq x \leq k, 0 \leq y \leq x^2\}.$$

Thus we have a function

$$f_2 \colon \mathbf{R}^+ \to \mathbf{R}^+,$$
$$: k \mapsto A.$$

In Chapter 2 we got a formula for this function:

$$k \mapsto \tfrac{1}{3}k^3$$

for every $k \geq 0$.

9) Given the graph of $y = x^n$, for $x \geq 0$.

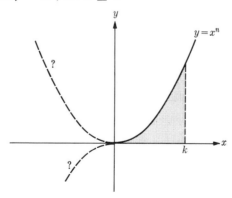

(The rest of the graph goes upward when n is even and downward when n is odd.) To each $k \geq 0$ there corresponds a number A which measures the area of the shaded region

$$R_n = \{(x, y) \mid 0 \leq x \leq k, 0 \leq y \leq x^n\}.$$

Thus for each n we have a function

$$f_n: R^+ \to R^+,$$

$$: k \mapsto A.$$

Only for the cases $n = 1$ and $n = 2$ do we know how to calculate the values of A. But for $n = 3$, we nevertheless have a well-defined function f_3. Later in this chapter, you will see how this function can be calculated.

Given a function $f: A \to B$, for each a in A we denote by $f(a)$ the element of B which corresponds to a. For example, if f is the function which squares things $(x \mapsto x^2)$, then

$$f(1) = 1, \quad f(2) = 4, \quad f(3) = 9, \quad f(\sqrt{2}) = 2;$$

and

$$f(\sqrt{x}) = x \quad \text{for every } x \geqq 0.$$

In Example 9 above, $f_3(1)$ is the area under the graph of $y = x^3$, from 0 to 1; and so on.

If the domain A and the range B are sets of real numbers, then we can draw pictures of the function. The *graph* of a function $f: A \to B$ is the set of all points of the coordinate plane that have the form $(x, f(x))$. In other words, to draw the graph of the function, we plot the point $(x, f(x))$ for each x in A.

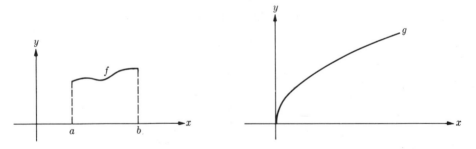

In the case shown in the left-hand figure above, the domain is a closed interval $[a, b]$. Consider, next, the function g in Example 4, which extracts nonnegative square roots:

$$g: R^+ \to R,$$

$$: x \mapsto \sqrt{x}.$$

The graph of g (the right-hand figure above) is the graph of the equation $y = \sqrt{x}$. To see that this graph is approximately right, observe that

$$y = \sqrt{x} \iff x \geqq 0, y \geqq 0, x = y^2.$$

We get $x = y^2$ by interchanging x and y in the equation $y = x^2$. Therefore the graph of $x = y^2$ is a parabola with directrix $x = -\frac{1}{4}$ and focus $(\frac{1}{4}, 0)$. And the graph of $y = \sqrt{x}$ is the upper half of this graph.

A curve which is the graph of a function is called a *function-graph*. It is easy to see what sort of curve is a function-graph: *A set of points in a coordinate plane is a function-graph if it intersects every vertical line in at most one point.*

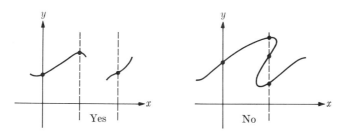

Ordinarily, we make no distinction between a function-graph in a coordinate plane and the corresponding function.

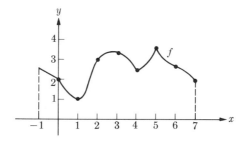

For example, in the figure above, f is a set of points, and is a function-graph. We use the same symbol f for the corresponding function. Thus we say that the domain of f is the closed interval $[-1, 7]$, and the range of f is **R**. (Obviously some smaller set could be used as the range, but it is not obvious from the figure just what the smallest possible range is.) We write $f(0) = 2$, $f(1) = 1$, $f(2) = 3$, and so on, because $0 \mapsto 2$, $1 \mapsto 1$, $2 \mapsto 3$, under the action of the function f.

Given a function

$$f: A \to B.$$

If b is $= f(a)$ for some a in A, we say that b is a *value* of the function. For example, 4 is a value of the function $x \mapsto x^2$, but -1 is not. The set of all values of a function is called the *image*. If you reexamine Examples 1 through 6 above, you will find that in 1 and 2 the image is all of **R**, and in the remaining cases the image is **R**+. (You should check these cases.)

Similarly, for

$$f: [0, 1] \to \mathbf{R}, \qquad x \mapsto \sqrt{1 - x^2}.$$

Here the graph is a quadrant of a circle, as shown on the left below, and the image is the closed interval [0, 1].

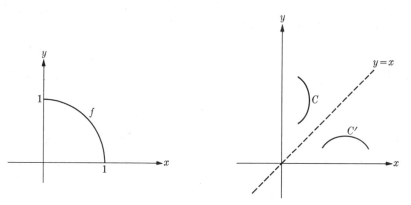

A word of caution: it often happens that a figure looks like a function-graph if you look at it sidewise. In the right-hand figure above, C is not a function-graph; but sidewise it looks like one. More precisely, if you reflect C across the line $y = x$ you get a curve which really is a function-graph. We often use this device, to study various curves C for which the reflection C' is a function-graph. But this does not mean that C was a function-graph in the first place. Therefore, in the following problem set, when you are asked whether certain curves are function-graphs, you must look at the curves right side up. For the curve C shown in the above figure, the answer is "No," even though for C' the answer is "Yes."

In some of the problems below, you are asked to find the image. In some cases, the image is not an interval; and you may find it convenient to use the notation

$$\{a, b, c, \ldots\}$$

for the set whose elements are $a, b, c, \ldots$. Thus

$$\mathbf{N} = \{1, 2, 3, \ldots\}$$

is the set of all positive integers;

$$\mathbf{Z} = \{\ldots, -2, -1, 0, 1, 2, \ldots\}$$

is the set of all integers; and

$$\{0, 1\}$$

is the set whose only elements are 0 and 1.

PROBLEM SET 3.1

1. Given $f(x) = x^2 + x + 1$, for every x. Find $f(0), f(1)$, and $f(2)$.
2. Given $f(x) = 2x^2 - x + 3$. Find $f(-1), f(0)$, and $f(2)$.
3. Given f as in Problem 2. Get a general formula for $f(2 + h)$.
4. For what positive integers n (if any) is the graph of $y = x^n$ a function-graph? For each such case (if any), what are the domain and the range?
5. Same question, for the graphs of the equations $x = y^n$.

6. For what positive integers n (if any) is the graph of $y = |x|^n$ a function-graph? For each such case (if any), what are the domain and range?

7. Same problem, for the graphs of the equations $|y|^n = x$.

*8. Same question as 6, for $y^3 + ny = x$.

9. Is the graph of $x = \sqrt{\overline{y}}$ a function-graph? If so, what are the domain and the image? Sketch.

10. Same question, for $y = |x|/x$.

11. Same question, for $|y| = x$.

12. Same question, for $y = |x| + x$.

13. Same question, for $y = x^2 + x + 1$. (Here, of course, the only trouble is in finding the image. The image is an interval, and should therefore be described in the interval notation.)

14. The postage rate for airmail letters within the United States is now (1971) ten cents per ounce *or fraction thereof*. Thus we have a function

$$\text{amp: } \mathbf{R}^+ \to \mathbf{R}^+,$$

where amp x is the airmail postage (in cents) for a letter of weight x (in ounces.) Thus amp $\frac{1}{2} = 10$, amp $1 = 10$, amp $\pi = 40$, amp $0 = 0$, and so on. Sketch the graph of this function. What is the image?

15. The *roundoff function* $r: \mathbf{R} \to \mathbf{R}$ assigns to each number the *nearest* integer (with a half-integer assigned to the next highest integer). Thus $r(2) = 2$, $r(2\frac{1}{4}) = 2$, $r(2\frac{1}{2}) = 3$, $r(2\frac{3}{4}) = 3$. Sketch the graph of this function from 0 to 3. What is its image?

16. Under what conditions is a semicircle a function-graph?

17. Under what conditions is a parabola a function-graph? (To solve this one, you will need a theorem from a problem in Chapter 2.)

3.2 THE DERIVATIVE OF A FUNCTION, INTUITIVELY CONSIDERED

In Section 2.7 we solved the tangent problem for parabolas. Given the graph of

$$y = ax^2 + bx + c,$$

we found that for each x_0, the slope of the tangent at the point (x_0, y_0) of the graph was

$$S_{x_0} = 2ax_0 + b.$$

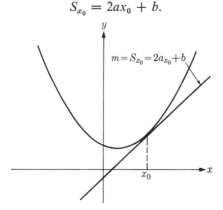

Obviously a parabola with its axis vertical is a function-graph; its equation expresses y in terms of x. Thus we have a function

$$f : \mathbf{R} \to \mathbf{R}$$
$$: x \mapsto ax^2 + bx + c.$$

Now at each point of the graph of f there is one and only one tangent; and this tangent has a certain slope. Thus we have another function

$$f' : \mathbf{R} \to \mathbf{R}$$
$$: x \mapsto S_x = 2ax + b.$$

For each x, $f'(x)$ is the slope of the tangent to the graph of f at the point $(x, f(x))$. To see how this works, consider the simplest example, in which

$$f(x) = x^2.$$

Here the parabola is the graph of the function

$$f : x \mapsto x^2,$$

and the line is the graph of the function

$$f' : x \mapsto 2x.$$

For each x, the value of f' is the slope of the tangent to the graph of f. For example, at the point where $x = 1$, the slope of the tangent to f is 2; and $f'(1) = 2 \cdot 1 = 2$. At $x = \frac{2}{3}$, we get $f'(\frac{2}{3}) = \frac{4}{3}$; and $\frac{4}{3}$ is the slope of the tangent to the parabola. Where $x = -1$, the slope of the tangent to the parabola is -2; and $f'(-1) = 2(-1) = -2$.

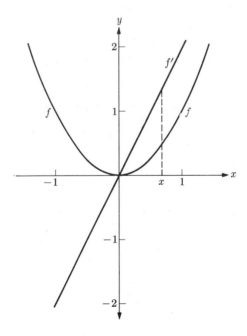

In general, suppose that we have a function

$$f: \mathbf{R} \to \mathbf{R}.$$

If the graph of f has a nonvertical tangent at each point $(x, f(x))$, we let $f'(x)$ be the slope of this tangent. This gives a new function

$$f': \mathbf{R} \to \mathbf{R}.$$

The new function f' is called the *derivative* of f. Consider another example.

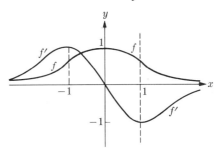

A careful inspection of the figure above indicates that f' is (at least approximately) the derivative of f. Thus, at $x = 0$, the tangent to f is horizontal; and $f'(0) = 0$, as it should be. At $x = 1$, the tangent to f seems to have slope $= -1$; and $f'(1) = -1$. At $x = -1$, the tangent to f has slope $= 1$; and $f'(-1) = 1$. For $x > 0$, the tangent to f has negative slope; and $f'(x) < 0$ for $x > 0$. For $x < 0$, the tangent to f has positive slope; and $f'(x) > 0$ for $x < 0$.

It may be that at some points f has no tangent. At such points, f' is not defined. Thus, in some cases, the domain of f' is a smaller set than the domain of f. Consider, for example, the function $f: x \mapsto |x|$.

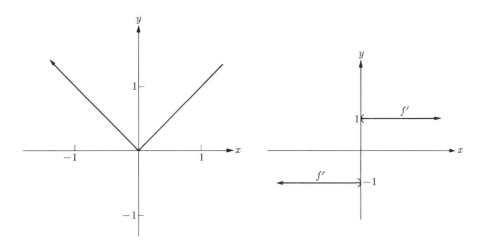

For every $x > 0$, the slope of the tangent is 1; and for every $x < 0$, the slope of the tangent is -1. Therefore the graph of f' looks like the figure on the right above.

Drawing both f and f' on the same set of axes, we get the left-hand figure below. You should carefully inspect the figure on the right below, to convince yourself that f' is the derivative of f, at least approximately. Here f has a tangent at $x = 0$, but the tangent is vertical, and therefore there is no such thing as $f'(0)$. When $x > 0$ and x is small, then $f'(x)$ is large, because f is rising steeply. When $x > 1$, $f'(x)$ is small. It looks as if $f'(2) = 0$; and the graph of f has a horizontal tangent at the point $(2, f(2))$.

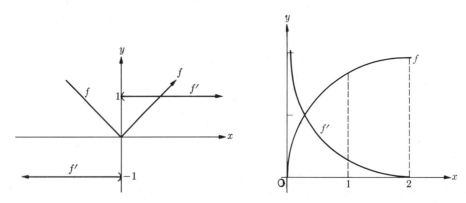

A function which has a derivative at every point of its domain is called *differentiable*. The following theorem describes a fundamental property of differentiable functions:

The Mean-Value Theorem. Every chord of a differentiable function is parallel to the tangent at some intermediate point.

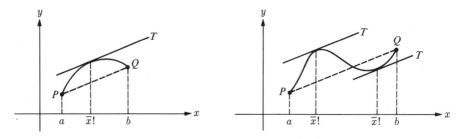

Here by a *chord* we mean a segment joining two points of the graph. The theorem says that if f is differentiable on $[a, b]$, then there is some $\bar{x}$ between a and b at which the slope of the tangent is the same as the slope of the chord. As indicated in the right-hand figure above, there may be more than one such point.

The situation with regard to this theorem is awkward. It is geometrically obvious. Also it is important and we shall need it soon. On the other hand, the proof of the theorem is hard, and involves ideas which belong in the later portion of a calculus course. We shall therefore postpone the proof, but use the theorem whenever we need it.

The theorem can be stated in a form which looks more algebraic. If f is defined on an interval $[a, b]$, then the slope of the chord joining the endpoints is

$$\frac{f(b) - f(a)}{b - a},$$

and the slope of the tangent at $\bar{x}$ is $f'(\bar{x})$. Thus the theorem states that

$$f'(\bar{x}) = \frac{f(b) - f(a)}{b - a}$$

for some $\bar{x}$ between a and b. In this style we can restate the theorem as follows:

The Mean-Value Theorem. Suppose that f is differentiable on the closed interval $[a, b]$. Then for some $\bar{x}$ between a and b we have

$$f'(\bar{x}) = \frac{f(b) - f(a)}{b - a}.$$

Note that, if we merely required that the graph have a tangent at every point, the theorem would become false. The graph shown below has a tangent at every point, but one of these tangents is vertical. Therefore the function f is not differentiable on $[a, b]$. And no tangent line is parallel to the chord from P to Q.

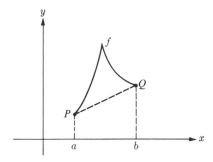

Hereafter, the mean-value theorem will be referred to as MVT.

PROBLEM SET 3.2

In each of the figures below, a function-graph f is given. Do a tracing of each graph on a sheet of writing paper, and then draw a plausible sketch of the graph of f'. Obviously your sketch of f' cannot be exact. But f' should be $= 0$ at points where the tangent to f is horizontal; $f'(x)$ should be > 0 where the original graph slopes upward; $f'(x)$ should be < 0 where f slopes downward; and so on. In some cases, you may find that the values of f' are so large that there is no room for them on the paper. In such cases, draw as much of the graph of f' as space permits.

Some but not all of the functions shown below satisfy the conditions of the mean-value theorem. For each such function, draw the chord between the endpoints of the graph, draw a tangent line which is parallel to this chord, and drop a dashed line from the point of tangency to the point $\bar{x}$ on the x-axis. (See the figures in the text, illustrating MVT.)

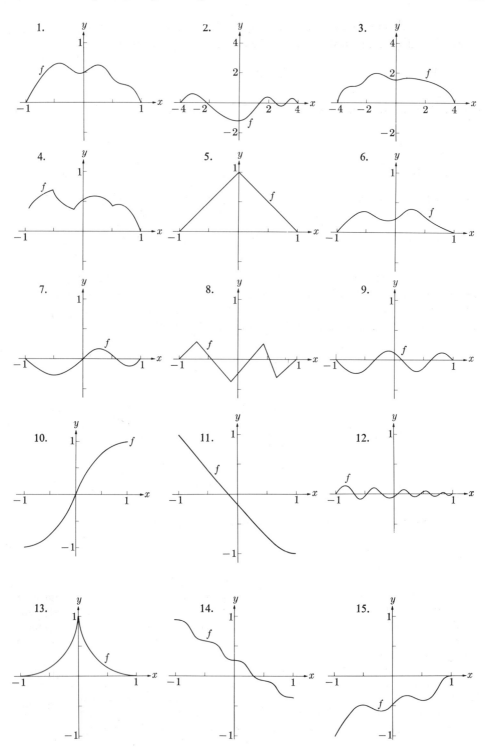

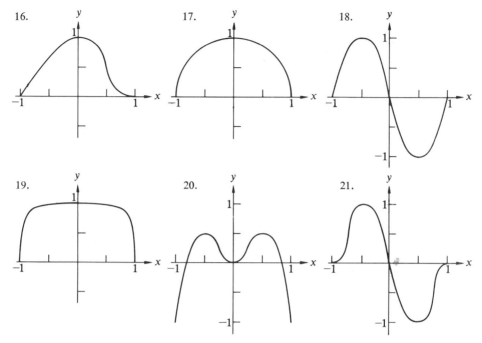

3.3 CONTINUITY AND LIMITS

Let *f* be a function defined on an interval, or defined for every *x*. Roughly speaking, *f* is *continuous* if we can draw the graph without lifting the pencil from the paper. For example, the function $f(x) = \sqrt{1 - x^2}$ is continuous, because it is the upper half of the circle $x^2 + y^2 = 1$. Most functions that arise naturally are of this type, and in this book we will rarely deal with any other kind of function. But some very simple functions are not continuous. Consider, for example, the airmail postage function defined in Problem 14 of Problem Set 3.1. The graph looks like this:

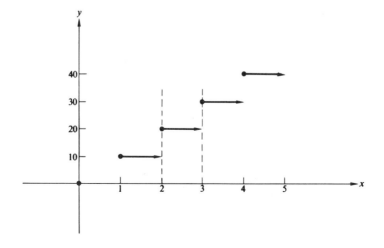

Here $y = \text{amp } x$, where amp x is the airmail postage on a letter weighing x ounces. The values of this function make sudden jumps at integral values: the graph cannot be drawn without lifting the pencil from the paper, and so the function is discontinuous.

Functions of this kind are used in physics. For example, the so-called *Heaviside function* is defined by the conditions

$$h(x) = \begin{cases} 0 & \text{if } x < 0 \\ 1 & \text{if } x \geq 0. \end{cases}$$

The graph looks like the figure below. It makes a sudden jump at $x = 0$.

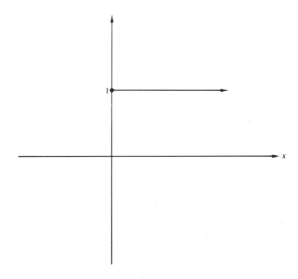

We shall now make the idea of continuity more exact, in several stages. Given a point x_0, in the domain of f, we want to explain what it means to say that f is continuous at x_0. First we try the following:

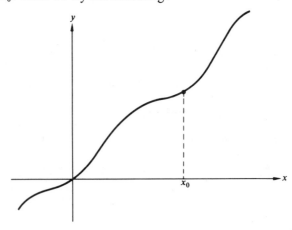

1)   Whenever x is close to x_0, $f(x)$ is close to $f(x_0)$. In symbols,

$$x \approx x_0 \quad \Rightarrow \quad f(x) \approx f(x_0).$$

This is the idea, but it is not good enough; the question is *how* close things are supposed to be to each other. As x gets very close to x_0, $f(x)$ is supposed to become very close to $f(x_0)$. This suggests:

2)   We can make $f(x)$ *as close as we please* to $f(x_0)$, by taking x *sufficiently close* to x_0.

This is better, but it can be improved. We measure the closeness of two numbers by taking the absolute value of their difference. Thus if ϵ is a positive number, and

$$|f(x) - f(x_0)| < \epsilon,$$

then we say that $f(x)$ is ϵ-close to $f(x_0)$. In these terms, we can restate (2) as follows:

3)   For each $\epsilon > 0$, $f(x)$ is ϵ-close to $f(x_0)$ whenever x is sufficiently close to x_0.

If $\delta > 0$ and $|x - x_0| < \delta$, then we say that x is δ-close to x_0. The idea of "sufficiently close" can be described by taking a positive number δ. This gives:

4)   For each $\epsilon > 0$, there is a $\delta > 0$ such that $f(x)$ is ϵ-close to $f(x_0)$ whenever x is δ-close to x_0.

We can now draw a picture.

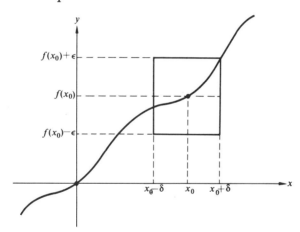

In the figure, the solid rectangular region is called an $\epsilon\delta$-*box* for the function f at the point $(x_0, f(x_0))$. When we call it a *box* for the function, we mean that no point of the graph lies above the box or below it. If the function is continuous, then for every positive number ϵ, no matter how small, we can find a $\delta > 0$ that gives an $\epsilon\delta$-box. We now restate (4) as follows.

Definition. Let x_0 be a point in the domain of the function f. Suppose that for every $\epsilon > 0$ there is a $\delta > 0$ such that

$$|x - x_0| < \delta \quad \Rightarrow \quad |f(x) - f(x_0)| < \epsilon.$$

Then f is *continuous* at x_0.

This definition applies very simply to the function $f(x) = 2x$, at the point $(1, 2)$.

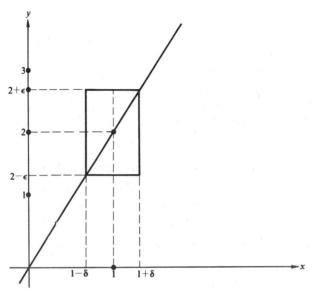

Given any $\epsilon > 0$, we can find an $\epsilon\delta$-box, as shown in the figure; we simply take $\delta = \epsilon/2$. Then, algebraically,

$$
\begin{aligned}
|x - 1| < \delta \quad &\Rightarrow \quad |x - 1| < \epsilon/2 \\
&\Rightarrow \quad |2x - 2| < \epsilon \\
&\Rightarrow \quad |f(x) - f(1)| < \epsilon;
\end{aligned}
$$

and this is what we need, to show that the function f is continuous at the point $x = 1$. Of course, the function is also continuous at all other points, and we can show in exactly the same way that the definition of continuity applies, taking $\delta = \epsilon/2$. (What would we do for $f(x) = 3x$, at any point x_0?)

We shall now apply the definition of continuity to the function $f(x) = x^2$, at the point $(1, 1)$.

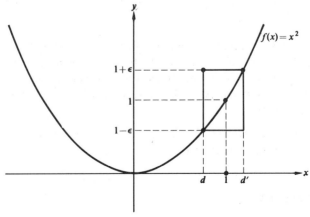

Given an $\epsilon > 0$, we want to find an $\epsilon\delta$-box for f, at $(1, 1)$, so that

$$|x - 1| < \delta \Rightarrow |x^2 - 1| < \epsilon.$$

To find the desired number δ directly requires clumsy calculations, but there is an easier way. Let d and d' be the numbers such that

$$f(d) = d^2 = 1 - \epsilon; \qquad f(d') = d'^2 = 1 + \epsilon.$$

(See the lower figure on page 78.) The graph rises from left to right. Therefore,

$$
\begin{aligned}
d < x < d' \;&\Rightarrow\; f(d) < f(x) < f(d') \\
&\Leftrightarrow\; d^2 < x^2 < d'^2 \\
&\Leftrightarrow\; 1 - \epsilon < x^2 < 1 + \epsilon \\
&\Leftrightarrow\; |x^2 - 1| < \epsilon.
\end{aligned}
$$

Thus the dotted rectangle in the figure boxes in the graph, in the same way that an $\epsilon\delta$-box does. We call such a rectangle a *dd'-box*. Obviously a *dd'*-box is just as good as an $\epsilon\delta$-box. And, in fact, given a *dd'*-box, we can always get an $\epsilon\delta$-box that lies in it. Let δ be the smaller of the positive numbers $1 - d$ and $d' - 1$. (In fact, $d' - 1$ is the smaller, but we don't need to use this.) Then

$$|x - 1| < \delta \;\Rightarrow\; d < x < d',$$

and so

$$|x - 1| < \delta \;\Rightarrow\; |x^2 - 1| < \epsilon,$$

which is what we wanted. We are going to use this method again, and so we record it as a theorem.

Theorem 1. Let x_0 be a point in the domain of the function f. Suppose that for every $\epsilon > 0$ there are numbers d and d' such that $d < x_0 < d'$ and

$$d < x < d' \;\Rightarrow\; f(x_0) - \epsilon < f(x) < f(x_0) + \epsilon.$$

Then f is continuous at x_0.

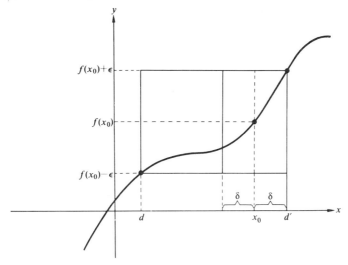

Proof. Let δ be the smaller of the numbers $x_0 - d$ and $d' - x_0$. (In the figure, $\delta = d' - x_0$.) Then

$$|x - x_0| < \delta \quad \Rightarrow \quad d < x < d' \quad \Rightarrow \quad f(x_0) - \epsilon < f(x) < f(x_0) + \epsilon$$

$$\Leftrightarrow \quad |f(x) - f(x_0)| < \epsilon.$$

We shall now reexamine the idea of a limit, which we used in defining the slope of the tangent to the graph of a function.

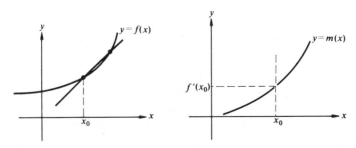

To find the slope of the tangent at the point $(x_0, f(x_0))$, we let $m(x)$ be the slope of the secant line through the points $(x_0, f(x_0))$ and $(x, f(x))$, where $x \neq x_0$. Thus the slope of the secant is a function, and we are now describing it in functional notation. By definition, the slope of the tangent is

$$f'(x_0) = \lim_{x \to x_0} m(x),$$

if such a limit exists. We shall now give a definition of the limit. The idea is that $\lim_{x \to x_0} m(x) = L$ if the function m becomes continuous at x_0 when we insert the value L as the value $m(x_0)$. Thus we want to use L as $m(x_0)$ in the definition of continuity. This gives the following:

Definition. Let m be a function defined at each point of an interval I, except at the point x_0. Suppose that for every $\epsilon > 0$ there is a $\delta > 0$ such that

$$0 < |x - x_0| < \delta \quad \Rightarrow \quad |m(x) - L| < \epsilon.$$

Then

$$\lim_{x \to x_0} m(x) = L.$$

For example, for $f(x) = x^2$, $x_0 = 1$, we have

$$m(x) = \frac{x^2 - 1}{x - 1} = x + 1 \qquad (x \neq 1).$$

When we insert the point $(1, 2)$ on the graph of the function m, we get a continuous function (which is equal to $x + 1$ for every x). Thus $\lim_{x \to x_0} m(x) = 2$, not just intuitively but also in terms of our definition of a limit.

There is one more problem to consider. What if $f(x_0)$ is defined? In this case, what do we mean by $\lim_{x \to x_0} f(x)$? The answer is that we ignore the value of f at x_0, and investigate how the rest of the graph behaves. To be exact:

Definition. Let f be a function defined on an interval I, except perhaps at the point x_0. Suppose that for every $\epsilon > 0$ there is a $\delta > 0$ such that

$$0 < |x - x_0| < \delta \;\;\Rightarrow\;\; |f(x) - L| < \epsilon.$$

Then

$$\lim_{x \to x_0} f(x) = L.$$

Note that here we have simply copied the preceding definition, using f for m: all along, the value $x = x_0$ was ruled out by the condition $0 < |x - x_0|$. The left-hand figure, showing the $\epsilon\delta$-box, looks the same as before, except that there is no point in plotting $f(x_0)$ (which may not be defined, and which will not in any case be used).

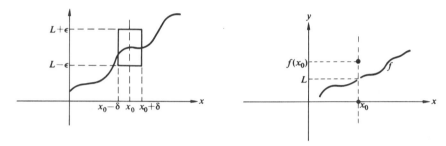

The definition of a limit applies in peculiar ways to certain peculiar functions. In the right-hand figure above, $\lim_{x \to x_0} f(x) = L$, but $L \neq f(x_0)$. The following theorem shows that the strange situation shown in the figure cannot happen if the function is continuous.

Theorem 2. f is continuous at x_0 if and only if

$$\lim_{x \to x_0} f(x) = f(x_0).$$

Here the displayed formula says three things at once:

1) $f(x_0)$ is defined. That is, x_0 is in the domain of f.
2) $\lim_{x \to x_0} f(x)$ exists. That is, f approaches a limit, as $x \to x_0$.
3) $f(x_0)$ and $\lim_{x \to x_0} f(x)$ are the same number.

PROBLEM SET 3.3

1. How close to 3 does x need to be, for $2x$ to be within 0.001 of 6? (Answer in the form $|x - 3| < \cdots \Rightarrow |2x - 6| < 0.001$. Sketch the graph of $f(x) = 2x$, and sketch your $\epsilon\delta$-box ($\epsilon = 0.001$).)

2. Find numbers d and d' such that $d < x < d' \Rightarrow |x^2 - 3^2| < 0.0001$. Sketch the graph of $f(x) = x^2$ and sketch your dd'-box. In your sketch, you will have to distort the scale grossly, because of the small size of your ϵ.

3. Show that the function $f(x) = x^2$ is continuous at the point $x_0 = 3$. Use the method that was used in the text for the same function at the point $x_0 = 1$ and apply Theorem 1. Thus your answer will include statements in the form: "Let $d = \ldots$, and let $d' = \ldots$. Then $d < x < d' \Rightarrow 3^2 - \epsilon < x^2 < 3^2 + \epsilon$." Sketch the function, showing your dd'-box.

Answer as in Problem 3 for the following functions, at the following points x_0.

4. $f(x) = 2x^2$, $x_0 = 1$.

5. $f(x) = \sqrt{x}$, $x_0 = 4$.

6. $f(x) = \sqrt[3]{x}$, $x_0 = 8$.

7. $f(x) = x^3$, $x_0 = 2$.

8. $f(x) = x^n$, where n is any positive integer and x_0 is any number.

9. $f(x) = \sqrt{x}$, x_0 is any positive number.

*10. A function f is called *Lipschitzian* if there is a number $m > 0$ such that for every two points x, x_0 we have

$$|f(x) - f(x_0)| \leqq m\,|x - x_0|.$$

Show that every Lipschitzian function is continuous.

3.4 THEOREMS ON LIMITS

In this section we shall give the elementary rules that we use in dealing with limits. These rules are much easier to learn and to use than they are to prove, and so many of the proofs are omitted from this section. (You will find the missing proofs in Appendix B.) But some of the proofs are easy, and they throw some light on the idea of a limit.

Theorem 1. If $\lim_{x \to x_0} f(x) = L$, then $\lim_{x \to x_0} [-f(x)] = -L$.

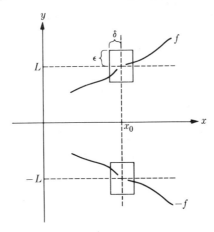

Proof. To get $-f$ from f, we flip the graph of f across the x-axis. We know that for every $\epsilon > 0$, f has an $\epsilon\delta$-box at (x_0, L). If we flip the box across the x-axis, in the same way that we flipped the graph, this gives us a box for $-f$ at $(x_0, -L)$.

This theorem can also be proved algebraically. The hypothesis means that: (1) for every $\epsilon > 0$ there is a $\delta > 0$ such that

$$0 < |x - x_0| < \delta \quad \Rightarrow \quad |f(x) - L| < \epsilon;$$

the conclusion means that: (2) for every $\epsilon > 0$ there is a $\delta > 0$ such that

$$0 < |x - x_0| < \delta \quad \Rightarrow \quad |-f(x) - (-L)| < \epsilon.$$

Since $|-f(x) - (-L)| = |f(x) - L|$, it is obvious that $(1) \Rightarrow (2)$, and thus the theorem holds.

Theorem 2. If $\lim_{x \to x_0} f(x) = L$, then $\lim_{x \to x_0} [f(x) - L] = 0$.

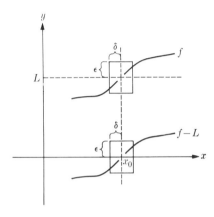

Proof. Given f, we get $f - L$ by moving the graph up or down a certain distance (down or up, according as L is positive or negative). We move the box along with the graph; and this gives a box for the function $f - L$.

Theorem 3. If $\lim_{x \to x_0} [f(x) - L] = 0$, then $\lim_{x \to x_0} f(x) = L$.

To prove this, merely use the previous proof in reverse; move the box along with the graph.

Theorem 4. If $\lim_{x \to x_0} f(x) = L$, and k is any number, then $\lim_{x \to x_0} kf(x) = kL$.

That is, the limit of a constant times a function is the same constant times the limit of the function.

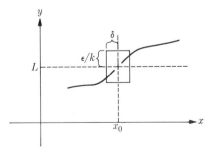

Proof.

1) For $k = 0$, this is easy: $kf(x) = 0$ for every x, and so $\lim_{x \to x_0} kf(x) = 0 = 0 \cdot L$.

2) Suppose that $k > 0$. For every $\epsilon > 0$, the graph of f has an $\epsilon\delta$-box at (x_0, L). Therefore, for every $\epsilon > 0$, the graph of f has an (ϵ/k) δ-box at (x_0, L). Thus

$$0 < |x - x_0| < \delta \;\Rightarrow\; |f(x) - L| < \epsilon/k.$$

Multiplying by k, in the inequality on the right, we get:

$$0 < |x - x_0| < \delta \;\Rightarrow\; |kf(x) - kL| < \epsilon,$$

and so $\lim_{x \to x_0} kf(x) = kL$, which was to be proved.

3) Proof for $k < 0$?

These proofs illustrate the way we work with boxes, to prove things about limits. The following theorems will be proved later (unless you find a way to prove them for yourself).

Theorem 5. If $\lim_{x \to x_0} f(x) = L$ and $\lim_{x \to x_0} g(x) = L'$, then

$$\lim_{x \to x_0} [f(x) + g(x)] = L + L'.$$

That is, if each of the functions f and g has a limit, as $x \to x_0$, then the sum also has a limit, and the limit of the sum is the sum of the limits.

Theorem 6. If $\lim_{x \to x_0} f(x) = L$ and $\lim_{x \to x_0} g(x) = L'$, then

$$\lim_{x \to x_0} [f(x)g(x)] = LL'.$$

Theorem 7. If $\lim_{x \to x_0} f(x) = L$ and $\lim_{x \to x_0} g(x) = L'$, and $L' \neq 0$, then

$$\lim_{x \to x_0} \frac{f(x)}{g(x)} = \frac{L}{L'}.$$

Caution: The preceding theorem says nothing about what happens when $L' = 0$. And in fact, for $L' = 0$ *anything* can happen, even in very simple cases. If

$$f(x) = 2x, \qquad g(x) = x,$$

then

$$\frac{f(x)}{g(x)} = \frac{2x}{x}, \qquad \text{and} \qquad \lim_{x \to 0} \frac{f(x)}{g(x)} = 2.$$

And any number k can be used in place of 2. Therefore, if $f(x) \to 0$ and $g(x) \to 0$, the quotient f/g can approach any number whatever as a limit. This should not surprise us, because every time we calculate a derivative we are finding the limit of a quotient

$$\frac{f(x) - f(x_0)}{x - x_0}$$

whose numerator and denominator are approaching 0.

In the preceding section we showed that f is continuous at x_0 if and only if

$$\lim_{x \to x_0} f(x) = f(x_0).$$

(This was Theorem 2.) Hereafter, we shall regard the above formula as interchangeable with the definition of continuity. Thus every theorem on limits automatically gives us a theorem on continuous functions. Some of these are as follows:

Theorem 8. If f is continuous at x_0, and k is any number, then kf is continuous at x_0.

Proof. We are given that

$$\lim_{x \to x_0} f(x) = f(x_0).$$

By Theorem 4,

$$\lim_{x \to x_0} kf(x) = kf(x_0),$$

and this means that kf is continuous at x_0.

Theorem 9. If f and g are continuous at x_0, then so also are $f + g$ and fg.

Proof? (Use Theorems 5 and 6.)

Theorem 10. If f and g are continuous at x_0, and $g(x_0) \neq 0$, then f/g is continuous at x_0.

(Use Theorem 7.)

Most of the time we shall apply these results not just at one point x_0 but throughout the domain of the functions f and g. For these cases, we can state our theorems more briefly as follows:

Theorem 11. Let f and g be functions with the same domain. If f and g are continuous, then so also are kf, $f + g$, and fg. And f/g is continuous at every point x_0 where $g(x_0) \neq 0$.

Thus, for example, given that $f(x) = x^2 + 1$ and $g(x) = x^4 + 4$ are continuous, on the entire real number system, we can infer immediately that kf, $f + g$, fg, and f/g have the same property. Here $g(x) \neq 0$ for every x. Given

$$h(x) = x^2 - 1,$$

we can infer that $f + h$ and fh are continuous everywhere, and that f/h is continuous except at 1 and -1. Of course, at $x = 1$ and $x = -1$ it is not just continuity that breaks down: the quotient function is not even defined at these points, because the denominator of f/h becomes 0.

Finally, a trivial observation.

Theorem 12. Every constant function is continuous.

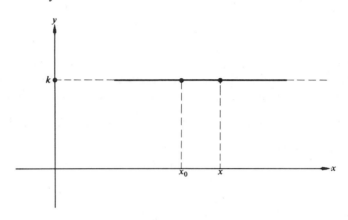

Proof. Given $f(x) = k$, for every x in a certain domain. We need to show that for every $\epsilon > 0$ there is a $\delta > 0$ such that

$$|x - x_0| < \delta \quad \Rightarrow \quad |f(x) - k| < \epsilon.$$

Obviously any positive number can be used as δ.

PROBLEM SET 3.4

In solving the following problems, you need not base your work directly on the definition of continuity or the definition of a limit; you are free to use all the theorems stated in this section. Note that the later problems are not based on this section at all; they are extensions of the theory.

1. Show that if $f(x) = kx$, then f is continuous.
2. Same, for $f(x) = kx^2$.
3. Same, for $f(x) = x^n$ (n a positive integer).
4. Same, for $f(x) = x^3 - 3x^2 + x - 4$.
5. A *polynomial of degree n* is a function of the form

$$f(x) = a_n x^n + a_{n-1} x^{n-1} + \cdots + a_1 x + a_0,$$

where $a_n \neq 0$. (The function which is $= 0$ for every x is a polynomial of degree 0.) Show that every polynomial is continuous.

6. Show that if

$$f(x) = \frac{x^n}{1 + x^2},$$

then f is continuous.

A function f is *bounded* if there is a number M such that $-M \leq f(x) \leq M$ for every x in the domain of M. If the above inequalities hold, then we say that M is a *bound* of f. Obviously, to show that a function f is bounded, you have to name a bound of f; and to show that a function is unbounded, you have to show that no number M is a bound of f.

Find out which of the following functions are bounded, on the given domains, and justify your answers:

7. $f(x) = \dfrac{1}{1 + x^2}$, $-\infty < x < \infty$

8. $f(x) = x^2$, $-\infty < x < \infty$

9. $f(x) = x^2$, $0 \leq x \leq 1$

10. $f(x) = \dfrac{x^2}{1 + x^2}$, $-\infty < x < \infty$

11. $f(x) = x^3$, $0 \leq x \leq 2$

12. $f(x) = \dfrac{x}{1 + x^2}$, $0 \leq x \leq 1$

13. $f(x) = \dfrac{x}{1 + x^2}$, $1 \leq x < \infty$

14. $f(x) = \dfrac{x}{1 + x^2}$, $-\infty < x < -1$

15. $f(x) = \dfrac{x}{1 + x^2}$, $-\infty < x < \infty$

16. $f(x) = \dfrac{1}{1 + x^3}$, $1 < x < \infty$

17. $f(x) = \dfrac{1}{1 + x^3}$, $-1 < x < 1$

18. $f(x) = \dfrac{x^4}{1 + x^3}$, $0 < x < \infty$

19. $f(x) = \dfrac{x + 1}{1 + x^3}$, $-1 < x < 1$

20. Show that if f is bounded, then so also is kf for every k.

21. Show that if f and g are bounded (on the same domain), then so also is $f + g$.

*22. Show that if f and g are bounded (on the same domain), then so also is fg. You may find it convenient to write the condition for boundedness in the form $|f(x)| \leq M$. Can you infer also that f/g is bounded? Why or why not?

*23. a) Show that if f is bounded and

$$\lim_{x \to x_0} g(x) = 0,$$

then

$$\lim_{x \to x_0} [f(x)g(x)] = 0.$$

(First try proving this for the case $M = 1$.)

b) Show that if

$$\lim_{x \to x_0} f(x) = 0,$$

then

$$\lim_{x \to x_0} [f(x) \sin x] = 0.$$

(Here it makes no difference whether degrees or radians are being used.)

c) In Problem 23(a), can we get along without the hypothesis that f is bounded? That is, if

$$\lim_{x \to x_0} g(x) = 0,$$

does it follow that

$$\lim_{x \to x_0} [f(x)g(x)] = 0,$$

no matter what kind of function f may be? Why or why not?

d) Show that if

$$\lim_{x \to 0} f(x) = 0,$$

then

$$\lim_{x \to 0} \left[f(x) \sin \frac{1}{x} \right] = 0.$$

(*Query:* Can Theorem 6 be applied to this problem?)

e) Show that

$$\lim_{x \to 0} x^2 \cos \frac{1}{x} = 0.$$

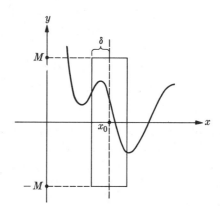

*24. a) A function f is *locally bounded* at x_0 if there are positive numbers M and δ such that

$$0 < |x - x_0| < \delta \Rightarrow |f(x)| < M.$$

b) If f is bounded, does it follows that f is locally bounded at each point of its domain?

c) Conversely, if f is locally bounded at each point of its domain, does it follow that f is bounded?

d) If f is locally bounded at each point of the open interval $(0, 1)$, does it follow that f is bounded on $(0, 1)$? Why or why not?

e) Show that if

$$\lim_{x \to x_0} f(x) = L,$$

then f is locally bounded at x_0. (This result does not require that x_0 be in the domain of f. If you draw a picture of what you *have*, and a picture of what you *want*, and compare the two, this proof may become obvious.)

f) Show that if f is locally bounded at x_0, and

$$\lim_{x \to x_0} g(x) = 0,$$

then

$$\lim_{x \to x_0} f(x)g(x) = 0.$$

3.5 THE PROCESS OF DIFFERENTIATION

The theorems in the preceding section tell us enough about limits to give us some information about derivatives.

To make some formulas easier to write, we introduce an alternative notation for the derivative: we write Df to mean the derivative of f. Thus

$$Df = f',$$

by definition. Similarly, if

$$h(x) = f(x) + g(x)$$

for every x in a certain domain, then

$$D(f + g) = h'.$$

Similarly, when we write

$$D(x^2 + 2x + 5)$$

we mean the derivative of the function

$$x \mapsto (x^2 + 2x + 5).$$

We know already what this derivative is:

$$D(x^2 + 2x + 5) = 2x + 2.$$

Here we are merely rewriting the result which we got quite a while ago: for each x, the slope of the tangent to the graph of

$$y = ax^2 + bx + c$$

is given by the formula

$$S_x = 2ax + b.$$

We recall that a function f is *differentiable* at a point x_0 if it has a derivative at x_0. When we say that f is *differentiable*, we mean that it has a derivative at every point of its domain. For example, if $f(x) = |x|$, then f is differentiable at 1, but not at 0. But if $f(x) = x^2$, then f is differentiable (without qualification).

Theorem 1. The derivative of a constant function is 0.

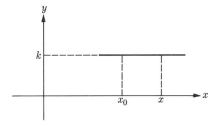

Here by a constant function we mean a function f for which $f(x) = k$ for every x. Obviously the slope of the tangent is 0 everywhere. Algebraically,

$$f'(x_0) = \lim_{x \to x_0} \frac{f(x) - f(x_0)}{x - x_0}$$

$$= \lim_{x \to x_0} \frac{k - k}{x - x_0} = \lim_{x \to x_0} 0 = 0.$$

Theorem 2. If f is differentiable, then so also is kf for every k, and

$$D(kf) = k\,Df.$$

Proof. Take any x_0 in the domain of f. Then

$$\lim_{x \to x_0} \frac{f(x) - f(x_0)}{x - x_0} = f'(x_0),$$

by the definition of $f'(x_0)$. Therefore, by Theorem 4 of Section 3.4,

$$\lim_{x \to x_0} \frac{kf(x) - kf(x_0)}{x - x_0} = kf'(x_0).$$

Therefore, at each point x_0, the derivative of kf is k times the derivative of f. We have seen an example of this:

$$Dx^2 = 2x,$$

and

$$D(kx^2) = 2kx,$$

as it should be.

Theorem 3. If f and g are differentiable, then so also is $f + g$, and

$$D(f + g) = Df + Dg.$$

Proof. Given an x_0 in the domain of f and g, we have

$$\lim_{x \to x_0} \frac{f(x) - f(x_0)}{x - x_0} = f'(x_0)$$

and

$$\lim_{x \to x_0} \frac{g(x) - g(x_0)}{x - x_0} = g'(x_0).$$

We want to prove that

$$\lim_{x \to x_0} \frac{[f(x) + g(x)] - [f(x_0) + g(x_0)]}{x - x_0} = f'(x_0) + g'(x_0).$$

Since we know by Theorem 5 of Section 3.4 that the limit of the sum is the sum of the limits, the result follows immediately; the big fraction in the third formula is the sum of the two fractions in the preceding two formulas.

We shall now show that

$$Dx^3 = 3x^2.$$

For each x, let

$$f(x) = x^3;$$

and take any x_0. Then

$$f'(x_0) = \lim_{x \to x_0} \frac{f(x) - f(x_0)}{x - x_0} = \lim_{x \to x_0} \frac{x^3 - x_0^3}{x - x_0}$$

$$= \lim_{x \to x_0} \frac{(x - x_0)(x^2 + x_0 x + x_0^2)}{x - x_0}$$

$$= \lim_{x \to x_0} (x^2 + x_0 x + x_0^2) \qquad (x \neq x_0)$$

$$= x_0^2 + x_0^2 + x_0^2,$$

by Theorems 6, 4, and 5 of Section 3.4. (Why?) Therefore

$$f'(x_0) = 3x_0^2$$

for every x_0, and

$$Dx^3 = 3x^2,$$

which is what we wanted.

To extend this result to $f(x) = x^n$, for every positive integer n, we merely need to know the general factorization formula

$$x^n - x_0^n = (x - x_0)(x^{n-1} + x^{n-2}x_0 + \cdots + xx_0^{n-2} + x_0^{n-1}).$$

This is easy to check by multiplication.

Theorem 4. $Dx^n = nx^{n-1}$, for every positive integer n.

Proof. Let $f(x) = x^n$ for every x, and take any x_0. Then

$$\lim_{x \to x_0} \frac{f(x) - f(x_0)}{x - x_0} = \lim_{x \to x_0} \frac{x^n - x_0^n}{x - x_0}$$

$$= \lim_{x \to x_0} \frac{(x - x_0)(x^{n-1} + x^{n-2}x_0 + \cdots + xx_0^{n-2} + x_0^{n-1})}{x - x_0}$$

$$= \lim_{x \to x_0} (x^{n-1} + x^{n-2}x_0 + \cdots + xx_0^{n-2} + x_0^{n-1})$$

$$= x_0^{n-1} + x_0^{n-1} + \cdots + x_0^{n-1} + x_0^{n-1} \qquad \text{(to } n \text{ terms)}$$

$$= nx_0^{n-1}.$$

Thus $f'(x_0) = nx_0^{n-1}$ for every x_0, and so

$$Dx^n = nx^{n-1},$$

which was to be proved.

The preceding theorems, in combination, enable us to differentiate any polynomial. For example,

$$D(17x^{29} + \pi x^{17} - 7x^5) = D17x^{29} + D\pi x^{17} - D7x^5,$$

because the derivative of the sum is the sum of the derivatives. This is

$$= 17 \cdot 29x^{28} + 17\pi x^{16} - 35x^4.$$

Theorem 5. If f is differentiable at x_0, then f is continuous at x_0.

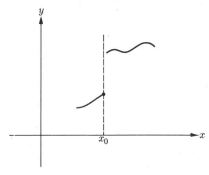

This is easy to see, as a matter of common sense. By definition, $f'(x_0)$ is the slope of the tangent. Therefore f must have a nonvertical tangent at x_0. But if f "made a jump" at x_0, there surely couldn't be a nonvertical tangent. The secant lines would get too steep for their slopes to approach a limit. Pictures are useful, but it is hard to be sure that the pictures we draw allow for all possibilities. In any case, the theorem is easy to prove algebraically. Given

$$\lim_{x \to x_0} \frac{f(x) - f(x_0)}{x - x_0} = f'(x_0),$$

then

$$\lim_{x \to x_0} [f(x) - f(x_0)] = \lim_{x \to x_0} \left[\frac{f(x) - f(x_0)}{x - x_0} \cdot (x - x_0) \right]$$

$$= f'(x_0) \cdot 0 = 0.$$

(Theorem needed, for the second step?) By Theorem 3 of Section 3.4, $\lim_{x \to x_0} f(x) = f(x_0)$, which was to be proved.

The differential calculus would be simpler if the derivative of the product were equal to the product of the derivatives; but this is not so. For example, take

$$f(x) = x^3, \qquad g(x) = x^2.$$

Then

$$f'(x) \cdot g'(x) = 3x^2 \cdot 2x = 6x^3,$$

which is not the same as

$$Dx^5 = 5x^4.$$

A correct formula can be derived as follows. Take any x_0, and suppose, as usual, that f and g are differentiable. Then

$$\lim_{x \to x_0} \frac{f(x) - f(x_0)}{x - x_0} = f'(x_0) \tag{1}$$

and

$$\lim_{x \to x_0} \frac{g(x) - g(x_0)}{x - x_0} = g'(x_0). \tag{2}$$

To find the derivative of the product, we need to find

$$\lim_{x \to x_0} \frac{f(x)g(x) - f(x_0)g(x_0)}{x - x_0}. \tag{3}$$

In a similar situation, when we were finding the derivative of $f + g$, there was no problem: we looked at the fraction whose limit we wanted to find, and observed that it was the sum of the two fractions

$$\frac{f(x) - f(x_0)}{x - x_0}, \qquad \frac{g(x) - g(x_0)}{x - x_0},$$

whose limits we knew. If these fractions appeared in (3), then we could use the fact that their limits are given. Since neither of them appears, we use a trick: we simply *put* one of them there, fix up the rest of the fraction so that its value is unchanged, and hope for the best:

$$\frac{f(x)g(x) - f(x_0)g(x_0)}{x - x_0} = \frac{f(x)g(x) - f(x_0)g(x) + f(x_0)g(x) - f(x_0)g(x_0)}{x - x_0}$$

$$= \frac{f(x) - f(x_0)}{x - x_0} g(x) + \frac{g(x) - g(x_0)}{x - x_0} f(x_0).$$

Now we can see what happens as $x \to x_0$:

$$\lim_{x \to x_0} \frac{f(x) - f(x_0)}{x - x_0} = f'(x_0),$$

$$\lim_{x \to x_0} g(x) = g(x_0),$$

$$\lim_{x \to x_0} \frac{g(x) - g(x_0)}{x - x_0} = g'(x_0).$$

Therefore, by our theorems on limits of sums and products, we have

$$\lim_{x \to x_0} \frac{f(x)g(x) - f(x_0)g(x_0)}{x - x_0} = f'(x_0)g(x_0) + f(x_0)g'(x_0).$$

In words:

Theorem 6. The derivative of the product of two differentiable functions is the derivative of the first, times the second, plus the first times the derivative of the second.

More briefly:

$$D(fg) = f'g + g'f.$$

Let us try this one out for

$$f(x) = x^3, \qquad g(x) = x^2.$$

Now

$$f'(x) = 3x^2, \qquad g'(x) = 2x.$$

Therefore

$$f'(x)g(x) + f(x)g'(x) = 3x^2 \cdot x^2 + x^3 \cdot 2x = 5x^4,$$

as it should be.

Next we want to find the derivative of the reciprocal $g = 1/f$ of a function f. As usual, we take a fixed x_0; and we assume that f is differentiable at x_0. We must have $f(x_0) \neq 0$, or $g(x_0)$ would not be defined. Now

$$g'(x_0) = \lim_{x \to x_0} \frac{g(x) - g(x_0)}{x - x_0} = \lim_{x \to x_0} \frac{1/f(x) - 1/f(x_0)}{x - x_0},$$

if such a limit exists. Algebraically,

$$\frac{1/f(x) - 1/f(x_0)}{x - x_0} = \frac{f(x_0) - f(x)}{f(x)f(x_0)(x - x_0)} = \frac{f(x) - f(x_0)}{x - x_0} \cdot \frac{-1}{f(x)f(x_0)}.$$

As $x \to x_0$, the first fraction approaches $f'(x_0)$, and the second fraction approaches $-1/[f(x_0)]^2$, because $f(x_0) \neq 0$. Therefore

$$g'(x_0) = \frac{-f'(x_0)}{[f(x_0)]^2},$$

and so

$$D\left(\frac{1}{f}\right) = \frac{-f'}{f^2},$$

at every point x where $f(x) \neq 0$. In words:

Theorem 7. The derivative of the reciprocal of a differentiable function is equal to minus the derivative of the function, divided by the square of the function (wherever the function is different from zero).

Using the preceding two theorems, we get

$$D\left(\frac{f}{g}\right) = D\left(f \cdot \frac{1}{g}\right) = f' \cdot \frac{1}{g} + f D\left(\frac{1}{g}\right) = \frac{f'}{g} + f \cdot \frac{-g'}{g^2} = \frac{f'g - g'f}{g^2}.$$

In words:

Theorem 8. The derivative of the quotient of two differentiable functions is equal to the denominator times the derivative of the numerator, minus the numerator times the derivative of the denominator, all divided by the square of the denominator (wherever the denominator is not $= 0$).

More briefly, at every point x where $g(x) \neq 0$, we have

$$D\left(\frac{f}{g}\right) = \frac{gf' - fg'}{g^2}.$$

Let us try this one out in the case

$$f(x) = x^4, \qquad g(x) = x^2,$$

where we already know the answer. By Theorem 8 we get

$$D\left(\frac{x^4}{x^2}\right) = \frac{x^2 \cdot 4x^3 - x^4 \cdot 2x}{(x^2)^2} = \frac{2x^5}{x^4} = 2x.$$

This is right, because $Dx^2 = 2x$.

For convenience of reference, we list our differentiation theorems as short formulas:

> (i) $Dk = 0,$
>
> (ii) $D(kf) = kDf,$
>
> (iii) $D(f + g) = Df + Dg,$
>
> (iv) $Dx^n = nx^{n-1}$ (n a positive integer),
>
> (v) $D(fg) = f'g + g'f,$
>
> (vi) $D\left(\dfrac{1}{f}\right) = -\dfrac{f'}{f^2}$ (wherever $f \neq 0$),
>
> (vii) $D\left(\dfrac{f}{g}\right) = \dfrac{gf' - fg'}{g^2}$ (wherever $g \neq 0$).

Theorem 5 did not involve a formula. It said that if a function is differentiable at a point, then it is continuous at the same point.

Finally, some remarks on the notation used for derivatives. When a function is named by a letter, such as f, then the notation Df is unambiguous: it means f'. But when we describe functions by formulas, it is not always obvious what function we mean. When we speak of "the function $x^2 - x + 1$," it is obvious that we mean $x \mapsto x^2 - x + 1$; and when we speak of "the function $t^2 + 2t - 3$," we mean $t \mapsto t^2 + 2t - 3$. But if we speak of the "function"

$$t^2 - tx^2 + x,$$

we might have either of two things in mind:

a) x is regarded as a constant; and our function is

$$f : t \mapsto t^2 - tx^2 + x.$$

b) t is regarded as a constant, and our function is

$$g: x \mapsto t^2 - tx^2 + x.$$

In such a case, it would hardly do to indicate a derivative by writing

$$(?) \quad D(t^2 - tx^2 + x) \quad (?),$$

because nobody could tell whether we meant f' or g'. To eliminate the ambiguity, we write D_t or D_x to indicate which letter does *not* represent a constant. Thus

$$D_t(t^2 - tx^2 + x) = f'(t) = 2t - x^2,$$

while

$$D_x(t^2 - tx^2 + x) = g'(x) = -2tx + 1.$$

Similarly,

$$D_x(ax^3z + z^2) = 3ax^2z,$$

$$D_z(ax^3z + z^2) = ax^3 + 2z,$$

$$D_a(ax^3z + z^2) = x^3z.$$

PROBLEM SET 3.5

All of the following are differentiation problems. Most of them can be worked by the standard formulas that we have just derived. But in some cases you will need to start with the definition of $f'(x_0)$ and then use various algebraic strategems.

1. $D(7x^{10} - x^8)$

2. $D\dfrac{1}{x + 1}$

3. $D\dfrac{x}{x + 1}$

4. $D\dfrac{1}{x^2 + 1}$

5. $D\dfrac{y}{y^3 - 3}$

6. $D(7y^4 - y^2 + \pi)$

7. $D\left(\dfrac{1}{x - 1}\right)$

8. $D\left(\dfrac{1}{2 - x}\right)$

9. $D(1 + x)^3$

10. a) $D_x(x^3y + ay^3 + xy^2)$ b) $D_y(x^3y + ay^3 + xy^2)$ c) $D_a(x^3y + ay^3 + xy^2)$

11. a) $D_x(3axy + x^2 + a^3)$ b) $D_a(3axy + x^2 + a^3)$

12. $D[(x^2 - x + 1)(x^2 + x + 1)]$

13. $D\dfrac{x^2 - x + 1}{x^2 + x + 1}$

14. $D\dfrac{x + 1}{x^3 - x}$

15. $D(x^2 + x)^2$

16. If you worked Problem 9 of Section 3.3, you know that $f(x) = \sqrt{x}$ is continuous in its entire domain $\mathbf{R}^+$. Assuming, in any case, that this is true, find f'. [*Hint:* Set up the fraction whose limit is $f'(x_0)$, rationalize the numerator and hope for the best.]

17. Given $f(x) = \sqrt{x + 1}$ $(x \geqq -1)$, find f'. Here you may assume that f is continuous. But you should mention this fact, at the stage where you need it.

18. Given $f(x) = \sqrt{x^2 + 1}$, find f'. (Assume that f is continuous.)

19. Given $f(x) = \sqrt{1 - x^2}$ $(-1 \leqq x \leqq 1)$, find f'.

20. Given $g(x) = x^2\sqrt{1 - x^2}$, find g'.   21. Find $D(1/\sqrt{x})$ $(x > 0)$.

22. Find $D(1/\sqrt{1 - x^2})$.   23. Find $D(x/\sqrt{1 - x^2})$ $(-1 < x < 1)$.

24. Now solve Problem 19 by the methods of Chapter 2, without using limits or differentiation formulas.

Find out whether the following formulas are correct, and give your reasons.

25. $D(x^2 + 1)^2 = 2(x^2 + 1)$ (?)   26. $D(x^2 + 1)^2 = 2(x^2 + 1) \cdot 2x$ (?)

27. $D(x^2 + 1)^3 = 3(x^2 + 1)$ (?)   28. $D(x^2 + 1)^3 = 3(x^2 + 1) \cdot 2x$ (?)

29. $D(x^2 + 1)^{500} = 500(x^2 + 1)^{499}$ (?)   [*Hint:* In fact, this formula is wrong. And it is possible to *prove* that it is wrong, without finding out what the derivative of the given function really is.]

30. Prove by induction that $Dx^n = nx^{n-1}$, without using either a factorization formula or the binomial theorem.

31. Same problem, for $Dx^{-n} = -nx^{-n-1}$.

32. Find $D\sqrt{x^2 + x + 1}$.

3.6 THE PROCESS OF DIFFERENTIATION: ROOTS AND POWERS OF FUNCTIONS

Some of the answers that you got in the preceding problem set deserve to be regarded as standard differentiation formulas. For example, you found that for

$$f(x) = \sqrt{x}$$

we have

$$f'(x) = \frac{1}{2\sqrt{x}} \, .$$

This problem is going to come up again. We had, therefore, better add it to the list of formulas at the end of Section 3.5:

$$\text{(viii)} \ D\sqrt{x} = \frac{1}{2\sqrt{x}} \, .$$

You found also that

$$D\sqrt{x + 1} = \frac{1}{2\sqrt{x + 1}} \qquad (x > -1),$$

$$D\sqrt{x^2 + 1} = \frac{x}{\sqrt{x^2 + 1}} \, ,$$

$$D\sqrt{1 - x^2} = \frac{-x}{\sqrt{1 - x^2}} \qquad (-1 < x < 1).$$

In each of these cases, we have the problem of finding the derivative of the positive square root $\sqrt{f}$ of the differentiable positive function f. If we can solve this problem in the general case, then we can get a formula

$$D\sqrt{f} = ?;$$

and we can then apply the formula hereafter. Suppose, then, that we are given

$$g(x) = \sqrt{f(x)},$$

on a domain where $f(x) > 0$; and suppose that f has a derivative at a certain point x_0 of the domain. We want to find $g'(x_0)$. By definition,

$$g'(x_0) = \lim_{x \to x_0} \frac{g(x) - g(x_0)}{x - x_0} = \lim_{x \to x_0} \frac{\sqrt{f(x)} - \sqrt{f(x_0)}}{x - x_0}$$

$$= \lim_{x \to x_0} \left[\frac{\sqrt{f(x)} - \sqrt{f(x_0)}}{x - x_0} \cdot \frac{\sqrt{f(x)} + \sqrt{f(x_0)}}{\sqrt{f(x)} + \sqrt{f(x_0)}} \right]$$

$$= \lim_{x \to x_0} \left[\frac{f(x) - f(x_0)}{x - x_0} \cdot \frac{1}{\sqrt{f(x)} + \sqrt{f(x_0)}} \right]$$

$$= f'(x_0) \lim_{x \to x_0} \frac{1}{\sqrt{f(x)} + \sqrt{f(x_0)}},$$

provided that the latter limit exists. It is easy to see that this limit exists, provided that

$$\lim_{x \to x_0} \sqrt{f(x)} = \sqrt{f(x_0)}. \tag{1}$$

Since the limit of the sum is the sum of the limits, it then follows that

$$\lim_{x \to x_0} \left[\sqrt{f(x)} + \sqrt{f(x_0)} \right] = 2\sqrt{f(x_0)};$$

and since the limit of the quotient is the quotient of the limits, we get:

Theorem 1. If f is positive and differentiable, then

$$D\sqrt{f} = \frac{f'}{2\sqrt{f}}.$$

Let us try this on the function $x \mapsto \sqrt{x + 1}$. Here $f(x) = x + 1$, and $f'(x) = 1$ for every x. Therefore

$$D\sqrt{x + 1} = \frac{1}{2\sqrt{x + 1}},$$

which is the right answer. For $x \mapsto \sqrt{x^2 + 1}$, we have

$$f(x) = x^2 + 1,$$
$$f'(x) = 2x,$$

$$D\sqrt{x^2 + 1} = D\sqrt{f(x)} = \frac{f'(x)}{2\sqrt{f(x)}} = \frac{2x}{2\sqrt{x^2 + 1}} = \frac{x}{\sqrt{x^2 + 1}},$$

which is the right answer. Formula (viii), of course, depends on

$$\lim_{x \to x_0} \sqrt{f(x)} = \sqrt{f(x_0)},\tag{1}$$

which we haven't proved. We postpone the proof, observing meanwhile that (1) is reasonable. Since f is continuous, we have

$$x \approx x_0 \;\Rightarrow\; f(x) \approx f(x_0).$$

Since $\sqrt{}$ is continuous, we have

$$f(x) \approx f(x_0) \;\Rightarrow\; \sqrt{f(x)} \approx \sqrt{f(x_0)}.$$

Fitting these two statements together, we get

$$x \approx x_0 \;\Rightarrow\; \sqrt{f(x)} \approx \sqrt{f(x_0)},$$

which is what we want.

Consider now the following function:

$$g(x) = (x^{50} - x^{17} + 1)^{247}.$$

If you want to find $g'(x)$, it is not helpful to observe that g is a polynomial, of degree 12,350, or to recall the binomial theorem. In fact, the right approach is to solve first a more general problem. Given a function g which is a power of a function f. Thus

$$g(x) = f^n(x),$$

where by $f^n(x)$ we mean $[f(x)]^n$. Then

$$g'(x_0) = \lim_{x \to x_0} \frac{f^n(x) - f^n(x_0)}{x - x_0}$$

$$= \lim_{x \to x_0} \left(\frac{[f(x) - f(x_0)][f^{n-1}(x) + f^{n-2}(x)f(x_0) + \cdots + f^{n-1}(x_0)]}{x - x_0} \right)$$

$$= \lim_{x \to x_0} \frac{f(x) - f(x_0)}{x - x_0} \lim_{x \to x_0} [f^{n-1}(x) + f^{n-2}(x)f(x_0) + \cdots + f^{n-1}(x_0)].$$

We see that this limit is

$$g'(x_0) = nf^{n-1}(x_0)f'(x_0),\tag{2}$$

provided that in the brackets on the right we have

$$\lim_{x \to x_0} f^k(x) = f^k(x_0)\tag{3}$$

for each positive integer k. Equation (3) states that the kth power of a continuous function is always continuous. And this is true: given that

$$\lim_{x \to x_0} f(x) = f(x_0),$$

it follows that

$$\lim_{x \to x_0} f^2(x) = \lim_{x \to x_0} [f(x)f(x)] = f(x_0)f(x_0) = f^2(x_0),$$

because the limit of the product is the product of the limits. For the same reason,

$$\lim_{x \to x_0} f^3(x) = \lim_{x \to x_0} f^2(x)f(x) = f^2(x_0)f(x_0) = f^3(x_0).$$

In $k - 1$ such steps, we get Eq. (3). Therefore Eq. (2) is correct; and we have:

Theorem 2. If f is differentiable, and n is a positive integer, then

$$Df^n = nf^{n-1}f'.$$

Let us try this on our polynomial of degree 12,350:

$$D(x^{50} - x^{17} + 1)^{247} = Df^{247} = 247f^{246}f'$$
$$= 247(x^{50} - x^{17} + 1)^{246}(50x^{49} - 17x^{16}).$$

Note that our use of the shortcut formula $Df^n = nf^{n-1}f'$ has two advantages over the method based on the binomial expansion. First, the calculation is *possible*, as a practical matter. Second, it gives the answer not merely in a correct form, but also in a factored form, which is easier to handle than the binomial expansion of the derivative.

Since we know how to differentiate fractions, we know how to differentiate functions of the form

$$f(x) = \frac{1}{x^k},$$

where k is a positive integer. We have

$$f'(x) = \frac{-Dx^k}{(x^k)^2} = \frac{-kx^{k-1}}{x^{2k}} = -\frac{k}{x^{k+1}},$$

where $k + 1 = 2k - (k - 1)$. If we express $1/x^k$ as x^{-k}, and make the same change in the formula for f', then we get

$$Dx^{-k} = -kx^{-k-1}.$$

This has the same form as our previous formula $Dx^n = nx^{n-1}$, with $n = -k$. What is needed, to take care of all such cases, is the following:

Theorem 3. If n is a positive integer, and f is differentiable, then

$$Df^n = nf^{n-1}f'.$$

If n is a negative integer, then the same formula holds at every point x where $f(x) \neq 0$.

The last condition is necessary: if $n < 0$, then $f(x)$ appears in the denominator of $f^n(x) = 1/f^{-n}(x)$, and f^n is therefore not defined at points where $f(x) = 0$.

Theorem 3 has already been proved for the case in which n is a positive integer. For the case in which n is a negative integer, $= -k$, the proof is as follows:

$$Df^n = Df^{-k} = D\frac{1}{f^k} = \frac{-kf^{k-1}f'}{f^{2k}}$$
$$= -kf^{-k-1}f' = nf^{n-1}f'.$$

For convenience of reference, we list all the differentiation formulas that we have so far:

$$\text{(i) } Dk = 0, \qquad\qquad\qquad \text{(ii) } D(kf) = kDf,$$

$$\text{(iii) } D(f + g) = Df + Dg, \qquad \text{(iv) } Dx^n = nx^{n-1} \qquad (n \neq 0),$$

$$\text{(v) } D(fg) = f'g + g'f, \qquad\qquad \text{(vi) } D\left(\frac{1}{f}\right) = -\frac{f'}{f^2},$$

$$\text{(vii) } D\left(\frac{f}{g}\right) = \frac{gf' - fg'}{g^2}, \qquad \text{(viii) } D\sqrt{x} = \frac{1}{2\sqrt{x}},$$

$$\text{(ix) } D\sqrt{f} = \frac{f'}{2\sqrt{f}}, \qquad\qquad \text{(x) } Df^n = nf^{n-1}f' \qquad (n \neq 0).$$

Each of these formulas holds for every x for which its right-hand member is defined.

PROBLEM SET 3.6

Find, by any method:

1. $D\sqrt{(x + 1)(x + 2)}$

2. $D\dfrac{1}{(x^2 + x + 1)^2}$

3. $D\dfrac{x}{(x^2 + 2x + 1)^2}$

4. $D\sqrt{x^4 + 5x^2 + 2}$

5. $D(x^3 + x^2 - x + 7)^{712}$

6. $D\dfrac{x^2 + 1}{x^2 - 1}$

7. $D\sqrt{x(x - 2)}$

8. $D\sqrt{x^3 + 2x + 1}$

9. $D\dfrac{\sqrt{x - 1}}{x^2 + 1}$

10. $D\dfrac{x^2 - 1}{x^2 + 1}$

11. $D\sqrt{\dfrac{1 - x}{1 + x}}$

12. $D\sqrt{\sqrt{x}}$

13. $D\sqrt{x^2}$ (*Warning:* Don't "simplify" yourself into a wrong answer.)

14. $D_x\sqrt{x^2 + a^2}$

15. $D_y(x^3y^2 - x^2y^3)^3$

16. $D_x\dfrac{x^3y^2}{(x^2 + y^2)^2}$

17. Recall that for $x > 0$,

$$x^{p/q} = \sqrt[q]{x^p},$$

by definition. Show that

$$\sqrt[q]{x^p} = (\sqrt[q]{x})^p \qquad (x > 0).$$

[*Hint:* By definition, $\sqrt[q]{x^p}$ is the number which, when raised to the qth power, gives x^p. Therefore you need to show that the qth power of the right-hand side of the above equation is x^p.]

    This is an instance of a frequent phenomenon: often, a problem becomes easy if we rewrite it, using the *definitions* of the ideas that the problem involves.

18. Find $Dx^{3/2}$, and write the answer in a form which brings out the analogy with the formula $Dx^n = nx^{n-1}$. You may assume that $x^{3/2}$ is continuous.

19. Find $D_x[\sqrt{x^3 + x}(x^3 + x)]$.

20. Find $D_x(x^4 + 2)^{3/2}$.

*21. Get a general formula for $Df^{3/2}$, where f is any positive function (differentiable, of course). The answer should be written in such a form as to bring out the analogy with the formula $Df^n = nf^{n-1}f'$.

*22. Find a formula for $Df^{5/2}$, where f is differentiable and positive.

23. Find $D_x(x^2 + 3x + 1)^{5/2}$.

24. Find a formula for $Dx^{-3/2}$ and write it in a form which brings out the analogy with the formula $Dx^n = nx^{n-1}$.

25. a) Simplify

$$\frac{a - b}{a^3 - b^3}.$$

(Obviously, the word "simplify" can mean many different things. In this case, it means to get rid of the numerator, so as to get a formula which will be useful to you in solving the next part of the problem.)

b) Find $D\sqrt[3]{x}$, assuming that $\sqrt[3]{x}$ is continuous.

26. a) Simplify, as in Problem 25,

$$\frac{a - b}{a^q - b^q}.$$

(Here q is a positive integer.)

b) Find $D\sqrt[q]{x}$, assuming that $\sqrt[q]{x}$ is continuous.

*27. Given positive integers p, q. Find a formula for $Dx^{p/q}$, valid for $x > 0$, and write the answer in a form which brings out the analogy with the formula $Dx^n = nx^{n-1}$.

3.7 THE INTEGRAL OF A NONNEGATIVE FUNCTION

Given the graph of $y = kx^2$. In Section 2.10, we calculated the area A_h of the shaded region

$$R = \{(x, y) \mid 0 \le x \le h, 0 \le y \le kx^2\}.$$

We found that

$$A_h = \frac{k}{3} h^3.$$

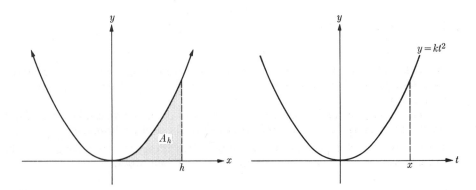

In this situation, we regarded h as a fixed number. On the other hand, it is plain that h can be any positive number, and that when h is named, A_h is determined. Thus A_h can be regarded as a function.

To discuss A_h as a function, without confusing the notation, let us relabel our horizontal axis as the t-axis, as shown on the right. Thus our parabola becomes the graph of the equation $y = kt^2$. (Here t is acting like x.) For each $x \geq 0$, let $F(x)$ be the area of the region under the parabola, from 0 to x. Thus $F(1)$ is the area under the parabola from 0 to 1; this is

$$F(1) = \frac{k}{3} \cdot 1^3 = \frac{k}{3}.$$

$F(3)$ is the area under the parabola from 0 to 3; this is

$$F(3) = \frac{k}{3} \cdot 3^3 = 9k.$$

And so on. Thus we have a function $F: \mathbf{R}^+ \to \mathbf{R}^+$. And we have a formula giving the values of F:

$$F(x) = \frac{k}{3} x^3.$$

Here we have replaced h by x in the area formula

$$A_h = \frac{k}{3} h^3.$$

Thus, starting with the nonnegative function

$$f: x \mapsto kx^2,$$

we have defined a new function

$$F: x \mapsto \frac{k}{3} x^3.$$

For each x, the *value* of the new function is the *area* under the graph of the old one.

We can generalize this scheme in the following way.

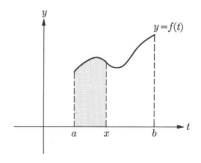

Given a continuous nonnegative function f, defined on an interval, $[a, b]$. As before, we label the horizontal axis as the t-axis, because we want to use the symbol x for another purpose. For each number x on the interval $[a, b]$, let R_x be the shaded region. Thus

$$R_x = \{(t, y) \mid a \leqq t \leqq x, 0 \leqq y \leqq f(t)\}.$$

And let $F(x)$ be the area of the region R_x.

This is the scheme that we used for the parabola. For the parabola, we had $f(t) = kt^2$, and we used $a = 0$. But we can go through the same proceeding starting with any number a and any continuous function f. Let us look at some more examples. Consider

$$f(t) = t + 1, \qquad t \geqq 1.$$

This is a function

$$f: [1, \infty) \to [2, \infty).$$

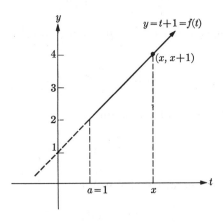

For every x on the infinite interval $[1, \infty)$, let $F(x)$ be the area under the graph, from 1 to x. We now have a new function

$$F: [1, \infty) \to [0, \infty).$$

In this case, it is easy to write a formula for F. For $x = 1$, the area is 0. Therefore $F(1) = 0$. For $x > 1$, $F(x)$ is the area of a trapezoid lying on its side, with its "bases" vertical. The altitude is $h = x - 1$, and the lengths of the bases are $b_1 = 2$ and $b_2 = x + 1$. Therefore

$$F(x) = \tfrac{1}{2}(b_1 + b_2)h = \tfrac{1}{2}(2 + x + 1)(x - 1)$$
$$= \tfrac{1}{2}(x + 3)(x - 1) = \tfrac{1}{2}(x^2 + 2x - 3).$$

Let us now try the parabola $y = t^2$, taking $a = -2$:

$$y = f(t) = t^2, \qquad t \geqq -2.$$

Here we have

$$f: [-2, \infty) \to [0, \infty).$$

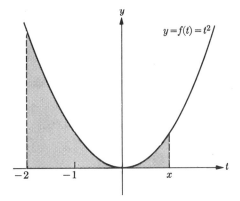

For each $x \geqq -2$, $F(x)$ is the area under this graph, from -2 to x. By our old formula,

$$F(x) = \tfrac{1}{3}x^3 - \tfrac{1}{3}(-2)^3 = \tfrac{1}{3}x^3 + \tfrac{8}{3}.$$

In all these situations, $F(x)$ is determined by (a) the given function f, (b) the number a, and (c) the number x. All this is conveyed by the notation

$$F(x) = \int_a^x f(t)\, dt.$$

That is, if f is continuous and nonnegative, and $a \leqq x$, then

$$\int_a^x f(t)\, dt$$

is the area under the graph of f, from $t = a$ to $t = x$.

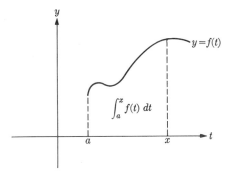

Here it should be understood that f is continuous on an interval containing a and x. The expression

$$\int_a^x f(t)\, dt$$

is called the *integral from a to x* of the function f. The number a is called the *lower limit of integration* (or, briefly, the *lower limit*) and x is called the *upper limit of integration* (or simply the *upper limit*). The function f is called the *integrand*. The notation for the integral may look formidable at first, but it is not hard to learn and is convenient: it shows at a glance that we are taking the integral, of a certain function, between certain limits.

We proceed to generalize these ideas in two ways.

a) Suppose that f is negative, for some values of t. In this case, areas below the t-axis are counted *negatively*. For example, in the left-hand figure below, A_1 and A_2 are positive numbers, representing the areas of the two shaded regions. We count A_1 positively; it is the area of a region above the t-axis. We count A_2 negatively; it is the area of a region below the t-axis.

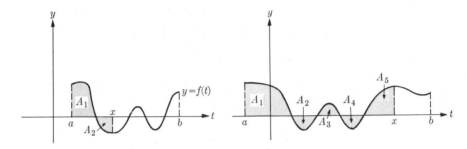

Thus we have

$$\int_a^x f(t)\, dt = A_1 - A_2.$$

Similarly, in the figure on the right,

$$\int_a^x f(t)\, dt = A_1 - A_2 + A_3 - A_4 + A_5.$$

b) So far, we have required $a < x$. If $a > x$, we first find

$$\int_x^a f(t)\, dt,$$

and then reverse the sign. Thus, in the figure on the left below,

$$\int_a^{x_2} f(t)\, dt = -A_3 + A_4,$$

under our old definition. And

$$\int_a^{x_1} f(t)\, dt = -\int_{x_1}^a f(t)\, dt = -(A_1 - A_2) = A_2 - A_1,$$

under our new definition.

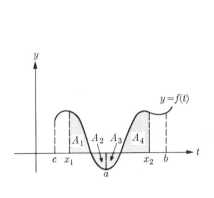

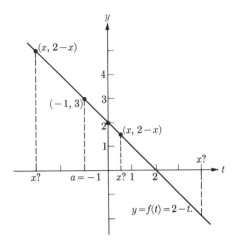

Consider, for example, $f(t) = 2 - t$, defined for every t (in the right-hand figure above). Take $a = -1$. Then, for $-1 \leqq x \leqq 2$ we have

$$\int_{-1}^{x} f(t)\, dt = \int_{-1}^{x} (2 - t)\, dt = \tfrac{1}{2}(x + 1)(3 + 2 - x) = \tfrac{1}{2}(x + 1)(5 - x),$$

by the formula for the area of a trapezoid.

For $x \geqq 2$ we have

$$\int_{-1}^{x} (2 - t)\, dt = \tfrac{1}{2} \cdot 3 \cdot 3 - \tfrac{1}{2}(x - 2)(x - 2)$$

$$= \tfrac{9}{2} - \tfrac{1}{2}(x - 2)^2.$$

For $x \leqq -1$ we have

$$\int_{-1}^{x} (2 - t)\, dt = -\int_{x}^{-1} (2 - t)\, dt$$

$$= -[\tfrac{1}{2}(-1 - x)(2 - x + 3)]$$

$$= \tfrac{1}{2}(x + 1)(5 - x).$$

You should check that these are the right answers for the three cases. In each case, we have computed areas of triangles and trapezoids by elementary area formulas, and then attached the correct sign to the area of each region.

PROBLEM SET 3.7

1. a) Consider $f(t) = |t|$. Get a formula for

$$\int_{0}^{x} |t|\, dt,$$

valid when $x \geqq 0$. Sketch.

b) Now get a formula for

$$\int_0^x |t|\, dt,$$

valid when $x \leq 0$. Sketch.

c) Now get *one* formula for the same integral, valid for every x.

d) Let

$$F(x) = \int_0^x |t|\, dt.$$

Get a formula for $F'(x)$, valid when $x > 0$.

e) Now get a formula for $F'(x)$, valid when $x < 0$.

f) Finally, get a formula for $F'(x)$, valid for every x.

2.   Do the same six things for

$$\int_0^x (t^2 + 1)\, dt.$$

3.   Consider the function defined by the graph below.

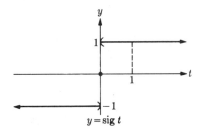

$$y = \text{sig } t$$

This function is called the *signum*. Algebraically,

$$\text{sig } t = \begin{cases} -1 & \text{when } t < 0, \\ 0 & \text{when } t = 0, \\ 1 & \text{when } t > 0. \end{cases}$$

Obviously sig is not continuous at 0. But we define

$$\int_0^x \text{sig } t\, dt$$

in the same way as for continuous functions. For example,

$$\int_0^1 \text{sig } t\, dt = 1;$$

here the integral is the area of a square of edge 1. Similarly,

$$\int_0^{-1} \text{sig } t\, dt = -(-1) = 1,$$

$$\int_0^3 \text{sig } t\, dt = 3,$$

$$\int_{-1}^{1} \operatorname{sig} t \, dt = -1 + 1 = 0;$$

and so on.

Do the same six things for $f(t) = \operatorname{sig} t$ that you did for $f(t) = |t|$ and $f(t) = t^2 + 1$.

4. Do the same six things, for $f(t) = t \, |t|$.

5. a) Explain why

$$\int_{-1}^{1} t^3 \, dt = 0.$$

b) Explain why

$$\int_{-1}^{1} t^{273} \, dt = 0.$$

c) Let f be a cubic polynomial, and suppose that

$$f(-a) = f(0) = f(a) = 0.$$

Show that

$$\int_{-a}^{a} f(t) \, dt = 0.$$

6. Explain why

$$0 < \int_{1}^{3} \frac{1}{t} \, dt < \frac{14}{12}.$$

7. Explain why

$$0 < \int_{1}^{3} \frac{1}{1 + t^2} \, dt < \frac{3}{5}.$$

3.8 THE DERIVATIVE OF THE INTEGRAL

In Problems 1 and 2 above, you found that if

$$F(x) = \int_{a}^{x} f(t) \, dt,$$

then

$$F'(x) = f(x)$$

for every x. That is, at each point x the *derivative* of the integral function F is simply the *value* of the integrand function f.

In fact, it is not hard to convince ourselves that if f is a continuous function, this is what always happens. Consider first the case in which f is positive.

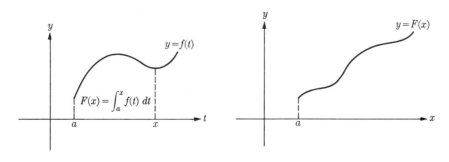

Take a fixed x_0. By definition,

$$F'(x_0) = \lim_{x \to x_0} \frac{F(x) - F(x_0)}{x - x_0}.$$

Now

$$F(x) = \int_a^x f(t)\, dt \quad \text{and} \quad F(x_0) = \int_a^{x_0} f(t)\, dt.$$

Therefore

$$F(x) - F(x_0) = \int_a^x f(t)\, dt - \int_a^{x_0} f(t)\, dt.$$

For $x_0 < x$, as in the figure below,

$$F(x) - F(x_0) = \int_{x_0}^x f(t)\, dt.$$

Since f is continuous,

$$F(x) - F(x_0) \approx (x - x_0) f(x_0).$$

Here "$\approx$" means "is approximately equal to." We are claiming that the area under the curve, from x_0 to x, is closely approximated by the area of a rectangle with base $x - x_0$ and altitude $f(x_0)$. Therefore

$$\frac{F(x) - F(x_0)}{x - x_0} \approx f(x_0);$$

and the approximation gets better as x gets closer to x_0.

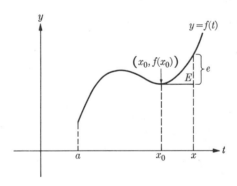

For the situation shown in the figure, in which the graph rises to the right of x_0, it is easy to see why the approximation is good. Here we have

$$F(x) - F(x_0) = (x - x_0) f(x_0) + E,$$

where E is the area of the little curvilinear triangle at the upper right. Now

$$E < e(x - x_0) \quad \text{and} \quad \frac{E}{x - x_0} < e,$$

where e is the altitude of the curvilinear triangle. Thus, when we write

$$F(x) - F(x_0) \approx (x - x_0)f(x_0),$$

the error in the approximation is E; and when we write

$$\frac{F(x) - F(x_0)}{x - x_0} \approx f(x_0),$$

the error in the approximation is $E/(x - x_0)$, which is less than e.

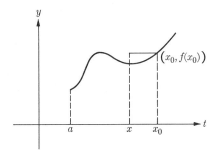

If $x < x_0$, the same approximation formula holds, although the reasons are slightly different. Here the area under the curve from x to x_0 is

$$F(x_0) - F(x);$$

the area of the rectangle is

$$(x_0 - x)f(x_0);$$

these are approximately equal. Changing the sign of each, we get

$$F(x) - F(x_0) \approx (x - x_0)f(x_0), \quad \text{and} \quad \frac{F(x) - F(x_0)}{x - x_0} \approx f(x_0),$$

as before.

But the fraction on the left is the slope of the secant line to the graph of F. The limit of this fraction is $F'(x_0)$; and since the fraction is close to $f(x_0)$ when x is close to x_0, we ought to have

$$F'(x_0) = f(x_0).$$

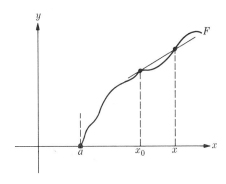

If this is true, then we have:

Theorem 1. If f is continuous on an interval containing a, then

$$D_x \int_a^x f(t)\, dt = f(x)$$

at each point x of the interval.

In fact this is true, and can be proved by a more careful use of the ideas that we have just been describing informally. But let us postpone the proof until the end of this chapter, and see, in the meantime, what the theorem is good for. Consider the following problem:

Problem 1. Calculate the area under the graph of $y = x^4$, from $x = 0$ to $x = 1$.

To solve this problem, the first step is to realize that whoever proposed the problem has asked the wrong question: the answer to his question is a *number*, and there is nothing about this number that is easy to see.

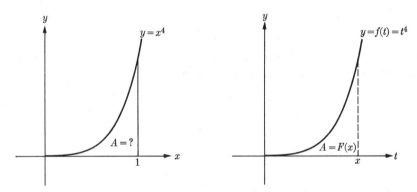

The easiest way to solve Problem 1 is to consider instead the following:

Problem 2. Find a formula for the function

$$F(x) = \int_0^x t^4\, dt.$$

It might seem that Problem 2 must be harder, but this is not true. The point is that, while information about the number $F(1)$ is hard to come by, Theorem 1 tells us something about the *function F*, namely,

$$F'(x) = x^4.$$

We now ask ourselves what sort of function has x^4 as its derivative. We get powers of x by differentiating powers of x, using the formula

$$Dx^n = nx^{n-1}.$$

Thus

$$Dx^5 = 5x^4.$$

This is like $F'(x)$, except for the factor 5. But the 5 is easy to get rid of: we divide x^5 by 5, getting

$$D(\tfrac{1}{5}x^5) = \tfrac{1}{5} \cdot 5x^4 = x^4.$$

Thus we have found a function

$$G(x) = \tfrac{1}{5}x^5,$$

which resembles the given function

$$F(x) = \int_0^x t^4 \, dt.$$

To be exact:

$$G'(x) = F'(x) \qquad \text{for every } x, \tag{1}$$
$$G(0) = F(0). \tag{2}$$

To see why (2) holds, we observe that

$$G(0) = \tfrac{1}{5}0^5 = 0$$

and that

$$F(0) = \int_0^0 t^4 \, dt = 0.$$

Equations (1) and (2) ought to guarantee that

$$G(x) = F(x) \qquad \text{for every } x;$$

that is,

$$G = F.$$

The functions F and G start with the same value, at $x = 0$. And (1) tells us that F and G always change at the same rate. This suggests the following:

Theorem 2. (*The uniqueness theorem.*) Let F and G be differentiable functions, defined on the same interval I, and let a be a point of I. If

$$F(a) = G(a) \tag{3}$$

and

$$F'(x) = G'(x) \qquad \text{for every } x \text{ in } I, \tag{4}$$

then

$$F(x) = G(x) \qquad \text{for every } x \text{ in } I. \tag{5}$$

Here we call the interval I because we want to allow intervals of *all* kinds, including $[a, b]$, $[a, b)$, $[a, \infty)$, $(-\infty, a]$, and so on. We also allow the case $I = (-\infty, \infty)$. This is the case for the functions

$$F(x) = \int_0^x t^4 \, dt, \qquad G(x) = \tfrac{1}{5}x^5$$

that we have been discussing in the last few pages.

The uniqueness theorem is a consequence of the mean-value theorem (MVT) of Section 3.2. The proof is as follows.

Suppose that $F(b) \neq G(b)$ for some b on the interval I. For each x on $[a, b]$, let

$$H(x) = F(x) - G(x).$$

Then $H(a) = 0$, but $H(b) \neq 0$.

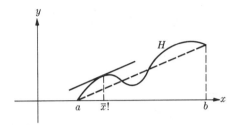

The slope of the chord joining the endpoints of the graph of H is

$$\frac{H(b) - H(a)}{b - a} \neq 0.$$

By MVT there is an $\bar{x}$ between a and b such that

$$H'(\bar{x}) = \frac{H(b) - H(a)}{b - a}.$$

Therefore $H'(\bar{x}) \neq 0$. But this is impossible: for every x we have

$$H'(x) = F'(x) - G'(x) = 0.$$

Theorems 1 and 2, in combination, enable us to solve some difficult area problems.

Example 1.   Find the area of the region above the x-axis and below the graph of $f(x) = 1 - x^2$.

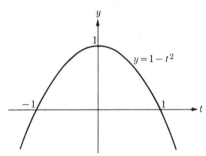

Obviously this area is

$$\int_{-1}^{1} (1 - t^2)\, dt.$$

Let

$$F(x) = \int_{-1}^{x} (1 - t^2)\, dt.$$

Then
$$F'(x) = 1 - x^2,$$
by Theorem 1. It is easy to find another function with this derivative, namely,
$$G(x) = x - \tfrac{1}{3}x^3.$$
Now $F(-1) = 0$, and
$$G(-1) = -1 + \tfrac{1}{3} = -\tfrac{2}{3}. \quad (?)$$
But this is easy to fix: we change our minds and write
$$G(x) = x - \tfrac{1}{3}x^3 + \tfrac{2}{3}.$$
Then
$$G(-1) = 0,$$
as it should be; and by the uniqueness theorem it follows that
$$G(x) = F(x)$$
for every x. Therefore
$$\int_{-1}^{1} (1 - t^2)\, dt = F(1) = G(1) = 1 - \tfrac{1}{3} + \tfrac{2}{3} = \tfrac{4}{3}.$$

Example 2. Find the area under the graph of $y = x^2 + x + 2$, from $x = -1$ to $x = 2$.

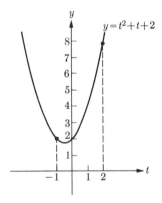

Here the area is
$$\int_{-1}^{2} (t^2 + t + 2)\, dt.$$
Let
$$F(x) = \int_{-1}^{x} (t^2 + t + 2)\, dt.$$
Then
$$F'(x) = x^2 + x + 2.$$
As our first guess, let $G(x) = \tfrac{1}{3}x^3 + \tfrac{1}{2}x^2 + 2x$, so that $G'(x) = F'(x)$. We find that

$G(-1) = -\frac{1}{3} + \frac{1}{2} - 2 = -\frac{11}{6}$. Therefore we really want $G(x) = \frac{1}{3}x^3 + \frac{1}{2}x^2 + 2x + \frac{11}{6}$, so that $G(-1) = 0$. This gives the answer in the form

$$\int_{-1}^{2} (t^2 + t + 2)\, dt = F(2) = G(2)$$

$$= \frac{1}{3}\cdot 8 + \frac{1}{2}\cdot 4 + 2\cdot 2 + \frac{11}{6} = \frac{63}{6} = \frac{21}{2}$$

The same scheme can be used to calculate integrals in which the integrand is negative and hence does not represent an area. Given

$$F(x) = \int_{a}^{x} f(t)\, dt,$$

we first find (if we can) another function G such that

$$G'(x) = F'(x) = f(x).$$

We then arrange for $G(a)$ to be 0, by adding a suitable constant to the first G that we tried. We then know that $G(x) = F(x)$ for every x. Therefore

$$\int_{a}^{b} f(t)\, dt = F(b) = G(b),$$

in the same way as for positive functions, and for the same reasons.

PROBLEM SET 3.8

1. By the methods of this section, find the area under the graph of $y = x - x^3$, from $x = 0$ to $x = 1$, and sketch. (Here, and hereafter in this problem set, you should explain what functions you are using as the functions f, F, and G.)

2. Find the area of the region lying below the x-axis and above the graph of $y = x^4 - 1$. Note that here the function f is negative, so that the area and the integral are different; the area A is positive, and

$$\int_{-1}^{1} f(t)\, dt = -A.$$

3. a) Find the area under the graph of $y = x^{10}$, from 0 to b.

   b) Find the area under the graph of $y = x^{10}$, from a to b. (*Query:* Do you need to give separate discussions for the cases $0 < a < b$, $a < 0 < b$, and so on?)

   c) Same as Problem 3(b), for the graph of $y = x^{100}$.

   d) Now find a general formula for

$$\int_{a}^{x} t^n\, dt,$$

   valid for every positive integer n.

4. Find

   a) $\displaystyle\int_{0}^{1} (t^2 + 2t + 5)\, dt$ b) $\displaystyle\int_{0}^{1} (x^2 + 2x + 5)\, dx$ c) $\displaystyle\int_{0}^{1} (z^2 + 2z + 5)\, dz$

d) A general formula for

$$F(x) = \int_0^x (t^2 + 2t + 5) \, dt.$$

5. a) Find

$$\int_0^2 (t^3 + t - 1) \, dt.$$

b) Get a general formula for

$$F(x) = \int_0^x (t^3 + t - 1) \, dt.$$

6. Get a general formula for

$$\int_a^x (t^5 - 2t^3 + 1) \, dt.$$

7. a) Find the area under the graph of

$$y = \frac{x}{\sqrt{x^2 + 1}},$$

from 1 to 2. Unless you happen to remember a function whose derivative is $x/\sqrt{x^2 + 1}$, you are going to have to figure out how this function might arise as the answer to a differentiation problem. The radical in the denominator suggests that somebody has been using the formula

$$D\sqrt{f} = \frac{f'}{2\sqrt{f}}.$$

b) Find

$$\int_{-1}^1 \frac{t}{\sqrt{t^2 + 1}} \, dt.$$

Then sketch the graph of

$$y = f(t) = \frac{t}{\sqrt{t^2 + 1}},$$

as well as you can, and explain how the numerical value that you got for the integral could have been predicted, without any calculations at all.

8. Let

$$F(x) = \int_0^x (1 + \sqrt{t})^5 \, dt.$$

Express $F'(x)$ by an elementary formula (that is, by a formula not involving integrals or differentiations). Note that you are *not* being asked to express $F(x)$ by an elementary formula.

9. Same as Problem 8, for

$$F(x) = \int_0^x (1 + \sqrt{t})^{500} \, dt.$$

10. Same as Problem 8, for

$$F(x) = \int_0^x (1 + t^{10})^{100} \, dt.$$

11. Find the area under the graph of $y = \dfrac{1}{\sqrt{x+1}}$ (and above the x-axis) from 0 to 1.

12. Same, for the function $y = \dfrac{1}{\sqrt{2-x}}$.

13. Find

$$\int_0^1 (1 + t^4)^3 t^3 \, dt.$$

[*Hint:* This problem is easier if you forget the binomial theorem.]

14. Find

$$\int_0^1 (1 + x^3)^4 x^2 \, dx.$$

15. Find

$$\int_1^2 \frac{x^2}{(1+x^3)^2} \, dx.$$

16. For $n \neq 0$, we have

$$Dx^n = nx^{n-1}.$$

Since $n \neq 0$, the function

$$f(x) = x^{-1} = \frac{1}{x}$$

never appears as the derivative of a power of x. If we allowed $n = 0$, then $x^n = x^0 = 1$ for $x > 0$; the derivative is 0; and $1/x$ still does not appear. Thus $f(x) = 1/x$ is not the derivative of any integral power of x.

   Question: Is there any function at all which has $f(x) = 1/x$ as its derivative, say, for $x > 0$? If so, what function?

*17. Consider

$$\int_{-1}^1 \frac{1}{t^2} \, dt.$$

If you attempt to evaluate this integral by applying the methods of this section in a mechanical sort of way, you will get an "answer." If you try to interpret your answer geometrically, you will see that your answer cannot possibly be right. What went wrong? (Evidently we must have been trying to apply a theorem in a case in which its hypothesis is not satisfied. The question is what theorem and what hypothesis.)

*18. In Theorem 1, suppose that we had omitted the hypothesis that f is continuous. Give an example to show that the resulting theorem would not have been true. [*Hint:* You have already seen cases in which a function of the type

$$F(x) = \int_a^x f(t) \, dt$$

fails to have a derivative at some point x_0; and surely we cannot have

$$F'(x_0) = f(x_0) \quad (?)$$

if there is no such thing as $F'(x_0)$.]

3.9   UNIFORMLY ACCELERATED MOTION

Suppose that a particle is moving, according to some given law, along a line. If we think of the line of motion as the y-axis, then the motion can be described by a function

$$f: I \to \mathbf{R};$$

for each time t on the interval, $f(t)$ is the y-coordinate of the moving particle at time t. Thus, for example, in the figure below, the total time interval I is the closed interval $[t_1, t_4]$. The figure tells us that, at the start of the motion, the time is t_1 and the particle is at the point $y = 1$; in the time interval $[t_1, t_2]$, the particle rises from 1 to 3; in the time interval $[t_2, t_3]$, the particle falls from 3 to -1; and in the time interval $[t_3, t_4]$, the particle rises from -1 to 4.

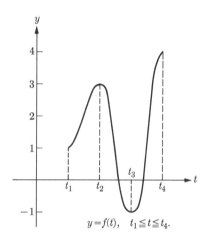

$$y = f(t), \quad t_1 \le t \le t_4.$$

The figure shows a finite time interval $I = [t_1, t_4]$. More generally, the function f may be defined on an infinite time interval $I = [t_1, \infty)$ or $I = \mathbf{R} = (-\infty, \infty)$. But most of the time, on or near the earth, the motion begins at some time t_0, and eventually the motion stops. The *velocity* is the function

$$v = f': I \to \mathbf{R},$$

provided that f is differentiable. The *acceleration* is the function

$$a = v': I \to \mathbf{R},$$

provided that v is differentiable. Thus the acceleration is the derivative of the derivative of f. We call this the *second derivative* of f, and denote it by f''. Thus we can sum up:

$$v = f', \quad a = v' = f'' \qquad \text{(by definition)}.$$

Finally, there is a fourth function associated with the motion. This is the function

$$F: I \to \mathbf{R}$$

which gives, for each time t, the *force* $F(t)$ acting on the body at time t.

We shall now see what form these functions take when f describes the motion of a freely falling body. Before we can work mathematically on the problem, we have to state our physical assumptions in mathematical form.

1) Newton's second law asserts that acceleration is proportional to force divided by mass; that is,

$$a(t) = k_1 \frac{F(t)}{m} \qquad (k_1 = \text{const}).$$

2) For a freely falling body (or a body projected vertically upward), the force is the resultant of the *weight* (which acts downward) and the air resistance (which acts upward when the body is falling and downward when the body is rising). If the speed is moderate, then the air resistance can be neglected. Hereafter, we shall assume that the weight is the only force, so that

$$F(t) = W(t) < 0 \qquad \text{for every } t \text{ on } [c, d],$$

where $W(t)$ is the weight at time t. $W(t) < 0$ because weight pulls things downward.

3) Evidently, the weight will not change merely with the passage of time, but it will depend on the altitude; the greater the altitude, the less the force of gravitation. But if the altitude is not very great, then the weight will be very nearly constant. We shall assume hereafter that the weight of a freely falling body is constant. Therefore $F(t) = W(t) = k_2 < 0$. Therefore

$$a(t) = k_1 \cdot \frac{k_2}{m} < 0 \qquad \text{for every } t;$$

and

$$a(t) = \frac{k_3}{m} < 0.$$

This last equation says that *for each falling body there is a constant which is equal to the acceleration, independently of the time.*

4) There remains, however, a question: is there one constant which works for all falling bodies, or does the constant acceleration depend on what sort of body is falling? Conceivably, the law governing the free fall of heavy bodies (such as cannon balls) might be different from the law governing the fall of light bodies (such as BB shots). In fact, until the time of Galileo, everybody thought that heavy bodies fell faster. The story goes that Galileo proved them wrong by dropping two iron balls of different sizes off the leaning tower of Pisa: they hit the ground at the same time.

Since k_3/m is independent of m, there is a constant $-g = k_3/m$ which gives the acceleration of every freely falling body, regardless of its mass. The number

$$g = -\frac{k_3}{m}$$

is called the *acceleration of gravity*. If distance is measured in feet and time in seconds, then numerically

$$g \approx 32, \qquad \text{measured in ft/sec}^2.$$

The above discussion can be summed up as follows:

> If we neglect air resistance and neglect the variation of weight with altitude, then the acceleration of a freely falling body is given by the formula
>
> $$a(t) = -g,$$
>
> where g is a constant and
>
> $$g \approx 32 \text{ ft/sec}^2.$$

We now consider the problem of finding the functions that satisfy the equation

$$a(t) = f''(t) = -g.$$

Problem. The function

$$f: \mathbf{R} \to \mathbf{R}$$

has the following properties:

a) $f''(t) = -g$ for every t,
b) $f'(0)$ is a given number v_0,
c) $f(0)$ is a given number y_0.

What is f?

Using the notation v for f' and a for $v' = f''$, we write these conditions in the form:

a) $a(t) = -g$ for every t,
b) $v(0) = v_0$,
c) $f(0) = y_0$.

Thus our data consist of (a) the constant acceleration $-g$, (b) the initial velocity $v_0 = v(0)$, and (c) the initial position $y_0 = f(0)$. The solution is as follows:

a)   We know that

$$v'(t) = a(t) = -g$$

for every t. The function

$$u(t) = -gt \quad (?)$$

has $-g$ as its derivative; the only trouble is that $u(0)$ is 0 instead of v_0. But this is easy to fix: we change our minds and let

$$u(t) = -gt + v_0.$$

Our function u then has the same derivative as v, and has the same value at $t = 0$. By the uniqueness theorem, u and v are the same function, and so

$$v(t) = -gt + v_0.$$

b)   We know now that

$$f'(t) = v(t) = -gt + v_0.$$

We want to find f. Now the function

$$z(t) = -\frac{g}{2}t^2 + v_0 t \quad (?)$$

has $-gt + v_0$ as its derivative; the only trouble is that $z(0)$ is 0 instead of y_0. But this is easy to fix: we change our minds and let

$$z(t) = -\frac{g}{2}t^2 + v_0 t + y_0.$$

The function z then has the same derivative as f, and has the same value at $t = 0$. By the uniqueness theorem, f and z are the same function, and so

$$f(t) = -\frac{g}{2}t^2 + v_0 t + y_0.$$

This completes the solution. We sum up in the following theorem:

Theorem 1. Let f be a function $\mathbf{R} \to \mathbf{R}$. If

a) $f''(t) = -g$ for every t,
b) $f'(0) = v_0$, and
c) $f(0) = y_0$,

then

d) $f(t) = (-g/2)t^2 + v_0 t + y_0$ for every t.

Thus the mathematical problem defined by (a), (b), and (c) has only one solution. This fact is important in applications, because, if our mathematical problem had two solutions, we would have to find out which of the two solutions applied to the physical situation that we started out to investigate. But, if $f''(t) = -g$, $f'(0) = v_0 = 10$, and $f(0) = v_0 = 5$, then f *must* be the function $f(t) = (-g/2)t^2 + 10t + 5$.

Theorem 1 can be stated in a more general form. If I is any time interval whatever (finite or infinite) and t_0 is any point of I, then we can consider a function

$$f: I \to \mathbf{R},$$

such that

a) $f''(t) = -g$ for every t,
b) $f'(t_0) = v_0$,
c) $f(t_0) = y_0$.

The uniqueness theorem applies to our problem in exactly the same way as before, and the algebra is only slightly more complicated. We are given

$$v'(t) = -g.$$

We try

$$u(t) = -gt \quad (?);$$

we observe that $u(t_0) = -gt_0$ instead of v_0; to fix this, we let

$$u(t) = -gt + gt_0 + v_0.$$

Now u has the same derivative as the unknown function v, and has the same value at $t = t_0$. By the uniqueness theorem, $u(t) = v(t)$ for every t, and so

$$v(t) = -gt + gt_0 + v_0.$$

This solves half of our problem.

We know that $f'(t) = v(t)$. We therefore try

$$z(t) = -\frac{g}{2}t^2 + gt_0 t + v_0 t \ (?);$$

we observe that $z(t_0) = (g/2)t_0^2 + v_0 t_0$ instead of y_0; and we fix this by letting

$$z(t) = -\frac{g}{2}t^2 + gt_0 t + v_0 t - \frac{g}{2}t_0^2 - v_0 t_0 + y_0$$

$$= -\frac{g}{2}(t - t_0)^2 + v_0(t - t_0) + y_0.$$

Then z has the same derivative as f, and has the same value at t_0. By the uniqueness theorem it follows that $z(t) = f(t)$ for every t. Therefore

$$f(t) = -\frac{g}{2}(t - t_0)^2 + v_0(t - t_0) + y_0.$$

None of these formulas should be learned. What you need to learn is the process by which they were derived; if you remember the method, you can use it. For example:

Problem. Given $f''(t) = 3, f'(3) = 1$, and $f(3) = 2$, what is f?

Solution. Let $v = f'$. Then $v'(t) = 3$. This suggests that $v(t) = 3t$. Adding the appropriate constant, to get $v(3) = 1$, we obtain

$$v(t) = 3t - 8.$$

Now

$$f'(t) = 3t - 8.$$

This suggests $f(t) = \frac{3}{2}t^2 - 8t$. Adding the appropriate constant, to get $f(3) = 2$, we have

$$f(t) = \tfrac{3}{2}t^2 - 8t - \tfrac{3}{2} \cdot 3^2 + 8 \cdot 3 + 2 = \tfrac{3}{2}t^2 - 8t + \tfrac{25}{2}.$$

This is the answer. (Two differentiations verify that it is *an* answer; and two applications of the uniqueness theorem tell us that it is the only answer.)

PROBLEM SET 3.9

Find formulas for the unknown functions, under each of the following sets of conditions. In all but one of these problems, the conditions are enough to determine the function. In three cases, however, there are infinitely many possibilities; and in these cases you should try to explain what the possible functions are.

1. $f'(t) = 3t + 4, f(0) = 4$ 2. $f'(x) = x^3 - 7x + 5, f(0) = -1$
3. $f''(t) = -1, f'(0) = 2, f(0) = 3$ 4. $f''(x) = 3x^2, f'(1) = 0, f(1) = 1$

5. $f''(t) = t^3, f'(0) = 1, f(1) = 0$ 6. $f'(x) = \dfrac{1}{x^2}, f(1) = 2$

7. $g'(x) = \dfrac{x}{\sqrt{1 - x^2}}, g(0) = -1$ 8. $g'(x) = x(x^2 + 1)^2, g(3) = 1$

9. $f'(t) = \dfrac{1}{\sqrt{t}}, f(2) = 5$

10. $f'(t) = t^2(1 + t^3)^{10}, f(0) = 2$ (By all means, do not use the binomial theorem on this one.)

11. $f'(t) = t^2 + 1, f(1) = 2$ 12. $f''(x) = x, f(1) = 0, f'(1) = 1$

13. $f'(t) = \dfrac{t^2}{(1 + t^3)^2}, f(1) = 1$ 14. $g''(t) = \dfrac{1}{(t + 1)^3}, g(0) = 1, g(1) = 1.$

15. A "theoretical projectile" is fired vertically upward, from the surface of the earth, at time 0, with initial velocity 10 ft/sec. When will it hit the ground again? For what time interval is its motion described by the condition $a(t) = -g$? (Following the advice given at the end of this section, you should solve this problem with your book closed, using the methods but not the results given in the text.)

16. A "theoretical projectile" is fired vertically upward, from the surface of the earth, and hits the ground again ten seconds later. What was the initial velocity?

17. A "theoretical projectile" is fired vertically downward from the top of a 200-foot building and hits the ground 2 seconds later. What was the initial velocity?

18. We state this problem in a nonmilitary form. A billiard ball is raised to a certain height y_0 and simply dropped, so that it begins its free fall at velocity $v_0 = 0$. Five seconds later it hits the ground. What was y_0?

19. Free fall near the surface of the moon works the same way as free fall near the surface of earth, except that the constant acceleration $-g_L$ (L for lunar) is different; the smaller mass of the moon makes the difference. Suppose you went to the moon, dropped a billiard ball as in Problem 18, and found that it dropped 3 feet in one second. What could you conclude about g_L?

*3.10 PROOF OF THE FORMULA FOR THE DERIVATIVE OF THE INTEGRAL

We shall now prove Theorem 1 of Section 3.8. We have a continuous function f; we let

$$F(x) = \int_a^x f(t) \, dt;$$

we take a point x_0; and we want to show that

$$F'(x_0) = f(x_0).$$

Let ϵ be any positive number. Since f is continuous, we know that the graph of f has an $\epsilon\delta$-box at the point $(x_0, f(x_0))$.

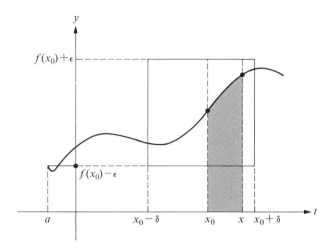

Thus

$$|x - x_0| < \delta \quad \Rightarrow \quad |f(x) - f(x_0)| < \epsilon$$
$$\Leftrightarrow \quad f(x_0) - \epsilon < f(x) < f(x_0) + \epsilon.$$

We are going to use these inequalities to get information about the function

$$m(x) = \frac{1}{x - x_0} [F(x) - F(x_0)].$$

Here m is the slope function for the function F, so that $\lim_{x \to x_0} m(x) = F'(x_0)$. Evidently

$$F(x) - F(x_0) = \int_a^x f(t) \, dt - \int_a^{x_0} f(t) \, dt$$

$$= \int_{x_0}^x f(t) \, dt,$$

and so

$$m(x) = \frac{1}{x - x_0} [F(x) - F(x_0)] = \frac{1}{x - x_0} \int_{x_0}^x f(t) \, dt.$$

If f is positive and $x_0 < x$, as in the figure, then $F(x) - F(x_0)$ is the area of the shaded region.

Case 1. Suppose that $x_0 < x < x_0 + \delta$, as in the figure. Then

$$f(x_0) - \epsilon < f(t) < f(x_0) + \epsilon \qquad (x_0 \leqq t \leqq x).$$

Therefore

$$\int_{x_0}^x [f(x_0) - \epsilon] \, dt < \int_{x_0}^x f(t) \, dt < \int_{x_0}^x [f(x_0) + \epsilon] \, dt,$$

and so

$$[f(x_0) - \epsilon](x - x_0) < \int_{x_0}^{x} f(t)\, dt < [f(x_0) + \epsilon](x - x_0).$$

Dividing by the positive number $x - x_0$, we get

$$f(x_0) - \epsilon < m(x) < f(x_0) + \epsilon.$$

Thus we have shown that

$$x_0 < x < x_0 + \delta \;\Rightarrow\; |f(x_0) - m(x)| < \epsilon. \tag{1}$$

Case 2. Suppose that $x_0 - \delta < x < x_0$. Then

$$f(x_0) - \epsilon < f(t) < f(x_0) + \epsilon \qquad (x \leqq t \leqq x_0),$$

just as in Case 1. Therefore

$$\int_{x}^{x_0} [f(x_0) - \epsilon]\, dt < \int_{x}^{x_0} f(t)\, dt < \int_{x}^{x_0} [f(x_0) + \epsilon]\, dt.$$

(We are integrating from left to right.) Therefore

$$[f(x_0) - \epsilon](x_0 - x) < \int_{x}^{x_0} f(t)\, dt < [f(x_0) + \epsilon](x_0 - x).$$

Since $x_0 - x > 0$ in Case 2, we can divide by $x_0 - x$, preserving these inequalities. This gives

$$f(x_0) - \epsilon < \frac{1}{x_0 - x} \int_{x}^{x_0} f(t)\, dt < f(x_0) + \epsilon.$$

When we interchange x and x_0, this changes the sign of each of the factors in the middle of this expression. Therefore

$$f(x_0) - \epsilon < \frac{1}{x - x_0} \int_{x_0}^{x} f(t)\, dt < f(x_0) + \epsilon.$$

To sum up:

$$x_0 - \delta < x < x_0 \;\Rightarrow\; f(x_0) - \epsilon < m(x) < f(x_0) + \epsilon \;\Rightarrow\; |m(x) - f(x_0)| < \epsilon,$$

$$\tag{2}$$

exactly as in Case 1. Fitting together our results in Cases 1 and 2, we get

$$0 < |x - x_0| < \delta \;\Rightarrow\; |m(x) - f(x_0)| < \epsilon.$$

Therefore

$$\lim_{x \to x_0} m(x) = \lim_{x \to x_0} \frac{1}{x - x_0} [F(x) - F(x_0)]$$

$$= f(x_0),$$

which was to be proved.

This proof is not easy, but it might have been worse. It was made simpler by the fact that for each $\epsilon > 0$, the $\delta > 0$ that we *get* from the hypothesis $\lim_{x \to x_0} f(x) = f(x_0)$ is precisely the δ that we *need*, to conclude that $\lim_{x \to x_0} m(x) = f(x_0)$.

PROBLEM SET 3.10

Find the first and second derivatives of the following functions.

1. $f(x) = \int_5^x t^2\, dt$

2. $g(x) = \int_\pi^x (t^3 + 1)\, dt$

3. $h(x) = \int_{2\pi}^x (t^4 - t)\, dt$

4. $f(x) = \int_0^x \dfrac{t^2}{t^4 + 1}\, dt$

5. $g(x) = \int_{-4}^x \sqrt{1 + t^8}\, dt$

6. $h(x) = \int_1^x \sqrt{t^2 + 1}\, dt$

7. $f(x) = \int_{4\pi}^x \sqrt{t}\, dt$

8. $g(x) = \int_x^3 (1 + t^3)^{100}\, dt$

9. $h(x) = \int_0^x \dfrac{1}{1 + t^2}\, dt$

10. $f(x) = \int_{-2}^x \sqrt{2 + t}\, dt$

11. $g(x) = \int_x^1 \sqrt{t}\,(t^2 + 1)\, dt$

12. $h(x) = \int_{-x}^x \dfrac{1}{1 + t^4}\, dt$

13. $\displaystyle\int_x^{} \dfrac{dt}{t}$

14. $\displaystyle\int_1^x \dfrac{dt}{t^3 + 1}$

15. $\displaystyle\int_2^x \dfrac{dt}{t^4 + 1}$

16. $\displaystyle\int_\pi^x \dfrac{t^2 + 1}{1 + t^{10}}\, dt$

17. $\displaystyle\int_{-1}^x \sqrt{\dfrac{1 + t}{1 - t}}\, dt$

18. $\displaystyle\int_3^x \sqrt{\dfrac{1 + t^2}{1 + t^4}}\, dt$

19. If you know that

$$D_x \int_a^x f(t)\, dt = f(x),$$

for every continuous function f, this does not immediately enable you to find the derivative of

$$f(x) = \int_0^{2x} \sqrt{1 + t^8}\, dt.$$

But find the answer f', by any method.

*20. Find $g'(x)$, given

$$g(x) = \int_0^{x^2} \sqrt{1 + t^8}\ dt.$$

Trigonometric and Exponential Functions

4.1 DIRECTED ANGLES. TRIGONOMETRIC FUNCTIONS OF ANGLES AND NUMBERS

In elementary geometry, when we speak of an angle we simply mean a geometric figure, that is, a set of points:

If $\overrightarrow{AB}$ and $\overrightarrow{AC}$ are rays which have the same endpoint A, but do not lie on the same line, then their union is the angle $\angle BAC$. (In the figure, the arrowheads remind us that the sides of an angle are rays rather than segments.) Some authors define the word angle in such a way as to allow "zero angles" and "straight angles."

In any case, in elementary geometry the idea of an angle does not include the idea of order; the sides of an angle are not arranged in an order, any more than the sides of a triangle are.

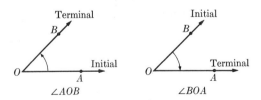

In trigonometry, however, the order of the sides of an angle makes a difference. Henceforth, whenever we speak of an *angle* we shall mean a *directed* angle. Thus, in the figures above, $\angle AOB$ is an *ordered pair* of rays $(\overrightarrow{OA}, \overrightarrow{OB})$; the ray $\overrightarrow{OA}$ is the *initial side*, and $\overrightarrow{OB}$ is the *terminal side*. Thus $\angle AOB$ is different from $\angle BOA$.

We suppose that a coordinate system is given in the plane. The counterclockwise direction is the direction from the positive *x*-axis to the positive *y*-axis, as shown below. The counterclockwise direction in a coordinate plane is regarded as positive; and the clockwise direction (running the other way) is regarded as the negative direction.

A new coordinate system is called *right-handed* if it gives the same counterclockwise direction; otherwise it is called *left-handed*. In the figure below, the right-handed coordinate systems are marked R, and the left-handed ones are marked L.

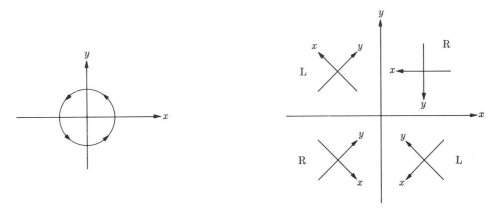

We can now define the trigonometric functions of an angle $\angle AOB$. The procedure is as follows. We set up a right-handed coordinate system, in which the initial side $\overrightarrow{OA}$ is the positive half of the *x*-axis. On the terminal side $\overrightarrow{OB}$ we choose a point $P \neq O$. *P* has coordinates (x, y), in the coordinate system that we have set up, and the distance *OP* is a positive number *r*.

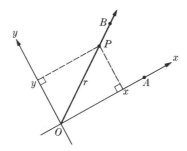

It is easy to show (by similar triangles) that the ratios x/r, y/r, y/x, x/y, r/x, r/y are independent of the choice of *P*; they depend only on the angle that we started with. Thus we can define the trigonometric functions of $\angle AOB$ as follows:

$$\sin \angle AOB = y/r, \qquad\qquad \cos \angle AOB = x/r,$$
$$\tan \angle AOB = y/x \quad \text{(for } x \neq 0\text{)}, \qquad \cot \angle AOB = x/y \quad \text{(for } y \neq 0\text{)},$$
$$\sec \angle AOB = r/x \quad \text{(for } x \neq 0\text{)}, \qquad \csc \angle AOB = r/y \quad \text{(for } y \neq 0\text{)}.$$

We have defined six functions. Note that the domains of these functions are not sets of numbers, but sets of angles.

Consider now the unit circle C, with center at the origin, in the xy-plane.

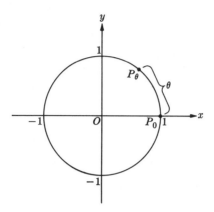

Let P_0 be the point $(1, 0)$, as in the figure. To each real number θ there corresponds a point P_θ of C, under the following rules:

1) Given $\theta > 0$, we start at P_0 and move around C in the counterclockwise direction until we have traced out a path whose total length is θ. The point where our path ends is P_θ.

2) Given $\theta < 0$, we start at P_0 and move around C in the *clockwise* direction, until we have traced out a path whose total length is $|\theta|$. The point where our path ends is P_θ.

These rules define a function

$$w: \mathbf{R} \to C$$
$$: \ \theta \mapsto P_\theta = w(\theta),$$

under which to each real number θ there corresponds a point of C. The function w is called the *winding function*. Note that the values of the function w are points rather than numbers. Note also that

$$P_{\theta+2\pi} = P_\theta,$$

for every θ. The reason is that when we add 2π to θ, this merely means that we take another round trip around the circle, ending at the same point P_θ where we began. Similarly,

$$P_{\theta-2\pi} = P_\theta \quad \text{and} \quad P_{\theta+2n\pi} = P_\theta,$$

for every integer n, positive, negative, or zero.

We shall use the winding function to define trigonometric functions of *numbers*, in the following way. For each *number* θ, let

$$\angle\theta = \angle P_0 O P_\theta.$$

The symbol $\angle \theta$ is pronounced "angle θ"; $\angle \theta$ is the angle which corresponds to the number θ. We now define

$$\sin \theta = \sin \angle \theta = \sin \angle P_0 O P_\theta,$$
$$\cos \theta = \cos \angle \theta = \cos \angle P_0 O P_\theta,$$

and so on.

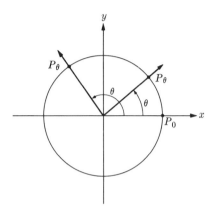

We have defined these functions in terms of $\angle \theta$ because we want to emphasize their geometric meaning. But for some purposes, it is simpler to forget about angles, and merely use the coordinates of P_θ. If

$$P_\theta = (x_\theta, y_\theta),$$

then

$$\sin \theta = y_\theta,$$
$$\cos \theta = x_\theta,$$
$$\tan \theta = \frac{y_\theta}{x_\theta} \qquad \text{(whenever } x_\theta \neq 0\text{)},$$
$$\cot \theta = \frac{x_\theta}{y_\theta} \qquad \text{(whenever } y_\theta \neq 0\text{)},$$
$$\sec \theta = \frac{1}{x_\theta} \qquad \text{(whenever } x_\theta \neq 0\text{)},$$
$$\csc \theta = \frac{1}{y_\theta} \qquad \text{(whenever } y_\theta \neq 0\text{)}.$$

Using these definitions, we can derive the usual formulas. Since P_θ is on the unit circle C, we know that $OP_\theta = 1$. Therefore

$$x_\theta^2 + y_\theta^2 = 1,$$

and we have:

Theorem 1. For every θ,

$$\cos^2 \theta + \sin^2 \theta = 1.$$

If the sign of θ is changed, this sends us around the circle C in the opposite direction. Therefore the points P_θ and $P_{-\theta}$ are symmetric across the x-axis, as in the figure.

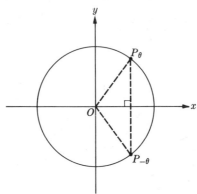

Therefore

$$x_{-\theta} = x_\theta, \qquad y_{-\theta} = -y_\theta.$$

This gives:

Theorem 2. For every θ,

$$\sin(-\theta) = -\sin\theta, \qquad \cos(-\theta) = \cos\theta.$$

Plotting the points P_0, $P_{\pi/2}$, and P_π, we get the following:

Theorem 3.

$$\sin 0 = 0, \qquad \cos 0 = 1,$$

$$\sin\frac{\pi}{2} = 1, \qquad \cos\frac{\pi}{2} = 0,$$

$$\sin\pi = 0, \qquad \cos\pi = -1.$$

Theorem 4. For each θ,

$$\sin(\pi + \theta) = -\sin\theta, \qquad \cos(\pi + \theta) = -\cos\theta.$$

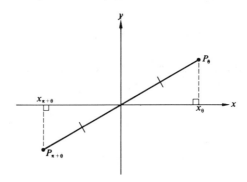

Proof. For each θ, the points P_θ and $P_{\pi+\theta}$ are symmetric across the origin. This holds in all quadrants. Therefore

$$x_{\pi+\theta} = -x_\theta, \qquad y_{\pi+\theta} = -y_\theta,$$

and the theorem follows, by definition of the sine and cosine.

In the kind of trigonometry that we are dealing with now, the relation between angles and numbers is a little tricky. If θ is known, then P_θ is determined, and so $\angle\theta$ is determined; $\angle\theta$ is $\angle P_0 O P_\theta$.

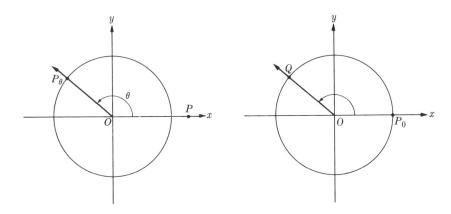

But if the angle is known, the number θ is *not* determined. In the figure on the right, $\angle P_0 O Q$ is given, but for this angle we may have

$$\theta = \tfrac{3}{4}\pi, \quad \text{or} \quad \theta = 2\pi + \tfrac{3}{4}\pi = \tfrac{11}{4}\pi, \quad \text{or} \quad \theta = \tfrac{3}{4}\pi - 2\pi = -\tfrac{5}{4}\pi.$$

In fact, for every integer n, positive, negative, or zero, we may have

$$\theta = \tfrac{3}{4}\pi + 2n\pi.$$

If an angle $\angle AOB$ corresponds to a number θ, under the rules that we have been giving, then we shall say that $\angle AOB$ *has measure* θ, and we shall write, for short,

$$\angle AOB = \angle\theta.$$

(We have seen that every angle $\angle AOB$ has infinitely many measures θ. For this reason, it would be misleading to speak of "*the* measure of an angle.")

So far, we have used the notation $\angle\theta$ only for angles "in standard position," that is, angles with the positive half of the x-axis as initial side. But it will be convenient to use the same shorthand for angles in general. Thus

$$\angle P_0 O Q = \angle\frac{\pi}{4}, \qquad \text{and} \qquad \angle P_0 O S = \angle\frac{3\pi}{4}.$$

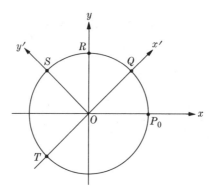

But if we set up new axes x', y', we can also say that

$$\angle QOS = \angle \frac{\pi}{2}, \quad \text{and} \quad \angle QOT = \angle \pi.$$

Using other axes, not shown in the figure, we see that

$$\angle SOP_0 = \angle \left(-\frac{3\pi}{4}\right), \quad \angle SOQ = \angle \left(-\frac{\pi}{2}\right),$$

and so on.

PROBLEM SET 4.1

Derive the trigonometric identities given or suggested below. The derivations should be based on the definitions and theorems given in this section of the text.

1. $\dfrac{1}{\sin \theta} =$ 2. $\dfrac{1}{\cos \theta} =$ 3. $\dfrac{1}{\tan x} =$ 4. $\dfrac{1}{\cot x} =$ 5. $\dfrac{1}{\sec y} =$

6. $\dfrac{1}{\csc z} =$ 7. $\dfrac{\sin \theta}{\cos \theta} =$ 8. $\dfrac{\cos \theta}{\sin \theta} =$ 9. $\dfrac{\sec \theta}{\csc \theta} =$ 10. $\dfrac{\csc \theta}{\sec \theta} =$

11. $1 + \tan^2 \theta =$ 12. $\cot^2 \theta + 1 =$ 13. $\dfrac{\csc x}{\sec x} =$ 14. $\sec^2 \theta - 1 =$

15. $\tan (-\theta) =$ 16. $\cot (-\theta) =$ 17. $\sec (-\theta) =$ 18. $\csc (-\theta) =$
19. $\tan (\pi + \theta) =$ 20. $\cot (\pi + \theta) =$ 21. $\sec (\pi + \theta) =$ 22. $\csc (\pi + \theta) =$
23. $\sin (\pi - \theta) =$ 24. $\cos (\pi - \theta) =$ 25. $\tan (\pi - \theta) =$ 26. $\cot (\pi - \theta) =$
27. $\sec (\pi - \theta) =$ 28. $\csc (\pi - \theta) =$
29. a) Show that for every θ, θ_0, we have

$$|\sin \theta - \sin \theta_0| \le |\theta - \theta_0|.$$

b) Show that the sine is a continuous function.

30. a) Show that for every θ, θ_0, we have

$$|\cos \theta - \cos \theta_0| \le |\theta - \theta_0|.$$

b) Show that the cosine is a continuous function.

4.2 THE LAW OF COSINES AND THE ADDITION FORMULAS

In the figure on the left below, we have $x_\theta = \cos\theta$, and $y_\theta = \sin\theta$, by definition of the sine and cosine.

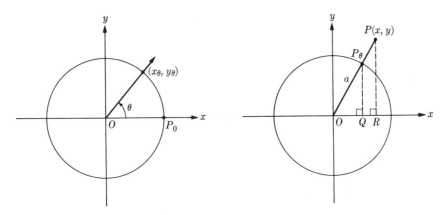

More generally, we have:

Theorem 1. Let P be any point of $\overrightarrow{OP_\theta}$, and let $OP = a$. Then the coordinates of P are

$$x = a\cos\theta, \qquad y = a\sin\theta.$$

Proof. By similar triangles,

$$\frac{|x|}{a} = \frac{|x_\theta|}{1}, \qquad \text{and} \qquad \frac{|y|}{a} = \frac{|y_\theta|}{1}.$$

Therefore

$$|x| = a\,|x_\theta|, \qquad |y| = a\,|y_\theta|.$$

In these equations x and x_θ also agree in sign, and similarly for y and y_θ. Therefore

$$x = ax_\theta = a\cos\theta, \qquad y = ay_\theta = a\sin\theta,$$

which was to be proved. From this we get immediately:

Theorem 2. (*The law of cosines*). If $\angle ACB = \angle\theta$, then

$$c^2 = a^2 + b^2 - 2ab\cos\theta.$$

(The notation is that of the following figure.)

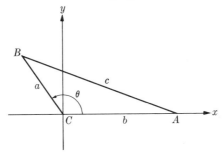

Proof. By the preceding theorem,

$$B = (a \cos \theta, a \sin \theta).$$

And obviously

$$A = (b, 0).$$

Therefore, by the distance formula,

$$c^2 = (a \cos \theta - b)^2 + (a \sin \theta - 0)^2$$
$$= a^2 \cos^2 \theta - 2ab \cos \theta + b^2 + a^2 \sin^2 \theta$$
$$= a^2 (\cos^2 \theta + \sin^2 \theta) + b^2 - 2ab \cos \theta$$
$$= a^2 + b^2 - 2ab \cos \theta,$$

which was to be proved.

Theorem 3. For every θ and ϕ,

$$\cos (\theta + \phi) = \cos \theta \cos \phi - \sin \theta \sin \phi.$$

Proof. Let $A = P_0$, $B = P_\theta$, and $C = P_{\theta+\phi}$.

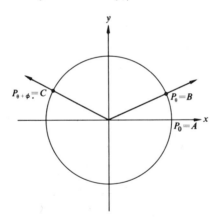

Then

$$A = (1, 0),$$
$$C = (\cos (\theta + \phi), \sin (\theta + \phi)),$$

and so by the distance formula

$$AC^2 = [\cos (\theta + \phi) - 1]^2 + \sin^2 (\theta + \phi)$$
$$= \cos^2 (\theta + \phi) - 2 \cos (\theta + \phi) + 1 + \sin^2 (\theta + \phi)$$
$$= 2 - 2 \cos (\theta + \phi).$$

We now set up a new coordinate system, with $\overrightarrow{OP_\theta}$ as the positive x'-axis.

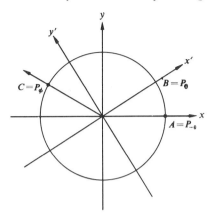

In the new coordinate system,

$$A = P_{-\theta} = \left(\cos(-\theta), \sin(-\theta)\right) = (\cos\theta, -\sin\theta),$$
$$C = P_\phi = (\cos\phi, \sin\phi).$$

Therefore, by the distance formula,

$$AC^2 = (\cos\theta - \cos\phi)^2 + (-\sin\theta - \sin\phi)^2$$
$$= \cos^2\theta - 2\cos\theta\cos\phi + \cos^2\phi + \sin^2\theta + 2\sin\theta\sin\phi + \sin^2\phi$$
$$= 2 - 2(\cos\theta\cos\phi - \sin\theta\sin\phi).$$

But the distance AC is independent of the coordinate system. Therefore

$$2 - 2\cos(\theta + \phi) = 2 - 2(\cos\theta\cos\phi - \sin\theta\sin\phi),$$

and

$$\cos(\theta + \phi) = \cos\theta\cos\phi - \sin\theta\sin\phi,$$

which was to be proved.

Once we have the addition formula for the cosine, it is easy to get similar formulas for the other trigonometric functions.

Theorem 4. For every θ and ϕ,

$$\cos(\theta - \phi) = \cos\theta\cos\phi + \sin\theta\sin\phi.$$

Proof. Using $-\phi$ for ϕ in the preceding theorem, we get

$$\cos(\theta - \phi) = \cos\theta\cos(-\phi) + \sin\theta\sin(-\phi).$$

But we know that

$$\cos(-\phi) = \cos\phi, \quad \text{and} \quad \sin(-\phi) = -\sin\phi.$$

Using these, we get the desired formula for $\cos(\theta + \phi)$.

Theorem 5. For every θ,

$$\cos\left(\frac{\pi}{2} - \theta\right) = \sin\theta, \quad\text{and}\quad \sin\left(\frac{\pi}{2} - \theta\right) = \cos\theta.$$

Proof. By Theorem 4,

$$\cos\left(\frac{\pi}{2} - \theta\right) = \cos\frac{\pi}{2}\cos\theta + \sin\frac{\pi}{2}\sin\theta$$

$$= 0\cdot\cos\theta + 1\cdot\sin\theta,$$

by Theorem 3 of Section 4.1. Therefore

$$\cos\left(\frac{\pi}{2} - \theta\right) = \sin\theta.$$

Using $\pi/2 - \theta$ for θ, we get

$$\cos\left[\frac{\pi}{2} - \left(\frac{\pi}{2} - \theta\right)\right] = \sin\left(\frac{\pi}{2} - \theta\right).$$

Therefore

$$\sin\left(\frac{\pi}{2} - \theta\right) = \cos\theta.$$

(The name of the cosine is a reference to this theorem; the word *cosine* is from the Latin *complementi sinus*, meaning *sine of the complement*.)

Theorem 6. For every θ and ϕ, $\sin(\theta + \phi) = \sin\theta\cos\phi + \cos\theta\sin\phi$.

Proof.

$$\sin(\theta + \phi) = \cos\left[\frac{\pi}{2} - (\theta + \phi)\right]$$

$$= \cos\left[\left(\frac{\pi}{2} - \theta\right) - \phi\right]$$

$$= \cos\left(\frac{\pi}{2} - \theta\right)\cos\phi + \sin\left(\frac{\pi}{2} - \theta\right)\sin\phi$$

$$= \sin\theta\cos\phi + \cos\theta\sin\phi.$$

PROBLEM SET 4.2

1. $\tan(A + B) = \dfrac{\tan A + \tan B}{1 - \tan A\tan B}$

2. $\tan(A - B) =$

3. $\cot(\theta + \phi) = \dfrac{\cot\theta\cot\phi - 1}{\cot\theta + \cot\phi}$

4. $\cot(A - B) =$

5. $\sin 2\theta = 2\sin\theta\cos\theta$

6. $\cos 2\theta = 2\cos^2\theta - 1$

7. $\cos 2\theta = 1 - 2\sin^2\theta$

8. $\cot(\theta - \phi) =$

9. a) $\sin \dfrac{3\pi}{2} =$ b) $\sin \left(\dfrac{3\pi}{2} + \theta\right) =$ 10. a) $\cos \dfrac{3\pi}{2} =$ b) $\cos \left(\dfrac{3\pi}{2} + \theta\right) =$

11. a) $\tan \dfrac{3\pi}{2} =$ b) $\tan \left(\dfrac{3\pi}{2} + \theta\right) =$

12. $2 \sin \dfrac{\theta}{2} \cos \dfrac{\theta}{2} =$ 13. $2 \cos^2 \dfrac{\theta}{2} - 1 =$ 14. $\sqrt{\dfrac{1 + \cos 2\theta}{2}} =$

15. $\sqrt{\dfrac{1 + \cos \theta}{2}} =$ 16. $\sqrt{\dfrac{1 - \cos 2\theta}{2}} =$ 17. $\sqrt{\dfrac{1 - \cos \theta}{2}} =$

18. $\tan \dfrac{\theta}{2} = \dfrac{\sin \theta}{1 + \cos \theta}$ [*Hint:* Let $\phi = \theta/2$, so that $\theta = 2\phi$, and rewrite the formula in terms of ϕ. *Then* prove it.]

19. $\tan \dfrac{\theta}{2} = \dfrac{1 - \cos \theta}{\sin \theta}$

*20. Show geometrically (without using any of the theory developed in this section) that the formula in Problem 18 holds whenever θ is between 0 and π. Discuss the problem of extending the formula from this special case to the general case.

21. Show that there is no formula which expresses $\sin (\theta/2)$ in terms of $\sin \theta$. That is, show that $\sin (\theta/2)$ is not determined if only $\sin \theta$ is known.

22. Find a formula which expresses $|\sin (\theta/2)|$ in terms of $\cos \theta$.

23. Show that there is no formula which expresses $\sin \theta$ in terms of $\tan \theta$. That is, show that $\tan \theta$ does not determine $\sin \theta$.

24. Show that there is no formula which expresses $\sin (\theta/2)$ in terms of $\sin \theta$ and $\cos \theta$.

25. Show that if P_θ is known, then $P_{3\theta}$ is determined. [*Hint:* If $P_\theta = P_\phi$, what is the relation between θ and ϕ? In this case, what is the relation between 3θ and 3ϕ? Between $P_{3\theta}$ and $P_{3\phi}$?]

26. It is a consequence of Problem 25 that, if $\sin \theta$ and $\cos \theta$ are known, then $P_{3\theta}$ is determined, and therefore $\sin 3\theta$ is determined. How? That is, find a formula which expresses $\sin 3\theta$ in terms of $\sin \theta$ and $\cos \theta$.

27. Can $\cos 3\theta$ be expressed in terms of $\cos \theta$? If so, derive such a formula. If not, explain how you know that no such formula exists.

4.3 THE DERIVATIVES OF THE TRIGONOMETRIC FUNCTIONS; THE DIFFERENCES Δx AND Δf; THE SQUEEZE PRINCIPLE

If we try, in a straightforward way, to find the derivative of the sine, we get into trouble. By definition,

$$f'(x_0) = \lim_{x \to x_0} \frac{f(x) - f(x_0)}{x - x_0},$$

if the indicated limit exists. For $f(x) = \sin x$, this definition says that

$$\sin' x_0 = \lim_{x \to x_0} \frac{\sin x - \sin x_0}{x - x_0},$$

if the indicated limit exists. In fact, the limit does exist. But it is not obvious what we ought to do to this expression

$$\frac{\sin x - \sin x_0}{x - x_0}$$

in order to find its limit. For functions f which were defined algebraically, we found ways to cancel out $x - x_0$ in fractions of the form

$$\frac{f(x) - f(x_0)}{x - x_0},$$

using various algebraic tricks. Evidently some new device is needed for the sine. It is as follows. Let

$$\Delta x = x - x_0.$$

The symbol Δx is all one symbol. It is pronounced "delta x," and the Greek delta is supposed to remind us that Δx is the difference in x. Obviously, $x = x_0 + \Delta x$. Similarly, let

$$\Delta f = f(x) - f(x_0).$$

Here Δf is the difference in f, as we pass from x_0 to x.

Geometrically, the use of the differences Δx and Δf is indicated by a new set of axes, with the new origin at the point $(x_0, f(x_0))$.

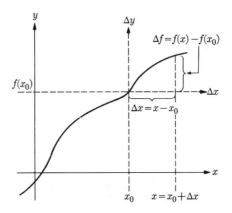

Geometrically, it is easy to see that the expressions

$$\lim_{x \to x_0} \frac{f(x) - f(x_0)}{x - x_0} = f'(x_0)$$

and

$$\lim_{\Delta x \to 0} \frac{\Delta f}{\Delta x} = \lim_{\Delta x \to 0} \frac{f(x_0 + \Delta x) - f(x_0)}{\Delta x} = f'(x_0)$$

are merely two different ways of describing the same limit $f'(x_0)$.

The point of this procedure, in finding the derivative of the sine, is that it enables us to apply the addition formula for the sine. For $f(x) = \sin x$, we have

$$f(x_0 + \Delta x) = \sin (x_0 + \Delta x)$$
$$= \sin x_0 \cos \Delta x + \cos x_0 \sin \Delta x.$$

Thus we have

$$f'(x_0) = \sin' x_0 = \lim_{\Delta x \to 0} \frac{\sin (x_0 + \Delta x) - \sin x_0}{\Delta x}$$

$$= \lim_{\Delta x \to 0} \frac{\sin x_0 \cos \Delta x + \cos x_0 \sin \Delta x - \sin x_0}{\Delta x}$$

$$= \lim_{\Delta x \to 0} \left[\sin x_0 \frac{\cos \Delta x - 1}{\Delta x} \right] + \lim_{\Delta x \to 0} \left[\cos x_0 \frac{\sin \Delta x}{\Delta x} \right].$$

We are going to show that

$$\lim_{\Delta x \to 0} \frac{\cos \Delta x - 1}{\Delta x} = 0 \tag{1}$$

and

$$\lim_{\Delta x \to 0} \frac{\sin \Delta x}{\Delta x} = 1. \tag{2}$$

It will then follow that

$$\sin' x_0 = \cos x_0,$$

and

$$D \sin x = \cos x.$$

The unknown limits (1) and (2) have curious forms. Since $\cos 0 = 1$, the first limit has the form

$$\lim_{\Delta x \to 0} \frac{\cos (0 + \Delta x) - \cos 0}{\Delta x} = \cos' 0. \tag{3}$$

And since $\sin 0 = 0$, the second limit is

$$\lim_{\Delta x \to 0} \frac{\sin (0 + \Delta x) - \sin 0}{\Delta x} = \sin' 0. \tag{4}$$

Thus we have found that if $\cos' 0 = 0$ and $\sin' 0 = 1$, then $\sin' x = \cos x$ for every x.

To simplify the notation, in the theorems that follow, we use θ in place of Δx, and state the theorems that we need in the following way:

Theorem 1. $\displaystyle \lim_{\theta \to 0} \frac{\sin \theta}{\theta} = 1.$

Theorem 2. $\displaystyle\lim_{\theta \to 0} \frac{\cos \theta - 1}{\theta} = 0.$

Theorem 1 is the hard part; given that Theorem 1 holds, Theorem 2 follows from it. To see this, we first observe that

$$\lim_{\theta \to 0} \frac{\cos \theta - 1}{\theta} = \lim_{\theta \to 0} \left[\frac{\cos \theta - 1}{\theta} \cdot \frac{\cos \theta + 1}{\cos \theta + 1} \right].$$

Simplifying on the right, we express this as

$$\lim_{\theta \to 0} \left[\frac{\cos^2 \theta - 1}{\theta(\cos \theta + 1)} \right] = \lim_{\theta \to 0} \left[\frac{-\sin^2 \theta}{\theta(\cos \theta + 1)} \right].$$

The last formula can be factored into three parts, giving

$$\lim_{\theta \to 0} \frac{\cos \theta - 1}{\theta} = \left[-\lim_{\theta \to 0} \frac{\sin \theta}{\theta} \right] \left[\lim_{\theta \to 0} \sin \theta \right] \left[\lim_{\theta \to 0} \frac{1}{\cos \theta + 1} \right].$$

Given that Theorem 1 holds, this gives

$$\lim_{\theta \to 0} \frac{\cos \theta - 1}{\theta} = -1 \cdot 0 \cdot \tfrac{1}{2} = 0.$$

(*Query:* How do we know that $\lim_{\theta \to 0} \sin \theta = 0$, and that $\lim_{\theta \to 0} \cos \theta = 1$?)

It remains to prove Theorem 1. First we observe that only positive values of θ need to be considered, because when we replace θ by $-\theta$, the value of the fraction $(\sin \theta)/\theta$ is unchanged. Thus if $(\sin \theta)/\theta \to 1$ as θ approaches 0 through positive values, it follows that $(\sin \theta)/\theta \to 1$ as $\theta \to 0$ through negative values.

We shall show that for $0 < \theta < \pi/2$ we have

$$\sin \theta \leqq \theta \leqq \tan \theta.$$

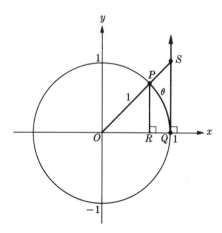

In the figure, θ is the length of the arc from Q to P. Since

$$RP = \sin \theta \quad \text{and} \quad QS = \tan \theta,$$

the inequalities that we want take the form

$$RP \leq \theta \leq QS.$$

To prove this, we have to go back to the definition of arc length.

The figure below shows a broken line inscribed in the arc from Q to P, with n segments of equal length $a_1 = a_2 = \cdots = a_n$. Thus the length of the broken line is

$$A_n = a_1 + a_2 + \cdots + a_n.$$

(In the figure, $n = 3$.) We extend the radii of the circle until they intersect the vertical line through Q; and for each segment of our broken line we let b_i be the length of the corresponding segment on the vertical line through Q.

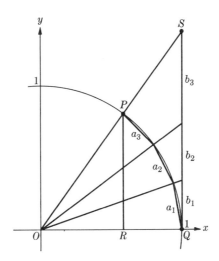

It is a matter of elementary geometry to check that

$$RP < A_n,$$

and that

$$a_i < b_i \quad \text{for each } i.$$

Therefore

$$RP < A_n < QS,$$

and so

$$\sin \theta < A_n < \tan \theta.$$

As $n \to \infty$, $A_n \to \theta$. In fact, this is the definition of the length of a circular arc. Therefore

$$\sin \theta \leq \theta \leq \tan \theta.$$

(When we pass to a limit, a "weak inequality" $a \leq A_n$ or $A_n \leq b$ is always preserved, but a "strong inequality" $a < A_n$ or $A_n < b$ is not necessarily preserved. For example,

$$\frac{n+1}{n} > 1 \qquad \text{for every } n,$$

but we cannot conclude that

$$(?) \lim_{n \to \infty} \frac{n+1}{n} > 1. \ (?)$$

In fact, the limit is 1, which is ≥ 1, but not >1. Hence the overcautious weak inequalities that we have written above. The strong inequalities $\sin \theta < \theta < \tan \theta$ always hold for $0 < \theta < \pi/2$, but we are not stopping to prove it.) Therefore

$$1 \leq \frac{\theta}{\sin \theta} \leq \frac{1}{\cos \theta}.$$

As $\theta \to 0$, $\cos \theta \to \cos 0 = 1$. (You proved this in Problem 30(b) of Section 4.1.) Therefore $1/\cos \theta \to 1$, because the limit of the reciprocal is the reciprocal of the limit. Thus the picture must look something like the figure below.

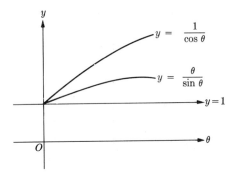

That is, the graph of $y = \theta/\sin \theta$ is "squeezed into 1," and

$$\lim_{\theta \to 0} \frac{\theta}{\sin \theta} = 1.$$

(This is an instance of a general "squeeze principle," to be discussed further at the end of this section.) Therefore

$$\lim_{\theta \to 0} \frac{\sin \theta}{\theta} = 1,$$

because the limit of the reciprocal is the reciprocal of the limit. As we have seen, this means that:

Theorem 3.

$$D \sin x = \cos x.$$

Once we know how to find $D \sin x$, the derivatives of the other trigonometric functions are easy.

$$\cos' x_0 = \lim_{\Delta x \to 0} \frac{\cos (x_0 + \Delta x) - \cos x_0}{\Delta x}$$

$$= \lim_{\Delta x \to 0} \frac{\cos x_0 \cos \Delta x - \sin x_0 \sin \Delta x - \cos x_0}{\Delta x}$$

$$= \cos x_0 \lim_{\Delta x \to 0} \frac{\cos \Delta x - 1}{\Delta x} - \sin x_0 \lim_{\Delta x \to 0} \frac{\sin \Delta x}{\Delta x}$$

$$= (\cos x_0) \cdot 0 - (\sin x_0) \cdot 1 = -\sin x_0.$$

Thus:

Theorem 4.

$$D \cos x = -\sin x.$$

By simpler methods, we get

$$D \tan x = \sec^2 x,$$
$$D \cot x = -\csc^2 x,$$
$$D \sec x = \sec x \tan x,$$
$$D \csc x = -\csc x \cot x.$$

You will be asked to derive these, in the problem set below.

In finding the limit of $\theta/\sin \theta$, we used the following idea:

Theorem 5. (*The squeeze principle*). Let f and g be functions defined at every point of the interval I, except perhaps at the point x_0. If

$$\lim_{x \to x_0} f(x) = L,$$

and for each x, $g(x)$ is between $f(x)$ and L, then

$$\lim_{x \to x_0} g(x) = L.$$

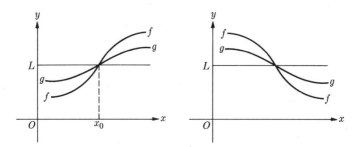

Two illustrations of the theorem are shown above. The theorem is geometrically clear, and is also easy to prove. The point is that since $g(x)$ is between $f(x)$ and L, any box for f at (x_0, L) is automatically a box for g at (x_0, L). Since f has an $\epsilon\delta$-box at (x_0, L), for every $\epsilon > 0$, it follows that g does also. Therefore

$$\lim_{x \to x_0} g(x) = L,$$

by definition of a limit.

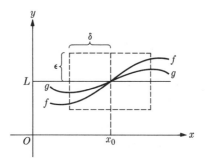

The same idea also works when two functions approach the same limit, and a third function lies between them.

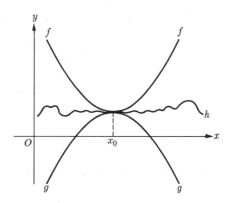

If

$$g(x) \leqq h(x) \leqq f(x),$$

and

$$\lim_{x \to x_0} f(x) = \lim_{x \to x_0} g(x) = L,$$

then it follows that

$$\lim_{x \to x_0} h(x) = L.$$

Similarly for the following situation:

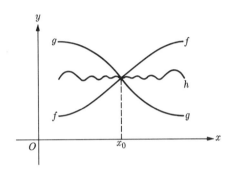

All of these ideas are very closely related, and we shall refer to all of them as the squeeze principle.

PROBLEM SET 4.3

Derive formulas for the following:

1. $D \tan x$ 2. $D \cot x$ 3. $D \sec x$ 4. $D \csc x$

5. $D\sqrt{1 - \sin^2 x}$ [*Warning:* It is very easy to get a wrong answer to this one.]

6. $D\sqrt{1 - \cos^2 x}$ [Same warning.]

7. $D \cos^2 x$ 8. $D(\cos^2 \theta + \sin^2 \theta)$

9. $D\, 2 \sin x \cos x$ 10. $D\sqrt{1 + \tan^2 \theta}$

11. $D(\csc^2 \theta - \cot^2 \theta)$ 12. $D \dfrac{\sin x}{1 + \cos x}$

13. $D \dfrac{\cos x}{1 + \sin x}$ 14. $D(x^2 \sin x)$

Show that the following differentiation formulas are correct:

15. $D \sin 2x = (\cos 2x)2$ 16. $D \cos 2x = (-\sin 2x)2$

17. $D \tan 2x = 2 \sec^2 2x$ 18. $D \sin (-x) = [\cos (-x)](-1)$

19. $D \cos (-x) = [-\sin (-x)](-1)$ 20. $D \cot 2x = -2 \csc 2x \cot 2x$

21. $D \tan (-x) = [\sec^2 (-x)](-1)$ 22. $D \sin 3x = (\cos 3x)3$

23. $D \cos 3x = (-\sin 3x)3$ 24. $D \tan 3x = 3 \sec^2 3x$

*25. Make a plausible guess for $D_x \sin ax$, and verify it if you can.

*26. Same, for $D_x \sin \sqrt{x}$.

*27. If $f(x) = \sin x$ and $g(x) = \cos x$, then

(a) $f' = g$,   (b) $g' = -f$,   (c) $f(0) = 0$,   (d) $g(0) = 1$.

Is it possible that there is another pair of functions satisfying the same four conditions? [*Hint:* Suppose that the pairs f_1, g_1 and f_2, g_2 satisfy (a) through (d). Consider the function

$$F = (f_1 - f_2)^2 + (g_1 - g_2)^2.$$

What sort of function is F? From what you learn about F, what can you conclude about f_1, f_2, g_1, and g_2?]

The answer to this problem has a rather curious significance: it means that all properties of the sine and cosine are contained, implicitly, in conditions (a) through (d). That is, the sine and cosine are completely described by the conditions

$$\sin' = \cos, \qquad \cos' = -\sin, \qquad \sin 0 = 0, \qquad \cos 0 = 1.$$

4.4 THE APPROXIMATION OF DIFFERENCES BY DIFFERENTIALS

We recall, from the preceding section, the apparatus which we set up in order to calculate the derivative of the sine. Given a function

$$f: I \to \mathbf{R},$$

where I is an interval, and a fixed point x_0 of I. For each point x of I, we let $\Delta x = x - x_0$, so that $x = x_0 + \Delta x$.

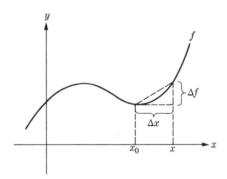

We let

$$\Delta f = f(x) - f(x_0) = f(x_0 + \Delta x) - f(x_0).$$

In the old notation,

$$f'(x_0) = \lim_{x \to x_0} \frac{f(x) - f(x_0)}{x - x_0},$$

by definition. In the new notation, this takes the form

$$f'(x_0) = \lim_{\Delta x \to 0} \frac{f(x_0 + \Delta x) - f(x_0)}{\Delta x} = \lim_{\Delta x \to 0} \frac{\Delta f}{\Delta x}.$$

When Δx is small, $\Delta f/\Delta x$ is close to $f'(x_0)$. Thus

$$\frac{\Delta f}{\Delta x} \approx f'(x_0) \qquad \text{when } \Delta x \approx 0,$$

where $\approx$ stands for the phrase "is approximately equal to." This ought to mean that

$$\Delta f \approx f'(x_0)\,\Delta x \qquad \text{when } \Delta x \approx 0.$$

Let us interpret this last statement geometrically.

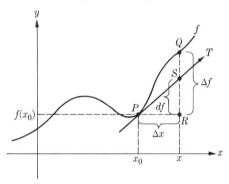

In the figure, the line T is the tangent to the graph of f at the point $(x_0, f(x_0))$. Thus the slope of T is $f'(x_0)$. If $S = (x, y)$, then

$$\frac{y - f(x_0)}{x - x_0} = f'(x_0),$$

because the slope of the segment from P to S is the slope of the line T. This gives

$$y - f(x_0) = f'(x_0)\,\Delta x.$$

This quantity is called the *differential of f at* x_0, and is denoted by *df*. (See the label in the figure.) To repeat:

$$df = f'(x_0)\,\Delta x,$$

by definition. Since x_0 is regarded as fixed, throughout this discussion, *df* is a function, whose value is determined when Δx is named. The differential is often convenient for purposes of numerical approximation. We have observed that

$$\Delta f \approx f'(x_0)\,\Delta x \qquad \text{when } \Delta x \approx 0.$$

In our new notation, this says that

$$\Delta f \approx df \qquad \text{when } \Delta x \approx 0.$$

Let us try this on some numerical examples, and see how good the approximation looks.

Example 1. Let
$$f(x) = \sqrt{x} \qquad (x \geq 0);$$
and take
$$x_0 = 25, \qquad \Delta x = 0.4.$$

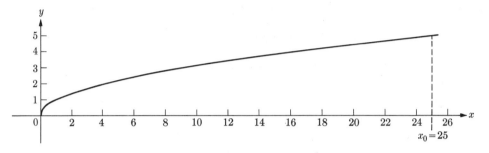

Then

$$f(x_0) = \sqrt{25} = 5,$$

$$f'(x) = \frac{1}{2\sqrt{x}} \quad (x > 0),$$

$$f'(x_0) = \frac{1}{2\sqrt{25}} = \frac{1}{10},$$

and
$$df = \tfrac{1}{10} \cdot \Delta x = \tfrac{1}{10} \cdot (0.4) = 0.04.$$

The approximation formula
$$df \approx \Delta f$$
suggests that

$$\sqrt{25.4} = f(x_0 + \Delta x) = f(x_0) + \Delta f \approx f(x_0) + df = \sqrt{25} + 0.04;$$

$$\sqrt{25.4} \approx 5.04.$$

The actual value of $\sqrt{25.4}$, correct to six decimal places, is

$$\sqrt{25.4} = 5.039841.$$

Thus the error in our approximation is 0.000159, which is not bad. Note also that the approximation $\Delta f \approx df$ wasn't supposed to be good except when Δx is small; and $\Delta x = 0.4$ is not very small. Using $\Delta x = 0.1$, we get

$$df = \frac{1}{2\sqrt{25}} (0.1) = 0.01;$$

$$\sqrt{25.1} \approx f(5) + df = 5.01.$$

The correct value is

$$\sqrt{25.1} = 5.00999,$$

so that our error is 0.00001, which looks better. Using $\Delta x = 0.01$, we get

$$df = \tfrac{1}{10}(0.01) = 0.001;$$
$$\sqrt{25.01} \approx 5.001.$$

Using five-place common logarithms, we get

$$\sqrt{25.01} \approx 5.0010.$$

Thus, in this case, the differential is as accurate as five-place tables.

It is natural to ask why the approximation

$$\Delta f \approx df = f'(x_0)\,\Delta x$$

should be as good as it is. The reason is as follows. We know that

$$f'(x_0) = \lim_{\Delta x \to 0} \frac{\Delta f}{\Delta x}\,.$$

On this basis, we wrote

$$f'(x_0) \approx \frac{\Delta f}{\Delta x} \qquad \text{when } \Delta x \approx 0. \tag{1}$$

Multiplying by Δx, we got

$$\Delta f \approx f'(x_0)\,\Delta x \qquad \text{when } \Delta x \approx 0. \tag{2}$$

The second of these approximations is *much* better than the first. The point is that if

$$\frac{\Delta f}{\Delta x} - f'(x_0) \approx 0 \qquad \text{and} \qquad \Delta x \approx 0,$$

then the product

$$\left[\frac{\Delta f}{\Delta x} - f'(x_0)\right]\Delta x \approx 0;$$

when you multiply two numbers each of which is small, the product is even smaller. We shall now express these ideas in a more exact form. For each Δx, let

$$E(\Delta x) = \frac{f(x_0 + \Delta x) - f(x_0)}{\Delta x} - f'(x_0) \qquad (\Delta x \neq 0).$$

Then

$$\lim_{\Delta x \to 0} E(\Delta x) = 0,$$

because

$$\lim_{\Delta x \to 0} \frac{f(x_0 + \Delta x) - f(x_0)}{\Delta x} = f'(x_0).$$

Thus the graph of the function E looks like the figure on the left below.

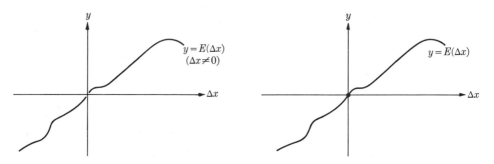

To this graph we add the origin. That is, we define

$$E(0) = 0.$$

The graph of the extended function E is shown on the right above. We now have

$$\lim_{\Delta x \to 0} E(\Delta x) = E(0) = 0.$$

Note that E is defined on some open interval containing 0.

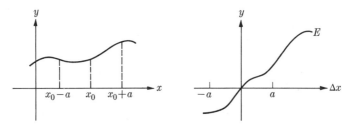

The reason is that if the open interval $(x_0 - a, x_0 + a)$ lies in the domain of f, then the open interval $(-a, a)$ lies in the domain of E. An open interval containing a given point will be called a *neighborhood* of the given point. In this language, we can sum up the above discussion in the following theorem.

Theorem 1. Let f be a function defined in a neighborhood of x_0, and suppose that f is differentiable at x_0. Then there is a function E, defined in a neighborhood of 0, such that

i) $\Delta f = f'(x_0) \, \Delta x + E(\Delta x) \, \Delta x$, and

ii) $\lim_{\Delta x \to 0} E(\Delta x) = E(0) = 0$.

(*Proof.* Use the function E which we have just defined.)

This theorem explains why

$$df = f'(x_0) \, \Delta x$$

is a good approximation of

$$\Delta f = f(x_0 + \Delta x) - f(x_0)$$

when $\Delta x \approx 0$. The reason is that

$$\Delta f - df = \Delta f - f'(x_0)\,\Delta x = E(\Delta x)\,\Delta x.$$

Thus when $\Delta x \approx 0$, the error in the approximation $\Delta f \approx df$ is doubly small, being a small multiple of the small number Δx.

In many cases, it is possible to estimate the largest possible error that can result when you use df as an approximation for Δf. For a discussion of this problem, see Appendix D.

PROBLEM SET 4.4

Following is a partial table of the sine and cosine functions, for ready reference in solving some of the following problems:

x	$\sin x$	$\cos x$
0.0814	0.0814	0.9967
0.1222	0.1219	0.9925
0.2094	0.2079	0.9781
0.3840	0.3746	0.9272

1. Find $\sin 0.1251$ approximately ($\sin 0.1251 = 0.1248$, correct to four decimal places).

2. Find $\cos 0.0844$ approximately ($\cos 0.0844 = 0.9964$).

3. Find $\sin 0.2123$ approximately ($\sin 0.2123 = 0.2108$).

4. Find $\cos 0.3869$ approximately ($\cos 0.3869 = 0.9261$).

5. How do you account for the first two entries in the first line of the above table?

6. Without the use of tables of any kind, get the best approximation you can for $\sin 0.5235988$. (This is a trick question.)

7. Same question, for $\cos(-6.2832)$.

Without using tables of any kind, get numerical approximations for the following. The answers given are the "right answers," correct to the indicated number of decimal places; it is not to be expected that an approximation process based on the differential will give them exactly.

8. $\sqrt[3]{27.1}$ [*Answer:* 3.004]   9. $\sqrt{25.2}$ [*Answer:* 5.0200]

10. $\sqrt[4]{16.3}$ 11. $\sqrt[3]{-7.9}$

12. One of the standard approximation formulas used in mathematical physics says that

$$\sin x \approx x \qquad \text{when } x \approx 0.$$

Explain how this formula is related to the ideas in this section of the text. [*Hint:* Consider the general approximation formula

$$\Delta f \approx df \qquad \text{when } \Delta x \approx 0.$$

What form does this take, for $f(x) = \sin x$, $x_0 = 0$?]

13. Same question, for

$$\cos(1 - x) \approx 1 \qquad \text{when } x \approx 1.$$

14. Another standard approximation formula says that

$$(1 + x)^n \approx 1 + nx \qquad \text{when } x \approx 0.$$

Interpret this in terms of the theory that we have been developing, and justify it. [*Hint:* Surely the given formula is equivalent to

$$(1 + \Delta x)^n - 1 \approx n\,\Delta x \qquad \text{when } \Delta x \approx 0.$$

Here, what is f? What is x_0? What are Δf and df?]

15. Same question, for the formula

$$\sqrt{1 + x} \approx 1 + \frac{x}{2} \qquad \text{when } x \approx 0.$$

16. Same question, for the formula

$$\frac{1}{\sqrt[3]{1 + x}} \approx 1 - \frac{x}{3} \qquad \text{when } x \approx 0.$$

17. Without using calculus at all, justify the approximation formula

$$\frac{1}{1 + x} \approx 1 - x \qquad \text{when } x \approx 0.$$

Is this a "doubly good" approximation in the same sense in which $\Delta f \approx df$ is "doubly good"? Why or why not?

4.5 COMPOSITION OF FUNCTIONS

In calculating derivatives, we have often found it convenient to regard one function as a power of another. For example, given

$$\phi(x) = (x^2 + 3x + 5)^5,$$

we let

$$g(x) = x^2 + 3x + 5,$$

so that

$$\phi = g^5.$$

We can then get ϕ' in the form

$$\phi' = 5g^4 g', \qquad \phi'(x) = 5(x^2 + 3x + 5)^4(2x + 3).$$

Similarly, we have found it convenient to regard one function as the positive square root of another. For example, given

$$\phi(x) = \sqrt{x^2 + 1},$$

we let

$$g(x) = x^2 + 1.$$

We then get ϕ' in the form

$$\phi' = \frac{g'}{2\sqrt{g}}, \qquad \phi'(x) = \frac{2x}{2\sqrt{x^2 + 1}} = \frac{x}{\sqrt{x^2 + 1}}.$$

The idea that we have been using is that of composition of functions. In the first case, the action of ϕ is described by

$$\phi: x \mapsto (x^2 + 3x + 5)^5.$$

We split this operation into two steps, like this:

$$x \mapsto x^2 + 3x + 5 \mapsto (x^2 + 3x + 5)^5.$$

The first of these steps represents the action of the function

$$g: x \mapsto x^2 + 3x + 5.$$

The second step raises things to the fifth power. It can thus be described by the function

$$f: u \mapsto u^5.$$

In this situation, g is called the *inside* function; it represents the first step. The function f is called the *outside* function; it represents the second step. And ϕ is called the *composition* of f and g. The reason for the use of the terms *inside* and *outside* is that we can write

$$\phi(x) = f(g(x)).$$

To get $\phi(x)$, we should substitute g in the formula for f.

Diagrammatically:

$$x \overset{g}{\mapsto} x^2 + 3x + 5 \overset{f}{\mapsto} (x^2 + 3x + 5)^5.$$
$$\phi = f(g)$$

Our second example fits the same pattern. We have

$$\phi(x) = \sqrt{x^2 + 1}, \qquad g(x) = x^2 + 1, \qquad f(u) = \sqrt{u},$$
$$\phi(x) = f(g(x)) = \sqrt{g(x)} = \sqrt{x^2 + 1}.$$

Diagrammatically:

$$x \overset{g}{\mapsto} x^2 + 1 \overset{f}{\mapsto} \sqrt{x^2 + 1}.$$
$$\phi = f(g)$$

Algebraically, to get the values of the composite function $\phi = f(g)$, we substitute $g(x)$ for u in the formula for $f(u)$. This is why we described the "square-root function" f by the formula

$$f(u) = \sqrt{u}$$

instead of the equally logical formula $f(x) = \sqrt{x}$. We want to form the composite function by setting

$$u = g(x) = x^2 + 1,$$

and it would hardly make sense to set $(?) \, x = g(x) = x^2 + 1 \, (?)$.

We sum all this up in the following definition:

Definition. Given two functions

$$g: A \to B, \qquad f: B \to C,$$

the *composition*

$$f(g): A \to C$$

is the function whose values are given by the formula

$$f(g)(x) = f(g(x)).$$

Here, for each x, $f(g)(x)$ denotes the value of the function $f(g)$ at the point x. Diagrammatically:

$$A \overset{g}{\mapsto} B \overset{f}{\mapsto} C.$$
$$\underbrace{}_{f(g)}$$

Let us consider some more examples.

Example 1. Let

$$f(u) = \sin u, \qquad g(x) = 2x + 1.$$

Then

$$f(g(x)) = \sin g(x) = \sin (2x + 1).$$

(In this example, what is A? What are B and C?)

Example 2. Let

$$f(u) = u^2 + u + 1, \qquad g(x) = \sqrt{x}.$$

Then

$$f(g(x)) = (\sqrt{x})^2 + \sqrt{x} + 1 = x + \sqrt{x} + 1.$$

(What are A, B, and C?)

Example 3. Let

$$f(u) = \sin u, \qquad g(x) = x^2 - 5.$$

Then

$$f(g(x)) = \sin (g(x)) = \sin (x^2 - 5).$$

Example 4. Let

$$f(u) = \int_0^u (t^4 - 1)\, dt, \qquad g(x) = x^2.$$

Then

$$f(g(x)) = \int_0^{g(x)} (t^4 - 1)\, dt = \int_0^{x^2} (t^4 - 1)\, dt.$$

Thus, for example,

$$f(g(3)) = \int_0^{3^2} (t^4 - 1)\, dt = \int_0^9 (t^4 - 1)\, dt.$$

In Examples 1 through 4 above, we supposed that f and g were given, and we then proceeded to form the composite function $\phi = f(g)$. More often, however, we are given a function ϕ, and in order to investigate the function ϕ, we express it as the composition of two other functions, each of which is simpler than ϕ. For example, to investigate the function

$$\phi(x) = \int_0^{x^2} (t^4 - 1)\, dt,$$

we first observe that it has the form

$$\phi(x) = \int_0^{g(x)} (t^4 - 1)\, dt,$$

where

$$g(x) = x^2.$$

Thus

$$\phi = f(g),$$

where

$$f(u) = \int_0^u (t^4 - 1)\, dt, \qquad g(x) = x^2.$$

Similarly, in the preceding three examples, if ϕ is given by the final formula, we shall for many purposes need to set up a pair of functions f and g in such a way that $\phi = f(g)$.

The derivative of a function is also a function; and so we can form composite functions of the type $f'(g)$ and $f(g')$. Consider, for example,

$$f(u) = u^3, \qquad g(x) = \sin x.$$

Then

$$f'(u) = 3u^2, \qquad g'(x) = \cos x.$$

Therefore

$$f(g(x)) = \sin^3 x, \qquad f'(g(x)) = 3 \sin^2 x, \qquad f'(g(x))g'(x) = 3 \sin^2 x \cos x.$$

These formulas are significant, because it will turn out that

$$Df(g(x)) = D \sin^3 x = f'(g(x))g'(x) = 3 \sin^2 x \cos x.$$

Similarly,

$$g(f(u)) = \sin u^3, \qquad g'(f(u)) = \cos u^3, \qquad g'(f(u))f'(u) = (\cos u^3)3\, u^2,$$

which will turn out to be the derivative of $\cos u^3$. (Here $\cos u^3$ is the cosine of u^3, not the cube of $\cos u$.) Let us try one more example:

$$f(u) = \cos u, \qquad g(x) = \sqrt{x}, \qquad g'(x) = \frac{1}{2\sqrt{x}},$$

$$f(g(x)) = \cos x, \qquad f'(g(x)) = -\sin \sqrt{x},$$

$$f'(g(x))g'(x) = (-\sin \sqrt{x})\frac{1}{2\sqrt{x}}.$$

In dealing with composite functions, we shall need the following:

Theorem 1. The composition of two continuous functions is continuous. That is, if

$$\lim_{x \to x_0} g(x) = g(x_0) = u_0$$

and

$$\lim_{u \to u_0} f(u) = f(u_0),$$

then

$$\lim_{x \to x_0} f(g(x)) = f(g(x_0)).$$

The idea here is that

$$x \approx x_0 \implies g(x) \approx g(x_0) \implies f(g(x)) \approx f(g(x_0)).$$

In Appendix E it is shown that this idea can be used to get a proof.

PROBLEM SET 4.5

For each of the functions ϕ, given in the problems below, find formulas for functions f and g, such that $\phi = f(g)$. Then get formulas for $f', g', f'(g)$, and ϕ'.

1. $\phi(x) = \sin^2 x$ 2. $\phi(x) = \cos^2 x$ 3. $\phi(x) = (\sin x + \cos x)^2$
4. $\phi(x) = \sin 2x$ 5. $\phi(x) = \tan 2x$ 6. $\phi(x) = \cos 2x$

7. $\phi(x) = \sqrt{1 - x^2}$ 8. $\phi(x) = \sin^6 x$ 9. $\phi(x) = \sqrt[3]{1 + x}$

10. $\phi(x) = \displaystyle\int_0^{\sin x} (t^2 + 1)\, dt$ 11. $\phi(x) = \displaystyle\int_0^{\cos x} (t^2 + 1)\, dt$

(Note that the function

$$f(u) = \int_0^u (t^2 + 1)\, dt$$

can be expressed without the use of integral signs; f can be calculated as a polynomial.)

12. a) Find $\displaystyle\lim_{u \to u_0} \frac{\sin u - \sin u_0}{u - u_0}$.   b) Find $\displaystyle\lim_{x \to x_0} \frac{\sin x^2 - \sin x_0^2}{x^2 - x_0^2}$.

(It is not hard to see a very plausible answer to Problem 12(b). To prove, in an orderly way, that your answer is right, you should express the function

$$\phi(x) = \frac{\sin x^2 - \sin x_0^2}{x^2 - x_0^2}$$

as the composition $f(g)$ of two functions f and g, and then apply Theorem 1.)

13. Find $\displaystyle\lim_{x \to x_0} \frac{\sin x^3 - \sin x_0^3}{x - x_0}$.

14. Given $\phi(x) = \sin x^2$, proceed as in Problems 1 through 11.

15. Do the same, for $\phi(x) = \sin x^3$.

16. Do the same, for $\phi(x) = \sin \sqrt{x}$.

17. Given

$$\phi(x) = \int_0^{\sin x} \sqrt{1 + t^2}\, dt.$$

On the basis of the theory that you know so far, you are in no position to calculate

$$f(u) = \int_0^u \sqrt{1 + t^2}\, dt.$$

And you have, so far, no general formula for

$$D[f(g)] = ?$$

On the other hand, you ought by this time to be able to make a good guess about $D[f(g)]$, and then use your guess to write *some* kind of formula for

$$\phi'(x) = D \int_0^{\sin x} \sqrt{1 + t^2}\, dt.$$

[*Hint:* As a start, what is $f'(u)$?]

18. Find $\lim\limits_{x \to \pi/2} \dfrac{\sin x - 1}{x - \pi/2}$.

[*Hint:* If you can figure out what the geometric meaning of this limit is, it will then be easy to find its numerical value.]

19. Find $\lim\limits_{x \to \pi} \dfrac{\cos x + 1}{x - \pi}$.   [Same hint.]   20. Find $\lim\limits_{x \to \pi/4} \dfrac{\tan x - 1}{x - \pi/4}$.

21. Find $\lim\limits_{x \to \pi/4} \dfrac{\sin 2x - 1}{x - \pi/4}$.   22. Find $\lim\limits_{x \to 0} \dfrac{\sec x - 1}{x}$.

4.6 THE CHAIN RULE

You may have observed, in the preceding problem set, that the formula

$$Df(g) = f'(g)g'$$

held in a number of cases. For example, if

$$f(u) = u^n,$$

then

$$f'(u) = nu^{n-1};$$

and

$$Df(g) = Dg^n = ng^{n-1}g' = f'(g)g'.$$

Similarly, if

$$f(u) = \sqrt{u},$$

then

$$f'(u) = \frac{1}{2\sqrt{u}},$$

$$Df(g) = D\sqrt{g} = \frac{1}{2\sqrt{g}} \cdot g' = f'(g)g'.$$

The same formula seems to hold for

$$f(u) = \sin u,$$

at least in the cases where we can test the formula by calculating $Df(g) = D \sin g$. For example, it turns out that

$$D \sin 2x = 2 \cos 2x,$$

and this has the form

$$Df(g) = f'(g)g',$$

where

$$f(u) = \sin u, \qquad f'(u) = \cos u,$$
$$g(x) = 2x, \qquad g'(x) = 2,$$
$$f'(g)g' = (\cos 2x) \cdot 2 = 2 \cos 2x.$$

The formula

$$Df(g) = f'(g)g'$$

is called the *chain rule*. In fact, it always holds, whenever the right-hand side has a meaning, that is, whenever $f'(g)$ and g' are defined. We shall prove this at the end of this section. First, we give some illustrations of its use.

Example 1. Consider

$$\phi(x) = \sin (3x + 1).$$

This is a composite function

$$\phi(x) = f(g(x)),$$

with

$$f(u) = \sin u, \qquad f'(u) = \cos u,$$
$$g(x) = 3x + 1, \qquad g'(x) = 3.$$

By the chain rule,

$$\phi'(x) = D \sin (3x + 1) = [\cos (3x + 1)]D(3x + 1)$$
$$= 3 \cos (3x + 1).$$

Example 2. Consider

$$\phi(x) = \sin (k + x).$$

By the chain rule,

$$\phi'(x) = [\cos (k + x)]D(k + x) = \cos (k + x).$$

Note that if the chain rule is known, and the formula

$$D \sin = \cos$$

is known, we can find $D \sin (k + x)$ without using the addition formula.

To give new applications of the chain rule, we should not be talking about cases where the outside function f is u^n or $\sqrt{u}$. For these outside functions, we have known for a long time that the chain rule held. After the trigonometric functions, the next outside functions to consider are integrals:

Example 3. Consider

$$\phi(x) = \int_1^{kx} \frac{1}{t}\, dt \qquad (k, x > 0).$$

Here

$$\phi(x) = f\big(g(x)\big), \qquad f(u) = \int_1^u \frac{1}{t}\, dt \quad (u > 0),$$

$$f'(u) = \frac{1}{u}, \qquad g(x) = kx, \qquad f'\big(g(x)\big) = \frac{1}{kx}, \qquad g'(x) = k,$$

$$D\phi = Df(g) = f'(g)g' = \frac{1}{kx}\, k = \frac{1}{x}.$$

This is a curious result:

$$D\int_1^{kx} \frac{1}{t}\, dt = \frac{1}{x} = D\int_1^{x} \frac{1}{t}\, dt.$$

What does it tell us about the functions?

Example 4. The chain rule can be applied several times in the same problem. For example, we know that

$$D \sin g = (\cos g)g',$$

whatever g may be. We can then apply the formula in cases where g' itself needs to be calculated by the chain rule:

$$D \sin \sin x = (\cos \sin x)D \sin x$$
$$= (\cos \sin x)\cos x.$$

Here $\sin \sin x$ is the sine of the sine of x, which is different from $\sin^2 x$.
    Therefore

$$D \sin \sin \sin x = (\cos \sin \sin x)D \sin \sin x$$
$$= (\cos \sin \sin x)(\cos \sin x)\cos x.$$

Example 5. Similarly,

$$D\{[(x^3 + 1)^2 + 1]^2 + 1\}^3 = 3\{[(x^3 + 1)^2 + 1]^2 + 1\}^2 D\{[(x^3 + 1)^2 + 1]^2 + 1\}$$
$$= 3\{\ \}^2 \cdot 2[(x^3 + 1)^2 + 1]D[(x^3 + 1)^2 + 1]$$
$$= 3\{\ \}^2 \cdot 2 \cdot [\ ] \cdot 2(x^3 + 1)D(x^3 + 1)$$
$$= 3\{\ \}^2 \cdot 2[\ ] \cdot 2(\ ) \cdot 3x^2.$$

Here we have left braces, brackets, and parentheses empty, in the intermediate stages, to make the steps easier to follow. The final answer is

$$3\{[(x^3 + 1)^2 + 1]^2 + 1\}^2 \cdot 2[(x^3 + 1)^2 + 1] \cdot 2(x^3 + 1) \cdot 3x^2,$$

which can be simplified slightly by collecting constants.

We shall now prove the chain rule. Given

$$\phi(x) = f(g(x)),$$

we want to show that for each x_0 we have

$$\phi'(x_0) = f'(g(x_0))g'(x_0).$$

Obviously, we must assume that

a) g has a derivative $g'(x_0)$ at x_0, and
b) f has a derivative $f'(g(x_0))$ at $g(x_0)$.

A differentiable function is continuous. Therefore

$$\lim_{x \to x_0} g(x) = g(x_0).$$

For convenience of reference later, we write this in the form

c) $$\lim_{\Delta x \to 0} [g(x_0 + \Delta x) - g(x_0)] = 0.$$

By definition,

$$\phi'(x_0) = \lim_{x \to x_0} \frac{\phi(x) - \phi(x_0)}{x - x_0}$$

$$= \lim_{\Delta x \to 0} \frac{\phi(x_0 + \Delta x) - \phi(x_0)}{\Delta x}$$

$$= \lim_{\Delta x \to 0} \frac{f(g(x_0 + \Delta x)) - f(g(x_0))}{\Delta x}.$$

Let

$$\Delta u = g(x_0 + \Delta x) - g(x_0) = g(x_0 + \Delta x) - u_0,$$

so that

$$g(x_0 + \Delta x) = u_0 + \Delta u.$$

Then

$$\phi'(x_0) = \lim_{\Delta x \to 0} \frac{f(u_0 + \Delta u) - f(u_0)}{\Delta x}.$$

Here the numerator is a difference

$$\Delta f = f(u_0 + \Delta u) - f(u_0)$$

between two values of the function f.

Now comes the crucial idea: we apply to f the theorem stated at the end of Section 4.4. We need to change the notation of the theorem, using u in place of x, to fit the notation of the present discussion. The theorem then says that there is a function E, defined in a neighborhood of 0, such that

$$\Delta f = f'(u_0) \Delta u + E(\Delta u) \Delta u,$$

and

$$\lim_{\Delta u \to 0} E(\Delta u) = E(0) = 0.$$

Therefore

$$\frac{\Delta f}{\Delta x} = f'(u_0)\frac{\Delta u}{\Delta x} + E(\Delta u)\frac{\Delta u}{\Delta x}$$

$$= f'(g(x_0))\frac{g(x_0 + \Delta x) - g(x_0)}{\Delta x} + E(\Delta u)\frac{g(x_0 + \Delta x) - g(x_0)}{\Delta x}.$$

It is now easy to see what the limit is. By definition of $g'(x_0)$, we have

$$\lim_{\Delta x \to 0}\frac{g(x_0 + \Delta x) - g(x_0)}{\Delta x} = g'(x_0).$$

As $\Delta x \to 0$, $\Delta u \to 0$. (Remember the definition of Δu, and recall condition (c), at the beginning of the proof.) Therefore, by Theorem 1 of Section 4.5, we have

$$\lim_{\Delta x \to 0} E(\Delta u) = 0.$$

This gives

$$\phi'(x_0) = \lim_{\Delta x \to 0}\frac{\Delta f}{\Delta x} = f'(g(x_0))g'(x_0) + 0 \cdot g'(x_0).$$

We therefore have:

Theorem 1. Let f and g be functions. Then

$$Df(g) = f'(g)g',$$

at every point x_0 at which the right-hand member has a meaning.

That is, the formula holds at every point x_0 such that (a) g is differentiable at x_0 and (b) f is differentiable at $g(x_0)$. These conditions illustrate the normal pattern of theorems involving differentiation formulas: the equation holds whenever the quantities mentioned in the right-hand member exist.

PROBLEM SET 4.6

In this problem set, your main job is to learn to use the chain rule. In each odd-numbered problem, from 1 to 19, you should indicate the logic of your work by writing formulas for $f, g, f', f'(g)$, and g', before writing the answer in the form $D[f(g)] = f'(g)g'$. For example, given the function

$$\phi(x) = \sin (x^2 + 1),$$

your solution should be written in the form

$$\begin{aligned} f(u) &= \sin u, & g(x) &= x^2 + 1, \\ f'(u) &= \cos u, & f'(g(x)) &= \cos (x^2 + 1), \\ g'(x) &= 2x, & \phi'(x) &= [\cos (x^2 + 1)]2x = 2x \cos (x^2 + 1). \end{aligned}$$

If you go through this routine for one day, you are less likely hereafter to omit the factor g' following $f'(g)$ in calculating $Df(g)$.

The parentheses and brackets in the expression $[\cos (x^2 + 1)]2x$ look clumsy, but to eliminate the brackets we have to change the order of the factors, as in the last expression above. It would have been simpler to write

$$(?) \quad \phi'(x) = \cos(x^2 + 1)2x \quad (?)$$

but this is the wrong answer: the function on the right is the function whose value, for each x, is the cosine of $2x^3 + 2x$. If you write this formula for ϕ', you are relying on the reader to remember what the problem was and to realize that you must not mean what you are saying.

In some cases you may not feel sure whether brackets are necessary. When in doubt, use them.

Now find the derivatives of the following functions:

1. $\sin x^2$ 2. $\sin^2 x$ 3. $\cos x^3$ 4. $\cos^3 x$ 5. $\tan(t^2 + 1)$ 6. $\tan t^2 + 1$

7. $\sin(x^3 + x)$ 8. $\sin x^3 + x$ 9. $\cos \sqrt{x}$ 10. $\sqrt{\cos x}$ 11. $\tan \dfrac{x - 1}{2}$

12. $\dfrac{\tan x - 1}{2}$ 13. $(\sqrt{x})^2$ 14. $|x|^2$ 15. a) $\sec^2 x$ b) $\sec x^2$

16. a) $\tan^2 x$ b) $\tan x^2$ 17. $\cos^4 x - \sin^4 x$ 18. $\cos 2x$ 19. $\cos^2 x - \sin^2 x$

20. $\sec 2x$ 21. $\sin x \cos^2 x + \sin^3 x$ 22. $\tan \dfrac{x}{2}$ 23. $\dfrac{\sin x}{1 + \cos x}$

24. $\sec \sqrt{x^2 + 1}$ 25. $\cos \sqrt{x^2 + 1}$ 26. a) $\sqrt{\tan x}$ b) $\tan \sqrt{x}$

27. a) $\sin \sqrt[3]{x}$ b) $\sqrt[3]{\sin x}$ 28. $\sin \sin x$ 29. $\sin^2 \sin x + \cos^2 \sin x$

30. $\cos \cos x$ 31. $\sin \cos x$ 32. $\sin \sin \sin t$ 33. $\tan \sin x$

34. $\sin(1 + \sin x)$ 35. $\displaystyle\int_0^{\cos x} t^3 \, dt$ 36. $\displaystyle\int_0^{x^3} \cos t \, dt$

37. Let k be any positive number; and for each positive number x, let

$$\phi(x) = \int_1^{kx} \frac{1}{t} \, dt - \int_1^x \frac{1}{t} \, dt - \int_1^k \frac{1}{t} \, dt.$$

Find the simplest possible formula for $\phi'(x)$. Then do the same, for the functions $\phi(x)$ defined by the following formulas.

38. $\displaystyle\int_1^{x^2} \frac{1}{t} \, dt \quad (x > 0)$ 39. $\displaystyle\int_1^{x^2} \frac{1}{t} \, dt - 2 \int_1^x \frac{1}{t} \, dt$ 40. $\displaystyle\int_1^{x^3} \frac{1}{t} \, dt$

41. $\displaystyle\int_1^{x^3} \frac{1}{t} \, dt - 3 \int_1^x \frac{1}{t} \, dt$ 42. $\displaystyle\int_1^{\sqrt{x}} \frac{1}{t} \, dt$ 43. $\displaystyle\int_1^{\sqrt{x}} \frac{1}{t} \, dt - \frac{1}{2} \int_1^x \frac{1}{t} \, dt$

44. $\displaystyle\int_1^{\sin x} \frac{1}{t} \, dt \quad (0 < x < \pi)$

45. For each $x > 0$, let

$$f(x) = \int_1^x \frac{1}{t} \, dt.$$

Show that for every pair of positive numbers a and b, we have

$$f(ab) = f(a) + f(b).$$

[*Hint:* When we try to attack this problem by the methods of calculus, the obvious trouble is that the problem does not appear to involve any functions. Therefore our first step should be to *introduce* a function into the problem.]

46. Let $\phi(x) = f(x^n)$, where $x > 0$ and f is as in the preceding problem. Find $\phi'(x)$.

*47. Given

$$D \sin = \cos, \qquad D \cos = -\sin,$$
$$\sin 0 = 0, \qquad \cos 0 = 1,$$

and given no other information whatever about the sine and cosine, prove that

$$\sin (k + x) = \sin k \cos x + \cos k \sin x,$$
$$\cos (k + x) = \cos k \cos x - \sin k \sin x,$$

for every k and x. [*Hint:* Let f be the function which is $= 0$ if the first equation holds; let g be the function which is $= 0$ if the second equation holds, and investigate the function

$$F = f^2 + g^2.]$$

This result tends to confirm a claim that was made in Problem *27 of Problem Set 4.3. The claim was that all properties of the sine and cosine are contained, implicitly, in the properties that we have just used to prove the addition formulas. Later we shall find further confirmation of this.

*48. Let f be a function, defined for every x, such that

   (a) $f'' = -f$,   (b) $f(0) = 0$,   (c) $f'(0) = 1$.

Show that $f(x) = \sin x$ for every x.

*49. Let g be a function, defined for every x, such that

   (a) $g'' = -g$,   (b) $g(0) = 1$,   (c) $g'(0) = 0$.

Show that $g(x) = \cos x$ for every x.

4.7 INVERTIBLE FUNCTIONS. THE INVERSE TRIGONOMETRIC FUNCTIONS

A function f is called *invertible* if its graph intersects every horizontal line in at most one point. Thus $f(x) = x^3$ is invertible, but $f(x) = x^2$ is not.

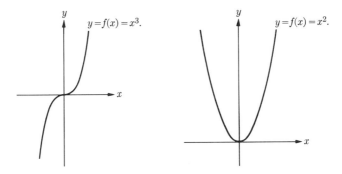

If f is invertible, then for each number y in the image of f there is exactly one number x in the domain of f such that $f(x) = y$.

Thus to every invertible function f there corresponds a new function f^{-1}, called the *inverse* of f. (This is pronounced f *inverse*. The symbol -1 is not an exponent, really; and f^{-1} is not $1/f$.) The inverse is defined by the condition that

$$f^{-1}(x) = y \qquad \text{if } f(y) = x.$$

If f is invertible, this condition defines a function, because for each x in the image of f there is exactly one such y. It is not hard to see what this relation between f and f^{-1} means geometrically. The point (x, y) is on the graph of f^{-1} if the point (y, x) is on the graph of f. Therefore, to get the graph of f^{-1} from the graph of f, we should reflect the graph of f across the line $y = x$.

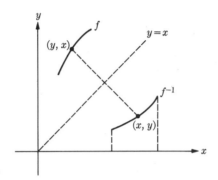

Let us see what this means algebraically. Consider

$$f(x) = x^3.$$

The graph of f is the graph of the equation

$$y = x^3. \tag{1}$$

The graph of f^{-1} is the graph of the equation

$$x = y^3. \tag{2}$$

Here we have simply interchanged x and y in Eq. (1). Now (2) is equivalent to

$$y = \sqrt[3]{x}.$$

Thus

$$f^{-1}(x) = \sqrt[3]{x},$$

as we would expect: the inverse of cubing is the extraction of cube roots.

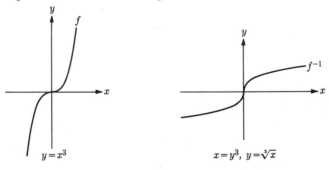

Theorem 1. Let f be an invertible function. Then

$$f(f^{-1}(x)) = x,$$

for every x.

Proof. For each x, let $y = f^{-1}(x)$. Then $f(y) = x$, by definition of f^{-1}. Therefore, $f(f^{-1}(x)) = f(y) = x$.

We can use this idea to calculate the derivatives of inverse functions, assuming that the inverse function has a derivative.

Example 1. The function $f(x) = x^3$ is invertible, and its inverse is $f^{-1}(x) = \sqrt[3]{x}$. Thus

$$(\sqrt[3]{x})^3 = x.$$

We take the derivative on each side of this equation, using the chain rule for the composite function on the left. This gives:

$$3(\sqrt[3]{x})^2 \, D\sqrt[3]{x} = 1,$$

$$D\sqrt[3]{x} = \frac{1}{3\sqrt[3]{x^2}} \qquad (x \neq 0).$$

You may have calculated this by another method, in Problem Set 3.6, but the present method is easier.

Example 2. A function of the form $f(x) = x^q$ (where q is a positive integer) is not necessarily invertible; in fact, it never is when q is even. We therefore restrict x to positive values. This gives an inverse function

$$f^{-1}(x) = \sqrt[q]{x}.$$

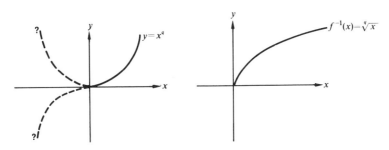

We calculate $D\sqrt[q]{x}$ in three steps, as follows:

1) $$(\sqrt[q]{x})^q = x,$$

2) $$q(\sqrt[q]{x})^{q-1} \, D\sqrt[q]{x} = 1,$$

3) $$D\sqrt[q]{x} = \frac{1}{q\sqrt[q]{x^{q-1}}} \qquad (x > 0).$$

When we use this method, the equations that we write have the following general form:

1)
$$f(f^{-1}(x)) = x,$$

2)
$$f'(f^{-1}(x))Df^{-1}(x) = 1,$$

3)
$$Df^{-1}(x) = \frac{1}{f'(f^{-1}(x))} \qquad (f'(f^{-1}(x)) \neq 0).$$

(You should check this against the preceding examples.) The method assumes that our problem has an answer, that is, that f^{-1} has a derivative. Thus we need to show that this holds, in every case in which the fraction at the last stage has a meaning. This is easy to see. Consider f, f^{-1}, as in the figure below, with

$$y_1 = f^{-1}(x_1), \qquad x_1 = f(y_1),$$

as the labels indicate.

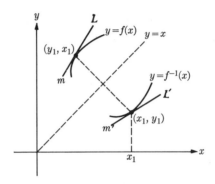

If f has a tangent line L, at (y_1, x_1), then f^{-1} has a tangent line L', at (x_1, y_1): to get this, we reflect both the graph and the tangent line across the line $y = x$. The slope of L is

$$m = f'(y_1) = f'(f^{-1}(x_1)).$$

If $m \neq 0$, then L is not horizontal. Therefore L' is not vertical, and f^{-1} has a derivative at x_1. Thus we have completed the proof of the following theorem.

Theorem 2.

$$Df^{-1}(x) = \frac{1}{f'(f^{-1}(x))},$$

wherever the fraction on the right has a meaning.

In most cases, the method used in deriving this formula is easier to use than the formula itself. To find Df^{-1}, we write $f(f^{-1}(x)) = 1$, differentiate, and solve for Df^{-1}, as in Examples 1 and 2.

We shall now discuss the so-called "inverse trigonometric functions." This involves a slight difficulty, because the fact is that no trigonometric function is invertible. The reason is that every trigonometric function satisfies the identity

$$f(x + 2\pi) = f(x),$$

for every x for which the trigonometric function $f(x)$ is defined at all. Therefore every value that a trigonometric function takes on at all is taken on for infinitely many values of x. For example, the graph of $f(x) = \sin x$ looks something like this:

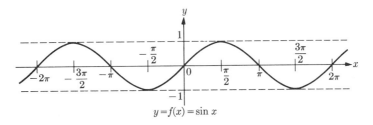

$$y = f(x) = \sin x$$

If we restrict x to the interval $[-\pi/2, \pi/2]$, then we get a new function whose graph includes some, but not all, of the original graph. This new function is denoted by Sin, and the graph of $y = \text{Sin } x$ looks like the left-hand figure below.

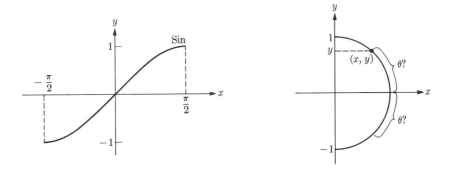

The graph looks as if Sin ought to be invertible; and in fact this is not hard to see. In the right-hand figure above, we have switched the notation to fit the definition of the sine, so that $y = \sin \theta$. Every point of the semicircle corresponds to exactly one θ on the interval $[-\pi/2, \pi/2]$; and every horizontal line intersects the semicircle in exactly one point.

As always for inverse functions, we get the graph of Sin^{-1} by reflecting the graph of Sin across the line $y = x$. Therefore the graph of Sin^{-1} looks like this:

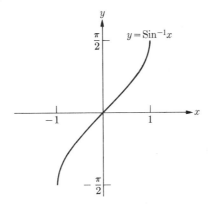

Similarly, we define Cos x to be equal to cos x, on the interval $[0, \pi]$, and we show that Cos is invertible. The graphs of Cos and Cos^{-1} look like this:

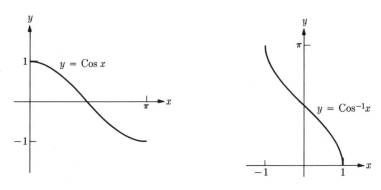

To find the derivative of Sin^{-1}, we write

$$\sin \text{Sin}^{-1} x = x,$$

$$(\cos \text{Sin}^{-1} x)D \,\text{Sin}^{-1} x = 1,$$

$$D \,\text{Sin}^{-1} x = \frac{1}{\cos \text{Sin}^{-1} x}.$$

We want to simplify the expression $\cos \text{Sin}^{-1} x$ on the right, and, while we are at it, we shall get a formula for $\sin \text{Cos}^{-1} x$. Since

$$\cos^2 u + \sin^2 u = 1,$$

we can now solve, getting

$$\cos u = \pm\sqrt{1 - \sin^2 u}, \tag{1}$$

$$\sin u = \pm\sqrt{1 - \cos^2 u}. \tag{2}$$

For

$$u = \text{Sin}^{-1} x,$$

this gives

$$\sin u = \sin \text{Sin}^{-1} x = x,$$

and so from (1) we get

$$\cos \text{Sin}^{-1} x = \pm\sqrt{1 - x^2}. \tag{3}$$

Similarly, for

$$u = \text{Cos}^{-1} x$$

we have

$$\cos u = \cos \text{Cos}^{-1} x = x,$$

and so from (2) we get

$$\sin \text{Cos}^{-1} x = \pm\sqrt{1 - x^2}. \tag{4}$$

Formulas (3) and (4) are correct, but they are not good enough for our purposes. In fact, *the double signs can be omitted, and the formulas still hold:*

Theorem 3.

$$\cos \text{Sin}^{-1} x = \sqrt{1 - x^2},$$

$$\sin \text{Cos}^{-1} x = \sqrt{1 - x^2}.$$

To see this, we merely need to remember that

$$-\frac{\pi}{2} \leqq \text{Sin}^{-1} x \leqq \frac{\pi}{2}.$$

On this interval, the cosine is $\geqq 0$. Therefore, in (3), it must be the plus sign that applies. Similarly,

$$0 \leqq \text{Cos}^{-1} x \leqq \pi.$$

On this interval, the sine is $\geqq 0$. Therefore, in (4), it must be the plus sign that applies.

We now substitute $\sqrt{1 - x^2}$ for $\cos \text{Sin}^{-1} x$, in the formula that we got for $D \text{ Sin}^{-1} x$. This gives:

Theorem 4. $D \text{ Sin}^{-1} x = 1/\sqrt{1 - x^2}$ $(-1 < x < 1)$.

Note that $D \text{ Sin}^{-1} x$ is always > 0, just as the graph suggests that it ought to be. At the endpoints of the graph, the tangents are vertical.

The proof of the following theorem is like that of the preceding one:

Theorem 5. $D \text{ Cos}^{-1} x = -1/\sqrt{1 - x^2}$ $(-1 < x < 1)$.

Note that $D \text{ Cos}^{-1} x$ is always < 0, as it should be.

For tan x, the process is simpler. The graph of $y = f(x) = \tan x$ looks something like this:

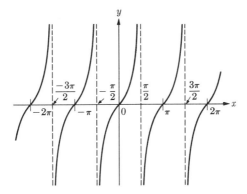

To get an invertible function Tan, we take the portion of the graph that lies between $x = -\pi/2$ and $x = \pi/2$. We could verify by brute force than Tan is invertible, but it is easier to prove first the following theorem:

Theorem 6.   Let f be a differentiable function on an interval I. If $f'(x) \neq 0$ for every x in I, then f is invertible.

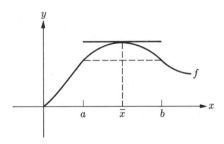

The proof is based on the mean-value theorem. If f is not invertible, then the graph intersects some horizontal line in more than one point. Thus

$$f(a) = f(b),$$

for some a and b in I. Therefore the graph has a horizontal chord. By MVT, this means that the graph has a horizontal tangent; that is, $f'(\bar{x}) = 0$ for some $\bar{x}$, which contradicts the hypothesis for f.

Now the domain of Tan is an open interval $(-\pi/2, \pi/2)$. On this interval, $\text{Tan}'\, x = \sec^2 x \neq 0$. Therefore Tan is invertible.

The graphs of Tan and Tan^{-1} look like this:

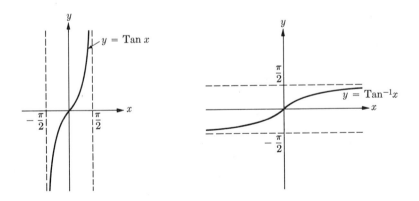

Theorem 7. $D \, \text{Tan}^{-1} x = 1/(1 + x^2)$.

The derivation is easier than the preceding ones, because it turns out that there are no double signs to be eliminated.

For the secant, the situation is trickier, and some handbooks contain formulas that are wrong. The reason is that the graph of the secant looks like this:

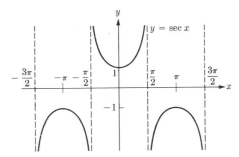

(Remember that $\sec x = 1/\cos x$ wherever $\cos x \neq 0$.)   This graph consists of infinitely many connected pieces, but none of these connected pieces is the graph of an invertible function. We therefore cannot use all of any one of the pieces. Everybody agrees that we ought to use the part of the graph where $0 \leqq x < \pi/2$, but there is no general agreement on what else we ought to use. To be safe, we define Sec x only for $0 \leqq x < \pi/2$. (See the graphs below.)

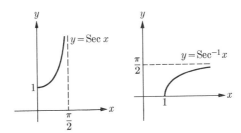

(*Query:* How do you know that the secant never takes on the same value twice, on the interval $[0, \pi/2)$?)

On the basis of the definition of Sec^{-1}, it is plain that the equation

$$y = \text{Sec}^{-1} x$$

means two things:

$$x = \sec y \tag{5}$$

and

$$0 \leqq y < \frac{\pi}{2}. \tag{6}$$

We now calculate the derivative. We have

$$\text{Sec Sec}^{-1} x = x, \tag{7}$$

and

$$\text{Sec}' u = \text{Sec } u \text{ Tan } u,$$

for every u from 0 to $\pi/2$. (Why are we justified in using capital letters on the right?) Therefore, by the chain rule,

$$(\text{Sec Sec}^{-1} x)(\text{Tan Sec}^{-1} x)D \text{ Sec}^{-1} x = 1,$$

and

$$x(\text{Tan Sec}^{-1} x)D \text{ Sec}^{-1} x = 1. \tag{8}$$

Therefore

$$D \text{ Sec}^{-1} x = \frac{1}{x \text{ Tan Sec}^{-1} x}. \tag{9}$$

We now need a formula for $\text{Tan Sec}^{-1} x$, analogous to the formulas for $\sin \text{Cos}^{-1} x$ and $\cos \text{Sin}^{-1} x$. We know that

$$1 + \tan^2 u = \sec^2 u$$

for every u. Therefore

$$\tan u = \pm\sqrt{\sec^2 u - 1}.$$

For $u = \text{Sec}^{-1} x$, this says that

$$\tan \text{Sec}^{-1} x = \pm\sqrt{\sec^2 \text{Sec}^{-1} x - 1} = \pm\sqrt{x^2 - 1}.$$

Since $0 \leqq \text{Sec}^{-1} x < \pi/2$, we have

$$\tan \text{Sec}^{-1} x = \text{Tan Sec}^{-1} x \geqq 0,$$

and so it must be the plus sign that applies on the right. Therefore

$$\text{Tan Sec}^{-1} x = \sqrt{x^2 - 1},$$

and we have:

Theorem 8. $D \text{ Sec}^{-1} x = 1/x\sqrt{x^2 - 1}.$

For convenience of reference, we repeat these differentiation formulas:

$$D \text{ Sin}^{-1} x = \frac{1}{\sqrt{1 - x^2}}, \qquad D \text{ Cos}^{-1} x = \frac{-1}{\sqrt{1 - x^2}},$$

$$D \text{ Tan}^{-1} x = \frac{1}{1 + x^2}, \qquad D \text{ Sec}^{-1} x = \frac{1}{x\sqrt{x^2 - 1}}.$$

We now have a new set of functions arising as derivatives: none of these four functions has appeared before as a result of differentiation. This means, for one thing, that we can use our new functions to solve certain area problems that we couldn't solve before. Later we shall see that the process by which we find a function whose derivative is a given function has many other applications.

You will also need to remember

$$\cos \text{Sin}^{-1} x = \sqrt{1 - x^2}, \qquad \sin \text{Cos}^{-1} x = \sqrt{1 - x^2}.$$

PROBLEM SET 4.7

For each of the following functions, calculate the derivative.

1. $\mathrm{Sin}^{-1}(x - 1)$
2. $\mathrm{Cos}^{-1}(x - 1)$
3. $\mathrm{Tan}^{-1}(x + 1)$
4. $\mathrm{Sec}^{-1}(x + 1)$
5. $\mathrm{Sin}\,\mathrm{Sin}^{-1}(x + 1)$
6. $\mathrm{Cos}\,\mathrm{Sin}^{-1}x$
7. $\mathrm{Sin}\,\mathrm{Sin}^{-1}x^2$
8. $\mathrm{Cos}\,\mathrm{Sin}^{-1}(2x)$
9. $\mathrm{Sin}^{-1}\sqrt{1 - x^2}$
10. $\mathrm{Cos}^{-1}\sqrt{1 - x^2}$
11. $\mathrm{Tan}^{-1}(x^2 + 1)$
12. $\mathrm{Tan}^{-1}(\sec^2 x - 1)$
13. $\mathrm{Sec}^{-1}x^2$
14. $\mathrm{Sec}^{-1}(1 + \tan^2 x)$
15. $\mathrm{Tan}^{-1}\dfrac{\sqrt{1 - x^2}}{x}$
16. $\mathrm{Sec}^{-1}\dfrac{1}{x}$
17. $\mathrm{Cos}^{-1}\dfrac{1}{x}$
18. $\mathrm{Sin}^{-1}\dfrac{1}{x}$
19. $\mathrm{Sec}\,\mathrm{Tan}^{-1}x$
20. $\mathrm{Tan}\,\mathrm{Sec}^{-1}x$
21. $\mathrm{Sin}\,\mathrm{Tan}^{-1}x$
22. $\mathrm{Cos}\,\mathrm{Tan}^{-1}x$
23. $\mathrm{Tan}\,\mathrm{Sin}^{-1}x$
24. $\mathrm{Tan}\,\mathrm{Cot}^{-1}x$
25. $\mathrm{Sin}^{-1}(2x + 1)$
26. $\mathrm{Tan}^{-1}(1 - x)$
27. $\mathrm{Sin}^{-1}x^2$

28. Show that

$$\mathrm{Sec}^{-1}x = \mathrm{Cos}^{-1}\frac{1}{x},$$

for every x on a certain interval. What interval?

29. Show that

$$\mathrm{Sin}^{-1}x + \mathrm{Cos}^{-1}x = \frac{\pi}{2},$$

for every x on the interval $[-1, 1]$. (A very short proof is possible. Remember the uniqueness theorem of Section 3.8.)

30. Find $\displaystyle\int_{-1}^{1}\frac{1}{1 + t^2}\,dt.$ Sketch.   31. Find $\displaystyle\int_{0}^{\sqrt{3}}\frac{1}{1 + t^2}\,dt.$ Sketch.

32. Find $\displaystyle\int_{0}^{1/2}\frac{1}{\sqrt{1 - t^2}}\,dt.$ 33. Find $\displaystyle\int_{0}^{1/\sqrt{2}}\frac{t}{\sqrt{1 - t^2}}\,dt.$

34. Try to get the right answer for the area under the graph of $y = 1/(1 + x^2)$, on the whole interval $(-\infty, \infty)$. You need not justify your answer, so long as it is right.

35. Given

$$f(x) = \sqrt{1 - x^2}, \qquad 0 \leqq x \leqq 1,$$

find a formula for $f^{-1}(x)$. Then explain how your answer might have been predicted without a calculation.

36. Find $\displaystyle\int_{2/\sqrt{3}}^{2}\frac{1}{t\sqrt{t^2 - 1}}\,dt.$ 37. Find $\displaystyle\int_{1}^{2}\frac{1}{x\sqrt{x^2 - 1}}\,dx.$

38. Find $\displaystyle\int_{0}^{1/\sqrt{3}}\frac{t}{\sqrt{t^2 + 1}}\,dt.$

39. In Theorem 6 we required that $f'(x)$ be different from 0 everywhere on the interval I. This hypothesis was satisfied by Tan on the open interval $(-\pi/2, \pi/2)$, and so we could conclude that Tan is invertible. But Theorem 6, as it stands, does not apply to Sin on $[-\pi/2, \pi/2]$ or to Sec on $[0, \pi/2)$, because the derivatives of these functions vanish at the endpoints $-\pi/2$, $\pi/2$, and 0. To take care of such cases, we need the following:

    Theorem. If f is differentiable on an interval I, and $f'(x) \neq 0$ at every interior point of I, then f is invertible.

    Here by an interior point of I we mean a point of I which is not an endpoint.

    Reread the proof of Theorem 6 and see whether it proves this more general theorem. If so, say so and explain. If not, furnish whatever additional reasoning is necessary.

40. It might also be convenient to have the following generalized form of the uniqueness theorem (of Section 3.8). Here we require that $F'(x) = G'(x)$ at all *interior* points of the interval I.

    Theorem (?). Let F and G be differentiable functions, defined on the same interval I, and let a be a point of I. If (i) $F(a) = G(a)$ and (ii) $F'(x) = G'(x)$ for every interior point x of I, then (iii) $F(x) = G(x)$ for every point x of I.

    Reexamine the proof of Theorem 2 of Section 3.8, and see whether it proves the more general theorem above. (If not, complete the proof.) Then name a case in which the more general theorem is more convenient to use.

4.8 SIMPSON'S RULE. THE COMPUTATION OF π

In Section 3.8 we developed a method for evaluating definite integrals. To find

$$\int_a^b f(x)\, dx,$$

where f is continuous, we first set up the function

$$F(x) = \int_a^x f(t)\, dt.$$

Then $F'(x) = f(x)$, for every x. We find another function G, such that $G' = f$. Then F and G have the same derivative f; and by adding a constant to G, we get a function, say H, such that $H' = G' = f$ and $H(a) = 0$. Since $F(a) = 0$, we know by the uniqueness theorem that $F(x) = H(x)$ for every x. Therefore

$$\int_a^b f(t)\, dt = H(b).$$

It is possible to write a theorem which sums this up very briefly:

Theorem 1. If f is continuous, and

$$G'(x) = f(x) \qquad (a \leqq x \leqq b),$$

then

$$\int_a^b f(x)\, dx = G(b) - G(a).$$

Proof. For each x, let

$$F(x) = \int_a^x f(t)\, dt.$$

Then

$$F'(x) = f(x), \quad \text{and} \quad F(a) = 0.$$

Let

$$H(x) = G(x) - G(a).$$

Then

$$H'(x) = G'(x) = f(x), \quad \text{and} \quad H(a) = 0.$$

Therefore $F(x) = H(x)$ for every x, and so $F(b) = H(b)$. Therefore

$$\int_a^b f(t)\, dt = H(b) = G(b) - G(a).$$

The proof reproduces the procedure that we have been using all along. G is the first G that we try, with $G' = f$; and H is the function that we get when we adjust the constant.

But in many cases it is hard to find a known function which has a given function f as its derivative. For example, if we had never heard of tan, Tan, or Tan^{-1}, then we would have had no chance at all of finding a known function G such that

$$G'(x) = \frac{1}{1 + x^2}.$$

Later, we shall learn more and better methods for attacking such problems. But no method, and no system of methods, works all the time. Therefore we often need to use numerical methods, to calculate definite integrals approximately.

One way is the following. Suppose that we didn't know anything about derivatives, but we needed to find $\int_0^1 (1 - x^3)\, dx$ approximately. We might divide the interval $[0, 1]$ into 10 subintervals of length 0.1, and add the areas of the circumscribed rectangles.

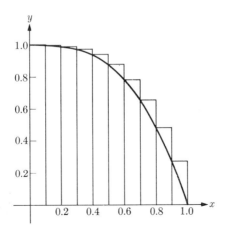

$i =$	$x_i =$	$y_i =$	$a_i =$
0	0	1	0.1
1	0.1	0.999	0.0999
2	0.2	0.992	0.0992
3	0.3	0.973	0.0973
4	0.4	0.936	0.0936
5	0.5	0.875	0.0875
6	0.6	0.784	0.0784
7	0.7	0.657	0.0657
8	0.8	0.488	0.0488
9	0.9	0.271	0.0271
			0.7975

Here the areas of the ten circumscribed rectangles are

$$a_0, a_1, \ldots, a_9;$$

and their total area is 0.7975. This gives

$$A = \int_0^1 (1 - x^3)\, dx \approx 0.7975 = A_1.$$

The approximation $A \approx A_1$ is not very good: by an easy calculation based on Theorem 1, we get the exact answer

$$\int_0^1 (1 - x^3)\, dx = 0.7500.$$

We might also have used *inscribed* rectangles. Their total area would be

$$0.7975 - 0.1000 = 0.6975 = A_2.$$

(Why?) The approximation $A \approx A_2$ is not very good either. But their *average* is considerably better.

$$A_3 = \tfrac{1}{2}(A_1 + A_2) = 0.7475 \approx 0.7500.$$

The sum A_3 has a geometric meaning: it is the sum of the areas of the inscribed trapezoids.

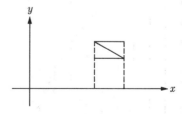

Over each of the little intervals, the area of the trapezoid is the average of the areas of the inscribed and circumscribed rectangles; and it is not hard to check that the same is true of the sums. This helps to explain why the approximation $A \approx A_3$ is reasonably good; we have approximated the graph of f by an inscribed broken line, and used the area under the broken line as an approximation of the integral.

In practice, however, nobody uses the approximation $A \approx A_3$, because there is another method which gives better results without any extra work. This method is *Simpson's rule*. The scheme is as follows.

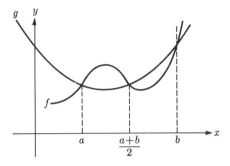

Suppose that we have a function f, whose values we can compute, on an interval $[a, b]$. We find a quadratic function

$$g(x) = Ax^2 + Bx + C$$

which agrees with f at a, at b, and at the midpoint of $[a, b]$; and we use the approximation

$$\int_a^b f(x)\, dx \approx \int_a^b g(x)\, dx.$$

Here by a *quadratic function* we mean a function given by a formula $Ax^2 + Bx + C$. We allow the case $A = 0$, and so the graph may turn out to be a line instead of a parabola. In any case, the integral on the right is easy to calculate: if

$$G(x) = \frac{A}{3}x^3 + \frac{B}{2}x^2 + Cx,$$

then

$$G' = g,$$

and so

$$\int_a^b g(x)\, dx = G(b) - G(a).$$

In the figure above, the approximation looks good, because the errors on the two halves of $[a, b]$ seem to cancel each other out. Most of the time, we cut $[a, b]$ into a certain number of little intervals $[a_i, a_{i+1}]$; we then use Simpson's rule on each of the little intervals, and add the results.

We shall now develop a shortcut formula for Simpson's rule, in a special case.

Theorem 2. Let g be a quadratic function, and let k be a positive number. Then

$$\int_{-k}^k g(x)\, dx = \frac{k}{3}(y_0 + 4y_1 + y_2),$$

where $y_0 = g(-k)$, $y_1 = g(0)$, and $y_2 = g(k)$.

Before proving that this formula is true, let us first check it, in a simple case, to make sure that it is not absurd. One of the possibilities is that $g(x) = 1$ for every x. In this case, the integral on the left is equal to $2k$. Thus our form la says that

$$2k = \frac{k}{3}(1 + 4 + 1),$$

which is correct. Any time you wonder whether you have remembered Simpson's rule correctly, you should check by this method; the check uncovers the most common errors in recollection.

We proceed to the proof. We have

$$g(x) = Ax^2 + Bx + C.$$

Let

$$G(x) = \frac{A}{3} x^3 + \frac{B}{2} x^2 + Cx,$$

so that $G' = g$. Then

$$\int_{-k}^{k} g(x)\, dx = G(k) - G(-k) = \tfrac{2}{3}Ak^3 + 2Ck.$$

(The algebra here is straightforward.) We need to express A and C in terms of y_0, y_1, y_2, and k. Evidently C is no problem:

$$C = g(0) = y_1.$$

To find A, we use

$$y_0 = Ak^2 - Bk + C, \qquad y_2 = Ak^2 + Bk + C,$$
$$y_0 + y_2 = 2Ak^2 + 2C, \qquad y_0 + y_2 = 2Ak^2 + 2y_1.$$

We can now solve for A:

$$A = \frac{1}{2k^2}(y_0 - 2y_1 + y_2).$$

Our expressions for A and C now give

$$\int_{-k}^{k} g(x)\, dx = \tfrac{2}{3}Ak^3 + 2Ck = \frac{k}{3}(y_0 - 2y_1 + y_2) + 2ky_1$$

$$= \frac{k}{3}(y_0 + 4y_1 + y_2),$$

which was to be proved.

Let us try Simpson's rule on the function

$$f(x) = \frac{1}{x + 2}, \qquad -1 \leq x \leq 1.$$

Here we have

$$k = 1, \qquad y_0 = 1, \qquad y_1 = \tfrac{1}{2}, \qquad y_2 = \tfrac{1}{3}.$$

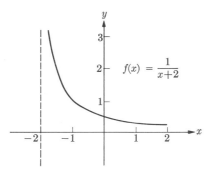

The rule gives

$$\int_{-1}^{1} \frac{dx}{x+2} \approx \tfrac{1}{3}(1 + 2 + \tfrac{1}{3}) \approx 1.11.$$

Later, we shall find ways to calculate this integral as exactly as we please. It will then turn out that the right answer, correct to four decimal places, is 1.0986. In this case, the approximation is good, in spite of the length of the interval $[-1, 1]$, because the portion of the graph of f that we are dealing with is very close to its approximating parabola.

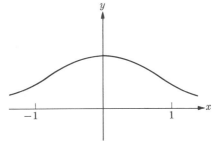

Let us now try

$$f(x) = \frac{1}{1 + x^2}, \qquad -1 \leq x \leq 1.$$

Here we have

$$k = 1, \qquad y_0 = \tfrac{1}{2}, \qquad y_1 = 1, \qquad y_2 = \tfrac{1}{2}.$$

The rule gives

$$\int_{-1}^{1} \frac{dx}{1 + x^2} \approx \tfrac{1}{3}(\tfrac{1}{2} + 4 + \tfrac{1}{2}) = \tfrac{5}{3} \approx 1.67.$$

Since

$$D \, \mathrm{Tan}^{-1} x = \frac{1}{1 + x^2},$$

the right answer is

$$\int_{-1}^{1} \frac{dx}{1 + x^2} = \mathrm{Tan}^{-1} 1 - \mathrm{Tan}^{-1}(-1) = \frac{\pi}{4} - \left(-\frac{\pi}{4}\right)$$

$$= \frac{\pi}{2} \approx 1.57.$$

Here the error is about 0.10, which is not very bad. To get better results, we need to cut up our intervals into smaller pieces. The first step in deriving the necessary formulas is to generalize Theorem 2, to take care of the case in which the origin is not necessarily the midpoint of the interval over which we are integrating.

Theorem 3. Let g be a quadratic function, and let k be a positive number. Then

$$\int_a^{a+2k} g(x)\, dx = \frac{k}{3}(y_0 + 4y_1 + y_2),$$

where

$$y_0 = g(a), \qquad y_1 = g(a+k), \qquad y_2 = g(a+2k).$$

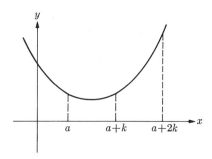

The easiest way to see this is to move the graph $a + k$ units to the left, so that the point $(a + k, 0)$ falls on the origin. When a parabola (or a line) is moved in this way, it is still a parabola (or a line); the integral does not change, and neither do the numbers k, y_0, y_1, and y_2. Therefore Theorem 3 is a consequence of Theorem 2.

Consider now a function f, on an interval $[a, b]$. We cut up the interval $[a, b]$ into an even number $2n$ of little intervals, each of length

$$k = \frac{b - a}{2n}.$$

The division points are $x_0, x_1, \ldots, x_{2n}$, as shown in the figure for $n = 2$.

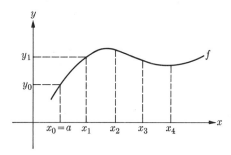

On the interval $[x_0, x_2] = [a, a + 2k]$, Simpson's rule gives

$$\int_a^{a+2k} f(x)\, dx \approx \frac{k}{3}(y_0 + 4y_1 + y_2),$$

where $y_i = f(x_i)$, for each i. On the interval $[x_2, x_4] = [a + 2k, a + 4k]$ we get

$$\int_{a+2k}^{a+4k} f(x)\, dx \approx \frac{k}{3}(y_2 + 4y_3 + y_4).$$

For the $2n$ little intervals we have

$$\int_a^b f(x)\, dx \approx \frac{k}{3}(y_0 + 4y_1 + 2y_2 + 4y_3 + 2y_4 + \cdots + 4y_{2n-1} + y_{2n}).$$

This formula is the final form of Simpson's rule. Let us try it, with $k = 0.2$, to get a better approximation of

$$\int_{-1}^{1} \frac{dx}{x + 2}.$$

The computation looks like this:

$i =$	$x_i =$	$y_i =$		
0	−1.0	1.0000	1	1.0000
1	−0.8	0.8333	4	3.3332
2	−0.6	0.7143	2	1.4286
3	−0.4	0.6250	4	2.5000
4	−0.2	0.5555	2	1.1110
5	0	0.5000	4	2.0000
6	0.2	0.4545	2	0.9090
7	0.4	0.4167	4	1.6668
8	0.6	0.3846	2	0.7692
9	0.8	0.3571	4	1.4285
10	1.0	0.3333	1	0.3333
				16.4795

This gives

$$\int_{-1}^{1} \frac{dx}{x + 2} \approx \frac{0.2}{3}(16.4795) \approx 1.0986.$$

This answer is actually correct, to the fourth decimal place. Obviously, however, we must have been lucky: Simpson's rule is not supposed to be exact, and, besides, we were carrying only four decimal places in the calculation.

   When you use Simpson's rule, it is a good idea to use a table like the one shown above. Make sure that the last entry in the fourth column of your table is 1 and not 2.

   We have postponed until now the presentation of Simpson's rule, because this is the first point at which we can do something interesting with it. The interesting thing is as follows. We know that

$$\int_0^1 \frac{dx}{1 + x^2} = \mathrm{Tan}^{-1}\, 1 - \mathrm{Tan}^{-1}\, 0 = \frac{\pi}{4}.$$

Therefore

$$\pi = 4 \int_0^1 \frac{dx}{1 + x^2}.$$

Applying Simpson's rule, we can thus get a numerical approximation of π. This is Problem 1 below.

PROBLEM SET 4.8

1. Apply Simpson's rule to the function

$$f(x) = \frac{1}{1 + x^2} \qquad (0 \leqq x \leqq 1),$$

with $k = \frac{1}{4}$. Check your answer against what people have been telling you about π. If you want to use $k = 0.1$, to get a more exact approximation, it might occur to you to use a slide rule to calculate the y_i's. Would this be a good idea? Why or why not? How about five-place log tables?

2. Apply Simpson's rule to the function

$$f(x) = 3x^3 - 5x^2 + 1 \qquad (-2 \leqq x \leqq 2),$$

with $k = 2$. Then calculate $\int_{-2}^{2} f(x)\, dx$ exactly, and compute the error in your approximation.

3. Apply Simpson's rule to the function

$$f(x) = x^3 + x^2 - 17 \qquad (-100 \leqq x \leqq 100),$$

with $k = 100$. Then calculate the integral exactly, and compute the error in the approximation.

4. Apply Simpson's rule to

$$f(x) = x^3 - 2x + 3, \qquad (-1 \leqq x \leqq 1),$$

with $k = \frac{1}{2}$. Compute the error.

5. Apply Simpson's rule to

$$f(x) = x^4 - 2x + 3,$$

over the same interval as in Problem 4, using the same k, and compute the error.

6. There ought to be a theorem which accounts for some of the results that you have been getting. State and prove the theorem.

7. Apply Simpson's rule to the function

$$f(x) = 1 - x^3,$$

on the interval $[0, 1]$, using $k = 0.1$. (This is the integral which we investigated in the text above, using inscribed rectangles, circumscribed rectangles, and finally trapezoids.)

*8. Given a positive number k and numbers y_0, y_1, and y_2, write an explicit formula for a quadratic function g such that $g(-k) = y_0$, $g(0) = y_1$, and $g(k) = y_2$. That is, write an expression of the form

$$g(x) = Ax^2 + Bx + C,$$

in which the coefficients A, B, and C are expressed algebraically in terms of k, y_0, y_1, and y_2.

*9. Does the theorem that you proved in Problem 6 hold only on intervals of the type $[-k, k]$ or does it hold on any interval $[a, b]$? Proof or refutation?

After finishing Problem 1, you may want to try a smaller k, to get a better approximation of π. As a check,

$$\pi = 3.14159265,$$

correct to eight decimal places.

In Appendix F, at the end of the book, you will find a theorem which enables us, under some conditions, to set a limit on the error in Simpson's rule.

4.9 EXPONENTIALS AND LOGARITHMS

For the case in which the exponents are positive integers, exponentials are part of elementary algebra. We begin with:

Definition. For each positive integer n,

$$x^n = xxx \cdots x \qquad \text{(to } n \text{ factors)}.$$

It is then easy to see, simply by counting factors on the left and on the right that the familiar laws of exponents hold:

$$x^m x^n = x^{m+n}, \tag{A}$$

$$(x^m)^n = x^{mn}. \tag{B}$$

If n is a negative integer, then $-n > 0$, and for $x \neq 0$ we define

$$x^n = \frac{1}{x^{-n}}.$$

Thus, for example, for $n = -3$ we have

$$x^{-3} = \frac{1}{x^{-(-3)}} = \frac{1}{x^3}.$$

For $x \neq 0$, we define

$$x^0 = 1.$$

It can be shown that, if $x \neq 0$, then formulas (A) and (B) hold for all integers m and n.

When the exponents are allowed to range over all real numbers, exponentials cease to be part of elementary algebra. In this section we shall state the facts about exponentials and logarithms, but will make no attempt to verify them. (In the following two sections, we shall see how these facts fit together to make a logical theory.) We begin with a positive base and a rational exponent.

1) Suppose that $a > 0$, and that x is a rational number p/q (where p and q are integers and $q > 0$). We want to define $a^x = a^{p/q}$ in such a way that (A) and (B) will continue to hold. For (B) to hold, we must have

$$(a^{p/q})^q = a^p.$$

That is, $a^{p/q}$ must be the qth root of a^p. Hence the following:

Definition. If $a > 0$ and $q > 0$, then

$$a^{p/q} = \sqrt[q]{a^p}.$$

Here we cannot allow the case $a < 0$. For $a = -1$, we would get

$$(-1)^{1/3} = \sqrt[3]{-1} = -1,$$
$$(-1)^{2/6} = \sqrt[6]{(-1)^2} = \sqrt[6]{1} = 1.$$

Thus, for $a < 0$, $a^{p/q}$ would depend not merely on the *number* that we use as an exponent but also on the *notation* in which the number is expressed. This would lead to nothing but trouble.

2)   It is a fact that for $a > 0$, and x and y rational, the following laws hold:

$$a^x \cdot a^y = a^{x+y}, \tag{A}$$
$$(a^x)^y = a^{xy}, \tag{B}$$
$$a^0 = 1. \tag{C}$$

3)   The rational numbers on the x-axis do not fill up the x-axis, because every interval, however short, contains irrational numbers. Therefore the set **Q** of all rational numbers forms a sort of infinitely dotted line. So far, the function $f(x) = a^x$ has been defined only for rational values of x. Therefore f is a function $\mathbf{Q} \to \mathbf{R}^+$, and the graph is an infinitely dotted curve, as in the figure below. Note that $f(x) > 0$ for every x, because $a^x = a^{p/q}$, which is the positive qth root of the positive number a^p.

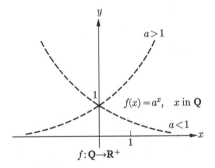

It is a fact that the definition of this function can be extended so as to give a new function:

$$f: \mathbf{R} \to (0, \infty)$$
$$x \mapsto a^x > 0,$$

defined for every x, such that f is continuous and satisfies (A), (B), and (C). For $a = 1$, we have $f(x) = 1^x = 1$ for every x. But for $a > 0$ and $a \neq 1$, f is invertible.

4)   We now define $\log_a$ as the inverse of $f(x) = a^x$ ($a \neq 1$). That is,

$$y = \log_a x \iff a^y = x,$$

by definition. The image of the exponential function includes all positive numbers.

Therefore the domain of its inverse includes all positive numbers, and we have a function

$$\log_a : (0, \infty) \to \mathbf{R}.$$

Logarithms to the base a obey the following laws:

$$\log_a xy = \log_a x + \log_a y, \tag{A'}$$

$$\log_a b^x = x \log_a b \qquad (b > 0, b \neq 1), \tag{B'}$$

$$\log_a 1 = 0. \tag{C'}$$

In fact, these are derivable from (A), (B), and (C).

Since the logarithm and exponential are inverses of each other, we have

$$\log_a a^x = x,$$

$$a^{\log_a x} = x.$$

And the graph of either of these functions is the reflection of the graph of the other across the line $y = x$.

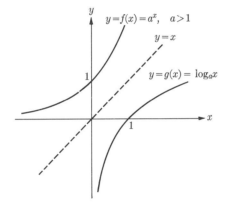

5)   We now want to calculate the derivative of the logarithm.   By definition,

$$\log'_a x_0 = \lim_{\Delta x \to 0} \frac{\log_a (x_0 + \Delta x) - \log_a x_0}{\Delta x}.$$

Using the laws (A'), (B'), and (C'), we express this in the form

$$\lim_{\Delta x \to 0} \frac{1}{\Delta x} \log_a \frac{x_0 + \Delta x}{x_0} = \lim_{\Delta x \to 0} \log_a \left(1 + \frac{\Delta x}{x_0}\right)^{1/\Delta x}$$

$$= \lim_{\Delta x \to 0} \left[\frac{1}{x_0} \log_a \left(1 + \frac{\Delta x}{x_0}\right)^{x_0/\Delta x}\right].$$

Let

$$h = \frac{\Delta x}{x_0}.$$

Since x_0 is fixed, $h \to 0$ as $\Delta x \to 0$. This gives

$$\log_a' x_0 = \frac{1}{x_0} \lim_{h \to 0} \log_a (1 + h)^{1/h}.$$

If $\log_a$ has a derivative, then the limit on the right-hand side exists, and conversely. Suppose that $(1 + h)^{1/h}$ approaches a limit, and call this limit e. Thus

$$e = \lim_{h \to 0} (1 + h)^{1/h}.$$

Suppose that $\log_a$ is continuous, so that the limit of the logarithm is the logarithm of the limit. Then

$$\log_a' x_0 = \frac{1}{x_0} \log_a \lim_{h \to 0} (1 + h)^{1/h} = \frac{1}{x_0} \log_a e.$$

Thus

$$D \log_a x = \frac{1}{x} \log_a e.$$

Since $e^1 = e$, we have

$$\log_e e = 1;$$

and so for $a = e$ our differentiation formula takes the form

$$D \log_e x = \frac{1}{x}.$$

Considering the complications of the preceding discussion, the simplicity of this formula is surprising; it is also important (see the next section).

6)   On the basis of the preceding formulas, it is easy to find Da^x, for $a \neq 1$. (For $a = 1$, we have no problem.) Let

$$f(x) = \log_a x,$$

so that

$$f^{-1}(x) = a^x.$$

The general formula

$$f(f^{-1}(x)) = x$$

thus takes the form

$$\log_a a^x = x.$$

Since

$$D_u \log_a u = \frac{1}{u} \log_a e,$$

the chain rule gives:

$$\left(\frac{1}{a^x} \log_a e \right) Da^x = 1.$$

Therefore

$$Da^x = \frac{1}{\log_a e} a^x.$$

In particular, for $a = e$ we have

$$De^x = e^x.$$

A final simplification: we assert that

$$\frac{1}{\log_a e} = \log_e a.$$

Proof. Let

$$x = \log_a e, \qquad y = \log_e a.$$

Then

$$a^x = e, \qquad e^y = a,$$

by the definition of the logarithm. Therefore

$$(a^x)^y = e^y, \qquad \text{and} \qquad a^{xy} = a.$$

This holds when $xy = 1$. Since the exponential function is invertible, it cannot take on the same value twice. Therefore the equation can hold *only* when $xy = 1$. Therefore

$$\frac{1}{x} = y,$$

which was to be proved. Thus we can write

$$Da^x = a^x \log_e a.$$

This is better, not just because it avoids a fraction, but also because e is one of the two bases for which tables of logarithms are published.

Throughout the following problem set you may assume that the statements made in this section are true. (They will be proved in the following two sections.) For convenience of reference, we give a summary.

Laws of Exponentials $(a > 0)$

a) $a^x \cdot a^y = a^{x+y}$,   b) $(a^x)^y = a^{xy}$,

c) $a^0 = 1$,   d) $a^x > 0$ for every x,

e) $Da^x = a^x \log_e a$,   f) $e = \lim_{h \to 0} (1 + h)^{1/h}$.

Laws of Logarithms $(a > 0, a \neq 1)$

g) $\log_a xy = \log_a x + \log_a y$,   h) $\log_a b^x = x \log_a b$ $(b > 0)$,

i) $\log_a 1 = 0$,   j) $D \log_a x = 1/(x \log_e a)$,

k) $\log_a a^x = x$,   l) $a^{\log_a x} = x$.

PROBLEM SET 4.9

Find the first and second derivatives of the following functions.

1. $x \log_e x$ 2. xe^x 3. xe^{2x}

4. $e^x \cos x$ 5. $e^x \sin x$ 6. $\frac{1}{2}e^x(\sin x + \cos x)$

7. $\log_e x^2$ 8. $[\log_e x]^2$ 9. $\log_e x^{500}$

10. $e^{\sin x}$ 11. 10^x 12. $\log_e e^x$

13. $[\log_e e]^x$ 14. $e^{\log_e x}$ 15. $x^2 e^x$

16. e^{x^2} 17. xe^{x^2} 18. $\log_e (x^3)$

19. $\log_e (1 - x)$ 20. $\log_e (e^x + 1)$ 21. e^{x-1}

22. e^{1-x} 23. $\log_e \sec x$ 24. e^{ex}

25. $\log_e \sin x$ 26. $\log_e \cos x$ 27. $\log_e (\sec x + \tan x)$

28. $\log_e \sqrt{\dfrac{1 + x}{1 - x}}$ 29. $\log_e (\csc x + \cot x)$ 30. $\log_e (x + \sqrt{x^2 + 1})$

31. Show that for every $x > 0$,

$$\log_e x = \int_1^x \frac{1}{t}\, dt.$$

32. Show that if a and b are positive and different from 1, then

$$\log_b a = \frac{1}{\log_a b}.$$

33. Show that, under the same conditions,

$$\log_b x = (\log_a x)(\log_b a).$$

34. Show that, if a and b are positive and different from 1, then

$$b^x = a^{x \log_a b}.$$

[*Hint:* What is $a^{\log_a b}$, and why?]

35. The function

$$f(x) = e^x$$

has the property of being its own derivative. But it is not the only such function, because for every $k, g(x) = ke^x$ has the same property. We have, however, the following theorem.

Theorem. If $g'(x) = g(x)$, on $(-\infty, \infty)$, then there is a constant k such that

$$g(x) = ke^x \qquad \text{for every } x.$$

That is, $g(x)/e^x$ is a constant.

Prove this.

36. Show that the function $f(x) = e^x$ is completely described by the conditions

$$f'(x) = f(x) \qquad (-\infty < x < \infty), \tag{1}$$
$$f(0) = 1. \tag{2}$$

That is, show that (1) and (2) imply that $f(x) = e^x$ for every x.

37. Show that the function $f(x) = e^{-x}$ is completely determined by the conditions

$$f'(x) = -f(x) \qquad (-\infty < x < \infty) \tag{1}$$
$$f(0) = 1. \tag{2}$$

That is, show that (1) and (2) imply that $f(x) = e^{-x}$ for every x.

4.10 THE FUNCTIONS ln AND exp

In the preceding section, we gave a sketch of the way that logarithms and exponentials ought to behave, postponing both the proofs and also the basic definitions. We shall now fill these gaps.

If you review the formulas of the preceding section, you will see that after considerable complications in the middle, we got a formula that looked simple:

$$D \log_e x = \frac{1}{x}.$$

This enables us to write a formula for $\log_e$:

$$\log_e x = \int_1^x \frac{1}{t}\, dt.$$

If the theory works, then this formula must be right: the functions on the two sides of the equation have the same derivative (namely, $1/x$), and they have the same value at $x = 1$ (namely, 0); and so it follows by the uniqueness theorem that they are the same function.

We shall use the function $\int_1^x (1/t)\, dt$ as the foundation of the theory of exponentials and logarithms. The scheme is to investigate the function $\int_1^x (1/t)\, dt$, learn its properties, and then define all our other functions in terms of it. Thus, at the beginning, we shall investigate $\int_1^x (1/t)\, dt$ without assuming that we know anything about logarithms, or about exponentials, or about the number e. To emphasize that we are starting afresh, we give the function a new name ln. (Here ln is suggested by *natural logarithm*.) And the official theory begins with the definition of ln in terms of an integral:

Definition. For each $x > 0$,

$$\ln x = \int_1^x \frac{dt}{t}.$$

Soon we shall show that every real number y is equal to $\ln x$ for some x. For this purpose we shall need:

Theorem 1 (*The no-jump theorem*). If f is continuous on $[x_1, x_2]$, then f takes on every value between $f(x_1)$ and $f(x_2)$.

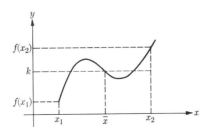

That is, if $f(x_1) < k < f(x_2)$, then there is an $\bar{x}$, between x_1 and x_2, such that $f(\bar{x}) = k$. And if $f(x_2) < k < f(x_1)$, then the same conclusion follows. This theorem will be proved in the next chapter.

Our first few theorems on the function ln are easy.

Theorem 2. $D \ln x = 1/x$.

Proof. This follows from the definition of ln and the formula for the derivative of the integral.

Theorem 3. $\ln 1 = 0$.

This is obvious.

Theorem 4. For every k, $x > 0$, $D_x \ln kx = 1/x$.

Proof. By the chain rule,

$$D_x \ln kx = \frac{1}{kx} \cdot k = \frac{1}{x}.$$

Theorem 5. For every a, $b > 0$, $\ln ab = \ln a + \ln b$.

Proof. The trouble with this theorem is that it does not appear to involve any functions. To prove it, we first restate it, using k for a and x for b. It then says that for every k, $x > 0$,

$$\ln kx = \ln k + \ln x.$$

The proof is now as follows. Let

$$f(x) = \ln kx, \qquad g(x) = \ln k + \ln x.$$

Then

$$f'(x) = \frac{1}{x} = g'(x), \qquad f(1) = \ln k,$$

and

$$g(1) = \ln k + \ln 1 = \ln k + 0 = \ln k.$$

By the uniqueness theorem, $f(x) = g(x)$ for every x.

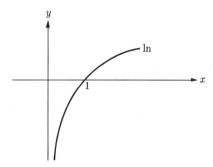

We now want to show that the graph of ln looks approximately like the drawing above. The figure suggests that $\ln 1 = 0$ and $\ln' 1 = 1$. These things we already know. Other things suggested by the figure are conveyed by some of the following theorems.

Theorem 6. ln is invertible.

Proof. We know that $\ln' x = 1/x$; and $1/x \neq 0$ for every x. Therefore $\ln' x \neq 0$ for every x. By Theorem 6 of Section 4.7, ln is invertible.

Theorem 7. For every $x > 0$, and every positive integer n,

$$\ln x^n = n \ln x.$$

Proof. Obviously this formula holds when $n = 1$. And if it holds for any particular integer n, then it holds for the next integer $n + 1$. Proof:

$$\ln x^{n+1} = \ln (x \cdot x^n) = \ln x + \ln x^n$$
$$= \ln x + n \ln x = (n + 1) \ln x.$$

Therefore, by induction, we have $\ln x^n = n \ln x$ for every x, which was to be proved.

A number M is called an *upper bound* for a function f if

$$f(x) \leq M \qquad \text{for every } x.$$

If there is such a number M, then we say that f is *bounded above*. (For example, the sine is bounded above, because $\sin x \leq 1$ for every x.) If no such number M exists, then we say that f is *unbounded above*. (For example, if $f(x) = x$, for every x, then f is unbounded above.)

Theorem 8. ln is unbounded above.

Proof. In Theorem 7, take $x = 2$.

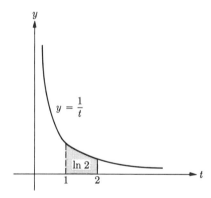

Then

$$\ln 2^n = n \ln 2,$$

for every n. And $\ln 2 > 0$, because $\ln 2$ is the area of a region. Therefore ln cannot have an upper bound; no number M is greater than or equal to all of the numbers $n \ln 2$, because

$$n \ln 2 > M \qquad \text{whenever } n > \frac{M}{\ln 2}.$$

Theorem 9. $\ln(1/x) = -\ln x.$

Proof.

$$\ln \frac{1}{x} + \ln x = \ln \left(\frac{1}{x} \cdot x \right) = \ln 1 = 0;$$

and from this the theorem follows.

Theorem 10. $\ln 2^{-n} = -n \ln 2.$

Proof. By Theorems 7 and 9.

A number m is called a *lower bound* of a function f if $m \leq f(x)$ for every x. If there is such a number m, then we say that f is *bounded below*. (For example, the sine is bounded below, because $-1 \leq \sin x$ for every x.) If no such number m exists, then we say that f is *unbounded below*. (For example, if $f(x) = x$, then f is unbounded below.)

Theorem 11. The function ln is unbounded below.

Because no number m is less than or equal to all the numbers $\ln 2^{-n} = -n \ln 2$.

Theorem 12. Every real number is a value of the function ln. That is, every number y is equal to $\ln \bar{x}$ for some $\bar{x} > 0$.

Proof. Since ln is unbounded both above and below, it follows that every number y lies *between* two values of ln. If $\ln x_1 < y < \ln x_2$, then it follows by Theorem 1 that $y = \ln \bar{x}$ for some $\bar{x}$. Thus the image of the function ln is the entire interval $\mathbf{R} = (-\infty, \infty)$.

We know by Theorem 6 that ln is invertible. Its inverse will be denoted by exp. That is:

Definition. $\exp = \ln^{-1}.$

Since $\ln x$ will turn out to be $\log_e x$, this means that $\exp x$ will turn out to be e^x. But we should not use the notation e^x, at this stage, because we have not yet defined e in the present treatment.

The graphs of ln and exp are shown in the figure opposite. Since exp and ln are inverses of each other, we have:

Theorem 13. $\exp \ln x = x.$

Theorem 14. $\ln \exp x = x.$

These are instances of the general rule

$$f^{-1}(f(x)) = f(f^{-1}(x)) = x.$$

As always, for functions which are inverses of one another, the image of exp is the domain of ln. Therefore

Theorem 15. $\exp x > 0$ for every x.

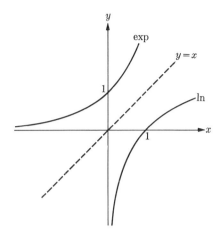

This theorem is also easy to see graphically, in the figure above. The graph of ln lies to the right of the y-axis. Reflecting this graph across the line $y = x$, we get the graph of exp. Therefore the graph of exp lies above the x-axis.

Theorem 16. $\exp 0 = 1$.

Because $\ln 1 = 0$.

Theorem 17. $\exp (k + x) = (\exp k)(\exp x)$ for every k and x.

Proof. Both sides of the equation have the same ln:

$$\ln \exp (k + x) = k + x,$$

because ln and exp are inverses of each other. And

$$\ln [(\exp k)(\exp x)] = \ln \exp k + \ln \exp x = k + x.$$

Since ln never takes on the same value twice, the theorem follows.

Theorem 18. $\exp' = \exp$.

Proof. We know that $\ln \exp x = x$. Since $\ln' u = 1/u$, the chain rules gives

$$(\ln' \exp x) \exp' x = 1, \qquad \frac{1}{\exp x} \exp' x = 1, \qquad \exp' x = \exp x,$$

for every x. Therefore $\exp' = \exp$, which was to be proved.

We now have functions $\ln x$ and $\exp x$ which have the properties that the functions $\log_e x$ and e^x are expected to have. A natural next step is to find a number e, and define the exponential function e^x, in such a way that $\exp x = e^x$. We shall do this in the next section. Meanwhile we give a quick summary of this one.

Definitions

a) $\ln x = \displaystyle\int_1^x \frac{dt}{t}$ $(x > 0)$,   b) $\exp = \ln^{-1}$.

Laws for ln

c) $\ln 1 = 0$,

d) $\ln x^n = n \ln x$ $(x > 0)$,

e) $\ln kx = \ln k + \ln x$ $(k, x > 0)$,

f) $\ln' x = 1/x$ $(x > 0)$.

Laws for exp

g) $\exp 0 = 1$,

h) $(\exp k)(\exp x) = \exp (k + x)$,

i) $\exp x > 0$ for every x,

j) $\exp \ln x = x$ $(x > 0)$,

k) $\ln \exp x = x$ for every x,

l) $\exp' = \exp$.

PROBLEM SET 4.10

Some of the problems below are to be solved by any method that works, including methods based on the unproved results of Section 4.9. Some, however, are supposed to be worked strictly on the basis of the theory developed in this section; and these are stated in the notation of ln and exp. Thus, if the problem uses the notation a^x, $\log_a x$, then the solution may use the theory in Section 4.9; but if the problem uses ln, exp, then the solution should also.

Find the derivatives of the following functions:

1. $\ln^2 x$ 2. $\ln \ln x$ 3. $\ln (x^2 + 1)$ 4. $\ln x^2 + 1$

5. $\exp x^2$ 6. $[\exp x]^2$ 7. $\exp (2 \ln x)$ 8. $\ln (\exp x^2)$

9. $\exp \sin x$ 10. $\sin (\exp x)$ 11. $\exp (x \ln x)$ 12. $e^x \log_e x$

13. $\ln \sin x$ 14. $\sin \ln x$ 15. x^x $(x > 0)$ 16. $(\sin x)^{\sin x}$ $(\sin x > 0)$

17. We found that the function

$$\ln x = \int_1^x \frac{dt}{t}$$

was unbounded above. Is this true also for

$$f(x) = \int_1^x \frac{dt}{\sqrt{t}} \ ?$$

Why or why not?

18. How about the function

$$g(x) = \int_1^x \frac{dt}{t^2} \ ?$$

19. Given

$$h(x) = \int_1^{x^2} \frac{dt}{\sqrt{t}} \quad (0 < x < \infty),$$

find $h'(x)$, by any method. Note, however, that you are not being asked to calculate h; you are being asked only to calculate h'.

20. Given

$$f(x) = \int_0^{\sin x} \sqrt{1 + t^2} \, dt,$$

find $f'(x)$.

21. Given

$$f(x) = \int_0^{\tan x} \sqrt{1 + t^2}\, dt,$$

find $f'(x)$.

22. Given

$$g(x) = \int_1^{e^x} \frac{dt}{t},$$

find $g'(x)$.

23. Given

$$h(x) = \exp\left(\int_0^x \frac{1}{t}\, dt\right),$$

find $h'(x)$.

24. Find

$$\lim_{x \to 3\pi/2} \frac{\sin x + 1}{x - 3\pi/2}.$$

(By far the easiest way to solve this problem is to think of a geometric meaning for it.)

25. Find $\displaystyle\lim_{x \to \pi/4} \frac{\tan x - 1}{x - \pi/4}$.   26. Find $\displaystyle\lim_{x \to 1} \frac{\ln x}{x - 1}$.   27. Find $\displaystyle\lim_{x \to 1} \frac{\ln x^2}{x - 1}$.

28. Find $\displaystyle\lim_{x \to -\pi/4} \frac{\tan x + 1}{x + \pi/4}$.   29. Find $\displaystyle\lim_{x \to 0} \frac{\exp(2x) - 1}{x}$.

30. Using Simpson's rule, compute an approximation of $\ln 2$. To four decimal places, the right answer is 0.6931; and if you cut up the interval $[1, 2]$ into ten parts, you get a good approximation.

31. Show that for $x \geqq 1$,

$$\ln x \leqq x - 1.$$

Show that for $0 < x < 1$, the same inequality holds.

32. Given $f(x) = x - 1$, find a formula for $f^{-1}(x)$, and sketch both functions on the same set of axes.

Show that $\exp x \geqq x + 1$, for every x. [*Hint:* Try to use a known property of $\ln$.]

33. Let $k = \ln^{-1} 1$. Show that $k > 2$.

34. Show that $k < 4$.

4.11 EXPONENTIALS AND LOGARITHMS. THE EXISTENCE OF e

In Section 4.9, we wanted to define e as the limit of the function

$$f(h) = (1 + h)^{1/h},$$

as $h \to 0$. To investigate this limit, we first need a proper definition of the *function* $(1 + h)^{1/h}$. Since the exponent $1/h$ varies continuously through real values, we need a definition of the exponential a^x, where $a > 0$ and x is not necessarily rational. The right definition is not hard to find. We know that if n is a positive integer, then

$$\ln a^n = n \ln a.$$

Therefore

$$a^n = \exp(n \ln a).$$

If the laws of logarithms hold for all exponents, then

$$\ln a^x = x \ln a,$$

which gives

$$a^x = \exp(x \ln a).$$

We take this last equation as our definition of the exponential function a^x. Thus:

Definition. For $a > 0$, $a^x = \exp(x \ln a)$.

This gives:

Theorem 1. $\ln a^x = x \ln a$.

We are now ready to show that $(1 + h)^{1/h}$ approaches a limit, as $h \to 0$. Let

$$f(x) = (1 + x)^{1/x}.$$

Then

$$\ln f(x) = \frac{1}{x} \ln(1 + x).$$

We now replace x by Δx, and observe that

$$\ln f(\Delta x) = \frac{1}{\Delta x} \ln(1 + \Delta x)$$

$$= \frac{\ln(1 + \Delta x) - \ln 1}{\Delta x}.$$

This last fraction is the fraction whose limit is $\ln' 1$, by definition of the derivative. Therefore

$$\lim_{\Delta x \to 0} \ln f(\Delta x) = \ln' 1 = \tfrac{1}{1} = 1,$$

and

$$\lim_{\Delta x \to 0} f(\Delta x) = \lim_{\Delta x \to 0} \exp \ln f(\Delta x) = \exp \lim_{\Delta x \to 0} \ln f(\Delta x)$$

$$= \exp 1 = \ln^{-1} 1.$$

Replacing Δx by h again, we have

$$\lim_{h \to 0} (1 + h)^{1/h} = \exp 1 = \ln^{-1} 1.$$

Now that we know the limit exists, we can use it as a definition of e:

Definition. $e = \lim_{h \to 0} (1 + h)^{1/h}$.

And we know:

Theorem 2. $e = \exp 1 = \ln^{-1} 1$.

This theorem has a geometric interpretation.

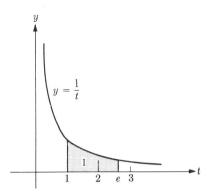

That is, e is the number such that

$$\int_1^e \frac{1}{t}\, dt = 1.$$

Later we will find an efficient method of calculating e. In fact,

$$e = 2.7182818,$$

correct to seven decimal places. It will turn out that

$$e = 1 + \frac{1}{1!} + \frac{1}{2!} + \cdots + \frac{1}{n!} + \cdots,$$

where

$$n! = 1 \cdot 2 \cdot 3 \cdots n.$$

The series on the right is infinite, but the terms diminish so rapidly as n increases that we get good numerical approximations by using the first few terms.

We expected $\exp x$ to be e^x. We can now show that this is true:

$$e^x = \exp (x \ln e),$$

by definition of e^x; and since $\ln e = 1$, we have:

Theorem 3. $e^x = \exp x$, for every x.

As before, we define the logarithm as the inverse of the exponential. That is

$$y = \log_a x \Leftrightarrow a^y = x.$$

Since e^x and $\exp x$ are the same function, they have the same inverse. Therefore we have:

Theorem 4. $\log_e x = \ln x$,   for every $x > 0$.

Thus $\exp$ really is an exponential, and $\ln$ really is a logarithm. Once we know the laws governing e^x and $\log_e x$, it is easy to derive the laws governing other positive bases. The first step is to express $\log_a$ in terms of $\ln$. We recall that for $a > 0$,

$$a^y = \exp (y \ln a),$$

by definition. Therefore
$$\ln a^y = \ln \exp (y \ln a),$$
and
$$\ln a^y = y \ln a.$$

Since a^x and $\log_a x$ are inverses of each other,

$$a^{\log_a x} = x.$$

In this equation, we take the ln of each side, getting

$$(\log_a x) \ln a = \ln x.$$

This gives:

Theorem 5. For every $x > 0$,

$$\log_a x = \frac{\ln x}{\ln a}.$$

Thus the function $\log_a$ is a constant times the function $\ln$; and this means that the extension of the theory from $\ln$ to $\log_a$ is easy.

Theorem 6. For every $a > 0, \neq 1$,

$$\log_a xy = \log_a x + \log_a y \qquad (x, y > 0), \tag{g}$$

$$\log_a b^x = x \log_a b, \tag{h}$$

$$\log_a 1 = 0, \tag{i}$$

$$\log_a' x = \frac{1}{x \log_e a}, \tag{j}$$

$$\log_a a^x = x, \qquad \text{for every } x, \tag{k}$$

$$a^{\log_a x} = x, \qquad \text{for } x > 0. \tag{l}$$

Here the formula designations are those of the summary at the end of Section 4.9. The proofs are as follows.

Proof. For $a = e$, the first three formulas are known to hold, because in this case $\log_a = \log_e = \ln$. If we divide every term by $\ln a$, then we get $\log_a$ throughout, and the equations still hold. To get Eq. (j), we observe that

$$\log_a' x = D \log_a x = D \frac{\ln x}{\ln a} = \frac{1}{x \ln a}.$$

Equations (k) and (l) merely remind us that $\log_a x$ and a^x are inverses of one another. Similarly, we get the laws governing the exponential a^x by using the fact that

$$a^x = \exp (x \ln a) = e^{x \ln a}.$$

Theorem 7. For every $a > 0$,

$$a^x \cdot a^y = a^{x+y}, \tag{a}$$

$$(a^x)^y = a^{xy}, \tag{b}$$

$$a^0 = 1, \tag{c}$$

$$a^x > 0 \qquad \text{for every } x, \tag{d}$$

$$Da^x = a^x \ln a. \tag{e}$$

The proofs are as follows.

a) We have

$$\ln (a^x \cdot a^y) = \ln a^x + \ln a^y = x \ln a + y \ln a$$
$$= (x + y) \ln a = \ln a^{x+y}.$$

Since $a^x \cdot a^y$ and a^{x+y} have the same ln, they must be the same; ln is invertible, and so ln never takes on the same value twice.

b) By definition, $b^y = \exp (y \ln b)$. Therefore

$$(a^x)^y = \exp (y \ln a^x) = \exp (yx \ln a)$$
$$= \exp (xy \ln a) = a^{xy}.$$

c) $a^0 = \exp (0 \ln a) = \exp 0 = 1.$

d) $a^x = \exp (x \ln a) > 0.$

e) $Da^x = D \exp (x \ln a) = [\exp (x \ln a)] \ln a$, by the chain rule. Therefore $Da^x = a^x \ln a$.

This completes the program that was sketched in Section 4.9. There are, however, some things that we still need to check. In the elementary theory, we stated:

Definition 1. For every positive integer n, and every real number x,

$$(x^n) = xxx \cdots x \qquad \text{(to n factors)}.$$

In the new theory, we stated:

Definition 2. If x is positive, and n is any real number, then

$$[x^n] = \exp (n \ln x).$$

We have used ( ) and [ ] to distinguish the two definitions. If n is a positive integer, and $x > 0$, then both definitions apply; and we need to know that they give the same answer. In fact, this is true:

$$\ln (x^n) = n \ln x,$$

by Theorem 7 of Section 4.10. Therefore

$$(x^n) = \exp [\ln (x^n)] = \exp (n \ln x) = [x^n],$$

by definition of $[x^n]$.

Similarly, we now have two definitions of $a^{p/q}$.

Definition 1. If $a > 0$, and x is a rational number p/q, then

$$a^x = a^{p/q} = \sqrt[q]{a^p}.$$

Definition 2. If $a > 0$, and $x = p/q$, then

$$a^x = a^{p/q} = \exp(x \ln a) = \exp\left(\frac{p}{q} \ln a\right).$$

These definitions agree if it is true that

$$\sqrt[q]{a^p} = \exp\left(\frac{p}{q} \ln a\right).$$

The proof is as follows. Let

$$y = \sqrt[q]{a^p}.$$

Then

$$y^q = a^p, \qquad \ln y^q = \ln a^p,$$

and

$$q \ln y = p \ln a.$$

Therefore

$$\ln y = \frac{p}{q} \ln a, \qquad \text{and} \qquad y = \exp\left(\frac{p}{q} \ln a\right),$$

which was to be proved.

We have found that the differentiation formula

$$Dx^k = kx^{k-1}$$

holds true in certain cases. We first proved it for the case in which k was a positive integer. Later we found that it held true when k was a negative integer. For $k = \frac{1}{2}$, and $x > 0$, it says that

$$Dx^{1/2} = \tfrac{1}{2}x^{1/2-1} = \tfrac{1}{2}x^{-1/2} = \frac{1}{2\sqrt{x}},$$

which is correct. We can now prove the following:

Theorem 8. For every $x > 0$, and every real number k,

$$Dx^k = kx^{k-1}.$$

Proof. By definition,

$$x^k = \exp(k \ln x).$$

Therefore

$$Dx^k = D[\exp(k \ln x)] = [\exp(k \ln x)] D[k \ln x]$$

$$= x^k \cdot k \cdot \frac{1}{x} = kx^{k-1}.$$

In this section we have presented no new results, except for Theorem 8; we have merely furnished proofs for the theory sketched in Section 4.9. You therefore have no new material for problem work. Hence we give the definitions of a new set of functions, the *hyperbolic* functions, and list various identities which they satisfy. In the following problem set you will be asked to derive these. The theory is simpler than the theory of trigonometric functions. In fact, once you know about the exponential function, most of the following formulas have straightforward derivations.

The functions are called the hyperbolic sine, hyperbolic cosine, hyperbolic tangent, and so on.

Definitions

$$\sinh x = \frac{e^x - e^{-x}}{2}.$$

$$\cosh x = \frac{e^x + e^{-x}}{2}.$$

$$\tanh x = \frac{e^x - e^{-x}}{e^x + e^{-x}} = \frac{\sinh x}{\cosh x}.$$

$$\coth x = \frac{e^x + e^{-x}}{e^x - e^{-x}} = \frac{\cosh x}{\sinh x}.$$

$$\operatorname{sech} x = \frac{2}{e^x + e^{-x}} = \frac{1}{\cosh x}.$$

$$\operatorname{csch} x = \frac{2}{e^x - e^{-x}} = \frac{1}{\sinh x}.$$

Identities

$$\sinh(-x) = -\sinh x. \tag{1}$$

$$\cosh(-x) = \cosh x. \tag{2}$$

$$\tanh(-x) = -\tanh x. \tag{3}$$

$$\cosh^2 x - \sinh^2 x = 1. \tag{4}$$

$$1 - \tanh^2 x = \operatorname{sech}^2 x. \tag{5}$$

$$\coth^2 x - 1 = \operatorname{csch}^2 x. \tag{6}$$

$$\sinh(x + y) = \sinh x \cosh y + \cosh x \sinh y. \tag{7}$$

$$\cosh(x + y) = \cosh x \cosh y + \sinh x \sinh y. \tag{8}$$

$$\tanh(x + y) = \frac{\tanh x + \tanh y}{1 + \tanh x \tanh y}. \tag{9}$$

$$\sinh 2x = 2 \sinh x \cosh x. \tag{10}$$

$$\cosh 2x = \cosh^2 x + \sinh^2 x. \tag{11}$$

$$e^x = \cosh x + \sinh x. \tag{12}$$

$$e^{-x} = \cosh x - \sinh x. \tag{13}$$

Derivatives

$$\sinh' x = \cosh x. \tag{14}$$

$$\cosh' x = \sinh x. \tag{15}$$

$$\tanh' x = \operatorname{sech}^2 x. \tag{16}$$

$$\coth' x = -\operatorname{csch}^2 x. \tag{17}$$

$$\operatorname{sech}' x = -\operatorname{sech} x \tanh x. \tag{18}$$

$$\operatorname{csch}' x = -\operatorname{csch} x \coth x. \tag{19}$$

PROBLEM SET 4.11

Verify the following. (The numbers in parentheses refer to the numbered formulas above in the text.)

1. (12)   2. (13)   3. (1)   4. (2)
5. (3)   6. (14)   7. (15)   8. (16)
9. (17)   10. (18)   11. (19)
12. Find the derivative of
$$F(x) = \cosh^2 x - \sinh^2 x.$$

13. Verify (4).   14. Verify (5).   15. Verify (6).
16. Let
$$A = \sinh (x + y) - \sinh x \cos y - \cosh x \sinh y,$$
$$B = \cosh (x + y) - \cosh x \cosh y - \sinh x \sinh y.$$

Show that $A + B = 0$. (It is not necessary to go back to the definitions to show this. Try Identities (12) and (13).)

17. Let A and B be as in Problem 16. Show that $A - B = 0$.

Now verify the following.

18. (7)   19. (8)   20. (9)   21. (10)   22. (11)
23. Express $\cosh 3x$ in terms of $\cosh x$.
24. Show that
$$x > 0 \Rightarrow \sinh x > 0.$$
25. Show that
$$x < 0 \Rightarrow \sinh x < 0.$$

26. Show that, on the interval $[0, \infty)$, cosh is increasing.

27. Show that, on the interval $(-\infty, 0]$, cosh is decreasing.

28. Show that
$$\cosh x \geqq 1 \qquad \text{for every } x.$$

29. Show that sinh is invertible.

30. Show that
$$\cosh x = \sqrt{1 + \sinh^2 x},$$

for every x. Note that there is no double sign in this formula; if your derivation leads to a "$\pm$" sign, you must find a way to get rid of it.

31. Find $\cosh \sinh^{-1} x$.

32. Find $D \sinh^{-1} x$.

33. Find $D \sinh^{-1} 2x$.

34. Show that cosh is not an invertible function.

35. Let

$$\text{Cosh } x = \cosh x \qquad (0 \leq x).$$

(Compare with the definition of Cos: $\text{Cos } x = \cos x \ (0 \leq x \leq \pi)$.) Show that Cosh is invertible.

36. Show that

$$\sinh x = \begin{cases} \sqrt{\cosh^2 x - 1}, & \text{for } x \geq 0, \\ -\sqrt{\cosh^2 x - 1}, & \text{for } x < 0. \end{cases}$$

37. Show that

$$\sinh \text{Cosh}^{-1} x = \sqrt{x^2 - 1}.$$

38. Find $D \text{ Cosh}^{-1} x$.

39. Find $D \text{ Cosh}^{-1} x^2$.

40. Show that tanh is invertible.

41. Show that

$$\text{sech } x = \sqrt{1 - \tanh^2 x},$$

without any double sign in front of the radical.

42. Find $D \tanh^{-1} x$.

43. Solve for x:

$$e^{2x} + e^x - 6 = 0,$$

and explain why this equation has only one root.

44. Solve for x:

$$e^x + 2 - 35e^{-x} = 0.$$

45. Solve for x, in terms of y:

$$e^x + y - 6y^2 e^{-x} = 0.$$

46. Find a formula which expresses $\sinh^{-1} x$ as the logarithm of an algebraic expression. *Hint:* The graph of sinh is the graph of the equation

$$y = \tfrac{1}{2}(e^x - e^{-x}). \tag{1}$$

Therefore the graph of $\sinh^{-1}$ is the graph of the equation

$$x = \tfrac{1}{2}(e^y - e^{-y}). \tag{2}$$

Here we have reflected the graph across the line $y = x$, by interchanging x and y in Eq. (1). Now solve for y in (2), getting

$$y = (\cdots).$$

Then

$$\sinh^{-1} x = (\cdots).$$

47. Analogously, get a formula for $\text{Cosh}^{-1} x$.

48. Analogously, get a formula for $\tanh^{-1} x$.

5 The Variation of Continuous Functions

5.1 INTERVALS ON WHICH A FUNCTION INCREASES, OR DECREASES

The function f is *increasing* if
$$x < x' \implies f(x) < f(x').$$
Similarly, f is *decreasing* if
$$x < x' \implies f(x) > f(x').$$

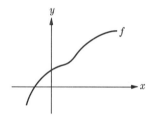 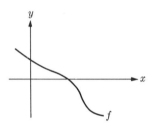

Here x and x' are any points in the domain of f. Some simple functions are neither increasing nor decreasing. For example, $f(x) = x^2$ satisfies neither of the above conditions.

Often, however, we can get a good description of a function by cutting up its domain into subintervals, in such a way that on each subinterval the function is either increasing or decreasing. For example, the domain might be a closed interval $[a, b]$, and the graph might look like this:

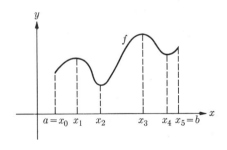

This function is *increasing* on the interval $[x_2, x_3]$. Similarly, f is increasing on $[x_0, x_1]$ and $[x_4, x_5]$, and is *decreasing* on the interval $[x_1, x_2]$.

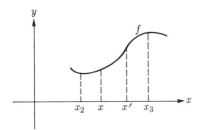 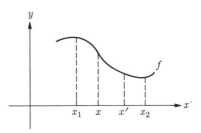

Similarly, f is decreasing on $[x_3, x_4]$.

If a function is differentiable, then we can find out where it is increasing or decreasing by examining the derivative.

Theorem 1. If $f'(x) > 0$ at every interior point of I, then f is increasing on I.

We recall that an *interior* point of an interval is any point which is not an endpoint. Theorem 1 is a consequence of the mean-value theorem (MVT).

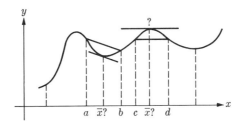

If we had

$$(?)\ a, b \text{ in } I, \qquad a < b, \qquad f(a) > f(b)\ (?),$$

as on the left of the graph above, then the slope of the chord would be

$$\frac{f(b) - f(a)}{b - a} < 0,$$

and this would give $f'(\bar{x}) < 0$ for some x between a and b. This is impossible, because such an $\bar{x}$ would be an interior point of I. If we had

$$(?)\ c, d \text{ in } I, \qquad c < d, \qquad f(c) = f(d)\ (?),$$

as on the right, then the chord would be horizontal, and we would have $f'(\bar{x}) = 0$ at an interior point $\bar{x}$ of I.

In Theorem 1, we allow the possibility that I is an infinite interval. Consider

$$f(x) = x^2, \qquad I = [0, \infty).$$

Here

$$f'(x) = 2x,$$

and so $f'(x) > 0$ at every interior point of I. Therefore f is increasing on I. Note that when we allow the possibility that $f'(x) = 0$ at endpoints of I, we are not splitting hairs; if we required that $f'(x)$ be >0 everywhere in I, then the theorem would not apply in the simple case $f(x) = x^2$, $I = [0, \infty)$, or to the case $f(x) = \cos x$, $I = [\pi, 2\pi]$. We don't need theorems to be as general as possible, but we want them to be general enough to be usable. And it is not unusual to find that $f'(x) = 0$ at an endpoint; in fact, this is what usually happens, when we break up the domain of our function into the largest possible subintervals on which the derivative does not change sign. Here f is increasing on I_1 and I_3, and decreasing on I_2; and the derivative vanishes at the endpoints x_1 and x_2.

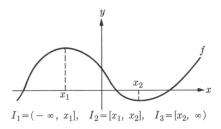

$$I_1 = (-\infty,\, x_1], \quad I_2 = [x_1,\, x_2], \quad I_3 = [x_2,\, \infty)$$

Consider another example:

$$f(x) = x^2 - x \qquad (0 \leqq x \leqq 2).$$

Here

$$f'(x) = 2x - 1, \qquad f'(\tfrac{1}{2}) = 0, \qquad \text{and} \qquad f'(x) > 0 \text{ on } (\tfrac{1}{2}, 2].$$

In the left-hand figure below, we have used this information, and have plotted $f(\tfrac{1}{2})$ and $f(2)$. Obviously $f(1) = 0$, and so we have plotted this point exactly also.

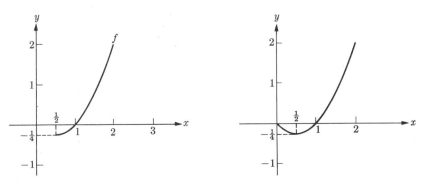

The same principle works in the opposite case:

Theorem 2. If $f'(x) < 0$ at every interior point of I, then f is decreasing on I.

(*Proof.* Apply Theorem 1 to the function $g = -f$. By Theorem 1, g is increasing on I. Therefore f is decreasing on I.)

To apply this theorem to the same function $f(x) = x^2 - x$, on the interval $I = [0, \frac{1}{2}]$, we observe that f is decreasing on this interval, because

$$f'(x) = 2x - 1 < 0 \qquad \text{for } 0 < x < \tfrac{1}{2}.$$

We use this information to complete our sketch.

This example doesn't look impressive, because we already knew how to sketch parabolas. It is not so obvious, however, how to sketch the graph of a cubic function taken at random, say,

$$f(x) = x^3 + 2x^2 - 3x - 4, \qquad -2 \leqq x \leqq 2.$$

This is not a put-up job; it is a "real-life" problem, and nothing is going to come out even. We need to find out where $f' > 0$ and where $f' < 0$. Now

$$f'(x) = 3x^2 + 4x - 3,$$

so that

$$f'(x) = 0 \quad \text{when} \quad x = \frac{-2 \pm \sqrt{13}}{3}.$$

Since $\sqrt{13} \approx 3.6$, the roots of the equation $f'(x) = 0$ are

$$x_1 \approx 0.5, \qquad x_2 \approx -1.9.$$

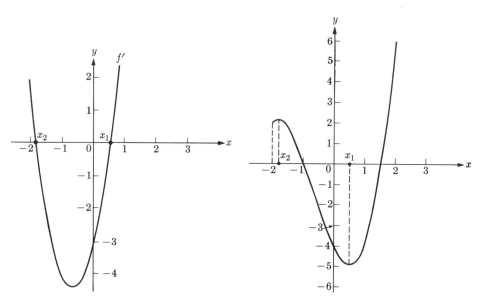

Since the graph of f' is a parabola opening upward, it must look like the drawing on the left above. Thus

$$f'(x) > 0 \quad \text{when} \quad x < x_2,$$
$$f'(x) < 0 \quad \text{when} \quad x_2 < x < x_1,$$
$$f'(x) > 0 \quad \text{when} \quad x > x_1.$$

Therefore f is increasing on $[-2, x_2]$; f is decreasing on $[x_2, x_1]$; and f is increasing on $[x_1, 2]$. We calculate

$$f(-2) = 2, \qquad f(x_2) \approx 2.1,$$
$$f(x_1) \approx -4.9, \qquad f(2) = 6.$$

This gives us our sketch on the right. (The problems in the following problem set are not this awkward.)

To apply this method, you need to know how the derivative behaves; and we may use the same method in investigating the derivative. For example, in the preceding problem we had

$$f'(x) = 3x^2 + 4x - 3.$$

If we let

$$g(x) = f'(x) = 3x^2 + 4x - 3,$$

then

$$g'(x) = 6x + 4.$$

Therefore g is increasing for $x > -\frac{2}{3}$, and is decreasing for $x < -\frac{2}{3}$. Plotting g exactly, at the points -2, x_2, 0, and x_1, we get the sketch of f' which is given above. We know that $f'(x) > 0$ for $x_1 < x < 2$, because f' increases, starting at the value $f'(x_1) = 0$. Similarly, $f'(x) > 0$ for $-2 < x < x_2$, because on the interval $[-2, x_2]$, f' decreases toward $f'(x_2) = 0$. Similarly in the middle interval $[x_1, x_2]$. This idea is simple enough, but it is so useful that we had better record it as a theorem:

Theorem 3. If f is increasing on $[x_1, x_2]$, then $f(x) > f(x_1)$ for every x on $(x_1, x_2]$.

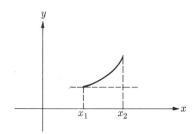

We recall that $(x_1, x_2]$ is a half-open interval;

$$(x_1, x_2] = \{x \mid x_1 < x \leq x_2\}.$$

We have been talking about the case $f'(x_1) = 0$.

Theorem 4. If f is decreasing on $[x_1, x_2]$, then $f(x) < f(x_1)$ for every x on $(x_1, x_2]$.

PROBLEM SET 5.1

For each function given, state on what intervals the function is increasing, and on what intervals it is decreasing; and sketch the graph.

1. $f(x) = \sin x,$ $\qquad\qquad\qquad -\pi \leqq x \leqq \pi$

2. $f(x) = \text{Sin}^{-1} x,$ $\qquad\qquad\quad -1 \leqq x \leqq 1$

3. $f(x) = \dfrac{1}{1 + x^2},$ $\qquad\qquad\quad -2 \leqq x \leqq 2$

4. $f(x) = \dfrac{x}{1 + x^2},$ $\qquad\qquad\quad -2 \leqq x \leqq 2$

5. $f(x) = x^3 - 3x,$ $\qquad\qquad\quad -2 \leqq x \leqq 2$

6. $f(x) = x^3 + 3x^2 - 2,$ $\qquad\qquad -1 \leqq x \leqq 3$

7. $f(x) = (\sin x + \cos x)^2 - 1,$ $\qquad -\pi \leqq x \leqq \pi$

8. $f(x) = x^3 + 6x^2 + 9x + 3,$ $\qquad -1 \leqq x \leqq 2$

9. $f(x) = e^{x^2},$ $\qquad\qquad\qquad\quad 0 \leqq x \leqq 1$

10. $f(x) = x \ln x,$ $\qquad\qquad\qquad 1 \leqq x \leqq 5$

11. $f(x) = \cos x,$ $\qquad\qquad\qquad 0 \leqq x \leqq 2\pi$

12. $f(x) = \sin 2x$ $\qquad\qquad\qquad\; 0 \leqq x \leqq \pi$

13. $f(x) = -x^4 + 2x^2$ $\qquad\qquad\; -2 \leqq x \leqq 2$

14. $f(x) = xe^{-x}$ $\qquad\qquad\qquad\; -1 \leqq x \leqq 2$

15. $f(x) = \dfrac{1}{1 + x^4}$ $\qquad\qquad\quad -1 \leqq x \leqq 1$

16. $f(x) = \dfrac{x}{1 + x^4}$ $\qquad\qquad\quad -1 \leqq x \leqq 1$

17. $f(x) = x \cos x - \sin x$ $\qquad\; -\pi \leqq x \leqq \pi$

18. $f(x) = x/2 + \sin x,$ $\qquad\qquad 0 \leqq x \leqq 2\pi$

19. $f(x) = e^x - 2x,$ $\qquad\qquad\quad\; 0 \leqq x \leqq 2$

(Here you are not going to be able to get answers in an exact numerical form. The figure should indicate plausible approximations.)

20. Investigate the converse of Theorem 1. That is, find out whether the following statement is true:

Theorem(?). If (i) f is continuous on $[a, b]$, (ii) f is differentiable on (a, b), and (iii) f is increasing on $[a, b]$, then (iv) $f'(x) > 0$ for every x of (a, b).

21. Is the following true?

Theorem(?). If f is differentiable at x_0 and $f'(x_0) > 0$, then some chord of the graph of f has a positive slope.

22. Investigate:

Theorem(?). Let f be a function satisfying (i), (ii), and (iii) of Problem 20. Then (iv') $f'(x) \geqq 0$ for every x of (a, b).

5.2 LOCAL MAXIMA AND MINIMA, DIRECTION OF CONCAVITY, INFLECTION POINTS

Again we consider a continuous function f, defined on a closed interval $[a, b]$. In the figure, $f(x_2) = M$; and M is the largest value of f.

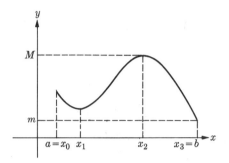

We say that f *has a maximum* at x_2; and we say that M is the *maximum value* of f. Similarly, $f(x_3) = m$; and m is the smallest value of f. We say that f *has a minimum* at x_3; and we say that m is the *minimum value* of f.

Here when we speak of maxima and minima, we mean maxima and minima on the *whole domain* of the function f; in this case the domain is $[a, b]$. Before you know what is a maximum or minimum, you must first know the domain of the function.

In the figure above, $f(x_1)$ is not a minimum value, because $f(x_3) < f(x_1)$. But $f(x_1)$ is the smallest value that f takes on when x is close to x_1. We say that f has a *local minimum* at x_1. This is abbreviated as LMin. Local minima can occur in three ways:

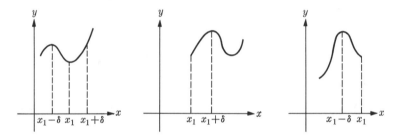

1) x_1 may lie on an open interval $(x_1 - \delta, x_1 + \delta)$, in the domain of f; and $f(x_1)$ may be the smallest value of the function on the interval $(x_1 - \delta, x_1 + \delta)$. In this case, we say that f has an *interior local minimum* at x_1. This is abbreviated as ILMin.

2) x_1 may be the left-hand endpoint of the domain of f; and $f(x_1)$ may be the smallest value of f on an interval $[x_1, x_1 + \delta)$.

3) x_1 may be the right-hand endpoint of the domain of f; and $f(x_1)$ may be the smallest value of f on an interval $(x_1 - \delta, x_1]$.

Thus, for the function f whose graph is sketched at the beginning of this section, we have local minima at x_1 and x_3. Note that every minimum is automatically a local minimum, just as the tallest man in the world is automatically the tallest in his own neighborhood.

Local maxima are defined similarly. *Local maximum* is abbreviated as LMax. A *local maximum* can occur in three ways:

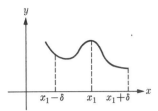 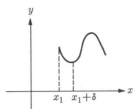 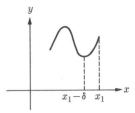

In the figure on the left, f has an *interior local maximum* at x_1. This is abbreviated ILMax.

There are simple conditions under which a function has an ILMax or an ILMin at a given point.

Theorem 1. If f is increasing on an interval $[x_1 - \delta, x_1]$ and decreasing on an interval $[x_1, x_1 + \delta]$, then f has an ILMax at x_1.

Theorem 2. If f is decreasing on an interval $[x_1 - \delta, x_1]$, and increasing on an interval $[x_1, x_1 + \delta]$, then f has an ILMin at x_1.

In applying these, we use the derivative. If $f' > 0$ on $(x_1 - \delta, x_1)$ and $f' < 0$ on $(x_1, x_1 + \delta)$, we can apply Theorem 1. Similarly for Theorem 2. In fact, if you find out where a function is increasing and where it is decreasing, it is always obvious where the interior local maxima and minima are; they are at the turning points, where the graph stops behaving in one way and starts behaving in the other way.

Most of the time, for functions defined on a closed interval, the endpoints of the interval give either local maxima or local minima. Therefore, if we are investigating a function for local maxima and minima, we always investigate the endpoints. Of course, interior local maxima and minima may occur anywhere in the interior of the interval. In searching for them, we use the theorem suggested by the figure below. If the function is differentiable, then at an interior local maximum the derivative must be $= 0$.

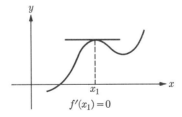

Theorem 3. If f has an ILMax at x_1, and f is differentiable at x_1, then $f'(x_1) = 0$.

This is geometrically obvious, and a logical proof is also easy. Let

$$m(x) = \frac{f(x) - f(x_1)}{x - x_1},$$

so that

$$\lim_{x \to x_1} m(x) = f'(x_1).$$

1) Suppose that $f'(x_1) > 0$. Then the function $m(x)$ must be >0 when $x \approx x_1$.

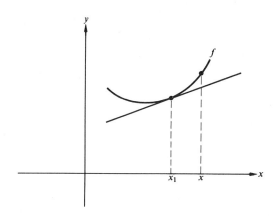

Take $x \approx x_1$, with $x > x_1$. Then

$$x \approx x_1 \quad \text{and} \quad x > x_1 \;\Rightarrow\; m(x) > 0 \quad \text{and} \quad x - x_1 > 0$$
$$\Rightarrow\; m(x)(x - x_1) > 0 \;\Rightarrow\; f(x) - f(x_1) > 0$$
$$\Rightarrow\; f(x) > f(x_1),$$

which is impossible, because f has an ILMax at x_1.

2) Suppose that $f'(x_1) < 0$. Then the function $m(x)$ must be <0 when $x \approx x_1$.

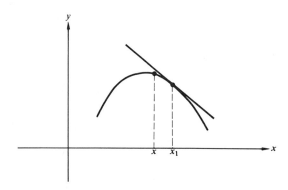

Take $x \approx x_1$, with $x < x_1$. Then

$$x \approx x_1 \quad \text{and} \quad x < x_1 \;\Rightarrow\; m(x) < 0 \quad \text{and} \quad x - x_1 < 0$$
$$\Rightarrow\; m(x)(x - x_1) > 0 \;\Rightarrow\; f(x) - f(x_1) > 0$$
$$\Rightarrow\; f(x) > f(x_1),$$

which is impossible.

Since (1) and (2) are both impossible, it follows that $f'(x_1) = 0$, which was to be proved.

This is the standard method for finding an ILMax. Given a differentiable function f, we find the points x where $f'(x) = 0$. Usually there are only a finite number of such points. *These are the only possible places where interior local maxima can occur.* Therefore we have only a finite number of values of x to investigate; and when we are done, our list of interior local maxima is complete.

Note, however, that the converse of Theorem 3 is false: if $f'(x_1) = 0$, it does *not* follow that f has a local maximum (or a local minimum) at x_1. For example, if $(fx) = x^3$, $-1 \leq x \leq 1$, then $f'(0) = 0$, but f is increasing on the whole interval $[-1, 1]$

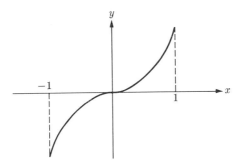

We have a similar theorem for interior local minima:

Theorem 4. If f has an ILMin at x_1, and f is differentiable at x_1, then $f'(x_1) = 0$.

Proof. Let
$$g(x) = -f(x).$$
Then g has an ILMax at x_1. Therefore $g'(x_1) = 0$. Therefore $f'(x_1) = -g'(x_1) = -0 = 0$, which was to be proved.

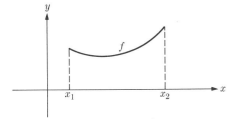 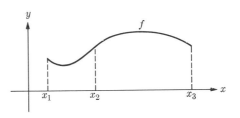

If f' is increasing, on an interval $[x_1, x_2]$, then f is *concave upward* on $[x_1, x_2]$. (You ought to be able to convince yourself that this is a reasonable use of language.) If f' is decreasing on $[x_2, x_3]$, then f is *concave downward* on $[x_2, x_3]$. In the figure on the right, x_2 is the point at which the direction of concavity changes. Such a point is called an *inflection point*. Of course, the direction of concavity can change from up to down or from down to up. Hence:

Definition. An *inflection point* of a function f is a point at which f' has either an ILMax or an ILMin.

Note the way in which these definitions fit together. If you know how to investigate (a) increasing, (b) decreasing, (c) interior local maxima, and (d) interior local minima, then automatically you know how to investigate direction of concavity and inflection points. The reason is that f', once you get it, is a function, and can be investigated in the same way as any other function, with the aid of *its* derivative f''. Where f' increases, f is concave upward; where f' decreases, f is concave downward; and where f' has an interior local maximum or minimum, f has an inflection point.

Most of the time, we investigate local maxima and local minima because we want to find the maxima and minima. We find the maxima and minima, on the whole domain, by looking to see which local maximum value is the largest and which local minimum value is the smallest.

Finally, we observe that a function may easily have a local maximum or minimum at an endpoint at which it is not differentiable. For example, the function $f(x) = x^{2/3}$ $(0 \leq x < \infty)$ has a minimum (and hence a local minimum) at $x = 0$. The theory takes care of this case. Since the derivative $\frac{2}{3}x^{-1/3}$ is positive in the interior of the interval $[0, \infty)$, it follows that the function is increasing, and so it has a minimum at the left-hand endpoint.

PROBLEM SET 5.2

1 through 19. For each of the functions described in Problems 1 through 19 of the preceding problem set, find the local maxima, the local minima, the maximum, the minimum, the inflection points (if any), and the image. (The image will always turn out to be a closed interval.) Tell where each of the functions is concave upward and where it is concave downward.

20. Consider the function defined by the following conditions:

$$f(x) = x \sin \frac{1}{x} \quad \text{for } 0 < x \leq \frac{1}{\pi},$$

$$f(0) = 0.$$

An exact sketch is not practical, because the ILMax and ILMin points are hard to calculate. Give a rough sketch, however, indicating as well as you can how the function behaves. Is it continuous at 0? Does it have a local maximum or minimum at 0? Is it differentiable at 0?

*21. Suppose that f is both continuous and differentiable on $[0, 1]$. Does it follow that f has an LMax or an LMin at 0? Why or why not?

5.3 THE BEHAVIOR OF FUNCTIONS AT INFINITY

So far, we have been discussing functions on closed intervals. In this section, we shall consider larger domains, including infinite intervals, such as $(-\infty, \infty)$, $[0, \infty)$, and so on, and also intervals with holes in them. For example, the domain of the tangent is

$$D = \{x \mid x \neq \pi/2 + n\pi\};$$

and this is an infinite interval $(-\infty, \infty)$ with infinitely many holes in it.

Most of the ideas that we shall be investigating are illustrated by a simple function, whose domain has a hole in it at 0.

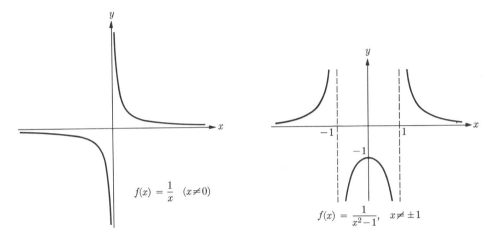

$$f(x) = \frac{1}{x} \quad (x \neq 0)$$

$$f(x) = \frac{1}{x^2-1}, \quad x \neq \pm 1$$

A careful inspection of the left-hand graph above will give you an idea of the meanings of the following statements:

$$\lim_{x \to \infty} f(x) = 0, \tag{1}$$

$$\lim_{x \to -\infty} f(x) = 0, \tag{2}$$

$$\lim_{x \to 0^+} f(x) = \infty, \tag{3}$$

$$\lim_{x \to 0^-} f(x) = -\infty. \tag{4}$$

Definitions will come later. Meanwhile, let us look at another example. The function whose graph is shown on the right above has the following properties:

i) f has an interior local maximum at 0.   (At $x = 0$, the denominator is -1. Everywhere else near 0, $x^2 > 0$, $x^2 - 1 > -1$, and $1/(x^2 - 1) < -1$.)

ii) $\lim_{x \to 1^-} f(x) = -\infty.$

iii) $\lim_{x \to 1^+} f(x) = \infty.$

iv) $\lim_{x \to -1^+} f(x) = -\infty.$

v) $\lim_{x \to -1^-} f(x) = \infty.$

Here statements (ii) through (v) mean the things that the figure suggests. An examination of the formula shows why the figure is right. For example, if $x > 1$, and $x \approx 1$, then $x^2 - 1 > 0$, and $x^2 - 1 \approx 0$. Therefore $1/(x^2 - 1)$ is positive and very large. This is shown in the figure and stated by (iii). Similarly, if $x < 1$ and $x \approx 1$, then $x^2 - 1 < 0$ and $x^2 - 1 \approx 0$. Therefore $1/(x^2 - 1)$ is negative and is *numerically* very large. This is shown in the figure and stated by (ii).

Let us now make this precise, by stating definitions that we can work with.

Definition. $\lim_{x \to \infty} f(x) = L$ means that for every $\epsilon > 0$ there is an M such that

$$x > M \implies L - \epsilon < f(x) < L + \epsilon.$$

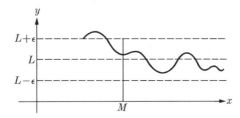

This is like the definition of $\lim_{x \to x_0} f(x) = L$. Roughly,

$$\lim_{x \to x_0} f(x) = L \qquad \text{means} \qquad x \approx x_0 \implies f(x) \approx L,$$

and

$$\lim_{x \to \infty} f(x) = L \qquad \text{means} \qquad x \approx \infty \implies f(x) \approx L.$$

In the definitions, the condition $x \approx x_0$ is expressed by $0 < |x - x_0| < \delta$, and the condition $x \approx \infty$ is expressed by $x > M$.

Let us see how our definition of $\lim_{x \to \infty}$ applies to the function

$$f(x) = \frac{1}{x} \qquad (x \neq 0).$$

We claim that

$$\lim_{x \to \infty} \frac{1}{x} = 0.$$

Under the definition, given $\epsilon > 0$, we are supposed to find an M such that

$$-\epsilon < \frac{1}{x} < \epsilon \qquad \text{whenever } x > M.$$

This is trivial: take $M = 1/\epsilon$. When $x > 1/\epsilon$, obviously $0 < 1/x < \epsilon$. Similarly:

Definition. $\lim_{x \to -\infty} f(x) = L$ means that for every $\epsilon > 0$ there is an M such that

$$x < M \implies L - \epsilon < f(x) < L + \epsilon.$$

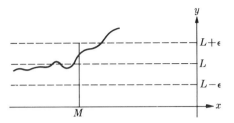

In the same spirit:

Definition. $\lim_{x \to x_0^+} f(x) = \infty$ means that for every M there is a $\delta > 0$ such that

$$x_0 < x < x_0 + \delta \implies f(x) > M.$$

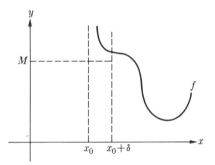

That is, you can make $f(x)$ as big as you want (i.e., $> M$) by taking x to the right of x_0 and very close to x_0 (i.e., between x_0 and $x_0 + \delta$.)

We need to talk about one-sided limits (as $x \to x_0^+$ or $x \to x_0^-$) because these often turn out to be different. In some cases, however, the one-sided limits have the same value. In such cases, $\lim_{x \to x_0} f(x)$ must exist, and must be their common value. Thus

$$\lim_{x \to 0} \frac{1}{x^2} = \infty.$$

The following two theorems justify the remarks that were made above about $f(x) = 1/(x^2 - 1)$.

Theorem 1. Suppose that $f(x) > 0$ on an interval (x_0, x_1). If $\lim_{x \to x_0^+} f(x) = 0$, then

$$\lim_{x \to x_0^+} \frac{1}{f(x)} = \infty.$$

Proof. Given $M > 0$. We need to find a $\delta > 0$ such that $1/f(x) > M$ whenever $x_0 < x < x_0 + \delta$. Let $\epsilon = 1/M$. By the definition of the statement $\lim_{x \to x_0} f(x) = 0$, we know that there is a $\delta > 0$ such that

$$f(x) < \epsilon \qquad \text{whenever} \qquad x_0 < x < x_0 + \delta.$$

This is the δ that we want: when $x_0 < x < x_0 + \delta$, we have

$$f(x) < \frac{1}{M},$$

and hence

$$\frac{1}{f(x)} > M.$$

(Remember that $M > 0$, and $f(x) > 0$ for the values of x that we are interested in.) Similarly, we have:

Theorem 2. Suppose that $f(x) < 0$ on an interval (x_0, x_1). If $\lim_{x \to x_0^+} f(x) = 0$, then

$$\lim_{x \to x_0^+} \frac{1}{f(x)} = -\infty.$$

Proof. Given $M < 0$. Let $\epsilon = -1/M > 0$. Let δ be a positive number such that

$$-\epsilon < f(x) \quad \text{whenever} \quad x_0 < x < x_0 + \delta.$$

When $x_0 < x < x_0 + \delta$, we have

$$f(x) > -\epsilon, \quad 1 < \frac{-\epsilon}{f(x)}, \quad -\frac{1}{\epsilon} > \frac{1}{f(x)}, \quad \frac{1}{f(x)} < M.$$

(Here we have been reversing inequalities, because we have been dividing by negative numbers.)

Following the analogy of the above definitions, you ought to be able to write your own definitions of the following statements:

$$\lim_{x \to \infty} f(x) = \infty, \qquad \lim_{x \to -\infty} f(x) = -\infty,$$

$$\lim_{x \to \infty} f(x) = -\infty, \qquad \lim_{x \to -\infty} f(x) = \infty.$$

Consider now the question of

$$(?) \ \lim_{x \to \infty} \left(1 + \frac{1}{x}\right)^x \ (?)$$

We use question marks, because it is not obvious that the indicated limit exists at all: as $x \to \infty$, $1/x \to 0$, and $1 + 1/x \to 1$. Therefore we have an "indeterminate form of the type 1^∞." We recall, however, a similar situation before where we got an answer like this:

$$\lim_{h \to 0} (1 + h)^{1/h} = e = \ln^{-1} 1.$$

This was also of the form "1^∞." And the two are related: if we let

$$f(u) = (1 + u)^{1/u} \quad \text{and} \quad g(x) = \frac{1}{x},$$

then

$$\left(1 + \frac{1}{x}\right)^x = f(g(x)),$$

and we want to find

$$\lim_{x \to \infty} f(g(x)).$$

This is like the situation in Section 4.5. There we found:

Theorem. If $\lim_{x \to x_0} g(x) = u_0 = g(x_0)$ and $\lim_{u \to u_0} f(u) = f(u_0)$, then

$$\lim_{x \to x_0} f(g(x)) = f(u_0).$$

For the case in which $x \to \infty$, instead of $x \to x_0$, this theorem is still true, and the proof is virtually the same. That is:

Theorem 3. If $\lim_{x \to \infty} g(x) = u_0$ and $\lim_{u \to u_0} f(u) = L$, then

$$\lim_{x \to \infty} f(g(x)) = L.$$

Roughly speaking, the reason is that

$$x \approx \infty \quad \Rightarrow \quad g(x) \approx u_0 \quad \Rightarrow \quad f(g(x)) \approx L.$$

In fact, the same result holds if $\lim_{x \to \infty} g(x) = \infty$.

Theorem 4. If $\lim_{x \to \infty} g(x) = \infty$ and $\lim_{u \to \infty} f(u) = L$, then

$$\lim_{x \to \infty} f(g(x)) = L.$$

These theorems give quick answers to some rather hard-looking problems. Returning to our discussion of

$$f(u) = (1 + u)^{1/u}, \qquad g(x) = \frac{1}{x}, \qquad f(g(x)) = \left(1 + \frac{1}{x}\right)^x,$$

we get immediately:

Theorem 5. $\lim_{x \to \infty} (1 + 1/x)^x = e$.

This limit is used as a definition of e, in some treatments of exponentials and logarithms. In such a treatment, the formula $e = \lim_{h \to 0} (1 + h)^{1/h}$ appears as a theorem.

PROBLEM SET 5.3

Investigate the following functions for maxima, minima, local maxima, local minima, direction of concavity, and inflection points. Then investigate for limits of the sort defined in this section.

1. $f(x) = \dfrac{1}{x(x - 2)}$ $\qquad$ $(x \neq 0, x \neq 2)$

2. $f(x) = \dfrac{1}{(x-1)(x-3)}$ $(x \neq 1,\ x \neq 3)$

3. $f(x) = \dfrac{1}{x^2 - x - 6}$ $(x \neq -2,\ x \neq 3)$

4. $f(x) = \dfrac{1}{x^2 + 1}$

5. $f(x) = \dfrac{1}{x^2}$ $(x \neq 0)$

6. $f(x) = \dfrac{x}{x^2 + 1}$

7. $f(x) = \dfrac{x^2}{x^2 + 1}$

8. $f(x) = \dfrac{1}{x^3 + 1}$ $(-1 \neq x)$

9. $f(x) = \dfrac{x+1}{x^3 + 1}$ $(-1 \neq x)$

10. $f(x) = \dfrac{1}{x^3 - x}$ $(x \neq 0,\ x \neq 1,\ x \neq -1)$

11. $f(x) = \dfrac{x^3}{1 + x^3}$ $(-1 \neq x)$

Investigate:

12. $\lim\limits_{x \to \infty} \left(1 + \dfrac{1}{x^2}\right)^{x^2}$ 13. $\lim\limits_{x \to \pi/2} (1 + \cos x)^{\sec x}$ 14. $\lim\limits_{x \to 0^+} (1 + \sqrt{x})^{1/\sqrt{x}}$

15. $\lim\limits_{x \to 0} (1 + x^4)^{1/x^4}$ 16. $\lim\limits_{x \to 1} (x)^{1/(x-1)}$

Investigate the following, for $\lim\limits_{x \to \infty}$.

17. $f(x) = \left(1 + \dfrac{1}{2x}\right)^{2x}$ 18. $f(x) = \left(1 + \dfrac{2}{x}\right)^{x/2}$ 19. $f(x) = \left(1 + \dfrac{1}{2x}\right)^{x}$

20. $f(x) = \left(1 + \dfrac{2}{x}\right)^{x}$ 21. $f(x) = (1 + e^{-x})^{e^x}$ 22. $\left(1 + \dfrac{1}{\ln x}\right)^{\ln x}$

23. $\left(1 + \dfrac{3}{x}\right)^{x/2}$

24. Discuss as in Problems 1 through 11

$$f(x) = (\ln x)/x   (x > 0).$$

(Here the sticky point is $\lim_{x \to \infty}$. You ought to be able to figure out what this limit is, and convince yourself that your answer must be right. But to *prove* that the answer is right is an unreasonably hard problem, at this stage.)

25. Find $\lim_{x \to 0^+} x \ln (1/x)$. You need not prove that your answer is right.
26. Is there such a thing as $\lim_{x \to \infty} \sin x$? Why or why not?
27. Is there such a thing as $\lim_{x \to \infty} (1/x) \sin x$? Why or why not?
28. Prove the following:

Theorem (*The squeeze principle*). If

$$f(x) \leq g(x) \leq h(x)   (x \geq a),$$

and

$$\lim_{x \to \infty} f(x) = \lim_{x \to \infty} h(x) = L,$$

then $\lim_{x \to \infty} g(x) = L$.

29. If you borrow a dollar for a year, at 100% simple interest, then at the end of the year you owe 2 dollars. (A certain Marcus Junius Brutus lent money at this rate, in the first century B.C. He was also an assassin.) If interest is compounded semiannually, then at the end of the year you owe

$$(1 + \tfrac{1}{2})^2 = \tfrac{9}{4} = \$2.25.$$

If the interest is compounded n times a year, then you owe

$$\left(1 + \frac{1}{n}\right)^n.$$

Suppose now that interest is compounded *continuously*: the bank passes to the limit, as n increases without limit, and at the end of the year they charge you the limit. How much do you owe?

30. Suppose that the basic interest rate is 6%, but interest is compounded continuously, as in Problem 29. How much do you owe? (To get a numerical answer to this one, you will need to use one of the tables at the end of the book.)

5.4 THE INTRODUCTION OF FUNCTIONS INTO GEOMETRIC PROBLEMS; THE USE OF EXISTENCE THEOREMS AS SHORTCUTS

On several occasions already we have been confronted with problems which did not appear to involve functions, and have solved them by introducing functions.

For example, in Section 3.7 we wanted to find the area under the graph of $y = x^4$, from 0 to 1.

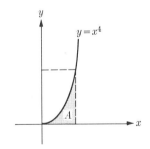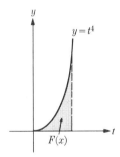

We solved this problem by attacking the more general problem of calculating the *function*

$$F(x) = \int_0^x t^4 \, dt.$$

We found that

$$F(x) = \frac{x^5}{5},$$

and then set $x = 1$ to get the answer $\tfrac{1}{5}$.

Similarly, in Section 4.10 we wanted to show that

$$\ln(ab) = \ln a + \ln b,$$

for every pair of positive numbers a, b. To use the methods of calculus, we had to introduce functions into the problem. Given $k > 0$, we set

$$f(x) = \ln kx \qquad\qquad (x > 0),$$
$$g(x) = \ln k + \ln x \qquad (x > 0).$$

We then found that $f'(x) = g'(x)$ for every x, and $f(1) = g(1)$. It followed that $f = g$; and this proved our theorem.

We use the same kind of method to attack problems in maxima and minima which may be stated in geometric or physical terms. Consider some examples.

Problem 1. A segment of length 1 has its endpoints on the sides of a right angle. What position for the segment gives maximum area for the resulting triangle?

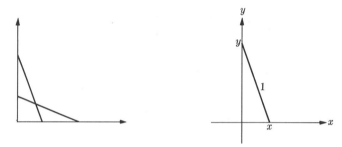

The first step is to introduce a coordinate system, as shown on the right above. The endpoints of the segment now lie on the positive ends of the axes.

Let x be the x-coordinate of the endpoint that lies on the x-axis; and let the other endpoint be $(0, y)$. When x is named, y is determined. Thus there is a function f which gives y in terms of x. Since

$$x^2 + y^2 = 1,$$

we have

$$f(x) = \sqrt{1 - x^2} \qquad (0 \leqq x \leqq 1).$$

And for each x, the area enclosed is

$$A(x) = \tfrac{1}{2}xf(x) = \tfrac{1}{2}x\sqrt{1 - x^2}.$$

We need to investigate the function A for maxima. Now

$$A'(x) = \frac{1}{2}\left[x\,\frac{-x}{\sqrt{1 - x^2}} + \sqrt{1 - x^2} \right] = \frac{1}{2}\,\frac{-x^2 + (1 - x^2)}{\sqrt{1 - x^2}}$$

$$= -\frac{1}{2}\,\frac{2x^2 - 1}{\sqrt{1 - x^2}} \qquad (0 \leqq x \leqq 1).$$

Therefore $A'(x) = 0$ when $x = \pm\sqrt{2}/2$. Since we are concerned only with numbers on the interval $[0, 1]$, only $x = \sqrt{2}/2$ is of interest to us. Here $A = \tfrac{1}{4}$. Any maximum of A is surely an ILMax, because $A(0) = A(1) = 0$, and $A(x) > 0$ for $0 < x < 1$.

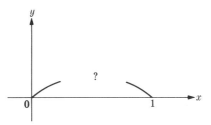

Therefore our problem is solved, with $A = \frac{1}{4}$, if we know the following theorem:

Theorem 1 (*Existence of maxima*). If f is continuous on $[a, b]$, then f has a maximum value on $[a, b]$.

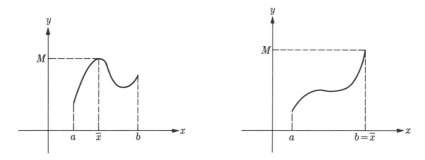

The maximum may be an ILMax, as on the left above, or it may be at an endpoint, as on the right. But in many cases, like the one we have just been discussing, it is plain that the second of these possibilities does not arise. In such cases, we can infer immediately that the maximum is an ILMax. If the derivative vanishes at only one point, then this point must be the maximum.

We shall prove Theorem 1 in Section 5.6. Meanwhile, let us look at some more applications of it. In the preceding example, there are other functions that we might equally well have introduced.

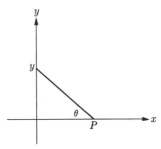

If the angle at P has measure θ ($0 \leqq \theta \leqq \pi/2$), then

$$y = \sin \theta, \qquad x = \cos \theta,$$

and

$$A(\theta) = \tfrac{1}{2}xy = \tfrac{1}{2} \sin \theta \cos \theta = \tfrac{1}{4} \sin 2\theta.$$

Therefore

$$A'(\theta) = \tfrac{1}{4}(\cos 2\theta) \cdot 2 = \tfrac{1}{2} \cos 2\theta.$$

The only point θ on the interval $[0, \pi/2]$ where $A'(\theta) = 0$ is the point where

$$2\theta = \frac{\pi}{2}, \qquad \theta = \frac{\pi}{4}.$$

We claim, without further investigation of derivatives, that this must be where the maximum occurs. (As in the previous discussion, there must be a maximum somewhere; this is not at an endpoint 0 or $\pi/2$; it is therefore an interior local maximum; at an ILMax, $A'(\theta) = 0$; and $\theta = \pi/4$ is the only point of the interval at which $A'(\theta) = 0$.)

Setting $\theta = \pi/4$, we get the maximum value of A as

$$A\!\left(\frac{\pi}{4}\right) = \tfrac{1}{4} \sin\left(2 \cdot \frac{\pi}{4}\right) = \tfrac{1}{4} \sin \frac{\pi}{2} = \tfrac{1}{4},$$

as before.

On reflection, you may find a way to solve this problem by purely geometrical methods, without taking any derivatives or even introducing any functions. The geometric method is easier if you think of it. Even in cases where elementary methods can be made to work, however, calculus does the same job methodically.

Problem 2. In a coordinate plane, let $A = (0, 1)$ and $B = (3, 2)$, as shown in the figure. What is the length of the shortest path from A to the x-axis to B? And where should the path touch the x-axis, for this minimum to be attained?

In other words, for what choice of $P = (x, 0)$ is the sum of the distances AP and PB as small as possible?

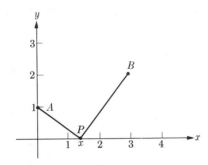

Solution. Let

$$f(x) = AP + PB$$

$$= \sqrt{1^2 + x^2} + \sqrt{(3 - x)^2 + 2^2}$$

$$= \sqrt{1 + x^2} + \sqrt{x^2 - 6x + 13}.$$

Then

$$f'(x) = \frac{x}{\sqrt{1+x^2}} + \frac{x-3}{\sqrt{x^2-6x+13}}$$

$$= \frac{x\sqrt{x^2-6x+13} + (x-3)\sqrt{1+x^2}}{\sqrt{1+x^2} \cdot \sqrt{x^2-6x+13}}$$

Therefore $f'(x) = 0$ when

$$x^2(x^2-6x+13) = (x^2-6x+9)(x^2+1),$$

or

$$x^4 - 6x^3 + 13x^2 = x^4 - 6x^3 + 9x^2 + x^2 - 6x + 9,$$

or

$$3x^2 + 6x - 9 = 0,$$

or

$$x^2 + 2x - 3 = 0,$$

or

$$(x+3)(x-1) = 0.$$

To examine second derivatives looks hard. Let us try to use reasoning instead.

1) When x decreases past 0, AP increases, and so does PB. The same is true when x increases past 3. Therefore, in searching for a minimum, we can restrict the search to, say, the interval $[-1, 4]$.

2) Suppose that we know that the function *has* a minimum, somewhere on the interval $[-1, 4]$. Then the minimum must be an ILMin, at which $f'(x) = 0$. There is only one such point on our interval, namely, $x = 1$. Therefore the minimum must be at $x = 1$.

Obviously, to complete this discussion, we need the following theorem:

Theorem 2 (*Existence of minima*). If f is continuous on $[a, b]$, then f has a minimum value on $[a, b]$.

The proof is easy, granted that Theorem 1 is true. Since $-f$ is continuous, it has a maximum; and any maximum of $-f$ is a minimum of f.

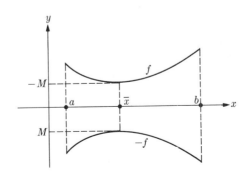

Here again, once the problem is solved, you may be able to think of a simpler attack on it. But the methods of calculus work in any case.

Problem 3. Find the right circular cylinder of largest volume, inscribed in a sphere of radius 1.

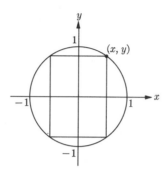

To avoid a difficult drawing problem, we show not the three-dimensional figure but merely a plane cross section of it. One way to introduce a function into this problem is to express the volume of the cylinder as

$$V(x) = \pi r^2 h = \pi x^2 \cdot 2y = 2\pi x^2 \sqrt{1 - x^2} \qquad (0 \leq x \leq 1).$$

This gives

$$V'(x) = 2\pi \left(2x \cdot \sqrt{1 - x^2} + x^2 \cdot \frac{-x}{\sqrt{1 - x^2}} \right)$$

$$= 2\pi \cdot \frac{2x(1 - x^2) - x^3}{\sqrt{1 - x^2}} .$$

Therefore

$$V'(x) = 0 \iff 2x - 3x^3 = 0 \iff x(3x^2 - 2) = 0.$$

Since x must lie on [0, 1], we find that $V'(x) = 0$ only when $x = 0$ or $x = \sqrt{\tfrac{2}{3}}$. Now V *has* a maximum, because V is continuous on [0, 1]. And this must be an ILMax, because $V(0) = 0 = V(1)$, and $V(x) > 0$ everywhere else on [0, 1]. Therefore $V'(x) = 0$ at the maximum. Therefore the maximum occurs at $x = \sqrt{\tfrac{2}{3}}$. Hence the maximum volume is

$$V\left(\sqrt{\tfrac{2}{3}} \right) = 2\pi \cdot \frac{2}{3} \sqrt{1 - \frac{2}{3}} = \frac{4\pi}{3} \cdot \frac{\sqrt{3}}{3} = \frac{4\pi\sqrt{3}}{9} .$$

There is another function that we might have used to solve the same problem. We might have written

$$V(y) = \pi r^2 h = \pi x^2 \cdot 2y = \pi(1 - y^2)2y = 2\pi \cdot (y - y^3).$$

This would give

$$V'(y) = 2\pi(1 - 3y^2),$$

so that

$$V'(y) = 0 \iff 3y^2 = 1 \iff y = \sqrt{\tfrac{1}{3}} \ \text{ or } \ y = -\sqrt{\tfrac{1}{3}}.$$

Here again only the positive number applies, because y must be on the interval $[0, 1]$. As before, we conclude that the maximum value occurs at $y = \sqrt{\tfrac{1}{3}}$;

$$V\left(\sqrt{\tfrac{1}{3}}\right) = \pi\left(1 - \tfrac{1}{3}\right)2 \cdot \sqrt{\tfrac{1}{3}} = \frac{4\pi}{3}\sqrt{\tfrac{1}{3}} = \frac{4\pi\sqrt{3}}{9}.$$

The second method is simpler. This sort of thing happens often. It is therefore a good idea to have a quick look at all of the functions that it seems natural to try, before doing any hard work with any one of them. If the first function that you try looks simple, there is no point in examining others.

Our third problem shows a danger which should be remembered hereafter. We might have supposed that the inscribed cylinder attains its maximum volume at the stage where the inscribed rectangle (in the cross section) attains its maximum area. But this is false: it is easy to show that the inscribed rectangle of maximum area is a square; and the cross section of the maximal cylinder is a rectangle of base $2\sqrt{\tfrac{2}{3}}$ and altitude $2\sqrt{\tfrac{1}{3}}$. Therefore *we should never assume without proof that two maximum-minimum problems are equivalent*.

A further word of caution: In establishing that a certain x_0 gives a maximum or minimum, you may use the theorems of the preceding sections. Under certain conditions, you may avoid these theorems (and the calculations that they require) by the sort of reasoning that we have used in the problems above. But in any case, you must use *either* the theorems of the preceding sections *or* a reasoning process which justifies your conclusions. To find a point x_0 where a derivative vanishes and hence infer that your problem is solved is a mistake. For one thing, x_0 may give a minimum when you were looking for a maximum, or vice versa. For another thing, x_0 may give a point of inflection.

PROBLEM SET 5.4

1. Find the area of the largest rectangle than can be inscribed in a semicircle of radius a.
2. Find the area of the largest rectangle that can be inscribed in an equilateral triangle whose sides have length a.
3. Find the area of the triangle with the smallest area which contains a square with side a.
4. Find the perimeter of the triangle with the smallest perimeter which contains a square with side a.
5. A rectangular field has one side along a river and a fence along the other three sides. If the total length of the fence is k, what is the maximum possible area of the field?
6. Given a rectangular field with one side along a river, as in Problem 5. If the area of the field is A, what is the minimum possible length of the fence?
7. If a rectangular wooden beam is supported horizontally at its ends, then the maximum weight that it can support at its midpoint is proportional (at least approximately) to its width, and to the square of its thickness. That is, $W = k \cdot x \cdot y^2$, where x is the width,

y is the thickness, and k is a constant depending on the wood (and on the units of length and weight).

Suppose that such a beam is to be cut from a cylindrical log of radius a, in such a way as to maximize W. What should be the width and the thickness?

8. An open pan is to be made out of a square metal sheet, by cutting out the square pieces from the corners of the sheet and folding up the sides of the metal that is left. (The square pieces are to be thrown away.) If the sheet has edges of length a, what is the volume of the pan of largest volume that can be made in this way?

9. An open pan, of the sort described in the preceding problem, has a total surface area of 128 sq. in. What is the largest possible volume?

10. Find the closed circular cylinder with volume 10 cu. in. and surface area as small as possible.

11. Solve the same problem, given that the cylinder is open at one end.

12. Solve the same problem, given that the cylinder is open at *both* ends. (It sits on a flat table and holds flour.)

13. A piece of sheet metal, n feet long and w feet wide, is to be bent so as to form a trough n feet long, with open top, open ends, and triangular cross sections. What is the greatest possible cross sectional area?

14. A trough is to be made with isosceles right triangles as endpieces and congruent rectangles as sides, as shown in the figure. If the total surface area is to be 100 sq. in., what is the maximum volume?

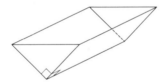

15. In a rectangular parallelepiped, with a square base, the total length of the edges is k. What is the largest possible volume?

16. A rectangle is to be inscribed in the region above the x-axis and below the graph of $y = 1 - x^2$. Find the area of the rectangle of maximum area.

17. Same problem, for $y = 1 - x^4$.

18. Find the rectangle of maximum area contained in the region above the line $y = \frac{1}{2}$, to the right of the line $x = 1$, and under the graph of $y = 1/x$.

19. A rectangle is inscribed in the region $R = \{(x, y) \mid |x| + |y| \le 1\}$, in such a way as to maximize the area. Find the area of the rectangle.

Find the values of x at which the following functions take on their maximum values, and justify your answers. You need not find the maximum values of the functions.

20. $f(x) = \dfrac{x}{1 + x^2}$

21. $g(x) = \dfrac{x}{1 + x^4}$

22. $h(x) = \displaystyle\int_{-1}^{x} \mathrm{Sin}^{-1} t \, dt$

23. $\phi(x) = \displaystyle\int_{0}^{\sin x} \sqrt{1 + t^8} \, dt$

Problems 24 through 27. Investigate the preceding four functions for *minimum* values.

28. An isosceles triangle has base d and altitude h. Find the area of the rectangle of largest area that can be inscribed in it.

29. Given a triangle with angles of $30°$, $60°$, and $90°$, there are three plausible ways of inscribing in it a rectangle of maximum area; the rectangle may have a side lying along any one of the three sides of the triangle. Show that all three of these "maximal" rectangles are really maximal; that is, show that they all have the same area.

30. Show that there are some triangles for which the conclusion in Problem 29 does not hold.

31. Show, however, that the conclusion of Problem 29 holds for a class of triangles which includes more than the $30°$–$60°$–$90°$ triangles.

32. Consider the curve which is the graph of the equation $x^2 + 4y^2 = 4$. Find the area of the rectangle of largest area that can be inscribed in this curve.

33. A right circular cone has a base of diameter d, and altitude h. Find the volume of the largest right circular cylinder that can be inscribed in it.

34. Find the area of the isosceles triangle of maximum area that can be inscribed in a circle of radius r.

35. Find the volume of the right circular cylinder of maximum volume that can be inscribed in a sphere of radius r.

36. Suppose that in Problem 34 the word "isosceles" is omitted. Is the solution of the resulting problem the same as before?

37. Similarly, discuss the problem obtained by omitting the word "right" in Problem 35.

38. Find the length of the longest ladder than can be carried (in a horizontal position) around the corner shown on the left below. The segment from P to Q shows a possible position of the ladder.

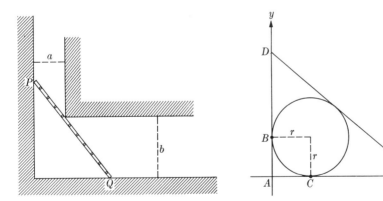

39. In the right-hand figure above, the circle (of radius r) is inscribed in the right angle $\angle BAC$. What is the minimum possible area of $\triangle ADE$?

**40. Suppose that in Problem 39 we do not require that $\angle BAC$ be a right angle. Given that $\angle BAC$ has measure α, find the minimum possible area of $\triangle ADE$, in terms of r and α. (This is *much* harder than Problem 39.)

5.5 THE USE OF FUNCTIONAL EQUATIONS AS SHORTCUTS

In the preceding section, we found that under some conditions we could locate maximum and minimum values merely by finding a point where the derivative vanishes. We shall now see that in some cases we can locate maximum and minimum values without calculating the function. Consider first a simple problem, from Section 5.4.

Problem 1. A segment of length 1 has its endpoints on the sides of a right angle. What position for the segment gives maximum area for the resulting triangle?

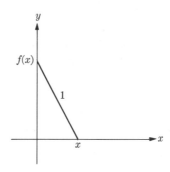

As in Section 5.4, we set up the axes as shown. Let x be the x-coordinate of the lower endpoint of the segment; and for each x from 0 to 1, let $f(x)$ be the y-coordinate of the other endpoint. Note that we are entitled to use functional notation: $f(x)$ really is determined when x is named. And for each x, we have

$$x^2 + [f(x)]^2 = 1^2,$$

because $x^2 + [f(x)]^2$ is the square of the length of the segment. Therefore the function f satisfies the equation

$$x^2 + f^2 = 1 \qquad (0 \leqq x \leqq 1). \tag{1}$$

The area of the triangle is

$$A(x) = \tfrac{1}{2}x \cdot f(x). \tag{2}$$

Now in (1), the left-hand member is a function, whose derivative is $2 \cdot x + 2 \cdot ff'$. But this function is known to be a constant, equal to 1 for every x from 0 to 1. Therefore

$$x + ff' = 0 \qquad (0 < x < 1). \tag{1'}$$

Here, of course, we are assuming that f *has* a derivative, for $0 < x < 1$, but this must be true, because the graph of f is a quadrant of a circle. Obviously

$$A'(x) = \tfrac{1}{2} \cdot x \cdot f'(x) + \tfrac{1}{2} \cdot f(x). \tag{2'}$$

The maximum of $A(x)$ must be an ILMax; and so, at the maximum of $A(x)$, we have

$$xf' + f = 0. \tag{2''}$$

We now know:

$$f' = -\frac{x}{f} \qquad \text{on } (0, 1),$$

$$f' = -\frac{f}{x} \qquad \text{at the maximum.}$$

Therefore, at the maximum, both of these equations hold, and

$$\frac{x}{f} = \frac{f}{x}, \qquad x^2 = f^2, \qquad \text{and} \qquad x = f(x).$$

That is, *the maximum is achieved when the triangle is isosceles.*

This discussion has been long, because ideas needed to be explained; but once the ideas are understood, the calculations are simple:

$$x^2 + f^2 = 1, \qquad 2 \cdot x + 2 \cdot ff' = 0, \qquad f' = -\frac{x}{f};$$

$$A(x) = \tfrac{1}{2}x \cdot f(x), \qquad A'(x) = \tfrac{1}{2}x \cdot f'(x) + \tfrac{1}{2}f(x);$$

and hence

$$A' = 0 \iff f' = -\frac{f}{x}.$$

Therefore, at the maximum,

$$-\frac{x}{f} = -\frac{f}{x}, \qquad \text{and} \qquad x = f(x).$$

In this case, of course, it was not much trouble to find a formula for f and use it. But in many cases, equations like

$$x^2 + f^2 = 1$$

are more convenient than formulas for the function f. These are called *functional equations.* Obviously every trigonometric identity is a functional equation. Usually, however, we use the word *identity* when the function is known, and the term *functional equation* when the equation itself is being used as a working definition of the function.

Consider another example, Problem 3 in Section 5.4.

Problem 3. Find the right circular cylinder of largest volume, inscribed in a sphere of radius a.

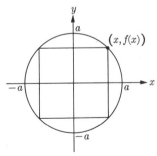

As before, we show a vertical cross section of the figure. Let x be the radius of the inscribed cylinder, and let $f(x)$ be half the altitude. Then

$$x^2 + f^2 = a^2, \qquad 2 \cdot x + 2 \cdot ff' = 0,$$

and

$$f' = -\frac{x}{f} \qquad (0 < x < a). \tag{3}$$

Now the volume is

$$V(x) = \pi x^2 \cdot 2f(x),$$

so that

$$V' = 2\pi \cdot (x^2 f' + 2xf).$$

At the maximum, $V' = 0$, and so

$$f' = -\frac{2xf}{x^2} = -\frac{2f}{x} \qquad \text{(at Max).} \qquad (4)$$

Therefore, at the maximum, both our formulas for f' must hold, and so

$$-\frac{x}{f} = -\frac{2f}{x},$$

and

$$f^2 = \tfrac{1}{2}x^2,$$

$$f = \frac{\sqrt{2}}{2} \cdot x. \qquad (5)$$

For $a = 1$, this tells us that

$$\sqrt{1 - x^2} = \frac{\sqrt{2}}{2} x, \qquad 1 - x^2 = \tfrac{1}{2}x^2, \qquad x = \sqrt{\tfrac{2}{3}},$$

$$V = \pi \cdot \tfrac{2}{3} \cdot 2 \cdot \frac{\sqrt{3}}{3} = \frac{4\pi\sqrt{3}}{9},$$

as before.

   Note, however, that in a way the most natural answer to a problem like this is a *shape*, rather than a size. And the solution based on the functional equation ordinarily gives the answer in the form of a shape, that is, in the form of a *ratio* between two measurements. For example, in the preceding problem the constant a, which determines the *size* of the whole configuration, disappeared immediately when we differentiated in the equation $x^2 + f^2 = a^2$. Our final equation (5) means that at the maximum,

$$2f(x) = \sqrt{2}\, x,$$

that is, the altitude of the maximum cylinder is equal to $\sqrt{2}$ times the radius of its base.

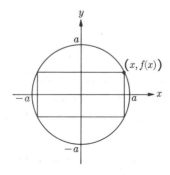

The answer is also a shape when the problem is to find the rectangle of maximum area in a given circle:

$$x^2 + f^2 = a^2, \qquad 2x + 2ff' = 0,$$

$$f' = -\frac{x}{f} \qquad (0 < x < a);$$

$$A(x) = (2x) \cdot 2f(x) = 4 \cdot x \cdot f,$$

$$A'(x) = 4(xf' + f),$$

$$A'(x) = 0 \iff f' = -\frac{f}{x}.$$

Therefore at the maximum,

$$-\frac{x}{f} = f' = -\frac{f}{x}, \qquad x^2 = f^2, \qquad \text{and} \qquad x = f,$$

because x and f are both positive. This is a qualitative answer, as it should be: it says that the maximum rectangle is a square. The constant a has disappeared, because the shape of the maximum rectangle is the same for all circles.

In the following problem set, you will find more cases in which maxima and minima can most conveniently be found by using functional equations. Meanwhile let us look carefully at what happens when we take the derivative on each side of a functional equation. The ideas here are illustrated by a simple case. When we write

$$x^2 + f^2 = a^2 \tag{6}$$
$$\Rightarrow 2 \cdot x + 2 \cdot ff' = 0, \tag{7}$$

we are claiming that *every differentiable function which satisfies Eq. (6) also satisfies Eq. (7)*. It often happens that there is more than one such function f. For example, consider

$$f_1(x) = \sqrt{a^2 - x^2}, \qquad f_2(x) = -\sqrt{a^2 - x^2}.$$

Here

$$f_1'(x) = \frac{-x}{\sqrt{a^2 - x^2}} = \frac{-x}{f_1(x)},$$

and

$$f_2'(x) = \frac{x}{\sqrt{a^2 - x^2}} = \frac{-x}{-\sqrt{a^2 - x^2}} = \frac{-x}{f_2(x)}.$$

Therefore

$$f_1(x)f_1'(x) = -x, \qquad \text{and} \qquad f_2(x)f_2'(x) = -x.$$

Therefore

$$\begin{cases} 2 \cdot x + 2 \cdot f_1 f_1' = 0, \\ 2 \cdot x + 2 \cdot f_2 f_2' = 0. \end{cases} \tag{8}$$

That is, both f_1 and f_2 satisfy (7). A figure makes it obvious what is going on here.

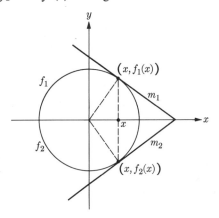

At each of the labeled points, we have

$$f_i'(x) = \frac{-1}{f_i(x)/x} = \frac{-x}{f_i(x)} ,$$

because the tangent is perpendicular to the radius.

The same sort of thing goes on in more complicated cases. The graph of

$$y = x^3 - x \qquad (9)$$

looks like the left-hand figure below.

Therefore the graph of

$$x = y^3 - y \qquad (10)$$

looks like the right-hand drawing below.

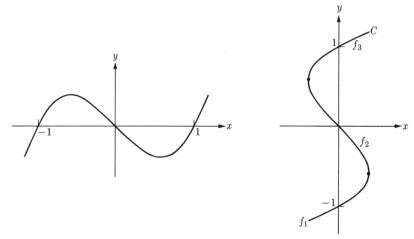

We have interchanged x and y in Eq. (9), and reflected the graph across the line

$y = x$. This gives the curve C which is the graph of (10). C is not a function-graph. But C is the union of the graphs of three functions f_1, f_2, f_3, as indicated in the figure. And each of the functions f_1, f_2, and f_3 satisfies the functional equation

$$x = f^3 - f.$$

Therefore each of these functions satisfies the differential equation

$$1 = 3 \cdot f^2 f' - f'.$$

This is what we are claiming when we differentiate the functional equation, and write

$$x = f^3 - f \;\Rightarrow\; 1 = 3 \cdot f^2 f' - f'.$$

PROBLEM SET 5.5

In Problems 1 through 10 below, the notation 5.4.n refers to Problem n of Problem Set 5.4. In each of these cases, the indicated ratio is to be found by the method based on functional equations.

1. In 5.4.1, find altitude/base, at the maximum.

2. In 5.4.2, same.

3. In 5.4.5, same, using the side parallel to the river as base.

4. In 5.4.7, find y/x, at the maximum.

5. In 5.4.14, let l be the length of the rectangular side and let w be the width. Find w/l at the maximum.

6. In 5.4.15, let h be the altitude and let e be the length of each edge of the base. Find h/e, at the maximum.

7. In 5.4.16, find altitude/base, at the maximum.

8. In 5.4.28, same.

9. In 5.4.33, let a be the altitude of the cylinder, and let r be the radius of the base. Find a/r, at the maximum.

10. In 5.4.34, let h be the altitude, and let a be *half* the length of the base. Find h/a, at the maximum.

11. In 5.4.35, let h be the altitude and let a be the radius of the base. Find h/a, at the maximum.

12. We know of a function f, with domain $[-1, 1]$, which is a solution of the functional equation $\sin f(x) = x$. (Our "known function," of course, is $f(x) = \operatorname{Sin}^{-1} x$.) What other continuous solutions of the equation have the entire interval $[-1, 1]$ as domain? Draw a figure.

13. Write a differential equation which is satisfied by all solutions of the functional equation

$$x^4 + [f(x)]^4 = 1.$$

14. a) Let $n = 10^{10^{10}}$. Sketch the graph of $x^n + y^n = 1$. [*Hint:* A commonly used drawing instrument will give you an excellent sketch.]
    b) Let $n = 10^{10^{10}} + 1$. Sketch the graph of $x^n + y^n = 1$. [Same hint.]

15. Find the functions f which satisfy the differential equation

$$x + ff' = 0.$$

(You need not show that the solutions that you describe are the only ones.)

16. Given that f and f' are continuous, let

$$F(x) = \int_0^x f(t)f'(t)\,dt.$$

Calculate $F(x)$, in terms of f.

*17. Now show that your list of solutions, in Problem 15, is complete.

18. Let f be the function whose graph is the union of (a) the lower left-hand quadrant of the circle with center at $(0, 1)$ and radius 1 and (b) the upper right-hand quadrant of the circle with center at $(0, -1)$ and radius 1. Show that f is a solution of the differential equation

$$[f'(x)]^2 = [x + f(x)f'(x)]^2,$$

except, of course, at the endpoints $x = \pm 1$, where the tangent lines are vertical and the function has no derivative. As a start, observe that at $x = 0$, the tangent to the graph is horizontal and the equation is satisfied: $0^2 = [0 + 0 \cdot 0]^2$.

*19. Consider the family of quadratic functions represented by the formula

$$f(x) = (x - a)^2. \tag{1}$$

Differentiating, we get

$$f'(x) = 2(x - a),$$

and squaring, we get $[f'(x)]^2 = 4(x - a)^2$, and

$$[f'(x)]^2 = 4 \cdot f(x). \tag{2}$$

Evidently (1) $\Rightarrow$ (2). But the converse is false.

a) Show that one of the solutions of (2) is a *linear* function f.

b) Show that (2) has some solutions which are neither quadratic nor linear; that is, the differential equation has solutions whose total graphs are neither lines nor parabolas.

5.6 THE COMPLETENESS OF **R** AND THE EXISTENCE OF MAXIMA

In Section 5.4 and later, we have used the fact that, if f is continuous on $[a, b]$, then f has a maximum value on $[a, b]$. In Section 5.4 this theorem was used as a shortcut in *finding* maximum values, but this is only one of the uses of the theorem. In fact, the theorem is part of the foundation of the calculus, as we shall see.

In proving it, we shall need to use, for the first time, the fact that the number line has no holes in it. As a guide in giving an exact description of this property of the number system, let us consider what happens when you remove a point from the number line, thus getting a system which really does have a hole in it.

Let A be the set of all negative numbers, and let B be the set of all positive numbers. We mean strictly positive and strictly negative, so that 0 belongs neither to A nor to B. Then

1) *B has no least element.*

The reason is that if x is a positive number, then so is $x/2$, and $x/2 < x$. Therefore no positive number x is less than all other positive numbers, and so (1) holds. Similarly,

2) *A has no greatest element.*

For if $x < 0$, then $x/2 < 0$, and $x/2 > x$.

Now let

$$K = A \cup B = \{x \mid x \neq 0\}.$$

Then obviously:

3) K is the union of two nonempty sets A and B, such that (a) every number in A is less than every number in B, but (b) A has no greatest element, and (c) B has no least element.

Evidently this situation could not have arisen if we had not excluded 0: if we put 0 in A, then 0 would be the greatest element of A; and if we put 0 in B, then 0 would be the least element of B. Thus the situation described in (3) can arise only in a number system with a hole in it, and so the following statement conveys the idea that there are no holes in **R**:

The Dedekind Cut Postulate (DCP). Suppose **R** is expressed as the union of two nonempty sets A and B, such that every element of A is less than every element of B. Then either A has a greatest element or B has a least element.

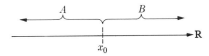

In the figure, x_0 must belong either to A or to B. Therefore x_0 is either the greatest element of A or the least element of B.

We have stated DCP as our first description of the completeness of **R**, because it is the best known description, and in some ways the most natural. But for some purposes, the following idea is easier to use. Given a sequence

$$[a_1, b_1], [a_2, b_2], \dots$$

of closed intervals. If every interval in the sequence contains the next, then we say that the sequence is *nested*. Algebraically, this means that

$$a_i \leqq a_{i+1} < b_{i+1} \leqq b_i \qquad \text{for every } i.$$

For example, if

$$[a_i, b_i] = \left(-\frac{1}{i}, \frac{1}{i} \right) \qquad \text{for every } i,$$

then the sequence is nested. This sequence "closes down on 0." That is, 0 lies in each of the intervals in the sequence, and 0 is the only number that lies in all of them. A more important example is as follows. Given a circle of radius 1, let p_n be the perimeter of an inscribed regular $(n + 2)$-gon, and let q_n be the perimeter of a circumscribed regular $(n + 2)$-gon. Evidently

$$p_1 < p_2 < p_3 < \cdots, \qquad q_1 > q_2 > q_3 > \cdots,$$

and

$$p_i < q_i \qquad \text{for each } i.$$

Thus we have a nested sequence

$$[p_1, q_1], \, [p_2, q_2], \, \ldots$$

of closed intervals. And this sequence "closes down on 2π." That is, 2π lies in all of the intervals in the sequence, and no other number lies in all of them.

The following postulate says that every nested sequence of intervals closes down on at least one point.

The Nested Interval Postulate (NIP). For every nested sequence of closed intervals there is a number $\bar{x}$ which lies in every interval in the sequence.

This conveys the idea that the number system is complete. Suppose, for example, that 2π were missing, so that the number system had a hole in it where 2π ought to be. Then no number at all would lie on all of the intervals $[p_1, q_1]$, $[p_2, q_2]$, $\ldots$ that we have just discussed. Similarly, if $\sqrt{2}$ were missing, then there would be a nested sequence of closed intervals closing down on no number whatever. (We could use $[\sqrt{2} - 1/i, \sqrt{2} + 1/i]$ as the ith interval in the sequence.)

Using the nested interval postulate (NIP), we shall prove the following theorem:

Theorem 1. If f is continuous on $[a, b]$, then f has an upper bound on $[a, b]$.

That is, there is a number M such that $f(x) \leqq M$ for each x of $[a, b]$.

Lemma. If f is unbounded above on an interval $[c, d]$, then f is unbounded above on at least one of the halves of $[c, d]$.

By the halves of $[c, d]$ we mean the intervals $[c, (c + d)/2]$ and $[(c + d)/2, d]$. The proof of the lemma is immediate: if f has an upper bound M_1 on $[c, (c + d)/2]$ and has an upper bound M_2 on $[(c + d)/2, d]$, then f has an upper bound on $[c, d]$. We merely use the larger of the bounds M_1 and M_2.

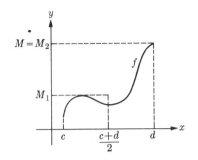 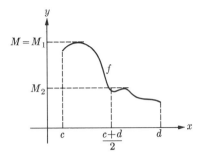

We proceed to prove the theorem. For short, we say that an interval is *good* if f is bounded above on the interval; and we say that an interval is *bad* if it is not good. Thus we need to prove that $[a, b]$ is good. We start by supposing that $[a, b]$ is bad, and we shall show this assumption leads to a contradiction.

If $[a, b]$ is bad, then it follows that at least one of the halves of $[a, b]$ must be bad. Let $[a_1, b_1]$ be a bad half of $[a, b]$. For the same reason, $[a_1, b_1]$ must have a bad half. Let $[a_2, b_2]$ be a bad half of $[a_1, b_1]$. Continuing this process to infinity, we get a sequence

$$[a_1, b_1], [a_2, b_2], \dots$$

of closed intervals, all of which are bad, and each of which is a half of the preceding one. Therefore

$$b_{i+1} - a_{i+1} = \tfrac{1}{2}(b_i - a_i),$$

and so

$$b_i - a_i = \frac{1}{2^i}(b - a).$$

By NIP, there is an $\bar{x}$ such that $a_i \leq \bar{x} \leq b_i$ for each i.

But f is continuous at $\bar{x}$. Thus, for every $\epsilon > 0$, f has an $\epsilon\delta$-box at the point $(\bar{x}, f(\bar{x}))$.

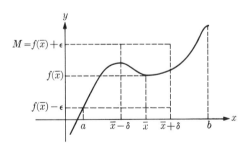

Thus

$$|x - \bar{x}| < \delta \;\; \Rightarrow \;\; f(\bar{x}) - \epsilon < f(x) < f(\bar{x}) + \epsilon,$$

and so $f(\bar{x}) + \epsilon$ is an upper bound for f on the interval $(\bar{x} - \delta, \bar{x} + \delta)$. But since

$$\lim_{i \to \infty} (b_i - a_i) = 0,$$

we have

$$b_i - a_i < \delta$$

for some i. For such an i, the closed interval $[a_i, b_i]$ lies inside the open interval $(\bar{x} - \delta, \bar{x} + \delta)$. That is,

$$\bar{x} - \delta < a_i < b_i < \bar{x} + \delta,$$

as shown in the figure.

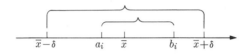

(This is easy to see geometrically, because $[a_i, b_i]$ contains the midpoint $\bar{x}$ of the open interval, and is less than half as long.)

But this situation is impossible, because f is bounded above on $(\bar{x} - \delta, \bar{x} + \delta)$ and is not bounded above on the smaller interval $[a_i, b_i]$. This contradiction completes the proof of the theorem.

One of the ideas that we have just used is going to be useful later. We therefore record it as a theorem:

Theorem 2. Suppose that

$$a_i \leqq \bar{x} \leqq b_i$$

for each i, and

$$\lim_{i \to \infty} (b_i - a_i) = 0.$$

Then every interval $(\bar{x} - \delta, \bar{x} + \delta)$ contains some interval $[a_i, b_i]$.

This was proved in the preceding discussion. If $(\bar{x} - \delta, \bar{x} + \delta)$ contains $[a_i, b_i]$, then of course it follows that $(\bar{x} - \delta, \bar{x} + \delta)$ contains all of the later intervals

$$[a_{i+1}, b_{i+1}], [a_{i+2}, b_{i+2}], \ldots$$

Given that a function f is bounded, it does not follow that f has a maximum or a minimum. Consider, for example,

$$f = \mathrm{Tan}^{-1}.$$

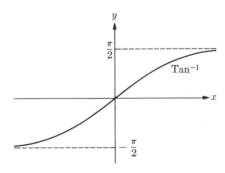

When x is far to the right, $\text{Tan}^{-1} x$ is close to $\pi/2$, but $\text{Tan}^{-1} x$ is never actually *equal* to $\pi/2$ for any x. Similarly, when x is far to the left, $\text{Tan}^{-1} x$ is close to $-\pi/2$, but $-\pi/2$ is not one of the values of the function. On the other hand, it is easy to see that the numbers $\pi/2$ and $-\pi/2$ are related to the function Tan^{-1} in a special way: $\pi/2$ is an upper bound of the function; and of all upper bounds of the function, $\pi/2$ is the smallest. We express this by writing

$$\frac{\pi}{2} = \sup \text{Tan}^{-1}.$$

Here *sup* is pronounced *supremum*. To be exact:

Definition. If k is an upper bound of a function f, and k is smaller than every other upper bound of f, then k is called the *supremum* of f, and we write

$$k = \sup f.$$

More generally, we define the supremum for any set of numbers:

Definition. Let B be a set of numbers. If $x \leq k$, for every x in B, then k is an *upper bound* of B. If k is smaller than every other upper bound of B, then k is called the *supremum* of B, and we write

$$k = \sup B.$$

Consider, for example, the case where B is an open interval (a, b). Every number $k \geq b$ is an upper bound of B. Thus the upper bounds of B form an interval $[b, \infty)$.

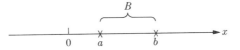

Here b is an upper bound of B, and b is smaller than all other upper bounds of B. Therefore

$$b = \sup B.$$

Consider now

$$B = \left\{ \frac{1}{2}, \frac{2}{3}, \frac{3}{4}, \ldots, \frac{n-1}{n}, \ldots \right\}.$$

Here the upper bounds of B are the points of the interval $[1, \infty)$, and $\sup B = 1$.

In each of these cases, starting with a nonempty set B which is bounded above, we have found that the upper bounds form an interval of the type $[k, \infty)$, and $k = \sup B$. The following postulate says that this is what always happens:

The Least Upper Bound Postulate (LUBP). Let B be a nonempty set of numbers. If B has an upper bound, then B has a supremum.

Using the least upper bound postulate, we shall show that no continuous function can behave like Tan^{-1} if its domain is a closed interval:

Theorem 3 (*Existence of maxima*). If f is continuous on $[a, b]$, then f has a maximum value on $[a, b]$.

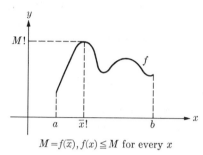

$M = f(\bar{x}), f(x) \leqq M$ for every x

Proof. We know by Theorem 1 that f is bounded. Let

$$k = \sup f.$$

Then $f(x) \leqq k$ for every x on $[a, b]$. We need to show that $f(x) = k$ for some x. Suppose not, and let

$$g(x) = \frac{1}{k - f(x)} \qquad (a \leqq x \leqq b).$$

Then g is continuous. But g is unbounded. For suppose that

$$g(x) \leqq M \qquad \text{for } a \leqq x \leqq b.$$

Then

$$\frac{1}{k - f(x)} \leqq M, \qquad \frac{1}{M} \leqq k - f(x),$$

and

$$f(x) \leqq k - \frac{1}{M} \qquad \text{for } a \leqq x \leqq b.$$

This is impossible, because k is the *least* of the upper bounds of f.

Thus, if f has no maximum, there is a continuous function g which is unbounded on $[a, b]$. This contradicts Theorem 1, and so completes the proof of Theorem 3.

We have already observed, in Section 5.4, that the existence of maxima implies the existence of minima. Therefore

Theorem 4 (*Existence of minima*). If f is continuous on $[a, b]$, then f has a minimum value on $[a, b]$.

(This was Theorem 2 of Section 5.4.)

PROBLEM SET 5.6

1. Let B be the set of all rational numbers p/q for which $p^2/q^2 < 2$. What is sup B?

2. Consider a circle of radius 1. For each polygon P inscribed in the circle, let $k(P)$ be the perimeter of P. Let B be the set of all numbers $k(P)$. What is sup B?

3. Consider the graph of $f(x) = \sin x$, $0 \leqq x \leqq \pi$. Suppose that we cut up the interval $[0, \pi]$ into little intervals, in any way, using subdivision points $0 = x_1 < x_2 < \cdots < x_i < x_{i+1} < \cdots < x_n = \pi$. Over each little interval $[x_i, x_{i+1}]$ we set up the tallest possible inscribed rectangle with $[x_i, x_{i+1}]$ as base. Let s be the sum of the areas of the rectangles. Let B be the set of all numbers s which are obtainable in this way. What is sup B? (A numerical answer is called for here.)

4. Let B be any set of numbers. If $b \in B$, and b is larger than every other element of B, then b is called the *greatest* element of B, and we write $b = $ Max B. *Question:* If B has an upper bound, does it follow that B has a Max?

5. Suppose that we had defined bounds and suprema in the following way:

"Let B be a set of numbers, and let k be a number. If $x < k$, for every x in B, then k is a *strict upper bound* of B. If k is a strict upper bound of B, and is smaller than every other strict upper bound of B, then $k = $ sup B."

a) What is the difference between this "definition" and the usual definition of upper bounds and suprema?

Under the new "definition" of "*supremum*," which if any of the following statements are true?

b) Every finite set has a "supremum."
c) No finite set has a "supremum."
d) Every open interval has a "supremum."
e) No open interval has a "supremum."
f) Every closed interval has a "supremum."
g) No closed interval has a "supremum."

6. If B is a set of numbers, then $-B$ denotes the set obtained when we replace every element x of B by its negative $-x$. That is,

$$-B = \{-x \mid x \in B\}.$$

For example, if $B = [1, 2]$, then $-B = [-2, -1]$; if $B = [-1, \infty)$, then $-B = (-\infty, 1]$, and so on. Prove the following:

Theorem. If (a) k is an upper bound of B, then (b) $-k$ is a lower bound of $-B$. And conversely, (b) implies (a).

(This is easy; don't try to make it hard.)

7. If k is a lower bound of the set B, and k is greater than every other lower bound of B, then k is called the infimum of B, and we write $k = $ inf B. Show that if a set B is bounded below, then B has an infimum.

8. Let B be a set which is bounded below, and let K be the set of all lower bounds of B. Describe K in the interval notation.

*9. Let $[a_1, b_1]$, $[a_2, b_2]$, ... be a nested sequence, and let $A = \{a_1, a_2, \ldots\}$, $B = \{b_1, b_2, \ldots\}$. Show that (a) every number b_i is an upper bound of A. Let $\bar{x} = $ sup A. Then show that (b) $a_i \leqq \bar{x} \leqq b_i$ for every i.

This result means that the least upper bound postulate (LUBP) implies the nested interval postulate (NIP).

*10. Let K be a (nonempty) set of numbers, bounded above. Let A be the set of all numbers a which are *not* upper bounds of K. That is, $a \in A$ if $a < k$ for some k in K. Show that A cannot contain a greatest element.

*11. Show that the Dedekind cut postulate (DCP) implies the least upper bound postulate (LUBP).

The results of Problems 9 and 11 mean that

$$DCP \Rightarrow LUBP \Rightarrow NIP.$$

Thus our only really new assumption, in this section, is DCP.

5.7 THE MEAN-VALUE THEOREM AND THE NO-JUMP THEOREM

The mean-value theorem was stated in Chapter 3, and we have been using it ever since. We are now finally in a position to prove it. We need one preliminary result.

Rolle's Theorem. If f is continuous on the closed interval $[a, b]$ and differentiable on the open interval (a, b), and $f(a) = f(b) = 0$, then $f'(\bar{x}) = 0$ for some $\bar{x}$ between a and b.

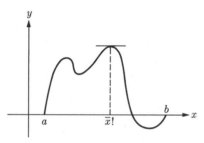

Proof. There are three cases to consider:

1) Suppose that $f(x) = 0$ for every x on $[a, b]$. Then any number $\bar{x}$ between a and b gives $f'(\bar{x}) = 0$.

2) Suppose that $f(x) > 0$ for some x on $[a, b]$. Now f has a maximum at some $\bar{x}$, and $\bar{x}$ is not a or b. Therefore f has an ILMax at $\bar{x}$. By Theorem 3 of Section 5.2 it follows that $f'(\bar{x}) = 0$.

3) If $f(x) < 0$ for some x, then the minimum of f is an ILMin. By Theorem 4 of Section 5.2 we know that at an ILMin the derivative vanishes.

The proof of MVT is now easy.

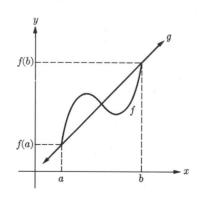

Given that f is continuous on $[a, b]$ and differentiable on (a, b), let g be the linear function which agrees with f at a and at b. Thus

$$g(a) = f(a), \qquad g(b) = f(b).$$

We could write a formula for g, in the form $g(x) = mx + k$, if we needed to, but we don't need to. Since the derivative of a linear function is simply the slope of the line which is its graph, we know that

$$g'(x) = \frac{f(b) - f(a)}{b - a},$$

for every x. For each x of $[a, b]$, let

$$\phi(x) = f(x) - g(x).$$

Then ϕ is continuous on $[a, b]$ (because f and g are), and ϕ is differentiable on (a, b), with

$$\phi'(x) = f'(x) - g'(x) = f'(x) - \frac{f(b) - f(a)}{b - a}$$

Since $\phi(a) = \phi(b) = 0$, we can apply Rolle's theorem. Therefore $\phi'(\bar{x}) = 0$ for some $\bar{x}$. Thus

$$f'(\bar{x}) - \frac{f(b) - f(a)}{b - a} = 0,$$

and

$$f'(\bar{x}) = \frac{f(b) - f(a)}{b - a}$$

for some $\bar{x}$, which was to be proved.

The no-jump theorem is harder. To prove it, we need to go back to first principles, and we need some preliminary results.

Lemma 1. Let f be a continuous function, on an open interval containing x_0. If $f(x_0) > 0$, then there is a $\delta > 0$ such that

$$x_0 - \delta < x < x_0 + \delta \quad \Rightarrow \quad f(x) > 0.$$

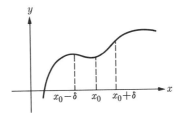

Proof. Since f is continuous, we know that

$$\lim_{x \to x_0} f(x) = f(x_0).$$

In the definition of a limit, we take $\epsilon = f(x_0) > 0$. There is a $\delta > 0$ such that

$$x_0 - \delta < x < x_0 + \delta \;\Rightarrow\; f(x_0) - \epsilon < f(x) < f(x_0) + \epsilon \;\Rightarrow\; 0 < f(x),$$

because $f(x_0) - \epsilon = 0$. Therefore the δ that we have is the δ that we wanted.

Lemma 2. Let f be a continuous function, on an interval containing x_0. If $f(x_0) < 0$, then there is a $\delta > 0$ such that

$$x_0 - \delta < x < x_0 + \delta \;\Rightarrow\; f(x) < 0.$$

Proof? (The proof of Lemma 1 can be adapted, to give a proof of Lemma 2. But it is quicker to derive Lemma 2 from the *statement* of Lemma 1.)

A function f *changes sign*, on an interval I, if $f(x) > 0$ for some x in I and $f(x') < 0$ for some x' in I.

Lemma 3. If f is continuous, on an interval containing x_0, and $f(x_0) \neq 0$, then there is a $\delta > 0$ such that f does not change sign on the interval $(x_0 - \delta, x_0 + \delta)$.

Proof. For $f(x_0) > 0$, this follows from Lemma 1. For $f(x_0) < 0$, it follows from Lemma 2.

We are now ready to prove the following convenient special case of the no-jump theorem.

Theorem 1. If f is continuous on $[a, b]$, and f changes sign on $[a, b]$, then

$$f(x_0) = 0$$

for some x_0 in $[a, b]$.

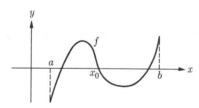

The proof is based on Lemma 3 and the nested interval postulate (NIP). We suppose that $f(x) \neq 0$ for every x in $[a, b]$. We shall show that this assumption leads to a contradiction.

Given that f changes sign on $[a, b]$ and that $f(x)$ is never $= 0$, it follows that f changes sign on one of the halves of $[a, b]$. We recall, from Section 5.6, that the halves of $[a, b]$ are $[a, (a + b)/2]$ and $[(a + b)/2, b]$. Let $[a_1, b_1]$ be half of $[a, b]$, such that f changes sign on $[a_1, b_1]$. Similarly, let $[a_2, b_2]$ be half of $[a_1, b_1]$, such that f changes sign on $[a_2, b_2]$. Proceeding to infinity in this way, we get a nested sequence

$$[a_1, b_1], [a_2, b_2], \ldots$$

of closed intervals, such that f changes sign on each of them. Evidently

$$b_i - a_i = 2^{-i}(b - a),$$

and so

$$\lim_{i \to \infty} (b_i - a_i) = 0,$$

as in the proof of Theorem 1 of Section 5.6. By NIP, there is an x_0 which lies on all of the intervals in the nested sequence. That is,

$$a_i \leqq x_0 \leqq b_i \qquad \text{for every } i.$$

By Lemma 3 there is a $\delta > 0$ such that f does not change sign on the interval $(x_0 - \delta, x_0 + \delta)$. By Theorem 2 of Section 5.6, there is an i for which $[a_i, b_i]$ lies in $(x_0 - \delta, x_0 + \delta)$, as indicated in the figure.

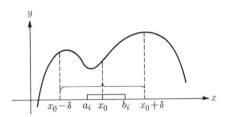

This is impossible, because f changes sign on $[a_i, b_i]$, but does not change sign on $(x_0 - \delta, x_0 + \delta)$. This contradiction completes the proof of Theorem 1.

It is now easy to prove the no-jump theorem.

Theorem 2 (*The no-jump theorem*). If f is continuous on $[x_1, x_2]$, then f takes on every value between $f(x_1)$ and $f(x_2)$.

Proof. Suppose first that

$$f(x_1) < k < f(x_2),$$

and let

$$g(x) = f(x) - k.$$

Then g changes sign on $[x_1, x_2]$. Therefore $g(x_0) = 0$ for some x_0 on $[x_1, x_2]$. This gives $f(x_0) - k = 0$, and $f(x_0) = k$.

If $f(x_2) < k < f(x_1)$, then the same function g still changes sign, and so the proof is exactly the same.

This completes our reexamination of the foundations of calculus. It now appears that the idea of a continuous function is adequately described by the $\epsilon\delta$-definition of a limit and that the completeness of the number system **R**, in the sense of "no holes," is adequately described by the Dedekind cut postulate (which implies the least upper bound postulate and the nested interval postulate).

The *theorems* in this section and the preceding one are not news; it was obvious at the outset that these theorems ought to be true. But the fact that these theorems can be proved, on the basis of a single simple assumption DCP, is significant. It means that mathematics hangs together in a special way.

Nobody expects that a doctor will write down a definition of the word *man* and then write a few assumptions about *men*, in such a way that all medical science can be derived by logical reasoning from the definition and from the assumptions. Medicine is an empirical science: it depends on observations of fact, not just at the outset but continually. Mathematics is different.

Moreover, in your study of mathematics you have already passed the point where the truth can be relied upon to be obvious and where obvious things can be relied on to be true. From now on, logic is going to be an important part of your mathematical equipment. This is partly due to recent developments. As late as 1800, calculus was illogical, and very few people cared. In the last century, however, mathematical ideas which require careful logical analysis have become more important, in pure research and also in applications.

5.8 THE DERIVATIVE OF ONE FUNCTION WITH RESPECT TO ANOTHER

Let f and g be differentiable functions. Take a point x_0, and form the differences

$$\Delta f = f(x_0 + \Delta x) - f(x_0),$$
$$\Delta g = g(x_0 + \Delta x) - g(x_0).$$

If $\Delta f/\Delta g$ approaches a limit, as $\Delta x \to 0$, then this limit is called the *derivative of f with respect to g*, and is denoted by df/dg. That is,

$$\lim_{\Delta x \to 0} \frac{\Delta f}{\Delta g} = \frac{df}{dg},$$

by definition. In fact, the limit always exists, whenever $g'(x_0) \neq 0$.

Theorem 1.

$$\frac{df}{dg} = \frac{f'}{g'},$$

wherever $g'(x) \neq 0$.

Proof.

$$\lim_{\Delta x \to 0} \frac{\Delta f}{\Delta g} = \lim_{\Delta x \to 0} \frac{\Delta f/\Delta x}{\Delta g/\Delta x} = \frac{f'(x_0)}{g'(x_0)}.$$

For the case in which $g(x) = x$ for every x, the derivative of f with respect to g reduces to an ordinary derivative:

Theorem 2. If $g(x) = x$ for every x, then

$$\frac{df}{dg} = \frac{df}{dx} = f'(x).$$

Obviously,

$$\frac{df}{dg} = \lim_{\Delta x \to 0} \frac{\Delta f}{\Delta x} = f'(x_0),$$

for each x_0.

Some examples are as follows:

$$\frac{d \sin x}{d \cos x} = \frac{\cos x}{-\sin x} = -\cot x, \qquad (\text{wherever } \cos x \neq 0)$$

$$\frac{d \sin x}{dx} = \cos x,$$

$$\frac{de^x}{dx^2} = \frac{e^x}{2x}. \qquad (\text{wherever } x \neq 0)$$

We often write

$$\frac{d}{dx} f(x) \qquad \text{for} \qquad \frac{df}{dx}.$$

Thus every derivative can be written in the form

$$f'(x) = \frac{d}{dx} f(x),$$

$$f' = \frac{df}{dx}.$$

The notation df/dx for derivatives is widely used, especially in physics, and it is natural to use it when you are continually dealing with the derivative df/dg of one function with respect to another. It has a disadvantage, however: there is no convenient way to write the value of the derivative at a particular point x_0. Sometimes we denote this by

$$\frac{df}{dx}\Big|_{x=x_0},$$

but the notation $f'(x_0)$ is more convenient.

We now want to prove a sort of cancellation law

$$\frac{df}{dg} \cdot \frac{dg}{dh} = \frac{df}{dh}.$$

We can derive this from the equation

$$\frac{\Delta f}{\Delta g} \cdot \frac{\Delta g}{\Delta h} = \frac{\Delta f}{\Delta h},$$

taking the limit as $\Delta x \to 0$. Thus we need

$$\frac{\Delta f}{\Delta g} \to \frac{df}{dg}, \qquad \frac{\Delta g}{\Delta h} \to \frac{dg}{dh},$$

as $\Delta x \to 0$. This requires

$$g'(x_0) \neq 0, \qquad h'(x_0) \neq 0,$$

as in Theorem 1. Hence the conditions in the following theorem:

Theorem 3. If f, g, and h are differentiable, then

$$\frac{df}{dg} \cdot \frac{dg}{dh} = \frac{df}{dh},$$

wherever $g' \neq 0$ and $h' \neq 0$.

Theorem 4.

$$\frac{dg}{df} = \frac{1}{df/dg},$$

wherever $df/dg \neq 0$.

(The limit of the quotient is the quotient of the limits.)

We shall now find a short-cut for calculating derivatives of the type df/dg. Consider

$$\frac{d \sin^2 x}{d \sin x} = \frac{2 \sin x \cos x}{\cos x} = 2 \sin x \qquad (\cos x \neq 0).$$

This has the form

$$\frac{du^2}{du} = 2u.$$

This is like

$$f(x) = x^2, \qquad \frac{df}{dx} = f'(x) = 2x.$$

That is, to find du^2/du (where u is a function), we treat u as if it were a dummy variable x and differentiate in one step. This is an example of the following situation.

Definition. Let f and g be functions. If there is a function ϕ such that $f = \phi(g)$, then we say that f is a function of g.

For example, $\sin^2 x$ is a function of $\sin x$, with $\phi(u) = u^2$. And $\cos^2 x - 2 \cos x$ is a function of $\cos x$, with $\phi(u) = u^2 - 2u$. The easiest way to calculate df/du, in each of these cases, is to write

$$\frac{d \sin^2 x}{d \sin x} = \frac{du^2}{du} = 2u = 2 \sin x,$$

$$\frac{d(\cos^2 x - 2 \cos x)}{d \cos x} = \frac{d(u^2 - 2u)}{du} = 2u - 2$$

$$= 2 \cos x - 2.$$

This procedure is justified by the following theorem.

Theorem 5. Let f be a function of g, $= \phi(g)$, where all the functions are differentiable. Then

$$\frac{df}{dg} = \phi'(g),$$

wherever $g' \neq 0$.

Proof.

$$\frac{df}{dg} = \frac{f'}{g'} = \frac{\phi'(g)g'}{g'} = \phi'(g).$$

Using this theorem, we can write immediately

$$\frac{d \tan^2 x}{d \tan x} = 2 \tan x,$$

instead of using Theorem 1 and writing

$$\frac{d \tan^2 x}{d \tan x} = \frac{2 \tan x \sec^2 x}{\sec^2 x} = 2 \tan x;$$

there is no point in writing $\sec^2 x = D \tan x$ in both numerator and denominator, since we are about to cancel it out in any case.

PROBLEM SET 5.8

Calculate df/dg, given:

1. $f(x) = e^{x^2}, g(x) = x^2$ 2. $f(x) = e^{x^2}, g(x) = 2x$
3. $f(x) = e^x, g(x) = \text{Tan } x$ 4. $f(x) = e^{x \sin x}, g(x) = x$
5. $f(x) = e^{x \sin x}, g(x) = x^3$ 6. $f(x) = e^{x^3}, g(x) = x^2$
7. $f(x) = e^{x^3}, g(x) = x^4$ 8. $f(x) = \sin x, g(x) = \cos x$
9. $f(x) = x^3, g(x) = \text{Tan } x$

Problems 10 through 14. In Problems 1 through 5, first calculate the function ϕ such that $f = \phi(g)$. (Answer in the form $\phi(u) = \cdots$.) Then calculate $\phi'(u) = \cdots$. Finally, calculate $\phi'(g)$, and compare it with your previous formula for df/dg. (Or, if you worked the problems this way in the first place, work them by the other method, and check.)

Calculate as for Problems 1 through 9.

15. $f(t) = \sin t, g(t) = e^t$ 16. $f(t) = \cos t, g(t) = \text{Tan } t$
17. $f(t) = t^6, g(t) = t^3$ 18. $f(t) = t^6, g(t) = \text{Tan } t$
19. $f(x) = \ln x, g(x) = e^x$

Problems 20 through 24. Solve the preceding five problems by another method.

25. Given $f^2 + t^2 + 1 = 0$, find df/dt.

26. Given $f^3 + t^3 = 1$, find df/dt. Then calculate $f = f(t)$, find $f'(t)$, and compare the result with df/dt.

27. Same, for $f^4 + t^4 = 1$. (Here there are *two* functions $f = f(t)$ to be considered.)

28. Now try to check your answer to Problem 25 in the same way that you checked your answers to Problems 26 and 27. (It often happens that a formal process gives "answers" in cases where there never was a question.)

The Technique
6 of Integration

6.1 INTRODUCTION

In Section 3.7 we found a way to solve certain types of area problem. To find the area under the graph of a continuous function f, from a to b, we introduce the area function

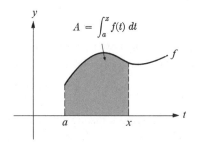

$$A = \int_a^x f(t)\, dt$$

We know that

$$F'(x) = D\int_a^x f(t)\, dt = f(x) \qquad \text{for every } x.$$

To calculate the area function, we find another function G such that

$$G' = f.$$

We then know that

$$G' = F'.$$

If it happens that $G(a) = 0$, then we have $G(x) = F(x)$ for every x. If not, we let

$$H(x) = G(x) - G(a).$$

Then

$$H'(x) = G'(x) = F'(x) \qquad \text{and} \qquad H(a) = 0.$$

Therefore

$$H(x) = F(x) \qquad \text{for every } x,$$

and so

$$\int_a^b f(t)\, dt = F(b) = H(b) = G(b) - G(a).$$

To sum up:

$$G' = f \quad \Rightarrow \quad \int_a^b f(t)\, dt = G(b) - G(a).$$

The notations t and G were introduced for the sake of the derivation. Once we have the answer, it is natural to use x and F, and write:

$$F' = f \quad \Rightarrow \quad \int_a^b f(x)\, dx = F(b) - F(a).$$

More formally:

Theorem 1 (*The fundamental theorem of integral calculus*). If f is continuous on $[a, b]$, and $F' = f$, then

$$\int_a^b f(x)\, dx = F(b) - F(a).$$

To apply the theorem, of course, we need to find F when f is given. This process is called *antidifferentiation*. We shall see later that the method of antidifferentiation enables us to solve not only the sort of area problems that we have used it on so far, but also a variety of problems which, offhand, don't look like area problems at all. But these applications should be postponed. The point is that, to apply the method, we need to know how to calculate a function F whose derivative is a given function f; up to now we have been finding such functions F only by hit-or-miss procedures, in simple cases; and it would not be good to reduce various problems to problems in antidifferentiation, when we are unable to solve the antidifferentiation problems. We should therefore first learn better methods for calculating functions when their derivatives are given.

6.2 INDEPENDENT VARIABLES AND INDEFINITE INTEGRALS

The usual way of defining a function is to write an expression which gives the value of the function for every number in the domain. For example, we may define functions f and g by writing

$$f(x) = x^2 \quad (-\infty < x < \infty), \qquad g(x) = \sqrt{x} \quad (x \geq 0).$$

In these formulas, the letter "x" is called the *independent variable*. It is simply a dummy letter, marking the places where numbers are to be inserted. Logically speaking, it makes no difference what letter we use as a dummy. For example, we could have defined exactly the same functions by writing

$$f(t) = t^2 \quad (-\infty < t < \infty), \qquad g(t) = \sqrt{t} \quad (t \geq 0).$$

If we have decided to use, say, x as the dummy, then we say that f is *a function of x*. Thus, when we write $g(t) = \cos^2 t$, we are describing g as a function of t; $h(\alpha) = \alpha^3 - 1$ is a function of α; and so on.

We return now to the problem of antidifferentiation. We found long ago, from the uniqueness theorem, that if two functions have the same derivative, on an interval, then they differ by a constant. Thus, if

$$f'(x) = x^2 \qquad (-\infty < x < \infty),$$

then f must be a function of the form

$$f(x) = \frac{x^3}{3} + C,$$

where C is a constant. The converse is trivial: for every C, $D(x^3/3 + C) = x^2$. Therefore

$$\{F \mid F' = x^2\} = \left\{ \frac{x^3}{3} + C \right\}.$$

The set of all functions F for which $F' = f$ is commonly denoted by

$$\int f(x)\, dx.$$

This is called the *indefinite integral of f*. Thus

$$\int x^4\, dx = \{F \mid F'(x) = x^4\} = \{ \tfrac{1}{5}x^5 + C \},$$

$$\int \cos x\, dx = \{F \mid F'(x) = \cos x\} = \{ \sin x + C \},$$

and so on. Any other dummy letter would have done as well:

$$\int t^4\, dt = \{F \mid F'(t) = t^4\},$$

and

$$\int \cos t\, dt = \{F \mid F'(t) = \cos t\}.$$

In each case, the braces on the right indicate that we are talking about the set of all functions of the form given inside. The symbol dx (or dt) merely reminds us that x (or t) is the dummy letter used in describing the function. In the examples above, the reminder may seem unnecessary. Similarly, when we write

$$\int (3x^3 + 2x^4)\, dx = \left\{ \frac{3x^4}{4} + \frac{2x^5}{5} + C \right\},$$

we might have gotten along without the "dx," because the only constants involved are the numerical constants 2 and 3. On the other hand, if we write

$$\int (\alpha x^3 y + \beta x^2 y^2)\, dx,$$

the "dx" is needed; it tells us that α, β, and y are to be regarded as constants, and that the function which we are dealing with is

$$f(x) = \alpha x^3 y + \beta x^2 y^2.$$

When the problem is understood in this sense, it is plain that the answer is

$$\int (\alpha x^3 y + \beta x^2 y^2)\, dx = \left\{ \frac{\alpha x^4 y}{4} + \frac{\beta x^3 y^2}{3} + C \right\}. \tag{i}$$

This should be compared with

$$\int (\alpha x^3 y + \beta x^2 y^2)\, dy = \left\{ \frac{\alpha x^3 y^2}{2} + \frac{\beta x^2 y^3}{3} + C \right\}, \tag{ii}$$

$$\int (\alpha x^3 y + \beta x^2 y^2)\, d\alpha = \left\{ \frac{\alpha^2 x^3 y}{2} + \beta x^2 y^2 \alpha + C \right\}, \tag{iii}$$

$$\int (\alpha x^3 y + \beta x^2 y^2)\, d\beta = \left\{ \alpha x^3 y \beta + \frac{\beta^2 x^2 y^2}{2} + C \right\}. \tag{iv}$$

In (ii), α, β, and x are constants, and the function is

$$g(y) = \alpha x^3 y + \beta x^2 y^2.$$

In (iii), x, y, and β are constants, and the function is

$$h(\alpha) = \alpha x^3 y + \beta x^2 y^2.$$

Similarly for (iv).

The process of calculating indefinite integrals is called *indefinite integration*, or briefly, *integration*. Given a differentiation formula, we can get a corresponding integration formula merely by writing the given formula "backwards," with minor adjustments in some cases to take care of constants. In each case below, the formula on the right follows from the formula or formulas on the left.

$$Dx^n = nx^{n-1} \quad (n \neq 0),$$

$$Dx^{n+1} = (n + 1)x^n \quad (n \neq -1),$$

$$D\left(\frac{x^{n+1}}{n + 1} \right) = x^n \quad (n \neq -1) \quad \Rightarrow \quad \int x^n\, dx = \left\{ \frac{1}{n + 1} x^{n+1} + C \right\} \quad (n \neq -1),$$

$$D\sqrt{x} = \frac{1}{2\sqrt{x}},$$

$$D(2\sqrt{x}) = \frac{1}{\sqrt{x}} \qquad \Rightarrow \quad \int \frac{1}{\sqrt{x}}\, dx = \{ 2\sqrt{x} + C \},$$

$$D \sin x = \cos x \qquad \Rightarrow \quad \int \cos x\, dx = \{ \sin x + C \},$$

$$D \cos x = -\sin x,$$

$$D(-\cos x) = \sin x \qquad \Rightarrow \quad \int \sin x\, dx = \{ -\cos x + C \},$$

$$D \ln x = \frac{1}{x} \quad (x > 0) \quad \Rightarrow \quad \int \frac{1}{x} dx = \{\ln x + C\} \quad (x > 0),$$

$$De^x = e^x \qquad \qquad \Rightarrow \quad \int e^x \, dx = \{e^x + C\}.$$

We know many more differentiation formulas than this, and so we could have written many more integration formulas. But we postpone the complete list until we can write it in a better form, which we shall now explain.

Given a function f, if u is another function, then $f(u)$ is a composite function. By the chain rule,

$$Df(u) = f'(u)u'.$$

It follows that

$$\int f'(u)u'(x) \, dx = \{f(u(x)) + C\}.$$

For example, if

$$f(u) = \sin u, \qquad u(x) = x^2 + 1,$$

then

$$D[f(u(x))] = D[\sin(x^2 + 1)] = f'(u)u'(x) = (\cos u)2x$$
$$= [\cos(x^2 + 1)]2x.$$

Therefore

$$\int [\cos(x^2 + 1)]2x \, dx = \{\sin(x^2 + 1) + C\}.$$

More generally,

$$D \sin u(x) = [\cos u(x)]u'(x),$$

and so

$$\int [\cos u(x)]u'(x) \, dx = \{\sin u(x) + C\}.$$

This works for any functions. If

$$F' = f,$$

so that

$$\int f(x) \, dx = \{F(x) + C\},$$

then

$$D[F(u(x))] = F'(u(x))u'(x) = f(u(x))u'(x),$$

so that

$$\int f(u(x))u'(x) \, dx = \{F(u(x)) + C\}.$$

In such formulas, we abbreviate $u'(x) \, dx$ by the symbol du. Thus we write

$$\int \cos u \, du = \{\sin u + C\},$$

which means that for every differentiable function $u(x)$, we have

$$\int [\cos u(x)]u'(x)\, dx = \{\sin u(x) + C\}.$$

Similarly, we write

$$\int e^u\, du = \{e^u + C\},$$

which means that if u is any differentiable function, then

$$\int e^{u(x)}u'(x)\, dx = \{e^{u(x)} + C\}.$$

This is true, because

$$De^{u(x)} = e^{u(x)}u'(x).$$

Using different dummy letters, we can convert the above formula to any of the forms

$$\int e^{u(\theta)}u'(\theta)\, d\theta, \qquad \text{or} \qquad \int e^{u(t)}u'(t)\, dt,$$

and so on. More often, however, we start with an integral described in the long notation and observe that it is convertible to a short form. For example,

$$\int e^{x^2+1}2x\, dx$$

has the form

$$\int e^{u(x)}u'(x)\, dx,$$

where $u(x) = x^2 + 1$. Therefore

$$\int e^{x^2+1}2x\, dx = \int e^u\, du = \{e^u + C\} = \{e^{x^2+1} + C\}.$$

Similarly, $\int [\sin(t^2 + 1)]2t\, dt$ has the form $\int \sin u\, du$. Therefore

$$\int [\sin(t^2 + 1)]2t\, dt = \int \sin u\, du = \{-\cos u + C\} = \{-\cos(t^2 + 1) + C\}.$$

Note that the solution is not finished in the third formula above, because u is a function. To complete the solution, we need to express the function u in terms of the dummy letter t. To sum up:

$$F' = f \quad \Rightarrow \quad D[F(u)] = f(u)u'.$$

Therefore

$$\int f(x)\, dx = \{F(x) + C\} \quad \Rightarrow \quad \int f(u(x))u'(x)\, dx = \{F(u) + C\}.$$

In the abbreviated form, using du for $u'(x)\, dx$, we have

$$\int f(x)\, dx = \{F(x) + C\} \quad \Rightarrow \quad \int f(u)\, du = \{F(u) + C\}.$$

Using this general idea, we can write all of our old integration formulas in the more general form. The first few look like this:

$$\int u^n\,du = \left\{\frac{u^{n+1}}{n+1} + C\right\}, \qquad\qquad \int \frac{1}{\sqrt{u}}\,du = \{2\sqrt{u} + C\},$$

$$\int \cos u\,du = \{\sin u + C\}, \qquad\qquad \int \sin u\,du = \{-\cos u + C\},$$

$$\int \frac{1}{u}\,du = \{\ln u + C\} \quad (u > 0), \qquad \int e^u\,du = \{e^u + C\}.$$

And of course we have

$$\int [f(x) + g(x)]\,dx = \int f(x)\,dx + \int g(x)\,dx,$$

$$\int kf(x) = k\int f(x)\,dx, \qquad k \neq 0,$$

because

$$D[f + g] = Df + Dg \qquad \text{and} \qquad D(kf) = k\,Df.$$

Let us now consider how to apply such formulas as these, as a practical matter.

Example 1. Consider

$$\int (x^2 + 1)^7 x\,dx.$$

This is almost, but not quite, in the form

$$\int u^7\,du.$$

If we take $u(x) = x^2 + 1$, then

$$du = u'(x)\,dx = 2x\,dx.$$

We therefore have

$$\int (x^2 + 1)^7 x\,dx = \int \tfrac{1}{2}(x^2 + 1)^7 2x\,dx = \int \tfrac{1}{2}u^7\,du$$

$$= \{\tfrac{1}{2}\cdot\tfrac{1}{8}u^8 + C\} = \{\tfrac{1}{16}(x^2 + 1)^8 + C\}.$$

This checks:

$$D[\tfrac{1}{16}(x^2 + 1)^8] = \tfrac{1}{16}\cdot 8(x^2 + 1)^7\cdot 2x = x(x^2 + 1)^7.$$

Example 2. Consider

$$\int \frac{\cos\sqrt{x}}{\sqrt{x}}\,dx \qquad (x > 0).$$

The only form that might fit this integral is the form $\int \cos u\,du$. Thus we would have

$$u(x) = \sqrt{x}, \qquad du = u'(x)\,dx = \frac{1}{2\sqrt{x}}\,dx.$$

The only difference between what we have and what we want is a multiplicative constant. Therefore

$$\int \frac{\cos \sqrt{x}}{\sqrt{x}}\, dx = \int (\cos \sqrt{x})\, \frac{1}{\sqrt{x}}\, dx = 2\int (\cos \sqrt{x})\, \frac{1}{2\sqrt{x}}\, dx$$

$$= 2\int \cos u\, du = \{2\sin u + C\} = \{2\sin \sqrt{x} + C\}.$$

Example 3. Consider

$$\int e^{\cos x}\, \sin x\, dx.$$

So far we have only one integration formula involving the exponential function:

$$\int e^u\, du = \{e^u + C\}.$$

If our problem fits this form, we must have

$$u(x) = \cos x, \qquad du = u'(x)\, dx = -\sin x\, dx.$$

Here again the multiplicative constant causes no trouble:

$$\int e^{\cos x}\, \sin x\, dx = \int -e^{\cos x}(-\sin x)\, dx = \int -e^u\, du$$

$$= \{-e^u + C\} = \{-e^{\cos x} + C\}.$$

Below we shall give a list of all the integration formulas that we can write, at this stage, on the basis of the differentiation formulas that we know. Special explanations are needed, however, in connection with the formula for $\int (1/u)\, du$. Given a function u, defined on a domain where $u(x) > 0$ for every x, we know that

$$D \ln u(x) = \frac{1}{u(x)}\, Du(x).$$

We need to know that $u(x) > 0$ on the domain under consideration, because only positive numbers have logarithms. Therefore we write

$$\int \frac{1}{u}\, du = \{\ln u + C\} \qquad (u > 0).$$

But even where $u(x) < 0$, it makes sense to write

$$\int \frac{1}{u}\, du\,;$$

that is, it makes sense to ask what functions f have $(1/u)u'$ as their derivatives. The answer is easy: if $u(x) < 0$, then $-u(x) > 0$. Therefore $-u(x)$ has a logarithm, and

$$D[\ln (-u(x))] = \frac{1}{-u(x)}\, D[-u(x)] = \frac{1}{-u(x)}(-u'(x)) = \frac{1}{u(x)}\, u'(x).$$

This gives us

$$\int \frac{1}{u}\, du = \{\ln (-u) + C\} \qquad (u < 0).$$

Hence the two formulas for $\int (1/u)\, du$ in the list below.

$$\int kf(x)\, dx = k\int f(x)\, dx \qquad (k \neq 0) \tag{1}$$

$$\int [f(x) + g(x)]\, dx = \int f(x)\, dx + \int g(x)\, dx \tag{2}$$

$$\int u^n\, du = \left\{\frac{u^{n+1}}{n+1} + C\right\} \qquad (n \neq -1) \tag{3}$$

$$\int \frac{1}{u}\, du = \{\ln u + C\} \qquad (u > 0) \tag{4}$$

$$\int \frac{1}{u}\, du = \{\ln(-u) + C\} \qquad (u < 0) \tag{5}$$

$$\int \cos u\, du = \{\sin u + C\} \tag{6}$$

$$\int \sin u\, du = \{-\cos u + C\} \tag{7}$$

$$\int \sec^2 u\, du = \{\tan u + C\} \tag{8}$$

$$\int \csc^2 u\, du = \{-\cot u + C\} \tag{9}$$

$$\int \sec u \tan u\, du = \{\sec u + C\} \tag{10}$$

$$\int \csc u \cot u\, du = \{-\csc u + C\} \tag{11}$$

$$\int e^u\, du = \{e^u + C\} \tag{12}$$

$$\int a^u\, du = \left\{\frac{a^u}{\ln a} + C\right\} \qquad (a > 0, a \neq 1) \tag{13}$$

$$\int \frac{du}{\sqrt{1 - u^2}} = \{\operatorname{Sin}^{-1} u + C\} \qquad (|u| < 1) \tag{14}$$

$$\int \frac{du}{1 + u^2} = \{\operatorname{Tan}^{-1} u + C\} \tag{15}$$

$$\int \frac{du}{u\sqrt{u^2 - 1}} = \{\operatorname{Sec}^{-1} u + C\} \qquad (u > 1) \tag{16}$$

To solve the following problems, you will start by expressing the given integral in the form $\int f(u)\,du$. In each such case, you should (a) say what u and du are and (b) state the general formula that you are applying. It is natural to write down the original integral first, and after this it would be awkward to interrupt the solution with the formulas for $u(x)$ and $du = u'(x)\,dx$. But u and du can be filled in on the right, like this, for example:

$$\int (x^3 + 1)^{10} x^2\,dx = \int \tfrac{1}{3}(x^3 + 1)^{10} 3x^2\,dx$$

$$\begin{cases} u(x) = x^3 + 1 \\ du = 3x^2\,dx \end{cases}$$

$$= \int \tfrac{1}{3}u^{10}\,du = \{\tfrac{1}{3} \cdot \tfrac{1}{11}u^{11} + C\}$$

$$= \{\tfrac{1}{33}(x^3 + 1)^{11} + C\}.$$

This form of the solution shows what we have in mind; writing formulas of the type $u(x) = x^3 + 1$, $du = 3x^2\,dx$, $\int \tfrac{1}{3}u^{10}\,du = \{\tfrac{1}{3} \cdot \tfrac{1}{11}u^{11} + C\}$ will help you to avoid mistakes. For example, you might write hastily

$$(?) \int (x^3 + 1)^{10}x^2\,dx = \{\tfrac{1}{11}(x^3 + 1)^{11} + C\},$$

as if it were true that for $u(x) = x^3 + 1$, $du = x^2\,dx$. When we write formulas for u and du, we uncover such errors.

Similarly for the following wrong solution:

$$(?) \int (x^2 + 1)^2\,dx = \{\tfrac{1}{3}(x^2 + 1)^3 + C\}.$$

In full, the solution would begin like this:

$$(?) \int (x^2 + 1)^2\,dx$$

$$\begin{cases} u = x^2 + 1 \\ du = dx(?!) \end{cases}$$

The error is obvious, and so we start over again:

$$\int (x^2 + 1)^2\,dx = \int (x^4 + 2x^2 + 1)\,dx = \{\tfrac{1}{5}x^5 + \tfrac{2}{3}x^3 + x + C\}.$$

PROBLEM SET 6.2

Calculate the following integrals, and check by differentiation in each case. Some of these problems fit together in sequences, in which the answer to one problem helps in the solution of another; you should watch for such patterns.

1. $\displaystyle\int (1 + x^2)^3 x\,dx$

2. $\displaystyle\int (1 + t^3)t^2\,dt$

3. $\displaystyle\int (2 + u^2)3u\,du$

4. $\displaystyle\int (t^4 + 1)t^3\,dt$

5. $\displaystyle\int (x^2 + t^2)^3\,dx$

6. $\displaystyle\int (x^2 + t^2)^3 tx\,dx$

7. $\int (x^2 + t^2)^3 tx \, dt$ 8. $\int (1 + \sqrt{x})^3 \frac{1}{\sqrt{x}} \, dx$ 9. $\int (t^{3/2} + 5)^{10} \sqrt{t} \, dt$

10. $\int (t^{5/2} - 1)t^{3/2} \, dt$ 11. $\int (1 + \sin x)^2 \cos x \, dx$ 12. $\int (1 + \tan x)^{3/2} \sec^2 x \, dx$

13. $\int \sqrt{\cos x} \sin x \, dx$ 14. $\int (e^x + 2)^4 e^x \, dx$ 15. $\int (e^x - 2)^3 e^{-x} \, dx$

16. $\int (e^x + e^{-x})^2 (e^x - e^{-x}) \, dx$ 17. $\int (e^x + e^{-x})^3 \, dx$ 18. $\int \frac{x}{(1 + x^2)^2} \, dx$

19. $\int \frac{x^2}{(1 + x^3)^3} \, dx$ 20. $\int \frac{x}{1 + x^2} \, dx$

21. $\int \frac{x^2}{1 + x^3} \, dx$ (There are two intervals to be considered in this problem.)

22. $\int \frac{1}{x} \ln x \, dx$ 23. $\int \frac{1}{x} \ln (x^2) \, dx$ (Same comment.)

24. $\int \sin x \cos x \, dx$ 25. $\int \sin^2 x \cos x \, dx$ 26. $\int \sin^3 x \cos x \, dx$

27. $\int \sin^{101} x \cos x \, dx$ 28. $\int \cos^2 x \sin x \, dx$ 29. $\int \cos^3 x \sin x \, dx$

30. $\int \cos^{57} x \sin x \, dx$ 31. $\int (1 + \tan^2 \theta) \, d\theta$ 32. $\int \tan^2 \theta \, d\theta$

33. $\int (\cot^2 \theta + 1) \, d\theta$ 34. $\int \cot^2 \theta \, d\theta$ 35. $\int \frac{\sin \theta}{\cos^2 \theta} \, d\theta$

36. $\int \frac{\cos \theta}{\sin^2 \theta} \, d\theta$ 37. $\int \sin 2\theta \, d\theta$ 38. $\int \cos 2\theta \, d\theta$

39. $\int (\cos^2 \theta - \sin^2 \theta) \, d\theta$ 40. $\int (\cos^2 \theta + \sin^2 \theta) \, d\theta$ 41. $\int (2 \cos^2 \theta - 1) \, d\theta$

42. $\int \cos^2 \theta \, d\theta$ 43. $\int (1 - 2 \sin^2 \theta) \, d\theta$ 44. $\int (2 \sin^2 \theta - 1) \, d\theta$

45. $\int \sin^2 \theta \, d\theta$ 46. $\int \sin^2 2\theta \, d\theta$ 47. $\int \sin^2 \theta \cos^2 \theta \, d\theta$

48. $\int \cos^2 \theta \sin \theta \, d\theta$ 49. $\int \sin \theta (1 - \sin^2 \theta) \, d\theta$ 50. $\int \sin^3 \theta \, d\theta$

51. $\int \cos (\theta/2) \, d\theta$ 52. $\int \sqrt{\frac{1 - \cos \theta}{2}} \, d\theta$ 53. $\int \sqrt{1 - \cos \theta} \sin \theta \, d\theta$

54. $\int xe^{-x^2} \, dx$ 55. $\int t^2 e^{t^3} \, dt$ 56. $\int xe^{x^2} \, dx$

57. $\int e^{2x} \, dx$ 58. $\int e^{5t} \, dt$ 59. $\int e^{t^2} t \, dt$

60. $\int e^{t^2+3} t \, dt$

61. $\int e^{\ln \sec^2 x} \, dx$

62. $\int e^{\sin x} \cos x \, dx$

63. $\int e^{\cos t} \sin t \, dt$

64. $\int 2^{x+1} \, dx$

65. $\int 10^{x^2} x \, dx$

66. $\int (10^x)^2 \, dx$

67. $\int (2 + t)^{-3/2} \, dt$

68. $\int t(2 + t^2)^{-3/2} \, dt$

69. $\int (2 + t^{-3/2}) \, dt$

70. $\int \dfrac{t \, dt}{\sqrt{1 - t^2}}$

71. $\int \dfrac{dt}{\sqrt{1 - t^2}} \, dt$

72. $\int \dfrac{t}{(\sqrt{1 - t^2})^3} \, dt$

73. $\int \dfrac{t^2}{\sqrt[3]{1 + t^3}} \, dt$

74. $\int \dfrac{dt}{\sqrt{4 - (2t)^2}}$

75. $\int \dfrac{t^3}{1 + t^4} \, dt$

76. $\int \dfrac{e^x}{1 + e^x} \, dx$

77. $\int \dfrac{e^x}{\sqrt{1 - e^x}} \, dx$

78. $\int \dfrac{e^x}{\sqrt{1 - e^{2x}}} \, dx$

(There are different intervals to consider in Problems 79 through 84.)

79. $\int \dfrac{\cos x}{\sin x} \, dx$

80. $\int \dfrac{\sin x}{\cos x} \, dx$

81. $\int \tan x \, dx$

82. $\int \dfrac{\sec^2 x + \sec x \tan x}{\sec x + \tan x} \, dx$

83. $\int \dfrac{\csc^2 x + \csc x \cot x}{\csc x + \cot x} \, dx$

84. $\int \sec x \, dx$

6.3 INTEGRALS LEADING TO THE LOGARITHM AND THE INVERSE SECANT. ALGEBRAIC DEVICES

In the preceding section, we got two formulas for $\int du/u$, for the intervals $(0, \infty)$ and $(-\infty, 0)$.

$$\int \frac{du}{u} = \{\ln u + C\} \qquad (u > 0) \tag{4}$$

$$\int \frac{du}{u} = \{\ln (-u) + C\} \qquad (u < 0). \tag{5}$$

Since $|u| = u$ when $u > 0$ and $|u| = -u$ when $u < 0$, these two formulas can be combined into one:

$$\int \frac{du}{u} = \{\ln |u| + C\} \qquad (\text{on } (0, \infty) \text{ or } (-\infty, 0)). \tag{17}$$

Here the expression in parentheses on the right reminds us that the formula can be used on an interval where $u > 0$, *or* on an interval where $u < 0$; it cannot be used on an interval where u takes on the value 0. When $u = 0$, there is no such thing as the "$1/u$" on the left or the "$\ln |u|$" on the right. Thus, whenever we apply formula (17), we might have used formula (4) or (5). The advantage of (17) is that it is easier to use. Consider

$$\int_{-2}^{-1} \frac{dx}{x} .$$

In the fundamental theorem of integral calculus, we take

$$f(x) = \frac{1}{x}, \qquad F(x) = \ln |x|.$$

Then $F' = f$. Therefore

$$\int_{-2}^{-1} \frac{dx}{x} = F(-1) - F(-2) = \ln |-1| - \ln |-2| = 0 - \ln 2 = -\ln 2.$$

This is negative, as it should be; the integrand is negative, and we are integrating from left to right. The calculation might be confusing if we used formula (5):

$$\int_{-2}^{-1} \frac{dx}{x} = [\ln (-x)]_{-2}^{-1} = \ln [-(-1)] - \ln [-(-2)] = 0 - \ln 2 = -\ln 2.$$

Hereafter, we shall use the following shorthand for this kind of calculation:

$$\int_{-2}^{-1} \frac{dx}{x} = [\ln |x|]_{-2}^{-1} = \ln |-1| - \ln |-2|.$$

In general

$$[F(x)]_a^b = F(b) - F(a),$$

by definition. Sometimes, where no confusion could result, we may omit the opening bracket on the left. Thus

$$\frac{x^3}{3}\bigg]_1^2 = \frac{8}{3} - \frac{1}{3} = \frac{7}{3}.$$

We can convert various integrals to the form $\int du/u$. For example,

$$\int \tan u \; du = \int \frac{\sin u}{\cos u} \; du.$$

Except for sign, this has the form

$$\int \frac{dv}{v}, \qquad v = \cos u, \qquad dv = -\sin u \; du.$$

Since

$$\int \frac{dv}{v} = \{\ln |v| + C\} \qquad (v > 0 \text{ or } v < 0),$$

we have

$$\int \tan u \; du = -\int \frac{-\sin u}{\cos u} \; du = -\int \frac{dv}{v}$$

$$= \{-\ln |v| + C\} = \{-\ln |\cos u| + C\};$$

$$\int \tan u \; du = \{\ln |\sec u| + C\} \qquad (\sec u > 0 \text{ or } \sec u < 0). \tag{18}$$

This is a standard formula.

Similarly,

$$\int \cot u \, du = \int \frac{\cos u}{\sin u} \, du.$$

This gives

$$\int \cot u \, du = \{\ln |\sin u| + C\} \qquad (\sin u > 0 \text{ or } \sin u < 0). \qquad (19)$$

By an ingenious device, we can find

$$\int \sec x \, dx.$$

We multiply and divide by $\sec x + \tan x$, getting

$$\int \sec x \, dx = \int \frac{\sec^2 x + \sec x \tan x}{\sec x + \tan x} \, dx.$$

Since

$$D \sec x = \sec x \tan x$$

and

$$D \tan x = \sec^2 x,$$

the integral has the form

$$\int \frac{du}{u},$$

where

$$u = \sec x + \tan x,$$
$$du = (\sec x \tan x + \sec^2 x) \, dx.$$

Therefore

$$\int \sec x \, dx = \{\ln |u| + C\} = \{\ln |\sec x + \tan x| + C\}.$$

As always, the chain rule gives us a more general formula for $\int \sec u \, du$:

$$\int \sec u \, du = \{\ln |\sec u + \tan u| + C\} \qquad (\sec u + \tan u > 0 \text{ or } < 0). \quad (20)$$

Similarly,

$$\int \csc x \, dx = \int \frac{\csc x(\csc x + \cot x)}{\csc x + \cot x} \, dx = \{-\ln |\csc x + \cot x| + C\};$$

and this gives

$$\int \csc u \, du = \{-\ln |\csc u + \cot u| + C\} \qquad (\csc u + \cot u > 0 \text{ or } < 0). \quad (21)$$

Consider now the formula

$$D \operatorname{Sec}^{-1} x = \frac{1}{x\sqrt{x^2 - 1}} \qquad (x > 1).$$

The graph of Sec^{-1} looks like this:

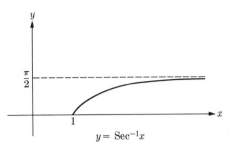

$$y = \mathrm{Sec}^{-1}x$$

(See Section 4.7.) Thus Sec^{-1} is defined on the interval $[1, \infty)$. But at 1 its tangent is vertical; and so the differentiation formula holds only for $x > 1$. It gives

$$\int \frac{dx}{x\sqrt{x^2 - 1}} = \{\mathrm{Sec}^{-1} x + C\} \qquad (x > 1),$$

and more generally

$$\int \frac{du}{u\sqrt{u^2 - 1}} = \{\mathrm{Sec}^{-1} u + C\} \qquad (u > 1). \tag{16}$$

Notice, however, that the integral

$$\int_{-3}^{-2} \frac{1}{x\sqrt{x^2 - 1}}\, dx$$

makes sense. We therefore need an integration formula which will apply to this integrand on the interval $(-\infty, -1)$. On this domain,

$$\int \frac{1}{x\sqrt{x^2 - 1}}\, dx = \int \frac{1}{-x\sqrt{(-x)^2 - 1}}\,(-1)\, dx = \int \frac{1}{u\sqrt{u^2 - 1}}\, du,$$

where

$$u(x) = -x, \qquad du = (-1)\, dx.$$

Therefore, for $x < -1$,

$$\int \frac{dx}{x\sqrt{x^2 - 1}} = \{\mathrm{Sec}^{-1} u + C\} = \{\mathrm{Sec}^{-1}(-x) + C\}$$

$$= \{\mathrm{Sec}^{-1} |x| + C\} \qquad (x < -1),$$

because $|x| = -x$ when $x < 0$. Fitting our two formulas together, and passing to the general case (with a function u instead of x), we get

$$\int \frac{du}{u\sqrt{u^2 - 1}} = \{\mathrm{Sec}^{-1} |u| + C\} \qquad (u > 1 \text{ or } u < -1). \tag{22}$$

There is a rough rule to help you decide which of our present list of formulas to apply to a given problem: *look in the integrand for functions which are the derivatives of other functions.* The point is that all our formulas have left-hand members of the form $\int f(u)\, du$; and we need to decide, in each case, what u is.

Example 1.

$$\int \frac{\ln^3 x}{x}\, dx.$$

Is there anything here that is the derivative of something else? Yes:

$$D \ln x = \frac{1}{x}.$$

Taking $u(x) = \ln x$, we have

$$du = u'(x)\, dx = \frac{1}{x}\, dx.$$

Thus our integral has the form

$$\int \frac{\ln^3 x}{x}\, dx = \int (\ln^3 x) \frac{1}{x}\, dx = \int u^3\, du \qquad\qquad \left\{ \begin{array}{l} u = \ln x \\[2mm] du = \dfrac{1}{x}\, dx \end{array} \right.$$

$$= \left\{ \frac{u^4}{4} + C \right\} = \{\tfrac{1}{4} \ln^4 x + C\}.$$

Example 2.

$$\int \frac{x\, dx}{(1 + x^2)^7}.$$

Looking for functions which are derivatives of other functions, we observe that

$$D(1 + x^2) = 2x.$$

Multiplicative constants are no trouble:

$$\int \frac{x\, dx}{(1 + x^2)^7} = \frac{1}{2} \int \frac{2x\, dx}{(1 + x^2)^7} = \frac{1}{2} \int u^{-7}\, du,$$

where

$$u(x) = 1 + x^2, \qquad du = u'(x)\, dx = 2x\, dx.$$

Therefore the answer is

$$\left\{ \frac{1}{2} \cdot \frac{1}{-7 + 1}\, u^{-7+1} + C \right\} = \left\{ \frac{-1}{12}\, u^{-6} + C \right\} = \left\{ \frac{-1}{12}\, (1 + x^2)^{-6} + C \right\}.$$

Example 3. Sometimes we have to hunt harder:

$$\int \frac{x\, dx}{\sqrt{1 - x^4}}.$$

There is no hope that $1/\sqrt{1 - x^4}$ is part of du. Either the problem is hard or du must be $x\,dx$, or a constant multiple of $x\,dx$. Now $2x = Dx^2$; and x^2 is what gets *squared* under the radical sign in the denominator. This suggests

$$u = x^2,$$

$$\int \frac{x\,dx}{\sqrt{1 - x^4}} = \frac{1}{2} \int \frac{2x\,dx}{\sqrt{1 - (x^2)^2}} = \frac{1}{2} \int \frac{du}{\sqrt{1 - u^2}}$$

$$= \{\tfrac{1}{2} \operatorname{Sin}^{-1} u + C\} = \{\tfrac{1}{2} \operatorname{Sin}^{-1} x^2 + C\}.$$

Example 4. Some obscure-looking integrals may be calculated algebraically:

$$\int \frac{x\,dx}{x - 1} = \int \left(1 + \frac{1}{x - 1}\right) dx.$$

(Here we have divided the denominator into the numerator, getting a quotient and a remainder.) Therefore

$$\int \frac{x\,dx}{x - 1} = \{x + \ln |x - 1| + C\}.$$

Example 5. Sometimes we need to find other algebraic devices, for such problems as this:

$$\int \frac{dx}{1 + e^{-x}}.$$

As it stands, this is hopeless: nothing in the integrand is the derivative of anything else. But

$$\int \frac{dx}{1 + e^{-x}} = \int \frac{e^x\,dx}{e^x + 1} = \int \frac{du}{u} \qquad u = e^x + 1$$

$$= \{\ln |u| + C\} = \{\ln (1 + e^x) + C\}.$$

(No absolute-value signs are needed, because $1 + e^x > 1$ for every x.)

Example 6. Sometimes the same devices appear in more complicated forms:

$$\int \frac{dx}{e^x + e^{-x}} = \int \frac{e^x\,dx}{1 + e^{2x}} = \int \frac{e^x\,dx}{1 + (e^x)^2} = \int \frac{du}{1 + u^2} \qquad (u = e^x,\ du = e^x\,dx)$$

$$= \{\operatorname{Tan}^{-1} u + C\} = \{\operatorname{Tan}^{-1} e^x + C\}.$$

Here we have used, in combination, the methods that worked in Examples 3 and 5.

Example 7. Often we need routine algebra and arithmetic:

$$\int \frac{dx}{4 + x^2} = \int \frac{1}{4} \cdot \frac{dx}{1 + (x/2)^2} = \int \frac{1}{4} \cdot 2 \cdot \frac{\tfrac{1}{2}\,dx}{1 + (x/2)^2}.$$

Here $u = x/2$, $du = \tfrac{1}{2}dx$. This gives

$$\left\{\tfrac{1}{2} \operatorname{Tan}^{-1} \frac{x}{2} + C\right\}.$$

Similarly,

$$\int \frac{dx}{\sqrt{3 - x^2}} = \int \frac{1}{\sqrt{3}} \frac{dx}{\sqrt{1 - (x/\sqrt{3})^2}} = \int \frac{(1/\sqrt{3})\, dx}{\sqrt{1 - (x/\sqrt{3})^2}}$$

$$= \left\{ \text{Sin}^{-1} \frac{x}{\sqrt{3}} + C \right\}.$$

There was nothing special about the numbers $4 = 2^2$ and $3 = (\sqrt{3})^2$. In the same way, we get

$$\int \frac{dx}{a^2 + x^2} = \left\{ \frac{1}{a} \text{Tan}^{-1} \frac{x}{a} + C \right\} \qquad (a > 0),$$

$$\int \frac{dx}{\sqrt{a^2 - x^2}} = \left\{ \text{Sin}^{-1} \frac{x}{a} + C \right\} \qquad (a > 0).$$

Passing from x to any differentiable function u, we get two more standard formulas:

$$\int \frac{du}{a^2 + u^2} = \left\{ \frac{1}{a} \text{Tan}^{-1} \frac{u}{a} + C \right\} \qquad (a > 0), \tag{23}$$

$$\int \frac{du}{\sqrt{a^2 - u^2}} = \left\{ \text{Sin}^{-1} \frac{u}{a} + C \right\} \qquad (a > 0). \tag{24}$$

PROBLEM SET 6.3

Calculate the following integrals, and check by differentiation in each case.

1. $\displaystyle\int \frac{x}{\sqrt{1 - x^4}}\, dx$ 2. $\displaystyle\int \frac{t}{\sqrt{1 - 4t^2}}\, dt$ 3. $\displaystyle\int \frac{y}{\sqrt{1 - 4y^4}}\, dy$

4. $\displaystyle\int \frac{y}{\sqrt{4 - y^4}}\, dy$ 5. $\displaystyle\int \frac{x}{1 + 9x^4}\, dx$ 6. $\displaystyle\int \frac{x^3}{1 + 9x^4}\, dx$

7. $\displaystyle\int \frac{x^3}{9 + x^4}\, dx$ 8. $\displaystyle\int \frac{t}{5 + t^4}\, dt$ 9. $\displaystyle\int \frac{z^2}{1 + z^6}\, dz$

10. $\displaystyle\int \frac{z^2}{2 + z^6}\, dz$ 11. $\displaystyle\int \frac{z^2}{1 + 2z^6}\, dz$ 12. $\displaystyle\int \frac{z^5}{1 + z^6}\, dz$

13. $\displaystyle\int \frac{z^5}{5 + z^6}\, dz$ 14. $\displaystyle\int \frac{z^5}{1 + 5z^6}\, dz$ 15. $\displaystyle\int \frac{x^7}{\sqrt{1 - x^8}}\, dx$

16. $\displaystyle\int \frac{x^3}{\sqrt{1 - x^8}}\, dx$ 17. $\displaystyle\int \frac{x^7}{1 + x^8}\, dx$ 18. $\displaystyle\int \frac{x^3}{1 + x^8}\, dx$

19. $\displaystyle\int \frac{e^z}{1 + e^z}\, dz$ 20. $\displaystyle\int \frac{e^t}{1 + e^{2t}}\, dt$ 21. $\displaystyle\int \frac{1}{e^x + e^{-x}}\, dx$

22. $\displaystyle\int \frac{e^x + e^{-x}}{e^x - e^{-x}}\, dx$ 23. $\displaystyle\int \frac{e^x - e^{-x}}{e^x + e^{-x}}\, dx$ 24. $\displaystyle\int (e^x + e^{-x})(e^x - e^{-x})^2\, dx$

25. $\displaystyle\int (e^x + e^{-x})(e^{2x} + e^{-2x})\, dx$ 26. $\displaystyle\int \frac{x^2}{\sqrt{2 - x^3}}\, dx$ 27. $\displaystyle\int \frac{x^2}{\sqrt{2 - x^6}}\, dx$

28. $\int \dfrac{x^2}{\sqrt{1-2x^3}}\,dx$ 29. $\int \dfrac{x^2}{\sqrt{1-2x^6}}\,dx$ 30. $\int \dfrac{1}{x\ln^2 x}\,dx$

31. $\int \dfrac{\text{Sin}^{-1} x}{\sqrt{1-x^2}}\,dx$ 32. $\int \dfrac{\text{Tan}^{-1} x}{1+x^2}\,dx$ 33. $\int (xe^x + e^x)\,dx$

34. $\int xe^x\,dx$ 35. $\int \ln e^{x^2}\,dx.$ 36. $\int \ln^2 e^x\,dx$

37. $\int (x\cos x + \sin x)\,dx$ 38. $\int x\cos x\,dx$ 39. $\int (x\sin x - \cos x)\,dx$

40. $\int x\sin x\,dx$ 41. $\int (2x\ln x + x)\,dx$ 42. $\int x\ln x\,dx$

43. $\int (3x^2 \ln x + x^2)\,dx$ 44. $\int x^2 \ln x\,dx$ 45. $\int \dfrac{\ln^3 x}{x}\,dx$

46. $\int \dfrac{1}{x(1+\ln^2 x)}\,dx$ 47. $\int \dfrac{(x+1)\ln(x^2+2x)}{x^2+2x}\,dx$ 48. $\int \dfrac{1}{t\sqrt{t^2-1}}\,dt$

49. $\int \dfrac{x}{x^2\sqrt{x^4-1}}\,dx$ 50. $\int \dfrac{1}{\sqrt{e^{2x}-1}}\,dx$ 51. $\int \left(\text{Sin}^{-1} x + \dfrac{x}{\sqrt{1-x^2}}\right)dx$

52. $\int \dfrac{x}{\sqrt{1-x^2}}\,dx$ 53. $\int \text{Sin}^{-1} x\,dx$ 54. $\int \left(\text{Cos}^{-1} x - \dfrac{x}{\sqrt{1-x^2}}\right)dx$

55. $\int \text{Cos}^{-1} x\,dx$ 56. $\int \text{Cos}^{-1}(2x)\,dx$ 57. $\int \dfrac{e^{2u}}{1+e^{2u}}\,du$

58. $\int \dfrac{e^{2u}}{1+e^{4u}}\,du$ 59. $\int \dfrac{e^{4u}}{1+e^{2u}}\,du$ 60. $\int \left(\text{Tan}^{-1} x + \dfrac{x}{1+x^2}\right)dx$

61. $\int \dfrac{x}{1+x^2}\,dx$ 62. $\int \text{Tan}^{-1} x\,dx$ 63. $\int \sqrt{1-z^2}\,dz$

64. We know that

$$D\frac{1}{x} = Dx^{-1} = -1x^{-2} = -\frac{1}{x^2}.$$

Consider

$$\int_{-1}^{1} \frac{1}{x^2}\,dx.$$

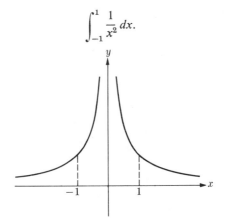

In the fundamental theorem of integral calculus, we take

$$f(x) = \frac{1}{x^2}, \qquad F(x) = -\frac{1}{x}.$$

Then $F' = f$. Therefore

$$\int_{-1}^{1} \frac{1}{x^2}\, dx = F(1) - F(-1) = -\frac{1}{1} - \left(-\frac{1}{-1}\right) = -2.$$

Now we interpret the problem geometrically. We seem to have proved that the region under a positive function has negative area.

a) What went wrong?

b) Show that the area in question is not only positive but infinite. (This does *not* follow from the mere fact that the region is unbounded. Some unbounded regions have finite areas.)

65. Let R be the region under the graph of $f(x) = 1/\sqrt{x}$, from $x = 0$ to $x = 1$. Show that R has finite area.

6.4 INTEGRATION BY PARTS

By differentiation, we get
$$D[x \sin x] = x \cos x + \sin x.$$
Since
$$D \cos x = -\sin x,$$
we have
$$D[x \sin x + \cos x] = x \cos x + \sin x - \sin x = x \cos x.$$
Therefore

$$\int x \cos x\, dx = \{x \sin x + \cos x + C\}.$$

Thus, working backward, we have found the solution of an integration problem which might have looked hard if we had approached it forward, starting with the unknown integral $\int x \cos x\, dx$. We shall now describe a general method of solving problems of this kind.

The formula for the derivative of a product is
$$D[u(x)v(x)] = u(x)v'(x) + u'(x)v(x).$$
Therefore

$$\int [u(x)v'(x) + u'(x)v(x)]\, dx = \{u(x)v(x) + C\}.$$
Therefore

$$\int u(x)v'(x)\, dx = u(x)v(x) - \int v(x)u'(x)\, dx.$$

Here we have dropped the constant, C, because each of the indefinite integrals on the two sides of the equation carries its own constant with it; what the equation says is that the two sides of the equation represent the same class of functions. Using the short notation

$$du = u'(x)\, dx, \qquad dv = v'(x)\, dx,$$

we get the formula

$$\int u \, dv = uv - \int v \, du.$$

This is the formula for *integration by parts*; the word *parts* refers to the functions $u(x)$ and $v'(x)$ in the integral on the left. Any time we apply the formula, we replace one integral by another. The method is useful when the new integral is easier to calculate than the old one.

Let us first try the method on

$$\int x \cos x \, dx.$$

Let
$$u = x, \qquad dv = \cos x \, dx,$$
so that
$$du = dx \qquad \text{and} \qquad v = \sin x.$$

(We need not allow for a constant here; any function v whose derivative is $\cos x$ will work. We will return to this point in a moment.) By the basic formula, we get

$$\int x \cos x \, dx = \int u \, dv = uv - \int v \, du = x \sin x - \int \sin x \, dx$$

$$= \{x \sin x + \cos x + C\}.$$

If we had used the seemingly more general

$$v = \sin x + c,$$

we would have

$$\int u \, dv = x(\sin x + c) - \int (\sin x + c) \, dx$$

$$= \{x \sin x + cx + \cos x - cx + C\} = \{x \sin x + \cos x + C\},$$

exactly as before. The same happens in general:

$$u(v + c) - \int (v + c) \, du = uv + uc - \int v \, du - uc = uv - \int v \, du.$$

In applying the basic formula, we made what may seem to be an arbitrary choice of u and dv. We might have taken

$$u = \cos x, \qquad du = -\sin x \, dx,$$

$$dv = x \, dx, \qquad v = \frac{x^2}{2}.$$

This would have given

$$\int x \cos x \, dx = \int u \, dv = uv - \int v \, du = \frac{x^2}{2} \cos x + \tfrac{1}{2} \int x^2 \sin x \, dx.$$

This is true, but is worthless as a method of finding $\int x \cos x \, dx$, because the new integral is harder to calculate than the old one.

An equally bad choice would be

$$u = x \cos x, \qquad du = \cos x - x \sin x,$$
$$dv = dx, \qquad v = x,$$

which gives

$$\int x \cos x \, dx = \int u \, dv = uv - \int v \, du$$

$$= x^2 \cos x - \int (x \cos x - x^2 \sin x) \, dx.$$

Here again the new integral is harder than the old one. We remember also that no term of the form $x^2 \cos x$ appears in the right answer. Therefore the term $x^2 \cos x$ cannot be the beginning of the solution, as we might hope: it must be a blind alley.

These examples indicate that integration by parts can be either a good or a bad method, according to the skill with which we choose the parts. Practice is a help, but there are general rules which help us to decide what choices are promising:

1) dv has *got* to be something that we know how to integrate. (If it isn't, we can't apply the method at all.)

2) We want $\int v \, du$ to be an easier integral than $\int u \, dv$. Therefore we want du to be simpler than u. At least, we don't want du to be more complicated than u.

3) For the same reason, we want v to be simpler than dv; at least we don't want it to look worse than dv.

These rules are not infallible, but they are a help. Let us try them on

$$\int x e^x \, dx.$$

We can integrate both x and e^x. Therefore (1) gives us no guidance. Rule (2) suggests that $u = x$ and $u = e^x$ are both acceptable, but that $u = x$ is to be preferred. ($De^x = e^x$, which is no worse than e^x, but $Dx = 1$, which looks good.) We therefore use

$$u = x, \qquad du = dx, \qquad dv = e^x \, dx, \qquad v = e^x.$$

This gives

$$\int x e^x \, dx = \int u \, dv = uv - \int v \, du = x e^x - \int e^x \, dx = \{x e^x - e^x + C\}.$$

Note that (3) advises us not to try

$$u = e^x, \qquad dv = x \, dx,$$

because we would then get $v = x^2/2$, which looks worse than dv. In fact, this choice won't work. Consider next

$$\int x^2 e^x \, dx.$$

Rule (3) tells us that we had better take $dv = e^x \, dx$. We therefore take

$$u = x^2, \qquad du = 2x \, dx, \qquad dv = e^x \, dx, \qquad v = e^x.$$

This looks good under rule (2), and acceptable under rule (3). We get

$$\int x^2 e^x \, dx = \int u \, dv = uv - \int v \, du = x^2 e^x - 2 \int x e^x \, dx,$$

$$= \{x^2 e^x - 2x e^x + 2e^x + C\},$$

by the result of the previous problem. If we hadn't known the answer to the previous problem, it would still be easy to see that we had made progress, in replacing $\int x^2 e^x \, dx$ by $\int x e^x \, dx$; we would then attack the new problem by the same method.

It sometimes happens that integration by parts gives us not an expression for the integral that we started with, but an equation that can be solved for this integral. Consider

$$\int e^x \sin x \, dx.$$

We take

$$u = e^x, \qquad du = e^x \, dx, \qquad dv = \sin x \, dx, \qquad v = -\cos x,$$

$$\int e^x \sin x \, dx = -e^x \cos x + \int e^x \cos x \, dx.$$

We repeat the process, taking

$$u = e^x, \qquad du = e^x \, dx, \qquad dv = \cos x \, dx, \qquad v = \sin x.$$

For short, we write

$$I = \int e^x \sin x \, dx.$$

We then have

$$I = -e^x \cos x + \int e^x \cos x \, dx = -e^x \cos x + e^x \sin x - \int e^x \sin x \, dx.$$

Here the last integral is simply the one we started with. Therefore

$$2I = \{e^x \sin x - e^x \cos x + C\},$$

and

$$\int e^x \sin x \, dx = \{\tfrac{1}{2}[e^x \sin x - e^x \cos x] + C\}.$$

Sometimes we need to make a strange choice, in which u is the whole integrand and dv is merely dx. This is what we need, to find

$$\int \ln x \, dx.$$

Here we use

$$u = \ln x, \qquad du = \frac{1}{x}\, dx, \qquad dv = dx, \qquad v = x.$$

When we replace 1 by x, we seem to have lost somewhat, but the profit in passing from $\ln x$ to $1/x$ more than makes up for it. In fact, this scheme works:

$$\int \ln x\, dx = uv - \int v\, du = x \ln x - \int x \cdot \frac{1}{x}\, dx$$

$$= x \ln x - \int 1 \cdot dx = \{x \ln x - x + C\}.$$

PROBLEM SET 6.4

Evaluate the following integrals. Each of them can be calculated by the method of integration by parts. You should try to work these problems with the smallest possible number of false starts. In each case, survey the situation and try to arrive at a conclusion on the question of what choice of u and dv is most promising. If you do this carefully, you ought to be able to solve each of the problems below on the first try.

Each answer should be checked by differentiation.

1. $\displaystyle\int \ln^2 x\, dx$

2. $\displaystyle\int \ln(x^2)\, dx$

3. $\displaystyle\int (ax)e^x\, dx$

4. $\displaystyle\int xe^{\,ax}dx$

5. $\displaystyle\int x \sin ax\, dx$

6. $\displaystyle\int x \cos ax\, dx$

7. $\displaystyle\int e^{ax} \sin x\, dx$

8. $\displaystyle\int e^{ax} \cos x\, dx$

9. $\displaystyle\int e^{ax} \sin bx\, dx$

10. $\displaystyle\int e^{ax} \cos bx\, dx$

11. $\displaystyle\int x^2 \sin x\, dx$

12. $\displaystyle\int x^2 \cos x\, dx$

13. $\displaystyle\int x^3 e^x\, dx$

14. $\displaystyle\int x^2 \ln x\, dx$

15. $\displaystyle\int x^2 \ln^2 x\, dx$

16. $\displaystyle\int x^3 \ln x\, dx$

17. $\displaystyle\int \mathrm{Sin}^{-1} x\, dx$

18. $\displaystyle\int \mathrm{Sin}^{-1}(2x)\, dx$

19. $\displaystyle\int \mathrm{Tan}^{-1} x\, dx$

20. $\displaystyle\int x\, \mathrm{Tan}^{-1} x\, dx$

21. $\displaystyle\int xe^x \sin x\, dx$

22. Derive and check:

$$\int \ln^n x\, dx = x \ln^n x - n \int \ln^{n-1} x\, dx.$$

Formulas of this kind are called *reduction* formulas. By $n - 1$ applications of the formula, we can calculate the integral on the left.

23. Find $\displaystyle\int \ln^3 x\, dx$.

24. Derive a reduction formula for $\int x^n \sin x\, dx$.

25. Derive a reduction formula for $\int x^n e^x \, dx$.

26. Derive a reduction formula which reduces n in $\int x^k \ln^n x \, dx$.

27. $\int \sin \ln x \, dx$. (Here you should survey the situation, decide on the most promising procedure, and then proceed with faith.)

28. $\int \cos \ln x \, dx$. (Same comment as for the preceding problem.)

6.5 INTEGRATION OF POWERS OF TRIGONOMETRIC FUNCTIONS

We shall find how to calculate integrals of the form

$$\int \sin^n x \cos^m x \, dx,$$

where n and m are any integers, positive, negative, or zero, and integrals of the forms

$$\int \sec^n x \tan^m x \, dx,$$

and

$$\int \csc^n x \cot^m x \, dx,$$

where $n \geq 0$ and $m \geq 0$. We shall discuss the various cases in the order of increasing difficulty.

$$\int \sin^n x \cos^m x \, dx, \qquad m \text{ odd and positive.} \tag{1}$$

For example, we might have

$$\int \sin^2 x \cos^3 x \, dx.$$

The method in such cases is as follows. Since

$$\cos^2 x + \sin^2 x = 1, \qquad \cos^2 x = 1 - \sin^2 x,$$

and

$$\int \sin^2 x \cos^3 x \, dx = \int \sin^2 x (1 - \sin^2 x) \cos x \, dx$$

$$= \int \sin^2 x \cos x \, dx - \int \sin^4 x \cos x \, dx$$

$$= \{\tfrac{1}{3} \sin^3 x - \tfrac{1}{5} \sin^5 x + C\}.$$

This method works whenever m is odd. (In this case n need not be an integer; it may be any real number.) For $m = 2k + 1$, our integral has the form

$$\int \sin^n x \cos^{2k+1} x \, dx = \int \sin^n x (\cos^2 x)^k \cos x \, dx$$

$$= \int \sin^n x (1 - \sin^2 x)^k \cos x \, dx.$$

We expand $(1 - \sin^2 x)^k$ by the binomial theorem. This gives us a sum of integrals of the form

$$\int u^m \, du \qquad (u = \sin x, \quad du = \cos x \, dx).$$

We integrate these one at a time and add the results.

$$\int \sin^n x \cos^m x \, dx, \qquad n \text{ odd and positive.} \tag{2}$$

This is like the preceding case. Here $n = 2k + 1$, and the integral has the form

$$\int \cos^m x (\sin^2 x)^k \sin x \, dx = -\int \cos^m x (1 - \cos^2 x)^k (-\sin x) \, dx.$$

Expanding by the binomial formula, we get a sum of integrals of the type

$$\int \cos^j x (-\sin x) \, dx;$$

we evaluate each of these by the formula for $\int u^j \, du$, and add.

$$\int \sin^n x \cos^m x, \qquad m \text{ and } n \geq 0 \text{ and even.} \tag{3}$$

To handle this one, we recall that

$$\cos 2x = \cos^2 x - \sin^2 x = \cos^2 x - (1 - \cos^2 x) = 2 \cos^2 x - 1.$$

Solving for $\cos^2 x$, we get

$$\cos^2 x = \frac{1 + \cos 2x}{2}.$$

Similarly,

$$\cos 2x = (1 - \sin^2 x) - \sin^2 x = 1 - 2 \sin^2 x,$$

and

$$\sin^2 x = \frac{1 - \cos 2x}{2}.$$

Making these substitutions in the integrand, we get a form in which the exponents are divided by 2. For example,

$$\int \sin^2 x \cos^2 x \, dx = \int \frac{1 - \cos 2x}{2} \cdot \frac{1 + \cos 2x}{2} \, dx$$

$$= \tfrac{1}{4} \int (1 - \cos^2 2x) \, dx = \tfrac{1}{4} \int \sin^2 2x \, dx.$$

We now make the same sort of substitution again, getting

$$\frac{1}{4}\int \frac{1 - \cos 4x}{2}\, dx = \tfrac{1}{8}x - \tfrac{1}{8}\int \cos 4x\, dx = \{\tfrac{1}{8}x - \tfrac{1}{32}\sin 4x + C\}.$$

When the exponents are large, this method is tedious, but at least we know that it will work.

$$\int \tan^n x\, dx \qquad (n\ \text{positive}). \tag{4}$$

For $n = 1$, we know that

$$\int \tan x\, dx = \{-\ln |\cos x| + C\} \qquad (\cos x > 0\ \text{or}\ \cos x < 0). \tag{4a}$$

For $n = 2$,

$$\int \tan^2 x\, dx = \int (\sec^2 x - 1)\, dx.$$

(Remember that $1 + \tan^2 x = \sec^2 x$.) This gives

$$\int \tan^2 x\, dx = \{\tan x - x + C\}. \tag{4b}$$

For $n > 2$, we have

$$\int \tan^n x\, dx = \int \tan^{n-2} x(\sec^2 x - 1)\, dx,$$

and so

$$\int \tan^n x\, dx = \frac{1}{n-1}\tan^{n-1} x - \int \tan^{n-2} x\, dx. \tag{4c}$$

This is called a *reduction* formula. By repeated applications of it, we can reduce the integral to one of the forms (4a) and (4b).

$$\int \cot^n x\, dx \qquad (n\ \text{positive}). \tag{5}$$

This is like (4). For $n = 1$,

$$\int \cot x = \int \frac{\cos x}{\sin x}\, dx = \{\ln |\sin x| + C\} \qquad (\sin x > 0\ \text{or}\ \sin x < 0). \tag{5a}$$

For $n = 2$,

$$\int \cot^2 x\, dx = \int (\csc^2 x - 1)\, dx.$$

(Remember that $\cot^2 x + 1 = \csc^2 x$.) Thus

$$\int \cot^2 x\, dx = \{-\cot x - x + C\}. \tag{5b}$$

For $n > 2$,

$$\int \cot^n x \, dx = \int \cot^{n-2} x \cot^2 x \, dx = \int \cot^{n-2} x (\csc^2 x - 1) \, dx;$$

and so

$$\int \cot^n x \, dx = -\frac{1}{n-1} \cot^{n-1} x - \int \cot^{n-2} x \, dx. \tag{5c}$$

By repeated applications of (5c), we can reduce our integral to one of the forms (5a) and (5b).

$$\int \sec^n x \, dx, \qquad n \text{ even and positive.} \tag{6}$$

For $n = 2$, $\int \sec^2 x \, dx = \{\tan x + C\}$. For $n = 2k$, $k > 1$,

$$\int \sec^n x \, dx = \int \sec^{2k} x \, dx = \int \sec^{2k-2} x \sec^2 x \, dx$$

$$= \int (1 + \tan^2 x)^{k-1} \sec^2 x \, dx.$$

When we expand $(1 + \tan^2 x)^{k-1}$ by the binomial formula, we get a sum of integrals of the form

$$\int \tan^j x \sec^2 x \, dx = \int u^j \, du.$$

We integrate each of these by the power formula and add the results. For example,

$$\int \sec^6 x \, dx = \int (\sec^2 x)^2 \sec^2 x \, dx = \int (1 + \tan^2 x)^2 \sec^2 x \, dx$$

$$= \int (1 + 2 \tan^2 x + \tan^4 x) \sec^2 x \, dx$$

$$= \{\tan x + \tfrac{2}{3} \tan^3 x + \tfrac{1}{5} \tan^5 x + C\}.$$

$$\int \csc^n x \, dx, \qquad n \text{ even and positive.} \tag{7}$$

This is like (6). For $n = 2$,

$$\int \csc^2 x \, dx = \{-\cot x + C\}.$$

For $n = 2k$, $k > 1$,

$$\int \csc^{2k} x \, dx = \int (\csc^2 x)^{k-1} \csc^2 x$$

$$= \int (\cot^2 x + 1)^{k-1} \csc^2 x \, dx$$

$$= -\int (\cot^2 x + 1)^{k-1} (-\csc^2 x \, dx).$$

When we expand the binomial, we get a sum of integrals of the form

$$\int \cot^j x(-\csc^2 x)\,dx = \int u^j\,du.$$

$$\int \sec^n x\,dx, \qquad n \text{ odd and positive.} \tag{8}$$

For $n = 1$, we found that

$$\int \sec x\,dx = \int \frac{\sec x(\sec x + \tan x)}{\sec x + \tan x}\,dx = \int \frac{du}{u},$$

where

$$u = \sec x + \tan x, \qquad du = (\sec x \tan x + \sec^2 x)\,dx.$$

Therefore

$$\int \sec x\,dx = \{\ln |\sec x + \tan x| + C\}.$$

For n odd and greater than 1, we have a problem. For example, in $\int \sec^3 x\,dx$ it does no good to write

$$\int \sec^2 x \sec x\,dx = \int (1 + \tan^2 x) \sec x\,dx,$$

because the second term fits no standard form. The solution is obtained by integrating by parts. We have

$$\int \sec^n x\,dx = \int \sec^{n-2} x \sec^2 x\,dx.$$

Let

$$u = \sec^{n-2} x, \qquad dv = \sec^2 x\,dx,$$

$$du = (n-2) \sec^{n-3} x \sec x \tan x\,dx = (n-2) \sec^{n-2} x \tan x\,dx,$$

$$v = \tan x.$$

This gives

$$\int \sec^n x\,dx = \int u\,dv = uv - \int v\,du$$

$$= \sec^{n-2} x \tan x - (n-2) \int \sec^{n-2} x \tan^2 x\,dx$$

$$= \sec^{n-2} x \tan x - (n-2) \int \sec^{n-2} x(\sec^2 x - 1)\,dx$$

$$= \sec^{n-2} x \tan x - (n-2) \int \sec^n x\,dx + (n-2) \int \sec^{n-2} x\,dx.$$

Thus, if I is the original integral, we have

$$I = \sec^{n-2} x \tan x - (n-2)I + (n-2)\int \sec^{n-2} x \, dx,$$

and

$$\int \sec^n x \, dx = \frac{1}{n-1} \sec^{n-2} x \tan x + \frac{n-2}{n-1} \int \sec^{n-2} x \, dx.$$

There is a similar reduction formula which works for odd powers of the cosecant:

$$\int \csc^n x \, dx = -\frac{1}{n-1} \csc^{n-2} x \cot x + \frac{n-2}{n-1} \int \csc^{n-2} x \, dx. \tag{9}$$

To derive this, we integrate by parts, taking $u = \csc^{n-2} x$, $dv = \csc^2 x$, and proceed as in the previous derivation to solve for $\int \csc^n x \, dx$.

By these formulas and methods we can integrate products of powers of trigonometric functions. *There is absolutely no need to memorize the formulas which we have just finished developing.* You can handle simple cases by remembering the methods. If you need to compute an integral, in one of the difficult cases—and this almost never happens to people, in real life—then you look up the appropriate reduction formula. Moreover, it isn't even safe to try to memorize complicated formulas: you are very likely to misremember them and get wrong answers.

PROBLEM SET 6.5

Before starting to work on these problems, you should read Section 6.5 carefully, until you understand what the methods are and why they work. In working the problems, you should refer to the text as seldom as possible. You should try to avoid looking up even the reduction formulas (8) and (9), unless a problem requires you to apply one of them more than once. If only one reduction is required, you should integrate by parts, instead of using the reduction formula. As you will see, the first few problems below are designed to remind you of the methods that we have been using. Check by differentiation in each case.

1. $\displaystyle\int \sin^2 x \cos^3 x \, dx$ 2. $\displaystyle\int \sin^3 x \cos^2 x \, dx$ 3. $\displaystyle\int \sin^2 x \, dx$

4. $\displaystyle\int \cos^2 x \, dx$ 5. $\displaystyle\int \sin^2 x \cos^2 x \, dx$ 6. $\displaystyle\int \tan^4 x \, dx$

7. $\displaystyle\int \cot^4 x \, dx$ 8. $\displaystyle\int \tan^5 x \, dx$ 9. $\displaystyle\int \cot^5 x \, dx$

10. $\displaystyle\int \sec^4 x \, dx$ 11. $\displaystyle\int \csc^4 x \, dx$ 12. $\displaystyle\int \sec^3 x \, dx$

13. $\displaystyle\int \csc^3 x\, dx$ 14. $\displaystyle\int \sec^5 x\, dx$ 15. $\displaystyle\int \csc^5 x\, dx$

16. $\displaystyle\int \sin x \sec x\, dx$ 17. $\displaystyle\int \cos x \csc x\, dx$ 18. $\displaystyle\int \sin x \sec^3 x\, dx$

19. $\displaystyle\int \sin^2 x \sin 2x\, dx$ 20. $\displaystyle\int \cos^3 2x\, dx$ 21. $\displaystyle\int \cos^2 2x\, dx$

22. $\displaystyle\int \tan 2x \sec 2x\, dx$ 23. $\displaystyle\int \csc x \tan x\, dx$ 24. $\displaystyle\int \frac{\sin x}{\cos^2 x}\, dx$

25. $\displaystyle\int \frac{\cos x}{\sin^2 x}\, dx$ 26. $\displaystyle\int \frac{1}{\cos^4 x}\, dx$ 27. $\displaystyle\int \frac{1}{\sin^4 x}\, dx$

28. $\displaystyle\int \frac{1}{\sec x \tan x}\, dx$ 29. $\displaystyle\int \frac{1}{\csc x \cot x}\, dx$

30. There is a reduction formula of the form

$$\int \sin^n x\, dx = A \sin^{n-1} x \cos x + B \int \sin^{n-2} x\, dx,$$

where A and B are constants, expressed, of course, in terms of n. Derive such a formula. [*Hint:* It is no use trying to do this merely by the use of the elementary trigonometric identities relating the sine and the cosine.]

31. There is a reduction formula of the form

$$\int \cos^n x\, dx = A \cos^{n-1} x \sin x + B \int \cos^{n-2} x\, dx.$$

Derive it.

6.6 INTEGRATION BY SUBSTITUTION

In Section 6.2 we found that there was a close connection between certain simple integrals and some more complicated ones. For example, if we know that

$$\int x^2\, dx = \{\tfrac{1}{3}x^3 + C\},$$

then we know that

$$\int \sin^2 \theta \cos \theta\, d\theta = \{\tfrac{1}{3} \sin^3 \theta + C\}.$$

(We are using a different dummy letter in the second problem, for reasons which will soon be clear.) Thus we have two related integration problems:

$$\int x^2\, dx \quad = \quad \{\tfrac{1}{3}x^3 + C\}$$

$$\Big\downarrow_{x \,\to\, \sin \theta} \qquad\qquad \Big\downarrow_{x \,\to\, \sin \theta}$$

$$\int \sin^2 \theta \cos \theta\, d\theta = \{\tfrac{1}{3} \sin^3 \theta + C\}.$$
$$\text{\small (!)}$$

The (!) at the bottom indicates that the equation in the bottom line is the final conclusion. The pattern here is the following:

$$\int f(x)\,dx \qquad\qquad = \quad \{F(x) + C\}$$

$$\left\downarrow\; x \to u(\theta) \right. \qquad\qquad\qquad \left\downarrow\; x \to u(\theta) \right.$$

$$\int f(u)\,du = \int f\big(u(\theta)\big)u'(\theta)\,d\theta \underset{(!)}{=} \{F\big(u(\theta)\big) + C\}.$$

Thus, if we know how to find $\int f(x)\,dx$, we can use the result to find $\int f(u)\,du$. It sometimes happens, however, that we want to move in the opposite direction; sometimes we can see how to calculate

$$\int f\big(u(\theta)\big)u'(\theta)\,d\theta,$$

and we want to use the result to calculate $\int f(x)\,dx$. So as to give ourselves a simple example to work with at first, let us suppose that we know about the functions Sin and Sin^{-1}, but do not know that $1/\sqrt{1-x^2}$ is the derivative of Sin$^{-1} x$. We then consider

$$\int \frac{dx}{\sqrt{1-x^2}}\,.$$

We observe that it does not fit any form that we know. But perhaps it would be manageable if we could extract the indicated square root. For $x = \text{Sin } \theta$, the square root can be extracted. (See below.) If we replace the dummy letter x by the function Sin θ, then dx becomes Sin$'\,\theta\,d\theta$, and we get the related integrals on the left in the following diagram:

$$\int \frac{dx}{\sqrt{1-x^2}} \quad = \; ?$$

$$\left\downarrow\; x \to \text{Sin } \theta \right.$$

$$\int \frac{\cos\theta\,d\theta}{\sqrt{1-\text{Sin}^2\,\theta}} \quad = \; ?$$

The trigonometric integral is easy:

$$\int \frac{\cos\theta\,d\theta}{\sqrt{1-\text{Sin}^2\,\theta}} \;=\; \int \frac{\cos\theta\,d\theta}{\sqrt{\cos^2\theta}} \;=\; \int 1\,d\theta = \{\theta + C\}.$$

(*Query:* Why is it true that

$$\sqrt{\cos^2\theta} = \cos\theta,$$

for the values of θ that we need to consider? What values of θ *do* we need to consider?)

The above calculation enables us to complete our diagram:

$$\int \frac{dx}{\sqrt{1 - x^2}} \underset{(!)}{=} \{\text{Sin}^{-1} x + C\}$$

$$\downarrow x \to \text{Sin } \theta \qquad \uparrow \theta \to \text{Sin}^{-1} x$$

$$\int \frac{\cos \theta \, d\theta}{\sqrt{1 - \text{Sin}^2 \theta}} = \{\theta + C\}$$

In this case, of course, the solution in the top line was known before we started. But the same scheme works in general, whenever we can calculate the new integral on the lower left:

$$\int f(x) \, dx \underset{(!)}{=} \{G(u^{-1}(x)) + C\}$$

$$\downarrow x \to u(\theta) \qquad \uparrow \theta \to u^{-1}(x)$$

$$\int f(u) \, du = \int f(u(\theta))u'(\theta) \, d\theta = \{G(\theta) + C\}$$

We shall prove, at the end of this section, that this procedure is valid, whenever the symbols u' and u^{-1} have a meaning; that is, whenever u has both a derivative and an inverse. Meanwhile, we shall show how the scheme is used to solve problems which would otherwise be hard.

Example 1.

$$\int \frac{dx}{x^2\sqrt{1 - x^2}} = ? \qquad (-1 < x < 1, \, x \neq 0).$$

As in the preceding case, it seems to be the radical that is causing the trouble; and so we get rid of it by the substitution

$$x \to \text{Sin } \theta \qquad \left(-\frac{\pi}{2} < \theta < \frac{\pi}{2}\right),$$

$$dx \to \text{Sin}' \, \theta \, d\theta = \cos \theta \, d\theta.$$

This gives

$$\int \frac{dx}{x^2\sqrt{1 - x^2}} \to \int \frac{\cos \theta \, d\theta}{\text{Sin}^2 \theta \cos \theta} = \int \csc^2 \theta \, d\theta = \{-\cot \theta + C\}.$$

(Throughout, $-\pi/2 < \theta < \pi/2$; on this interval, $\sin x = \text{Sin } x$, and the usual identities hold automatically.)

We now reverse the substitution, using $\theta \to \text{Sin}^{-1} x$. This gives

$$\int \frac{dx}{x^2\sqrt{1 - x^2}} = \{-\cot \text{Sin}^{-1} x + C\}.$$

A formula for $\cot \text{Sin}^{-1} x$ is easy to read off from a figure.

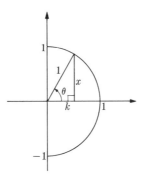

Here

$$-\frac{\pi}{2} \leqq \theta \leqq \frac{\pi}{2},$$

$$\text{Sin } \theta = x, \qquad \theta = \text{Sin}^{-1} x;$$

$$\cot \text{Sin}^{-1} x = \cot \theta = \frac{k}{x} = \frac{\sqrt{1 - x^2}}{x}.$$

Therefore

$$\int \frac{dx}{x^2\sqrt{1 - x^2}} = \left\{-\frac{\sqrt{1 - x^2}}{x} + C\right\}.$$

Note that all the trigonometry has cancelled out of the problem. Our answer checks:

$$D\frac{\sqrt{1 - x^2}}{x} = \frac{1}{x^2}\left(x \cdot \frac{-x}{\sqrt{1 - x^2}} - \sqrt{1 - x^2}\right)$$

$$= \frac{1}{x^2\sqrt{1 - x^2}}[-x^2 - (\sqrt{1 - x^2})^2] = \frac{-1}{x^2\sqrt{1 - x^2}}.$$

We can sum this up in a diagram as follows:

$$\int \frac{dx}{x^2\sqrt{1 - x^2}} \, (\qquad = \qquad \left\{-\frac{\sqrt{1 - x^2}}{x} + C\right\}$$

$$\downarrow x \to \text{Sin } \theta \qquad\qquad \uparrow \theta \to \text{Sin}^{-1} x$$

$$\int \csc^2 \theta \, d\theta \quad = \quad \{-\cot \theta + C\}$$

The substitution $x \to \text{Sin } \theta$ is the usual one to try, if the troublesome part of the integrand is $\sqrt{1 - x^2}$. In other cases, $x \to \text{Tan } \theta$ works in much the same way.

Example 2. Consider

$$\int \sqrt{1 + x^2} \, dx = \; ?$$

To get rid of the radical, we use

$$x \to \text{Tan } \theta \quad \left(-\frac{\pi}{2} < \theta < \frac{\pi}{2}\right), \quad dx \to \sec^2 \theta \, d\theta.$$

This gives

$$\int \sqrt{1 + x^2} \, dx \to \int \sqrt{1 + \text{Tan}^2 \theta} \, \sec^2 \theta \, d\theta.$$

The domain of Tan is the interval $(-\pi/2, \pi/2)$, on which $\sec \theta > 0$. Therefore $\sec \theta = \sqrt{1 + \text{Tan}^2 \theta}$, and

$$\int \sqrt{1 + \text{Tan}^2 \theta} \, \sec^2 \theta \, d\theta = \int \sec^3 \theta \, d\theta.$$

We now use one of the reduction formulas of Section 6.5:

$$\int \sec^3 \theta \, d\theta = \int \sec^n \theta \, d\theta \qquad\qquad (n = 3)$$

$$= \frac{1}{n-1} \sec^{n-2} \theta \tan \theta + \frac{n-2}{n-1} \int \sec^{n-2} \theta \, d\theta$$

$$= \tfrac{1}{2} \sec \theta \tan \theta + \tfrac{1}{2} \int \sec \theta \, d\theta$$

$$= \{\tfrac{1}{2} \sec \theta \tan \theta + \tfrac{1}{2} \ln |\sec \theta + \tan \theta| + C\}$$

$$= \{G(\theta) + C\}.$$

We complete the solution by letting $\theta \to \text{Tan}^{-1} x$. This gives

$$\int \sqrt{1 + x^2} \, dx = \{G(\text{Tan}^{-1} x) + C\}$$

$$= \{\tfrac{1}{2}(\sec \text{Tan}^{-1} x)(\tan \text{Tan}^{-1} x)$$

$$+ \tfrac{1}{2} \ln |\sec \text{Tan}^{-1} x + \tan \text{Tan}^{-1} x| + C\}.$$

Obviously $\tan \text{Tan}^{-1} x = x$. The formula for $\sec \text{Tan}^{-1} x$ can be read off from a figure.

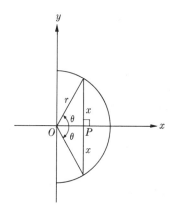

In the figure, $-\pi/2 < \theta < \pi/2$, but θ may be positive or negative. We take $OP = 1$. This gives

$$x = \text{Tan } \theta, \qquad \theta = \text{Tan}^{-1} x, \qquad r = \sec \theta = \sec \text{Tan}^{-1} x,$$

so that

$$\sec \text{Tan}^{-1} x = \sqrt{1 + x^2}.$$

Therefore the answer is

$$\int \sqrt{1 + x^2} \, dx = \{\tfrac{1}{2}x\sqrt{1 + x^2} + \tfrac{1}{2} \ln |\sqrt{1 + x^2} + x| + C\}.$$

This can be simplified slightly: since $\sqrt{1 + x^2} + x > 0$ for every x, we can omit the absolute value bars, getting:

$$\int \sqrt{1 + x^2} \, dx = \{\tfrac{1}{2}x\sqrt{1 + x^2} + \tfrac{1}{2} \ln (\sqrt{1 + x^2} + x) + C\}.$$

As before, we sum up in a diagram the process by which the problem was solved:

$$\int \sqrt{1 + x^2} \, dx \underset{(!)}{=} \{\tfrac{1}{2}x\sqrt{1 + x^2} + \tfrac{1}{2} \ln (\sqrt{1 + x^2} + x) + C\}$$

$$\downarrow x \to \text{Tan } \theta \qquad\qquad\qquad \uparrow \theta \to \text{Tan}^{-1} x$$

$$\int \sec^3 \theta \, d\theta \quad = \{\tfrac{1}{2} \sec \theta \tan \theta + \tfrac{1}{2} \ln |\sec \theta + \tan \theta| + C\}$$

Such diagrams are worth drawing, especially the first few times you use the substitution process; often the calculations are long, and it is easy to lose track of what the process means.

The answer in Example 2 suggests that no method would have made the problem seem easy. Note that the formulas of Section 6.5 are turning out to be useful in solving problems which do not appear, at first, to involve trigonometry at all.

We return to the general theory, to see why this method works. The pattern of our work is described by the diagram:

$$\int f(x) \, dx \quad \underset{(!)}{=} \{G(u^{-1}(x)) + C\}$$

$$\downarrow x \to u(\theta) \qquad\qquad \uparrow \theta \to u^{-1}(x)$$

$$\int f(u(\theta))u'(\theta) \, d\theta = \quad \{G(\theta) + C\}$$

What we are claiming, when we use the method of substitution, is that, if the second equation holds, so does the first. In terms of the definition of the indefinite integral, this means the following:

Theorem 1. If u is differentiable and invertible, then
$$G' = f(u)u' \implies D[G(u^{-1})] = f.$$
The proof is as follows. By the chain rule,
$$D[G(u^{-1})] = G'(u^{-1})\, Du^{-1}.$$
By hypothesis, $G' = f(u)u'$. Therefore
$$G'(x) = f(u(x))u'(x)$$
for every x. Therefore
$$G'(u^{-1}) = f(u(u^{-1}))u'(u^{-1}),$$
and
$$D[G(u^{-1})] = G'(u^{-1})\, Du^{-1} = f(u(u^{-1}))u'(u^{-1})\, Du^{-1}.$$
Now
$$f(u(u^{-1})) = f,$$
because $u(u^{-1}(x)) = x$ for every x. Therefore
$$f(u(u^{-1}))u'(u^{-1})\, Du^{-1} = f \cdot u'(u^{-1})\, Du^{-1} = f \cdot u'(u^{-1}) \cdot \frac{1}{u'(u^{-1})},$$
by the general formula for the derivative of the inverse of a function. Now $u'(u^{-1})$ cancels, and gives us $D[G(u^{-1})] = f$, which was to be proved.

PROBLEM SET 6.6

Calculate each of the following integrals, by any method. In most cases, but not all, the easiest method is to use a substitution of the form $x \to \operatorname{Sin} \theta$, $x \to \operatorname{Tan} \theta$, or $x \to \operatorname{Sec} \theta$. In each case where you do use the method of substitution, you should sum up the process of solution in a diagram as in Examples 1 and 2 in the text. Finally, check in each case by differentiation.

1. $\displaystyle\int (1 - x^2)^{-3/2}\, dx$

2. $\displaystyle\int \frac{dx}{\sqrt{x^2 + 1}}$

3. $\displaystyle\int \frac{dx}{\sqrt{x^2 - 1}}$

4. $\displaystyle\int \frac{dx}{x(1 + x^2)}\, dx$

5. $\displaystyle\int \frac{dx}{x\sqrt{x^2 - 1}}\, dx$

6. $\displaystyle\int x(1 - x^2)^{-3/2}\, dx$

7. $\displaystyle\int \frac{dx}{x^2(1 + x^2)}$

8. $\displaystyle\int \frac{x\, dx}{1 + x^2}$

9. $\displaystyle\int \sqrt{1 - x^2}\, dx$

10. $\displaystyle\int \frac{dx}{x^2\sqrt{x^2 - 1}}$

11. $\displaystyle\int \frac{x\, dx}{\sqrt{x^2 - 1}}$

12. $\displaystyle\int x^2\sqrt{1 - x^2}\, dx$

13. $\displaystyle\int \frac{x^2\, dx}{\sqrt{1 - x^2}}$

14. $\displaystyle\int \frac{x^3\, dx}{\sqrt{1 - x^2}}$

15. $\displaystyle\int \frac{x^2\, dx}{1 + x^2}$

16. $\displaystyle\int x^2\sqrt{1 + x^2}\, dx$

17. $\displaystyle\int x\sqrt{1 - x^2}\, dx$

18. $\displaystyle\int \frac{x}{(1 + x^2)^3}\, dx$

19. $\displaystyle\int x\sqrt{1 + x^2}\, dx$

20. $\displaystyle\int x^3\sqrt{1 + x^2}\, dx$

21. $\displaystyle\int \frac{x^3}{1 + x^2}\, dx$

22. $\displaystyle\int x^2\sqrt{x^2 - 1}\, dx$

23. a) Show that

$$\int_a^b f(u(\theta))u'(\theta)\, d\theta = \int_{u(a)}^{u(b)} f(x)\, dx,$$

   whether u is invertible or not.

   b) Show that if u is invertible, then

$$\int_c^d f(x)\, dx = \int_{u^{-1}(c)}^{u^{-1}(d)} f(u(\theta))u'(\theta)\, d\theta.$$

24. Obviously there is no point in writing this on a paper which is to be turned in and graded; but for your own benefit, reproduce the proof of the following, without reference to the text:

$$G' = f(u)u' \Rightarrow D[G(u^{-1})] = f.$$

6.7 ALGEBRAIC SUBSTITUTIONS

It is a good rule, if you have a problem which you don't see how to solve, to try to think of an easier problem that resembles it. If you can solve the easier problem, and bridge the gap between the two, then you have solved the problem which you started with. For example, consider

$$\int \frac{x^2}{\sqrt{2x+1}}\, dx.$$

This does not fit any of the standard forms that we know about. There is no reason to suppose that a trigonometric substitution would help; and in fact, none of them would. We note, however, that if the denominator were of the form $\sqrt{x}$, or $\sqrt{t}$, then the problem would become easier. Now

$$2x + 1 = t \Leftrightarrow x = \tfrac{1}{2}(t - 1).$$

We therefore try the substitution

$$\boxed{x \to u(t) = \tfrac{1}{2}(t - 1),}$$

$$dx \to u'(t)\, dt = \tfrac{1}{2}\, dt.$$

Under this substitution,

$$\int \frac{x^2\, dx}{\sqrt{2x+1}} \quad\to\quad \int \frac{\tfrac{1}{4}(t-1)^2 \cdot \tfrac{1}{2}\, dt}{\sqrt{t}}.$$

The latter integral is easy to calculate. It is

$$\frac{1}{8} \int \frac{t^2 - 2t + 1}{\sqrt{t}}\, dt = \tfrac{1}{8} \cdot \int (t^{3/2} - 2t^{1/2} + t^{-1/2})\, dt$$

$$= \{\tfrac{1}{8}(\tfrac{2}{5}t^{5/2} - 2 \cdot \tfrac{2}{3}t^{3/2} + 2t^{1/2}) + C\}$$

$$= \{\tfrac{1}{20}t^{5/2} - \tfrac{1}{6}t^{3/2} + \tfrac{1}{4}t^{1/2} + C\}$$

$$= \{G(t) + C\}.$$

To get the answer to the problem which we started with, we use the inverse substitution

$$t \to u^{-1}(x) = 2x + 1.$$

This gives

$$\int \frac{x^2\, dx}{\sqrt{2x+1}} = \{\tfrac{1}{20}(2x+1)^{5/2} - \tfrac{1}{6}(2x+1)^{3/2} + \tfrac{1}{4}(2x+1)^{1/2} + C\}$$

$$= \{G(u^{-1}(x)) + C\}.$$

The scheme here is the same as in the preceding section:

$$\int \frac{x^2}{\sqrt{2x+1}}\, dx \underset{(!)}{=} \{G(u^{-1}(x)) + C\}$$

$$\downarrow{\scriptstyle x\to u(t)} \qquad\qquad \uparrow{\scriptstyle t\to u^{-1}(x)}$$

$$\int \frac{t^2 - 2t + 1}{8\sqrt{t}}\, dt = \quad \{G(t) + C\}$$

The only differences are that (a) the functions u and u^{-1} are described algebraically, and (b) the formulas for $G(t)$ and $G(u^{-1}(x))$ are too long to be conveniently written in the diagram. In any case, we know that the method works: this follows from Theorem 1 of Section 6.6.

Often we can tell that a substitution is going to work, long before we know what the answer is. As soon as we wrote

$$\int \frac{x^2\, dx}{\sqrt{2x+1}} \quad\to\quad \int \frac{\tfrac{1}{4}(t-1)^2 \cdot \tfrac{1}{2}\, dt}{\sqrt{t}},$$

it was evident that the numerator was a polynomial. We can integrate the quotient of a polynomial and a power. Similarly, we know that we can integrate

$$\int \frac{(x^3 - 3x + 4)^2}{x^{2/3}}\, dx;$$

the calculation will be tedious, but the outcome is not in doubt.

If one algebraic substitution works, there are usually others that also work. In the preceding problem, we might have used

$$y = \sqrt{2x+1}, \qquad y^2 = 2x + 1, \qquad x = \tfrac{1}{2}(y^2 - 1).$$

Thus we use

$$x \to u(y) = \tfrac{1}{2}(y^2 - 1),$$

$$dx \to u'(y)\, dy = y\, dy,$$

$$x^2 \to \tfrac{1}{4}(y^2 - 1)^2, \qquad \sqrt{2x+1} \to y,$$

$$\int \frac{x^2\, dx}{\sqrt{2x+1}} \quad\to\quad \int \frac{\tfrac{1}{4}(y^2-1)^2 \cdot y\, dy}{y} = \tfrac{1}{4}\int (y^4 - 2y^2 + 1)\, dy$$

$$= \{\tfrac{1}{20}y^5 - \tfrac{1}{6}y^3 + \tfrac{1}{4}y + C\}.$$

As usual, we now reverse the substitution, using

$$y \to u^{-1}(x) = \sqrt{2x + 1}.$$

This gives the final answer

$$\{\tfrac{1}{20}(2x + 1)^{5/2} - \tfrac{1}{6}(2x + 1)^{3/2} + \tfrac{1}{4}(2x + 1)^{1/2} + C\},$$

as before.

There are no rules which tell us the best substitution to try in every case. The best approach is to look at the integrand, ask ourselves what feature of it is most troublesome, and then choose a substitution which seems likely to remove the troublesome feature. For example, if we want to calculate

$$\int \frac{dx}{\sqrt{x^2 + 1}},$$

we want to extract the square root; we can do this if

$$x \to \operatorname{Tan} \theta \quad \left(-\frac{\pi}{2} < \theta < \frac{\pi}{2}\right),$$

$$x^2 + 1 \to \sec^2 \theta.$$

This works:

$$\int \frac{dx}{\sqrt{x^2 + 1}} \quad \to \quad \int \frac{\sec^2 \theta \, d\theta}{\sec \theta} = \int \sec \theta \, d\theta,$$

which leads to a solution, as you found in Problem 2 of Problem Set 6.6.

We might also have tried

$$z = \sqrt{x^2 + 1}, \qquad z^2 = x^2 + 1,$$

$$x \to u(z) = \sqrt{z^2 - 1}$$

$$dx = \frac{z}{\sqrt{z^2 - 1}} \, dz.$$

This gives

$$\int \frac{dx}{\sqrt{x^2 + 1}} \quad \to \quad \int \frac{1}{z} \cdot \frac{z}{\sqrt{z^2 - 1}} \, dz = \int \frac{dz}{\sqrt{z^2 - 1}},$$

which gets us nowhere, unless we happen to remember the solution of Problem 3 of Problem Set 6.6.

Usually, to find out what algebraic substitution is going to work, we need to solve an algebraic equation. For example, given

$$\int \frac{dz}{1 + \sqrt{z}},$$

we wish that the denominator had merely the form t. To get

$$1 + \sqrt{z} \rightarrow t,$$

we need

$$\sqrt{z} \rightarrow t - 1, \qquad z \rightarrow (t - 1)^2.$$

We usually write this with "=" signs:

$$1 + \sqrt{z} = t, \qquad \sqrt{z} = t - 1, \qquad z = (t - 1)^2.$$

This gives

$$z \rightarrow u(t) = (t - 1)^2,$$

$$dz \rightarrow u'(t)\, dt = 2(t - 1)\, dt,$$

$$\int \frac{dz}{1 + \sqrt{z}} \quad \rightarrow \quad \int \frac{2(t - 1)\, dt}{t} = \int \left(2 - \frac{2}{t}\right) dt$$

$$= \{2t - 2 \ln |t| + C\}.$$

The reverse substitution

$$t \rightarrow u^{-1}(z) = 1 + \sqrt{z}$$

gives the final answer

$$\{2(1 + \sqrt{z}) - 2 \ln (1 + \sqrt{z}) + C\}.$$

(*Query:* Would it be all right to delete the "1" in the first parenthesis?)

   This is probably the most efficient solution. If we hadn't thought of it, we might have tried

$$\sqrt{z} = t, \qquad z = t^2,$$

which gives the substitution

$$z \rightarrow u(t) = t^2,$$

$$dz \rightarrow u'(t)\, dt = 2t\, dt,$$

$$\int \frac{dz}{1 + \sqrt{z}} \quad \rightarrow \quad \int \frac{2t\, dt}{1 + t}.$$

Dividing $1 + t$ into t, we get

$$\frac{t}{1 + t} = 1 - \frac{1}{1 + t}.$$

Therefore

$$2 \int \frac{t\, dt}{1 + t} = 2 \int \left(1 - \frac{1}{1 + t}\right) dt = \{2t - 2 \ln |1 + t| + C\}.$$

Finally we apply the inverse substitution

$$t \to u^{-1}(z) = \sqrt{z},$$

getting

$$\int \frac{dz}{1 + \sqrt{z}} = \{2\sqrt{z} - 2 \ln (1 + \sqrt{z}) + C\}.$$

(Is this really the same as the previous answer? Why or why not?)

We have used the substitutions $x \to \mathrm{Sin}\ \theta$; $x \to \mathrm{Tan}\ \theta$, and $x \to \mathrm{Sec}\ \theta$ to handle integrands involving the radicals

$$\sqrt{1 - x^2}, \qquad \sqrt{1 + x^2}, \qquad \text{and} \qquad \sqrt{x^2 - 1}.$$

Slight variations enable us to take care of more general cases, involving

$$\sqrt{a^2 - x^2}, \qquad \sqrt{a^2 + x^2}, \qquad \sqrt{x^2 - a^2}.$$

For example, to find

$$\int \sqrt{a^2 - x^2}\, dx \qquad (a > 0),$$

we use $x \to a\ \mathrm{Sin}\ \theta = u(\theta)$, so that

$$a^2 - x^2 = a^2(1 - \mathrm{Sin}^2\ \theta), \qquad \sqrt{a^2 - x^2} = a \cos \theta.$$

Here $x/a = \mathrm{Sin}\ \theta$, so that

$$\theta = \mathrm{Sin}^{-1} \frac{x}{a}.$$

Thus

$$\int \sqrt{a^2 - x^2}\, dx \quad \to \quad \int (a \cos \theta) a \cos \theta\, d\theta = a^2 \int \cos^2 \theta\, d\theta$$

$$= a^2 \int \tfrac{1}{2}(\cos 2\theta + 1)\, d\theta$$

$$= \left\{ \frac{a^2}{4} \sin 2\theta + \frac{a^2}{2} \theta + C \right\}.$$

Now

$$\sin 2\theta = 2 \sin \theta \cos \theta = 2 \cdot \frac{x}{a} \cdot \frac{1}{a} \sqrt{a^2 - x^2}.$$

This gives

$$\int \sqrt{a^2 - x^2}\, dx = \left\{ \tfrac{1}{2}x\sqrt{a^2 - x^2} + \frac{a^2}{2} \mathrm{Sin}^{-1} \frac{x}{a} + C \right\}.$$

In the same way, we use

$$x \to a\ \mathrm{Tan}\ \theta, \qquad \theta \to \mathrm{Tan}^{-1} \frac{x}{a}$$

to get rid of $\sqrt{a^2 + x^2}$; and we use

$$x \to a \operatorname{Sec} \theta, \qquad \theta \to \operatorname{Sec}^{-1} \frac{x}{a}$$

to get rid of $\sqrt{x^2 - a^2}$.

There are miscellaneous substitutions which work on miscellaneous problems. For example, in

$$\int \frac{dx}{x^2(x^2 + 1)},$$

the trouble seems to be that the integrand is concentrated in its own denominator. We ought to be able to correct this by letting

$$x \to u(t) = \frac{1}{t},$$

$$dx \to \frac{-1}{t^2} dt.$$

This gives

$$\int \frac{dx}{x^2(x^2 + 1)} \quad \to \quad \int \left(-1 + \frac{1}{1 + t^2} \right) dt = \{ -t + \operatorname{Tan}^{-1} t + C \}$$

$$\to \quad \left\{ -\frac{1}{x} + \operatorname{Tan}^{-1} \frac{1}{x} + C \right\}.$$

Here, in the last step, we have applied the inverse substitution

$$t \to u^{-1}(x) = \frac{1}{x}.$$

In writing up solutions of problems, in the following problem set, you need not draw diagrams of the form:

$$\int f(x)\, dx \quad \underset{(!)}{=} \quad \{ G(u^{-1}(x)) + C \}$$

$$\Big\downarrow_{x \to u(t)} \qquad\qquad \Big\uparrow_{t \to u^{-1}(x)}$$

$$\int f(u(t)) u'(t)\, dt = \quad \{ G(t) + C \}$$

But whenever you use a substitution, you should explain what you are doing, by writing formulas of the type

$$x \to u(t) = \cdots, \qquad t \to u^{-1}(x) = \cdots$$

PROBLEM SET 6.7

Calculate the following, by any method.

1. $\displaystyle\int \frac{dx}{(1 + \sqrt{x})^3}$ 2. $\displaystyle\int \frac{\sqrt{x}}{(1 + \sqrt{x})^3}\, dx$ 3. $\displaystyle\int (a^2 - x^2)^{-3/2}\, dx$

4. $\displaystyle\int (a^2 + x^2)^{-3/2}\, dx$ 5. $\displaystyle\int \frac{dx}{\sqrt{1 + e^x}}$ 6. $\displaystyle\int \frac{dx}{\sqrt{1 - e^x}}$

7. $\displaystyle\int \frac{dx}{1 + \sqrt[3]{x}}$ 8. $\displaystyle\int \frac{dx}{(1 - \sqrt[3]{x})^2}$ 9. $\displaystyle\int \frac{dx}{\sqrt{\sqrt{x} + 1}}$

10. $\displaystyle\int \frac{z^3\, dz}{\sqrt{z + 1}}$ 11. $\displaystyle\int \frac{z^3\, dz}{\sqrt{z^2 - 1}}$ 12. $\displaystyle\int \frac{dx}{\sqrt{1 + \sqrt[3]{x}}}$

13. $\displaystyle\int \frac{dx}{\sqrt[3]{1 + \sqrt{x}}}$ 14. $\displaystyle\int \frac{dx}{x^4(x - 1)}$ 15. $\displaystyle\int (1 - x^2)^4\, dx$

16. $\displaystyle\int (1 + \sqrt{x})^3 \sqrt{x}\, dx$ 17. $\displaystyle\int \frac{dx}{\sqrt{1 + e^{2x}}}$ 18. $\displaystyle\int \frac{dx}{(1 + e^x)^4}$

19. $\displaystyle\int \frac{dx}{\sqrt{1 + e^{2x}}}$ 20. $\displaystyle\int \mathrm{Sin}^{-1} x\, dx$ 21. $\displaystyle\int x \ln x\, dx$

22. $\displaystyle\int x\, \mathrm{Sin}^{-1} x\, dx$ 23. $\displaystyle\int \mathrm{Tan}^{-1} x\, dx$ 24. $\displaystyle\int x\, \mathrm{Tan}^{-1} x\, dx$

25. $\displaystyle\int \frac{1}{(1 + \sqrt{x})^2}\, dx$ 26. $\displaystyle\int \frac{1}{\sqrt{x}(1 + \sqrt{x})^2}\, dx$ 27. $\displaystyle\int \frac{1}{1 + \sqrt[4]{x}}\, dx$

28. $\displaystyle\int \frac{1}{\sqrt{1 + \sqrt[3]{x}}}\, dx$ 29. $\displaystyle\int \frac{1}{(1 + e^x)^2}\, dx$ 30. $\displaystyle\int \frac{dx}{\sqrt{1 + e^{3x}}}$

31. $\displaystyle\int \frac{dx}{x^3(1 - x)}$ 32. $\displaystyle\int x^2 \ln x\, dx$ 33. $\displaystyle\int x^2\, \mathrm{Tan}^{-1} x\, dx$

6.8 ALGEBRAIC DEVICES: COMPLETING
THE SQUARE AND PARTIAL FRACTIONS

In Section 6.6 we used trigonometric substitutions to calculate integrals involving $\sqrt{a^2 \pm x^2}$ and $\sqrt{x^2 - a^2}$. By completing the square, we can extend these methods so as to take care of expressions of the form $\sqrt{ax^2 + bx + c}$. For example,

$$x^2 + x + 1 = x^2 + x + \tfrac{1}{4} + \tfrac{3}{4} = (x + \tfrac{1}{2})^2 + \left(\frac{\sqrt{3}}{2}\right)^2.$$

Therefore

$$\int \frac{dx}{\sqrt{x^2 + x + 1}} = \int \frac{dx}{\sqrt{(x + \tfrac{1}{2})^2 + (\sqrt{3}/2)^2}},$$

which has the form

$$\int \frac{du}{\sqrt{u^2 + a^2}}.$$

We can calculate this by the substitution

$$u \to a \, \text{Tan} \, \theta.$$

This gives

$$\{\ln |\sqrt{u^2 + a^2} + u| + C\} = \{\ln |\sqrt{x^2 + x + 1} + x + \tfrac{1}{2}| + C\}.$$

Similarly,

$$x^2 + x = x^2 + x + \tfrac{1}{4} - \tfrac{1}{4} = (x + \tfrac{1}{2})^2 - (\tfrac{1}{2})^2,$$

and so

$$\int \frac{dx}{\sqrt{x^2 + x}} = \int \frac{dx}{\sqrt{(x + \tfrac{1}{2})^2 - (\tfrac{1}{2})^2}},$$

which has the form

$$\int \frac{du}{\sqrt{u^2 - a^2}}.$$

Here we would use

$$u \to a \, \text{Sec} \, \theta,$$

and proceed as in Section 6.6.

The following simple-looking problem has a curious solution:

$$\int \frac{dx}{x^2 - 1} = ?$$

We try

$$x \to \text{Sec} \, \theta \qquad (x > 1)$$

so that

$$dx \to \sec \theta \tan \theta \, d\theta,$$

giving

$$\int \frac{\sec \theta \tan \theta \, d\theta}{\tan^2 \theta} = \int \csc \theta \, d\theta = \{-\ln |\csc \theta + \cot \theta| + C\}$$

$$\to \left\{ -\ln \left(\frac{x}{\sqrt{x^2 - 1}} + \frac{1}{\sqrt{x^2 - 1}} \right) + C \right\}.$$

Now

$$\frac{x}{\sqrt{x^2 - 1}} + \frac{1}{\sqrt{x^2 - 1}} = \frac{x + 1}{\sqrt{(x + 1)(x - 1)}} = \sqrt{\frac{x + 1}{x - 1}} \qquad (x > 1).$$

This gives the answer

$$\int \frac{dx}{x^2 - 1} = \left\{ \tfrac{1}{2} \ln \left| \frac{x - 1}{x + 1} \right| + C \right\} \qquad (x > 1).$$

We check by differentiation:

$$D\left(\tfrac{1}{2} \ln \left| \frac{x - 1}{x + 1} \right| \right) = D(\tfrac{1}{2} \ln |x - 1| - \tfrac{1}{2} \ln |x + 1|)$$

$$= \frac{1}{2} \cdot \frac{1}{x - 1} - \frac{1}{2} \cdot \frac{1}{x + 1} = \frac{1}{2} \cdot \frac{(x + 1) - (x - 1)}{(x - 1)(x + 1)}$$

$$= \frac{1}{x^2 - 1}.$$

This shows that our answer was right. But it also shows that our use of trigonometry was unnecessary; the solution depends merely on the algebraic identity

$$\frac{1}{x^2 - 1} = \frac{\frac{1}{2}}{x - 1} + \frac{-\frac{1}{2}}{x + 1}.$$

This suggests that we should have a systematic method of breaking up rational functions into sums of simpler functions. We call this the method of *partial fractions*.

Theorem 1. If $a \neq b$, then there are numbers A and B such that

$$\frac{cx + d}{(x - a)(x - b)} = \frac{A}{x - a} + \frac{B}{x - b} \qquad (x \neq a, b).$$

Proof. The obvious method works:

$$\frac{cx + d}{(x - a)(x - b)} = \frac{A}{x - a} + \frac{B}{x - b}$$

$$\Leftrightarrow \quad cx + d = A(x - b) + B(x - a)$$

$$\Leftrightarrow \quad \begin{cases} A + B = c & \text{and} \\ Ab + Ba = -d. \end{cases}$$

We solve for A and B by any method, getting the solution

$$A = \frac{ac + d}{a - b}, \qquad B = \frac{bc + d}{b - a}.$$

These values satisfy both equations.

Note that since $a - b$ appears in the denominators, we really needed the hypothesis $a \neq b$. And for $a = b$, the theorem is false. That is, you cannot express $1/(x - a)^2$ in the form

$$\frac{A}{x - a} + \frac{B}{x - a} = \frac{A + B}{x - a}.$$

It might seem that we should have stated a stronger theorem, as follows:

Theorem 1. If $a \neq b$, then

$$\frac{cx + d}{(x - a)(x - b)} = \frac{ac + d}{a - b} \cdot \frac{1}{x - a} + \frac{bc + d}{b - a} \cdot \frac{1}{x - b}.$$

But nobody could remember this formula. The efficient way to handle such problems is the following. Given

$$\int \frac{dx}{(x - 2)(x - 5)} = ?$$

we know by Theorem 1 that there are numbers A and B such that

$$\frac{1}{(x - 2)(x - 5)} = \frac{A}{x - 2} + \frac{B}{x - 5}.$$

The only problem is to find out what they are, numerically. We first write

$$1 = A(x - 5) + B(x - 2).$$

Since this equation holds for every x, it must hold for $x = 2$ and for $x = 5$. Therefore

$$1 = A \cdot (-3), \qquad 1 = B \cdot 3;$$
$$A = -\tfrac{1}{3}, \qquad B = \tfrac{1}{3}.$$

This is another example of the efficiency of existence theorems: often, if you know in advance that a problem has a solution, you can use a simple procedure to find out what the answer is. Without Theorem 1, the shortcut calculation of

$$\frac{1}{(x - 2)(x - 5)} = -\frac{1}{3} \cdot \frac{1}{x - 2} + \frac{1}{3} \cdot \frac{1}{x - 5}$$

would not have been valid. To see this, consider the following analogous procedure:

"Problem." Find the numbers a and b such that

$$\sin x = ax + b.$$

"*Solution.*" Letting $x = 0$, we get

$$0 = a \cdot 0 + b.$$

Therefore $b = 0$, and

$$\sin x = ax.$$

Letting $x = \pi/2$, we get

$$1 = a \cdot \frac{\pi}{2},$$

and $a = 2/\pi$. Therefore

$$(?) \quad \sin x = \frac{2x}{\pi} \qquad \text{for every } x \quad (?)$$

This is wrong: in fact, our formula is correct for only three values of x, namely, $x = 0$ and $x = \pm\pi/2$. The fallacy was in assuming at the outset that the problem had a solution, when in fact it has none. What the above line of reasoning really proves is the following:

$$\sin x \text{ is a linear function} \tag{1}$$
$$\Rightarrow \quad \sin x \text{ is the linear function } 2x/\pi. \tag{2}$$

The statement $(1) \Rightarrow (2)$ is true, but it is not useful, because (1) is false.

The method that we used for quadratic denominators also works whenever the denominator can be factored into linear factors all of which are different.

$$\frac{2x^2 + 1}{(x + 1)(x + 2)(x + 3)} = \frac{A}{x + 1} + \frac{B}{x + 2} + \frac{C}{x + 3},$$
$$2x^2 + 1 = A(x + 2)(x + 3) + B(x + 1)(x + 3) + C(x + 1)(x + 2),$$
$$x = -1, \qquad 3 = 2A, \qquad A = \tfrac{3}{2};$$
$$x = -2, \qquad 9 = -B, \qquad B = -9;$$
$$x = -3, \qquad 19 = 2C, \qquad C = \tfrac{19}{2}.$$

This solution depends on the following existence theorem.

Theorem 2. If $p(x)$ is of degree ≤ 2, and a, b, and c are all different, then there are numbers A, B, and C such that

$$\frac{p(x)}{(x - a)(x - b)(x - c)} = \frac{A}{x - a} + \frac{B}{x - b} + \frac{C}{x - c}. \tag{3}$$

With our present equipment, we could give only a brute-force proof. But we know how to handle simple cases, and in the following problem set you will see how various more difficult problems of the same type can be solved.

PROBLEM SET 6.8

Find:

1. $\displaystyle\int \frac{dx}{x^2 + 2x + 5}$ 2. $\displaystyle\int \frac{dx}{\sqrt{x^2 + 2x + 5}}$ 3. $\displaystyle\int \frac{dx}{x^2 - 4}$

4. $\displaystyle\int \frac{dx}{x^2 + x - 4}$ 5. $\displaystyle\int \frac{dx}{\sqrt{x^2 + x - 4}}$ 6. $\displaystyle\int \frac{dx}{\sqrt{2 - x^2}}$

7. $\displaystyle\int \frac{dx}{\sqrt{2 - 2x - x^2}}$

8. $\displaystyle\int \frac{dx}{\sqrt{-2x - x^2}}$ (Is this an impossible problem at the outset?)

9. $\displaystyle\int \frac{dx}{x^2 + 6x + 10}$ 10. $\displaystyle\int \frac{dx}{\sqrt{-x^2 - 6x + 10}}$ 11. $\displaystyle\int \frac{dx}{x^2 - 6x + 10}$

12. $\displaystyle\int \frac{dx}{(x - 1)(x - 2)}$ 13. $\displaystyle\int \frac{x\,dx}{(x - 1)(x - 2)}$ 14. $\displaystyle\int \frac{dx}{x(x - 1)(x - 2)}$

15. $\displaystyle\int \frac{x\,dx}{x(x - 1)(x - 2)}$

Find the unknown coefficients A, B, C, ... which satisfy the following equations:

16. $\displaystyle\frac{1}{(x - 1)^2(x - 2)} = \frac{A}{(x - 1)^2} + \frac{B}{x - 1} + \frac{C}{x - 2}$ 17. Find $\displaystyle\int \frac{dx}{(x - 1)^2(x - 2)}$

18. $\displaystyle\frac{1}{x^2(x - 1)^2} = \frac{A}{x^2} + \frac{B}{x} + \frac{C}{(x - 1)^2} + \frac{D}{x - 1}$ 19. Find $\displaystyle\int \frac{dx}{x^2(x - 1)^2}$

20. $\displaystyle\frac{1}{(x + 1)(x - 2)^3} = \frac{A}{x + 1} + \frac{B}{(x - 2)^3} + \frac{C}{(x - 2)^2} + \frac{D}{(x - 2)}$

21. Find $\displaystyle\int \frac{dx}{(x + 1)(x - 2)^3}$ 22. $\displaystyle\frac{1}{x(x^2 + 1)} = \frac{A}{x} + \frac{Bx + C}{x^2 + 1}$

23. Find $\displaystyle\int \frac{dx}{x(x^2 + 1)}$ 24. $\displaystyle\frac{1}{x(x^2 + 1)^2} = \frac{A}{x} + \frac{Bx + C}{(x^2 + 1)^2} + \frac{Dx + E}{x^2 + 1}$

25. Find $\displaystyle\int \frac{dx}{x(x^2 + 1)^2}$ 26. Find $\displaystyle\int \frac{\sin \theta}{1 + \cos \theta}\,d\theta$

27. Given $\theta = 2 \operatorname{Tan}^{-1} x$, calculate $\sin \theta$ and $\cos \theta$, in terms of x.

28. Find $\int d\theta/(1 + \cos \theta)$. (One way to do this is to use the substitution

$$\theta \to u(x) = 2 \operatorname{Tan}^{-1} x, \qquad d\theta \to u'(x)\, dx = \frac{2\, dx}{1 + x^2},$$

$$\sin \theta \to ? \qquad \cos \theta \to ?$$

But when you see the answer, you may be able to think of a simpler method of solution.)
Find:

29. $\displaystyle\int \frac{d\theta}{1 + \sin \theta}$
30. $\displaystyle\int \frac{d\theta}{\cos \theta}$
31. $\displaystyle\int \frac{dx}{x^2 + 6x + 9}$

32. $\displaystyle\int \frac{d\theta}{\sin \theta + \cos \theta}$
33. $\displaystyle\int \frac{d\theta}{\sec \theta + \tan \theta}$
34. $\displaystyle\int \sec^3 \theta \cot \theta\, d\theta$

35. $\displaystyle\int \frac{d\theta}{1 - \sin \theta}$
36. $\displaystyle\int \frac{d\theta}{2 + \cos \theta}$
37. $\displaystyle\int \frac{d\theta}{\cos \theta - \sin \theta}$

The
Definite Integral

7.1 THE PROBLEM OF ARC LENGTH

Consider the parabolic arc which is the graph of $y = x^2$, $0 \leq x \leq 1$. We shall calculate its length. The ideas that we use to do this will apply to other curves, but the general problem is no harder than the special one; if we were interested in arc length only for parabolas, the ideas in this section would all be needed.

The length of an arc of a circle is defined (as in Section 4.3) as the limit of the lengths of the inscribed broken lines. More generally, suppose that f is a continuous function on a closed interval $[a, b]$.

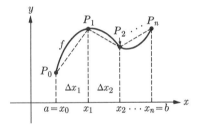

By a *net* over $[a, b]$ we mean an ascending sequence N of numbers

$$a = x_0 < x_1 < \cdots < x_i < x_{i+1} < \cdots < x_n = b.$$

For each i, let

$$y_i = f(x_i), \qquad P_i = (x_i, y_i).$$

We join successive points P_{i-1}, P_i with segments, getting a broken line as in the figure. Such a broken line is said to be *inscribed in* the graph of f. Its length is

$$P_0P_1 + P_1P_2 + \cdots + P_{n-1}P_n.$$

We denote this by $p(N)$. That is,

$$p(N) = \sum_{i=1}^{n} P_{i-1}P_i.$$

We use the functional notation $p(N)$, because when the net N is named, the broken line is determined, and so also is its length.

The graph of a continuous function on a closed interval may have infinite length. But if the length is finite, we ought to be able to approximate it by using a net N which cuts up $[a, b]$ into very small pieces. This idea is the basis of the following definitions.

Definition. Let

$$N = x_0, x_1, \ldots, x_n$$

be a net over $[a, b]$. The *mesh* of N is the largest of the numbers

$$\Delta x_i = x_i - x_{i-1}.$$

The mesh of N is denoted by $|N|$.

Definition. If $p(N)$ approaches a limit L, as $|N|$ approaches 0, then f is said to be *rectifiable*, and the number L is called its *length*.

We need, of course, to explain what is meant by the statement

$$\lim_{|N| \to 0} p(N) = L.$$

Intuitively, this means that $p(N) \approx L$ when $|N| \approx 0$. We define this idea by the same method that we used to define the limit of a function at a point. To make the analogy clearer, we write the old and new definitions in parallel.

Definition. Let f be a function, on an interval $[a, b]$. Let x_0 be a point of $[a, b]$.

Suppose that for every $\epsilon > 0$ there is a $\delta > 0$ such that if x is a point of $[a, b]$, then

$$0 < |x - x_0| < \delta$$
$$\Rightarrow \ |f(x) - L| < \epsilon.$$

Then

$$\lim_{x \to x_0} f(x) = L.$$

Definition. Let f be a function, on an interval $[a, b]$.

Suppose that for every $\epsilon > 0$ there is a $\delta > 0$ such that if N is a net over $[a, b]$, then

$$|N| < \delta$$
$$\Rightarrow \ |p(N) - L| < \epsilon.$$

Then

$$\lim_{|N| \to 0} p(N) = L.$$

To calculate the arc length, we first express the length $p(N)$ (of the inscribed broken line) in terms of things that we know how to handle. By definition,

$$p(N) = \sum_{i=1}^{n} P_{i-1}P_i.$$

(See the figure on p. 303.) The segment from P_{i-1} to P_i looks like this:

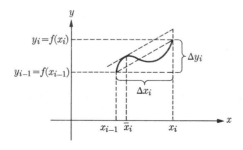

Thus
$$(P_{i-1}P_i)^2 = (x_i - x_{i-1})^2 + (y_i - y_{i-1})^2 = \Delta x_i^2 + \Delta y_i^2$$
$$= \left[1 + \left(\frac{\Delta y_i}{\Delta x_i}\right)^2\right] \Delta x_i^2.$$

Here the fraction $\Delta y_i/\Delta x_i$ is the slope of the chord from P_{i-1} to P_i; and the mean-value theorem says that this is the slope of the tangent line at some intermediate point. Thus we have

$$\frac{\Delta y_i}{\Delta x_i} = f'(\bar{x}_i) \qquad (x_{i-1} < \bar{x}_i < x_i).$$

Making this substitution, and extracting the square root, we get

$$P_{i-1}P_i = \sqrt{1 + [f'(\bar{x}_i)]^2} \cdot \Delta x_i.$$

Taking the sum from $i = 1$ to n, we get

$$p(N) = \sum_{i=1}^{n} P_{i-1}P_i = \sum_{i=1}^{n} \sqrt{1 + [f'(\bar{x}_i)]^2} \cdot \Delta x_i.$$

The problem is to find out what happens to the sum on the right as $|N| \to 0$. We can find this by giving a geometric interpretation to the sum.

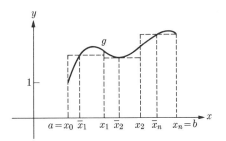 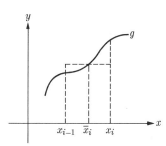

For each x on $[a, b]$, let
$$g(x) = \sqrt{1 + [f'(x)]^2}.$$

In the figure,
$$y = g(x), \qquad x_{i-1} < \bar{x}_i < x_i \qquad \text{for} \quad 1 \leq i \leq n.$$

On each little interval $[x_{i-1}, x_i]$, of length $\Delta x_i = x_i - x_{i-1}$, we have set up a rectangle with $[x_{i-1}, x_i]$ as base, and altitude $g(\bar{x}_i)$. The area of this rectangle is then

$$g(\bar{x}_i)\, \Delta x_i.$$

The sum of these areas is

$$\sum_{i=1}^{n} g(\bar{x}_i)\, \Delta x_i = \sum_{i=1}^{n} \sqrt{1 + [f'(\bar{x}_i)]^2}\, \Delta x_i = p(N).$$

If f' is continuous, then so is g; and $\sum g(\bar{x}_i)\, \Delta x_i$ ought to be close to the area under the graph of g, when the mesh of the net N is small. That is, we ought to have

$$|N| \approx 0 \quad \Rightarrow \quad \sum_{i=1}^{n} g(\bar{x}_i) \Delta x_i \approx \int_a^b g(x) \, dx,$$

which means that

$$\lim_{|N| \to 0} \sum_{i=1}^{n} g(\bar{x}_i) \Delta x_i = \int_a^b g(x) \, dx.$$

This gives us a formula for arc length:

$$L = \int_a^b \sqrt{1 + [f'(x)]^2} \, dx.$$

This holds whenever f' is continuous; and we will complete the proof later in this chapter. Meanwhile, consider some examples.

Example 1. Let

$$f(x) = 1, \qquad 0 \leq x \leq 1.$$

Then $f'(x) = 0$, and

$$L = \int_0^1 \sqrt{1 + 0^2} \, dx = 1,$$

which is the right answer.

Example 2. Let

$$f(x) = kx, \qquad 0 \leq x \leq 1.$$

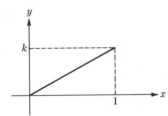

Then $f'(x) = k$, and

$$L = \int_0^1 \sqrt{1 + k^2} \, dx = \sqrt{1 + k^2},$$

which is the right answer, by the Pythagorean theorem.

Example 3. Let

$$f(x) = \sqrt{1 - x^2}, \qquad 0 \leq x \leq \frac{\sqrt{2}}{2}.$$

Then

$$f'(x) = \frac{-x}{\sqrt{1 - x^2}}, \qquad [f'(x)]^2 = \frac{x^2}{1 - x^2},$$

$$1 + [f'(x)]^2 = \frac{1 - x^2 + x^2}{1 - x^2} = \frac{1}{1 - x^2},$$

and

$$L = \int_0^{\sqrt{2}/2} \sqrt{1 + [f'(x)]^2}\, dx = \int_0^{\sqrt{2}/2} \frac{dx}{\sqrt{1 - x^2}} = [\text{Sin}^{-1} x]_0^{\sqrt{2}/2}$$

$$= \text{Sin}^{-1} \frac{\sqrt{2}}{2} = \frac{\pi}{4}.$$

This is the right answer, because L is one-eighth of the circumference of a circle of radius 1.

Example 4. We return to $f(x) = x^2$, $0 \leq x \leq 1$. Here

$$f'(x) = 2x, \qquad 1 + [f'(x)]^2 = 1 + 4x^2,$$

$$L = \int_0^1 \sqrt{1 + 4x^2}\, dx.$$

Now

$$\int \sqrt{1 + 4x^2}\, dx = \frac{1}{2} \int \sqrt{1 + u^2}\, du, \qquad u = 2x.$$

This is

$$\{\tfrac{1}{2} x \sqrt{1 + 4x^2} + \tfrac{1}{4} \ln |2x + \sqrt{1 + 4x^2}| + C\}.$$

Therefore, by the fundamental theorem of integral calculus, we have

$$L = \frac{\sqrt{5}}{2} + \tfrac{1}{4} \ln (2 + \sqrt{5}).$$

The answer suggests that no method would have made the problem look easy.

PROBLEM SET 7.1

Find the lengths of the graphs of the following functions, between the indicated limits.

1. $f(x) = x^{3/2}$, $\ 0 \leq x \leq 2$
2. $f(x) = \ln \cos x$, $\ 0 \leq x \leq \pi/4$
3. $f(x) = \ln \sin x$, $\ \pi/4 \leq x \leq \pi/2$
*4. $f(x) = \ln x$, $\ 1 \leq x \leq 3$
5. $f(x) = 1 + \tfrac{2}{3} x^{3/2}$, $\ 0 \leq x \leq 4$
*6. $f(x) = e^x$, $\ 0 \leq x \leq 1$
7. a) $f(x) = \tfrac{1}{2}(e^x + e^{-x})$, $0 \leq x \leq 1$ (You can solve this one, by an algebraic trick, without using any of the standard formulas for hyperbolic functions. But the problem is a little easier if you remember that $\sinh x = \tfrac{1}{2}(e^x - e^{-x})$, $\cosh x = \tfrac{1}{2}(e^x + e^{-x})$. For the definitions of the hyperbolic functions sinh and cosh, and the formulas governing them, see the end of Section 4.11.)

    b) $f(x) = \dfrac{1}{2a}(e^{ax} + e^{-ax})$, $\ 0 \leq x \leq 1$

8. Let f be a function with a continuous derivative, on an interval containing x_0.

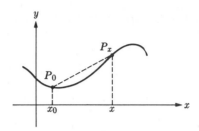

Let $r(x) = \overset{\frown}{P_0P_x}/P_0P_x$, that is, the ratio of the arc length to the length of the chord. Show that $\lim_{x \to x_0} r(x) = 1$.

To prove this, you will need to use the formula which expresses $\overset{\frown}{P_0P_x}$ as an integral. In many books you will see a "proof" of the integral formula for arc length, based on the assumption that $r(x) \to 1$. This is an example of a proof of infinite thinness: the hole in it is as big as the proof, because to fill the hole you must first prove the theorem itself by another method.

9. Consider the sequence of broken lines suggested by the figure below. Each broken line forms a stairway from P to Q. The nth stairway has n vertical segments and n horizontal segments. For each n, let B_n be the length of the nth broken line. Find $\lim_{n \to \infty} B_n$.

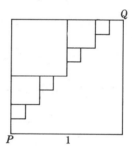

10. Let f be any function on $[a, b]$, and let x be any point of (a, b). Let m_1, m_2, and m be the slopes of the chords over the intervals $[a, x]$, $[x, b]$, and $[a, b]$. Show that m is between m_1 and m_2. (Unless, of course, $m_1 = m_2 = m$.)

More precisely,

$$m_1 = \frac{f(x) - f(a)}{x - a}, \qquad m_2 = \frac{f(b) - f(x)}{b - x},$$

and

$$m = \frac{f(b) - f(a)}{b - a}.$$

The theorem says that either (a) $m_1 \leqq m \leqq m_2$ or (b) $m_2 \leqq m \leqq m_1$.

7.2 THE DEFINITE INTEGRAL, DEFINED AS A LIMIT OF SAMPLE SUMS

In Section 3.7, we defined the definite integral in terms of area, with areas above the x-axis counted positively and those below counted negatively. In the preceding section, however, we regarded the integral as the limit of a sum:

$$\int_a^b g(x)\,dx = \lim_{|N|\to 0} \sum_{i=1}^n g(\bar{x}_i)\,\Delta x_i.$$

Most of the time hereafter, the definite integral will be used in this way, and so we shall redefine the integral, using the above formula as a definition. For this purpose, we need to investigate nets, and sums of the type

$$\sum_{i=1}^n g(\bar{x}_i)\,\Delta x_i.$$

Consider first an increasing continuous function f, on an interval $[a, b]$.

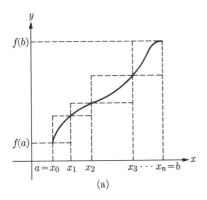

(a)

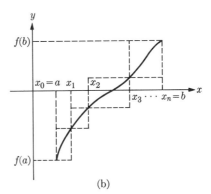

(b)

On the interval $[a, b]$ we form a net

$$N:\ a = x_0 < x_1 < \cdots < x_{i-1} < x_i < \cdots < x_n = b.$$

The points of the net N cut up the interval $[a, b]$ into little intervals $[x_{i-1}, x_i]$. For each i from 1 to n, let m_i be the minimum value of f on the ith interval $[x_{i-1}, x_i]$, and let M_i be the maximum value. Since f is increasing, we have $m_i = f(x_{i-1})$, $M_i = f(x_i)$. As usual, $\Delta x_i = x_i - x_{i-1}$, so that Δx_i is the length of the ith interval $[x_{i-1}, x_i]$. If f is positive, as in part (a) of the figure above, then the sum

$$s(N) = \sum_{i=1}^n m_i\,\Delta x_i$$

is the sum of the areas of the inscribed rectangles, and the sum

$$S(N) = \sum_{i=1}^n M_i\,\Delta x_i$$

is the sum of the areas of the circumscribed rectangles. For functions which may be negative, as in part (b) of the figure, $s(N)$ and $S(N)$ are sums of *signed* areas. In either case, $s(N)$ is called the *lower sum* of f over the net N, and $S(N)$ is called the *upper sum* of f over N.

On each interval $[x_{i-1}, x_i]$ we choose a sample point $\bar{x}_i$. Thus

$$x_{i-1} \leqq \bar{x}_i \leqq x_i \qquad (1 \leqq i \leqq n).$$

The sequence

$$X: \bar{x}_1, \bar{x}_2, \dots, \bar{x}_n$$

is called a *sample* of the net N. The sample gives a sum

$$\sum(X) = \sum_{i=1}^{n} f(\bar{x}_i) \Delta x_i.$$

This is called the *sample sum* of f over the sample X.

As in the preceding section, the *mesh* of N is the largest of the numbers Δx_i. The mesh is denoted by $|N|$. Thus

$$|N| = \max \{\Delta x_i\}.$$

Let R be the region between the graph of f and the x-axis, from a to b.

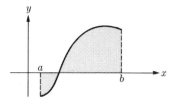

Theorem 1. If f is continuous, and N_1 and N_2 are any nets over $[a, b]$, then

$$s(N_1) \leqq S(N_2).$$

That is, every lower sum of f is less than or equal to every upper sum of f. For positive functions this is obvious, because in this case $s(N_1)$ is the area of an inscribed polygonal region (lying under the curve) and $S(N_2)$ is the area of a circumscribed polygonal region. In general,

$$s(N_1) = A - B, \qquad S(N_2) = C - D,$$

where A and D are areas of inscribed regions, and B and C are areas of circumscribed regions. (To see how this works, see the figure (b) above.) Therefore,

$$A \leqq C, \qquad B \geqq D, \qquad -B \leqq -D, \qquad A - B \leqq C - D,$$

and $s(N_1) \leqq S(N_2)$, as before.

Note that in Theorem 1 it is not required that f be an increasing function.

Theorem 2. If f is continuous and increasing, then

$$\lim_{|N| \to 0} [S(N) - s(N)] = 0.$$

That is, the upper sums are close to the corresponding lower sums, when the mesh is small. To prove this, we observe that the difference $S(N) - s(N)$ has a geometric interpretation.

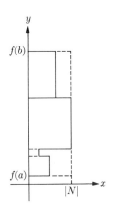
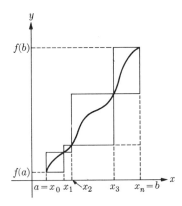

This difference is

$$S(N) - s(N) = \sum_{i=1}^{n} M_i \, \Delta x_i - \sum_{i=1}^{n} m_i \, \Delta x_i = \sum_{i=1}^{n} (M_i - m_i) \, \Delta x_i;$$

and this is the sum of the areas of the rectangles drawn solid in the figure. These rectangles can all be moved to the left and stacked up inside a rectangle of altitude $f(b) - f(a)$ and base $|N|$. (Remember that $|N|$ is the largest of the numbers Δx_i.) Therefore

$$S(N) - s(N) \leqq [f(b) - f(a)] \cdot |N|,$$

and so $S(N) - s(N) \to 0$ as $|N| \to 0$.

Theorem 3. If f is continuous, and

$$\lim_{|N| \to 0} [S(N) - s(N)] = 0,$$

then the sample sums $\sum (X)$ approach a limit, as $|N| \to 0$.

That is, there is a number k such that

$$\lim_{|N| \to 0} \sum (X) = k.$$

Proof. We have to start the proof by naming the number k. The numbers $s(N)$ are bounded above. (By Theorem 1, every upper sum is an upper bound of the lower sums.) Let

$$k = \sup \{s(N)\}.$$

Consider an interval $[x_{i-1}, x_i]$. For each i,

$$m_i \leqq f(\bar{x}_i) \leqq M_i.$$

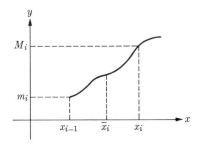

Since $\Delta x_i > 0$, this gives

$$m_i \, \Delta x_i \leqq f(\bar{x}_i) \, \Delta x_i \leqq M_i \, \Delta x_i.$$

Therefore the sums from 1 to n rank in the same order:

$$\sum_{i=1}^{n} m_i \, \Delta x_i \leqq \sum_{i=1}^{n} f(\bar{x}_i) \, \Delta x_i \leqq \sum_{i=1}^{n} M_i \, \Delta x_i,$$

so that

$$s(N) \leqq \sum (X) \leqq S(N).$$

That is, every sample sum lies between the lower sum and the upper sum.

We are now almost done. Given $\epsilon > 0$, we want a $\delta > 0$ such that

$$|N| < \delta \quad \Rightarrow \quad \left| \sum (X) - k \right| < \epsilon.$$

By Theorem 3 there is a $\delta > 0$ such that

$$|N| < \delta \quad \Rightarrow \quad S(N) - s(N) < \epsilon.$$

Thus when $|N| < \delta$, the interval from $s(N)$ to $S(N)$ has length less than ϵ.

This interval contains $\sum (X)$. (See the inequalities above.) And it also contains k: $s(N) \leqq k$, because k is *an* upper bound for the lower sums; and $k \leqq S(N)$, because k is the *least* upper bound of the lower sums. Therefore $\left| \sum (X) - k \right| < \epsilon$, because $\sum (X)$ and k are squeezed together: they both lie on the same short interval.

We can now give the new definition of the integral.

Definition. $\int_a^b f(x) \, dx = \lim_{|N| \to 0} \sum (X)$, if the indicated limit on the right exists. If the limit exists, then we say that f is *integrable* on $[a, b]$.

Theorems 2 and 3 fit together to give:

Theorem 4. *If f is continuous and increasing on $[a, b]$, then f is integrable on $[a, b]$.*

Later we shall see that all continuous functions are integrable, whether or not they are increasing.

Our calculations of definite integrals have been based on the differentiation formula

$$D \int_a^x f(t) \, dt = f(x),$$

where f is continuous. We need to know that this differentiation formula still holds, under the new definition of the integral. This is the purpose of the following theorem.

Theorem 5 (*The betweenness theorem for integrals*). *If f is integrable on $[a, b]$, and*

$$m \leqq f(x) \leqq M \qquad (a \leqq x \leqq b),$$

then

$$m(b - a) \leq \int_a^b f(x)\, dx \leq M(b - a).$$

Proof. Let N be any net over $[a, b]$, and let X be any sample of N. Then

$$m \leq f(\bar{x}_i) \leq M \qquad \text{for every } i.$$

Therefore

$$m\, \Delta x_i \leq f(\bar{x}_i)\, \Delta x_i \leq M \Delta x_i.$$

Forming the sample sum $\sum (X)$ by addition, we get

$$\sum_{i=1}^n m\, \Delta x_i \leq \sum (X) \leq \sum_{i=1}^n M\, \Delta x_i.$$

But

$$\sum_{i=1}^n m\, \Delta x_i = m \sum_{i=1}^n \Delta x_i; \qquad \sum_{i=1}^n M\, \Delta x_i = M \sum_{i=1}^n \Delta x_i,$$

and

$$\sum_{i=1}^n \Delta x_i = b - a.$$

(Why?) Therefore

$$m(b - a) \leq \sum (X) \leq M(b - a);$$

and this holds for every sample sum, over every net N. Therefore the same inequalities hold for $\lim \sum (X)$, and the integral lies between $m(b - a)$ and $M(b - a)$, which was to be proved.

If you review the proof of the formula

$$D \int_a^x f(t)\, dt = f(x),$$

in Section 3.10, you will find that in this proof, all that we needed to know about the integral was the information conveyed by the betweenness theorem for integrals. Therefore the differentiation formula continues to hold, under our new definition, wherever the integrand is continuous. It follows that the fundamental theorem of integral calculus still holds true.

At the end of this chapter we shall prove that every continuous function is integrable.

*PROBLEM SET 7.2

1. In Theorem 2 it was assumed that the function f is increasing. Does the same scheme of proof work, for a decreasing function? If so, draw a figure illustrating the proof for decreasing functions. If not, explain how the scheme breaks down, for the case of a decreasing function.

2. In Theorem 2 it was assumed that f is both continuous and increasing. Suppose we assume that f is increasing, but not that f is continuous. What changes (if any) do we then need to make in

   a) the definitions of m_i and M_i,   b) the definitions of $s(N)$ and $S(N)$, and
   c) the proof of Theorem 2?

3. Prove the following:

Theorem A   (*The mean-value theorem for integrals*). If f is continuous on $[a, b]$, then there is a point $\bar{x}$, between a and b, such that

$$\int_a^b f(x)\,dx = f(\bar{x})(b - a).$$

4. Consider the function f defined by the following conditions: $f(\tfrac{1}{3}) = \tfrac{1}{3}$, $f(\tfrac{1}{2}) = \tfrac{1}{2}$, $f(\tfrac{2}{3}) = \tfrac{1}{3}$; $f(x) = 0$ for every other x on $[0, 1]$. Is this function integrable? Why or why not?

*5. Consider the following function, on $[0, 1]$. If x is irrational, then $f(x) = 0$. If $x = p/q$, in lowest terms, then $f(x) = 1/q$. At what points (if any) is this function continuous? Is the function integrable? Why or why not?

6. Given a continuous function g, on $[a, b]$, and a net N over $[a, b]$. Show that there is a sample X of N such that $\sum (X) = \int_a^b g(x)\,dx$.

7. In Section 7.1 we showed that for every net N we could choose a sample X in such a way that the length of the inscribed broken line is equal to the sample sum $\sum (X)$, not just approximately but *exactly*. Is it always possible to choose a sample X' such that $\sum (X')$ is exactly equal to the arc length? (Here we are assuming, as usual, that f' is continuous.)

*8. The following remarks are a very sketchy indication of an amusing proof of an important theorem which is known to you in a slightly weaker form. Fill in the gaps, and state the theorem which is proved.

$F' = f$. f is known to be *integrable* on $[a, b]$, but is not necessarily continuous.

$$F(x_i) - F(x_{i-1}) = f(\bar{x}_i)\Delta x_i,$$

$$\sum_{i=1}^n [F(x_i) - F(x_{i-1})] = \sum_{i=1}^n f(\bar{x}_i)\,\Delta x_i.$$

As $|N| \to 0$, $\sum_{i=1}^n f(\bar{x}_i)\,\Delta x_i \to$?; but $\sum_{i=1}^n [F(x_i) - F(x_{i-1})]$ was something simple, all along.

*9. Let f be differentiable on $[a, b]$. Show that if

$$f'(a) < k < \frac{f(b) - f(a)}{b - a},$$

then $k = f'(\bar{x})$, for some $\bar{x}$ between a and b.

[*Hint:* Remember the definition of $f'(a)$. Do a sketch illustrating the definition.]

*10. **Theorem** (*The no-jump theorem for derivatives*). If f is differentiable on $[a, b]$, and k is between $f'(a)$ and $f'(b)$, then $k = f'(\bar{x})$ for some $\bar{x}$ between a and b.

Thus, for example, the function

$$g(x) = \begin{cases} \dfrac{x}{|x|} & \text{for } x \neq 0, \\[2mm] 0 & \text{for } x = 0 \end{cases}$$

cannot be the derivative of any other function f.

*11. **Theorem.** If f is differentiable at a, then

$$\lim_{\substack{x_1 \to a^- \\ x_2 \to a^+}} \left(\frac{f(x_2) - f(x_1)}{x_2 - x_1} \right) = f'(a).$$

More precisely, for every $\epsilon > 0$ there is a $\delta > 0$ such that

$$a - \delta < x_1 < a < x_2 < a + \delta$$

$$\Rightarrow \quad \left| \frac{f(x_2) - f(x_1)}{x_2 - x_1} - f'(a) \right| < \epsilon.$$

12. a) A function of f is *Lipschitzian* on $[a, b]$ if there is a number $k > 0$ such that for every x_1 and x_2 of $[a, b]$,

$$|f(x_1) - f(x_2)| \leqq k |x_1 - x_2|.$$

Show that $f(x) = \sin x$ is Lipschitzian on the interval $(-\infty, \infty)$.

b) Show that every Lipschitzian function is continuous.

c) Give an example to show that a continuous function is not necessarily Lipschitzian.

d) Show that if f' is continuous on $[a, b]$, then f is Lipschitzian on $[a, b]$.

e) Show that if f is Lipschitzian on $[a, b]$, then f is integrable.

   Since Theorem 3 is known, it will be sufficient to show that if f is Lipschitzian, then

$$\lim_{|N| \to 0} [S(N) - s(N)] = 0.$$

13. a) A function f is *uniformly continuous* on $[a, b]$ if for every $\epsilon > 0$ there is a $\delta > 0$ such that

$$|x - x'| < \delta \Rightarrow |f(x) - f(x')| < \epsilon.$$

(Here x and x' are any points of $[a, b]$.) Show that if f' is continuous, then f is uniformly continuous on $[a, b]$.

b) Show that every uniformly continuous function is integrable.

7.3 THE CALCULATION OF VOLUMES, BY THE METHOD OF DISKS

The volumes of various solids can be expressed as definite integrals. In this process, we shall assume that the following volume formulas are known.

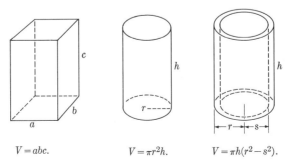

$V = abc.$ $V = \pi r^2 h.$ $V = \pi h(r^2 - s^2).$

The first of these solids is a rectangular parallelepiped; the second is a right circular cylinder; and the third is a *cylindrical shell*, that is, the portion of the larger cylinder that lies outside the smaller cylinder.

We get a coordinate system in space by setting up a z-axis, perpendicular to the xy-plane at the origin. Here, and throughout this chapter, we shall indicate only the positive half of each axis, thus getting a picture of only the "first octant," in which the points have nonnegative coordinates.

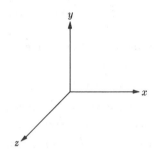

Consider now the function

$$f(x) = 1/x$$

on the interval $[1, 2]$. Let R be the region under the graph, in the xy-plane. We rotate the region R about the x-axis. This gives a solid S.

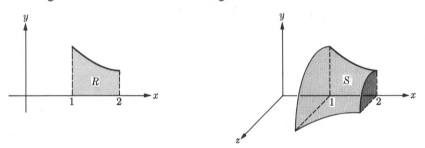

Let vS be the volume of S. We shall express vS as a definite integral. First we form a net

$$N = x_0, x_1, \ldots, x_n$$

over the interval $[1, 2]$. For convenience, we use equally spaced points, so that

$$x_i - x_{i-1} = \Delta x = 1/n$$

for each i. Over the intervals $[x_{i-1}, x_i]$ we set up the circumscribed rectangles. These form a region R_n which is an approximation of the region R. Then we rotate R_n about the x-axis. Each of our rectangles then gives a cylinder (lying on its side), and the cylinders form a solid S_n which is an approximation of S. In the figure on the right below we show only the ith cylinder. Its altitude is $\Delta x = x_i - x_{i-1}$, and the radius of its base is

$$r_i = 1/x_{i-1}.$$

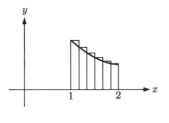 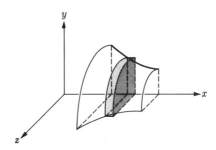

Therefore its volume is

$$v_i = \pi r_i^2 \cdot \Delta x = \pi \left(\frac{1}{x_{i-1}}\right)^2 \cdot \Delta x,$$

and so the total volume of the circumscribed solid S_n is

$$vS_n = \sum_{i=1}^{n} v_i = \sum_{i=1}^{n} \pi \left(\frac{1}{x_{i-1}}\right)^2 \cdot \Delta x.$$

This is a sample sum of the function

$$g(x) = \pi \frac{1}{x^2},$$

over the net

$$N:\ x_0, x_1, \ldots, x_n.$$

(In fact, it is the upper sum of g over the net N, because $g(x_{i-1})$ is the maximum value of g on the interval $[x_{i-1}, x_i]$.) The mesh of N is

$$|N| = \Delta x = \frac{1}{n},$$

and so $|N| \to 0$ as $n \to \infty$. Therefore

$$\lim_{n \to \infty} vS_n = \int_1^2 \frac{\pi}{x^2}\, dx = -\left. \frac{\pi}{x} \right]_1^2 = \frac{\pi}{2}. \qquad (1)$$

If we use inscribed rectangles, and rotate them about the x-axis, then we get an inscribed solid S_n', with volume

$$vS_n' = \sum_{i=1}^{n} \pi \left(\frac{1}{x_i}\right)^2 \cdot \Delta x.$$

This is also a sample sum, of the same function $g = \pi/x^2$. Therefore

$$\lim_{n \to \infty} vS_n' = \int_1^2 \frac{\pi}{x^2}\, dx. \qquad (2)$$

Therefore the volume vS of S is squeezed between the volumes of the inscribed and circumscribed solids:

$$vS_n' \leqq vS \leqq vS_n \qquad \text{for every } n,$$

$$\lim_{n \to \infty} vS_n' \leqq vS \leqq \lim_{n \to \infty} vS_n;$$

and so

$$vS = \int_1^2 \frac{\pi}{x^2}\,dx = \frac{\pi}{2}. \tag{3}$$

We shall now review this process and state the assumptions on which it is based. Not all solids are *measurable*, in the sense that they have volumes; but the solids that you are likely to encounter soon are measurable, and their volumes are governed by the following laws.

By an *elementary solid* we mean a right parallelpiped, cylinder, or cylindrical shell, as at the beginning of this section. We have been assuming that:

V.1. Elementary solids are measurable, and their volumes are given by the formulas $v = abc$, $v = \pi r^2 h$, $v = \pi h(r^2 - s^2)$.

Two solids are *nonoverlapping* if they have no solid in common. (They may have *surfaces* in common.)

V.2. If $s_1, s_2, \ldots, s_n$ are nonoverlapping elementary solids, and S_n is their union, then S_n is measurable, and

$$vS_n = vs_1 + vs_2 + \cdots + vs_n.$$

V.3. If S and S' are measurable, and S' lies in S, then $vS' \leqq vS$.

V.4 (*The squeeze principle*). If (a) $S_1, S_2, \ldots$ are measurable solids containing S, (b) $S_1', S_2', \ldots$ are measurable solids lying in S, and (c) $\lim_{n \to \infty} vS_n = L = \lim_{n \to \infty} vS_n'$, then S is measurable, and

$$vS = L.$$

Using V.1 through V.4, we can show that the method of disks, which we have used for the function

$$f(x) = 1/x,$$

works for every function f which is $\geqq 0$ and continuous.

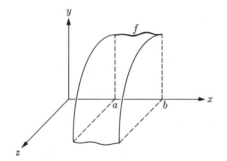
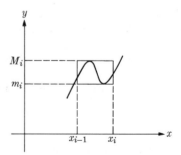

Given such a function f, on a closed interval $[a, b]$, let S be the solid of revolution of R, about the x-axis. Take a net N over $[a, b]$, with equal spacing. As usual, let m_i and M_i be the minimum and maximum values of f on the ith interval $[x_{i-1}, x_i]$.

If we rotate the inscribed rectangles about the x-axis, we get an inscribed solid S'_n, of volume

$$vS'_n = \sum_{i=1}^{n} \pi m_i^2 \Delta x.$$

If we rotate the circumscribed rectangles about the x-axis, we get a circumscribed solid S_n, of volume

$$vS_n = \sum_{i=1}^{n} \pi M_i^2 \Delta x.$$

V.3 says that

$$vS'_n \leqq vS_n \qquad \text{for every } n.$$

But vS_n is an upper sum of the function

$$g(x) = \pi f(x)^2,$$

and vS'_n is a lower sum of the same g. As $n \to \infty$, $|N| \to 0$;

$$vS_n \to \int_a^b \pi f(x)^2 \, dx, \qquad vS'_n \to \int_a^b \pi f(x)^2 \, dx.$$

By the squeeze principle it follows that S is measurable, and

$$vS = \int_a^b \pi f(x)^2 \, dx.$$

We also use this formula sidewise.

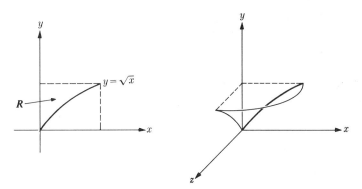

Suppose that the region R on the left is rotated about the y-axis. Sidewise, R can be regarded as the region under the graph of a function

$$x = f(y) = y^2, \qquad 0 \leqq y \leqq 1.$$

Therefore the volume is

$$\int_0^1 \pi [f(y)]^2 \, dy = \int_0^1 \pi y^4 \, dy = \frac{\pi}{5}.$$

PROBLEM SET 7.3

1. Obviously a right circular cone can be regarded as the solid of revolution of a right triangle about one of its legs. If we place the triangle in the xy-plane as shown in the figure, then the hypotenuse becomes the graph of a function f. Calculate f, and find the volume of the cone by the methods of this section.

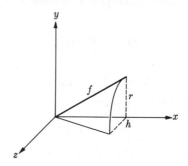

2. Similarly, a round ball of radius r can be regarded as the solid of revolution of a semi-circular region about its diameter. Find the volume, by the methods of this section.

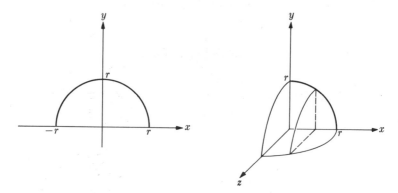

3. The region under the graph of $f(x) = \sqrt{x}$ $(0 \leq x \leq 1)$ is rotated about the x-axis. Find the volume of the resulting solid.

4. Same, for $f(x) = \sin x$, $0 \leq x \leq \pi$.

5. Same, for $f(x) = x^{3/2}$, $0 \leq x \leq 1$.

6. Same, for $f(x) = \cos x$, $-\pi/2 \leq x \leq \pi/2$.

7. Let $R = \{(x, y) \mid 0 \leq x \leq 1,\ x^2 \leq y \leq 1\}$ be rotated about the y-axis. Find the volume.

8. Same, for $R = \{(x, y) \mid 0 \leq x \leq 1,\ \operatorname{Sin}^{-1} x \leq y \leq \pi/2\}$.

9. Same, for $R = \{(x, y) \mid 0 \leq x \leq \pi/2,\ \sin x \leq y \leq 1\}$.

10. Same, for $R = \{(x, y) \mid 0 \leq x \leq 1,\ x^3 \leq y \leq 1\}$.

11. Same, for $R = \{(x, y) \mid 0 \leq x \leq \sqrt{2}/2,\ x \leq y \leq \sqrt{1 - x^2}\}$.

12. Find out whether the following is true:

    Theorem (?). Let T and T' be triangles each of which has a side on the x-axis. If T and T' have the same area, then when they are rotated about the x-axis, they give solids with the same volume.

13. a) For each x from 0 to 1, let T_x be the triangle whose vertices are $(0, 0)$, $(1, 0)$, and $(x, 1)$. What value or values of x give maximum volume, when T_x is rotated about the x-axis?

    b) Suppose that the triangles T_x are rotated about the y-axis (instead of the x-axis). Which value or values of x give maximum volume?

14. For each k from 0 to 1, let T_k be the triangle whose vertices are $(0, 0)$, $(k, 0)$, and $(0, \sqrt{1 - k^2})$. (Thus the hypotenuse of T_k has length 1.) T_k is rotated about the x-axis. What value of k gives maximum volume? What is the maximum volume?

15. a) Given $f(x) = 1/x$. Let R be the region under the graph of f, from 1 to ∞. Give a reasonable definition of the area of R. Is this area finite?

    b) The region R is rotated about the x-axis, giving a solid S. Give a reasonable definition of the volume of S. Is this volume finite?

16. a) The region R under the graph of $f(x) = 1/x^2$ from 1 to ∞ is rotated about the x-axis giving a solid S. Does S have finite volume?

    b) If the region R is rotated about the y-axis, do we obtain a solid with finite volume?

7.4 THE GENERAL METHOD OF CROSS SECTIONS, AND THE METHOD OF SHELLS

The method of disks can be generalized in the following way. Given a solid S in space. Suppose that we can calculate the areas of the cross sections perpendicular to the x-axis.

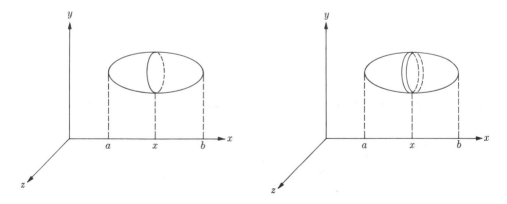

For each x from a to b we let $A(x)$ be the area of the cross section. This gives a function A which expresses the cross-sectional area in terms of x. (In our previous examples, the cross sections were all circular.) As before, we divide the interval $[a, b]$ into n equal parts, and we approximate the volume by cylinders. In the figure at the right, we show only the ith cylinder.

We then have

$$V_n = \sum_{i=1}^{n} A(x_i)\, \Delta x,$$

and the sum on the right-hand side is a sample sum of the function A. Therefore, as the mesh goes to 0,

$$V_n \to \int_a^b A(x)\, dx.$$

It is plausible to suppose that as the mesh approaches 0, V_n approaches the volume of S; and in fact this is true, for measurable solids, although we are not in a position to prove it. Thus

$$vS = \int_a^b A(x)\, dx.$$

By this method we can calculate volumes. For example, take the parabola $y = x^2$, for $0 \leq x \leq 1$. For each y from 0 to 1, we take the horizontal segment from $(0, y)$ to the point (x, y) of the parabola; and using this segment as an edge, we construct a horizontal square. Thus we get a solid, as shown in the figure.

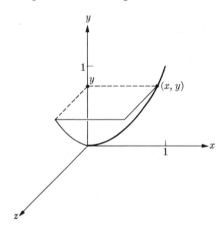

The cross-sectional areas perpendicular to the y-axis are given by the formula

$$A(y) = x^2 = y.$$

Therefore the volume is

$$V = \int_0^1 A(y)\, dy = \int_0^1 y\, dy = \left[\frac{y^2}{2}\right]_0^1 = \frac{1}{2}.$$

The general method of cross sections applies, in a sense, to every volume problem. That is, it is always *true* that

$$vS = \int_a^b A(x)\, dx.$$

But often this formula leads to difficult calculations.

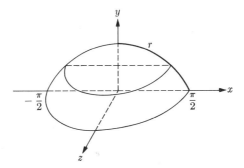

Consider, for example, the region R under the graph of

$$f(x) = \cos x, \qquad 0 \leqq x \leqq \frac{\pi}{2}.$$

We rotate R about the y-axis, getting a solid of revolution, of which only the front half is shown in the figure. We can find the volume by the cross-section method. We have

$$A(y) = \pi x^2 = \pi(\text{Cos}^{-1} y)^2.$$

Therefore

$$V = \int_0^1 A(y)\, dy = \int_0^1 \pi(\text{Cos}^{-1} y)^2\, dy.$$

We can calculate this by integrating by parts twice, but there is a better way. Instead of approximating the solid by thin cylinders, we approximate it by thin cylindrical shells.

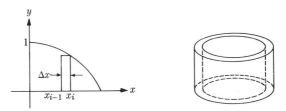

First we approximate the region R by rectangles with equal bases $\Delta x = \pi/2n$, as shown on the left. Then we rotate each of these rectangles about the y-axis, getting cylindrical shells, as shown on the right. The altitude of the ith shell is

$$f(x_i) = \cos x_i;$$

the outer radius is x_i; and the inner radius is $x_{i-1} = x_i - \Delta x$. Therefore the volume of the ith shell is

$$
\begin{aligned}
v_i &= \pi x_i^2 \cos x_i - \pi(x_i - \Delta x)^2 \cos x_i \\
&= \pi(x_i^2 - x_i^2 + 2x_i \Delta x - \Delta x^2) \cos x_i \\
&= 2\pi x \cos x_i \, \Delta x - \pi(\cos x_i)(\Delta x)^2.
\end{aligned}
$$

Therefore the total volume of the inscribed solid S_n is

$$vS_n = \sum_1^n v_i$$

$$= \sum_1^n 2\pi x_i \cdot \cos x_i \cdot \Delta x - \pi \Delta x \sum_1^n \cos x_i \Delta x.$$

We need to find out what happens as the mesh goes to 0. The first sum is a sample sum of the function $2\pi x \cos x$. Therefore

$$\sum_1^n 2\pi x_i \cos x_i \cdot \Delta x \to \int_0^{\pi/2} 2\pi x \cos x \, dx.$$

For the same reason,

$$\sum_1^n \cos x_i \cdot \Delta x \to \int_0^{\pi/2} \cos x \, dx.$$

Therefore

$$\pi \Delta x \cdot \sum \cos x_i \cdot \Delta x \to \pi \cdot 0 \cdot \int_0^{\pi/2} \cos x \, dx.$$

Thus the entire second sum, in the expression for vS_n, drops out when we pass to the limit. Therefore

$$vS_n \to \int_0^{\pi/2} 2\pi x \cos x \, dx,$$

and

$$V = \int_0^{\pi/2} 2\pi x \cos x \, dx = 2\pi [x \sin x + \cos x]_0^{\pi/2}$$

$$= 2\pi \left[\frac{\pi}{2} + 0 \right] - 2\pi[1] = \pi^2 - 2\pi.$$

The same method applies if we rotate a region lying to the right of the y-axis.

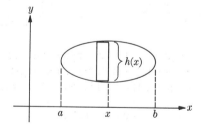

If the width of the region is given by a function $h(x)$, then the volume of the ith cylindrical shell is

$$v_i = \pi x_i^2 h(x_i) - \pi(x_i - \Delta x)^2 h(x_i)$$
$$= \pi h(x_i)(x_i^2 - x_i^2 + 2x_i \Delta x - \Delta x^2)$$
$$= 2\pi x_i h(x_i) \Delta x - \pi h(x_i) \Delta x^2.$$

Therefore

$$V_n = \sum_1^n v_i = 2\pi \sum_1^n x_i h(x_i)\, \Delta x - \pi \Delta x \sum_1^n h(x_i)\, \Delta x$$

$$\rightarrow 2\pi \int_a^b x h(x)\, dx - \pi \cdot 0 \cdot \int_a^b h(x)\, dx.$$

Therefore

$$V = 2\pi \int_a^b x h(x)\, dx.$$

Here again the second part of the sum drops out when we pass to the limit. Thus the sums behave as if v_i were given by the formula

$$v_i \approx 2\pi x_i h(x_i)\, \Delta x.$$

There is a simple reason why v_i is well approximated by $2\pi x_i h(x_i)\, \Delta x$.

 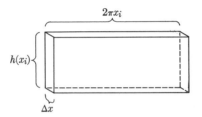

If we make a vertical cut in the ith cylindrical shell, and flatten it out, we get a rectangular prism. The length of the prism is the circumference of the outer circle in the base of the shell. This is $2\pi x_i$. The altitude and the thickness of the prism are the same as the altitude and the thickness of the shell; these are $h(x_i)$ and Δx. Therefore the volume of the prism is exactly $2\pi x_i h(x_i)\, \Delta x$; and this ought to be a good approximation to v_i when Δx is small, because when the shell is thin, we can flatten it out without distorting it very much. As we have seen, the error goes to zero as the mesh goes to zero.

The method of shells applies to the problem that we were discussing above. We know that the volume is

$$V = \int_0^{\pi/2} 2\pi x \cos x\, dx.$$

Integrating by parts, we get

$$\int x \cos x\, dx = \{x \sin x + \cos x + C\}.$$

Therefore

$$V = [2\pi(x \sin x + \cos x)]_0^{\pi/2} = \pi^2 - 2\pi.$$

The same method applies more generally. Consider the region

$$R = \{(x, y) \mid 2 \leqq x \leqq 3, 0 \leqq y \leqq -x^2 + 5x - 6\}.$$

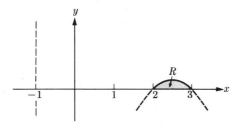

If R is rotated about the line $x = -1$, then by the shell method the volume of the resulting solid is

$$\int_2^3 2\pi(x + 1)(-x^2 + 5x - 6) \, dx = \int_2^3 2\pi(-x^3 + 4x^2 - x - 6) \, dx$$

$$= 2\pi\left[-\frac{x^4}{4} + \frac{4x^3}{3} - \frac{x^2}{2} - 6x\right]_2^3 = \tfrac{7}{6}\pi.$$

If the same region is rotated about the line $x = 6$, the volume is

$$\int_2^3 2\pi(6 - x)(-x^2 + 5x - 6) \, dx,$$

which is also equal to $7\pi/6$. (Why?)

In some cases there is little to choose between the cross-section method and the shell method. For example, suppose we take the region below the graph of $y = x^2$, $0 \leq x \leq 1$, and rotate it about the y-axis.

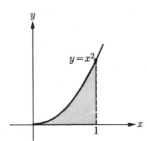

By the shell method,

$$V = \int_0^1 2\pi x \cdot x^2 \, dx = \frac{\pi}{2}.$$

The horizontal cross section at height y is the region between a circle of radius 1 and a circle of radius $x = \sqrt{y}$. Therefore

$$V = \int_0^1 A(y) \, dy = \int_0^1 [\pi \cdot 1^2 - \pi y] \, dy$$

$$= \pi - \pi \int_0^1 y \, dy = \frac{\pi}{2},$$

as before.

PROBLEM SET 7.4

1. Let R be the circular region with center at $(5, 0)$ and radius 2. R is rotated about the y-axis. Find the volume of the resulting solid.

2. A solid of the sort described in Problem 1 is called a *solid torus*. More generally, suppose we have given a circular region of radius a, and a line L in the same plane, such that the perpendicular distance from L to the center of R is b, with $b \geqq a$. When R is rotated about the line L, the result is a solid torus. Find its volume, in terms of a and b.

3. Let R be the square region with center at $(4, 0)$ and sides of length 2, parallel to the coordinate axes. R is rotated about the y-axis. Find the volume of the resulting solid.

4. Let T be the square region with center at $(4, 0)$ and sides of length 2, with diagonals parallel to the coordinate axes. Find the volume of the solid which results when T is rotated about the y-axis.

5. a) The region under the graph of

$$y = \ln x, \qquad 1 \leqq x \leqq e,$$

is rotated about the x-axis. Find the volume, by the method of disks.

   b) Now solve this problem by the method of shells.

6. For each x from 0 to 1, let R_x be the circular region perpendicular to the xy-plane, with center at the point (x, x^2) and radius 1. Let S be the solid formed by the regions R_x. Find the volume of S.

7. a) The region described in Problem 5a is rotated about the y-axis. Find the volume, by the shell method.

   b) Now solve Problem 7a by the method of cross sections.

8. a) The region under the graph of $y = e^x, 0 \leqq x \leqq 1$, is rotated about the y-axis. Find the volume by the method of shells.

   b) Now solve Problem 8a by the cross-section method.

9. Let C be the cylinder with the y-axis as its axis of symmetry, and radius 1. Let S be the sphere with center at the origin and radius 2. Find the volume of the solid which lies inside the sphere and outside the cylinder.

10. Let C_x be the cylinder of radius 1, with the x-axis as its axis of symmetry; and let C_y be the cylinder of radius 1 with the y-axis as its axis of symmetry. Find the volume of the solid which lies in both C_x and C_y.

11. Let S be the sphere of radius $\sqrt{2}$ with center at the origin. Let C be the cone with vertex at the origin, axis along the y-axis, and passing through the point $(1, 1)$. Find the volume of the solid which lies inside the sphere and inside the cone.

7.5 THE AREA OF A SURFACE OF REVOLUTION

Given a line and a curve, lying in the same plane and lying on one side of the given line. If the curve is rotated about the line, the resulting surface is called a *surface of revolution*. The area of such a surface can be expressed as an integral. We begin with the simplest case, in which a function-graph is rotated about the y-axis. Here the function f is defined on a closed interval $[a, b]$ on the positive half of the x-axis. We assume that f has a continuous derivative.

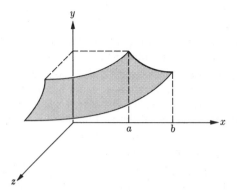

To calculate the area of the surface of revolution, we need the formula for the lateral surface of a right circular cone. Let s be the slant height of the cone, and let r be the radius of the base, so that the circumference of the base is $2\pi r$. We assert that the lateral surface is the same as the area of a circular sector of radius s, with boundary arc of length $2\pi r$. The reason is that we can make a straight cut in the cone, starting at the vertex, and then flatten out the surface, without changing its area, so that the resulting surface lies in a plane. The plane surface thus obtained is the sector shown below.

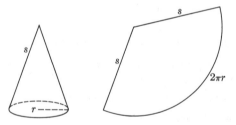

But the area of a circular sector is half the product of its radius and the length of its boundary arc. Therefore, for cones, we have

$$A = \pi r s.$$

(Note that for a "cone of altitude 0," that is, a disk, this formula gives the right answer $\pi r s = \pi r^2$.) From this we can get a formula for the lateral area of a frustum of a cone. If the larger cone (with slant height s_2) has area A_2, and the smaller cone

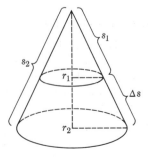

has area A_1, then the area of the frustum is

$$\Delta A = A_2 - A_1 = \pi r_2 s_2 - \pi r_1 s_1.$$

Evidently

$$\frac{s_2}{r_2} = \frac{s_1}{r_1}.$$

If $s_1 = kr_1$, $s_2 = kr_2$, then

$$\Delta A = \pi k r_2^2 - \pi k r_1^2 = \pi k (r_2 - r_1)(r_2 + r_1)$$

$$= \pi (s_2 - s_1)(r_2 + r_1) = 2\pi \cdot \frac{r_1 + r_2}{2} \Delta s,$$

and so

$$\Delta A = 2\pi \bar{r} \, \Delta s,$$

where

$$\bar{r} = \tfrac{1}{2}(r_1 + r_2).$$

That is, *the area of the frustum is equal to its "average circumference" $2\pi\bar{r}$ times its slant height Δs.*

Consider now the surface of revolution obtained by rotating the graph of f about the y-axis. We take a net

$$N: x_0, x_1, \ldots, x_{i-1}, x_i, \ldots x_n,$$

over the interval $[a, b]$, with equal subdivisions, so that

$$x_i - x_{i-1} = \Delta x = \frac{b - a}{n} = |N|$$

for each i. For each i, let P_i be the point $(x_i, f(x_i))$. These points determine a broken line B_n which is an approximation of the graph of f. When B_n is rotated about the y-axis, we get a surface S_n, with area A_n. By definition, the area of the surface of revolution of f is

$$A = \lim_{|N| \to 0} A_n,$$

if the limit exists. (This is like the definition of arc length.)

We shall now calculate A_n, and find its limit as $|N| \to 0$. Consider the ith segment, from P_{i-1} to P_i. When this segment is rotated, it gives a frustum whose area is

$$a_i = 2\pi \bar{x}_i \cdot P_{i-1}P_i;$$

where $\bar{x}_i = \tfrac{1}{2}(x_{i-1} + x_i).$

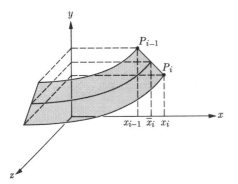
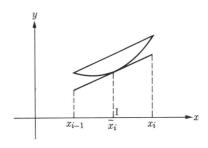

As in the calculation of arc length,

$$P_{i-1}P_i = \sqrt{\Delta x^2 + [f(x_i) - f(x_{i-1})]^2}$$

$$= \sqrt{1 + \left(\frac{f(x_i) - f(x_{i-1})}{\Delta x}\right)^2}\,\Delta x$$

$$= \sqrt{1 + f'(\bar{x}_i')^2}\,\Delta x,$$

where $x_{i-1} < \bar{x}_i' < x_i$, as shown in the last figure.

We now have a formula for the area A_n of the approximating surface:

$$A_n = \sum_{i=1}^{n} a_i = \sum_{i=1}^{n} 2\pi\bar{x}_i P_{i-1}P_i = \sum_{i=1}^{n} 2\pi\bar{x}_i\sqrt{1 + f'(\bar{x}_i')^2}\,\Delta x.$$

Here $\bar{x}_i$ is the midpoint of $[x_{i-1}, x_i]$, and $\bar{x}_i'$ is somewhere on the same interval.

If it were true that $\bar{x}_i' = \bar{x}_i$ for each i, then A_n would be a sample sum of the function

$$g(x) = 2\pi x\sqrt{1 + f'(x)^2},$$

and we would have no problem, because for

$$\sum_{i=1}^{n} a_i' = \sum_{i=1}^{n} 2\pi\bar{x}_i'\sqrt{1 + f'(\bar{x}_i')^2}\,\Delta x,$$

we know that

$$\lim_{n\to\infty} \sum_{i=1}^{n} a_i' = \int_a^b g(x)\,dx = \int_a^b 2\pi x\sqrt{1 + f'(x)^2}\,dx.$$

What we need to show, therefore, is that

$$\lim_{|N|\to 0} \left|\sum_{i=1}^{n} a_i - \sum_{i=1}^{n} a_i'\right| = 0.$$

It will then follow that

$$\lim_{|N|\to 0} \sum_{i=1}^{n} a_i = \lim_{|N|\to 0} \sum_{i=1}^{n} a_i' = \int_a^b 2\pi x\sqrt{1 + f'(x)^2}\,dx.$$

Now

$$|\bar{x}_i - \bar{x}_i'| \leq \frac{\Delta x}{2},$$

because $\bar{x}_i'$ lies on the interval $[x_{i-1}, x_i]$, whose midpoint is $\bar{x}_i$. Therefore

$$|a_i - a_i'| = |2\pi\bar{x}_i\sqrt{1 + f'(\bar{x}_i')^2}\,\Delta x - 2\pi\bar{x}_i'\sqrt{1 + f'(\bar{x}_i')^2}\,\Delta x|$$

$$= 2\pi\sqrt{1 + f'(\bar{x}_i')^2}\,\Delta x\,|\bar{x}_i - \bar{x}_i'|$$

$$\leq \pi\sqrt{1 + f'(\bar{x}_i')^2}\,\Delta x^2.$$

Therefore

$$\left| \sum_{i=1}^{n} a_i - \sum_{i=1}^{n} a'_i \right| \leq \sum_{i=1}^{n} |a_i - a'_i| \leq \sum_{i=1}^{n} \pi\sqrt{1 + f'(\bar{x}'_i)^2}\, \Delta x^2$$

$$= \Delta x \sum_{i=1}^{n} \pi\sqrt{1 + f'(\bar{x}'_i)^2}\, \Delta x.$$

Now

$$\lim_{|N| \to 0} \Delta x \sum_{i=1}^{n} \pi\sqrt{1 + f'(\bar{x}'_i)^2}\, \Delta x = 0 \cdot \int_a^b \pi\sqrt{1 + f'(x)^2}\, dx.$$

Therefore $|\sum a_i - \sum a'_i|$ is squeezed to 0, which was to be proved.

Let us try this formula on some problems to which we already know the answers.

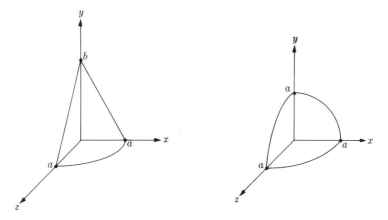

A cone is the surface of revolution of a segment. Here

$$f(x) = b - (b/a)x, \qquad f'(x) = -b/a,$$

and

$$A = \int_0^a 2\pi x \sqrt{1 + \frac{b^2}{a^2}}\, dx$$

$$= \frac{2\pi}{a} \sqrt{a^2 + b^2} \int_0^a x\, dx$$

$$= \pi a \sqrt{a^2 + b^2},$$

which is the right answer. Consider next a quadrant of a circle of radius a, rotated about the y-axis. In this case,

$$f(x) = \sqrt{a^2 - x^2} \qquad (0 \leq x \leq a),$$

$$f'(x) = \frac{-x}{\sqrt{a^2 - x^2}},$$

$$1 + f'(x)^2 = 1 + \frac{x^2}{a^2 - x^2} = \frac{a^2}{a^2 - x^2}.$$

Therefore the area is

$$A = \int_0^a 2\pi a \frac{x}{\sqrt{a^2 - x^2}}\, dx = 2\pi a\left[-\sqrt{a^2 - x^2}\right]_0^a = 2\pi a^2.$$

It follows that the total area of a sphere of radius a is $4\pi a^2$. This is the standard formula.

It is harder to find the area when we rotate a function-graph about the x-axis instead of the y-axis.

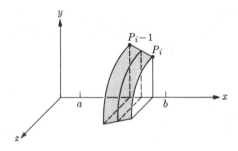

Given a nonnegative function f, on an interval $[a, b]$. As before, take a net N over $[a, b]$, with equal subdivisions, so that for each i,

$$x_i - x_{i-1} = \Delta x = \frac{b - a}{n} = |N|.$$

As before, we approximate the graph by a broken line B_n. Then we rotate B_n about the x-axis, getting a surface S_n, with area A_n. We define the area of the surface of revolution to be $\lim_{|N| \to 0} A_n$, if such a limit exists. We proceed to calculate:

$$A_n = \sum_{i=1}^n a_i,$$

where a_i is the area of the ith frustum, shown in the figure. As before,

$$P_{i-1}P_i = \sqrt{1 + f'(\bar{x}_i')^2}\, \Delta x,$$

where

$$x_{i-1} < \bar{x}_i' < x_i.$$

But when we rotate the chord from P_{i-1} to P_i about the x-axis, the "average circumference" is

$$2\pi \bar{r}_i = 2\pi \cdot \tfrac{1}{2}[f(x_{i-1}) + f(x_i)].$$

Obviously $\bar{r}_i$ is between $f(x_{i-1})$ and $f(x_i)$, because $\bar{r}_i$ is their average. By the no-jump theorem of Section 5.7,

$$\bar{r}_i = f(\bar{x}_i) \qquad (x_{i-1} \leq \bar{x}_i \leq x_i).$$

Therefore

$$A_n = \sum_{i=1}^n a_i = \sum_{i=1}^n 2\pi f(\bar{x}_i)\sqrt{1 + f'(\bar{x}_i')^2}\, \Delta x.$$

If it were true that $\bar{x}_i = \bar{x}'_i$ for each i, then the sum on the right-hand side would be a sample sum of the function

$$g(x) = 2\pi f(x)\sqrt{1 + f'(x)^2}.$$

As it stands, it is very close to being a sample sum. The idea is that

$$|N| = \Delta x \approx 0$$

$$\Rightarrow \quad \bar{x}_i \approx \bar{x}'_i \qquad \text{for each } i$$

$$\Rightarrow \quad f(\bar{x}_i) \approx f(\bar{x}'_i) \qquad \text{for each } i$$

$$\Rightarrow \quad \sum_{i=1}^{n} 2\pi f(\bar{x}_i)\sqrt{1 + f'(\bar{x}'_i)^2}\, \Delta x \approx \sum_{i=1}^{n} 2\pi f(\bar{x}'_i)\sqrt{1 + f'(\bar{x}'_i)^2}\, \Delta x$$

$$\approx \int_a^b 2\pi f(x)\sqrt{1 + f'(x)^2}\, dx.$$

At the end of the chapter, these ideas will be turned into a proof. Meanwhile let us look at some applications of the formula

$$A = \int_a^b 2\pi f(x)\sqrt{1 + f'(x)^2}\, dx.$$

1) If $f(x) = k$, on $[a, b]$, then the surface of revolution is a cylinder. By the integral formula,

$$A = \int_a^b 2\pi k\sqrt{1 + 0^2}\, dx = 2\pi k(b - a),$$

which is the right answer.

2) A sphere of radius a is the surface of revolution of a semicircle of radius a. Here

$$f(x) = \sqrt{a^2 - x^2}, \qquad x^2 + f^2 = a^2.$$

Hence

$$2x + 2ff' = 0, \qquad \text{and} \qquad f' = -\frac{x}{f}.$$

Therefore

$$1 + f'^2 = 1 + \frac{x^2}{f^2} = \frac{f^2 + x^2}{f^2} = \frac{a^2}{f^2},$$

and

$$A = \int_{-a}^a 2\pi f(x)\sqrt{\frac{a^2}{f(x)^2}}\, dx = \int_{-a}^a 2\pi a\, dx = 4\pi a^2.$$

PROBLEM SET 7.5

1. Let C_a be the circle with center at the origin and radius a, and let A be the arc of C_a lying above the interval $[-a/2, a/2]$. A is rotated about the x-axis. Find the area of the resulting surface. What proportion is this, of the total area of the sphere?

2. The entire circle C_a is rotated about the x-axis, giving a sphere of radius a. E_b and E_c are two planes, perpendicular to the x-axis, at $x = b$ and $x = c$; and S is the part of the sphere that lies between them. Find the area of S, in terms of a, b, and c. The form of your answer ought to suggest a somewhat surprising theorem which can be stated without the use of formulas. What is the theorem?

3. The circle with center at $(b, 0)$ and radius a, $a < b$, is rotated about the y-axis. The resulting surface is called a *torus*. Find its area.

4. The square with corners at the points $(a, 0)$, $(a + k, k)$, $(a + k, -k)$, and $(a + 2k, 0)$ is rotated about the y-axis. (Here $0 < k < a$.) Find the area of the resulting surface.

5. Find the volume of the solid obtained when the corresponding square *region* is rotated about the y-axis.

6. The same square is rotated about the line $x = a + 2k$. Find the surface area.

7. The square region is rotated about the line $x = a + 2k$. Find the volume.

8. The square with center at $(a, 0)$ and sides of length $2k$ parallel to the coordinate axes is rotated about the line $x = 2a$. Find the area of the resulting surface. (Here $0 < k < a$.)

9. Find the volume when the corresponding square *region* is rotated about the line $y = k/2$.

10. Consider the curve consisting of (a) the segment from $(0, 0)$ to $(a, 0)$, (b) the segment from $(0, 1)$ to $(a, 1)$, and (c) the semicircle, pointing outward, with endpoints at $(a, 0)$ and $(a, 1)$. This curve is to be rotated about the y-axis. For what value of a is it true that the total area of the resulting surface is equal to 15?

11. For each a, let S_a be the area of the surface described in Problem 10, and let V_a be the volume of the solid that it encloses. What value of a maximizes the ratio V_a/S_a?

12. The circle with center at (b, b) and radius a is rotated about the line

$$x + y = 1.$$

Here a and b are both positive, and the circle does not intersect the line. Find (a) the area of the resulting surface, and (b) the volume of the solid that it encloses.

13. Same question, for the circle with center at $(2, 10)$ and radius 1 and the line $x + y = 2$. (The only natural solutions of this, on the basis of the theory that we have so far, are rather clumsy. This suggests that some new ideas are needed.)

14. The graph of

$$y = 2x^2,$$

from $x = -1$ to $x = 1$, is rotated about the line $x = 5$. Find the area of the resulting surface.

15. If the same surface is rotated about the line $x = 4$, would the area of the resulting surface be greater, or would it be less, than the answer to Problem 14? Get a plausible answer to this, and justify it as well as you can.

16. The graph of $y = \frac{1}{2}(e^x + e^{-x})$, $0 \le x \le 1$, is rotated about the x-axis. Find the area of the resulting surface.

17. The same graph is rotated about the y-axis. Find the area of the resulting surface.

18. Let G be the graph of $f(x) = \sin x$, from $x = 0$ to $x = \pi/2$. G is first rotated about the line $x + y = 4$, and then about the line $x + y = 5$. Which of the resulting surfaces has the larger area? Why? (A right answer, with a plausibility argument, is acceptable

as an answer to this one. It is possible, however, to give a proof of the right answer, without calculating the area of either of the surfaces. That is, you can prove an inequality of the form $A < B$, without calculating either A or B.)

7.6 MOMENTS AND CENTROIDS. THE THEOREMS OF PAPPUS

The ideas in this section are mathematical descriptions of physical ideas. Given a finite set of "point masses" m_i, at the points $P_i = (x_i, y_i)$ in a coordinate plane, the *moment* (of the system) *about the y-axis* is

$$M_y = \sum_{i=1}^{n} x_i m_i.$$

The left-hand figure below shows the general case.

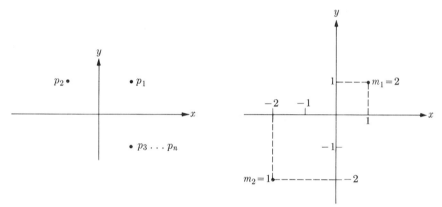

In the example on the right, we have:

$$M_y = x_1 m_1 + x_2 m_2 = 1 \cdot 2 + (-2) \cdot 1 = 0.$$

Physically speaking, this means that if the plane is horizontal, resting on a knife-edge along the y-axis, it will balance. The formula $\sum m_i x_i$ for M_y makes it plain that the effect of each point mass depends only on the product $m_i x_i$; if we divide m_i by 2, and double x_i, then the moment M_y is unchanged.

Similarly, the moment about the x-axis, of our finite system of point masses, is defined to be

$$M_x = \sum_{i=1}^{n} y_i m_i.$$

The total mass of all the particles in the system is denoted by m. That is,

$$m = \sum_{i=1}^{n} m_i.$$

The *centroid* of the system is defined to be the point
$$\bar{P} = (\bar{x}, \bar{y})$$
such that
$$M_y = \bar{x} m, \quad \text{and} \quad M_x = \bar{y} m.$$

Thus if we concentrate the entire mass of the system at $\bar{P}$, the moments about the x-axis and y-axis are unchanged.

For example, if we have $m_1 = 2$ at $P_1 = (1, 2)$ and $m_2 = 3$ at $P_2 = (2, 5)$, then

$$M_y = 2 + 6 = 8, \qquad M_x = 4 + 15 = 19,$$
$$m = \sum m_i = 5,$$
$$8 = \bar{x} \cdot 5, \qquad 19 = \bar{y} \cdot 5,$$
$$\bar{x} = \tfrac{8}{5}, \qquad \bar{y} = \tfrac{19}{5}.$$

The above discussion does not prove that M_y, M_x, and $\bar{P} = (\bar{x}, \bar{y})$ have any physical significance; only experiments can prove this. The fact, however, is that the physical conditions for equilibrium are described by moments and centroids.

Let us now consider how these ideas can be applied to a region R in the xy-plane. We shall think of R as a very thin sheet of homogeneous material, so that the mass per unit area is constant, say, $= 1$.

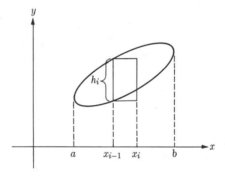

Suppose that we take a net over the interval $[a, b]$, as in the figure; for each x, we let $h(x)$ be the height of the cross section of R at x, and we let

$$h_i = h(x_i).$$

We use equal subdivisions, so that

$$x_i - x_{i-1} = \Delta x_i = \Delta x, \qquad \text{for each } i.$$

Then

$$m_i = h_i \, \Delta x$$

is the area of the rectangle in the figure. The rectangle is narrow, and so its moment about the y-axis should be approximately

$$x_i m_i = x_i h_i \, \Delta x.$$

If we approximate the region R by a finite set of such narrow rectangles, then the moment of R about the y-axis ought to be approximately

$$\sum_{i=1}^{n} x_i h_i \, \Delta x = \sum_{i=1}^{n} x_i h(x_i) \, \Delta x,$$

and the approximation ought to get better as the mesh Δx decreases. This is the idea of the following definition.

Definition. Let R be the region lying between the graphs of two continuous functions f_1 and f_2, on an interval $[a, b]$, with $f_1 \leqq f_2$, and let

$$h(x) = f_2(x) - f_1(x).$$

Then the *moment of R about the y-axis* is

$$M_y = \int_a^b x h(x)\, dx.$$

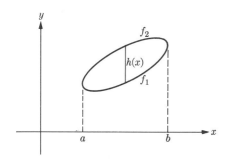 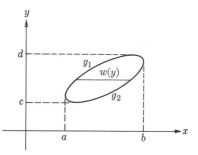

The definition of M_x is similar. Here (see the right-hand figure)

$$R = \{(x, y) \mid c \leqq y \leqq d \quad \text{and} \quad g_1(y) \leqq x \leqq g_2(y)\},$$
$$w(y) = g_2(y) - g_1(y),$$

and by definition,

$$M_x = \int_c^d y w(y)\, dy.$$

Since the total mass of R is its area

$$A = \int_a^b h(x)\, dx = \int_c^d w(y)\, dy,$$

it is natural to define the centroid of R as the point $\bar{P} = (\bar{x}, \bar{y})$ such that

$$M_y = \bar{x}A, \qquad M_x = \bar{y}A.$$

For example, consider a quadrant of a circle of radius a.

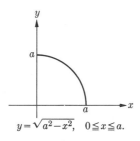

$$y = \sqrt{a^2 - x^2}, \quad 0 \leqq x \leqq a.$$

Here
$$M_y = \int_0^a x\sqrt{a^2 - x^2}\, dx = [-\tfrac{1}{2} \cdot \tfrac{2}{3}(a^2 - x^2)^{3/2}]_0^a = \tfrac{1}{3}a^3.$$

Obviously
$$A = \tfrac{1}{4}\pi a^2,$$

and so
$$\tfrac{1}{3}a^3 = \bar{x} \cdot \tfrac{1}{4}\pi a^2.$$

Therefore
$$\bar{x} = \frac{4}{3\pi}\, a.$$

By symmetry, interchanging x and y, we get:
$$\bar{y} = \bar{x} = \frac{4}{3\pi}\, a.$$

The moment about the line $x = x_0$ is defined to be
$$M_{x=x_0} = \int_a^b (x - x_0)h(x)\, dx,$$

and the moment about the line $y = y_0$ is
$$M_{y=y_0} = \int_c^d (y - y_0)w(y)\, dy.$$

It is now easy to see that:

Theorem 1. $M_{x=\bar{x}} = 0 = M_{y=\bar{y}}.$

Proof. For any x_0,
$$M_{x=x_0} = \int_a^b (x - x_0)h(x)\, dx = \int_a^b xh(x)\, dx - x_0\int_a^b h(x)\, dx = M_y - x_0 A.$$

This is 0 for $x_0 = \bar{x}$. The proof of the other half of the theorem is the same. In fact, the equation
$$M_{x=x_0} = M_y - x_0 A$$

shows that the converse of Theorem 1 is also true.

Theorem 2. If $M_{x=x_0} = 0$, then $x_0 = \bar{x}$; and if $M_{y=y_0} = 0$, then $y_0 = \bar{y}$.

Centroids are easy to find for regions which are *symmetric*, in the sense now to be defined. Two points P and P' are *symmetric across the line* L if L is the perpendicular bisector of the segment between them. In the left-hand figure below, we say that P' is the point symmetrically across L from P. A figure is *symmetric about a line* L if for each point P of the figure, P' also lies in the figure. The figure may be either a region or a curve. For example, a circle is symmetric about any line through its center, and the interior of a circle has the same property.

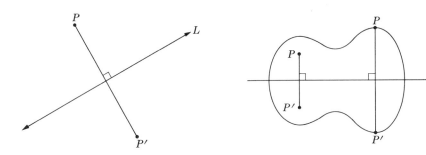

It is easy to see that if R is symmetric about the y-axis, then $\bar{x} = 0$. In the figure on the left, $h(x)$ is an *even* function, with $h(-x) = h(x)$. Therefore $xh(x)$ is an *odd* function, with $(-x)h(-x) = -[xh(x)]$. Therefore

$$M_y = \int_{-a}^{a} xh(x) = 0,$$

and $\bar{x} = 0$.

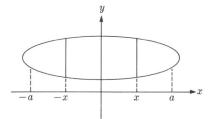 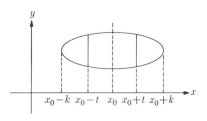

More generally, as in the right-hand figure, we have:

Theorem 3. If R is symmetric about the line $x = x_0$, then $\bar{x} = x_0$.

Proof.

$$M_{x=x_0} = \int_{x_0-k}^{x_0+k} (x - x_0)h(x)\, dx = \int_{x_0-k}^{x_0+k} \phi(x)\, dx.$$

By symmetry,

$$h(x_0 - t) = h(x_0 + t)$$

for every t; and so

$$\phi(x_0 - t) = [(x_0 - t) - x_0]h(x_0 - t) = -th(x_0 - t)$$
$$= -\phi(x_0 + t).$$

Therefore the graph of ϕ must be like the graph shown below.

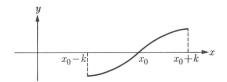

Therefore

$$\int_{x_0-k}^{x_0} \phi(x) \, dx = -\int_{x_0}^{x_0+k} \phi(x) \, dx,$$

and

$$M_{x=x_0} = \int_{x_0-k}^{x_0+k} \phi(x) \, dx = 0.$$

It follows that $x_0 = \bar{x}$.

In this proof, all that we have used is the assumption that

$$h(x_0 - t) = h(x_0 + t).$$

This condition may hold for regions which are not symmetric, as below.

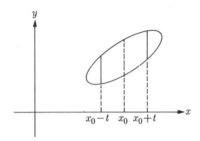

And interchanging x and y, we get the following theorem.

Theorem 4. If R is symmetric about the line $y = y_0$, then $\bar{y} = y_0$.

These ideas have the following geometric consequence:

Theorem 5 (*Pappus' theorem, for volumes*). If a region is rotated about a line not intersecting it, then the volume of the resulting solid is equal to the area of the region times the circumference of the circle described by the centroid.

That is, if the region below is rotated about the y-axis, then

$$V = 2\pi\bar{x}A.$$

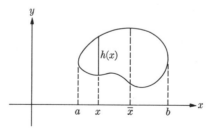

Proof. By the method of shells,

$$V = \int_a^b 2\pi x h(x) \, dx.$$

Therefore

$$V = 2\pi M_y = 2\pi\bar{x}A,$$

because $\bar{x}$ was defined by the equation

$$M_y = \bar{x}A.$$

Pappus' theorem can be applied in two ways. If we know $\bar{x}$ and A, we can compute $V = 2\pi\bar{x}A$; and if we know V and A, we can solve for $\bar{x} = V/2\pi A$. For example, consider a circular region R, of radius a, with center at the point $(b, 0)$, $b > a$. When R is rotated about the y-axis, we get a solid which is called a *solid torus*. (The surface of the solid is called a *torus*.) By Pappus' theorem, we get

$$V = 2\pi b \cdot \pi a^2 = 2\pi^2 a^2 b.$$

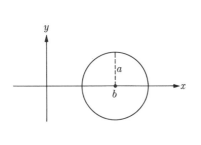 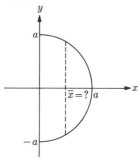

We can use the theorem in reverse to find the centroid of a semicircular region. If the region is rotated about the y-axis, we get a sphere of radius a, with volume

$$V = \tfrac{4}{3}\pi a^3.$$

Obviously

$$A = \tfrac{1}{2}\pi a^2.$$

Therefore

$$\tfrac{4}{3}\pi a^3 = 2\pi\bar{x} \cdot \tfrac{1}{2}\pi a^2,$$

and

$$\bar{x} = \frac{4}{3\pi} \cdot a.$$

These ideas apply also to arcs. We shall think of an arc as a thin homogeneous wire whose mass per unit length is constant, say, $= 1$. Suppose that the arc is the graph of a function f, on an interval $[a, b]$. As usual, we take a net over $[a, b]$, with equal subdivisions. The arc length over the interval $[x_{i-1}, x_i]$ is

$$s_i = \int_{x_{i-1}}^{x_i} \sqrt{1 + f'(x)^2} \, dx.$$

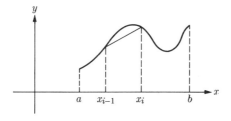

Now
$$s_i \approx \sqrt{1 + f'(x_i)^2}\, \Delta x;$$
the moment of this little arc about the y-axis ought to be approximately $x_i s_i$; and so the moment of the whole graph ought to be
$$M_y \approx \sum_{i=1}^{n} x_i \sqrt{1 + f'(x_i)^2}\, \Delta x.$$

Definitions. Given the function f, with a continuous derivative f', on $[a, b]$, the moment of the graph about the y-axis is
$$M_y = \int_a^b x\sqrt{1 + f'(x)^2}\, dx;$$
and the moment about the line $x = x_0$ is
$$M_{x=x_0} = \int_a^b (x - x_0)\sqrt{1 + f'(x)^2}\, dx.$$

Similarly, we state the following:

Definitions. The moment of the graph of f about the x-axis is
$$M_x = \int_a^b f(x)\sqrt{1 + f'(x)^2}\, dx,$$
$$M_{y=y_0} = \int_a^b (f(x) - y_0)\sqrt{1 + f'(x)^2}\, dx;$$
and the centroid of the graph is the point $\bar{P} = (\bar{x}, \bar{y})$ for which
$$M_y = \bar{x}L, \qquad M_x = \bar{y}L,$$
where L is the total arc length.

Our previous theorems for regions now have analogous forms for arcs, as follows.

Theorem 6. If the graph of f is symmetric about a line $x = x_0$ (or $y = y_0$) then this line contains the centroid.

Theorem 7. If the graph of f is rotated about a horizontal or vertical line not intersecting the graph, then the area of the resulting surface of revolution is equal to the length of the arc times the circumference of the circle described by the centroid.

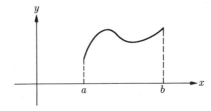

For example, if we rotate about the y-axis, then the area of the resulting surface is

$$S = \int_a^b 2\pi x \sqrt{1 + f'(x)^2}\, dx$$
$$= 2\pi M_y = 2\pi \bar{x} L,$$

by definition of $\bar{x}$. The proof of the theorem in the other cases is similar.

Throughout this section, we have used a fixed coordinate system to define and investigate moments and centroids of regions and arcs. It is a fact, however, that moments and centroids do not depend on the choice of a coordinate system; they depend only on the regions and the arcs. In particular, any line of symmetry (horizontal, vertical, or sloping) must contain the centroid. You may use this fact in the problem set below.

PROBLEM SET 7.6

1. Let A, B, and C be the points $(0, 0)$, $(a, 0)$, and (b, c), $a, c > 0$. At each of these points there is a particle of mass 1. Find the centroid of the resulting system.

2. Suppose at the points A, B, and C of Problem 1 there are particles of mass 1, 2, and 3 respectively. Find the centroid of the resulting system.

3. A *median* of a triangle is a segment between a vertex and the midpoint of the opposite side. Show that every median of the triangle described in Problem 1 passes through the centroid.

4. Now consider the triangular region R determined by the same points A, B, and C. Find the centroid of R.

5. The region R is rotated about the x-axis. Find the volume.

6. R is rotated about the y-axis. Find the volume.

7. The figure formed by the sloping sides of the triangle is rotated about the x-axis. Find the area of the resulting surface.

8. Same question, for rotation about the y-axis, assuming $b \geqq 0$.

9. A trapezoid has vertices $(0, 0)$, (a, c), $(b - a, c)$, $(b, 0)$ (with $b > a > 0$ and $c > 0$). Find the centroid of the region T bounded by this trapezoid.

10. The region T is rotated about the x-axis. Find the volume.

11. The region T is rotated about the y-axis. Find the volume.

12. The figure formed by the four sides of the trapezoid is rotated about the y-axis. Find the surface area.

13. The circle with center at $(b, 0)$ and radius a, with $0 < a < b$, is rotated about the y-axis. Find the area of the resulting torus.

14. Let the arc A be the portion of the circle with center at the origin and radius a which lies in the first quadrant. Find the centroid of A.

15. The square with corners at $(a, 0)$, $(a + k, k)$, $(a + k, -k)$, $(a + 2k, 0)$ is rotated about the y-axis. (Here $0 < k < a$.) Find the area of the resulting surface.

16. Find the volume of the solid obtained if the corresponding square *region* is rotated.

17. The same square is rotated about the line $x = a + 2k$. Find the surface area.

18. The square region is rotated about the line $x = a + 2k$. Find the volume.

19. Consider the curve consisting of: (a) the segment from $(0, 0)$ to $(a, 0)$; (b) the segment from $(0, 1)$ to $(a, 1)$; and (c) a semicircle, pointing outward, with endpoints at $(a, 0)$ and $(a, 1)$. This curve is to be rotated about the y-axis. For what value of a is it true that the total area of the resulting surface is equal to 15?

20. For each a let S_a be the area of the surface described in Problem 19, and let V_a be the volume of the solid that it encloses. What value of a maximizes the ratio V_a/S_a?

7.7 IMPROPER INTEGRALS

The definite integral is defined as a limit of sample sums as the mesh of the net approaches 0. This limit exists if the integrand f is continuous. But this definition of the integral does not apply to the function

$$f(x) = \frac{1}{\sqrt{x}}$$

on the half-open interval from 0 to 1.

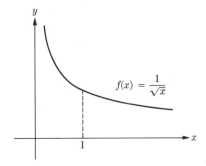

It really is the half-open interval

$$(0, 1] = \{x \mid 0 < x \leqq 1\}$$

that we are dealing with, because at $x = 0$ the function is not defined. On this half-open interval the function is unbounded. Therefore, for every net over $(0, 1]$ we can form a sample sum as large as we please, by taking the first sample point x_1' close to 0. Thus,

$$\sum_{i=1}^{n} f(x_i') \Delta x_i > f(x_1') \Delta x_1 = \frac{1}{\sqrt{x_1'}} \Delta x_1,$$

the sample sum is large when x_1' is small, and so the sample sums do not approach a limit as the mesh approaches 0.

Nevertheless, we can extend the definition of the integral in such a way that our problem has an answer. The function $f(x) = 1/\sqrt{x}$ is defined and continuous on every closed interval $[a, 1]$, where $a > 0$. Therefore $\int_a^1 (1/\sqrt{x})\, dx$ is well defined.

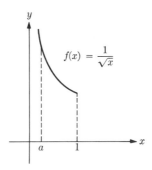

We define a new kind of integral by saying that

$$\int_0^1 \frac{dx}{\sqrt{x}} = \lim_{a \to 0^+} \int_a^1 \frac{dx}{\sqrt{x}},$$

if the indicated limit on the right exists. (We write $a \to 0^+$, because a takes on only positive values.) In the present case, the limit exists and is finite:

$$\int_a^1 \frac{dx}{\sqrt{x}} = \int_a^1 x^{-1/2}\, dx = [2x^{1/2}]_a^1 = 2 - 2\sqrt{a}.$$

Therefore

$$\lim_{a \to 0^+} \int_a^1 \frac{dx}{\sqrt{x}} = \lim_{a \to 0^+} [2 - 2\sqrt{a}] = 2.$$

There are similar-looking problems for which the limit is infinite. For example,

$$\int_0^1 \frac{dx}{x^2} = \lim_{a \to 0^+} \int_a^1 \frac{dx}{x^2},$$

if the limit exists. In this case,

$$\int_a^1 \frac{dx}{x^2} = \left[-\frac{1}{x} \right]_a^1 = -1 + \frac{1}{a} \qquad (a > 0),$$

and so

$$\lim_{a \to 0^+} \int_a^1 \frac{dx}{x^2} = \infty,$$

in the sense defined in Section 5.3.

We abbreviate this by writing

$$\int_0^1 \frac{dx}{x^2} = \infty.$$

The same test can be applied at any point where the function "blows up," as long as there is only one such point, at an endpoint of the interval. For example, consider

$$\int_0^{\pi/2} \sec x\, dx = \lim_{a \to \pi/2^-} \int_0^a \sec x\, dx,$$

where the minus sign means that $a \to \pi/2$ through values less than $\pi/2$.

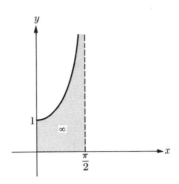

Now

$$\int_0^a \sec x \, dx = [\ln |\sec x + \tan x|]_0^a$$

$$= \ln |\sec a + \tan a| - \ln |1 + 0|$$

$$= \ln |\sec a + \tan a|.$$

As $a \to \pi/2^-$,

$$\sec a = \frac{1}{\cos a} \to \infty \quad \text{and} \quad \tan a = \frac{\sin a}{\cos a} \to \infty.$$

Therefore $\ln [\sec a + \tan a] \to \infty$, and

$$\lim_{a \to \pi/2^-} \int_0^a \sec x \, dx = \infty.$$

We use the same method to define and evaluate such integrals as

$$\int_1^\infty \frac{dx}{x^2} .$$

Here the integration is supposed to be carried out all the way to the right, starting at $x = 1$. Again our definition of the definite integral (as the limit of the sample sums) does not apply, and so we define the improper integral as a limit:

$$\int_1^\infty \frac{dx}{x^2} = \lim_{a \to \infty} \int_1^a \frac{dx}{x^2} ,$$

if the limit on the right exists. For the function

$$g(a) = \int_1^a \frac{dx}{x^2} ,$$

the limit exists and is finite:

$$\int_1^a \frac{dx}{x^2} = \left[-\frac{1}{x} \right]_1^a = -\frac{1}{a} + 1;$$

and so

$$\lim_{a \to \infty} \int_1^a \frac{dx}{x^2} = \lim_{a \to \infty} \left[1 - \frac{1}{a} \right] = 1.$$

Here we have a second example of an unbounded figure with a finite area.

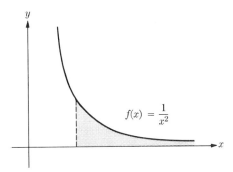

For very similar-looking functions, the integral from 1 to ∞ may be infinite. For example,

$$\int_1^\infty \frac{dx}{x} = \lim_{a \to \infty} \int_1^a \frac{dx}{x} = \lim_{a \to \infty} [\ln x]_1^a = \lim_{a \to \infty} \ln a = \infty.$$

Integrals of the type that we have been discussing here are called *improper integrals*, of the first and second kind respectively. The two kinds can occur in combination. For example,

$$\int_0^\infty \frac{dx}{\sqrt{x}(1 + x)}$$

is improper in two ways: the function blows up at the lower limit, and also the upper limit is ∞. Thus, for the integral to be finite, the two limits

$$\lim_{a \to 0^+} \int_a^1 \frac{dx}{\sqrt{x}(1 + x)}, \qquad \lim_{b \to \infty} \int_1^b \frac{dx}{\sqrt{x}(1 + x)}$$

must both be finite; if they are, then the integral from 0 to ∞ is their sum. (Any positive number k would have done equally well in place of 1.) In this case, both limits are finite.

Improper integrals may appear in a disguised form, and so we need to be careful. For example, a careless calculation gives

$$\int_{-1}^1 \frac{dx}{x^2} = \left[\frac{-1}{x} \right]_{-1}^1 = -1 - 1 = -2.$$

This is impossible, because the integrand is positive. The trouble here is that the function blows up at $x = 0$, and so we need to make separate investigations of the integrals

$$\int_{-1}^0 \frac{dx}{x^2}, \qquad \int_0^1 \frac{dx}{x^2}.$$

We find that both these limits are equal to ∞. Therefore the original integral is equal to ∞.

Note that when we got the "answer" -2 we had no right to complain that the theory was wrong, because the Fundamental Theorem of Integral Calculus does not apply to functions whose domains have holes in them; the theorem applies only to functions which are defined and continuous on the interval over which we are integrating.

An even worse example of the same kind is

$$\int_{-1}^{1} \frac{dx}{x} = [\ln |x|]_{-1}^{1} = \ln 1 - \ln 1 = 0 \quad (?).$$

Splitting this into two parts, we get

$$\int_{-1}^{0} \frac{dx}{x} = \lim_{a \to 0^-} [\ln |x|]_{-1}^{a} = \lim_{a \to 0^-} \ln |a| = -\infty,$$

$$\int_{0}^{1} \frac{dx}{x} = \lim_{a \to 0^+} [\ln |x|]_{a}^{1} = \lim_{a \to 0^+} [-\ln a] = \infty.$$

The limits $-\infty$ and ∞ do not combine to give a well-defined limit, either finite or infinite, and therefore the original integral is not defined at all. We also get no answer for

$$\int_{0}^{\infty} \sin x \, dx.$$

Here

$$\int_{0}^{a} \sin x \, dx = [-\cos x]_{0}^{a} = -\cos a + 1,$$

which oscillates forever between 0 and 2, and therefore does not approach a limit.

Thus, when we investigate an improper integral, there are three situations that we may encounter.

1) The integral may exist, as a finite limit. For example,

$$\int_{1}^{\infty} \frac{dx}{x^2} = \lim_{a \to \infty} \int_{1}^{a} \frac{dx}{x^2} = \lim_{a \to \infty} \left[-\frac{1}{x} \right]_{1}^{a} = 1.$$

2) The integral may exist, as an infinite limit ∞ or $-\infty$. For example,

$$\int_{1}^{\infty} \frac{dx}{x} = \lim_{a \to \infty} \int_{1}^{a} \frac{dx}{x} = \lim_{a \to \infty} \ln a = \infty.$$

3) Finally, the limit may not exist as *any* limit, finite or infinite. This is what we find for

$$(?) \quad \int_{-1}^{1} \frac{dx}{x} .$$

In the following problem set, when you are asked to "investigate" an improper integral, you should find out which of the above three cases it represents. If it is Case 1, you should find the limit, unless the contrary is stated.

PROBLEM SET 7.7

Investigate:

1. $\displaystyle\int_1^\infty \frac{dx}{1+x^2}$

2. $\displaystyle\int_1^\infty \frac{dx}{x^2+x}$

3. $\displaystyle\int_1^\infty \frac{dx}{x^3}$

4. $\displaystyle\int_0^\infty \frac{dx}{x^3}$

5. $\displaystyle\int_0^1 \frac{dx}{\sqrt{x}}$

6. $\displaystyle\int_1^\infty \frac{dx}{\sqrt{x}}$

7. $\displaystyle\int_1^\infty \frac{dx}{x^{1.0001}}$

8. $\displaystyle\int_1^\infty \frac{dx}{x^{0.9999}}$

9. $\displaystyle\int_2^\infty \frac{dx}{x \ln x}$

10. $\displaystyle\int_1^2 \frac{dx}{x \ln x}$

11. $\displaystyle\int_0^1 \frac{dx}{x^{1.0001}}$

12. $\displaystyle\int_0^1 \frac{dx}{x^{0.9999}}$

13. $\displaystyle\int_2^\infty \frac{1}{(x-1)^3}\, dx$

14. $\displaystyle\int_1^2 \frac{1}{(x-1)^3}\, dx$

15. $\displaystyle\int_0^\infty e^{-x}\, dx$

16. $\displaystyle\int_0^\infty xe^{-x}\, dx$

17. $\displaystyle\int_0^\infty x^2 e^{-x}\, dx$

18. $\displaystyle\int_0^\infty \frac{dx}{\sqrt{x}(1+x)}$

19. $\displaystyle\int_1^\infty \frac{x\, dx}{1+x^4}$

20. Show that $\int_0^\infty x^n\, dx$ is never finite, for any value of n, positive, negative, or zero. (The point is that something always goes wrong, either at 0 or at ∞.)

21. Consider the graph of $f(x) = 1/x$, $1 \leqq x < \infty$. Let R be the region under the graph; let S be the solid of revolution (about the x-axis); and let T be the *surface* of revolution. Investigate the improper integrals which represent (a) the area of R, (b) the volume of S, and (c) the area of T.

Investigate the following for existence. (That is, find out *whether* the integral represents a finite limit; but in the cases where it does, you need not calculate the limit.)

22. $\displaystyle\int_1^\infty \frac{dx}{1+e^x}\, dx$

23. $\displaystyle\int_2^\infty \frac{dx}{1+x\ln x}$

24. $\displaystyle\int_0^\infty \frac{dx}{1+\sqrt{x}(1+x)}$

25. $\displaystyle\int_1^\infty \frac{x^2 e^{-x}}{1+x}\, dx$

26. $\displaystyle\int_{-\infty}^\infty e^{-x^2}\, dx$

(To show that this is finite, it will be sufficient to show that $\int_1^\infty e^{-x^2}\, dx < \infty$. It will then follow by symmetry that $\int_{-\infty}^{-1} e^{-x^2} \times dx < \infty$. And obviously $\int_{-1}^1 e^{-x^2}\, dx < \infty$, because e^{-x^2} is continuous on $[-1, 1]$.)

27. $\displaystyle\int_{-\infty}^\infty e^{-x^3}\, dx$

28. $\displaystyle\int_1^\infty \frac{1}{1+x^4}\, dx$

29. $\displaystyle\int_2^\infty \frac{dx}{\sqrt{x}\ln x}$

30. $\displaystyle\int_{-\infty}^\infty e^{-x^4}\, dx$

31. $\displaystyle\int_1^\infty \left(\frac{\pi}{2} - \mathrm{Tan}^{-1} x\right) dx$

32. $\displaystyle\int_0^{\pi/2} \tan x\, dx$

33. $\displaystyle\int_0^{\pi/2} \csc x\, dx$

34. $\displaystyle\int_1^\infty x^2 |\sin x|\, dx$

35. $\displaystyle\int_\pi^\infty \frac{|\sin x|}{x^2}\, dx$

*36. $\displaystyle\int_\pi^\infty \frac{\sin x}{x^2}\, dx$

**37. a) Show that for each n,

$$\left| \int_{\sqrt{n\pi}}^{\sqrt{(n+1)\pi}} \sin x^2 \, dx \right| < \left| \int_{\sqrt{(n-1)\pi}}^{\sqrt{n\pi}} \sin x^2 \, dx \right| .$$

b) Investigate $\displaystyle\int_{\sqrt{\pi}}^{\infty} \sin x^2 \, dx$.

38. Let f be a decreasing function, with a continuous derivative, on the interval $[a, \infty)$. The graph (and the region under it) are rotated about the x-axis. Show that if the surface area is finite, then so also is the volume.

*7.8 THE INTEGRABILITY OF CONTINUOUS FUNCTIONS

Let f be a function which is continuous on an interval I. The function f is continuous if for each x, and each $\epsilon > 0$, the graph has an $\epsilon\delta$-box at the point $(x, f(x))$. The total height of an $\epsilon\delta$-box is $h = 2\epsilon$. The box is then called an *h-box* of every point of the graph that lies in it.

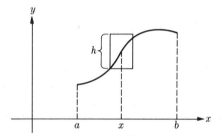

The definition of continuity applies to the points x of the interval, one at a time. It may appear, therefore, that if f is continuous at each point x of the interval, then we have to use infinitely many boxes (one for each x) in order to exhibit the fact. But if I is a closed interval, this is not so:

Theorem 1 (*The finite covering theorem*). Let f be a continuous function on the closed interval $[a, b]$, and let h be a positive number. Then there is a finite collection of h-boxes, covering the entire graph of f.

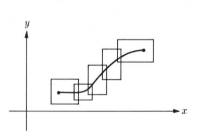

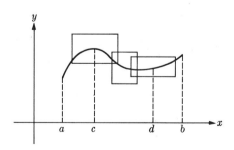

Proof. Let $[c, d]$ be a subinterval of $[a, b]$. If there is a finite collection of h-boxes, covering the part of the graph determined by $[c, d]$, then we shall say that $[c, d]$ is *good*.

If no such finite collection exists, then we say that $[c, d]$ is *bad*. We allow the case in which $[c, d]$ is all of $[a, b]$. Thus what we need to show is that $[a, b]$ is good. Suppose, then, that $[a, b]$ is bad. We shall show that this leads to a contradiction.

Let $[a_1, b_1] = [a, b]$. If the left-hand half of $[a_1, b_1]$ and the right-hand half of $[a_1, b_1]$ are both good, then it follows that $[a_1, b_1]$ is good; we can fit together two finite collections of boxes, getting another finite collection of boxes that covers the whole graph. Therefore one or both of the halves of $[a_1, b_1]$ must be bad. Let $[a_2, b_2]$ be a bad half of $[a_1, b_1]$. Similarly, one of the halves of $[a_2, b_2]$ must be bad. Let $[a_3, b_3]$ be a bad half of $[a_2, b_2]$.

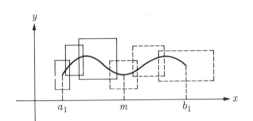 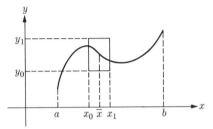

Proceeding to infinity in this way, we get a nested sequence

$$[a_1, b_1], [a_2, b_2], \ldots, [a_n, b_n], \ldots$$

of closed intervals, each of which is bad. By the nested interval postulate, there is a number $\bar{x}$ which lies on all of these intervals.

Now f is continuous at $\bar{x}$. Therefore f has an h-box at $(\bar{x}, f(\bar{x}))$. Suppose that the box is $\{(x, y) \mid x_0 < x < x_1, y_0 < y < y_1\}$. Since

$$b_n - a_n = \frac{b - a}{2^{n-1}},$$

the length of the nth interval $[a_n, b_n]$ approaches 0. Therefore we must have

$$x_0 < a_n < b_n < x_1 \qquad \text{for some } n.$$

This means that $[a_n, b_n]$ is good after all: *one h-box covers the part of the graph that lies above it;* and 1 is finite.

We continue now at the point where we left off in Section 7.2. There we defined *net, mesh* (of a net), *upper sum $S(N)$, lower sum $s(N)$,* and *sample sum $\sum (X)$.* Theorem 3 of Section 7.2 was as follows:

Theorem. If f is continuous, and

$$\lim_{|N| \to 0} [S(N) - s(N)] = 0,$$

then the sample sums approach a limit, as $|N| \to 0$.

Now $\int_a^b f(x)\, dx$ is defined to be $\lim_{|N| \to 0} \sum (X)$. Therefore what we need, to complete the proof that continuous functions are integrable, is the following theorem:

Theorem 2. If f is continuous on $[a, b]$, then

$$\lim_{|N| \to 0} [S(N) - s(N)] = 0.$$

Proof. Let $\epsilon > 0$ be given. We need to show that there is a $\delta > 0$ such that

$$|N| < \delta \quad \Rightarrow \quad S(N) - s(N) < \epsilon.$$

We know by the finite covering theorem that for every $h > 0$ there is a finite collection of h-boxes, covering the graph. (See the left-hand figure below. We have not yet decided what h we want to use.) The x-coordinates of the vertical sides of the boxes, together with a and b, form a net N_0 over $[a, b]$. Let δ be the length of the shortest interval in N_0. We assert that if N is any other net over $[a, b]$, with $|N| < \delta$, then every little interval $[x_{i-1}, x_i]$ in N lies under some one of our boxes. We illustrate

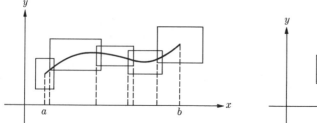
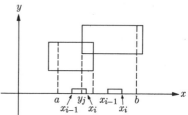

with the simpler figure on the right. If $[x_{i-1}, x_i]$ contains no point of N_0 (as on the right) this is evident. If $[x_{i-1}, x_i]$ contains a point y_j of N_0 (as on the left), then y_j lies on the *open* interval under one of our boxes, and so $[x_{i-1}, x_i]$ lies under the same box.

Now take a net N, with $|N| < \delta$. The difference $S(N) - s(N)$ is the sum of the areas of a collection of rectangles, like this:

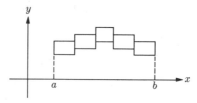

Each of these rectangles lies in one of our h-boxes. (Why?) Therefore each of them has height $\leq h$. Hence

$$S(N) - s(N) \leqq h(b - a).$$

Thus we want

$$h(b - a) < \epsilon,$$

and this will hold if

$$h < \frac{\epsilon}{b - a}.$$

This is the way we should choose h at the beginning of the proof. The resulting δ is the δ that we need.

Theorem 3. Every continuous function on a closed interval is integrable.

Proof. By the preceding theorem, $\lim [S(N) - s(N)] = 0$. By Theorem 3 of Section 7.2, f is integrable.

Some of the ideas in this proof are worth examining further. In Problem Set 7.2. we gave the following definition.

Definition. Let f be a function on an interval I. Suppose that for every $\epsilon > 0$ there is a $\delta > 0$ such that

$$|x - x'| < \delta \;\Rightarrow\; |f(x) - f(x')| < \epsilon,$$

where x and x' are any two points of I. Then f is *uniformly continuous* on I.

Note that while continuity is defined for one point x at a time, uniform continuity is defined for the graph as a whole. The difference between these ideas may be clarified by an analogy:

1) A man is *literate* if there is a language that he can read and write.
2) A group of men is *literate* if each of its members is literate.
3) A group of men is *uniformly literate* if there is one language that every member of the group can read and write.

Thus, *uniform literacy* is a property not of the individuals in a group but of the group as a whole; if each of the members of the group is literate, it follows that the group is literate [see (2)], but it does not follow that the group is uniformly literate.

The difference between continuity of a function f on an interval I and uniform continuity of f on I is analogous. For example, $f(x) = 1/x$ is continuous on the open interval $I = (0, 1)$, because f is continuous at every point x of I. But f is not uniformly continuous on I. (For every $\epsilon > 0$, we can find two points x, x', as close together as we please, such that $|f(x) - f(x')| > \epsilon$.)

But continuity implies uniform continuity when the domain of the function is a closed interval.

Theorem 4 (*the uniform continuity theorem*). If f is continuous on $[a, b]$, then f is uniformly continuous on $[a, b]$.

Proof. Let $\epsilon > 0$ be given. By the finite covering theorem there is a finite collection of boxes of height ϵ, covering the graph. (We are using ϵ as the h of Theorem 1.) Let N_0 be the corresponding net over $[a, b]$, as in the proof of Theorem 2. As before, let δ be the length of the shortest interval in N_0. It follows that if $|x - x'| < \delta$, then x and x' lie under some one of our boxes. Therefore $|f(x) - f(x')| < \epsilon$, which was to be proved.

This is the idea that we need, to complete the proof of the formula

$$A = \int_a^b 2\pi f(x)\sqrt{1 + f'(x)^2}\, dx$$

for the area of a surface of revolution about the x-axis. In Section 7.5, we knew that as $|N| \to 0$, the sample sum

$$\Sigma' = \sum_{i=1}^{n} 2\pi f(\bar{x}_i')\sqrt{1 + f'(\bar{x}_i')^2}\, \Delta x$$

approaches the integral. But the area of the approximating surface was

$$\Sigma = \sum_{i=1}^{n} 2\pi f(\bar{x}_i)\sqrt{1 + f'(\bar{x}_i')^2}\, \Delta x,$$

with two different sample points $\bar{x}_i$, $\bar{x}_i'$ used on each interval $[x_{i-1}, x_i]$. Thus we need to show that

$$\lim_{|N| \to 0} |\Sigma' - \Sigma| = 0.$$

We are assuming that f' is continuous. Therefore so also is $2\pi\sqrt{1 + f'(x)^2}$. Therefore the latter function is bounded. Let M be such that

$$2\pi\sqrt{1 + f'(x)^2} < M, \qquad \text{for every } x.$$

Let ϵ be any positive number. Then

$$\frac{\epsilon}{M(b - a)} > 0.$$

By the uniform continuity theorem, there is a $\delta > 0$ such that

$$|x - x'| < \delta \quad \Rightarrow \quad |f(x) - f(x')| < \frac{\epsilon}{M(b - a)}.$$

This is the δ that we need.

Proof.

$$|N| < \delta \quad \Rightarrow \quad |\bar{x}_i - \bar{x}_i'| < \delta \qquad \text{for each } i$$

$$\Rightarrow \quad |f(\bar{x}_i) - f(\bar{x}_i')| < \frac{\epsilon}{M(b - a)} \qquad \text{for each } i.$$

Now

$$\Sigma - \Sigma' = \sum_{i=1}^{n} 2\pi[f(\bar{x}_i) - f(\bar{x}_i')]\sqrt{1 + f'(\bar{x}_i')^2}\, \Delta x,$$

and so

$$|\Sigma - \Sigma'| \leq \sum_{i=1}^{n} |f(\bar{x}_i) - f(\bar{x}_i')| \cdot 2\pi\sqrt{1 + f'(\bar{x}_i')^2}\, \Delta x$$

$$< \sum_{i=1}^{n} M \cdot |f(\bar{x}_i) - f(\bar{x}_i')|\, \Delta x.$$

Therefore

$$|N| < \delta \quad \Rightarrow \quad |\Sigma - \Sigma'| < \sum_{i=1}^{n} M \cdot \frac{\epsilon}{M(b - a)} \Delta x = M \cdot \frac{\epsilon}{M(b - a)} \sum_{i=1}^{n} \Delta x$$

$$= \frac{\epsilon}{b - a} \cdot (b - a) = \epsilon,$$

which was to be proved.

*PROBLEM SET 7.8

Most of the questions below can be answered on the basis of a careful reexamination of the theorems and proofs in Sections 7.2 and 7.8. Some of them, however, require independent investigation. Naturally, all answers should be explained.

1. Suppose that f is known to be increasing on $[a, b]$, but is not known to be continuous. Does it follow that f is integrable?

2. Show that Tan^{-1} is uniformly continuous on $(-\infty, \infty)$.

3. Same, for $f(x) = x$ on $(-\infty, \infty)$.

4. Is it possible for a function to be uniformly continuous on an open interval (a, b)? Why or why not?

5. If a function is uniformly continuous on an interval I, does it follow that f is continuous on I? Why or why not?

6. Suppose that f is (a) continuous at a, (b) continuous at b, and (c) uniformly continuous on (a, b). Does it follow that f is uniformly continuous on $[a, b]$?

*7. Suppose that f is bounded and integrable, but not necessarily continuous, on $[a, b]$. For each x of $[a, b]$, let

$$F(x) = \int_a^x f(t)\, dt.$$

Show that F is continuous. (The betweenness theorem for integrals, which is Theorem 5 of Section 7.2, may be useful here.)

*8. For the definition of *Lipschitzian*, see Problem 12a of Problem Set 7.2. Show that if f is Lipschitzian on I, then f is uniformly continuous on I. (Here I may be any interval, open or closed, finite or infinite.)

9. Let $a_1, a_2, \dots, a_n$ be any finite sequence of numbers. Show that

$$\left| \sum_{i-1}^n a_i \right| \leq \sum_{i=1}^n |a_i|.$$

10. Let f be continuous on $[a, b]$. Show that

$$\left| \int_a^b f(x)\, dx \right| \leq \int_a^b |f(x)|\, dx.$$

8   The Conic Sections

8.1   TRANSLATION OF AXES

In Section 2.2 we stated the definition of a coordinate system on a line L.

A coordinate system on L is a one-to-one correspondence

$$L \leftrightarrow \mathbf{R}, \qquad P \leftrightarrow x,$$

between the points P of L and the real numbers x, such that the distance between any two points is the absolute value of the difference of the corresponding numbers. That is,

$$P_1 \leftrightarrow x_1, \quad P_2 \leftrightarrow x_2 \quad \Rightarrow \quad P_1 P_2 = |x_1 - x_2|.$$

If we subtract the same number from the coordinate of every point, we obtain another coordinate system on the line. If we subtract h from every x, then

$$P_1 \leftrightarrow x_1' = x_1 - h, \qquad P_2 \leftrightarrow x_2' = x_2 - h,$$

and so

$$P_1 P_2 = |(x_1 - h) - (x_2 - h)| = |x_1' - x_2'|.$$

Therefore the distance formula works, for the new coordinates $x' = x - h$.

This process is called *translation*. The origin is moved to the point h, and all the other number labels are moved with it.

Thus the old and new coordinates are related by the formulas

$$x' = x - h, \qquad x = x' + h.$$

Consider now a plane with a coordinate system.

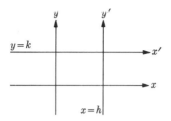

Suppose that we translate the coordinate system on the x-axis, subtracting h from every x-coordinate, and then translate the coordinate system on the y-axis, subtracting k from every y-coordinate. The effect is to move the origin to the point (h, k). Every point p now has a new pair of coordinates x', y', and these are related to the old ones by the formulas

$$x = x' + h, \qquad x' = x - h,$$
$$y = y' + k, \qquad y' = y - k.$$

These formulas are easy to remember; the only way you are likely to go wrong is to get them backwards (by writing $x' = x + k$ (?) or $y' = y + k$ (?)). It is easy to see, however, that the new origin must have old coordinates h, k and new coordinates 0, 0; and from this we can tell which way the formulas ought to go.

As usual, (x, y) denotes the point whose old coordinates are x and y. Thus the old origin is $(0, 0)$, and the new origin is (h, k). When we write $(a, b)'$ (with a prime outside the parentheses) we mean the point whose new coordinates are a and b. Thus the new origin is $(0, 0)'$, and the old origin is $(-h, -k)'$. More examples are given below:

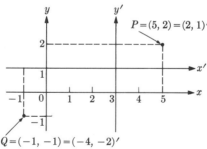

In the figure, $h = 3$ and $k = 1$. Two points have been labeled both ways. At the point P, we have

$$x = 5, \qquad y = 2, \qquad x' = 2, \qquad y' = 1,$$

so that the label

$$P = (5, 2) = (2, 1)'$$

is correct. Similarly, at Q we have

$$x = -1, \qquad y = -1, \qquad x' = -4, \qquad y' = -2,$$

so that
$$Q = (-1, -1) = (-4, -2)'.$$

When we write an equation to describe a figure in the plane, the equation depends on the choice of axes; and often one choice of axes gives a simpler equation than any other. If we didn't start with the axes in the best position, then we can simplify the equation by translation of axes. For example, consider the parabola with directrix $y = -1$ and focus $F = (3, 3)$.

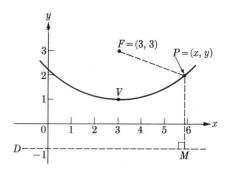

The parabola is the graph of the condition
$$FP = MP.$$
Algebraically, this says that

$$\sqrt{(x - 3)^2 + (y - 3)^2} = \sqrt{(y + 1)^2}$$
$$\Leftrightarrow x^2 - 6x + 9 + y^2 - 6y + 9 = y^2 + 2y + 1$$
$$\Leftrightarrow 8y = x^2 - 6x + 17$$
$$\Leftrightarrow y = \tfrac{1}{8}x^2 - \tfrac{3}{4}x + \tfrac{17}{8}.$$

We know, however, that a parabola with a horizontal directrix and vertex at the origin always has an equation of the form $y = ax^2$. The vertex of our parabola is halfway between the focus and the directrix, at the point $V = (3, 1)$. This means that we should translate the axes so that the new origin becomes the point
$$O' = (h, k) = (3, 1).$$
Relative to the new axes, the equation becomes
$$8(y' + 1) = (x' + 3)^2 - 6(x' + 3) + 17;$$
here we have replaced x by $x' + h = x' + 3$ and y by $y' + k = y' + 1$. This gives
$$8y' + 8 = x'^2 + 6x' + 9 - 6x' - 18 + 17,$$
or
$$8y' = x'^2, \quad \text{or} \quad y' = \tfrac{1}{8}x'^2.$$

This is in the standard form $y' = ax'^2$, where $a = 1/2p$ and p is the distance between the focus and the directrix.

Thus, by a translation of axes, we have eliminated the linear term in x and the constant term. Here we knew in advance where the origin ought to be for the equation to appear in a simple form. If we hadn't known this, we could still have investigated algebraically, to find out what sort of simplifications a translation could accomplish. To do this, we would regard h and k as unknown quantities, and make the substitution

$$x = x' + h, \qquad y = y' + k$$

in general form. This gives:

$$8(y' + k) = (x' + h)^2 - 6(x' + h) + 17,$$

or

$$x'^2 + (2h - 6)x' - 8y' + h^2 - 6h + 17 - 8k = 0.$$

Certain facts are now obvious: (1) We can't get rid of the term x'^2, by any choice of h and k, because h and k do not appear in the coefficient of x'^2. (2) For the same reason, we can't get rid of the linear term in y'. (3) The total coefficient of x' is $2h - 6$, and the total constant term is

$$h^2 - 6h + 17 - 8k.$$

We can therefore get rid of the x' term, by using $h = 3$. The constant term then becomes

$$9 - 18 + 17 - 8k, \qquad \text{or} \qquad 8 - 8k,$$

which is 0 when $k = 1$. Thus, translating the origin to the point $(h, k) = (3, 1)$, we get the equation in the form

$$8y' = x'^2 \qquad \text{or} \qquad y' = \tfrac{1}{8}x'^2,$$

as before. This is the process that you follow if you don't know the answer in advance

PROBLEM SET 8.1

1. Find a translation which eliminates both of the linear terms from the equation

$$xy - 5y - 6x - 30 = 0.$$

Then sketch the graph, showing both sets of axes.

2. Is there a translation which eliminates the xy-term from the above equation? Why or why not? How about the possibility of removing the constant term?

3. Find a translation which removes both linear terms from the equation

$$x^2 + y^2 + x + y - 2 = 0.$$

Then sketch the graph, showing both sets of axes.

4. Find a translation which removes both linear terms from the equation

$$2xy - x + 3y - 2 = 0.$$

5. Find a translation which removes both linear terms from the equation

$$x^2 + y^2 + 4x + 2y + 1 = 0.$$

6. Find a translation which removes both linear terms from the equation

$$x^2 + xy - 3x + 2 = 0.$$

7. Find a translation that eliminates both linear terms in the equation

$$x^2 + xy + y^2 + x + y + 5 = 0.$$

8. Find a translation which removes both linear terms from the equation

$$x^2 + xy + y^2 + x + y + 1 = 0.$$

9. Show that there is no translation which removes both linear terms from the equation

$$x^2 + 2xy + y^2 + x - y + 1 = 0.$$

10. Show that there is no translation which removes both linear terms from the equation

$$4x^2 + 4xy + y^2 + 2x + y + 8 = 0.$$

11. Consider the equation $x^2 + y^2 + x + y - 2 = 0$.   Under what conditions for h and k does this equation take the form

$$x'^2 + y'^2 + Ax' + By' = 0,$$

with possible linear terms but no constant term? (You may be able to think of a way to answer this question without doing any calculations at all.)

12. Show that if $ad - bc \neq 0$, then the linear system

$$ah + bk = e, \qquad ch + dk = f$$

always has a solution. (Simply start solving it; at some point, you will need to assume that $ad - bc \neq 0$.)

13. Consider an equation of the form

$$Ax^2 + Bxy + Cy^2 + Dx + Ey + F = 0.$$

Show that if $B^2 - 4AC \neq 0$, then there is always a translation that eliminates both of the linear terms. (The converse is not true; there are simple examples of equations where $B^2 - 4AC = 0$, but where the linear terms are absent to begin with. Examples?)

14. Sketch the graph of the equation $y^2 = x(x + 1)(x - 1)$.

15. Let C be the graph of an equation of the form

$$b_2 y^2 + b_1 y + b_0 = a_3 x^3 + a_2 x^2 + a_1 x + a_0.$$

Show that if the axes are translated, then C is the graph of an equation of the same form, in the new coordinates x' and y'.

8.2 THE ELLIPSE

Let F and F' be two points, let c be half the distance between them, so that

$$FF' = 2c,$$

and let a be a number greater than c. Let C be the graph of the equation

$$FP + F'P = 2a.$$

The curve C is called *the ellipse with foci F, F' and focal sum $2a$.*

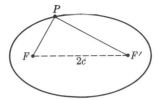

To draw an ellipse, we put two thumbtacks in a drawing board, at the foci F and F'. We tie the ends of a string to the thumbtacks, in such a way that the length of the string left free between the thumbtacks is $2a$. Then we put a pencil in the loop of string, placing the point so that the string is taut, and move the pencil around, keeping the string taut all the way. (We need to do this in two steps, on the two sides of the line through F and F'.)

In the definition of an ellipse, we really mean that F and F' are *two* points; that is, $F \neq F'$. (Thus a circle is not an ellipse.) Also, we really mean $a > c$. (For $a = c$, the graph of the condition $FP + F'P = 2a$ is the segment from F to F'; and for $a < c$, the graph is empty.)

Some things about ellipses are easily seen from the definition. For the definition of symmetry of a figure, about a line or a point, see Section 7.6.

Theorem 1. An ellipse is symmetric about the line through its foci.

Proof. In the left-hand figure below, P is on the ellipse, so that

$$FP + F'P = 2a.$$

And L is the perpendicular bisector of the segment from P to P'. By elementary geometry, $FP = FP'$, and $F'P = F'P'$. Therefore $FP' + F'P' = 2a$, and P' is on the ellipse.

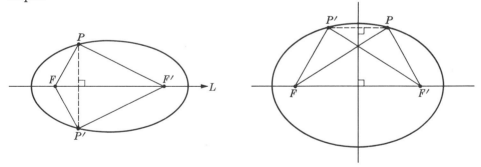

Theorem 2. An ellipse is symmetric about the perpendicular bisector of the segment between its foci.

Proof? (This is not quite as simple as the preceding theorem. See the right-hand figure above.)

Theorem 3. If a curve is symmetric about each of two perpendicular lines, then it is symmetric about their point of intersection.

Proof? (We need to show that $OP = OP''$.)

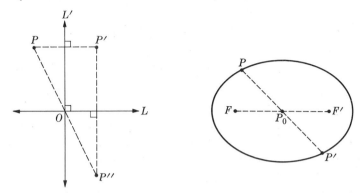

For ellipses, this gives us:

Theorem 4. Every ellipse is symmetric about the point midway between its foci.

P_0 is called the *center* of the ellipse. (See the right-hand figure above.)

These symmetry theorems convey nearly all that is easy to see about ellipses merely from the definition. Our next step is to set up a coordinate system, and describe our ellipses by equations. We take the origin at the center of the ellipse, and the foci on the *x*-axis. The ellipse is then said to be in *standard position*, relative to the axes.

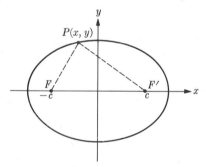

As indicated in the figure above, let F and F' be the points $(-c, 0)$ and $(c, 0)$. Then

$$FP + F'P = 2a$$
$$\Leftrightarrow \sqrt{(x + c)^2 + y^2} + \sqrt{(x - c)^2 + y^2} = 2a$$
$$\Leftrightarrow \sqrt{(x + c)^2 + y^2} = 2a - \sqrt{(x - c)^2 + y^2}$$
$$\Rightarrow x^2 + 2cx + c^2 + y^2 = 4a^2 - 4a\sqrt{(x - c)^2 + y^2} + x^2 - 2cx + c^2 + y^2$$
$$\Leftrightarrow a\sqrt{(x - c)^2 + y^2} = a^2 - cx$$
$$\Rightarrow a^2x^2 - 2a^2cx + a^2c^2 + a^2y^2 = a^4 - 2a^2cx + c^2x^2$$
$$\Leftrightarrow (a^2 - c^2)x^2 + a^2y^2 = a^2(a^2 - c^2)$$
$$\Leftrightarrow \frac{x^2}{a^2} + \frac{y^2}{a^2 - c^2} = 1.$$

Thus every point on the ellipse satisfies the final equation. It is possible to show, conversely, that every point (x, y) that satisfies the final equation lies on the ellipse. (See Problem 22 below.) Thus we have:

Theorem 5. The ellipse with foci at $(\pm c, 0)$ and focal sum $2a$ is the graph of the equation

$$\frac{x^2}{a^2} + \frac{y^2}{a^2 - c^2} = 1.$$

For example, for $a = 3$ and $c = 2$ we get

$$\frac{x^2}{9} + \frac{y^2}{5} = 1.$$

To sketch, we observe that for $y = 0$, $x = \pm 3$, and for $x = 0$, $y = \pm\sqrt{5}$. We then sketch an oval with these as its extreme points.

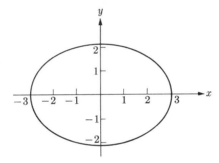

Given an equation

$$\frac{x^2}{a^2} + \frac{y^2}{b^2} = 1, \qquad b^2 < a^2,$$

the graph is always an ellipse. Since $a^2 - b^2 > 0$, it follows that $a^2 - b^2 = c^2$ for some $c > 0$. The graph is therefore the ellipse described in Theorem 5. Thus we have proved half of the following theorem.

Theorem 6. Given the equation

$$\frac{x^2}{a^2} + \frac{y^2}{b^2} = 1.$$

For $b^2 < a^2$, the graph is the ellipse with focal sum $2a$ and foci at $(\pm c, 0)$, where $c = \sqrt{a^2 - b^2}$. For $a^2 < b^2$, the graph is the ellipse with focal sum $2b$ and foci at $(0, \pm c)$, where $c = \sqrt{b^2 - a^2}$.

Proof of the second half of the theorem?

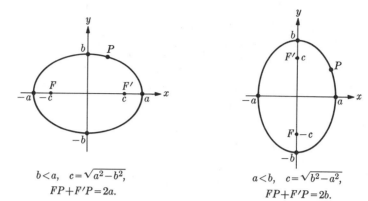

$$b < a, \quad c = \sqrt{a^2 - b^2},$$
$$FP + F'P = 2a.$$

$$a < b, \quad c = \sqrt{b^2 - a^2},$$
$$FP + F'P = 2b.$$

If the foci are not in either of the two positions shown above, then the equation of the ellipse is more complicated. In some cases, when the equation is given, we can simplify the equation by a translation of axes. Consider

$$4x^2 + 9y^2 - 8x + 18y - 23 = 0.$$

Making the substitutions $x = x' + h$, $y = y' + k$, we get

$$4x'^2 + 9y'^2 + (8h - 8)x' + (18k + 18)y' + 4h^2 + 9k^2 - 8h + 18k - 23 = 0.$$

Evidently we want $h = 1$, $k = -1$; and this gives the equation in the form

$$4x'^2 + 9y'^2 - 36 = 0, \quad \text{or} \quad \frac{x'^2}{9} + \frac{y'^2}{4} = 1.$$

The graph is the ellipse with foci at $(\pm\sqrt{5}, 0)'$ and focal sum 6; it intersects the x'- and y'-axes at the points $(\pm 3, 0)'$, $(0, \pm 2)'$. We can now sketch, showing both sets of axes

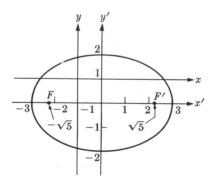

In doing such sketches, we start by drawing the new axes and the curve, in a convenient position on the paper, and then draw the old axes, in the position where they must have been.

PROBLEM SET 8.2

Write equations for the ellipses described by the following conditions and sketch.

1. Foci at $(\pm 1, 0)$; focal sum 4.   2. Foci at $(0, \pm 1)$; focal sum 4.

3. Foci at $(1, 2)$, $(1, 4)$; focal sum 4.   4. Foci at $(-1, -1)$, $(1, 1)$; focal sum 4.

5. Foci at $(-1, 1)$, $(1, -1)$; focal sum 4.

6. Foci at $(\pm 2, 0)$; focal sum 6.   7. Foci at $(0, \pm 2)$; focal sum 6.

8. Foci at $(-1, 1)$ and $(1, -1)$; focal sum 6.

Find the foci and the focal sum, and sketch, showing both sets of axes, in cases where more than one set is used.

9. $x^2/4 + y^2 = 1$ 10. $x^2 + 9y^2 - 2x + 36y + 28 = 0$

11. $9x^2 + y^2 + 36x - 2y + 28 = 0$

12. $x^2 + 2y^2 + 3x + 4y - 6 = 0$ (This one does not "come out even.")

13. $4x^2 + y^2 = 1$ 14. $x^2 + x + \dfrac{y^2}{9} + \dfrac{2y}{3} + 1 = 0$

15. Given an equation of the form

$$Ax^2 + By^2 + Cx + Ey + F = 0,$$

where A and B are both positive, show that the graph is (a) an ellipse, (b) a point, or (c) the empty set. (The same conclusion follows if A and B are both negative.)

16. A function f is *odd* if $f(-x) = -f(x)$ for every x. Show that the graph of an odd function is symmetric about the origin.

17. a) Let C be the graph of the sine function. Show that C is symmetric about the origin.

    b) Now show that C is also symmetric about infinitely many other points. (Thus it may happen that an unbounded figure has more than one "center." In fact, there is a simpler example: a line is symmetric about each of its points, and so every point of a line is "a center" of the line. For this reason, we ordinarily use the word *center* only for bounded figures.)

18. a) Show that the graph of the cosine is symmetric about infinitely many points.

    b) Show that the graph of the sine is symmetric about infinitely many *lines*.

19. Consider the infinite strip R between the lines $y = 1$ and $y = -1$. That is,

$$R = \{(x, y) \mid -\infty < x < \infty, \; -1 \leq y \leq 1\}.$$

Show that R is symmetric about infinitely many points, and find a simple description of the set C consisting of all points which are "centers" of R.

20. Show that every cubic curve is symmetric about its point of inflection. Here by a cubic curve we mean the graph of an equation $y = ax^3 + bx^2 + cx + d$, with $a \neq 0$.

21. Suppose that in Theorem 3 of this section we drop the hypothesis that the two lines of symmetry are perpendicular. Would the resulting theorem be true? Why or why not?

*22. Given $0 < c < a$, as in the definition of an ellipse. Let $F = (-c, 0)$, $F' = (c, 0)$. Let $P = (x, y)$ be a point satisfying the equation

$$\frac{x^2}{a^2} + \frac{y^2}{a^2 - c^2} = 1.$$

a) Show that

$$y^2 = \frac{a^2 - c^2}{a^2} (a^2 - x^2).$$

b) Show that

$$FP + F'P = \frac{1}{a}\sqrt{(a^2 + cx)^2} + \frac{1}{a}\sqrt{(a^2 - cx)^2}.$$

c) Show that $a^2 + cx > 0$.   (There are two cases to consider, $x \geq 0$ and $x \leq 0$. Remember that $0 < c < a$, and use the fact that $|x| \leq a$.)

d) Show that $a^2 - cx > 0$.

e) Show that $FP + MP = 2a$.

(This completes the proof of Theorem 5.)

8.3 THE HYPERBOLA

Given $0 < a < c$, and the points F and F', with $FF' = 2c$. Let C be the graph of the condition

$$FP - F'P = \pm 2a \qquad (a < c).$$

The curve C is called the *hyperbola with foci F, F' and focal difference 2a*. The figure

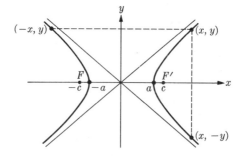

shows what a hyperbola looks like, but the reasons for this appearance of the graph are not obvious; the only thing that is easy to see, on the basis of the definition, is that the hyperbola is symmetric about each of the two perpendicular lines. The first step in our investigation of hyperbolas is to take the axes in a convenient position, as shown above, with $F = (-c, 0)$ and $F' = (c, 0)$, and get an equation for the curve.

$FP - F'P = \pm 2a$

$\Leftrightarrow FP = F'P \pm 2a$

$\Leftrightarrow \sqrt{(x + c)^2 + y^2} = \sqrt{(x - c)^2 + y^2} \pm 2a$

$\Rightarrow x^2 + 2cx + c^2 + y^2 = x^2 - 2cx + c^2 + y^2 \pm 4a\sqrt{(x - c)^2 + y^2} + 4a^2$

$\Leftrightarrow cx - a^2 = \pm a\sqrt{(x - c)^2 + y^2}$

$$\Rightarrow \quad c^2x^2 - 2a^2cx + a^4 = a^2x^2 - 2a^2cx + a^2c^2 + a^2y^2$$

$$\Leftrightarrow \quad (c^2 - a^2)x^2 - a^2y^2 = a^2(c^2 - a^2)$$

$$\Leftrightarrow \quad \frac{x^2}{a^2} - \frac{y^2}{c^2 - a^2} = 1.$$

Thus every point $P = (x, y)$ of the hyperbola satisfies the final equation. As in the case of the ellipse, it can be shown conversely that every point on the graph of the final equation is on the hyperbola. (See Problem 32 below.) Since $c^2 > a^2$, we may let

$$b^2 = c^2 - a^2.$$

This substitution gives the standard form of the equation:

$$\frac{x^2}{a^2} - \frac{y^2}{b^2} = 1.$$

And we can sum up as follows:

Theorem 1. The graph of the equation

$$\frac{x^2}{a^2} - \frac{y^2}{b^2} = 1$$

is the hyperbola with foci at $(\pm c, 0)$ (where $c = \sqrt{a^2 + b^2}$) and focal difference $2a$.

We shall use our equation to justify the sketch which we gave at the outset.

1) No point of the hyperbola lies between the lines $x = -a$ and $x = a$. The reason is as follows. Solving for y, we get

$$y = \pm \frac{b}{a} \sqrt{x^2 - a^2}.$$

Therefore the hyperbola is the union of the graphs of two functions

$$f(x) = \frac{b}{a} \sqrt{x^2 - a^2} \quad \text{and} \quad g(x) = -\frac{b}{a} \sqrt{x^2 - a^2},$$

and neither of these functions is defined for $-a < x < a$.

2) The curve is symmetric about each of the coordinate axes. This is easy to see algebraically. For each point (x, y), the symmetric point across the x-axis is $(x, -y)$; and if (x, y) is on the hyperbola, then so also is $(x, -y)$. Similarly, if (x, y) is on the curve, then so also is $(-x, y)$; and so the hyperbola is symmetric about the y-axis.

3) The hyperbola is unbounded in both the x- and y-directions. Obviously $f(x)$ and $g(x)$ are defined whenever $|x| \geq a$. As $x \to \infty$, $f(x) \to \infty$ and $g(x) \to -\infty$. And as $x \to -\infty$, $f(x) \to \infty$ and $g(x) \to -\infty$.

It remains to discuss the two lines which the curve seems to be getting close to when both x and y become numerically large. The behavior of the hyperbola relative to these lines seems to be similar to that of the curve

$$y = f(x) = \frac{1}{x},$$

relative to the coordinate axes. The coordinate axes are called *asymptotes* of this curve. By this we mean, roughly speaking, that points of the curve, far from the origin, in the appropriate directions, are close to the axes. We want to extend this idea to cases in which the asymptote is neither horizontal nor vertical.

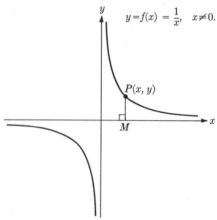

As $x \to \infty$, the distance from the point $P = (x, y)$ to the x-axis approaches 0. (This distance is $MP = |y| = |1/x|$.) We shall take this property as our definition of an asymptote. That is, a line L is an *asymptote* of a function-graph if the distance from the line to the point $P = (x, f(x))$ approaches 0 as $x \to \infty$, or as $x \to -\infty$. It is evident that the x-axis is an asymptote of the graph of $f(x) = 1/x$ under this definition. In fact, the x-axis is an asymptote in both the positive and negative directions. We also say that a line L is an asymptote of a curve C if C *contains* a function-graph which has L as an asymptote. In the case of $y = 1/x$, we also have $x = g(y) = 1/y$; thus the curve, looked at sidewise, is still a function-graph, and has the y-axis as an asymptote, in both the positive and negative directions. This is shown in the left-hand figure below.

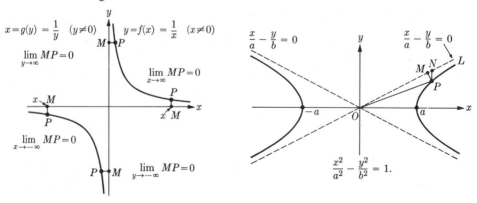

We return to our hyperbola. In the last figure on the preceding page, the slope
of the segment from the origin to the point $P = (x, y)$ is

$$m(x) = \frac{y}{x} = \frac{1}{x} \cdot \frac{b}{a} \sqrt{x^2 - a^2} = \frac{b}{a} \sqrt{1 - \frac{a^2}{x^2}}.$$

Obviously

$$\lim_{x \to \infty} m(x) = b/a,$$

and this suggests that the line $y = bx/a$, or $x/a - y/b = 0$, is an asymptote of the
part of the curve that lies in the first quadrant. If we show this, then it will follow
by symmetry that the lines $x/a \pm y/b = 0$ are asymptotes of the curve in all four
quadrants.

Thus we need to show that $\lim_{x \to \infty} MP = 0$. Since $MP < NP$, it will be sufficient
to show that $\lim NP = 0$. This can be done by an algebraic trick.

$$NP = \frac{b}{a} x - \frac{b}{a} \sqrt{x^2 - a^2} = \frac{b}{a}(x - \sqrt{x^2 - a^2})$$

$$= \frac{b}{a} \cdot (x - \sqrt{x^2 - a^2}) \cdot \frac{x + \sqrt{x^2 - a^2}}{x + \sqrt{x^2 - a^2}}$$

$$= \frac{b}{a} \cdot \frac{a^2}{x + \sqrt{x^2 - a^2}}.$$

Obviously $NP \to 0$ as $x \to \infty$. Therefore $MP \to 0$, which was to be proved. This
gives the following theorem.

Theorem 2. The lines

$$\frac{x}{a} \pm \frac{y}{b} = 0$$

are asymptotes of the hyperbola

$$\frac{x^2}{a^2} - \frac{y^2}{b^2} = 1.$$

You can sketch a hyperbola by drawing in the asymptotes and x-intercepts
exactly, and then filling in the curve freehand.

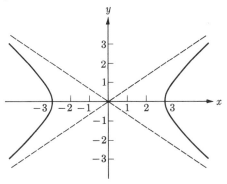

For example, consider

$$\frac{x^2}{9} - \frac{y^2}{4} = 1.$$

The x-intercepts are at $x = \pm 3$, and the asymptotes are the lines

$$\frac{x}{3} \pm \frac{y}{2} = 0, \quad \text{or} \quad y = \pm \tfrac{2}{3}x.$$

A hyperbola whose asymptotes are perpendicular is called *rectangular*.

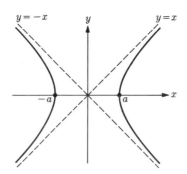

If such a hyperbola is in standard position, then the asymptotes must be the lines $x \pm y = 0$, and the equation must have the form

$$\frac{x^2}{a^2} - \frac{y^2}{a^2} = 1, \quad \text{or} \quad x^2 - y^2 = a^2.$$

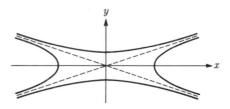

If the foci are on the y-axis, at the points $(0, \pm c)$, then the equation of the hyperbola takes the form

$$\frac{y^2}{a^2} - \frac{x^2}{c^2 - a^2} = 1.$$

It follows that the graph of the condition

$$\frac{x^2}{a^2} - \frac{y^2}{b^2} = \pm 1$$

is the union of two hyperbolas with the same asymptotes. These are called *conjugate* hyperbolas.

PROBLEM SET 8.3

Sketch the graphs of the following equations.

1. $x^2 - 4y^2 = 4$　　　　　2. $y^2 - 4x^2 = 4$　　　　　3. $x^2 - 4y^2 = -4$

4. $y^2 - 4x^2 = -4$　　　　5. $9x^2 - 4y^2 = 36$　　　6. $9x^2 - 4y^2 = -36$

7. $-x^2 + 9y^2 = 9$　　　　8. $-9x^2 + y^2 = 9$　　　　9. $25x^2 - 4y^2 = 100$

10. $25y^2 - 4x^2 = 100$

Derive equations for the hyperbolas determined by the following conditions, and sketch.

11. Foci at $(\pm 2, 0)$; focal difference 3.　　　12. Foci at $(\pm 2, 2)$; focal difference 3.

13. Foci at $(0, 0)$ and $(0, 4)$; focal difference 3.　　14. Foci at $(0, \pm 2)$; focal difference 3.

15. Foci at $(\pm 1, \pm 1)$; focal difference 2.

16. Foci at $(\pm 2, 0)$; passing through the point $(3, 4)$.

17. Foci at $(\pm 2, 0)$; focal difference 2.

18. Foci at $(\pm 3, 0)$; focal difference 4.

19. Foci at $(0, \pm 3)$; focal difference 4.

20. Foci at $(\pm 3, 0)$; passing through the point $(5, 5)$.

21. Given F, F', and a, as for a hyperbola in standard position. What is the graph of the condition $FP - F'P = 2a$? How about the graph of $FP - F'P = -2a$?

22. Find a rectangular hyperbola in standard position (with asymptotes $x + y = 0$ and $x - y = 0$) passing through the point $(5, 3)$.

Investigate the graphs of the following equations. In each case, find all asymptotes.

23. $(x^2 - y^2 - 1)^2 = 0$　　　　　24. $(x^2 - y^2)^2 = 1$

25. $x^2y^2 - xy + \frac{1}{4} = 0$　　　　26. $y = \dfrac{1}{(x - 1)(x - 2)}$

27. Let D be the line $x = -1$, let F be the origin, and for each point P let DP be the perpendicular distance between D and P. Let C be the graph of the condition

$$\frac{FP}{DP} = 2.$$

What sort of curve is this? Sketch.

28. Let F and D be as in the preceding problem, and let C' be the graph of the condition

$$\frac{FP}{DP} = \frac{1}{2}.$$

What sort of curve is this? Sketch.

29. Let G be the set of points P such that $CP = 2DP$, where C is the circle $x^2 + y^2 = 1$ and D is the line $x = 4$. What sort of figure is G? Discuss and sketch.

30. The following passage occurs in the U.S. Internal Revenue Act of 1964.

"... There shall be allowed as a deduction moving expenses paid ... in connection with the commencement of work by the taxpayer ... at a new principal place of work ... [However,] no deduction shall be allowed ... unless ... the taxpayer's new principal place of work ... is at least 20 miles farther from his former residence than was his former principal place of work ..."

Give a sketch, showing what this means. Your sketch should show (a) the former residence, (b) the former place of work, and (c) the region in which the new place of work must lie, for the moving expenses to be deductible. (The author is indebted, for this problem, to Dr. Henry Pollak, of the Bell Telephone Laboratories.)

31. The region between two conjugate hyperbolas stretches out infinitely far, in each of four directions. Find out whether the area of such a region is finite.

*32. Given $0 < a < c$, as in the definition of a hyperbola. Let $F = (-c, 0)$, $F' = (c, 0)$. Let $P = (x, y)$ be a point satisfying the equation

$$\frac{x^2}{a^2} - \frac{y^2}{c^2 - a^2} = 1.$$

a) Show that

$$y^2 = \frac{c^2 - a^2}{a^2} (x^2 - a^2).$$

b) Show that

$$FP - F'P = \frac{1}{a} \sqrt{(cx + a^2)^2} - \frac{1}{a} \sqrt{(cx - a^2)^2}.$$

c) Show that, if $x \geqq a$, then

$$cx + a^2 > 0, \qquad cx - a^2 > 0, \qquad \text{and} \qquad FP - F'P = 2a.$$

d) Show that if $x \leqq -a$, then

$$cx + a^2 < 0, \qquad cx - a^2 < 0, \qquad \text{and} \qquad FP - F'P = -2a.$$

(This completes the proof of Theorem 1.)

8.4 THE GENERAL EQUATION OF THE SECOND DEGREE. ROTATION OF AXES

An equation of the second degree in x and y is an equation of the form

$$Ax^2 + Bxy + Cy^2 + Dx + Ey + F = 0,$$

where at least one of the coefficients A, B, and C is different from zero. The latter condition is to guarantee that the degree of the equation really is 2, rather than 1 or 0. We have found that all conic sections are graphs of equations of this type; and we shall now investigate the converse. That is, we propose to find out what sort of figure can be the graph of a second-degree equation. The possibilities that we have already found are

a)   a *circle*,
b)   a *parabola*,
c)   an *ellipse*,
d)   a *hyperbola*.

There are other possibilities, which we noted as exceptional cases when we were studying the equation

$$x^2 + y^2 + Dx + Ey + F = 0,$$

in connection with the circle. The graph of

$$x^2 + y^2 = 0$$

is a point; and the graph of

$$x^2 + y^2 + 1 = 0$$

is empty. (See Theorem 2 of Section 2.3.) Our list of possible graphs of second-degree equations must therefore include

e) *a point*, and
f) *the empty set.*

And this is not all. The graph of

$$y^2 = 0$$

is a line, namely, the x-axis. And the graph of

$$xy = 0$$

is the union of two lines, namely, the two axes. Similarly, the graph of

$$x^2 - y^2 = 0$$

is the union of two lines. The reason is that $x^2 - y^2 = (x - y)(x + y)$. This is $= 0$ if and only if either $x - y = 0$ or $x + y = 0$. Therefore a point $P = (x, y)$ is on the graph of $x^2 - y^2 = 0$ if and only if (i) P is on the line $y = x$ or (ii) P is on the line $y = -x$.

In this example, the lines intersect, but we may get the union of two parallel lines. The equation

$$x^2 - x = 0$$

is equivalent to

$$x(x - 1) = 0;$$

and the graph is therefore the union of the two parallel lines $x = 0$ and $x = 1$. Thus the graphs, for the general equation of the second degree, include

g) *a line*, and
h) *the union of two lines*, either parallel or intersecting.

We shall show that the eight possibilities that we have just listed are the only possibilities. The method will be to reduce the equation to a recognizable form by moving the axes. In some cases, this cannot be done by translation; we may also have to use rotation of the axes.

Suppose that we rotate the axes through an angle of measure θ, getting a new pair of axes.

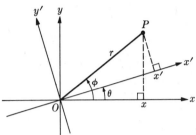

In the figure, r is the distance OP; P has coordinates x, y in the old coordinate system, and coordinates x', y' in the new coordinate system. Evidently

$$x = r \cos \phi, \qquad y = r \sin \phi,$$
$$x' = r \cos (\phi - \theta) = r \cos \phi \cos \theta + r \sin \phi \sin \theta,$$
$$y' = r \sin (\phi - \theta) = r \sin \phi \cos \theta - r \cos \phi \sin \theta.$$

Therefore the new coordinates are given in terms of the old ones by the formulas

$$x' = x \cos \theta + y \sin \theta, \qquad y' = -x \sin \theta + y \cos \theta.$$

If we rotate the new axes through an angle of measure $-\theta$ we are back where we started. Therefore the old coordinates are expressed in terms of the new ones by the formulas

$$x = x' \cos (-\theta) + y' \sin (-\theta), \qquad y = -x' \sin (-\theta) + y' \cos (-\theta).$$

These give

$$x = x' \cos \theta - y' \sin \theta, \qquad y = x' \sin \theta + y' \cos \theta.$$

Theorem 1. In any second-degree equation, the xy-term can be eliminated by a rotation of axes.

Before going into the proof, let us try a simple example:

$$xy = 1.$$

To rotate the axes through an angle θ, we should substitute

$$x = x' \cos \theta - y' \sin \theta, \qquad y = x' \sin \theta + y' \cos \theta. \tag{1}$$

The equation then becomes

$$(x' \cos \theta - y' \sin \theta)(x' \sin \theta + y' \cos \theta) = 1,$$

or

$$x'^2 \sin \theta \cos \theta + x'y'(\cos^2 \theta - \sin^2 \theta) - y'^2 \sin \theta \cos \theta = 1.$$

We want the $x'y'$-term to vanish. Thus we want

$$\cos^2 \theta - \sin^2 \theta = 0, \qquad \text{or} \qquad \cos 2\theta = 0;$$

and this will happen when

$$2\theta = \frac{\pi}{2} + n\pi, \qquad \theta = \frac{\pi}{4} + \frac{n\pi}{2}.$$

One value of θ is all we need, and so we take $\theta = \pi/4$, which gives

$$\sin\theta = \cos\theta = \frac{1}{\sqrt{2}}$$

and

$$\sin\theta\cos\theta = \tfrac{1}{2}.$$

Thus our new equation is

$$\frac{x'^2}{2} - \frac{y'^2}{2} = 1.$$

This is the equation of a rectangular hyperbola.

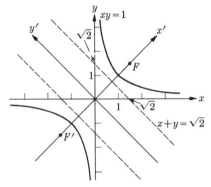

Let us now return to our general equation

$$Ax^2 + Bxy + Cy^2 + Dx + Ey + F = 0.$$

Making the usual substitution, to rotate the axes through θ, we get

$$A(x'\cos\theta - y'\sin\theta)^2 + B(x'\cos\theta - y'\sin\theta)(x'\sin\theta + y'\cos\theta)$$
$$+ C(x'\sin\theta + y'\cos\theta)^2 + D(x'\cos\theta - y'\sin\theta)$$
$$+ E(x'\sin\theta + y'\cos\theta) + F = 0.$$

When we collect coefficients for the terms of various types, we get a new equation of the same form, like this:

$$A'x'^2 + B'x'y' + C'y'^2 + D'x' + E'y' + F' = 0.$$

Algebraically,

$$A' = A\cos^2\theta + B\sin\theta\cos\theta + C\sin^2\theta,$$
$$B' = -2A\sin\theta\cos\theta + B(\cos^2\theta - \sin^2\theta) + 2C\sin\theta\cos\theta,$$
$$C' = A\sin^2\theta - B\sin\theta\cos\theta + C\cos^2\theta,$$
$$D' = D\cos\theta + E\sin\theta,$$
$$E' = -D\sin\theta + E\cos\theta,$$
$$F' = F.$$

For future reference, we have written down all of these, but for the moment, all we are interested in is B': we want to find a θ that makes $B' = 0$. Simplifying trigonometrically, we get

$$B' = (C - A)\sin 2\theta + B\cos 2\theta.$$

There are now two cases:

1) If $A = C$, then $B' = B\cos 2\theta$. We must have $B \neq 0$, or there wouldn't be any xy-term in the original equation. Therefore

$$B' = 0 \qquad \text{when} \qquad \cos 2\theta = 0,$$

and $\cos 2\theta = 0$ when

$$\theta = \frac{\pi}{4}.$$

Thus a rotation through $\pi/4$ eliminates the xy-term whenever $A = C$.

2) If $A \neq C$, then we can divide by $A - C$. Therefore $B' = 0$ when

$$B\cos 2\theta = (A - C)\sin 2\theta, \qquad \text{or} \qquad \frac{B}{A - C} = \tan 2\theta.$$

Thus, to get $B' = 0$, we take

$$\theta = \tfrac{1}{2}\operatorname{Tan}^{-1}\frac{B}{A - C}.$$

This proves the theorem. (The theorem did not say that the coefficients in the new equation were easy to compute.)

Theorem 2. The graph of a second-degree equation is (a) a circle, (b) a parabola, (c) an ellipse, (d) a hyperbola, (e) a point, (f) the empty set, (g) a line, or (h) the union of two lines (either parallel or intersecting).

Proof. By the preceding theorem, we can assume that there is no xy-term. The equation then has the form

$$Ax^2 + Cy^2 + Dx + Ey + F = 0.$$

We now need to discuss various cases.

1) Suppose that neither A nor C is $= 0$. We can then write

$$A\left(x^2 + \frac{D}{A}x\right) + C\left(y^2 + \frac{E}{C}y\right) = -F,$$

and complete the square twice to get

$$A\left(x + \frac{D}{2A}\right)^2 + C\left(y + \frac{E}{2C}\right)^2 = -F + \frac{D^2}{4A^2} + \frac{E^2}{4C^2},$$

which has the form

$$Ax'^2 + Cy'^2 = F'.$$

Here we have translated the axes letting

$$x' = x + \frac{D}{2A}, \qquad y' = y + \frac{E}{2C}.$$

Since $A \neq 0$, we can divide by A, getting

$$x'^2 + C'y'^2 = F'' \qquad (C' = C/A \neq 0).$$

There are six possibilities to be considered. For each of these cases, we have indicated on the right what sort of figure the graph is.

$$\begin{cases} C' > 0, \, F'' > 0 \\ C' > 0, \, F'' = 0 \\ C' > 0, \, F'' < 0 \end{cases} \qquad \begin{array}{l} \text{a circle or ellipse} \\ \text{a point} \\ \text{the empty set} \end{array}$$

$$\begin{cases} C' < 0, \, F'' > 0 \\ C' < 0, \, F'' = 0 \\ C' < 0, \, F'' < 0 \end{cases} \qquad \begin{array}{l} \text{a hyperbola} \\ \text{two intersecting lines} \\ \text{a hyperbola (with foci on the new } y\text{-axis)} \end{array}$$

2) Suppose that $C = 0$. The equation then has the form

$$Ax^2 + Dx + Ey + F = 0,$$

where $A \neq 0$, because the degree is 2. We divide by A, getting

$$x^2 + D'x + E'y + F' = 0;$$

and then we complete the square in x, so as to eliminate the linear term in x. This gives an equation of the form

$$x'^2 + F'' = E'y.$$

For $E' \neq 0$, this is a parabola. For $E' = 0$, the equation $x'^2 = -F''$ gives one line, two lines, or the empty set.

3) Suppose that $A = 0$. This is exactly like Case 2; we interchange x and y, and proceed as before. This completes the proof of the theorem.

It is easy to compute the new coefficients produced by a translation of axes. For a rotation, the new coefficients are expressed in terms of $\sin \theta$ and $\cos \theta$, and θ is defined by the equation

$$\theta = \tfrac{1}{2} \operatorname{Tan}^{-1} \frac{B}{A - C}.$$

Thus we want to express $\sin \theta$ and $\cos \theta$ in terms of $\tan 2\theta \, (= B/(A - C))$ for the case where

$$-\frac{\pi}{2} < 2\theta < \frac{\pi}{2}.$$

When 2θ is in the first or fourth quadrant, $\cos 2\theta > 0$, and $\sin 2\theta$ has the same sign as $\tan 2\theta$.

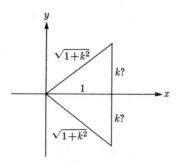

In the figure,

$$k = \tan 2\theta = \frac{B}{A - C}.$$

Therefore

$$\cos 2\theta = \frac{1}{\sqrt{1 + k^2}}.$$

The half-angle formulas are

$$\cos \frac{x}{2} = \pm\sqrt{\frac{1 + \cos x}{2}}, \qquad \sin \frac{x}{2} = \pm\sqrt{\frac{1 - \cos x}{2}}.$$

For the present case, these give

$$\cos \theta = \sqrt{\frac{1 + \cos 2\theta}{2}}, \qquad \sin \theta = \pm\sqrt{\frac{1 - \cos 2\theta}{2}},$$

where

$$\cos 2\theta = \frac{1}{\sqrt{1 + k^2}}$$

and where the sign in the formula for $\sin \theta$ is the same as the sign of $k = \tan 2\theta$.
    For example, consider

$$3x^2 + 2xy + y^2 = 1.$$

Here

$$A = 3, \qquad B = 2, \qquad C = 1,$$

and

$$k = \frac{B}{A - C} = \frac{2}{3 - 1} = 1.$$

Therefore

$$\cos 2\theta = \frac{1}{\sqrt{1 + k^2}} = \frac{1}{\sqrt{2}}.$$

Hence

$$\cos \theta = \sqrt{\frac{1 + 1/\sqrt{2}}{2}} = \sqrt{\frac{2 + \sqrt{2}}{4}},$$

and

$$\sin \theta = \sqrt{\frac{1 - 1/\sqrt{2}}{2}} = \sqrt{\frac{2 - \sqrt{2}}{4}}.$$

(In the second formula, $\sin \theta > 0$ because $k > 0$.) Therefore

$$\cos^2 \theta = \frac{2 + \sqrt{2}}{4}, \qquad \sin^2 \theta = \frac{2 - \sqrt{2}}{4}, \qquad \sin \theta \cos \theta = \frac{\sqrt{2}}{4}.$$

The new equation is

$$A'x'^2 + C'y'^2 = 1,$$

where

$$A' = A \cos^2 \theta + B \sin \theta \cos \theta + C \sin^2 \theta$$

$$= 3 \cdot \frac{2 + \sqrt{2}}{4} + 2 \cdot \frac{\sqrt{2}}{4} + \frac{2 - \sqrt{2}}{4} = \sqrt{2} + 2$$

and

$$C' = A \sin^2 \theta - B \sin \theta \cos \theta + C \cos^2 \theta$$

$$= 3 \cdot \frac{2 - \sqrt{2}}{4} - 2 \cdot \frac{\sqrt{2}}{4} + \frac{2 + \sqrt{2}}{4} = -\sqrt{2} + 2.$$

PROBLEM SET 8.4

In these problems, when you are asked to *investigate* an equation, you should find out what sort of figure the graph is, and sketch. If the graph is a conic section, you should also find the coefficients in the standard form.

1. Investigate

$$x - xy = 1.$$

(Here it is easier to translate first and rotate second. Sketch, showing all three sets of axes.)

2. Investigate

$$x^2 - xy = 1.$$

3. Investigate

$$xy - x - 2y = 0.$$

4. Investigate

$$2xy - y^2 + 2 = 0.$$

5. Investigate

$$x^2 + 2xy + y^2 + 2x + 2y + 1 = 0.$$

6. Investigate

$$x^2 + 4xy + 4y^2 + 4x + 8y + 3 = 0.$$

7. Show that, under a rotation of axes,

$$A' + C' = A + C \qquad \text{and} \qquad F' = F.$$

We express this by saying that $A + C$ and F are *invariant under rotation*.

8. a) Given the general equation of the second degree. Let A_θ, B_θ, C_θ, ... be the new coefficients, when the axes are rotated through an angle of measure θ. Thus A_θ, B_θ, C_θ, ... are the A', B', C', ... of the text; and so

$$A_\theta = A \cos^2 \theta + B \sin \theta \cos \theta + C \sin^2 \theta,$$
$$B_\theta = (C - A) \sin 2\theta + B \cos 2\theta,$$
$$C_\theta = A \sin^2 \theta - B \sin \theta \cos \theta + C \cos^2 \theta.$$

Show that the derivatives A'_θ, B'_θ, C'_θ satisfy the differential equations

$$A'_\theta = B_\theta, \qquad C'_\theta = -B_\theta, \qquad B'_\theta = 2(C_\theta - A_\theta).$$

b) Show that the function

$$f(\theta) = B_\theta^2 - 4A_\theta C_\theta$$

is a constant. Thus we say that $B^2 - 4AC$ is *invariant* under rotation of axes. It may be of some interest to check this, in the cases where we have computed the new coefficients.

9. Given $x^2 + 2xy + 3y^2 + 4x + 5y + 6 = 0$, the axes are rotated so as to eliminate the xy-term. What are the possibilities for the coefficients of x^2 and y^2, in the new equation?

10. Same question, for the equation

$$x^2 + 2xy + 5y^2 - 10 = 0.$$

11. Same question for

$$4x^2 + \sqrt{3}\, xy + y^2 + 2x + 3 = 0.$$

12. a) Let D be a line, let F be a point not on D, and let e be a positive number. Let G be the set of all points P such that

$$\frac{FP}{DP} = e,$$

where DP is the perpendicular distance from D to P. G is called *the conic section with directrix D, focus F, and eccentricity e*. Show that G is (a) an ellipse if $e < 1$, (b) a parabola if $e = 1$, and (c) a hyperbola if $e > 1$.

b) Is a circle a conic section, in the sense defined in Problem 12(a)? Why or why not?

9 Paths and Vectors in a Plane

9.1 MOTION OF A PARTICLE IN A PLANE

To describe the motion of a particle in a plane E, we need to explain where the particle is, at each time t on a certain interval I. Thus to each time t on the given interval I there corresponds a point $P(t)$; and the motion is described by a function

$$P: I \to E.$$

For the motion shown in the figure, I is the infinite interval $[0, \infty)$, and the initial point $P(0)$ is the point $(1, 1)$.

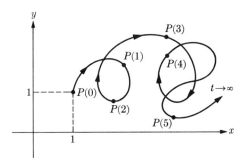

In general:

Definition. A *plane path* is a function

$$P: I \to E,$$

where I is an interval and E is a plane.

The same idea applies more generally: a path in space is a function $P: I \to S$, where I is an interval and S is space. In this chapter we shall be dealing only with plane paths, and so we shall refer to them for short simply as paths.

The *locus* of a path is the curve which is traced out by the moving point. More precisely:

Definition. Given a path

$$P: I \to E,$$

the *locus* of P is the set of all points Q which are $= P(t)$ for some t in I.

Briefly, the locus of P is the *image* of I under the function P. The locus is determined when the path is named, but given a locus, the path is not determined: the same curve can be traced out by a moving point in infinitely many ways.

We describe a path in a coordinate plane by defining two functions which give the coordinates of the moving point for each time t. For example, we might take

$$\left. \begin{array}{l} x = f(t) = 4t \\ y = g(t) = 8t^2 \end{array} \right\}$$

Here

$$I = (-\infty, \infty), \quad \text{and} \quad P(t) = (4t, 8t^2).$$

At $t = 0$, $P(t) = (0, 0)$. As t increases, starting from 0, both x and y increase, but y increases faster. In fact, the locus of this path is a parabola. To see this, we observe that from the first equation, $t = x/4$. Substituting in the second equation, we get

$$y = 8\left(\frac{x}{4}\right)^2 = \tfrac{1}{2}x^2.$$

Thus every point of the path lies on the graph of the equation

$$y = \tfrac{1}{2}x^2.$$

And it is easy to check, conversely, that every point (x, y) of the parabola is on the path.

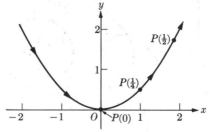

When a path is described by a pair of functions $x = f(t)$, $y = g(t)$, the two functions are called the *coordinate functions*, and t is called a *parameter*. Sometimes we can get a simple description of the locus of a path by writing an equation in x and y. We then say that we have *eliminated the parameter*, getting a *rectangular equation* of the locus. Often this process is useful: a path may trace out a simple figure, such as a segment or a circle, in a complicated way; and when this happens we want to know it.

The process of getting rectangular equations for loci is often tricky. Consider, for example, the path P described by the equations

$$x = f(t) = t^2, \qquad y = g(t) = t^4.$$

Every point of this path lies on the parabola

$$y = x^2.$$

But the converse is not true. On the path, we always have $x \geq 0$, because $t^2 \geq 0$. Therefore the locus of P is only half of the parabola.

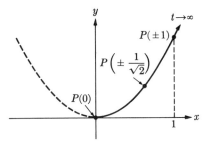

We are always free to regard a parameter as representing time, and in many physical problems, this is what the parameter means. But it often makes equally good sense to regard the parameter as the measure of an angle. Consider, for example,

$$x = \cos t, \qquad y = \sin t.$$

These functions describe uniform motion around a circle. Here we may regard the parameter as the measure of an angle, and write

$$x = \cos \theta, \qquad y = \sin \theta$$

to describe the same path.

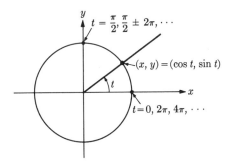

Somewhat similar looking paths have ellipses as their loci. For example, consider

$$x = a \cos \theta, \qquad y = b \sin \theta.$$

Here the locus of P is an ellipse:

$$\frac{x}{a} = \cos \theta, \qquad \frac{y}{b} = \sin \theta,$$

$$\frac{x^2}{a^2} + \frac{y^2}{b^2} = \cos^2 \theta + \sin^2 \theta = 1.$$

Investigating further, we see what values of θ correspond to what points of the elliptical locus. We draw circles of radii a and b, with centers at the origin, and construct $\angle \theta$ in standard position.

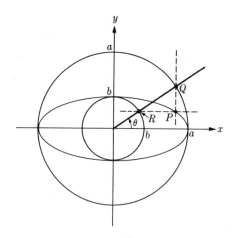

In the figure,

$$Q = (a \cos \theta, a \sin \theta),$$
$$R = (b \cos \theta, b \sin \theta).$$

Therefore

$$P = P(\theta) = (a \cos \theta, b \sin \theta).$$

Following the scheme of the above figure, using drawing instruments, you can plot as many points of the ellipse as you want to, without making any numerical calculations. The same idea is used in the construction of a drawing instrument called the ellipsograph, which can be adjusted so as to draw the ellipse with any pair of semiaxes a, b.

PROBLEM SET 9.1

Investigate the paths described by the following pairs of coordinate functions, sketch the loci, and label a few points as $P(0)$, $P(\pi/4)$, and so on, so as to indicate the way in which the moving point traverses the locus.

1. $x = \sec \theta, y = \tan \theta$ 2. $x = \cos \theta, y = \cos^2 \theta$ 3. $x = 2 \cos \theta, y = \sin \theta$

4. $x = \cos^2 \theta, y = \sin^2 \theta$ 5. $x = t^3, y = |t^3|$

(Check that not only $f(t) = t^3$ but also $g(t) = |t^3|$ have continuous derivatives. Thus a moving point can go smoothly around a sharp corner, if only it does so slowly enough.)

6. $x = \sec^2 \theta, y = \tan^2 \theta$ 7. $x = \sec \theta, y = \cos \theta$ 8. $x = \csc \theta, y = \cot \theta$

9. $x = \sin \theta, y = |\sin \theta|$ 10. $x = t^6, y = t^4$

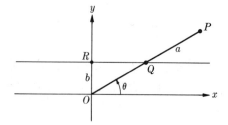

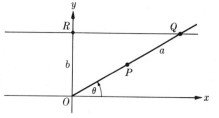

11. In the left-hand figure above, θ ranges over the open interval $(0, \pi)$, $OR = b$, and QP is a constant a. Find a parametric description of the path, and sketch the locus.

12. In the right-hand figure above, $OR = b$ as before, and QP is a constant a. Find a parametric description of the path, and sketch the loci, showing the three cases $a < b$, $a = b$, $a > b$.

13. A circle of radius a rolls without slipping inside a circle of radius $2a$. The initial position is shown on the left below; a later position is shown on the right. Observe that $RQ = 2a\theta$, $PQ = 2a\theta$, $PQ = a\phi$. Therefore $\phi = 2\theta$. Let

$$S = (h, k) = (a \cos \theta, a \sin \theta).$$

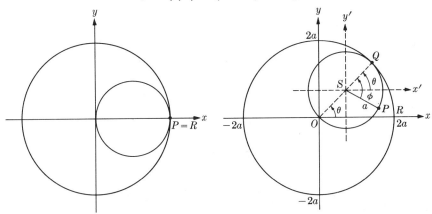

Then

$$x' = a \cos (\theta - \phi), \qquad y' = a \sin (\theta - \phi),$$
$$x = x' + h, \qquad\qquad y = y' + k.$$

Complete the discussion to get a parametric description of the path, and find out what the locus is. It will turn out that the figure on the right above is slightly misleading.

**14. If you solved the preceding problem correctly, you found that some of the machinery that you used was not necessary after all. But consider the case where the outer circle has radius a and the inner circle has radius $b = a/4$. Find parametric equations for the path, and eliminate the parameter to get the rectangular equation

$$\sqrt[3]{x^2} + \sqrt[3]{y^2} = \sqrt[3]{a^2}.$$

This curve is called a *four-cusped hypocycloid*.

*15. Show that if f is differentiable for $x \ne a$, and $\lim_{x \to a} f'(x) = k$, then f is differentiable at a, and $f'(a) = k$. [*Hint:* The theorem is conceptual, and the proof goes back to first principles. Start by writing out the hypothesis and conclusion in terms of the basic definitions of the statements (a) $\lim_{\bar{x} \to a} f'(\bar{x}) = k$ and (b) $f'(a) = k$.]

9.2 THE PARAMETRIC MEAN-VALUE THEOREM; L'HÔPITAL'S RULE

Given a path described parametrically by a pair of coordinate functions

$$x = f(t), \qquad y = g(t),$$

we may want to find the slope of the tangent line at the point corresponding to a particular t.

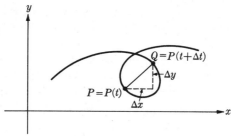

In the figure, we see the path; we want to find the slope of the tangent at P, if such a tangent exists. Suppose that P is the point corresponding to a certain t; and let Q be the neighboring point corresponding to $t + \Delta t$. Let

$$\Delta y = g(t + \Delta t) - g(t) \qquad \text{and} \qquad \Delta x = f(t + \Delta t) - f(t),$$

as indicated in the figure. Then the slope of the tangent at P is

$$m = \lim_{\Delta t \to 0} \frac{\Delta y}{\Delta x},$$

if such a limit exists. Suppose now that f and g are differentiable, and that $f'(t) \neq 0$. Then we can write

$$m = \lim_{\Delta t \to 0} \frac{\Delta y}{\Delta x} = \lim_{\Delta t \to 0} \frac{\Delta y/\Delta t}{\Delta x/\Delta t} = \lim_{\Delta t \to 0} \frac{[g(t + \Delta t) - g(t)]/\Delta t}{[f(t + \Delta t) - f(t)]/\Delta t}$$

$$= \frac{\lim_{\Delta t \to 0} \{[g(t + \Delta t) - g(t)]/\Delta t\}}{\lim_{\Delta t \to 0} \{[f(t + \Delta t) - f(t)]/\Delta t\}} = \frac{g'(t)}{f'(t)}.$$

Thus we get the formula

$$m = \frac{g'(t)}{f'(t)}.$$

This will be called the *parametric slope formula*. We have shown:

Theorem 1. Given a path defined by functions

$$x = f(t), \qquad y = g(t).$$

If f and g are differentiable at t, and $f'(t) \neq 0$, then the path has a tangent at the corresponding point P, and the slope of the tangent is given by the formula

$$m = m(t) = \frac{g'(t)}{f'(t)}.$$

An important case is the one in which $f'(t) \neq 0$ for $a < t < b$. Here $x = f(t)$ can never take on the same value twice, and so the locus of the path is the graph of a function ϕ.

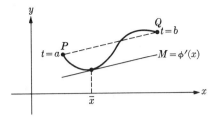

If P and Q are the endpoints of the graph, as in the figure, then the slope of the secant line through P and Q is

$$\frac{g(b) - g(a)}{f(b) - f(a)}.$$

By the mean-value theorem, there is a point $\bar{x}$ where the derivative $\phi'(\bar{x})$ is the slope of the secant line. Thus

$$\frac{g(b) - g(a)}{f(b) - f(a)} = \phi'(\bar{x}).$$

This number $\bar{x}$ must have come from somewhere. That is, there must be a $\bar{t}$ between a and b such that

$$\bar{x} = f(\bar{t}).$$

It follows that

$$\phi'(\bar{x}) = \frac{g'(\bar{t})}{f'(\bar{t})}.$$

Therefore

$$\frac{g(b) - g(a)}{f(b) - f(a)} = \frac{g'(\bar{t})}{f'(\bar{t})} \qquad (a < \bar{t} < b).$$

What we have just proved is a parametric form of the mean-value theorem. The idea is that, if a function-graph is presented parametrically, then we can rewrite the mean-value theorem parametrically, expressing both the slope of the secant and the slope of the tangent in terms of the parameter.

Theorem 2 (*The parametric mean-value theorem*). Given two continuous functions f and g, for $a \leqq t \leqq b$. If both functions are differentiable for $a < t < b$, and $f'(t) \neq 0$ for $a < t < b$, then

$$\frac{g(b) - g(a)}{f(b) - f(a)} = \frac{g'(\bar{t})}{f'(\bar{t})}$$

for some $\bar{t}$ between a and b.

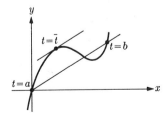

This theorem takes a simple form when

$$f(a) = g(a) = 0.$$

In this case, the theorem says that

$$\frac{g(b)}{f(b)} = \frac{g'(\bar{t})}{f'(\bar{t})},$$

for some $\bar{t}$ between a and b. This has the following consequence: if $g'(t)/f'(t)$ approaches a limit L, as $t \to a$, then $g(t)/f(t)$ approaches the same limit L. That is, if

$$f(a) = g(a) = 0, \qquad \text{and} \qquad \lim_{t \to a} \frac{g'(t)}{f'(t)} = L,$$

then

$$\lim_{t \to a} \frac{g(t)}{f(t)} = L.$$

This is called *l'Hôpital's rule*. Roughly, the reason why it holds true is as follows. Since $\bar{t}$ is between t and a, we know that $t \approx a \Rightarrow \bar{t} \approx a$. But

$$\bar{t} \approx a \quad \Rightarrow \quad \frac{g'(\bar{t})}{f'(\bar{t})} \approx L,$$

because $g'(t)/f'(t) \to L$. Therefore

$$t \approx a \Rightarrow \bar{t} \approx a \quad \Rightarrow \quad \frac{g(t)}{f(t)} = \frac{g'(\bar{t})}{f'(\bar{t})} \approx L.$$

Therefore

$$t \approx a \quad \Rightarrow \quad \frac{g(t)}{f(t)} \approx L,$$

and so

$$\lim_{t \to a} \frac{g(t)}{f(t)} = L.$$

It is very easy to express this idea in the form of an ϵ-δ-proof; all we do is to formalize our statements involving "$\approx$" in the following way:

1) *Hypothesis.* For every $\epsilon > 0$ there is a $\delta > 0$ such that

$$0 < |t - a| < \delta \quad \Rightarrow \quad \left| \frac{g'(t)}{f'(t)} - L \right| < \epsilon.$$

2) *Conclusion.* For every $\epsilon > 0$ there is a $\delta > 0$ such that

$$0 < |t - a| < \delta \quad \Rightarrow \quad \left| \frac{g(t)}{f(t)} - L \right| < \epsilon.$$

We need to show that (1) $\Rightarrow$ (2). Given $\epsilon > 0$, as in (2), we take the $\delta > 0$ furnished by (1). For each t, let $\bar{t}$ be the $\bar{t}$ furnished by the parametric mean-value theorem. This is the δ that we need:

$$0 < |t - a| < \delta \;\Rightarrow\; 0 < |\bar{t} - a| < \delta$$

$$\Rightarrow \left| \frac{g'(\bar{t})}{f'(\bar{t})} - L \right| < \epsilon$$

$$\Rightarrow \left| \frac{g(t)}{f(t)} - L \right| < \epsilon.$$

These fit together to give the desired conclusion.

In the above discussion, we assumed that $f(a)$ and $g(a)$ were both defined, and were $= 0$. It would have been sufficient, however, to suppose that

$$\lim_{t \to a} f(t) = 0 = \lim_{t \to a} g(t).$$

If these relations hold, and f and g are not defined at $x = a$, then we *define* $f(a)$ and $g(a)$ to be 0; f and g are then continuous, and the discussion of the limit of g/f goes through exactly as before.

Using x instead of t, we get our theorem in the following form:

Theorem 3 (*l'Hôpital's rule, first form*). If

$$\lim_{x \to a} f(x) = \lim_{x \to a} g(x) = 0 \quad \text{and} \quad \lim_{x \to a} \frac{g'(x)}{f'(x)} = L,$$

then

$$\lim_{x \to a} \frac{g(x)}{f(x)} = L.$$

Let us now look at some applications. Consider

$$\lim_{x \to 0} \frac{\sin x}{x}.$$

This satisfies the conditions of l'Hôpital's rule, with

$$g(x) = \sin x, \quad f(x) = x.$$

We investigate the quotient of the derivatives:

$$\lim_{x \to 0} \frac{\cos x}{1} = 1.$$

Therefore

$$\lim_{x \to 0} \frac{\sin x}{x} = 1.$$

This discussion does not supersede the geometric proof of the same statement, given in Section 4.2. The reason is that to apply l'Hôpital's rule, we had to know the derivative of the sine, and to find the derivative of the sine we needed to know that $\lim_{x \to 0} [(\sin x)/x] = 1$. Moreover, if you know the derivative of the sine, you can remind yourself of what $\lim [(\sin x/x]$ is, without using l'Hôpital's rule. The point is that

$$\lim_{x \to 0} \frac{\sin x}{x} = \lim_{x \to 0} \frac{\sin x - \sin 0}{x - 0} = \sin' 0,$$

by definition of $\sin' 0$. Since $\sin' = \cos$, and $\cos 0 = 1$, we get the answer immediately.

It is not an accident that in applying the first form of l'Hôpital's rule, we sometimes find that we are merely solving a differentiation problem. The reason is that the formula used in the definition of the derivative is always an instance of the rule, whenever the function is differentiable. By definition,

$$F'(x_0) = \lim_{x \to x_0} \frac{F(x) - F(x_0)}{x - x_0}.$$

The indicated limit on the right satisfies the conditions of Theorem 1, with

$$g(x) = F(x) - F(x_0) \to 0, \qquad f(x) = x - x_0 \to 0,$$

as $x \to x_0$. Thus every differentiation problem is a problem of the sort that l'Hôpital's rule deals with. The rule, of course, applies in many other cases; and it is the other cases that make it significant. For example,

$$\lim_{x \to 0} \frac{\sin^2 x + x}{e^x - 1} = \lim_{x \to 0} \frac{2 \sin x \cos x + 1}{e^x} = 1,$$

by the rule; and here the rule is needed.

Often, the application of l'Hôpital's rule requires the use of some preliminary device. For example, consider the possible limit

$$\lim_{x \to 0} x \cot x.$$

Here we should start by writing

$$\lim_{x \to 0} \frac{x \cos x}{\sin x}$$

and then use Theorem 3 (unless we can think of something simpler).

PROBLEM SET 9.2

Investigate the following indicated limits. (That is, calculate the ones which exist.)

1. $\displaystyle\lim_{x \to 0} x \cot x$

2. $\displaystyle\lim_{\theta \to 0} \frac{\sin^2 \theta}{\theta^2}$

3. $\displaystyle\lim_{\theta \to 0} \frac{\sin^3 \theta}{\theta^3}$

4. $\displaystyle\lim_{x \to 0} \frac{\cos^2 x - 1}{x^2}$

5. $\displaystyle\lim_{x \to 1} \frac{\ln x}{x - 1}$

6. $\displaystyle\lim_{x \to 1} \frac{x - 1}{e^x - 1}$

7. $\displaystyle\lim_{x \to 0} \frac{e^x - 1}{\ln (x + 1)}$

8. $\displaystyle\lim_{y \to 2} \frac{y^2 - 2y + 4}{y - 3}$

9. $\displaystyle\lim_{x \to \pi/2} \frac{\cos^3 x - 1}{x^3 - 1}$

10. $\displaystyle\lim_{x \to \pi/2} x^2 \sec^2 x$

11. $\displaystyle\lim_{x \to 0} x^2 \sin \frac{1}{x^2}$

12. $\displaystyle\lim_{x \to \pi} \frac{x^4 - 1}{\sin^2 x - 1}$

13. $\lim\limits_{x \to e} \dfrac{\ln^2 x - 1}{x^2 - 1}$

14. $\lim\limits_{x \to 0} \dfrac{\sin x \cos x - \tan x}{\sqrt{x}}$

15. $\lim\limits_{x \to 0} \dfrac{\tan x}{x}$

16. $\lim\limits_{\theta \to \pi/4} \dfrac{\sin \theta - \cos \theta}{\theta - \pi/4}$

17. $\lim\limits_{x \to 0} x^2 \csc^2 x$

18. $\lim\limits_{x \to 1} \dfrac{x^2 - 4x + 3}{x^2 - 3x + 2}$

19. $\lim\limits_{x \to 1} \dfrac{x^2 - 1}{\ln x}$

20. $\lim\limits_{x \to 0} \dfrac{3x}{e^{x^2} - 1}$

21. $\lim\limits_{t \to e} \dfrac{\ln (t) - 1}{t - e}$

22. $\lim\limits_{t \to \pi/2} \left[\sec t \displaystyle\int_{\pi/2}^{t} \sqrt{1 + \sin^3 t} \; dt \right]$

23. $\lim\limits_{x \to 1} \left[\dfrac{1}{x - 1} \displaystyle\int_{1}^{x} e^{t^3} \sqrt{t} \; dt \right]$

24. $\lim\limits_{x \to \pi} \left[\csc x \displaystyle\int_{\pi/2}^{x} \sqrt{1 + \sin^3 t} \; dt \right]$

*25. $\lim\limits_{x \to 0^+} x \ln x$

26. $\lim\limits_{x \to 0^+} x \ln \sin x$

*27. $\lim\limits_{x \to 0^+} x \ln (\sin x \cos x)$

*28. $\lim\limits_{x \to 0^+} x \ln (\cos^2 x \sin^2 x)$

29. A circle starts off tangent to the x-axis at the origin. The circle then rolls, without slipping, along the x-axis. The point P which started at the origin then traces out a path; this path is called a *cycloid*. (The same term *cycloid* is applied also to the locus of the path.) The parameter in the coordinate functions is the θ indicated in the figure. Sketch the locus, and calculate the coordinate functions of the path.

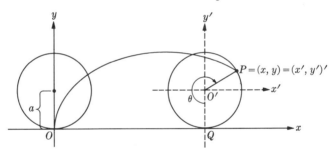

As the figure suggests, the easiest method is to use a "moving coordinate system," as in Problem 13 of Problem Set 9.1; we need to calculate the coordinates (h, k) of the "moving origin" O', and calculate x' and y' as $a \cos \phi$ and $a \sin \phi$. (What is ϕ?)

30. a) When a circle rolls on the inside of another circle, we get a *hypocycloid*. In the figure, the fixed circle has radius a and the rolling circle has radius b.

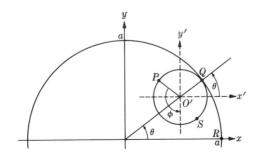

Calculate the coordinate functions, using θ as the parameter. The answer is

$$x = f(\theta)$$

$$= (a - b) \cos \theta + b \cos \frac{b - a}{b} \theta,$$

$$y = g(\theta)$$

$$= (a - b) \sin \theta + b \sin \frac{b - a}{b} \theta.$$

   b) Get a rectangular equation for the locus, for the case $b = a/4$. Sketch.

31. When one circle rolls around the *outside* of another, the figure traced out is called an *epicycloid*. Derive parametric equations for the epicycloid, using radius a for the fixed circle and radius b for the moving circle. Use the same parameter θ as in Problem 30(a).

32. Suppose that a railroad wheel rolls (without slipping) along a flat track. Find coordinate functions for the path traced out by a point at the outer edge of the flange on the wheel. In the figure below, the outer radius is b and the inner radius is a. Sketch the locus, bearing in mind that it is *not* a function-graph; it has loops in it.

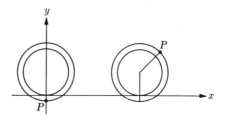

33. Make the same modification in the definition of the epicycloid, as suggested by the figure below, and sketch the curve. The fixed circle has radius a; the rolling wheel has inner radius b and outer radius c.

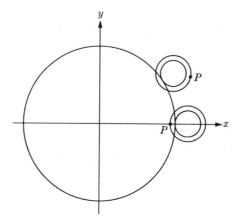

*34. A path is *regular* if the coordinate functions f, g are differentiable, and we never have $f'(t) = g'(t) = 0$ for the same t. Show that every chord of a regular path is parallel to the tangent at some intermediate point. Examples of this are as follows:

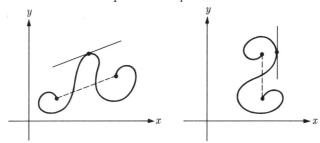

(Evidently the locus need not be a function-graph, and the chord may be vertical.)

*35. Given a path, with differentiable coordinate functions f, g. Show that, if the axes are rotated, then the coordinate functions F, G that work for the new set of axes are also differentiable. Show that if f' and g' never vanish simultaneously, then F' and G' never vanish simultaneously.

9.3 OTHER FORMS OF L'HÔPITAL'S RULE

The first form of l'Hôpital's rule says that if

$$\lim_{x \to a} f(x) = \lim_{x \to a} g(x) = 0, \quad \text{and} \quad \lim_{x \to a} \frac{g'(x)}{f'(x)} = L,$$

then

$$\lim_{x \to a} \frac{g(x)}{f(x)} = L.$$

This can be generalized in three ways.

1) If $x \to \infty$ or $x \to -\infty$, instead of $x \to a$, it doesn't matter; the rule still holds.

2) If $g'(x)/f'(x) \to \infty$ or $\to -\infty$, as $x \to a$, the rule still holds.

3) If $f(x) \to \infty$ and $g(x) \to \infty$, instead of $f(x) \to 0$, $g(x) \to 0$, the rule still holds. Similarly if $f(x) \to -\infty$ and $g(x) \to -\infty$.

Thus, in the most general form of l'Hôpital's rule, we have: (1) $x \to a$, $x \to \infty$, or $x \to -\infty$; (2) $g'(x)/f'(x) \to L$, $g'(x)/f'(x) \to \infty$, or $g'(x)/f'(x) \to -\infty$; and (3) $f(x), g(x) \to 0$, or $\to \infty$, or $\to -\infty$. Thus we have a grand total of 27 theorems, all of which are true. One of these has already been proved, and the only hard one among the others is the following.

Theorem 1 (*The Northeast Theorem*). If

$$\lim_{x \to \infty} f(x) = \lim_{x \to \infty} g(x) = \infty, \quad \text{and} \quad \lim_{x \to \infty} \frac{g'(x)}{f'(x)} = L,$$

then

$$\lim_{x \to \infty} \frac{g(x)}{f(x)} = L.$$

This is proved in Appendix H. Meanwhile we shall use it.

Example 1. To find

$$\lim_{x \to \infty} \frac{\ln x}{x},$$

we take derivatives and find

$$\lim_{x \to \infty} \frac{1/x}{1} = 0.$$

By the Northeast theorem,

$$\lim_{x \to \infty} \frac{\ln x}{x} = 0.$$

The case in which $g'(x)/f'(x) \to \infty$ causes no trouble.

Example 2. To find

$$\lim_{x \to \infty} \frac{e^x}{x},$$

we investigate

$$\lim_{x \to \infty} \frac{e^x}{1} = \infty.$$

It follows, by one form of l'Hôpital's rule, that

$$\lim_{x \to \infty} \frac{e^x}{x} = \infty.$$

The theorem being applied here is the following.

Theorem 2. If

$$\lim_{x \to \infty} f(x) = \lim_{x \to \infty} g(x) = \infty, \qquad \text{and} \qquad \lim_{x \to \infty} \frac{g'(x)}{f'(x)} = \infty,$$

then

$$\lim_{x \to \infty} \frac{g(x)}{f(x)} = \infty.$$

The proof is easy, on the basis of the Northeast theorem; we merely investigate reciprocals. Since $g'(x)/f'(x) \to \infty$, we have

$$\lim_{x \to \infty} \frac{f'(x)}{g'(x)} = 0.$$

By the Northeast theorem,

$$\lim_{x \to \infty} \frac{f(x)}{g(x)} = 0.$$

And $f(x)/g(x) > 0$ when x is large. Therefore

$$\lim_{x \to \infty} \frac{g(x)}{f(x)} = \infty.$$

Example 3. Consider

$$\lim_{x \to 0^+} \frac{\ln x}{x} = -\lim_{x \to 0^+} \frac{-\ln x}{x}.$$

Here the limit on the right takes the form ∞/∞. Taking derivatives, we find

$$\lim_{x \to 0^+} \frac{-1/x}{1} = 0.$$

By one form of l'Hôpital's rule, it follows that

$$\lim_{x \to 0^+} \frac{-\ln x}{x} = 0,$$

and so the answer to the original problem is $-0 = 0$. The theorem being used here is the following.

Theorem 3. If

$$\lim_{x \to a^+} f(x) = \lim_{x \to a^+} g(x) = \infty, \quad \text{and} \quad \lim_{x \to a^+} \frac{g'(x)}{f'(x)} = L,$$

then

$$\lim_{x \to a^+} \frac{g(x)}{f(x)} = L.$$

This is not quite as easy as Theorem 2.

Let

$$y = \frac{1}{x - a}, \quad x = a + \frac{1}{y},$$

so that $y \to \infty$ as $x \to a^+$ and $x \to a^+$ as $y \to \infty$. Then

$$\lim_{x \to a^+} \frac{g(x)}{f(x)} = \lim_{y \to \infty} \frac{g(a + 1/y)}{f(a + 1/y)}.$$

Taking derivatives, we find

$$\lim_{y \to \infty} \frac{g'(a + 1/y)(-1/y^2)}{f'(a + 1/y)(-1/y^2)} = \lim_{y \to \infty} \frac{g'(a + 1/y)}{f'(a + 1/y)}$$

$$= \lim_{x \to a^+} \frac{g'(x)}{f'(x)} = L.$$

The Northeast theorem now applies to

$$\lim_{y \to \infty} \frac{g(a + 1/y)}{f(a + 1/y)},$$

and tells us that this limit is L. Therefore

$$\lim_{x \to a^+} \frac{g(x)}{f(x)} = L.$$

We have now discussed all the troublesome cases of l'Hôpital's rule; once we have gotten this far, the rest of the derivations are routine. Hereafter, we shall use all forms of the rule without comment.

Sometimes we can apply the rule by taking logarithms. Consider

$$\lim_{x \to 0} x^x = ?$$

Let $\phi(x) = x^x$. Then

$$\ln \phi(x) = x \ln x = \frac{\ln x}{1/x} = \frac{g(x)}{f(x)}.$$

Now

$$\lim_{x \to 0^+} \frac{g'(x)}{f'(x)} = \lim_{x \to 0} \frac{1/x}{-1/x^2} = \lim_{x \to 0^+} (-x) = 0.$$

Therefore

$$\lim_{x \to 0^+} \ln \phi(x) = 0, \qquad \text{and} \qquad \lim_{x \to 0^+} x^x = e^0 = 1.$$

PROBLEM SET 9.3

1. $\lim_{x \to \infty} x^2 e^{-x}$

2. $\lim_{y \to \infty} y(\mathrm{Tan}^{-1} y - \pi/2)$

3. $\lim_{x \to \infty} (\ln \ln x)/\sqrt{x}$

4. $\lim_{x \to \infty} (x^x/e^x)$

5. $\lim_{x \to 0^+} e^{-1/x^2}$

6. $\lim_{x \to 0^+} [(1/x)e^{-1/x^2}]$

7. $\lim_{x \to \infty} [(\sin x)(\ln x)/x]$

8. $\lim_{\alpha \to \infty} (1 + \mathrm{Tan}^{-1} \alpha)^{\mathrm{Tan}^{-1}\alpha}$

9. $\lim_{x \to 0^+} (1 + \csc x)^{\sin^2 x}$

10. $\lim_{x \to 0} (1 + ax)^{1/x}$

11. $\lim_{x \to \infty} x^3 e^{-x}$

12. $\lim_{x \to \infty} x e^{-1/x}$

13. $\lim_{x \to \infty} (1 - 2x)^{1/x}$

14. $\lim_{x \to \pi/2} \left(1 + \dfrac{1}{\mathrm{Tan}^{-1}(x)}\right)^{\mathrm{Tan}^{-1} x}$

15. $\lim_{x \to 0^+} x^2 e^{-1/x}$

16. $\lim_{\theta \to 0^+} (\tan \theta)^{\tan \theta}$

17. $\lim_{t \to \infty} \left(\dfrac{1+t}{t}\right)^t$

18. $\lim_{x \to 0^+} x^2 \ln x$

19. $\lim_{x \to 0} (e^{-1/x^2}/x^2)$

20. $\lim_{x \to 0} (e^{-1/x^2}/x^3)$

21. $\lim_{x \to 0^+} x \ln x$

22. $\lim_{x \to 0^+} (1/x + \ln x)$

23. $\lim_{x \to 0^+} (1/x + n \ln x)$

24. $\lim_{x \to 0^+} (1/x^2 + n \ln x)$

25. Show that for every n, $\lim_{x \to 0} (e^{-1/x^2}/x^n) = 0$.

26. $\lim_{x \to 0^+} (\sin x)^{\sin x}$

27. $\lim_{y \to 0^+} (1 - \cos y)^{1 - \cos y}$

28. $\lim_{y \to 0^+} (1 + \cos y)^{1 - \cos y}$

29. $\lim_{x \to 0} (1 + \sin kx)^{\csc x}$

30. $\lim_{x \to 0} (1 - \sin kx)^{\csc x}$

31. $\lim_{v \to 0^+} (v^2)^v$

32. $\lim_{W \to 0^+} W^{W^2}$

33. $\lim_{\theta \to 0} (1 + \tan 3\theta)^{-\csc \theta}$

*34. The nth derivative of a function f is denoted by $f^{(n)}$. Let

$$f(x) = \begin{cases} e^{-1/x^2} & \text{for } x \neq 0, \\ 0 & \text{for } x = 0. \end{cases}$$

Show that for each $n > 0$, f has an nth derivative, for every x; and show that $f^{(n)}(0) = 0$ for every n. [*Hint:* You are not likely to find a manageable general formula for $f^{(n)}(x)$. But you ought to be able to show that for $x \neq 0$, $f^{(n)}(x)$ is always given by a formula of a certain *form*, involving certain constant coefficients; and you may be able to use this form, to show that $f^{(n)}(0) = 0$, without needing to determine the coefficients.]

9.4 POLAR COORDINATES

When we set up a rectangular coordinate system in a plane E, to every ordered pair (x, y) of numbers there corresponds a point P of E. Thus we have a function

$$\{(x, y)\} \to E, \qquad (x, y) \mapsto P.$$

And the correspondence also works in reverse: when P is named, x and y are determined. Thus a rectangular coordinate system gives us a one-to-one correspondence $\{(x, y)\} \leftrightarrow E$ between the ordered pairs of real numbers and the points of E.

We now consider another way of labeling points with pairs of numbers.

Given two numbers r and θ, we first draw the ray which starts at the origin and has direction θ. On the line containing this ray we set up a coordinate system, with the direction θ as the positive direction; and we let P be the point with coordinate r. (This is equivalent to saying that the directed distance $\overline{OP}$ is $= r$.) We then say that P has *polar coordinates* (r, θ).

For example, in the left-hand figure it looks as if P_1 has polar coordinates $(2, \pi/3)$, and P_2 has polar coordinates $(-2, \pi/3)$.

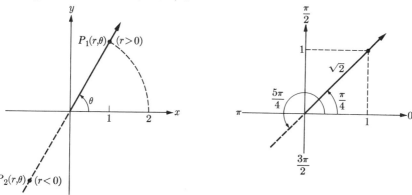

Thus to every pair (r, θ) of numbers there corresponds a point P. But the correspondence does not work uniquely the other way: every point P corresponds to infinitely many number pairs (r, θ). Thus, in the right-hand figure, the point P with rectangular coordinates $(1, 1)$ has polar coordinates $(\sqrt{2}, \pi/4)$. But P also has polar coordinates $(-\sqrt{2}, 5\pi/4)$. And this is not all; the possible polar coordinates for P are

$$(\sqrt{2}, \pi/4 + 2n\pi), \qquad (-\sqrt{2}, 5\pi/4 + 2n\pi),$$

where n is any integer (positive, negative, or zero).

Thus, when we set up a polar coordinate system we have a function

$$\{(r, \theta)\} \to E,$$

but we do not have a one-to-one correspondence; the polar coordinates of a point are not determined when the point is named. For this reason, graphs in polar coordinates can naturally be thought of as paths. Let us look at some examples.

1) Consider the graph of

$$r = \cos \theta \qquad (0 \leqq \theta \leqq 2\pi).$$

Since the cosine is periodic, with period 2π, we can get all of the locus by restricting θ to the interval $[0, 2\pi]$.

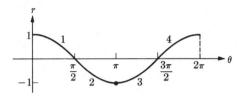

As an aid in sketching the polar graph, we first sketch the rectangular graph of the equation $r = \cos \theta$. We cut the curve into four parts, as indicated, and then sketch the portion of the polar graph corresponding to each of them. As θ increases from 0 to $\pi/2$, r decreases from 1 to 0. As θ increases from $\pi/2$ to π, r decreases from 0 to -1. Therefore the second part of the curve, in the *fourth* quadrant, comes from values of θ in the *second* quadrant. (See the figure on the left.)

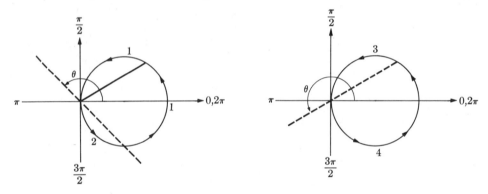

As θ continues to increase, from π to 2π, we trace out a curve shown on the right. This looks like the curve that we had already. And in fact it is exactly the same curve as before, because

$$\cos (\theta + \pi) = -\cos \theta.$$

Further investigation shows that the graph is a circle.

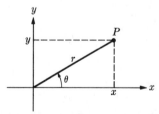

If P has polar coordinates (r, θ), then the rectanguar coordinates of P are

$$x = r \cos \theta, \qquad y = r \sin \theta.$$

(There are two cases to check. If $r > 0$, then these formulas follow from the definitions of the sine and cosine. Verification for $r = 0$ and $r < 0$?) Therefore

$$x^2 + y^2 = r^2 \cos^2 \theta + r^2 \sin^2 \theta = r^2.$$

This gives the three conversion formulas

$$x = r \cos \theta, \qquad y = r \sin \theta, \qquad x^2 + y^2 = r^2.$$

We note that the equation

$$r = \cos \theta$$

does not involve any of the three expressions $r \cos \theta$, $r \sin \theta$, r^2 which we know how to convert into rectangular coordinates. But we can multiply by r, on both sides, getting

$$r^2 = r \cos \theta;$$

and this means that

$$x^2 + y^2 = x.$$

This is the circle with center at the point $(\frac{1}{2}, 0)$ and radius $\frac{1}{2}$.

2) Consider

$$r = \sec \theta.$$

Here we might sketch the graph without using rectangular coordinates.

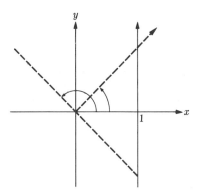

It is easier, however, to multiply both sides of the equation by $\cos \theta$. This gives

$$r \cos \theta = 1, \qquad \text{or} \qquad x = 1.$$

As θ increases from 0 to 2π (skipping $\pi/2$ and $3\pi/2$), this line is traversed twice. (It is worthwhile to figure out how.)

3) In these two examples, it was easy to work back to a rectangular equation. Consider, however,

$$r = 1 - \sin \theta \qquad (0 \leqq \theta \leqq 2\pi).$$

(As for $r = \cos\theta$, the interval $[0, 2\pi]$ gives us the entire locus of the path.) First we do a rectangular sketch as on the left. We then sketch the polar graph in four parts:

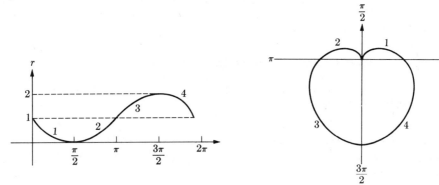

This curve is called, for obvious reasons, a *cardioid*.

It is possible to write a rectangular equation for the cardioid. First we observe that since

$$r = 1 - \sin\theta \gtrless 0, \tag{1}$$

we always have (on this particular curve)

$$r = \sqrt{r^2} = \sqrt{x^2 + y^2}.$$

We can therefore write

$$r^2 = r - r\sin\theta, \tag{2}$$

$$x^2 + y^2 = \sqrt{x^2 + y^2} - y. \tag{3}$$

4) To get the equation of a line L, in polar coordinates, we proceed as follows. Let N be the perpendicular to L through the origin. Rotate the axes through an angle of measure ϕ, choosing ϕ so that N becomes the x'-axis.

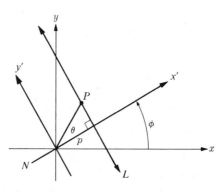

Then L is the graph of an equation

$$x' - p = 0,$$

where p is a constant. Since $x' = x \cos \phi + y \sin \phi$, this gives

$$x \cos \phi + y \sin \phi - p = 0.$$

Converting to polar form, we get

$$r \cos \theta \cos \phi + r \sin \theta \sin \phi - p = 0,$$

$$r \cos (\theta - \phi) - p = 0.$$

This is the standard form of the equation of a line in polar coordinates.

Some geometric problems are most conveniently attacked by introducing polar coordinates at the outset. To do this, we need a distance formula.

Theorem 1. Let d be the distance between the points with polar coordinates (r_1, θ_1) and (r_2, θ_2). Then

$$d^2 = r_1^2 + r_2^2 - 2r_1 r_2 \cos (\theta_1 - \theta_2).$$

Proof. The rectangular coordinates of the two points are

$$x_i = r_i \cos \theta_i, \qquad y_i = r_i \sin \theta_i$$

for $i = 1, 2$. Therefore

$$
\begin{aligned}
d^2 &= (x_1 - x_2)^2 + (y_1 - y_2)^2 \\
&= (r_1 \cos \theta_1 - r_2 \cos \theta_2)^2 + (r_1 \sin \theta_1 - r_2 \sin \theta_2)^2 \\
&= r_1^2(\cos^2 \theta_1 + \sin^2 \theta_1) + r_2^2(\cos^2 \theta_2 + \sin^2 \theta_2) \\
&\quad - 2r_1 r_2(\cos \theta_1 \cos \theta_2 + \sin \theta_1 \sin \theta_2) \\
&= r_1^2 + r_2^2 - 2r_1 r_2 \cos (\theta_1 - \theta_2),
\end{aligned}
$$

which was to be proved.

For $r_1 > 0, r_2 > 0, 0 < \theta_1 - \theta_2 < \pi$, this is simply the law of cosines.

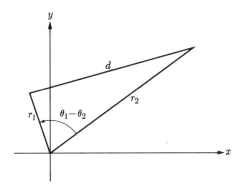

But the polar distance formula applies for any values of r_1, r_2, θ_1, and θ_2.

PROBLEM SET 9.4

Sketch the following, and convert to rectangular coordinates if possible.

1. $r = 2 \csc \theta$

2. $r = -2 \sec \theta$

3. $r \cos (\theta - \pi/4) = 1$

4. $r = 1 + \sin \theta$

5. $r = 1 - \cos \theta$

6. $r = \sin \theta \sec^2 \theta$

7. $r^2 = \sin \theta \sec^3 \theta$

8. $r = \dfrac{1}{\sin \theta + \cos \theta}$

9. $r = \sin 2\theta$

10. $r = \sin 3\theta$

11. $r = \sin 4\theta$

12. $r = 1 + r \cos \theta$

13. $r = \dfrac{1}{1 + \cos \theta}$

14. $r^2 = \dfrac{36}{4 \cos^2 \theta + 9 \sin^2 \theta}$

15. $r = 1 + \csc \theta$

16. $r \sin \theta = 1$

17. $r = 1 - \sin \theta$

18. $r = \cos 3\theta$

19. $r = \dfrac{1}{\sin \theta - \cos \theta}$

20. $r^2 = \sin^2 \theta$

21. $r^2 = \sin \theta$

22. $r = 2 + \cos \theta$

23. $\dfrac{r}{1 + r \cos \theta} = 2.$

24. $r = e^{\theta/\pi}$

25. The figure given in the text suggests that at the origin, the two sides of the cardioid have the same tangent, namely, the line $\theta = \pi/2$. Show that this is correct.

Discuss, as in Problems 1 through 24.

26. $r^2 = \cos^2 \theta$

27. $r^2 = \cos \theta$

28. $r^2 = \cos 2\theta$

29. $r^2 = a^2 \cos 2\theta$ (This curve is called a *lemniscate*.)

30. $r^2 = a^2 \sin 2\theta$

Find polar equations for the curves defined by the following conditions, and sketch. Identify the curve if possible.

31. The set of all points which are equidistant from the origin and the line $r = \csc \theta$.

32. The set of all points which are equidistant from the origin and the point $(2\sqrt{2}, \pi/4)$.

33. The set of all points P such that $PA = 2PB$, where A is the origin and $B = (2\sqrt{2}, \pi/4)$.

34. The circle with center at $(2, \pi/4)$ and radius 2.

Sketch:

35. $r = 2 - \sin \theta$

36. $r = 3 - 2 \sin \theta$

37. $r = 3 + 2 \cos \theta$

38. $\dfrac{r}{1 + r \cos \theta} = \dfrac{1}{2}$ (What sort of curve is this, and why?)

9.5   AREAS IN POLAR COORDINATES

Given

$$r = f(\theta) \geq 0, \qquad \alpha \leq \theta \leq \beta,$$

where f is continuous, and the length of the interval $[\alpha, \beta]$ is $\leq 2\pi$. Let R be the region between the polar graph and the origin.

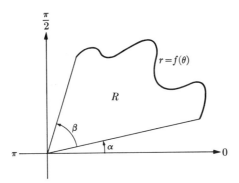

That is,

$$R = \{(r, \theta) \mid \alpha \le \theta \le \beta \quad \text{and} \quad 0 \le r \le f(\theta)\}.$$

Consider a subinterval $[\theta_{i-1}, \theta_i]$ of the interval $[\alpha, \beta]$. Let m_i be the minimum value of f on $[\theta_{i-1}, \theta_i]$, let M_i be the maximum value, and let ΔA_i be the area of the region between the origin and the part of the curve from $\theta = \theta_{i-1}$ to $\theta = \theta_i$.

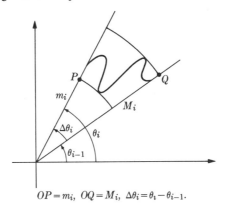

$$OP = m_i, \quad OQ = M_i, \quad \Delta\theta_i = \theta_i - \theta_{i-1}.$$

The area of the inner circular sector, with radius m_i, is

$$\tfrac{1}{2} m_i^2 \Delta\theta_i,$$

and the area of the outer sector, with radius M_i, is

$$\tfrac{1}{2} M_i^2 \Delta\theta_i.$$

Therefore

$$\tfrac{1}{2} m_i^2 \Delta\theta_i \le \Delta A_i \le \tfrac{1}{2} M_i^2 \Delta\theta_i.$$

Now take a net

$$N: \alpha = \theta_0, \theta_1, \ldots, \theta_{i-1}, \theta_i, \ldots, \theta_n = \beta$$

over the interval $[\alpha, \beta]$. The area of R is

$$A = \sum_{i=1}^{n} \Delta A_i.$$

The above inequalities hold for every i; and so by addition we get

$$\sum_{i=1}^{n} \tfrac{1}{2} m_i^2 \Delta \theta_i \leq A \leq \sum_{i=1}^{n} \tfrac{1}{2} M_i^2 \Delta \theta_i.$$

But the sum on the left is the lower sum $s(N)$ of the function

$$F(\theta) = \tfrac{1}{2} f(\theta)^2,$$

over the net N, and the sum on the right is the upper sum $S(N)$, of the same function F, over the net N. Thus

$$s(N) \leq A \leq S(N);$$

and so

$$\lim_{|N| \to 0} s(N) \leq A \leq \lim_{|N| \to 0} S(N).$$

Since

$$\lim_{|N| \to 0} s(N) = \int_{\alpha}^{\beta} \tfrac{1}{2} f(\theta)^2 \, d\theta = \lim_{|N| \to 0} S(N),$$

it follows that A is squeezed, and

$$A = \int_{\alpha}^{\beta} \tfrac{1}{2} f(\theta)^2 \, d\theta.$$

Thus we have:

Theorem 1. Let f be continuous and ≥ 0 on $[\alpha, \beta]$, with $\beta - \alpha \leq 2\pi$, and let R be the region between the origin and the polar graph of f. Then the area of R is

$$A = \int_{\alpha}^{\beta} \tfrac{1}{2} f(\theta)^2 \, d\theta.$$

Let us try this in some simple cases. For the circular region with radius a and center at the origin, the formula gives

$$A = \int_0^{2\pi} \tfrac{1}{2} a^2 \, d\theta = \tfrac{1}{2} a^2 \cdot 2\pi = \pi a^2,$$

which is the right answer. For

$$r = \frac{1}{\cos \theta + \sin \theta}, \qquad 0 \leq \theta \leq \pi/2,$$

we get

$$A = \int_0^{\pi/2} \frac{1}{2} \cdot \frac{1}{(\cos \theta + \sin \theta)^2} \, d\theta$$

$$= \frac{1}{2} \int_0^{\pi/2} \frac{1}{1 + \sin 2\theta} \, d\theta$$

$$= \frac{1}{2} \int_0^{\pi/2} \frac{1 - \sin 2\theta}{\cos^2 2\theta} \, d\theta$$

$$= \tfrac{1}{2} [\tfrac{1}{2} \tan 2\theta - \tfrac{1}{2} \sec 2\theta]_0^{\pi/2}$$

$$= \tfrac{1}{2}(0 + \tfrac{1}{2}) - \tfrac{1}{2}(0 - \tfrac{1}{2}) = \tfrac{1}{2}.$$

This is correct, because the region is a right triangle with legs of length 1.

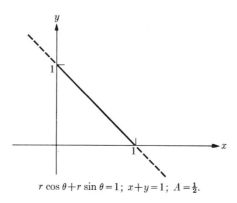

$r \cos \theta + r \sin \theta = 1$; $x + y = 1$; $A = \frac{1}{2}$.

PROBLEM SET 9.5

Find the areas of the regions enclosed by the following curves, and sketch.

1. $r = 1 - \sin \theta$
2. $r = 1 - \cos \theta$
3. $r = \cos \theta$
4. $r^2 = \cos^2 \theta$
5. $r^2 = \cos 2\theta$
6. $r = \sec \theta \tan \theta, 0 \leq \theta \leq \pi/4$
7. $r = \sec \theta \sqrt{\tan \theta}, 0 \leq \theta \leq \pi/4$
8. $r = \sin \theta \cos \theta$
9. $r^2 = 4 \cos 2\theta$
10. Find the area of the inside loop of the graph of $r = 1 - 2 \sin \theta$, and sketch.
11. $r = \dfrac{1}{|\cos \theta| + |\sin \theta|}$
12. $r = e^\theta, 0 \leq \theta \leq 2\pi$
13. $r = e^{2\theta}, 0 \leq \theta \leq 2\pi$
14. Given a polar graph defined by a differentiable function $r = f(\theta)$ ($\alpha \leq \theta \leq \beta$), derive a formula for the slope of the tangent, at a point (r_0, θ_0) $(r_0 = f(\theta_0))$. Here we really mean the slope, relative to a rectangular coordinate system superimposed on the polar coordinate system.

9.6 THE LENGTH OF A PATH

Roughly speaking, the length of a path is the total distance traversed by the moving point. For example, consider the path defined by the coordinate functions

$$f(t) = a \cos t, \qquad g(t) = a \sin t \qquad (0 \leq t \leq 4\pi).$$

The locus of this path is a circle with radius a and circumference $2\pi a$. But as t increases from 0 to 4π, this locus is traversed twice. Therefore the length of the *path* is $2 \cdot 2\pi a = 4\pi a$.

Lengths of paths are undirected; they are always positive (or zero, in trivial cases). Thus the length of the path

$$x = f(t) = \cos t, \qquad y = g(t) = 0 \qquad (0 \leq t \leq 2\pi)$$

is four, not zero; the two halves of the path do not cancel each other out.

To be exact, path length is defined as follows. Given a path

$$x = f(t), \qquad y = g(t) \qquad (a \leqq t \leqq b),$$

let

$$N: t_0, t_1, \ldots, t_{i-1}, t_i, \ldots, t_n$$

be a net over $[a, b]$; for each i from 0 to n, let

$$x_i = f(t_i), \qquad y_i = g(t_i), \qquad P_i = (x_i, y_i).$$

Then

$$\sum_{i=1}^{n} P_{i-1} P_i$$

is the length of an inscribed broken line.

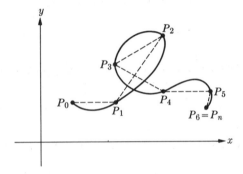

The path length, by definition, is

$$s = \lim_{|N| \to 0} \sum_{i=0}^{n} P_{i-1} P_i.$$

We can express the path length as an integral. For each i, let

$$P_i = (x_i, y_i), \qquad x_i = f(t_i), \qquad y_i = g(t_i),$$

$$\Delta x_i = x_i - x_{i-1}, \qquad \Delta y_i = y_i - y_{i-1},$$

as indicated in the figure.

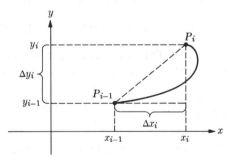

Then

$$P_{-1} P_i = \sqrt{\Delta x_i^2 + \Delta y_i^2}.$$

Since

$$\Delta x_i = f(t_i) - f(t_{i-1}) \quad \text{and} \quad \Delta y_i = g(t_i) - g(t_{i-1}),$$

we know by the mean-value theorem that

$$\Delta x_i = f'(\bar{t}_i) \, \Delta t_i,$$

for some $\bar{t}_i$ between t_{i-1} and t_i; and

$$\Delta y_i = g'(\bar{t}_i') \, \Delta t_i,$$

for some $\bar{t}_i'$ between t_{i-1} and t_i. (We do not know that $\bar{t}_i = \bar{t}_i'$, and this leads to trouble, as we shall see.) Therefore

$$P_{i-1}P_i = \sqrt{[f'(\bar{t}_i) \, \Delta t_i]^2 + [g'(\bar{t}_i') \, \Delta t_i]^2} = \sqrt{f'(\bar{t}_i)^2 + g'(\bar{t}_i')^2} \, \Delta t_i,$$

and so

$$\sum_{i=1}^{n} P_{i-1}P_i = \sum_{i=1}^{n} \sqrt{f'(\bar{t}_i)^2 + g'(\bar{t}_i')^2} \, \Delta t_i.$$

This is almost, but not quite, a sample sum of the function

$$\alpha(t) = \sqrt{f'(t)^2 + g'(t)^2};$$

it differs from a sample sum in that we have used two different sample points $\bar{t}_i, \bar{t}_i'$ on each interval $[t_{i-1}, t_i]$ of our net N. Since $|N| \to 0$, and g' is continuous, we ought to have

$$\sum_{i=1}^{n} \sqrt{f'(\bar{t}_i)^2 + g'(\bar{t}_i')^2} \, \Delta t_i \approx \sum_{i=1}^{n} \sqrt{f'(\bar{t}_i)^2 + g'(\bar{t}_i)^2} \, \Delta t_i = \sum_{i=1}^{n} \alpha(\bar{t}_i) \, \Delta t_i$$

$$\approx \int_a^b \sqrt{f'(t)^2 + g'(t)^2} \, dt$$

when $|N| \approx 0$. For a proof of this, see Appendix I. Meanwhile we shall state the following theorem and use it.

Theorem 1. If the coordinate functions f, g of a path have continuous derivatives f', g', then the length of the path is

$$s = \int_a^b \sqrt{f'(t)^2 + g'(t)^2} \, dt.$$

This formula can be converted to polar coordinates in the following way. Suppose that a polar path is described by a function

$$r = \phi(\theta) \qquad (a \leqq \theta \leqq b).$$

The *rectangular* coordinate functions of the path are then

$$x = f(\theta) = \phi(\theta) \cos \theta, \qquad y = g(\theta) = \phi(\theta) \sin \theta.$$

This gives

$$f'(\theta) = \phi'(\theta) \cos \theta - \phi(\theta) \sin \theta, \qquad g'(\theta) = \phi'(\theta) \sin \theta + \phi(\theta) \cos \theta.$$

For short, let us write ϕ' for $\phi'(\theta)$, c for $\cos\theta$, and s for $\sin\theta$. This gives

$$f'^2 = \phi'^2 c^2 - 2\phi\phi' cs + \phi^2 s^2, \qquad g'^2 = \phi'^2 s^2 + 2\phi\phi' cs + \phi^2 c^2,$$

and

$$f'^2 + g'^2 = \phi'^2 + \phi^2.$$

Thus we have the following theorem.

Theorem 2. Given a path defined in polar coordinates by a function

$$r = \phi(\theta) \qquad (a \leq \theta \leq b),$$

where ϕ' is continuous, the length of the path is

$$s = \int_a^b \sqrt{\phi^2 + \phi'^2}\, d\theta.$$

PROBLEM SET 9.6

It is hard to propose reasonable problems in the calculation of path length; sometimes the integral takes a troublesome but manageable form such as $\int \sqrt{1 + x^2}\, dx$, but most of the time, path length problems are either easy or impossible. Therefore, if some of the problems below look impossible, you should try to think of an approach that might make them easy.

Find the lengths of the following paths.

1. $r = \cos\theta, 0 \leq \theta \leq \pi$

2. $r = e^\theta, 0 \leq \theta \leq 2\pi$

3. $x = 0, y = \cos t, 0 \leq t \leq 3\pi/2$

4. $r = \dfrac{1}{\cos\theta + \sin\theta}, 0 \leq \theta \leq \pi/2$

5. $r = 2\sin\theta, 0 \leq \theta \leq \pi$

6. $r = \dfrac{2}{\cos\theta - \sin\theta}, \pi/2 \leq \theta \leq \pi$

7. $x = t^4, y = t^6, 0 \leq t \leq 1$

8. $r = \dfrac{1}{e^\theta}, 0 \leq \theta \leq 2\pi$

9. $r = \sec\theta\tan\theta, 0 \leq \theta \leq \pi/4$ (Remember the above remarks.)

10. $x = 1 + \sin t, y = 1 - \cos t, 0 \leq t \leq \pi/4$

11. $x = \cos^3 t, y = \sin^3 t, 0 \leq t \leq \pi/4$ (What sort of curve is the locus of this path?)

12. $x = t^3, y = |t^3|, -1 \leq t \leq 1$ (Do these coordinate functions satisfy the conditions of Theorem 1? That is, does $g(t) = |t^3|$ have a continuous derivative?)

*13. The proof of Theorem 1 would have been much easier if we had been able to use the following:

Theorem (?). For each i, there is a single point $\bar{t}_i$, between t_{i-1} and t_i, such that

$$P_{i-1}P_i = \sqrt{f'(\bar{t}_i)^2 + g'(\bar{t}_i)^2}\, \Delta t_i.$$

We could then have expressed $P_{i-1}P_i$ as a sample sum, and passed to the limit, as in Section 7.1. But the above theorem is false. Give an example of a path (with f' and g' continuous) for which the theorem fails. There is a very simple example of this kind.

9.7 VECTORS IN A PLANE

In Section 3.8, we found that the motion of a particle on a line could be described by a single function f, with real numbers as values, and that the velocity and acceleration functions were the first and second derivatives

$$v = f' \quad \text{and} \quad a = v' = f''.$$

As we remarked at the time, these ideas are not adequate to describe the motion of a particle in a plane (or in space). The motion of a particle in a plane E is described by a path, which is a function

$$P: I \to E$$
$$: t \mapsto P(t),$$

where I is an interval, and $P(t)$ is the location of the moving particle at time t. Velocity in this case is a "vector quantity," with both a magnitude and a direction, conveniently pictured by an arrow. At each point $P(t)$, the direction of the velocity vector is the direction of the motion, so that the arrow always lies on the tangent line, pointing in the appropriate direction on the tangent line; and the length of the velocity vector is the speed.

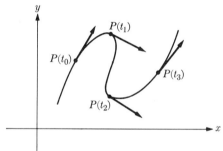

This is the idea. We need to express it in a mathematical form in which it can be used. The idea of a vector appears in a variety of forms. The simplest of these is as follows.

With each point P of the plane we associate the directed segment $\overrightarrow{OP}$, starting at the origin and ending at P. Such a directed segment $\overrightarrow{OP}$ will be called a *vector*.

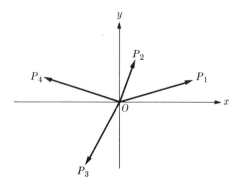

We allow the "degenerate segment" $\overrightarrow{OO}$; this is called the *zero vector*, and may be denoted simply by $\overrightarrow{O}$. Moreover, since all our directed segments, in this section, are going to start at the origin, we can denote the directed segment $\overrightarrow{OP}$ by the shorter symbol $\overrightarrow{P}$. Three operations can be performed, in this system:

Addition. Given $\overrightarrow{P_1}, \overrightarrow{P_2}$, with $P_1 = (x_1, y_1)$ and $P_2 = (x_2, y_2)$, the sum is defined to be

$$\overrightarrow{P_1} + \overrightarrow{P_2} = \overrightarrow{Q},$$

where

$$Q = (x_1 + x_2, y_1 + y_2).$$

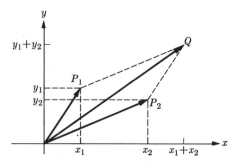

Vector addition is governed by the same formal laws that govern addition of real numbers, as follows.

A.1 *Associativity.* $(\overrightarrow{P_1} + \overrightarrow{P_2}) + \overrightarrow{P_3} = \overrightarrow{P_1} + (\overrightarrow{P_2} + \overrightarrow{P_3})$.

A.2 *Existence of $\overrightarrow{O}$.* There is a vector $\overrightarrow{O}$ such that $\overrightarrow{O} + \overrightarrow{P} = \overrightarrow{P} + \overrightarrow{O} = \overrightarrow{P}$ for every vector $\overrightarrow{P}$.

A.3 *Existence of negatives.* For each vector $\overrightarrow{P}$ there is a vector $-\overrightarrow{P}$ such that

$$\overrightarrow{P} + (-\overrightarrow{P}) = (-\overrightarrow{P}) + \overrightarrow{P} = \overrightarrow{O}.$$

A.4 *Commutativity.* $\overrightarrow{P_1} + \overrightarrow{P_2} = \overrightarrow{P_2} + \overrightarrow{P_1}$.

These follow from the corresponding laws for real numbers. For example, if

$$(\overrightarrow{P_1} + \overrightarrow{P_2}) + \overrightarrow{P_3} = \overrightarrow{Q}, \quad \text{and} \quad \overrightarrow{P_1} + (\overrightarrow{P_2} + \overrightarrow{P_3}) = \overrightarrow{Q'},$$

then

$$Q = ((x_1 + x_2) + x_3, (y_1 + y_2) + y_3)$$
$$= (x_1 + (x_2 + x_3), y_1 + (y_2 + y_3)) = Q',$$

and so $\overrightarrow{Q} = \overrightarrow{Q'}$. The existence of $\overrightarrow{O}$ is obvious: $\overrightarrow{O}$ is $\overrightarrow{OO}$. If $P = (x, y)$, then $-\overrightarrow{P} = \overrightarrow{Q}$, where $Q = (-x, -y)$. Similarly for A.4.

Scalar multiplication. When we are discussing vectors, we refer to real numbers as scalars. To multiply a vector $\vec{P}$ by a scalar α, we multiply the coordinates of P by α. That is,

$$\alpha\vec{P} = \vec{Q},$$

where

$$Q = (\alpha x, \alpha y).$$

We then have a kind of associative law.

M.1. $(\alpha\beta)\vec{P} = \alpha(\beta\vec{P})$.

Because $(\alpha\beta)x = \alpha(\beta x)$, and $(\alpha\beta)y = \alpha(\beta y)$. Multiplication is connected with vector addition by two distributive laws.

M.2. $(\alpha + \beta)\vec{P} = \alpha\vec{P} + \beta\vec{P}$.

M.3. $\alpha(\vec{P}_1 + \vec{P}_2) = \alpha\vec{P}_1 + \alpha\vec{P}_2$.

Zero and 1 work in the usual way:

M.4. $0 \cdot \vec{P} = \vec{O}$, for every $\vec{P}$.

M.5. $1 \cdot \vec{P} = \vec{P}$, for every $\vec{P}$.

M.6. $\alpha \cdot \vec{O} = \vec{O}$, for every α.

Let $\mathscr{V}$ be the set of all vectors $\vec{P}$. In $\mathscr{V}$ we have defined two operations (addition and scalar multiplication), and shown that they satisfy the laws A.1 through A.4 and M.1 through M.6; $\mathscr{V}$ is called a *vector space* (relative to these two operations). More generally, any collection $\mathscr{V}$ of objects is called a vector space if it is provided with two operations satisfying the above formal laws. There are many important vector spaces other than the one which we are now discussing. For example, we may consider the directed segments $\vec{P} = \vec{OP}$, starting from the origin in three-dimensional space, with the two operations defined in an analogous way.

Finally, we introduce another kind of multiplication for vectors, called the *dot product* or *inner product*. If $P_1 = (x_1, y_1)$ and $P_2 = (x_2, y_2)$, as before, then the inner product is a scalar, namely,

$$\vec{P}_1 \cdot \vec{P}_2 = x_1 x_2 + y_1 y_2.$$

The following properties of this operation are easy to check:

S.1. $\vec{P}_1 \cdot \vec{P}_2 = \vec{P}_2 \cdot \vec{P}_1$.

S.2. $(\alpha\vec{P}_1) \cdot \vec{P}_2 = \alpha(\vec{P}_1 \cdot \vec{P}_2)$.

S.3. $\vec{P}_1 \cdot (\vec{P}_2 + \vec{P}_3) = \vec{P}_1 \cdot \vec{P}_2 + \vec{P}_1 \cdot \vec{P}_3$.

S.4. $\vec{P} \cdot \vec{P} \geqq 0$, for every $\vec{P}$.

S.5. If $\vec{P} \cdot \vec{P} = 0$, then $\vec{P} = \vec{O}$.

(The last condition rules out trivial "dot products" for which $\vec{P}_1 \cdot \vec{P}_2$ is always 0, for every $\vec{P}_1, \vec{P}_2$.)

Thus, $\mathscr{V}$ is called an *inner product space* (relative to the three operations which have now been defined). More generally, any collection $\mathscr{V}$ is called an inner product space if it is provided with three operations (addition, scalar multiplication, and inner product) satisfying all the above laws.

As a matter of convenience, we have defined our three operations algebraically, using the coordinates (x, y) of the terminal points P of the vectors. But it is important to understand that all three of them have geometric meanings. We can add two vectors, geometrically, by completing a parallelogram, as shown on the left.

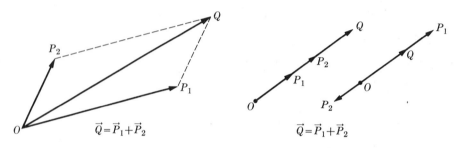

To do this, we don't need to know the directions of the x- and y-axes. Therefore the axes can be, and have been, omitted from the figure. If $\vec{P}_1$ and $\vec{P}_2$ are collinear, then the parallelogram collapses, but the idea is the same.

Geometrically, $-\vec{P}$ is the vector $\vec{Q}$ which has the same length as $\vec{P}$, but has the opposite direction.

To multiply a vector $\vec{P}$ by a positive scalar α, we draw a vector with the same direction as $\vec{P}$, but multiply the length by α. If $\alpha < 0$, we go in the opposite direction, and multiply the length by $|\alpha|$.

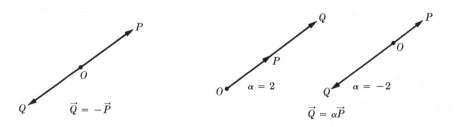

The geometric meaning of the inner product is less obvious. Algebraically,

$$\vec{P}_1 \cdot \vec{P}_2 = x_1 x_2 + y_1 y_2.$$

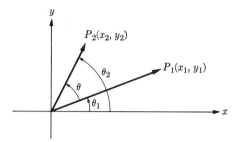

Under the conditions given in the figure,

$$\cos \theta = \cos (\theta_2 - \theta_1)$$

$$= \cos \theta_1 \cos \theta_2 + \sin \theta_1 \sin \theta_2.$$

Substituting $\cos \theta_1 = x_1/OP_1$, $\sin \theta_1 = y_1/OP_1$, $\cos \theta_2 = x_2/OP_2$, $\sin \theta_2 = y_1/OP_2$, we get

$$\cos \theta = \frac{x_1 x_2}{OP_1 \cdot OP_2} + \frac{y_1 y_2}{OP_1 \cdot OP_2},$$

so that

$$\vec{P}_1 \cdot \vec{P}_2 = OP_1 \cdot OP_2 \cos \theta.$$

Obviously $\cos \theta$ is independent of the directions of the axes, because θ measures the angle *between* the two vectors. Note that the length of the vector $\vec{P}$ can be expressed in terms of the dot product:

$$\vec{P} \cdot \vec{P} = x^2 + y^2 = OP^2.$$

The length of a vector $\vec{P}$ may also be denoted by $|\vec{P}|$. Thus

$$|\vec{P}| = \sqrt{\vec{P} \cdot \vec{P}}.$$

By a *linear combination* of two vectors $\vec{P}_1$, $\vec{P}_2$ we mean a vector $\vec{Q}$ which can be expressed in the form

$$\vec{Q} = \alpha \vec{P}_1 + \beta \vec{P}_2,$$

where α and β are scalars. In a coordinate plane, it is easy to find two vectors i and j such that every vector is a linear combination of them. If the vectors i and j are as in the left-hand figure below, and $P = (x, y)$, then

$$\vec{P} = x\mathrm{i} + y\mathrm{j} \qquad (\mathrm{i} = (1, 0), \mathrm{j} = (0, 1)).$$

This is an equation between *vectors*, not numbers. On the right, we have multiplied the vectors i and j by the scalars x and y, and added the resulting vectors.

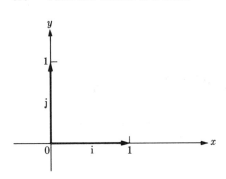
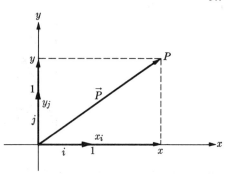

This section contains no new information, but quite a lot of new language. Learning a language takes practice. Therefore, while some of the problems below are genuine problems, many of them are merely exercises in the process of translation from the language of coordinate systems to the language of vectors and back again.

PROBLEM SET 9.7

Sketch the set of all points P satisfying the following conditions.

1. $\vec{P} = \alpha i \; (-\infty < \alpha < \infty)$ 2. $\vec{P} = \alpha j \; (-\infty < \alpha < \infty)$
3. $\vec{P} = \alpha i + \alpha j \; (-\infty < \alpha < \infty)$ 4. $\vec{P} = \alpha i - \alpha j \; (-\infty < \alpha < \infty)$
5. $\vec{P} \cdot i = 0$ 6. $\vec{P} \cdot j = 0$
7. $\vec{P} \cdot (i + j) = 1$ 8. $\vec{P} \cdot (i + j) = 0$
9. $\vec{P} \cdot \vec{P} = 1$ 10. $\vec{P} \cdot \vec{P} = 0$
11. $\vec{P} = \alpha i + 2\alpha j \; (0 \leqq \alpha \leqq \sqrt{3})$ 12. $\vec{P} \cdot i = \frac{1}{2}$
13. $\vec{P} \cdot j = \frac{1}{2}$ 14. $\vec{P} = \alpha i + \alpha^2 j \; (-\infty < \alpha < \infty)$
15. $\vec{P} = 2\alpha i - \alpha j \; (-\infty < \alpha < \infty)$ 16. $\vec{P} = \alpha i + \alpha^3 j \; (-\infty < \alpha < \infty)$
17. $\vec{P}(i + 2j) = 3$ 18. $\vec{P} = \alpha^2 i + \alpha^0 j \; (-\infty < \alpha < \infty)$
19. $\vec{P} = \alpha j + \alpha^2 i \; (-\infty < \alpha < \infty)$ 20. $\vec{P} \cdot (i + 2j) = 0$
21. $\vec{P} \cdot (2i + j) = 0$
22. Let $c = i + j$, $d = i - j$. Express i as a linear combination of c and d. (To do this you will need to calculate with vectors, by the same processes that you use with real numbers. This can be done; and this is why we stated and verified the laws A.1 through A.4 and M.1 through M.6.)
23. Express j as a linear combination of c and d.
24. Now show how any vector $\vec{P}$ can be expressed as a linear combination of c and d.
25. Let $e = i + 2j$, $f = 2i - j$.
    a) Express i as a linear combination of e and f.
    b) Express j as such a linear combination.
    c) Show how every vector $\vec{P}$ can be so expressed.
26. Same problem, for $e = i - 2j$, $f = 3i + 2j$.
27. The vectors g and h *span* the vector space $\mathscr{V}$ if every vector in $\mathscr{V}$ is a linear combination of g and h. (Thus in Problem 25 (c) you showed that e and f span $\mathscr{V}$.) Is it true that every pair of vectors in $\mathscr{V}$ span $\mathscr{V}$? Why or why not?

28. Let $P_1 = (2, 1)$, $P_2 = (1, 2)$. Sketch the set of all points P such that

$$\vec{P} = \alpha \overrightarrow{P_1} + (1 - \alpha)\overrightarrow{P_2} \qquad (0 \leqq \alpha \leqq 1).$$

29. Let $\overrightarrow{P_1}$ and $\overrightarrow{P_2}$ be any two vectors (by which we mean two *different* vectors). Sketch the set of all points P such that

$$\vec{P} = \alpha \overrightarrow{P_1} + (1 - \alpha)\overrightarrow{P_2} \qquad (0 \leqq \alpha \leqq 1).$$

Sketch the set of all points P satisfying the following conditions:

30. $\vec{P} \cdot \mathbf{i} \geqq 0$ 31. $\vec{P} \cdot \mathbf{j} \geqq 0$

32. $\vec{P} \cdot (\mathbf{i} + \mathbf{j}) \geqq 0$ 33. $\vec{P} \cdot (\mathbf{i} - \mathbf{j}) \geqq 0$

34. $\vec{P} = \alpha \mathbf{i} + \beta \mathbf{j} \; (\alpha \geqq 0, \beta \geqq 0)$ 35. $\vec{P} = \alpha \mathbf{i} + \beta(\mathbf{i} + \mathbf{j}) \quad (\alpha \geqq 0, \beta \geqq 0)$

36. $\vec{P} \cdot (\mathbf{i} + 2\mathbf{j}) \geqq 0$ 37. $\vec{P} \cdot (\mathbf{i} - 2\mathbf{j}) < 0$

*33. Let $\mathscr{W}$ be the set of all continuous functions on the interval $[-1, 1]$. State, for the functions in $\mathscr{W}$, definitions of (a) addition, (b) scalar multiplication, and (c) inner product, in such a way that $\mathscr{W}$ forms an inner product space. Verify that under your definitions, the inner product space laws are all satisfied. (There is only one reasonable definition for (a), and similarly for (b); but the "right" definition of the inner product is less obvious. *Hint:* The "$\cdot$" operation is supposed to assign a *number* $f \cdot g$ to each pair of functions f, g. Under what significant operation does a number correspond to *one* function? As a check on your definition, it should turn out that if $f(x) = x^3, g(x) = 1$, $h(x) = x$, then $f \cdot g = 0, g \cdot h = 0$, and $f \cdot h = \frac{2}{5}$.)

This inner product space has important uses, later in the theory of functions.

9.8   FREE VECTORS

In the last section, we defined a vector to be a directed segment $\overrightarrow{OP}$, starting at the origin. We shall now introduce a different form of the vector concept, which for some purposes is better.

By a *translation* of a coordinate plane we mean a correspondence of the form

$$x \leftrightarrow x + h, \qquad y \leftrightarrow y + k,$$

where h and k are constants. This is different from the idea of translation of axes, which we used in Chapter 8. Then, we were moving the *axes*, while now we are moving the *points* (x, y), with $(x, y) \mapsto (x + h, y + k)$.

Suppose that we have given two directed segments $\overrightarrow{PQ}$, $\overrightarrow{P'Q'}$, in a coordinate plane. If there is a translation under which $P \leftrightarrow P'$ and $Q \leftrightarrow Q'$, then we say that $\overrightarrow{PQ}$ and $\overrightarrow{P'Q'}$ are *equivalent*.

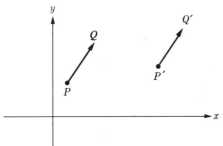

This idea is easy to describe in terms of coordinates. Let

$$P = (x_1, y_1), \qquad Q = (x_2, y_2),$$

$$P' = (x_1', y_1'), \qquad Q' = (x_2', y_2').$$

We can always move P onto P' by a translation

$$x \mapsto x + h, \qquad y \mapsto y + k,$$

where

$$h = x_1' - x_1, \qquad k = y_1' - y_1.$$

If it is true that

$$x_1' - x_1 = x_2' - x_2, \qquad y_1' - y_1 = y_2' - y_2,$$

then this translation also moves Q onto Q', and $\overrightarrow{PQ}$ and $\overrightarrow{P'Q'}$ are equivalent.

For each pair P, Q, the symbol $\overrightarrow{\mathbf{PQ}}$ denotes the set of all directed segments $\overrightarrow{P'Q'}$ that are equivalent to $\overrightarrow{PQ}$. Such a set of equivalent directed segments is called a *free vector* (or simply a *vector*, if the context makes it obvious what meaning is intended). Thus the figure on the left below is a partial picture of exactly one free vector. A free vector is called an *equivalence class* of directed segments; and any directed segment which belongs to such an equivalence class is called a *representative* of the class. Thus each of the arrows in the figure is a representative of the free vector $\overrightarrow{\mathbf{PQ}}$.

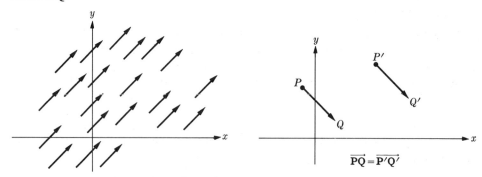

If two directed segments $\overrightarrow{PQ}$, $\overrightarrow{P'Q'}$ are equivalent, then they determine the same free vector, and $\overrightarrow{\mathbf{PQ}} = \overrightarrow{\mathbf{P'Q'}}$. And if $\overrightarrow{\mathbf{PQ}} = \overrightarrow{\mathbf{P'Q'}}$, then the segments $\overrightarrow{PQ}$ and $\overrightarrow{P'Q'}$ are equivalent. Therefore, when we write an equation of the form

$$\overrightarrow{\mathbf{PQ}} = \overrightarrow{\mathbf{P'Q'}},$$

we are saying that the segments $\overrightarrow{PQ}$, $\overrightarrow{P'Q'}$ are equivalent under a translation.

It is now easy to define, for free vectors, the operations of addition, scalar multiplication, and dot product. If

$$\overrightarrow{\mathbf{OP}} + \overrightarrow{\mathbf{OQ}} = \overrightarrow{\mathbf{OR}},$$

in the sense defined in the preceding section, then

$$\overrightarrow{OP} + \overrightarrow{OQ} = \overrightarrow{OR},$$

by definition. This definition is complete, because every free vector $\overrightarrow{ST}$ has exactly one representative segment which starts at the origin. Similarly, if $\alpha\overrightarrow{OP} = \overrightarrow{OQ}$, then

$$\alpha\overrightarrow{OP} = \overrightarrow{OQ},$$

by definition; and

$$\overrightarrow{OP} \cdot \overrightarrow{OQ} = \overrightarrow{OP} \cdot \overrightarrow{OQ},$$

by definition. The form of these definitions makes it clear that all the vector laws and inner product laws of the preceding section also hold true for free vectors. Since all we would need to do is rewrite them in the new notation (using $\overrightarrow{OP}$ for $\vec{P}$), it is not worth while to do so. The length of a free vector is the length of each of its representatives. That is,

$$|\overrightarrow{OP}| = |\overrightarrow{OP}| = OP.$$

The free vector of length 0 is denoted by $\mathbf{0}$.

It is convenient, in figures, to use the label $\overrightarrow{PQ}$ for any representative of the free vector $\overrightarrow{PQ}$. Thus different segments may have the same label, as in the left-hand figure below; and when they do, this means that the segments are equivalent.

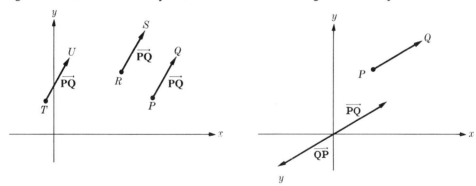

It is easy to see that the right-hand figure above is correctly labeled. Therefore:

Theorem 1. $\overrightarrow{PQ} + \overrightarrow{QP} = \mathbf{0}$, for every P, Q.

Similarly, the labels are correct in the parallelogram below. Since

$$\overrightarrow{OP} + \overrightarrow{OR} = \overrightarrow{OQ},$$

we have

$$\overrightarrow{OP} + \overrightarrow{PQ} = \overrightarrow{OQ}.$$

This has a geometric meaning: we can add free vectors by laying representative segments end to end. Solving for $\overrightarrow{PQ}$, we get $\overrightarrow{PQ} = \overrightarrow{OQ} - \overrightarrow{OP}$. And this gives:

Theorem 2. $\overrightarrow{PQ} + \overrightarrow{QR} + \overrightarrow{RP} = O$, for every P, Q, R.

Proof.

$$\overrightarrow{PQ} + \overrightarrow{QR} + \overrightarrow{RP} = \overrightarrow{OQ} - \overrightarrow{OP} + \overrightarrow{OR} - \overrightarrow{OQ} + \overrightarrow{OP} - \overrightarrow{OR} = O.$$

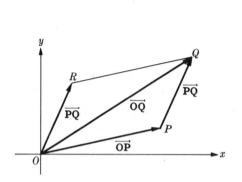 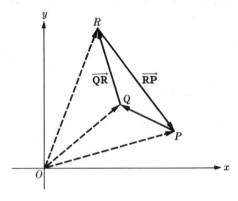

As a matter of convenience, we have defined equivalence of directed segments in terms of a coordinate system. But in fact this relation of equivalence is independent of the choice of the coordinate system. The directed segments $\overrightarrow{PQ}$ and $\overrightarrow{P'Q'}$ are equivalent under translation if (a) their lengths PQ and $P'Q'$ are the same and (b) the directions θ and θ' are the same.

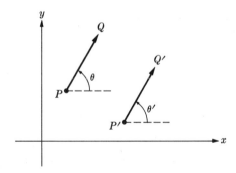

Note that while the directions θ, θ' depend on the directions of the axes, the equation $\theta = \theta'$ does not; if the equation holds, and the axes are rotated, then the equation continues to hold.

Thus we say that the relation of equivalence between directed segments, used in defining free vectors, is *invariant* under changes in the coordinate system.

It very often happens that we use coordinate systems in the study of things which are invariant under changes of coordinates. Thus the distance between two points is invariant, and so also is the question whether a given curve is a parabola. But we use coordinate systems in the study of parabolas, and similarly we use coordinate systems in the study of vectors. If $P = (x, y)$, then x and y are called the x- and y-*components* of $\overrightarrow{OP}$. In this case

$$\overrightarrow{P} = \overrightarrow{OP} = x\mathbf{i} + y\mathbf{j},$$

where $\overrightarrow{P}$, i, and j are as in the preceding section. Corresponding to the vectors i, j we have free vectors **i, j**; and $\overrightarrow{OP}$ is a linear combination of these free vectors:

$$\overrightarrow{OP} = x\mathbf{i} + y\mathbf{j},$$

as shown on the left below. And of course pictures of the new **i** and **j** can be drawn starting at any point that we want. In the right-hand figure, $\overrightarrow{PQ}$, **i**, and **j** are all free vectors. In general, if **V** and **T** are any vectors, with $\mathbf{T} \neq 0$, then the **T**-*component* of **V** is the number

$$\mathbf{V}_\mathbf{T} = |\mathbf{V}| \cos \theta,$$

where θ measures the angle between the direction of **T** and the direction of **V**.

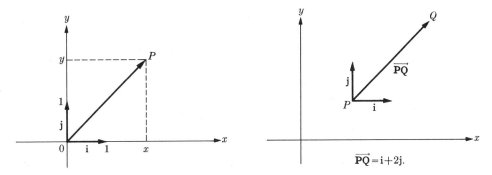

$$\overrightarrow{PQ} = \mathbf{i} + 2\mathbf{j}.$$

Thus, in the figure below $\mathbf{V}_\mathbf{T}$ is the directed distance $\overline{PQ}$, relative to the given positive direction on the line that contains $\overrightarrow{PR}$.

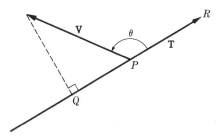

Since

$$\mathbf{V} \cdot \mathbf{T} = |\mathbf{V}| \cdot |\mathbf{T}| \cos \theta,$$

it is easy to express the **T**-component in terms of the dot product:

$$\mathbf{V}_\mathbf{T} = \frac{\mathbf{V} \cdot \mathbf{T}}{|\mathbf{T}|}$$

PROBLEM SET 9.8

In the figures below, we use tick marks to indicate that segments have the same length. Thus the tick marks in the figure below say that $AB = AC$.

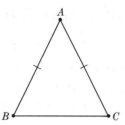

1. a) Calculate $\overrightarrow{OS}$ as a linear combination of $\overrightarrow{OR}$ and $\overrightarrow{OP}$. (The figure is a parallelogram.)

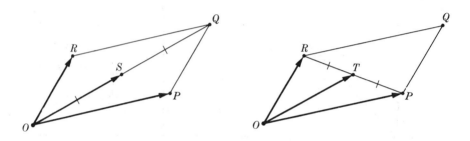

b) Calculate $\overrightarrow{OT}$ as a linear combination of $\overrightarrow{OR}$ and $\overrightarrow{OP}$, shown on the right above. (These two answers, in combination, give a vector proof that the diagonals of a parallelogram bisect each other.)

2. a) Calculate $\overrightarrow{SR}$ and $\overrightarrow{OT}$ as linear combinations of $\overrightarrow{OR}$ and $\overrightarrow{OS}$. (The figure is a rhombus, so $|\overrightarrow{OS}| = |\overrightarrow{OR}|$.)

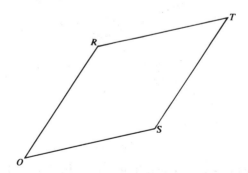

b) Show that, in a rhombus, $\overrightarrow{SR} \cdot \overrightarrow{OT} = 0$. (These two answers, in combination, give a vector proof that the diagonals of a rhombus are perpendicular.)

3. a) Calculate $\overrightarrow{OS}$ as a linear combination of $\overrightarrow{OP}$ and $\overrightarrow{OR}$ in the left-hand figure below.

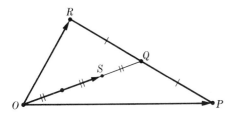 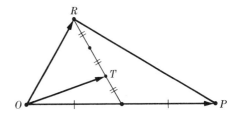

b) Calculate $\overrightarrow{OT}$ as a linear combination of $\overrightarrow{OP}$ and $\overrightarrow{OR}$, in the right-hand figure.

4. Do Problems 3a and 3b give a vector proof that the three medians of a triangle are concurrent? Or do you need to carry out a third calculation of the same kind, to complete the proof?

5. a) Show that

$$|V \cdot T| \leqq |V| \cdot |T|,$$

for every two free vectors V and T.

b) Show that for any real numbers a, b, x, y, we have

$$|ax + by| \leqq \sqrt{a^2 + b^2}\sqrt{x^2 + y^2}.$$

6. Show that if P, Q, R, and S are any four points of the plane, then

$$\overrightarrow{PQ} + \overrightarrow{QR} + \overrightarrow{RS} + \overrightarrow{SP} = 0.$$

7. Let V_0 be a fixed (free) vector. Show that if $V_0 \cdot V = 0$, for every V, then $V_0 = 0$.

8. Suppose that $V \cdot i = 0$, and $|V| = 1$. Is this information enough to determine V? If so, what is V? If not, give a figure, showing the possibilities for V.

9. Given that $V \cdot i = 0$ and $V \cdot j = 0$, discuss as in Problem 8.

10. Given $V \cdot i = 1$ and $V \cdot j = 1$, discuss as in Problem 8.

11. a) A set of vectors $V_1, V_2, \ldots, V_n$ are *linearly dependent* if there are scalars $\alpha_1, \alpha_2, \ldots, \alpha_n$, not all $= 0$, such that

$$\alpha_1 V_1 + \alpha_2 V_2 + \cdots + \alpha_n V_n = 0.$$

Show that for any V, the vectors i, j, and V are linearly dependent.

b) Show that if one of the vectors V_i is $= 0$, then the vectors $V_1, V_2, \ldots, V_n$ are linearly dependent.

c) Find a number a such that $2i + j$ and $7i + aj$ are linearly dependent.

12. a) A set of vectors $V_1, V_2, \ldots, V_n$ are *linearly independent* if they are not linearly dependent. Thus the V_i's are linearly dependent if

$$\sum_{i=1}^{n} \alpha_i V_i = 0 \Rightarrow \alpha_1 = \alpha_2 = \cdots = \alpha_n = 0.$$

Show that i and j are linearly independent.

b) Are i and $i + j$ linearly independent? Why or why not?

c) Given that i and $\overrightarrow{OP}$ are linearly dependent, what are the possibilities for P?

13. Show that if

$$|V_1| = |W_1|, \qquad |V_2| = |W_2|, \qquad \text{and} \qquad V_1 \cdot W_1 = V_2 \cdot W_2,$$

then

$$|V_1 - V_2| = |W_1 - W_2|.$$

(Remember that $|V|^2 = V \cdot V$, for every V.) Then draw a figure, and restate the theorem in the language of elementary geometry.

14. Explain how Problems 5a and 5b can be regarded as the same problem.

15. a) Consider the vector space which you were asked to define in the last problem of the preceding problem set. Let **1** be the constant function which is $= 1$ for each x on $[-1, 1]$. Find ten nonconstant functions $f_1, f_2, \ldots, f_{10}$ such that $\mathbf{1} \cdot f_i = 0$ for each i.
   b) Show that in the same vector space, $f \cdot f_0 = 0$ for every $f \Rightarrow f_0 = 0$.

9.9 VELOCITY, ACCELERATION, AND CURVATURE

We return to the discussion of moving particles in a plane. Suppose that the motion is described by a path

$$P: I \to E$$

$$: t \mapsto P(t),$$

where I is a time interval. Let the coordinate functions of the path be f and g, so that

$$P(t) = \big(f(t), g(t)\big) \qquad (t \text{ on } I).$$

We now regard the path as a function whose values are the vectors

$$\overrightarrow{P}_t = \overrightarrow{OP}_t,$$

where $\overrightarrow{P}_t$ is a vector in the sense of Section 9.7, and $P(t)$ is denoted by P_t, to fit it into the vector notation.

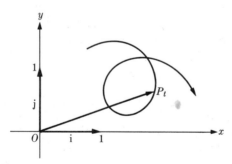

We then have

$$\overrightarrow{P}_t = f(t)\mathbf{i} + g(t)\mathbf{j}.$$

We can now define the velocity and acceleration. These are the *free* vectors

$$V_t = f'(t)\mathbf{i} + g'(t)\mathbf{j},$$
$$A_t = f''(t)\mathbf{i} + g''(t)\mathbf{j},$$

where **i** and **j** are the free vectors corresponding to i and j. Since $\mathbf{V}_t$ and $\mathbf{A}_t$ are free vectors, we can draw pictures of them in any position we want; and so we picture them by drawing arrows starting at the point P_t.

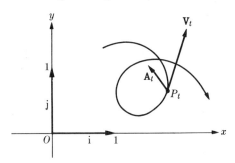

The picture then says that at time t, the moving particle is at the point P_t and has the indicated velocity and acceleration vectors $\mathbf{V}_t$ and $\mathbf{A}_t$. Note that $\mathbf{V}_t$ lies along the tangent line; and this is right. (This should be checked, for the various possible cases. (a) If $f'(t)$ and $g'(t)$ are both 0, then $\mathbf{V}_t = \mathbf{0}$, and there is nothing to prove. (b) If $f'(t) \neq 0$, then $\mathbf{V}_t$ and the tangent line both have slope $g'(t)/f'(t)$. (c) If $f'(t) = 0$ and $g'(t) \neq 0$, then $\mathbf{V}_t$ and the tangent line are both vertical.)

When we write $\mathbf{V}_t = f'(t)\mathbf{i} + g'(t)\mathbf{j}$, $\mathbf{A}_t = f''(t)\mathbf{i} + g''(t)\mathbf{j}$, we are describing each of the vectors $\mathbf{V}_t$ and $\mathbf{A}_t$ by a pair of numbers. Unfortunately, the numbers $f'(t)$, $g'(t)$, $f''(t)$, $g''(t)$ have no physical meaning, because they depend on the coordinate system. It is possible, however, to describe the acceleration by a pair of numbers which do have physical meanings. This is done in the following way. First we take a free vector $\mathbf{T}$, with the same direction as $\mathbf{V}_t$, but with length 1. $\mathbf{T}$ is called the *unit tangent vector* at the point P_t. (Here, and throughout the following discussion, we are assuming that the speed $|\mathbf{V}_t|$ is not zero. If the length of $\mathbf{V}_t$ is 0, then its direction is not determined, and so $\mathbf{T}$ is not determined either.)

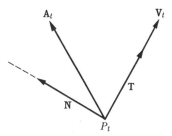

Next we take a free vector $\mathbf{N}$, with length 1, perpendicular to $\mathbf{T}$, and lying on the same side of $\mathbf{T}$ as $\mathbf{A}_t$. Then $\mathbf{A}_t$ must be expressible as a linear combination

$$\mathbf{A}_t = \alpha \mathbf{T} + \beta \mathbf{N}$$

of $\mathbf{T}$ and $\mathbf{N}$. Here α is the $\mathbf{T}$-component of $\mathbf{A}_t$, and β is the $\mathbf{N}$-component. These numbers are called the *tangential* and *normal* components of the acceleration. We shall now compute them.

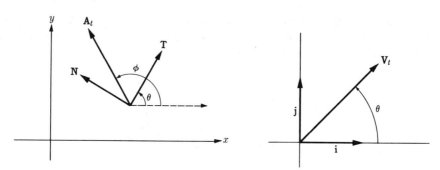

In the right-hand figure above, θ is the direction of $\mathbf{V}_t$. Since $\mathbf{V}_t = f'(t)\mathbf{i} + g'(t)\mathbf{j}$, we have

$$\cos\theta = \frac{f'(t)}{|\mathbf{V}_t|}, \qquad \sin\theta = \frac{g'(t)}{|\mathbf{V}_t|},$$

where

$$|\mathbf{V}_t| = \sqrt{f'(t)^2 + g'(t)^2}.$$

Similarly, ϕ is the direction of the acceleration, so that

$$\cos\phi = \frac{f''(t)}{|\mathbf{A}_t|}, \qquad \sin\phi = \frac{g''(t)}{|\mathbf{A}_t|}.$$

By definition of the T-component α of $\mathbf{A}_t$, we have

$$\alpha = |\mathbf{A}_t|\cos(\phi - \theta) = |\mathbf{A}_t|\cos\phi\cos\theta + |\mathbf{A}_t|\sin\phi\sin\theta$$

$$= f''(t)\cos\theta + g''(t)\sin\theta = \frac{f''(t)f'(t) + g''(t)g'(t)}{|\mathbf{V}_t|}$$

$$= \frac{f'(t)f''(t) + g'(t)g''(t)}{\sqrt{f'(t)^2 + g'(t)^2}}.$$

Theorem 1. The tangential component of acceleration is the derivative of the speed. That is,

$$\alpha = \mathbf{A}_\mathbf{T} = \frac{d}{dt}|\mathbf{V}_t| = |\mathbf{V}_t|'.$$

Once this has been observed, it is easy to check it, by differentiating the function $|\mathbf{V}_t| = \sqrt{f'(t)^2 + g'(t)^2}$.

The normal component β is computed as follows. If $\mathbf{N}$ is counterclockwise from $\mathbf{T}$, as in the figure on p. 423, then

$$\beta = |\mathbf{A}_t|\cos[(\theta + \pi/2) - \phi]$$

$$= |\mathbf{A}_t|\cos[(\theta - \phi) + \pi/2] = -|\mathbf{A}_t|\sin(\theta - \phi).$$

If the direction of $\mathbf{N}$ is reversed, the sign of $\sin(\theta - \phi)$ is also reversed. In any case, we want $\beta \geqq 0$, because $\mathbf{N}$ is taken on the same side of the tangent as $\mathbf{A}_t$. Therefore we must have

$$\beta = |\mathbf{A}_t| \cdot |\sin(\theta - \phi)|,$$

in all cases. Therefore

$$\beta = |\mathbf{A}_t| \cdot |\sin\theta\cos\phi - \cos\theta\sin\phi|$$

$$= \big||\mathbf{A}_t|\sin\theta\cos\phi - |\mathbf{A}_t|\cos\theta\sin\phi\big| = |f''(t)\sin\theta - g''(t)\cos\theta|$$

$$= \frac{|f''(t)g'(t) - g''(t)f'(t)|}{|\mathbf{V}_t|} = \frac{|f''(t)g'(t) - g''(t)f'(t)|}{\sqrt{f'(t)^2 + g'(t)^2}}.$$

This formula for β also has an interpretation, but its interpretation is harder to see, and requires the idea of the curvature of a path at a point.

For the sake of simplicity, we start with the idea of the curvature of the graph of a twice differentiable function at a point.

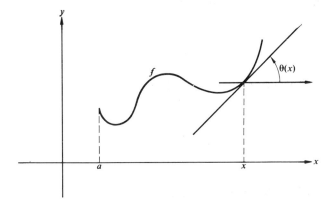

For each x, let $s(x)$ be the length of the graph from $t = a$ to $t = x$. Then

$$s(x) = \int_a^x \sqrt{1 + f'(t)^2}\, dt,$$

and

$$s'(x) = \sqrt{1 + f'(x)^2}.$$

For each x, let $\theta(x)$ be the direction of the tangent line, with $-\pi/2 < \theta(x) < \pi/2$. Since s is an increasing function, $\theta(x)$ is determined when $s(x)$ is known. Thus there is a function h such that

$$\theta(x) = h(s(x)),$$

and (in the language of Section 5.8)

$$\frac{d\theta}{ds} = h'(s(x)) = \frac{d\theta/dx}{ds/dx}.$$

The *curvature* κ is defined to be

$$\kappa = \left|\frac{d\theta}{ds}\right|.$$

This is easy to calculate. Since

$$\theta(x) = \text{Tan}^{-1} f'(x),$$

we have

$$\theta'(x) = \frac{d\theta}{dx} = \frac{1}{1 + f'(x)^2} \cdot f''(x).$$

Therefore

$$\kappa = \left| \frac{d\theta}{ds} \right| = \left| \frac{d\theta/dx}{ds/dx} \right| = \left| \frac{f''(x)}{1 + f'(x)^2} \cdot \frac{1}{\sqrt{1 + f'(x)^2}} \right|$$

$$= \left| \frac{f''(x)}{[1 + f'(x)^2]^{3/2}} \right|.$$

For future reference:

Theorem 2. The curvature of the graph of a twice differentiable function is given by the formula

$$\kappa = \left| \frac{f''(x)}{[1 + f'(x)^2]^{3/2}} \right|.$$

For paths, the idea is similar. Take a fixed t_0, and for each t, let $s(t)$ be the length of the path, from t_0 to t. Then

$$s(t) = \int_{t_0}^{t} \sqrt{f'(u)^2 + g'(u)^2} \, du,$$

and

$$s'(t) = \sqrt{f'(t)^2 + g'(t)^2}.$$

For each t, let $\theta(t)$ be the direction of the velocity vector at time t. We are working on a portion of the path where $|\mathbf{V}_t| = \sqrt{f'(t)^2 + g'(t)^2} \neq 0$. On such a portion of the path, s is an increasing function, and so $\theta(t)$ is determined when $s(t)$ is known. Thus there is a function h such that

$$\theta(t) = h(s(t)).$$

Therefore

$$\frac{d\theta}{ds} = h'(s(t)) = \frac{d\theta/dt}{ds/dt}.$$

But according to the definition of the curvature κ of the path,

$$\kappa = \left| \frac{d\theta}{ds} \right|.$$

In order to calculate κ we take first the case in which $\mathbf{V}_t$ is not vertical, so that

$$\tan \theta(t) = \frac{g'(t)}{f'(t)}.$$

Taking the derivative, we get

$$[\sec^2 \theta(t)]\theta'(t) = \frac{f'(t)g''(t) - g'(t)f''(t)}{f'(t)^2}.$$

Now

$$\sec^2 \theta(t) = 1 + \tan^2 \theta(t) = 1 + \frac{g'(t)^2}{f'(t)^2} = \frac{f'(t)^2 + g'(t)^2}{f'(t)^2}.$$

Therefore

$$\theta'(t) = \frac{f'(t)g''(t) - g'(t)f''(t)}{f'(t)^2 + g'(t)^2}.$$

This derivation works whenever the velocity vector is nonvertical. (*Query:* How would you derive the same formula, in the case where the velocity vector is vertical?) This gives

$$\kappa = \left| \frac{d\theta}{ds} \right| = \left| \frac{d\theta/dt}{ds/dt} \right| = \left| \frac{\theta'(t)}{s'(t)} \right|$$

$$= \left| \frac{f'(t)g''(t) - g'(t)f''(t)}{f'(t)^2 + g'(t)^2} \cdot \frac{1}{\sqrt{f'(t)^2 + g'(t)^2}} \right|$$

$$= \frac{|f'(t)g''(t) - g'(t)f''(t)|}{[f'(t)^2 + g'(t)^2]^{3/2}}.$$

Thus we have:

Theorem 3. The curvature of a twice differentiable path, at any point where the speed is not 0, is given by the formula

$$\kappa = \frac{|f'(t)g''(t) - g'(t)f''(t)|}{[f'(t)^2 + g'(t)^2]^{3/2}}.$$

Comparing this with the formula

$$\beta = \frac{|f''(t)g'(t) - g''(t)f'(t)|}{\sqrt{f'(t)^2 + g'(t)^2}},$$

we get:

Theorem 4. At any point where the speed is not zero, the normal component of acceleration is given by the formula

$$\beta = \mathbf{A_N} = \kappa \, |\mathbf{V}_t|^2.$$

In our discussion, we used the notation $f', g', \ldots$ for derivatives, most of the time, in order to connect our work with the preceding theory. We used the notation $d\theta/dt$, $d\theta/ds$, $\ldots$ only when we really needed to talk about the derivative of one function with respect to another, in defining and calculating curvature. In the literature of physics, however, the notation $f', g', \ldots$ is hardly used at all. The following notations are far more common:

$$\mathbf{V} = \frac{df}{dt}\mathbf{i} + \frac{dg}{dt}\mathbf{j}, \quad \mathbf{V} = \frac{dx}{dt}\mathbf{i} + \frac{dy}{dt}\mathbf{j}, \quad \mathbf{V} = \dot{x}\mathbf{i} + \dot{y}\mathbf{j}.$$

In the last expression the dots over x and y indicate differentiation with respect to *time*. Similarly,

$$\mathbf{A} = \frac{d^2f}{dt^2}\mathbf{i} + \frac{d^2g}{dt^2}\mathbf{j} = \frac{d^2x}{dt^2}\mathbf{i} + \frac{d^2y}{dt^2}\mathbf{j} = \ddot{x}\mathbf{i} + \ddot{y}\mathbf{j}.$$

In these notations,

$$\alpha = A_T = \frac{d}{dt}|V|,$$

and

$$\kappa = \frac{|\dot{x}\ddot{y} - \dot{y}\ddot{x}|}{[\dot{x}^2 + \dot{y}^2]^{3/2}} = \frac{|(dx/dt)(d^2y/dt^2) - (dy/dt)(d^2x/dt^2)|}{[(dx/dt)^2 + (dy/dt)^2]^{3/2}}.$$

There is a good reason, in physics, for the use of the "fractional" notation df/dx, $dy/dt, \ldots$ for derivatives. Most of the time, physical problems involve a large number of interrelated functions, and physicists need to talk about the derivative of one of these with respect to another. Therefore, the rest of the time, they use the same "fractional" notation df/dx for ordinary derivatives f'.

PROBLEM SET 9.9

1. Find the point of maximum curvature of the parabola $y = x^2$, and find the maximum value of κ.

2. Find the point of maximum curvature of the parabola $y = 2 + x + x^2$, and find the maximum value of κ.

3. Find the points of maximum and minimum curvature of the graph of $y = x^3$, and calculate the values of κ at these points.

4. Calculate the curvature of a circle of radius a.

5. Calculate the curvature at the points $(a, 0)$ and $(0, b)$ for the ellipse

$$\frac{x^2}{a^2} + \frac{y^2}{b^2} = 1.$$

6. Sketch the path

$$\vec{P}_t = i \cos t + j \sin t,$$

showing the velocity and acceleration vectors at several points.

7. Discuss (as in Problem 6)

$$\vec{P}_t = i \cos \frac{t}{2} + j \sin \frac{t}{2}.$$

8. Discuss $\vec{P}_t = 2i \cos t + j \sin t$.

9. Discuss $\vec{P}_t = it + jt^2$.

10. Discuss $\vec{P}_t = ti + (t - t^2)j$ $(0 \le t \le 1)$.

11. Discuss $\vec{P}_t = i \cos t^2 + j \sin t^2$.

12. Discuss $\vec{P}_t = it + jt^3$.

13. Discuss $\vec{P}_t = it^3 + jt^2$.

14. Discuss $\vec{P}_t = i(1 - t)^2 + j(t^3 - t)$.

15. Discuss $\vec{P}_t = (t \cos \alpha)i + (-\frac{1}{2}gt^2 + t \cos \alpha)j$. In the sketch, show the velocity at $t = 0$, and find the direction of this vector.

16. For a certain path, the velocity at time 0 has direction α and length 1. The initial point $\vec{P}_0$ is the origin. For each t, $\mathbf{A}_t = -g\mathbf{j}$. Express the path in the form $\vec{P}_t = f(t)\mathbf{i} + g(t)\mathbf{j}$.

17. Discuss as in Problems 6 through 15, and express the tangential and normal components of acceleration as functions of the time:

$$\vec{P}_t = 3\mathbf{i} \cos 2t + 3\mathbf{j} \sin 2t.$$

18. Discuss and sketch

$$\vec{P}_\theta = \mathbf{i} \cos^2 \theta + \mathbf{j} \sin \theta \cos \theta.$$

Describe this as a path in polar coordinates; find a rectangular equation for its locus, and identify the locus.

19. Discuss and sketch

$$\vec{P}_\theta = (\cos \theta - \cos^2 \theta)\mathbf{i} + (\sin \theta - \sin \theta \cos \theta)\mathbf{j}.$$

20. Discuss and sketch

$$\vec{P}_\theta = (3 \cos \theta + \cos 3\theta)\mathbf{i} + (3 \sin \theta - \sin 3\theta)\mathbf{j}.$$

21. Discuss and sketch

$$\vec{P}_\theta = \tfrac{1}{2}(\cos^2 \theta - \sin^2 \theta)\mathbf{i} + \sin \theta \cos \theta\mathbf{j}.$$

22. Discuss and sketch

$$\vec{P}_\theta = (\cos \theta - \cos \theta \sin \theta)\mathbf{i} + (\sin \theta - \sin^2 \theta)\mathbf{j}.$$

23. Is the following statement true? (Why or why not?)

Theorem (?). Given a path with coordinate functions f and g, on an interval $[a, b]$, such that f and g are differentiable, and the velocity is nowhere $= 0$, then there is a time t such that $\mathbf{V}_t$ has the same direction as $\overrightarrow{P_a P_b}$.

The figure indicates that in *some* cases, at least, there is such a time t.

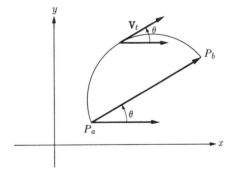

24. Given a path which has curvature κ at time t_0, suppose that the axes are rotated. Does the curvature change? Why or why not? [*Hint:* This problem does not require a calculation.]

25. Let $\mathbf{a} = \mathbf{i} + \mathbf{j}$, $\mathbf{b} = \mathbf{i} - \mathbf{j}$. Suppose we define an "inner product" $\mathbf{V}_1 * \mathbf{V}_2$, by agreeing that for

$$\mathbf{V}_1 = x_1\mathbf{a} + y_1\mathbf{b},$$
$$\mathbf{V}_2 = x_2\mathbf{a} + y_2\mathbf{b},$$

the $*$ product is

$$V_1 * V_2 = x_1 x_2 + y_1 y_2.$$

a) Does $*$ obey the same formal laws as the old inner product?

b) Is it true that $V_1 * V_2 = V_1 \cdot V_2$ for every V_1 and V_2? Why or why not? In any case, express the new operation $*$ in terms of the old.

9.10 CONCLUDING REMARKS ON VECTOR SPACES AND INNER PRODUCT SPACES

The treatment of vectors in this chapter has been brief, because so far we are working in a plane, and the main advantages of a vector approach appear in three-dimensional space, and in spaces of higher dimensions. Meanwhile we must bear in mind that vector ideas appear in many different forms.

1) *Free vectors.* Velocity and acceleration are vectors in this sense, as in Sections 9.8 and 9.9.

2) *Bound vectors.* These have not only length and direction, but also position. For example, if two forces act in opposite directions on the ends of a spring, then they may be regarded as bound vectors.

In the figure, the two forces have the same length and opposite directions, but they do not cancel each other out, as free vectors would; on the contrary, they compress the spring.

3) *Sequences of numbers, regarded as vectors.* For example, ordered quadruplets (w, x, y, z) can be regarded as vectors. We make the natural definitions

$$(w_1, x_1, y_1, z_1) + (w_2, x_2, y_2, z_2) = (w_1 + w_2, x_1 + x_2, y_1 + y_2, z_1 + z_2),$$
$$\alpha(w, x, y, z) = (\alpha w, \alpha x, \alpha y, \alpha z),$$
$$(w_1, x_1, y_1, z_1) \cdot (w_2, x_2, y_2, z_2) = (w_1 w_2 + x_1 x_2 + y_1 y_2 + z_1 z_2.$$

In fact, this is the usual way of describing a space of four dimensions.

4) *Systems of other kinds, regarded as vector spaces and inner product spaces.* Some of these are unexpected, but turn out to be useful. See, for example, Problem 33 of Problem Set 9.7, in which it appeared that a set of functions can be regarded as an inner product space, although functions may not seem like vectors when we look at them one at a time.

For this reason, when people speak of "vectors," we need to find out what kind of vectors they are talking about.

10   Infinite Series

10.1 LIMITS OF SEQUENCES

Most of the time, so far, we have dealt with limits of functions, as $x \to a$ or as $x \to \infty$. But often we have dealt with limits of sequences, as $n \to \infty$. For example, in Section 2.10, we wanted to find the area A, under the graph of $y = x^2$, from $x = 0$ to $a = h$. We expressed A as the limit of a sequence $A_1, A_2, \ldots$, where A_n is the area of a circumscribed polygonal region R_n. We calculated

$$A_n = \frac{h^3}{3}\left(1 + \frac{1}{n}\right)\left(1 + \frac{1}{2n}\right),$$

and we found that

$$\lim_{n \to \infty} A_n = \frac{h^3}{3}.$$

We are now going to use limits of sequences more extensively, as a way of dealing with infinite series. Given an infinite sum

$$\sum_{i=1}^{\infty} a_i = a_1 + a_2 + \cdots,$$

we define

$$A_n = \sum_{i=1}^{n} a_i,$$

and we call the A_n's the *partial sums* of $\sum_{i=1}^{\infty} a_i$. Thus the A_n's form a sequence

$$A_1, A_2, \ldots$$

If

$$\lim_{n \to \infty} A_n = A,$$

then we say that the infinite sum is *convergent*, and we write

$$\sum_{i=1}^{\infty} a_i = A.$$

We shall now examine limits of sequences more carefully, starting with the definition of the limit, and building up the theory that is needed.

Definition. Given a sequence $A_1, A_2, \ldots$ of numbers, and a number L. Suppose that for every $\epsilon > 0$ there is an integer N such that

$$n > N \quad \Rightarrow \quad |A_n - L| < \epsilon.$$

431

Then

$$\lim_{n \to \infty} A_n = L.$$

Note that this is like the definition of $\lim_{x \to \infty} f(x)$. A sequence which has a limit is called *convergent*. Here, as always, when we speak of a limit we mean a finite limit (unless the contrary is stated.)

Theorem 1. $\lim_{n \to \infty} \dfrac{1}{n} = 0$.

Proof. Here $L = 0$, and $|A_n - L| = |1/n - 0| = 1/n$. Thus we need to show that for every $\epsilon > 0$ there is an N such that

$$n > N \quad \Rightarrow \quad \frac{1}{n} < \epsilon.$$

Now

$$\frac{1}{n} < \epsilon \quad \Leftrightarrow \quad n > \frac{1}{\epsilon}.$$

If $1/\epsilon$ is an integer, let $N = 1/\epsilon$. In any case, there is an integer $N > 1/\epsilon$. Then

$$n > N \quad \Rightarrow \quad n > 1/\epsilon \quad \Rightarrow \quad 1/n < \epsilon,$$

which is what we wanted.

On the basis of the definition of $\lim_{n \to \infty} A_n$, we can prove the expected theorems on sums, products, and quotients. These are much like the corresponding theorems for limits of functions. In Appendix C they are listed in such an order that they became easy to prove. Meanwhile we shall state the main results and use them.

Theorem 2. If $\lim_{n \to \infty} A_n = A$ and $\lim_{n \to \infty} B_n = B$, then

$$\lim_{n \to \infty} (A_n + B_n) = A + B,$$

and

$$\mathrm{Lim}_{n \to \infty} A_n B_n = AB.$$

If $B \neq 0$, and $B_n \neq 0$ for each n, then

$$\lim_{n \to \infty} A_n / B_n = A/B.$$

These theorems justify the procedures that we have been using informally. For example, they give a proof that

$$\lim_{n \to \infty} \frac{h^3}{3} \left(1 + \frac{1}{n} \right) \left(1 + \frac{1}{2n} \right) = \frac{h^3}{3}.$$

The steps are as follows:

$$\lim_{n \to \infty} \frac{1}{n} = 0$$

$$\lim_{n \to \infty} \left(1 + \frac{1}{n}\right) = 1$$

$$\lim_{n \to \infty} \frac{1}{2n} = \frac{1}{2} \lim_{n \to \infty} \frac{1}{n} = 0$$

$$\lim_{n \to \infty} \left(1 + \frac{1}{2n}\right) = 1$$

$$\lim_{n \to \infty} \frac{h^3}{3} \left(1 + \frac{1}{n}\right) \left(1 + \frac{1}{2n}\right) = \frac{h^3}{3}.$$

(Justification for each of these steps?)

If we start with convergent sequences $A_1, A_2, \ldots, B_1, B_2, \ldots$, and so on, then Theorem 2 tells us that certain other sequences are convergent. But often we deal with sequences which are not built up out of convergent sequences as in Theorem 2. We then need the following ideas.

Definition. A sequence $A_1, A_2, \ldots$ is *increasing* if $A_n \leq A_{n|1}$ for every n. The sequence is *decreasing* if $A_n \geq A_{n+1}$ for every n. (If $A_n < A_{n+1}$ for every n, then the sequence is *strictly increasing*; and if $A_{n+1} < A_n$ for every n, then the sequence is *strictly decreasing*.)

Definition. If there is a number M such that $A_n \leq M$ for every n, then M is called an *upper bound* of the sequence $A_1, A_2, \ldots$, and we say that the sequence is *bounded above*. If there is a number m such that $m \leq A_n$ for every n, then m is called a *lower bound* of the sequence, and we say that the sequence is *bounded below*. If there is a $K > 0$ such that $|A_n| \leq K$ for every n, then the sequence is *bounded*.

Example: (1) If $A_n = \sqrt{n}$ for every n, then the sequence is increasing, and is bounded below but not above. (2) If $A_n = e^{-n}$, then the sequence is decreasing, and is bounded. (3) If $A_n = \sin n$, then the sequence is neither increasing nor decreasing, but is bounded, with $|\sin n| \leq 1$ for each n.

It is easy to see that if a sequence is bounded both above and below, then it is bounded. Given $m \leq A_n \leq M$ for every n, let K be the larger of the numbers $|m|$ and $|M|$.

Theorem 3. If a sequence is increasing, and is bounded above, then it is convergent.

That is, if

$$A_1 \leq A_2 \leq \cdots \leq A_n \leq A_{n+1} \leq \cdots \leq M,$$

then the sequence has a limit. The first application of this principle that you may have seen is in geometry. Given a circle of diameter 1, we inscribe in it a regular polygon of $2n$ sides. For each n, let A_n be the perimeter of our $2n$-gon. (Note that we had better start with $n = 2$.) It is a matter of elementary geometry to show that the sequence A_2, A_3, ... is increasing. Also, $A_n < 4$ for every n, because the perimeter of every inscribed polygon is less than the perimeter of the circumscribed square. (Draw a figure.) Therefore the sequence is convergent. Its limit, of course, is π.

We proceed to the proof. Let S be the set of all numbers A_n. That is,

$$S = \{A_n\}.$$

Then S has an upper bound. By the Least Upper Bound Postulate (LUBP), S has a least upper bound. (See Section 5.6.) This is called the *supremum* of S, and is denoted by sup S. Let

$$A = \text{sup } S.$$

We shall show that

$$\lim_{n\to\infty} A_n = A.$$

Let ϵ be any positive number. Then $A - \epsilon < A$. Therefore $A - \epsilon$ is *not* an upper bound of S. Therefore $A_N > A - \epsilon$ for some N. Since the sequence is increasing, this means that

$$n > N \implies A_n > A - \epsilon.$$

Since A is an upper bound of S, and $A + \epsilon > A$, it follows that $A + \epsilon$ is an upper bound of S.

Therefore

$$A_n < A + \epsilon \qquad \text{for every } n.$$

Therefore

$$n > N \implies |A_n - A| < \epsilon,$$

and $\lim_{n\to\infty} A_n = A$, which was to be proved.

We have a similar theorem for decreasing sequences:

Theorem 4. If a sequence is decreasing, and is bounded below, then it is convergent.

That is, if A_1, A_2, ... is decreasing, and $A_n \geqq K$ for every n, then the sequence has a limit.

Proof. For each n, let $B_n = -A_n$. Then B_1, B_2, ... is increasing, and is bounded above. Therefore it is convergent. Let $\lim_{n\to\infty} B_n = B$. Then $\lim_{n\to\infty} A_n = -B$.

Some simple sequences converge for reasons which are not covered by the preceding theorems. For example, given that

$$\lim_{n\to 0} \frac{1}{n} = 0,$$

it is obvious that

$$\lim_{n\to\infty} \frac{1}{n + 2\sqrt[3]{n}} = 0,$$

because the second sequence is smaller, term by term. This is the idea of the following theorem.

Theorem 5 (*The squeeze principle*). If $\lim_{n\to\infty} A_n = L$, $\lim_{n\to\infty} C_n = L$, and $A_n \leqq B_n \leqq C_n$ for every n, then $\lim_{n\to\infty} B_n = L$.

In many cases, it is easier to use this theorem than to do awkward calculations. Similarly, it ought to be true that

$$\lim_{n\to\infty} \frac{\cos n}{n} = 0,$$

because $|\cos n| \leqq 1$, and $1/n \to 0$. But we can't get this result from Theorem 2, because $\cos n$ does not approach a limit as $n \to \infty$. Hence we need the following:

Theorem 6 (*The annihilation theorem*). If $\lim_{n\to\infty} A_n = 0$, and $B_1, B_2, \ldots$ is bounded, then $\lim_{n\to\infty} A_n B_n = 0$.

Theorem 7. Every convergent sequence is bounded.

For increasing sequences, this is trivial: if $A_1, A_2, \ldots$ is increasing, and $\lim_{n\to\infty} A_n = A$, then $A_1 \leqq A_n \leqq A$ for every n. Similarly for decreasing sequences. For a proof in the general case, see Appendix C. This theorem has simple applications: if we show that a sequence is *not* bounded, then it follows that the sequence is not convergent.

The statements

$$\lim_{n\to\infty} A_n = \infty, \qquad \lim_{n\to\infty} A_n = -\infty$$

mean what you would expect. You should be able to state your own definitions of them, following, if you need to, the models of Section 5.3. Sequences like this are *not* called convergent. If $\lim_{n\to\infty} A_n = \infty$, then we say that the sequence *diverges to infinity*. And if $\lim_{n\to\infty} A_n = -\infty$, we say that the sequence *diverges to minus infinity*. We have to be careful about this: if convergence allowed the limits ∞ and $-\infty$, then Theorem 7 would become false, and Theorem 2 would be meaningless in many cases. (You can't perform algebraic operations on the "numbers" ∞ and $-\infty$.)

PROBLEM SET 10.1

Investigate the following indicated limits. That is, find out whether they exist, and find out, if possible, what they are.

1. $\lim_{n\to\infty} \dfrac{1}{n^2}$

2. $\lim_{n\to\infty} \dfrac{2 + 3n}{3n}$ (Try dividing the numerator and denominator by n.)

3. $\lim_{n\to\infty} \dfrac{1}{n^2 + n}$ (Try using one of the last theorems in this section.)

4. $\lim_{n\to\infty} \dfrac{1}{2n^3 + n}$ 5. $\lim_{n\to\infty} \dfrac{n}{n^3 + n^2 + \pi}$

6. $\lim_{n\to\infty} \dfrac{\sin(2n + 1)}{n}$ 7. $\lim_{n\to\infty} \dfrac{\cos(n - 1)}{n + 1}$

8. $\lim\limits_{n\to\infty} \left(1 + \dfrac{1}{n}\right)^n$

9. $\lim\limits_{n\to\infty} \left(1 - \dfrac{1}{n}\right)^{-1/n}$ [*Hint:* Surely you know $\lim_{x\to 0} (1 + x)^{1/x}$. Now find $\lim_{y\to\infty}$ $(1 + 1/y)^y$, and apply the result to the problem in hand.]

10. $\lim_{n\to\infty} B_n$, where B_n is the perimeter of a regular $2n$-gon circumscribed about a circle of radius 1. (You need not prove that your answer to this one is right.)

11. $\lim\limits_{n\to\infty} \ln n$

12. $\lim\limits_{n\to\infty} \ln (n^2)$

13. $\lim\limits_{n\to\infty} \ln 1/n$

14. $\lim\limits_{n\to\infty} \ln \left(\dfrac{1}{n^2}\right)$

15. $\lim\limits_{n\to\infty} \displaystyle\int_1^n \dfrac{dx}{x^2}$

16. $\lim\limits_{n\to\infty} \displaystyle\int_1^n \dfrac{dx}{x^3}$

17. $\lim\limits_{n\to\infty} \displaystyle\int_1^n \dfrac{dx}{x^{3/2}}$

18. $\lim\limits_{n\to\infty} \displaystyle\sum_{i=1}^n \dfrac{1}{i^2}$ (Investigate existence only. A geometric interpretation is useful.)

19. $\lim\limits_{n\to\infty} \displaystyle\int_1^n \dfrac{dx}{x}$

20. $\lim\limits_{n\to\infty} \displaystyle\sum_{i=1}^n \dfrac{1}{i^3}$ (Investigate existence only.)

21. $\lim\limits_{n\to\infty} \displaystyle\sum_{i=1}^n \dfrac{1}{i^{3/2}}$ (Investigate existence only.)

22. $\lim\limits_{n\to\infty} \displaystyle\sum_{i=1}^n \dfrac{1}{i}$ (Investigate existence only.)

23. $\lim\limits_{n\to\infty} \displaystyle\sum_{i=1}^n \left(\sin \dfrac{i\pi}{n}\right) \dfrac{\pi}{n}$ (Geometric interpretation?)

24. $\lim\limits_{n\to\infty} \displaystyle\sum_{i=1}^n \left(\cos \dfrac{i\pi}{2n}\right) \dfrac{\pi}{2n}$

25. $\lim\limits_{n\to\infty} \displaystyle\sum_{i=1}^n \dfrac{1}{1 + (i/n)} \left(\dfrac{1}{n}\right)$

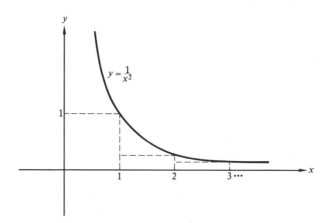

26. $\displaystyle\lim_{n\to\infty} \sum_{i=1}^{n} \frac{1}{1 + (i/2n)} \left(\frac{1}{n}\right)$

27. $\displaystyle\lim_{n\to\infty} \frac{1}{n} \sum_{i=1}^{n} e^{i/n}$

28. $\displaystyle\lim_{n\to\infty} \frac{1}{n} \sum_{i=1}^{n} e^{-i/n}$

29. $\displaystyle\lim_{n\to\infty} n \sin \frac{1}{n}$

30. $\displaystyle\lim_{n\to\infty} n^2 \sin \frac{1}{n^2}$

31. $\displaystyle\lim_{n\to\infty} n\left(1 - \cos \frac{1}{n}\right)$

32. $\displaystyle\lim_{n\to\infty} n \tan \frac{1}{n}$

33. $\displaystyle\lim_{n\to\infty} n \sec \frac{1}{n}$

34. $\displaystyle\lim_{n\to\infty} \left[\sum_{i=1}^{n} \frac{1}{i} - \ln n \right]$

(In fact, this limit exists; if you can find a geometric interpretation of the problem, you can prove it. The limit is known as Euler's constant. Nobody knows whether it is rational.)

35. $\displaystyle\lim_{n\to\infty} \sum_{i=1}^{n} \frac{1}{(2i + 1)^2}$ (Investigate existence only.)   36. $\displaystyle\lim_{n\to\infty} \sum_{i=1}^{n} \frac{1}{(3i^2 + 1)^2}$

10.2 INFINITE SERIES. CONVERGENCE. COMPARISON TESTS

By an *infinite series* we mean an indicated sum of the form

$$\sum_{i=1}^{\infty} a_i = a_1 + a_2 + \cdots + a_n + \cdots$$

We say "an indicated sum" because in many cases there is no such thing as the sum of infinitely many terms. For example, the series

$$1 + 1 + 1 + \cdots \quad \text{(to infinity)}$$

has no sum; and neither does the series

$$1 - 1 + 1 - 1 + 1 \cdots \quad \text{(to infinity)}.$$

In many cases, however, the "sum of infinitely many terms" can be defined, by a passage to a limit, in the following way.

    Given the series

$$\sum_{i=1}^{\infty} a_i,$$

for each n, let

$$A_n = \sum_{i=1}^{n} a_i = a_1 + a_2 + \cdots + a_n.$$

Then A_n is called the *nth partial sum* of the series. If

$$\lim_{n\to\infty} A_n = A,$$

where A is a (finite) number, then we say that the series is *convergent* and that A is its *sum*. We also say that the series *converges to A*. If the sequence $A_1, A_2, \ldots$ has no limit, then we say that the series is *divergent*. If

$$\lim_{n\to\infty} A_n = \infty,$$

then the series *diverges to infinity*; and if

$$\lim_{n \to \infty} A_n = -\infty,$$

then the series *diverges to minus infinity.* We may write these statements briefly as

$$\sum_{i=1}^{\infty} a_i = A, \qquad \sum_{i=1}^{\infty} a_i = \infty, \qquad \sum_{i=1}^{\infty} a_i = -\infty.$$

Probably the first example that you have seen of a convergent series is the geometric series

$$1 + r + r^2 + \cdots + r^n + \cdots \qquad (0 < r < 1).$$

Here

$$A_n = 1 + r^2 + \cdots + r^n = \frac{1 - r^{n+1}}{1 - r} = \frac{1}{1 - r} - \frac{r^{n+1}}{1 - r} = \frac{1}{1 - r} - r^n \frac{r}{1 - r}.$$

If we know that

$$\lim_{n \to \infty} r^n = 0 \qquad (0 < r < 1), \tag{1}$$

then it follows that

$$\lim_{n \to \infty} A_n = \frac{1}{1 - r} \qquad (0 < r < 1); \tag{2}$$

and this means that

$$\sum_{i=0}^{\infty} r^i = \frac{1}{1 - r} \qquad (0 < r < 1). \tag{3}$$

There are many ways of proving (1). The following proof is the easiest. Since $0 < r < 1$, we have

$$r^{n+1} < r^n \quad \text{for every } n.$$

Therefore the sequence $r, r^2, r^3, \ldots$ is decreasing. And it has a lower bound, namely 0. Therefore the sequence is convergent, to some limit L. Thus

$$\lim_{n \to \infty} r^n = L, \qquad \text{and} \qquad \lim_{n \to \infty} r^{n+1} - L.$$

(Why? What happens to the limit of a sequence, if you omit the first term?) Therefore

$$L = \lim_{n \to \infty} r^{n+1} = r \lim_{n \to \infty} r^n = rL,$$

and so

$$L = rL, \qquad \text{and} \qquad (1 - r)L = 0.$$

Since $1 - r \neq 0$, it follows that $L = 0$. Therefore

$$\lim_{n \to \infty} r^n = 0 \qquad (0 < r < 1).$$

In fact, the same conclusion holds for $-1 < r \leq 0$. We get this from the following observation:

Theorem 1. If $\lim_{n \to \infty} |a_n| = 0$, then $\lim_{n \to \infty} a_n = 0$, and conversely.

Proof? (If you rewrite these two statements, using the definitions of the statements $\lim_{n \to \infty} |a_n| = 0$ and $\lim_{n \to \infty} a_n = 0$, they hardly even look different.) Thus we get the following theorem.

Theorem 2. If $-1 < r < 1$, then

$$\lim_{n \to \infty} r^n = 0.$$

Algebraically, the formula

$$1 + r + r^2 + \cdots + r^n = (1 - r^{n+1})/(1 - r)$$

holds for every $r \ne 1$. We therefore have a more general result for geometric series:

Theorem 3. If $-1 < r < 1$, then

$$\sum_{i=0}^{\infty} r^i = \frac{1}{1 - r}.$$

This holds because $\lim_{n \to \infty} [-r^{n+1}/(1 - r)] = 0/(1 - r) = 0$. If the first term is any number a, rather than 1, then we have:

Theorem 4. If $-1 < r < 1$, then

$$\sum_{i=0}^{\infty} ar^i = \frac{a}{1 - r}.$$

The following theorem often makes it easy to see that a series *diverges*.

Theorem 5. If $\sum_{i=1}^{\infty} a_i$ is convergent, then $\lim_{n \to \infty} a_n = 0$.

Proof. For each n, let

$$A_n = \sum_{i=1}^{n} a_i.$$

Let $\lim_{n \to \infty} A_n = A$. Then $\lim_{n \to \infty} A_{n-1} = A$, where $n > 1$. Therefore

$$\lim_{n \to \infty} (A_n - A_{n-1}) = A - A = 0.$$

But $A_n - A_{n-1} = a_n$. Therefore $\lim_{n \to \infty} a_n = 0$, which was to be proved.

For example, the geometric series $\sum_{i=0}^{\infty} ar^i$ is divergent for $a \ne 0$ and $|r| \geq 1$. In this case, $|a_n| = |a| \cdot |r|^n \geq |a|$, and so a_n does not approach 0.

Warning. The converse of Theorem 5 is false. That is, the nth term of a series may approach 0, and the series may still diverge. The simplest example of this is the series

$$1 + \tfrac{1}{2} + \tfrac{1}{2} + \tfrac{1}{3} + \tfrac{1}{3} + \tfrac{1}{3} + \tfrac{1}{4} + \tfrac{1}{4} + \tfrac{1}{4} + \tfrac{1}{4} + \cdots$$

The next five terms are each equal to $\tfrac{1}{5}$; and so on. Here $a_n \to 0$, but the series diverges to infinity.

A more natural example of the same phenomenon is the *harmonic* series

$$\sum_{i=1}^{\infty} \frac{1}{i} = 1 + \frac{1}{2} + \frac{1}{3} + \cdots + \frac{1}{n} + \cdots$$

In fact, this diverges. The easiest way to see this is to draw a picture:

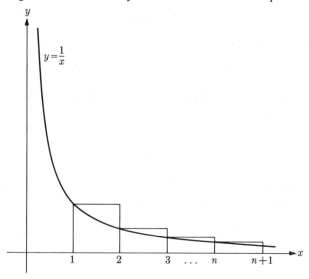

For each n, the area under the graph from $x = 1$ to $x = n + 1$ is less than the total area of the circumscribed rectangles. Therefore

$$A_n = 1 + \frac{1}{2} + \frac{1}{3} + \cdots + \frac{1}{n} > \int_1^{n+1} \frac{dx}{x}.$$

But this integral is $\ln (n + 1)$; and

$$\lim_{n \to \infty} \ln (n + 1) = \infty.$$

Therefore the partial sums A_n form an unbounded sequence, and the series must diverge to infinity. Briefly:

Theorem 6. $\sum_{i=1}^{\infty} (1/i) = \infty$.

The same sort of comparison scheme can be used for other series, to show that they converge. Consider, for example,

$$\sum_{i=1}^{\infty} \frac{1}{i^2}.$$

Here $a_i = 1/i^2$, and so $\lim_{i \to \infty} a_i = 0$. This does not, in itself, show that the series converges. But the algebraic pattern suggests that the series is related to the improper integral

$$\int_1^\infty \frac{dx}{x^2} = \lim_{a \to \infty} \int_1^a \frac{dx}{x^2} = \lim_{a \to \infty} \left[\frac{-1}{x} \right]_1^a = \lim_{a \to \infty} \left(\frac{-1}{a} + 1 \right) = 1.$$

Since $1/x^2 > 0$ for every x, the integral approaches its limit from below, and

$$\int_1^n \frac{dx}{x^2} < 1 \quad \text{for every } n.$$

And since the function $1/x^2$ is decreasing,

$$\int_{n-1}^{n} \frac{dx}{x^2} > \frac{1}{n^2} \qquad (n > 1).$$

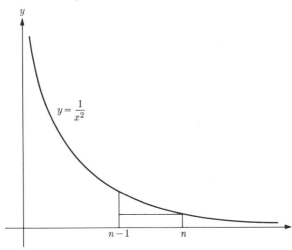

$$y = \frac{1}{x^2}$$

(Here the area of the rectangle is $1/n^2$.) Therefore

$$\frac{1}{2^2} < \int_{1}^{2} \frac{dx}{x^2}, \qquad \frac{1}{3^2} < \int_{2}^{3} \frac{dx}{x^2},$$

and

$$A_n = \sum_{i=1}^{n} \frac{1}{n^2} < 1 + \int_{1}^{n} \frac{dx}{x^2} < 2.$$

Obviously the sequence $A_1, A_2, \dots$ of partial sums is increasing; and we have just seen that it is bounded above. Therefore:

Theorem 7. $\sum_{i=1}^{\infty} (1/i^2)$ is convergent.

Some of the ideas that we have been using to get these results are useful in so many connections that they are worth recording as theorems.

Theorem 8 (*The comparison theorem*). Let $\sum_{i=1}^{\infty} a_i$ and $\sum_{i=1}^{\infty} b_i$ be series, with

$$0 \leqq a_i \leqq b_i \quad \text{for each } i.$$

Then (1) if $\sum_{i=1}^{\infty} b_i$ is convergent, then so also is $\sum_{i=1}^{\infty} a_i$; and (2) if $\sum_{i=1}^{\infty} a_i$ is divergent, then so also is $\sum_{i=1}^{\infty} b_i$.

Proof. For each n, let

$$A_n = \sum_{i=1}^{n} a_i, \qquad B_n = \sum_{i=1}^{n} b_i.$$

Then

$$A_n \leqq B_n \quad \text{for each } n.$$

(Why?) And each of the sequences $A_1, A_2, \ldots$ and $B_1, B_2, \ldots$ is increasing. An increasing sequence is convergent if it is bounded, and conversely. We can therefore prove (1) in the following steps:

$$\sum_{i=1}^{\infty} b_i \quad \text{is convergent}$$

$$\Rightarrow B_1, B_2, \ldots \text{ is convergent}$$

$$\Rightarrow B_1, B_2, \ldots \text{ is bounded}$$

$$\Rightarrow A_1, A_2, \ldots \text{ is bounded}$$

$$\Rightarrow A_1, A_2, \ldots \text{ is convergent}$$

$$\Rightarrow \sum_{i=1}^{\infty} a_i \quad \text{is convergent.}$$

(Reason for each of these implication signs?) Similarly, we can prove (2) in the following steps:

$$\sum_{i=1}^{\infty} a_i \quad \text{is divergent}$$

$$\Rightarrow A_1, A_2, \ldots \text{ is divergent}$$

$$\Rightarrow A_1, A_2, \ldots \text{ is unbounded}$$

$$\Rightarrow B_1, B_2, \ldots \text{ is unbounded}$$

$$\Rightarrow B_1, B_2, \ldots \text{ is divergent}$$

$$\Rightarrow \sum_{i=1}^{\infty} b_i \quad \text{is divergent.}$$

The comparison theorem gives us easy tests for some series. Consider for example,

$$\sum_{i=0}^{\infty} \frac{1}{i!} = 1 + \frac{1}{1} + \frac{1}{2!} + \frac{1}{3!} + \cdots$$

Here

$$n! = 1 \cdot 2 \cdot 3 \cdots n \qquad (n \geq 1)$$

and $0! = 1$, by definition. For each i, let

$$a_i = \frac{1}{i!}, \qquad b_i = \left(\frac{1}{2}\right)^{i-1}.$$

Then

$$a_i \leq b_i \quad \text{for each } i;$$

$$\frac{1}{0!} < \left(\frac{1}{2}\right)^{-1} \qquad (i = 0),$$

$$\frac{1}{1!} = \left(\frac{1}{2}\right)^{0} \qquad (i = 1),$$

and thereafter the strict inequality holds, with $1/n! < (1/2)^{n-1}$ for $n \geq 2$. Therefore our series is term by term less than the geometric series

$$\sum_{i=0}^{\infty} 2 \cdot \left(\frac{1}{2}\right)^i,$$

which is known to converge. Therefore:

Theorem 9. $\sum_{i=0}^{\infty} (1/i!)$ is convergent.

In fact,

$$\sum_{i=0}^{\infty} \frac{1}{i!} = e = \ln^{-1} 1 = \lim_{x \to 0} (1 + x)^{1/x}.$$

But we won't be able to prove this until we have developed the theory much further. The situation here is peculiar: the easiest way to get this special result is first to show that

$$e^x = \sum_{i=0}^{\infty} \frac{x^i}{i!} \quad \text{for every } x,$$

and then to set $x = 1$. (You have seen a situation like this before. The easiest way to find $\int_0^1 x^4 \, dx$ is first to calculate the function $\int_0^x t^4 \, dt$, and then to set $x = 1$.) Consider next

$$\sum_{i=2}^{\infty} \frac{1}{n - \sqrt{n}}.$$

Since

$$\frac{1}{n - \sqrt{n}} > \frac{1}{n} \quad \text{for every } n,$$

and

$$\sum_{i=2}^{\infty} \frac{1}{n} = \infty,$$

it follows that the given series diverges:

$$\sum_{i=2}^{\infty} \frac{1}{n - \sqrt{n}} = \infty.$$

While the comparison theorem tells us, under some conditions, that a series converges, it never tells us what the sum is. But such partial information may be useful. In fact, some of the most important uses of series are in cases where a number (or a function) can best be described by a series; in such cases, we use $\sum_{i=1}^{n} a_i$ (for some large n) to get an approximation of $\sum_{i=1}^{\infty} a_i$. For example, the approximation

$$\sum_{i=0}^{n} \frac{1}{i!} \approx \sum_{i=0}^{\infty} \frac{1}{i!}$$

is excellent, even for fairly small values of n; it gives by far the best way of computing e; and in fact, the series approaches its infinite sum so fast that e is much easier to compute than $\sqrt{2}$.

Therefore, when you are asked to show that a series has a sum, without finding out what the sum is, you should not consider that the problem is artificial.

PROBLEM SET 10.2

Find out which of the following series are convergent. If the series is geometric, calculate the sum.

1. $\displaystyle\sum_{i=1}^{\infty} \frac{1}{\sqrt{i}}$

2. $\displaystyle\sum_{i=1}^{\infty} \frac{1}{i^{3/2}}$

3. $\displaystyle\sum_{i=1}^{\infty} \frac{1}{\sqrt{i}+1}$

4. $\displaystyle\sum_{i=1}^{\infty} \frac{1}{i^{3/2}+2}$

5. $\displaystyle\sum_{i=1}^{\infty} \left(\frac{1}{3}\right)^{2i+1}$

6. $\displaystyle\sum_{i=1}^{\infty} \left(\frac{1}{4}\right)^{2i}$

7. $\displaystyle\sum_{i=1}^{\infty} (-1)^i \pi^{-i}$

8. $\displaystyle\sum_{i=1}^{\infty} (-2)^i e^{2i}$

9. $\displaystyle\sum_{i=1}^{\infty} \frac{\sin^2 (2i-1)}{i^2}$

10. $\displaystyle\sum_{i=1}^{\infty} \frac{\cos^3 (2i)}{i^2}$

11. $\displaystyle\sum_{i=1}^{\infty} \frac{1}{i^3}$

12. $\displaystyle\sum_{i=1}^{\infty} \frac{1}{i^7}$

13. $\displaystyle\sum_{i=1}^{\infty} \frac{1}{i^{1.1}}$

14. $\displaystyle\sum_{i=1}^{\infty} \frac{1}{i^{0.9}}$

15. $\displaystyle\sum_{i=2}^{\infty} \frac{1}{i \ln i}$

16. $\displaystyle\sum_{i=2}^{\infty} \frac{1}{i^2 \ln i}$

17. $\displaystyle\sum_{i=2}^{\infty} \frac{1}{i \ln^2 i}$

18. $\displaystyle\sum_{i=2}^{\infty} \frac{1}{i^2 \ln^2 i}$

19. $\displaystyle\sum_{i=2}^{\infty} \frac{1}{\ln^2 i}$

20. $\displaystyle\sum_{i=2}^{\infty} \frac{1}{i^3 \ln i}$

21. $\displaystyle\sum_{i=0}^{\infty} \frac{1}{(i!)^2}$

22. $\displaystyle\sum_{i=0}^{\infty} \frac{(i!-1)}{(i!)^3}$

23. $\displaystyle\sum_{i=1}^{\infty} \frac{1}{i(i+1)}$

24. $\displaystyle\sum_{i=2}^{\infty} \frac{1}{i(i-1)}$

25. $\displaystyle\sum_{i=1}^{\infty} \frac{1}{i(i+1)(i+2)}$

26. $\displaystyle\sum_{i=2}^{\infty} \frac{1}{(i-1)(i)(i+2)}$

27. $\displaystyle\sum_{i=1}^{\infty} \frac{i}{i+1}$

28. $\displaystyle\sum_{i=1}^{\infty} \frac{i+1}{2i}$

29. $\displaystyle\sum_{i=1}^{\infty} \frac{i}{i^2+1}$

30. $\displaystyle\sum_{i=1}^{\infty} \frac{2i}{i^2-1}$

31. If you think of Theorem 3 backwards, it says that

$$\frac{1}{1-r} = 1 + r + r^2 + \cdots$$

That is, $1/(1-r)$ can be expressed as the sum for an infinite series. Express $1/(1+x)$ as the sum of an infinite series. For what numbers x does your series converge?

32. Express $1/(1+x^2)$ as an infinite series. For what numbers x does the series converge?

33. Same question, for $1/(1+x^4)$.

*34. Suppose that $\sum_{i=0}^{\infty} a_i x^i$ converges for every x. The series then defines a function

$$f(x) = \sum_{i=0}^{\infty} a_i x^i.$$

It will turn out that functions which can be defined in this way are always differentiable, and that their derivatives can be calculated by differentiating the series a term at a time. That is,

$$f'(x) = \sum_{i=1}^{\infty} i a_i x^{i-1}.$$

(Don't try to prove this; you haven't got a chance.) Granted that all this is true, what must the a_i's be, if $f(0) = 1$ and $f'(x) = f(x)$ for every x? Comment on your result.

35. $\displaystyle\sum_{i=1}^{\infty} \frac{i}{i^3 + 1}$

36. $\displaystyle\sum_{i=1}^{\infty} \frac{2i^2}{i^4 + 2}$

*37. For which numbers α is the series $\sum_{i=1}^{\infty} (1/n^{\alpha})$ convergent?

*38. Prove the following.

Theorem A (*The Integral Test*). Let f be a positive decreasing continuous function, on the interval $[1, \infty)$. If

$$\int_1^{\infty} f(x)\, dx < \infty, \tag{1}$$

then

$$\sum_{i=1}^{\infty} f(i) < \infty, \tag{2}$$

and conversely.

10.3 ABSOLUTE CONVERGENCE. ALTERNATING SERIES

Given a series $\sum_{i=0}^{\infty} a_i$ (in which the terms may be positive, negative, or zero), we can form a new series by taking the absolute value $|a_i|$ of each term a_i. For example, if

$$\sum_{i=0}^{\infty} a_i = \sum_{i=0}^{\infty} (-1)^i r^i = 1 - r + r^2 - \cdots,$$

then

$$\sum_{i=0}^{\infty} |a_i| = \sum_{i=0}^{\infty} |r|^i = 1 + |r| + |r|^2 + \cdots$$

Given that $\sum a_i$ converges, it does not follow that $\sum |a_i|$ converges. For example, the series

$$\sum a_i = 1 - 1 + \tfrac{1}{2} - \tfrac{1}{2} + \tfrac{1}{3} - \tfrac{1}{3} + \cdots$$

is convergent, but the series

$$\sum |a_i| = 1 + 1 + \tfrac{1}{2} + \tfrac{1}{2} + \tfrac{1}{3} + \tfrac{1}{3} + \cdots$$

is not, because the harmonic series is not. The same sort of thing happens if we take absolute values in the series

$$\sum_{i=1}^{\infty} a_i = \sum_{i=1}^{\infty} (-1)^{i+1} \frac{1}{i} = 1 - \frac{1}{2} + \frac{1}{3} - \frac{1}{4} + \cdots$$

Here it is plain that $\sum |a_i|$ diverges, but it is not quite so easy to see that $\sum a_i$ is convergent. This is worth proving, however, because the idea used in the proof is useful in other connections.

Theorem 1. $\sum_{i=1}^{\infty} (-1)^{i+1}(1/i)$ is convergent.

Proof. Let

$$A_n = \sum_{i=1}^{n} (-1)^{i+1} \frac{1}{i} = 1 - \frac{1}{2} + \cdots + (-1)^{n+1} \frac{1}{n}.$$

If n is even, with $n = 2k$, then

$$A_n = A_{2k} = 1 - \frac{1}{2} + \frac{1}{3} - \cdots - \frac{1}{2k}$$

$$= \left(1 - \frac{1}{2}\right) + \left(\frac{1}{3} - \frac{1}{4}\right) + \cdots + \left(\frac{1}{2k-1} - \frac{1}{2k}\right).$$

Therefore the sequence $A_2, A_4, A_6, \ldots, A_{2k}, \ldots$ is *increasing*. And it has an upper bound, because

$$A_{2k} = 1 - \left(\frac{1}{2} - \frac{1}{3}\right) - \left(\frac{1}{4} - \frac{1}{5}\right) - \cdots - \left(\frac{1}{2k-2} - \frac{1}{2k-1}\right) - \frac{1}{2k} < 1.$$

Therefore the sequence $A_2, A_4, \ldots$ has a limit. Let

$$A = \lim_{k \to \infty} A_{2k}. \tag{1}$$

We shall show that A is the sum of the series. First we observe that

$$\lim_{k \to \infty} A_{2k+1} = \lim_{k \to \infty} [A_{2k} + a_{2k+1}] = \lim_{k \to \infty} A_{2k} + \lim_{k \to \infty} \frac{1}{2k + 1},$$

so that

$$\lim_{k \to \infty} A_{2k+1} = A + 0 = A. \tag{2}$$

Thus we see that (1) as $n \to \infty$ through even values, $A_n \to A$ and (2) as $n \to \infty$ through odd values, $A_n \to A$. It follows that (3) $\lim_{n \to \infty} A_n = A$.

Proof? (You need to show that for every $\epsilon > 0$ there is an N such that $|A_n - A| < \epsilon$ for every $n > N$. Given such an ϵ, you know from (1) that there is an N_1 such that $|A_{2k} - A| < \epsilon$ for every $k > N_1$; and you know from (2) that there is an N_2 such that $|A_{2k+1} - A| < \epsilon$ for every $k > N_2$. How can N be defined in terms of N_1 and N_2?)

The scheme that we used to prove Theorem 1 applies more generally. If you reexamine the proof, you will see that the only facts about the series

$$\sum_{i=1}^{\infty} (-1)^{i+1} \frac{1}{i} = \sum_{i=1}^{\infty} a_i$$

that were used were the following:

1) The series is *alternating*. That is, successive terms a_i, a_{i+1} have opposite signs.
2) $\lim_{n \to \infty} a_n = 0$.
3) The sequence $|a_1|, |a_2|, \ldots$ is decreasing.

We have therefore proved the following theorem.

Theorem 2 (*The alternating series test*). Given an alternating series $\sum_{i=1}^{\infty} a_i$. If $\lim_{n \to \infty} a_n = 0$, and the sequence $|a_1|, |a_2|, \ldots$ is decreasing, then the series converges.

(Strictly speaking, some of our formulas in the proof of Theorem 1 used the fact that the first term was positive instead of negative. If you know that the theorem holds in this case, how would you show that it also holds when $a_1 < 0$?)

We have seen that if $\sum a_i$ converges, it does not follow that $\sum |a_i|$ converges. But the reverse implication does hold:

Theorem 3. If $\sum_{i=1}^{\infty} |a_i|$ is convergent, then so also is $\sum_{i=1}^{\infty} a_i$.

If $\sum |a_i|$ is convergent, then $\sum a_i$ is said to be *absolutely convergent*. In this language, we can restate Theorem 3 as follows:

Every absolutely convergent series is convergent.

To prove this, we break up each partial sum

$$A_n = \sum_{i=1}^{n} a_i$$

into a sum of positive terms and a sum of negative terms. To do this, we let

$$a_i^+ = \begin{cases} a_i & \text{if } a_i \geqq 0, \\ 0 & \text{if } a_i < 0, \end{cases}$$

and let

$$a_i^- = \begin{cases} a_i & \text{if } a_i \leqq 0, \\ 0 & \text{if } a_i > 0. \end{cases}$$

Let

$$A_n^+ = \sum_{i=1}^{n} a_i^+, \qquad A_n^- = \sum_{i=1}^{n} a_i^-.$$

Then

$$A_n = A_n^+ + A_n^-,$$

for each n, because

$$a_i = a_i^+ + a_i^-$$

for each i. Obviously $A_1^+, A_2^+, \ldots$ is an increasing sequence, and $A_1^-, A_2^-, \ldots$ is a decreasing sequence. Let

$$k = \sum_{i=1}^{\infty} |a_i|.$$

Then

$$\sum_{i=1}^{n} |a_i| \leqq k \quad \text{for every } n.$$

Also

$$A_n^+ \leqq \sum_{i=1}^{n} |a_i|,$$

because A_n^+ is the sum of some (perhaps all) of the terms on the right-hand side.

Therefore $A_1^+, A_2^+, \ldots$ is convergent. Let

$$A^+ = \lim_{n \to \infty} A_n^+.$$

Similarly,

$$A_n^- \geqq \sum_{i=1}^{n} (-|a_i|),$$

because A_n^- is the sum of some (perhaps all) of the terms on the right-hand side; and if you omit negative terms, the sum becomes larger. Therefore the decreasing sequence $A_1^-, A_2^-, \ldots$ is bounded below. Therefore it has a limit. Let $A^- = \lim_{n \to \infty} A_n^-$. Then

$$\lim_{n \to \infty} A_n = \lim_{n \to \infty} (A_n^+ + A_n^-) = A^+ + A^-,$$

and $\sum_{i=1}^{\infty} a_i$ is convergent, which was to be proved. In fact, we can say a little more:

Theorem 4. If $\sum_{i=1}^{\infty} |a_i|$ is convergent, then

$$\left| \sum_{i=1}^{\infty} a_i \right| \leqq \sum_{i=1}^{\infty} |a_i|.$$

Proof. We know that

$$|a_1 + a_2| \leqq |a_1| + |a_2|.$$

By induction it follows that

$$\left| \sum_{i=1}^{n} a_i \right| \leqq \sum_{i=1}^{n} |a_i| \quad \text{for every } n.$$

Passing to the limit, we get the inequality that we wanted.

PROBLEM SET 10.3

Find out which of the following series are alternating, which are convergent, and which are absolutely convergent.

1. $\displaystyle\sum_{i=1}^{\infty} (-1)^i \pi^{-i}$ 2. $\displaystyle\sum_{i=1}^{\infty} (-\tfrac{3}{4})^i$ 3. $\displaystyle\sum_{i=1}^{\infty} (-1)^i \frac{i}{i+1}$

4. $\displaystyle\sum_{i=1}^{\infty} (-1)^i \frac{1}{i^4}$ 5. $\displaystyle\sum_{i=1}^{\infty} \frac{i^2 - 1}{i^2 + 1}$ 6. $\displaystyle\sum_{i=1}^{\infty} \frac{\cos \pi i}{i}$

7. $\displaystyle\sum_{i=1}^{\infty} (-2)^i \frac{1}{i!}$ 8. $\displaystyle\sum_{i=1}^{\infty} (-1)^{i!} \frac{1}{i^2}$ 9. $\displaystyle\sum_{i=2}^{\infty} (-\tfrac{1}{2})^{i^2(i-1)}$

10. $\displaystyle\sum_{i=1}^{\infty} \left(-\frac{1}{e}\right)^{2i}$ 11. $\displaystyle\sum_{i=1}^{\infty} \left(\sin \frac{\pi i}{2} + \cos \frac{\pi i}{2}\right) \frac{1}{i}$ 12. $\displaystyle\sum_{i=1}^{\infty} (-i)^{-i}$

10.4 ESTIMATES OF REMAINDERS

Given that a series converges, we often want to use a partial sum

$$A_n = \sum_{i=1}^{n} a_i$$

as an approximation of the limit

$$A = \lim_{n \to \infty} A_n = \sum_{i=1}^{\infty} a_i.$$

The approximation $A_n \approx A$ is used in some of the most important applications, and in all applications that use computers. As in all approximation processes, we are better off if we can set a limit on the error. We shall now find ways to do this.

Given that $\sum_{i=1}^{\infty} a_i$ converges to a sum A, let $R_n = A - A_n$. Then

$$R_n = \sum_{i=n+1}^{\infty} a_i,$$

and obviously

$$\lim_{n \to \infty} R_n = 0.$$

For alternating series, of the type treated in Theorem 2 of the preceding section, it is easy to get an estimate of R_n. Let the series be

$$\sum_{i=1}^{\infty} a_i = \sum_{i=1}^{\infty} (-1)^{i+1} b_i = b_1 - b_2 + b_3 - \cdots,$$

where $b_i = |a_i|$. Then

$$R_n = \sum_{i=n+1}^{\infty} a_i = \sum_{i=n+1}^{\infty} (-1)^{i+1} b_i.$$

If n is even, then

$$R_n = b_{n+1} - b_{n+2} + b_{n+3} - \cdots$$
$$= (b_{n+1} - b_{n+2}) + (b_{n+3} - b_{n+4}) + \cdots \geqq 0.$$

But we can also write

$$R_n = b_{n+1} - (b_{n+2} - b_{n+3}) - \cdots \leqq b_{n+1}.$$

Therefore

1) $0 \leqq R_n \leqq b_{n+1}$,   when n is even.

If n is odd, then

$$R_n = -b_{n+1} + b_{n+2} - b_{n+3} + \cdots$$
$$= -(b_{n+1} - b_{n+2}) - (b_{n+3} - b_{n+4}) - \cdots \leqq 0;$$
$$R_n = -b_{n+1} + (b_{n+2} - b_{n+3}) + (b_{n+4} - b_{n+5}) + \cdots \geqq -b_{n+1}.$$

Thus

2) $-b_{n+1} \leqq R_n \leqq 0$,   when n is odd.

Therefore

$$-b_{n+1} \leqq R_n \leqq b_{n+1} \quad \text{for every } n.$$

Since $b_{n+1} = |a_{n+1}|$, we have proved the following theorem.

Theorem 1. Given $\sum_{i=1}^{\infty} a_i$. If (1) the series is alternating, (2) $\lim_{n \to \infty} a_n = 0$, and (3) the sequence $|a_1|, |a_2|, \ldots$ is decreasing, then

$$|R_n| \leq |a_{n+1}| \quad \text{for every } n.$$

That is, when you stop after a finite number of terms, the error is numerically no larger than the first term that you omit. For example, take

$$\sum_{i=1}^{\infty} (-1)^{i+1} \frac{1}{i^2} = 1 - \frac{1}{2^2} + \frac{1}{3^2} - \cdots.$$

By the alternating series test, this series converges. Let A be its sum. Then

$$A \approx 1 - \frac{1}{2^2} + \frac{1}{3^2} - \cdots + \frac{1}{9^2};$$

and the error in the approximation is $\leq 1/10^2 = 0.01$. This series does not converge very fast. Next consider

$$\sum_{i=0}^{\infty} (-1)^i \frac{1}{i!} = 1 - 1 + \frac{1}{2!} - \frac{1}{3!} + \cdots.$$

This series converges to a sum A. (It will turn out that $A = 1/e$.) We have

$$A \approx \frac{1}{2!} - \frac{1}{3!} + \cdots + \frac{1}{10!};$$

and the error is less than $1/11!$. This series converges very rapidly:

$$11! = 39{,}916{,}800, \quad \text{and} \quad \frac{1}{11!} \approx 2.5052 \cdot 10^{-8} \approx 0.000000025052.$$

If you reexamine the proof of Theorem 1, you will see that the method that we used to get an estimate of the error was very much like the method that we used to establish convergence in the first place, in the proof of the alternating series test. This happens most of the time: that is, a proof of convergence usually gives an estimate of R_n. Consider, for example,

$$\sum_{i=1}^{\infty} \frac{1}{i^2}.$$

We let

$$A_n = \sum_{i=1}^{n} \frac{1}{i^2},$$

and we observe that the sequence $A_1, A_2, \ldots$ is increasing. To show that it is bounded above, we draw a picture and observe that

$$A_n = \sum_{i=1}^{n} \frac{1}{i^2} < 1 + \int_1^n \frac{dx}{x^2}.$$

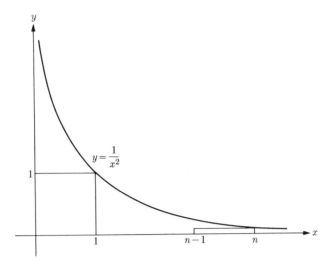

The same sort of reasoning tells us that

$$R_n = \sum_{i=n+1}^{\infty} \frac{1}{i^2} < \int_n^{\infty} \frac{dx}{x^2}.$$

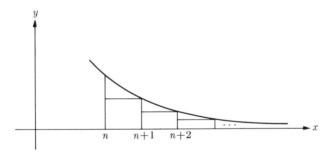

Since

$$\int_n^{\infty} \frac{dx}{x^2} = \left[\frac{-1}{x} \right]_n^{\infty} = \frac{1}{n},$$

we conclude that

$$R_n < \frac{1}{n} \quad \text{for every } n.$$

This is nowhere nearly so small as the estimate of error for the corresponding alternating series. In fact, the positive series $\sum (1/i^2)$ converges very slowly.

Similarly, Theorem 4 of Section 10.3 gives an estimate of the error for series which are absolutely convergent.

Theorem 2. Suppose that $\sum |a_i|$ is convergent. Let

$$R_n = \sum_{i=n+1}^{\infty} a_i.$$

Then

$$|R_n| \leq \sum_{i=n+1}^{\infty} |a_i|.$$

That is, the error in $\sum a_i$ is numerically no greater than the error in $\sum |a_i|$. To prove this, we apply Theorem 4 to the series

$$\sum_{i=n+1}^{\infty} a_i, \qquad \sum_{i=n+1}^{\infty} |a_i|.$$

If we use the comparison theorem of Section 10.2, to establish the convergence of a positive series, then any estimate of the remainder of the larger series automatically is an estimate of the remainder of the smaller one. For example, we have found that

$$\sum_{i=1}^{\infty} \frac{1}{i^2} < \infty,$$

with $R_n < 1/n$ for every n. Since $0 < 1/(i^2 + 1) < 1/i^2$ for every i, it follows by the comparison theorem that

$$\sum_{i=1}^{\infty} \frac{1}{i^2 + 1} < \infty.$$

It also follows, for the remainder R'_n in the new series, that

$$R'_n = \sum_{i=n+1}^{\infty} \frac{1}{i^2 + 1} < \sum_{i=n+1}^{\infty} \frac{1}{i^2} < \frac{1}{n},$$

and so

$$R'_n < \frac{1}{n}.$$

This scheme always works, whenever we establish convergence by means of the comparison theorem.

PROBLEM SET 10.4

Each of the following series is convergent. In each case, get an estimate of the remainder R_n, in the form $|R_n| \leq \cdots$

1. $\displaystyle\sum_{i=1}^{\infty} \left(\frac{1}{3}\right)^{2i+1}$

2. $\displaystyle\sum_{i=1}^{\infty} \left(-\frac{3}{4}\right)^i$

3. $\displaystyle\sum_{i=1}^{\infty} (-1)^i \pi^{-i}$

4. $\displaystyle\sum_{i=1}^{\infty} \frac{\cos \pi i}{i}$

5. $\displaystyle\sum_{i=1}^{\infty} \frac{\sin^2 (2i - 1)}{i^2}$

6. $\displaystyle\sum_{i=1}^{\infty} \frac{1}{i^3}$

7. $\displaystyle\sum_{i=1}^{\infty} \frac{(-1)^{i+1}}{i^{-4}}$

8. $\displaystyle\sum_{i=1}^{\infty} \frac{1}{i^{1.1}}$

9. $\displaystyle\sum_{i=1}^{\infty} \frac{(-1)^i}{i^{0.9}}$

10. $\displaystyle\sum_{i=2}^{\infty} \frac{1}{i \ln^2 i}$

11. $\displaystyle\sum_{i=0}^{\infty} \frac{1}{(i\,!)^2}$

12. $\displaystyle\sum_{i=1}^{\infty} \left(-\frac{1}{e}\right)^{2i}$

13. $\displaystyle\sum_{i=1}^{\infty} \frac{1}{i(i+1)}$

14. $\displaystyle\sum_{i=2}^{\infty} \frac{1}{i^2(1+i)}$

15. $\displaystyle\sum_{i=1}^{\infty} \frac{1}{i(i+1)(i+2)}$

16. $\displaystyle\sum_{i=1}^{\infty} (-i)^{-i}$

*17. Let $f_1, f_2, \ldots$ be a sequence of continuous functions defined on the same interval $[a, b]$. Let f be a function such that

$$\lim_{n\to\infty} f_n(x) = f(x),$$

for each x on $[a, b]$. Questions: (1) Does it follow that f is continuous? (2) If f is known to be continuous, does it follow that

$$(?) \lim_{n\to\infty} \int_a^b f_n(x)\,dx = \int_a^b f(x)\,dx?$$

*18. Consider $\sum_{i=1}^{\infty} (-1)^{i+1}(1/i)$. Show that by writing the terms of this series in a different *order* (using each term once and only once) you can get a series $\sum a_i$ whose sum is 10.

19. Now reexamine your solutions of Problems 1 through 16. If you used any method other than Theorem 1, in estimating the remainder in an alternating series, try using Theorem 1, and compare the new estimate with the old one. (The alternating series test usually gives a good estimate, in the cases where it applies at all.)

10.5 TERMWISE INTEGRATION OF SERIES. POWER SERIES FOR Tan⁻¹ AND ln

A *power series* is a series of the form

$$\sum_{i=0}^{\infty} a_i x^i = a_0 + a_1 x + a_2 x^2 + \cdots$$

(Here, as a matter of convenience, we are defining $0^0 = 1$, so that $a_0 x^0 = a_0$ for every x, including $x = 0$.) Thus every geometric series is a power series; writing x for r in the old formula, we get

$$\sum_{i=0}^{\infty} x^i = \frac{1}{1-x} \qquad (-1 < x < 1).$$

If a given series is convergent, for every x on an open interval $(-r, r)$, then the series defines a function f, on the same interval, and we write

$$f(x) = \sum_{i=0}^{\infty} a_i x^i \qquad (-r < x < r).$$

The following theorem is fundamental:

Theorem A. Given

$$f(x) = \sum_{i=0}^{\infty} a_i x^i \qquad (-r < x < r).$$

Then f is continuous and differentiable on $(-r, r)$, and the derivative of the sum is the sum of the derivatives. That is,

$$f'(x) = \sum_{i=1}^{\infty} i\, a_i x^{i-1} \qquad (-r < x < r).$$

The same idea applies to the integral.

Theorem B. Given

$$f(x) = \sum_{i=0}^{\infty} a_i x^i \qquad (-r < x < r).$$

Then the integral of the sum is the sum of the integrals. That is,

$$\int_0^x f(t)\, dt = \sum_{i=0}^{\infty} \int_0^x a_i t^i\, dt = \sum_{i=0}^{\infty} \frac{a_i}{i+1}\, x^{i+1}.$$

As you might expect, the proofs are hard; they will be postponed until the end of this chapter. But the theorems are easy to apply, and Theorem B gives the best method of finding series for many functions. The method is as follows.

We know that

$$1 + x + x^2 + \cdots + x^n + \cdots = \frac{1}{1-x} \qquad (-1 < x < 1).$$

Writing this backwards, we can express the function $1/(1-x)$ as a power series:

$$1/(1-x) = 1 + x + x^2 + \cdots \qquad (-1 < x < 1).$$

Replacing x by $-x$, we get

$$\frac{1}{1+x} = 1 - x + x^2 - \cdots + (-1)^n x^n + \cdots \qquad (-1 < x < 1);$$

and then, replacing x by x^2, we get

$$\frac{1}{1+x^2} = 1 - x^2 + x^4 - x^6 + \cdots + (-1)^n x^{2n} + \cdots \qquad (-1 < x < 1).$$

Theorem B says that the series on the right can be integrated a term at a time. Thus

$$\int_0^x \frac{dt}{1+t^2} = \int_0^x dt - \int_0^x t^2\, dt + \cdots + (-1)^n \int_0^x t^{2n}\, dt + \cdots \qquad (-1 < x < 1),$$

and this gives

$$\int_0^x \frac{dt}{1+t^2} = x - \frac{x^3}{3} + \cdots (-1)^n \frac{x^{2n+1}}{2n+1} + \cdots$$

$$= \sum_{i=0}^{\infty} (-1)^i \frac{x^{2i+1}}{2i+1} \qquad (-1 < x < 1).$$

The integral on the left is equal to $\mathrm{Tan}^{-1} x$. Thus we have:

Theorem 1.

$$\text{Tan}^{-1} x = \sum_{i=0}^{\infty} (-1)^i \frac{x^{2i+1}}{2i+1} \qquad (-1 < x < 1).$$

Granted that Theorem B is true, there is no need to test the convergence of the series on the right; Theorem B tells us not only that the series has a sum, but also that its sum is $\text{Tan}^{-1} x$. Note that the series includes only terms of odd degree. This could have been predicted, because Tan^{-1} is an odd function, with $\text{Tan}^{-1}(-x) = -\text{Tan}^{-1} x$ for every x.

The same method can be used to get a series for the natural logarithm.

Theorem 2.

$$\ln(1+x) = x - \frac{x^2}{2} + \frac{x^3}{3} - \cdots + (-1)^{i+1}\frac{x^i}{i} + \cdots$$

$$= \sum_{i=1}^{\infty} (-1)^{i+1} \frac{x^i}{i} \qquad (-1 < x < 1).$$

Proof. We know that

$$\ln(1+x) = \int_0^x \frac{dt}{1+t},$$

and we know that

$$\frac{1}{1+t} = 1 - t + t^2 - t^3 + \cdots + (-1)^i t^i + \cdots \qquad (-1 < t < 1).$$

By Theorem B,

$$\int_0^x \frac{dt}{1+t} = \int_0^x dt - \int_0^x t\,dt + \int_0^x t^2\,dt - \cdots + (-1)^i \int_0^x t^i\,dt + \cdots$$

$$= x - \frac{x^2}{2} + \frac{x^3}{3} - \cdots + (-1)^{i+1}\frac{x^i}{i} + \cdots$$

$$= \sum_{i=1}^{\infty} (-1)^{i+1}\frac{x^i}{i} \qquad (-1 < x < 1).$$

Note that this method cannot be used to calculate the integral from 0 to 2, because the series for $1/(1 + t)$ converges only for $|t| < 1$.

The method that we have just been using can be applied so as to give answers, in the form of series, for problems which up to now we could not have solved. Consider

$$\int_0^{-1/2} \frac{dx}{1+x^4}.$$

In Chapter 6, this would have been an impossible problem. But now we can solve it, by expressing the integrand as a power series, and integrating a term at a time. In the series for $1/(1+x)$, we replace x by x^4. This gives

$$\frac{1}{1+x^4} = 1 - x^4 + x^8 - \cdots + (-1)^i x^{4i} + \cdots$$

$$= \sum_{i=0}^{\infty} (-1)^i x^{4i}.$$

Therefore

$$\int_0^{-1/2} \frac{dx}{1+x^4} = \sum_{i=0}^{\infty} \int_0^{-1/2} (-1)^i x^{4i}\, dx$$

$$= \sum_{i=0}^{\infty} (-1)^i \frac{1}{4i+1} (-\tfrac{1}{2})^{4i+1}$$

$$= -\tfrac{1}{2} - \tfrac{1}{5}(-\tfrac{1}{2})^5 + \tfrac{1}{9}(-\tfrac{1}{2})^9 + \cdots$$

$$= -\frac{1}{2} + \frac{1}{5 \cdot 2^5} - \frac{1}{9 \cdot 2^9} + \cdots$$

This is an alternating series; the terms diminish numerically, and approach 0 as $n \to \infty$. Therefore, if we use the first three terms as an approximation of the integral, the error is less than the fourth term. This is

$$E = \tfrac{1}{13} \cdot (\tfrac{1}{2})^{13},$$

which is quite small: $2^{10} = 1024$, $2^{13} = 8192$, and so $E < 10^{-5}$.

PROBLEM SET 10.5

1. Calculate $\text{Tan}^{-1}\, 0.02$ to six decimal places, and explain how you know that the error in your approximation is less than $5 \cdot 10^{-7}$.

2. Calculate $\int_0^{1/10} \frac{1}{1+x^4}\, dx$ to five decimal places, and explain how you know that the error in your approximation is less than $5 \cdot 10^{-6}$.

3. Using the first term only, in the series for Tan^{-1}, we get the approximation formula

$$\text{Tan}^{-1} x \approx x \quad \text{for } x \approx 0.$$

How might you explain and justify this approximation formula if you knew nothing about infinite series?

4. Given

$$f(x) = \frac{1}{1+x^6}.$$

a) Express $f(x)$ as an infinite series.
b) Express

$$\int_0^{0.6} \frac{dx}{1+x^6}$$

as an infinite series.
c) Calculate numerically the sum of the first three terms of your series.
d) Get (by any method) an estimate of the error in the resulting approximation of the integral.

5. Do the same four things, starting with $f(x) = 1/(1 + \sqrt{x})$, on the interval $[0, 0.49]$. (Your infinite series will use powers of $\sqrt{x}$, but the same methods will apply, for the same reasons.)

6. Do the same four things, starting with $f(x) = 1/(1 + x)$, on the interval $[0, 0.2]$.

7. Do the same, starting with $f(x) = \dfrac{1}{1 + x^{5/2}}$, on the interval $[0, 0.25]$.

8. Do the same, starting with $f(x) = \dfrac{1}{1 + x^3}$, on $[0, 1/2]$.

9. Express in the form of a series:

$$\int_0^k \left[\sum_{i=0}^{\infty} \frac{x^i}{i + 1} \right] dx \qquad (0 < k < 1).$$

10. Using the first term only, in the series for ln, we get the approximation formula

$$\ln (1 + x) \approx x \quad \text{for } x \approx 0.$$

How might you explain and justify this formula, if you knew nothing about infinite series?

11. Consider the function $f(x)$ defined by the series

$$1 + x + \frac{x^2}{2} + \frac{x^3}{3!} + \cdots + \frac{x^n}{n!} + \cdots$$

a) Express $f'(x)$ as a series.
b) Express $\int_0^x f(t)\, dt$ as a series.
The results that you get ought to enable you to guess what the function is.

*12. For each n, let

$$f_n(x) = nx^n(1 - x^{2n}) \qquad (0 \leq x \leq 1).$$

a) Find $\lim_{n \to \infty} \int_0^1 f_n(x)\, dx$.
b) For each x on $[0, 1]$, let $f(x) = \lim_{n \to \infty} f_n(x)$. Get a formula for the function $f(x)$.
c) Find

$$\int_0^1 f(x)\, dx = \int_0^1 [\lim_{n \to \infty} f_n(x)]\, dx.$$

*13. Your answers in Problem 12 suggest that the functions f_n behave rather peculiarly. Investigate as follows:
a) For each n, let $\bar{x}_n$ be the point at which f_n takes on its maximum value. Get a formula for $\bar{x}_n$, and find $\lim_{n \to \infty} \bar{x}_n$.
b) For each n, let $\bar{y}_n = f_n(\bar{x}_n)$. Get a formula for $\bar{y}_n$, and find $\lim_{n \to \infty} \bar{y}_n$.
c) Draw a sketch showing what the graph of f_n looks like for $n = 1$, $n = 2$, and $n \approx \infty$. Your sketch will throw some light on the results that you got in Problem 12.

10.6 THE RATIO TEST FOR ABSOLUTE CONVERGENCE. APPLICATIONS TO POWER SERIES

Consider a series $\sum a_i$, in which the terms may be positive or negative, but not equal to 0. For each i, let

$$r_i = \left| \frac{a_{i+1}}{a_i} \right|,$$

so that

$$|a_{i+1}| = |a_i|\, r_i.$$

An examination of the sequence $r_1, r_2, \ldots$ gives us a convergence test which works very quickly, in the cases where it applies.

Theorem 1 (*The ratio test*). If

$$\lim_{i \to \infty} r_i = r < 1,$$

then $\sum_{i=0}^{\infty} a_i$ is absolutely convergent.

Proof. Let s be any number such that $r < s < 1$. Then there is an N such that

$$i \geqq N \;\Rightarrow\; r_i < s.$$

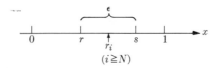

$$(i \geqq N)$$

(In the definition of $\lim_{i \to \infty} r_i$, take $\epsilon = s - r$, so that $r = s - \epsilon$.) It follows that

$$|a_{N+1}| = |a_N|\, r_N < |a_N|\, s,$$
$$|a_{N+2}| = |a_{N+1}|\, r_{N+1} < |a_N|\, ss = |a_N|\, s^2;$$

and in general, given

$$|a_{N+j}| < |a_N|\, s^j,$$

it follows that

$$|a_{N+j+1}| = |a_{N+j}|\, r_{N+j} < |a_N|\, s^j s = |a_N|\, s^{j+1}.$$

By induction,

$$|a_{N+j}| < |a_N|\, s^j \quad \text{for every } j.$$

Therefore

$$\sum_{j=0}^{\infty} |a_{N+j}| \leqq \sum_{j=0}^{\infty} |a_N|\, s^j = |a_N| \sum_{j=0}^{\infty} s^j = |a_N|\, \frac{1}{1-s} < \infty.$$

It follows that

$$\sum_{i=N}^{\infty} |a_i| < \infty,$$

and so

$$\sum_{i=0}^{\infty} |a_i| = \sum_{i=0}^{N-1} |a_i| + \sum_{i=N}^{\infty} |a_i| < \infty,$$

which was to be proved.

What we are really using here is a comparison test between the series $\sum |a_i|$ and a geometric series; the comparison does not necessarily work for the first few terms, but it does start working after a certain point; and this is good enough to tell us what we want to know.

In Section 10.2, Theorem 9, we showed by a comparison test that $\sum_{i=0}^{\infty} (1/i!)$ is convergent. The ratio test gives this result very quickly. We have

$$a_i = \frac{1}{i!},$$

$$r_i = \frac{a_{i+1}}{a_i} = \frac{i!}{(i+1)!} = \frac{1}{i+1},$$

and so $\lim_{i \to \infty} r_i = 0$. It follows that the series converges.

There are simple cases in which a series converges, but in which convergence cannot be established by the ratio test. Consider

$$\sum_{i=1}^{\infty} \frac{1}{i^2},$$

which is known to converge. Here

$$r_i = \frac{a_{i+1}}{a_i} = \frac{i^2}{(i+1)^2}.$$

Therefore, while $r_i < 1$ for each i, we have

$$\lim_{i \to \infty} r_i = \lim_{i \to \infty} \frac{1}{[1 + (1/i)]^2} = 1,$$

and so the ratio test does not apply. And Theorem 1 cannot be generalized to take care of these cases, because if $r_i \to 1$ it may easily happen that the series diverges. This happens for

$$\sum_{i=1}^{\infty} \frac{1}{i} = \infty, \qquad r_i = \frac{i}{i+1} \to 1.$$

An even simpler case is

$$\sum_{i=0}^{\infty} (-1)^i = 1 - 1 + 1 - 1 + \cdots$$

Here

$$r_i = |(-1)^{i+1}/(-1)^i| = 1 \quad \text{for every } i,$$

so that $r_i \to 1$ automatically, but the series diverges.

On the basis of the ratio test, we can derive a more general result for power series:

Theorem 2. Given the series

$$\sum_{i=0}^{\infty} a_i x^i,$$

where $a_i \neq 0$ for every i. Suppose that

$$\lim_{i \to \infty} \left| \frac{a_{i+1}}{a_i} \right| = L.$$

If $L = 0$, then the series is absolutely convergent for every x. If $L > 0$, then the series is absolutely convergent for

$$|x| < 1/L.$$

Proof. For $x = 0$, there is nothing to prove. For each $x \neq 0$, we have

$$r_i = \left| \frac{a_{i+1}x^{i+1}}{a_i x^i} \right| = |x| \cdot \left| \frac{a_{i+1}}{a_i} \right|.$$

Therefore

$$\lim_{i \to \infty} r_i = |x| \cdot L.$$

If $L = 0$, then $r_i \to 0$, no matter what x may be. If $L > 0$, then $\lim_{i \to \infty} r_i < 1$ whenever $|x| < 1/L$. In either case, the series converges absolutely.

By the first half of Theorem 2, we conclude that

$$\sum_{i=0}^{\infty} \frac{x^i}{i!}$$

converges absolutely for every x. By the second half of Theorem 2, we see that

$$\sum_{i=1}^{\infty} \frac{x^i}{i^2}$$

converges absolutely for $|x| < 1$. In each of these cases, the sum of the coefficients forms a convergent series. But the theorem also applies in cases where the sum of the coefficients diverges. Consider

$$\sum_{i=1}^{\infty} i\pi^i x^i.$$

Here

$$\lim_{i \to \infty} \left| \frac{a_{i+1}}{a_i} \right| = \lim_{n \to \infty} \frac{(i + 1)\pi^{i+1}}{i\pi^i} = \pi.$$

Therefore the series converges absolutely whenever $|x| < 1/\pi$.

If the ratio r_n approaches a limit which is greater than 1, then the series $\sum a_i$ always diverges. The reason is that in this case we have an N such that

$$i \geqq N \implies r_i > 1.$$

Therefore $|a_{i+1}| > |a_i|$ for $i \geqq N$, and so after a certain point the sequence $|a_1|$, $|a_2|, \ldots$ becomes an increasing sequence. Therefore a_i cannot approach 0. This observation enables us to add something to the conclusion of Theorem 2.

Theorem 3. Given the series $\sum_{i=0}^{\infty} a_i x^i$, with

$$\lim_{i \to \infty} \left| \frac{a_{i+1}}{a_i} \right| = L.$$

If $L = 0$, then the series converges absolutely for every x. If $L > 0$, then the series converges absolutely for $|x| < 1/L$ and diverges for $|x| > 1/L$.

This theorem can be adapted to take care of cases in which some terms of the series are equal to 0. For example,

$$\sum_{i=0}^{\infty} (-1)^i x^{2i}.$$

Setting $x^2 = y$, we get

$$\sum_{i=0}^{\infty} (-1)y^i,$$

which converges absolutely for $|y| < 1$ and diverges for $|y| > 1$. Therefore the given series converges absolutely for $|x| < 1$ and diverges for $|x| > 1$. Similarly,

$$\sum_{i=0}^{\infty} \frac{x^{2i+1}}{2^i} = x \sum_{i=0}^{\infty} \frac{x^{2i}}{2^i}.$$

Here

$$\lim_{i \to \infty} \left| \frac{a_{i+1}}{a_i} \right| = \frac{1}{2}.$$

Therefore the series converges absolutely for $x^2 < 2$ and diverges for $x^2 > 2$.

Some more observations about Theorem 3 are in order.

1) The theorem applies only to the case in which $|a_{i+1}/a_i|$ approaches a limit. This usually happens for series which are describable by simple formulas. But for series in general it should be regarded as a remarkable accident. Suppose, for example, that we start with

$$\sum_{i=0}^{\infty} x^i = 1 + x + x^2 + \cdots$$

Here $a_i = 1$ for every i, and so $r_i = 1$ for every i. We now divide x^i by $i!$ for every *even i*. This gives

$$\sum_{i=0}^{\infty} b_i x^i = 1 + x + \frac{x^2}{2!} + x^3 + \frac{x^4}{4!} + \cdots$$

The series still converges, for $|x| < 1$, but the ratio approaches no limit at all.

2) The theorem tells us that the series converges everywhere on the open interval $(-1/L, 1/L)$, but it tells us nothing about what happens at the endpoints of the interval. In fact, at the endpoints anything can happen. For example, $\sum_{i=1}^{\infty} (x^i/i^2)$ converges on $(-1, 1)$, and converges at both the endpoints. The series $\sum_{i=1}^{\infty} i x^i$ converges on the same interval, but converges at *neither* of the endpoints. The series $\sum_{i=1}^{\infty} (x^i/i)$ converges on $(-1, 1)$, and converges at $x = -1$, but diverges at $x = 1$. The series $\sum_{i=1}^{\infty} (-1)^i (x^i/i)$ converges on $(-1, 1)$, and converges at $x = 1$, but diverges at $x = -1$. For this reason, to tell where the series converges, we have to make separate tests at the endpoints.

3) Obviously every power series $\sum a_i x^i$ converges for $x = 0$; the sum is a_0. But sometimes 0 is the only value of x that gives convergence. Consider $\sum_{i=0}^{\infty} i! x^i$. For every $x \neq 0$, we have

$$r_i = |(i + 1)! x^{i+1}/i! x^i|$$
$$= (i + 1) |x| \to \infty.$$

Therefore the series converges only for $x = 0$.

4) Finally, the results that we have been getting for power series suggest a conjecture. In every case that we have investigated, the domain of convergence of $\sum a_i x^i$ has turned out to be of one of the following types:

  i) The entire interval $(-\infty, \infty)$.

  ii) An open interval $(-a, a)$, plus, perhaps, one or both of the endpoints.

  iii) The point 0 alone.

The question arises whether these are the only possibilities. For example, is there a series $\sum a_i x^i$ whose domain of convergence is an interval whose midpoint is not the origin? We shall see, as the theory develops, that the domain of convergence of $\sum a_i x^i$ is always a set of one of the forms

$$(-\infty, \infty), \qquad (-a, a), \qquad [-a, a), \qquad (-a, a], \qquad [-a, a], \qquad \{0\}.$$

PROBLEM SET 10.6

For each of the following series, find the domain of convergence, remembering, of course, to test the endpoints.

1. $\displaystyle\sum_{i=1}^{\infty} i^2 x^i$

2. $\displaystyle\sum_{i=1}^{\infty} i^3 x^i$

3. $\displaystyle\sum_{i=1}^{\infty} i^2 x^{2i-1}$

4. $\displaystyle\sum_{i=1}^{\infty} i x^{2i}$

5. $\displaystyle\sum_{i=1}^{\infty} \frac{x^i}{\sqrt{i}}$

6. $\displaystyle\sum_{i=1}^{\infty} \frac{x^{2i}}{\sqrt{i}}$

7. $\displaystyle\sum_{i=1}^{\infty} \frac{x^i}{\sqrt{2i+1}}$

8. $\displaystyle\sum_{i=2}^{\infty} \frac{(-x)^i}{\sqrt{i-1}}$

9. $\displaystyle\sum_{i=1}^{\infty} (3i)^3 x^i$

10. $\displaystyle\sum_{i=1}^{\infty} (3i) x^{2i}$

11. $\displaystyle\sum_{i=1}^{\infty} (3i)^2 x^{2i-1}$

12. $\displaystyle\sum_{i=1}^{\infty} (3i)^4 x^{3i}$

13. $\displaystyle\sum_{i=1}^{\infty} (-1)^{i+1} \frac{x^{2i-1}}{(2i-1)!}$

14. $\displaystyle\sum_{i=0}^{\infty} (-1)^i \frac{x^{2i}}{(2i)!}$

15. $\displaystyle\sum_{i=0}^{\infty} (-1)^i \frac{x^{4i}}{(2i)!}$

16. $\displaystyle\sum_{i=1}^{\infty} i^i x^i$

17. $\displaystyle\sum_{i=1}^{\infty} i^{-i} x^i$

18. $\displaystyle\sum_{i=1}^{\infty} (\mathrm{Tan}^{-1} i) x^i$

19. $\displaystyle\sum_{i=1}^{\infty} (-1)^i \frac{x^i}{i^2+1}$

20. $\displaystyle\sum_{i=1}^{\infty} i(2x-1)^i$ (Does the answer to this one contradict Theorem 3?)

21. $\displaystyle\sum_{i=1}^{\infty} \frac{e^i}{i^3} (x-4)^i$ (Same query as for Problem 20.)

22. $\displaystyle\sum_{i=1}^{\infty} \frac{(x-2)^i}{i}$ (Same query as for Problem 20.)

23. Show that $\sum_{i=1}^{\infty} (\sin i) x^i$ is absolutely convergent when $|x| < 1$.

24. Prove the following theorem:

   Theorem. If $\sum_{i=1}^{\infty} a_i$ is absolutely convergent, and $b_1, b_2, \ldots$ is a bounded sequence, then $\sum_{i=1}^{\infty} a_i b_i$ is absolutely convergent.

*25. Show that there are infinitely many integers i for which $\sin i > \frac{1}{2}$.

*26. Show that $\sum_{i=1}^{\infty} (\sin i)x^i$ is divergent when $|x| > 1$.

(The results of Problems 23 and 26 show that for this very irregular series, the domain of convergence is still of one of the types described by Theorem 3.)

*27. You may have noticed that the number 1 has come up very often as an endpoint of our domains of convergence. The following theorem helps to account for this:

Theorem. Let $p(i)$ and $q(i)$ be polynomials in i, of any degree, with $q(i)$ never equal to 0. If $a_i = p(i)/q(i)$, then

$$\sum_{i=0}^{\infty} a_i x^i$$

converges absolutely for $|x| < 1$, and diverges for $|x| > 1$.

Prove this theorem.

10.7 POWER SERIES FOR exp, sin, AND cos

Theorem A of Section 10.5 asserts that power series can be differentiated a term at a time. That is, if

$$f(x) = \sum_{i=0}^{\infty} a_i x^i \qquad (-r < x < r),$$

then

$$f'(x) = \sum_{i=1}^{\infty} i a_i x^{i-1} \qquad (-r < x < r).$$

We shall use this to find a series for the exponential function. We start by assuming that e^x can be expressed in *some* way as a power series, so that

$$f(x) = e^x = \sum_{i=0}^{\infty} a_i x^i = a_0 + a_1 x + \cdots,$$

for some sequence of coefficients $a_0, a_1, \ldots$. On any open interval where this works, we have

$$f'(x) = \sum_{i=1}^{\infty} i a_i x^{i-1} = a_1 + 2a_2 x + \cdots + i a_i x^{i-1} + (i + 1)a_{i+1} x^i + \cdots$$

It must be true that $f'(x) = f(x)$, and $f(0) = 1$; and so we want to find a sequence of coefficients $a_0, a_1, a_2, \ldots$ which gives these results for the series. This is easy: we want

$$(i + 1)a_{i+1} = a_i, \qquad a_{i+1} = \frac{a_i}{i + 1},$$

which gives $f'(x) = f(x)$; and we want $a_0 = 1$, which gives $f(0) = 1$. Thus

$$a_0 = 1, \qquad a_1 = a_0/1 = 1, \qquad a_2 = a_1/2 = \tfrac{1}{2}, \qquad a_3 = a_2/3 = 1/(2 \cdot 3);$$

and, in general,

$$a_i = 1/i!.$$

This can be checked by induction. For $i = 0, 1, 2$, the formula $a_i = 1/i!$ holds true. And

$$a_i = \frac{1}{i!} \quad \Rightarrow \quad a_{i+1} = \frac{a_i}{i+1} = \frac{1}{(i+1)i!} = \frac{1}{(i+1)!}.$$

This proceeding does *not* prove that

$$e^x = \sum_{i=0}^{\infty} \frac{x^i}{i!}, \tag{1}$$

because we started off with an unproved assumption that e^x had *some* power series expansion. But now that we know what series to examine, it is very easy to show that Eq. (1) holds. By the ratio test, the series on the right-hand side converges for every x. It therefore defines a function g. Thus

$$g(x) = \sum_{i=0}^{\infty} \frac{x^i}{i!} \quad (-\infty < x < \infty).$$

We chose the coefficients a_i in such a way that $g' = g$ and $g(0) = 1$. We need to show that $g(x) = e^x$ for every x. For each x, let $\phi(x) = g(x)/e^x$. Then

$$\phi'(x) = \frac{e^x g'(x) - g(x)e^x}{e^{2x}} = \frac{1}{e^x}[g'(x) - g(x)] = 0.$$

Therefore ϕ is a constant, and $\phi(x) = \phi(0)$ for every x. But $\phi(0) = 1$. Therefore

$$g(x)/e^x = 1, \quad \text{and} \quad g(x) = e^x,$$

which was to be proved.

What makes this scheme work is the fact that the function $f(x) = e^x$ is completely described by the conditions $f' = f$, $f(0) = 1$; no other function satisfies these conditions. Thus we have

Theorem 1. $e^x = \sum_{i=0}^{\infty} (x^i/i!)$.

Setting $x = 1$ we get

Theorem 2. $e = \sum_{i=0}^{\infty} (1/i!)$.

This series converges so fast that some people enjoy using it to calculate $e \approx 2.7182818$, correct in the seventh decimal place.

We now want to get a series for the sine. As before, we start by assuming that our problem has a solution, and then we try to find out what form the solution must take. For $f(x) = \sin x$ we have

$$f'(x) = \cos x, \quad f''(x) = -\sin x.$$

Therefore

$$f(0) = 0, \quad f'(0) = 1,$$

and

$$f''(x) = -f(x) \quad \text{for every } x.$$

Thus if

$$\sin x = \sum_{i=0}^{\infty} a_i x^i = a_0 + a_1 x + \cdots .$$

we must have $a_0 = 0$, and $a_1 = 1$. Now

$$f'(x) = \sum_{i=1}^{\infty} i a_i x^{i-1},$$

$$f''(x) = \sum_{i=2}^{\infty} i(i - 1)a_i x^{i-2}$$

$$= 2a_2 + 3 \cdot 2a_3 x + \cdots + i(i - 1)a_i x^{i-2}$$

$$+ (i + 1)i a_{i+1} x^{i-1} + (i + 2)(i + 1)a_{i+2} x^i + \cdots$$

To get $f'' = -f$, we want

$$(i + 2)(i + 1)a_{i+2} = -a_i, \quad \text{or} \quad a_{i+2} = -\frac{a_i}{(i + 1)(i + 2)} .$$

Since $a_0 = 0$, it follows that every even-numbered coefficient a_{2i} is also equal to 0. The odd-numbered coefficients are

$$a_1 = 1,$$

$$a_3 = -\frac{a_1}{(1 + 1)(1 + 2)} = -\frac{a_1}{2 \cdot 3} = -\frac{1}{3!},$$

$$a_5 = -\frac{a_3}{(3 + 1)(3 + 2)} = +\frac{1}{3! \cdot 4 \cdot 5} = \frac{1}{5!};$$

and in general

$$a_{2i+1} = (-1)^i \cdot \frac{1}{(2i + 1)!} .$$

To check this by induction, we note that

$$a_{2i+1} = (-1)^i \frac{1}{(2i + 1)!}$$

$$\Rightarrow \quad a_{(2i+1)+2} = -\frac{a_{2i+1}}{[(2i + 1) + 1] \cdot [(2i + 1) + 2]}$$

$$\Leftrightarrow \quad a_{2(i+1)+1} = (-1)^i(-1) \cdot \frac{1}{(2i + 1)!} \cdot \frac{1}{(2i + 2)(2i + 3)}$$

$$= (-1)^{i+1} \frac{1}{(2i + 3)!} = (-1)^{i+1} \frac{1}{[2(i + 1) + 1]!}$$

Therefore, if there is a series for the sine, the series must have the form

$$g(x) = \sum_{i=0}^{\infty} (-1)^i \frac{x^{2i+1}}{(2i + 1)!} .$$

We need to show that $g(x) = \sin x$ for every x.

Let $h(x) = g'(x)$. Then we know that

1) $g' = h$,
2) $h' = -g$,
3) $g(0) = 0$,
4) $h(0) = 1$.

It ought to be true that

$$g(x) = \sin x, \qquad h(x) = \cos x.$$

If so, the function

$$\phi(x) = [g(x) - \sin x]^2 + [h(x) - \cos x]^2$$

must be equal to 0 for every x. And conversely, if $\phi(x) = 0$ for every x, it follows that $g(x) = \sin x$ and $h(x) = \cos x$. Now

$$
\begin{aligned}
\phi'(x) &= 2[g(x) - \sin x][g'(x) - \cos x] + 2[h(x) - \cos x][h'(x) + \sin x] \\
&= 2[g(x) - \sin x][h(x) - \cos x] + 2[h(x) - \cos x][-g(x) + \sin x] \\
&= 0 \quad \text{for every } x.
\end{aligned}
$$

Therefore ϕ is a constant. But

$$\phi(0) = [0 - 0]^2 + [1 - 1]^2 = 0.$$

Therefore $\phi(x) = 0$ for every x, which was to be proved. Thus we have:

Theorem 3.

$$\sin x = \sum_{i=0}^{\infty} (-1)^i \frac{x^{2i+1}}{(2i+1)!} = x - \frac{x^3}{3!} + \frac{x^5}{5!} - \frac{x^7}{7!} + \cdots$$

By differentiation,

$$\cos x = \sum_{i=0}^{\infty} (-1)^i \frac{(2i+1)x^{2i}}{(2i+1)!} = \sum_{i=0}^{\infty} (-1)^i \frac{x^{2i}}{(2i)!}.$$

Thus:

Theorem 4.

$$\cos x = \sum_{i=0}^{\infty} (-1)^i \frac{x^{2i}}{(2i)!} = 1 - \frac{x^2}{2!} + \frac{x^4}{4!} - \frac{x^6}{6!} + \cdots$$

Obviously, the series that we have been developing in this section can be used for calculating the values of the corresponding functions. In fact, this is the way people arrived at the values that you find in the tables of exp, sin, and cos. And the series can be adapted, in simple ways, to handle a variety of related problems. For example, consider

$$\int_0^{0.5} e^{x^2}\, dx.$$

If we could get a simple formula for a function F such that

$$F'(x) = e^{x^2},$$

then the integral could be expressed as $F(0.5) - F(0)$. There is no such simple formula. But we can express such an F as an infinite series, in the following way. We know that

$$e^x = \sum_{i=0}^{\infty} \frac{x^i}{i!} = 1 + x + \frac{x^2}{2!} + \cdots$$

Therefore

$$e^{x^2} = \sum_{i=0}^{\infty} \frac{x^{2i}}{i!} = 1 + x^2 + \frac{x^4}{2!} + \cdots$$

Integrating a term at a time, we get a function

$$F(x) = x + \frac{x^3}{3} + \frac{x^5}{5 \cdot 2!} + \cdots = \sum_{i=0}^{\infty} \frac{x^{2i+1}}{i!(2i+1)}.$$

Evidently

$$F(0) = 0, \quad \text{and} \quad F'(x) = e^{x^2}.$$

Therefore

$$\int_0^x e^{t^2}\, dt = F(x),$$

and so, using the series for F, we can calculate $F(\frac{1}{2})$ approximately, with an error as small as we please.

PROBLEM SET 10.7

Find a series for each of the following functions. In each case, name the interval on which you know that your series converges to the given function.

1. $f(x) = x \ln (x + 1)$.

2. $f(x) = x^2 \ln (x^2 + 1)$.

3. $f(x) = x^2 \ln (x + 1)$.

4. $\displaystyle\int_0^x f(t)\, dt$, where f is as in
　　　　　　　　　　　　Problem 2.

5. $f(x) = \sin 2x$.

6. $f(x) = x \sin x$.

7. $f(x) = \sin (x/2)$.

8. $f(x) = \cos 2x$.

9. $f(x) = \cos \left(\dfrac{x}{3}\right)$.

10. $f(x) = \sin x \cos x$.

11. $f(x) = x^3 e^{x^3}$.

12. $f(x) = xe^x - x$.

13. $f(x) = \displaystyle\int_0^x t^3 e^{t^3}\, dt$.

14. $f(x) = \begin{cases} \dfrac{\text{Tan}^{-1} x}{x} & \text{for } x \neq 0 \\ 1 & \text{for } x = 0 \end{cases}$

15. $F(x) = \displaystyle\int_0^x f(t)\, dt$, where f is as in Problem 14.

16. $f(x) = \begin{cases} \dfrac{e^x - 1}{x} & \text{for } x \neq 0 \\ 1 & \text{for } x = 0 \end{cases}$

17. $F(x) = \displaystyle\int_0^x f(t)\, dt$, where f is as in Problem 16.

18. $f(x) = \begin{cases} \dfrac{\sin x}{x} & \text{for } x \neq 0 \\ 1 & \text{for } x = 0 \end{cases}$

19. $F(x) = \int_0^x f(t)\, dt$, where f is as in Problem 18.   20. $f(x) = \cos^2 x$.

21. $f(x) = x^3 \cos^2 x$.   22. $F(x) = \int_0^x f(t)\, dt$, where f is as in Problem 21.

23. $f(x) = \cos^2 x - \sin^2 x$ 24. $F(x) = x \cos^2 x + x \sin^2 x$.

25. Find a series for a function f such that (1) $f' = \frac{1}{2}f$ and (2) $f(0) = 1$. Either before or after finding the series, find an elementary formula for such a function f.

26. Find a series for a function f such that (1) $f'(x) = 2f(x)/x$ for every $x \neq 0$ and (2) $f(0) = 0$.

27. Is there only one function satisfying the conditions of Problem 26? Why or why not?

28. Get a formula for $D^i x^i$, where $D^i f$ denotes the ith derivative of the function f.

29. Get a formula for $D^i x^j$, valid for $i < j$.

30. Do the same, for the case $i > j$.

*31. Given $f(x) = \sum_{i=0}^{\infty} a_i x^i$. Get a formula for $f^{(i)}(0)$. (Here $f^{(i)}$ denotes the ith derivative of f.)

*32. Is it possible that there are two different power series for the same function, valid on the same open interval I? That is, given

$$f(x) = \sum_{i=0}^{\infty} a_i x^i = \sum_{i=0}^{\infty} b_i x^i \quad \text{on } I,$$

does it follow that $a_i = b_i$ for each i? Why or why not?

*33. A function f is called *real-analytic* on an interval I if f can be expressed as a power series $\sum_{i=0}^{\infty} a_i x^i$. Does there exist a real-analytic function f, on an interval $(-a, a)$, such that $f^{(i)}(0) = (i!)^2$ for each i? Why or why not?

10.8 THE BINOMIAL SERIES

It is possible to show, by induction, that if n is a positive integer, then

$$(a + b)^n = a^n + na^{n-1}b + \frac{n(n-1)}{2} a^{n-2}b^2 + \cdots$$

$$+ \frac{n(n-1)\cdots(n-i+1)}{i!} a^{n-i}b^i + \cdots + b^n.$$

Here the coefficient of $a^{n-i}b^i$ can be written more briefly as

$$\binom{n}{i} = \frac{n!}{i!(n-i)!} = \frac{n(n-1)\cdots(n-i+1)}{i!}.$$

The induction proof of the binomial theorem depends on the identity

$$\binom{n+1}{i} = \binom{n}{i} + \binom{n}{i-1}.$$

You may have seen this proved. In any case, we shall not stop to prove it now, because the elementary form of the binomial theorem is a corollary of a more general result which we shall prove presently.

We would like to generalize the familiar binomial formula

$$(a + b)^n = \sum_{i=0}^{n} \binom{n}{i} a^{n-i} b^i$$

to take care of the case in which n is not an integer. That is, we want a formula for $(a + b)^k$, where k is any real number. The following observations are obvious:

1) For $k = 0$, we have $(a + b)^k = 1$, and our problem is solved. We may therefore assume that

$$k \neq 0.$$

2) For the case of interest, in which k is not an integer, the exponential c^k is defined only for $c > 0$. (See Section 4.9.) Therefore we *must* assume that

$$a + b > 0.$$

3) For $a = b$, the problem has an immediate solution: $(a + b)^k = (2a)^k = 2^k a^k$. We therefore may assume hereafter that

$$a \neq b.$$

And we want to assume this, because the case $a = b$ does not fit the pattern that is going to emerge.

4) It is now a matter of notation to suppose that

$$a > b.$$

We let $x = b/a$, so that

$$a + b = a(1 + x), \quad \text{and} \quad (a + b)^k = a^k(1 + x)^k.$$

If we had $|x| = |b/a| \geq 1$, then either $b \geq a$ or $b \leq -a$; and these possibilities are ruled out by conditions (2) and (4). Therefore $|x| < 1$, and our problem takes the following form:

Problem. Given $k \neq 0$, and

$$f(x) = (1 + x)^k \qquad (|x| < 1).$$

Find a formula for $f(x)$, analogous to the binomial formula.

Our past experience with sin, cos, and exp suggests that we should investigate the relation between $f(x) = (1 + x)^k$ and its derivatives, and use the results in the investigation of the series. Now

$$f'(x) = k(1 + x)^{k-1}.$$

Therefore

$$(1 + x)f'(x) = kf(x).$$

If $f(x) = \sum_{i=0}^{\infty} a_i x^i$, then

$$f'(x) = \sum_{i=0}^{\infty} i a_i x^{i-1} = \sum_{i=1}^{\infty} i a_i x^{i-1}.$$

Therefore

$$xf'(x) = \sum_{i=1}^{\infty} i a_i x^i.$$

We want to express $(1 + x)f'(x)$ as a series, and so we need to express $f'(x)$ in the form $\sum b_i x^i$. For this purpose we use a trick. Let $j = i - 1$, so that $i = j + 1$. This gives

$$f'(x) = \sum_{j=0}^{\infty} (j + 1)a_{j+1} x^j = \sum_{i=0}^{\infty} (i + 1)a_{i+1} x^i.$$

The equation $(1 + x)f'(x) = kf(x)$ now takes the form

$$\sum_{i=0}^{\infty} [(i + 1)a_{i+1} + i a_i] x^i = \sum_{i=0}^{\infty} k a_i x^i.$$

Comparing coefficients of x^i, we get

$$(i + 1)a_{i+1} + i a_i = k a_i \iff (i + 1)a_{i+1} = (k - i)a_i$$

$$\iff a_{i+1} = \frac{k - i}{i + 1} a_i.$$

Obviously

$$a_0 = f(0) = (1 + 0)^k = 1.$$

Therefore

$$a_0 = 1;$$

$$a_1 = \frac{k - 0}{0 + 1} a_0 = k;$$

$$a_2 = \frac{k - 1}{1 + 1} a_1 = \frac{k(k - 1)}{2};$$

$$a_3 = \frac{k - 2}{2 + 1} a_2 = \frac{k(k - 1)(k - 2)}{3 \cdot 2} = \frac{k(k - 1)(k - 2)}{3!};$$

and in general, for $i > 1$,

$$a_i = \frac{k(k - 1) \cdots (k - i + 1)}{i!}.$$

We denote the fraction on the right by the symbol $\binom{k}{i}$, just as in the case where k is a positive integer. The above formula then takes the form

$$a_i = \binom{k}{i} \quad \text{for each } i \geq 0,$$

and the net result of the above discussion is that

$$\sum_{i=0}^{\infty} a_i x^i = (1 + x)^k \implies a_i = \binom{k}{i} \quad \text{for each } i.$$

That is, the series that we have found is the only series that might work. To know that our series does work, we need the following two theorems.

Theorem 1. The series $\sum_{i=0}^{\infty} \binom{k}{i} x^i$ is convergent for $|x| < 1$.

Proof. As in Section 10.6, let

$$r_i = \left| \frac{a_{i+1} x^{i+1}}{a_i x^i} \right| = \left| \binom{k}{i+1} x^{i+1} \middle/ \binom{k}{i} x^i \right|.$$

Then

$$r_i = \left| \frac{k(k-1) \cdots (k-i+1)(k-i)}{(i+1)!} \cdot \frac{i!}{k(k-1) \cdots (k-i+1)} \cdot x \right|$$

$$= \left| \frac{k-i}{i+1} \cdot x \right|.$$

Evidently

$$\lim_{i \to \infty} r_i = |x|.$$

Therefore, by the ratio test, the series converges absolutely for $|x| < 1$, and diverges for $|x| > 1$.

Theorem 2. For every $k \neq 0$, and every x between -1 and 1,

$$\sum_{i=0}^{\infty} \binom{k}{i} x^i = (1 + x)^k.$$

Proof. Let g be the function which is the sum of the series. We determined the coefficients in such a way that

1) $g(0) = 1$;
2) $(1 + x)g'(x) = kg(x)$.

We need to prove that

$$g(x) = f(x) = (1 + x)^k \qquad (-1 < x < 1).$$

For this purpose, we use the same device that worked for the exponential. Let

$$\phi(x) = \frac{g(x)}{(1 + x)^k}.$$

Then

$$\phi'(x) = \frac{(1 + x)^k g'(x) - g(x) k (1 + x)^{k-1}}{(1 + x)^{2k}}$$

$$= \frac{(1 + x)g'(x) - kg(x)}{(1 + x)^{k+1}} = 0,$$

so that ϕ is a constant. And the constant is 1, because $g(0) = f(0) = 1$. Therefore $g(x) = f(x)$ for every x, which was to be proved.

PROBLEM SET 10.8

1. Write a series for $\sqrt{x + 1}$, and find out how many terms of the series you would need to use, to calculate $\sqrt{1.1}$, correct to three decimal places.

2. Do the same, for $\sqrt[3]{x + 1}$.

3. Do the same, for $\sqrt[4]{x + 1}$.

4. Do the same, for $\sqrt{2x + 1}$.

5. Let n be a positive integer. Using the definition

$$\binom{n}{i} = \frac{n!}{(n - i)!i!},$$

write formulas for $\binom{n+1}{i}$ and $\binom{n}{i-1}$.

6. Using the apparatus of Problem 5, show that

$$\binom{n + 1}{i} = \binom{n}{i} + \binom{n}{i - 1}.$$

7. Using the result of Problem 6, show that

$$(1 + x)^n = \sum_{i=0}^{n} \binom{n}{i} x^i \quad \Rightarrow \quad (1 + x)^{n+1} = \sum_{i=0}^{n+1} \binom{n + 1}{i} x^i.$$

(The first and last terms on the right-hand side require a separate discussion. But note that $\binom{n}{0} = \binom{n+1}{0}$, because both are equal to 1; and similarly that $\binom{n}{n} = \binom{n+1}{n+1} = 1$.) Since obviously $(1 + x)^1 = \binom{1}{0} + \binom{1}{1}x^1$, this gives an induction proof of the elementary form of the binomial theorem.

Find a series for each of the following functions, and discuss for convergence. You need not test for convergence at the endpoints.

8. $f(x) = \dfrac{1}{\sqrt{1 + x}}$

9. $f(x) = \dfrac{x^2}{\sqrt{1 + x}}$

10. $f(x) = x\sqrt{1 + x}$

11. $f(x) = \dfrac{x}{\sqrt[3]{1 + x^2}}$

12. $f(x) = \displaystyle\int_0^x \sqrt{1 + t^{10}}\, dt$

13. $f(x) = (1 + x)^{3/2}$

14. $f(x) = \displaystyle\int_0^{x^2} \sqrt{1 + t^{10}}\, dt$

15. $f(x) = \sqrt[3]{2 + x}$

16. $f(x) = \sqrt{3 + x}$

17. $f(x) = \displaystyle\int_0^x \sqrt[3]{2 + t^2}\, dt$

18. $f(x) = (2 + x^2)^k \quad (k \neq 0)$

19. $f(x) = \displaystyle\int_0^x (2 + t^2)^k\, dt$

20. $f(x) = \dfrac{x}{\sqrt{1 + x^2}}$

21. $f(x) = \displaystyle\int_0^x \dfrac{t}{\sqrt{1 + t^2}}\, dt$

22. Find a function f such that (1) $(1 + x^2)f'(x) = f(x)$ and (2) $f(0) = 1$. Then show that the function that you found is the only function satisfying conditions (1) and (2).

23. Same question, for the conditions (1) $f'(x) \sec x = f(x)$ and (2) $f(0) = 1$.

10.9 TAYLOR SERIES

Obviously we cannot get a series of the type $\sum a_i x^i$, converging to $f(x) = 1/x$ on an open interval containing 0, because any such series is continuous at $x = 0$, while $\lim_{x \to 0^+} 1/x = \infty$. On the other hand, there is a series

$$\sum_{i=0}^{\infty} (-1)^i x^i = \frac{1}{1+x} \qquad (|x| < 1).$$

If we let $x' = 1 + x$, $x = x' - 1$, then the above equation takes the form

$$\sum_{i=0}^{\infty} (-1)^i (x' - 1)^i = \frac{1}{x'} \qquad (|x' - 1| < 1);$$

and dropping the prime we get

$$f(x) = \frac{1}{x} = \sum_{i=0}^{\infty} (-1)^i (x - 1)^i \qquad (|x - 1| < 1).$$

The series on the right-hand side is called a *Taylor series*, or a *Taylor expansion* of the function f about the point $x = 1$. A power series $\sum a_i x^i$, of the type that we have been discussing so far, is called a *Maclaurin series*. Thus every Maclaurin series is a Taylor series:

$$\sum_{i=0}^{\infty} a_i x^i = \sum_{i=0}^{\infty} a_i (x - 0)^i,$$

which is a Taylor series, with $a = 0$. In this language, we may say that $f(x) = 1/x$ has no Maclaurin series, but it does have a Taylor expansion about the point 1.

Similarly, $f(x) = \ln x$ cannot have a Maclaurin series, because at $x = 0$ the function approaches $-\infty$. But we do have a series for

$$g(x) = \ln (1 + x) = \sum_{i=1}^{\infty} (-1)^{i+1} \frac{x^i}{i} \qquad (|x| < 1).$$

Setting $x' = 1 + x$, $x = x' - 1$, we get

$$\ln x' = \sum_{i=1}^{\infty} (-1)^{i+1} \frac{(x' - 1)^i}{i} \qquad (|x' - 1| < 1);$$

and dropping the prime we get

$$\ln x = \sum_{i=1}^{\infty} (-1)^{i+1} \frac{(x - 1)^i}{i} \qquad (|x - 1| < 1).$$

This is a Taylor expansion of ln, about the point 1.

With the obvious modifications, all our theorems for Maclaurin series hold also for Taylor series; and to prove them in the general case, we merely translate the axes by the substitution $x = x' + a$, $x' = x - a$. For example, Theorems A and B of Section 10.5 take the following forms:

Theorem A′. Given

$$f(x) = \sum_{i=0}^{\infty} a_i (x - a)^i \qquad (a - r < x < a + r),$$

then f is continuous and differentiable on the interval $(a - r, a + r)$, and the derivative of the sum is the sum of the derivatives. That is

$$f'(x) = \sum_{i=1}^{\infty} ia_i(x - a)^{i-1} \qquad (a - r < x < a + r).$$

Theorem B′. Given

$$f(x) = \sum_{i=0}^{\infty} a_i(x - a)^i \qquad (a - r < x < a + r).$$

Then the integral of the sum is the sum of the integrals. That is,

$$\int_0^x f(t)\, dt = \sum_{i=0}^{\infty} \int_0^x a_i(t - a)^i\, dt = \sum_{i=0}^{\infty} \frac{a_i}{i + 1}(x - a)^{i+1}.$$

Our other theorems can be generalized in the same style; we treat $x - a$ in exactly the same way that we used to treat x. Another example is as follows. You found, in Problem 31 of Problem Set 10.7, that if $f(x) = \sum_{i=0}^{\infty} a_i x^i$, on an open interval containing 0, then

$$f^{(n)}(0) = n!a_n,$$

so that

$$a_n = \frac{f^{(n)}(0)}{n!}.$$

An analogous formula holds for Taylor series:

Theorem 1. If $f(x) = \sum_{i=0}^{\infty} a_i(x - a)^i$, on an interval $(a - r, a + r)$, then

$$a_n = \frac{f^{(n)}(a)}{n!} \quad \text{for every } n.$$

Proof. The nth derivative of a function described by a formula $\phi(x)$ will be denoted by $D^n\phi(x)$. We observe that

$$D^n(x - a)^i = 0 \quad \text{for} \quad i < n,$$
$$D^n(x - a)^n = n!,$$
$$D^n(x - a)^i = i(i - 1) \cdots (i - n + 1)(x - a)^{i-n} \quad \text{for} \quad i > n.$$

Therefore, for $f(x) = \sum a_i(x - a)^i$, we have

$$f^{(n)}(x) = a_n \cdot n! + \sum_{i=n+1}^{\infty} b_i(x - a)^{i-n}.$$

We don't care what form the b_i's have, because every term of the sum on the right-hand side has $(x - a)$ raised to a positive power, and we are about to set $x = a$. This gives

$$f^{(n)}(a) = a_n \cdot n!,$$

so that

$$a_n = \frac{f^{(n)}(a)}{n!},$$

which was to be proved.

We have found that for some functions the use of Taylor series in place of Maclaurin series is a necessity. For example, $1/x$ and $\ln x$ don't have any Maclaurin expansions. In other cases a Taylor series may be preferable, even though the Maclaurin expansion exists. The point is that $\sum a_i x^i$ usually converges rapidly when x is close to 0, and more slowly when x is larger. To take an extreme case, we know that $\sin 10{,}000\pi = 0$, because 10,000 is even. Therefore it must be true that

$$\sum_{i=0}^{\infty} (-1)^i \frac{1}{(2i+1)!} (10{,}000\pi)^{2i+1} = 0.$$

But in waiting for the partial sums to get close to 0, we had better not be impatient. In general, if we want to use a series to calculate a function numerically, we should choose the "base point" a as close as possible to the value of x that we want to substitute. Suppose, for example, that we have calculated $\ln 1.5 = 0.4055$. One way to calculate $\ln 1.6$ would be to take $x = 1.6$ in the series

$$\ln x = \sum_{i=1}^{\infty} (-1)^{i+1} \frac{(x-1)^i}{i}.$$

But the convergence of the series

$$\sum_{i=1}^{\infty} (-1)^{i+1} \frac{(0.6)^i}{i}$$

is slow. We therefore use the base point 1.5. Thus

$$\ln x = \sum_{i=0}^{\infty} a_i (x - 1.5)^i.$$

For $f(x) = \ln x$, we have

$$f'(x) = 1/x = x^{-1},$$

$$f''(x) = (-1)x^{-2}, \ldots,$$

$$f^{(i)}(x) = (-1)^{i+1}(i-1)! x^{-i},$$

$$a_i = \frac{f^{(i)}(1.5)}{i!} = \frac{(-1)^{i+1}}{i}(1.5)^{-i} \qquad (i > 0),$$

$$f^{(0)}(1.5) = 0.4055.$$

Therefore

$$\ln x = 0.4055 + \sum_{i=1}^{\infty} \frac{(-1)^{i+1}}{i} \left(\frac{x - 1.5}{1.5} \right)^i.$$

For $x = 1.6$, this gives

$$\ln 1.6 = 0.4055 + \sum_{i=1}^{\infty} \frac{(-1)^{i+1}}{i} \left(\frac{1}{15} \right)^i,$$

which converges much more rapidly.

Note that the above derivation tacitly assumes that $\ln$ *has* a Taylor expansion about the point $a = 1.5$; if it has, then the coefficients must be given by the formula

$$a_i = \frac{f^i(a)}{i!}.$$

It is a fact that if a function f has a Taylor expansion

$$f(x) = \sum_{i=0}^{\infty} a_i(x - a)^i,$$

converging on an interval $|x - a| < r$, then any other point b of the interval can also be used as a base point, giving an expansion

$$f(x) = \sum_{i=0}^{\infty} b_i(x - b)^i,$$

which converges on some interval containing b. But the proof would be hard, in the present context, and should be postponed until we can use the theory of functions of a complex variable.

PROBLEM SET 10.9

For some of the functions in the first twelve problems below, it is a practical proceeding to derive a general formula for $f^{(n)}(a)$, and use the formula to calculate the coefficients a_i in the series $\sum a_i(x - a)^i$. In each such case, calculate the coefficients by this method. In cases where the derivation of the general formula seems unreasonably difficult, merely calculate the first three terms of the series.

1. $f(x) = \sin x$, $a = 0$.
2. $f(x) = \tan x$, $a = 0$.
3. $f(x) = \tan x$, $a = \pi$.
4. $f(x) = \cos x$, $a = 2\pi$.
5. $f(x) = \text{Tan}^{-1} x$, $a = 0$.
6. $f(x) = \text{Tan}^{-1} x$, $a = 1$.
7. $f(x) = e^x$, $a = 0$.
8. $f(x) = e^x$, $a = 1$.
9. $f(x) = \ln(2 + x)$, $a = 0$.
10. $f(x) = e^x$, $a = $ any number.
11. $f(x) = \ln(1 + x^2)$, $a = 0$.
12. $f(x) = \sin x$, $a = 1$.

13. This is a separate problem, and it requires you to think of a trick. Given that $\ln 1.4 = 0.3365$, find a way, using series, to calculate $\ln 2$, correct to three decimal places. (To *four* decimal places, $\ln 2 = 0.6931$.)

14. Let $\sum_{i=1}^{\infty} a_i$ be any series. For each i, let

$$b_i = \tfrac{1}{2}(a_i + |a_i|),$$
$$c_i = \tfrac{1}{2}(a_i - |a_i|).$$

Show that if $\sum b_i$ and $\sum c_i$ both converge, then $\sum a_i$ converges absolutely.

15. Let $\sum a_i$, $\sum b_i$, $\sum c_i$ be as in Problem 14. Show that if $\sum a_i$ converges and $\sum |a_i| = \infty$, then

$$\sum b_i = \infty$$

and

$$\sum c_i = -\infty.$$

*16. Let $n_1, n_2, \ldots$ be a sequence of positive integers in which each positive integer appears exactly once. That is, the numbers $n_1, n_2, \ldots$ are the integers $1, 2, 3, \ldots$ arranged in some order. For each series $\sum_{i=1}^{\infty} a_i$, we can then form a "rearranged series" $\sum_{i=1}^{\infty} a_{n_i}$, in which the same terms appear in some order. The following theorem is a sort of "commutative law of addition" for positive series.

Theorem. If $a_i > 0$ for each i, and $\sum a_i = A$, then every rearrangement of $\sum a_i$ converges to the same sum.

Prove this.

*17. Show that

$$\sum_{i=1}^{\infty} (-1)^{i+1} \frac{1}{i}$$

has a rearrangement which converges to 0. (Thus the "commutative law for infinite sums" does not hold in general.)

*18. Show that for every number k there is a rearrangement of the above series which converges to k.

10.10 TAYLOR'S THEOREM. ESTIMATES OF REMAINDERS

In the preceding section, we showed that, if a function is expressible by a Taylor series, with

$$f(x) = \sum_{i=0}^{\infty} a_i(x - a)^i \qquad (a - r < x < a + r),$$

then the coefficients a_i are given by the formula

$$a_i = \frac{f^{(i)}(a)}{i!}.$$

Using the formula, we can write down a series. But there are three questions which it is natural to ask:

1) For what values of x does the series converge? (We recall that $\mathrm{Tan}^{-1} x$ is defined for every x, but its series converges only for $-1 < x \leq 1$.)

2) Does the series converge to the function f that we started with?

3) If we use a partial sum

$$S_n(x) = \sum_{i=0}^{n} \frac{f^{(i)}(a)}{i!} (x - a)^i$$

as an approximation of $f(x)$, what is the error? For this, we need an estimate of the "remainder function"

$$R_n(x) = f(x) - S_n(x)$$

$$= f(x) - \sum_{i=1}^{n} \frac{f^{(i)}(a)}{i!} (x - a)^i.$$

Partial answers to these questions are given by the following theorem.

Theorem 1 (*Taylor's theorem*). If f has $n + 1$ derivatives, on the interval $[a, x]$ or the interval $[x, a]$, then

$$R_n(x) = \frac{f^{(n+1)}(\bar{x})}{(n + 1)!} (x - a)^{n+1},$$

for some $\bar{x}$ between a and x.

The proof is artificial, and hard to remember. We regard x as a constant; and for each t we let

$$F(t) = f(x) - \sum_{i=0}^{n} \frac{f^i(t)}{i!}(x-t)^i.$$

Here we have simply replaced a by t in the formula for $R_n(x)$. For $t = x$ we have

$$F(x) = f(x) - \frac{f^{(0)}(x)}{0!} = f(x) - f(x) = 0.$$

For $t = a$ we have $F(a) = R_n(x)$. Since

$$F(t) = f(x) - \frac{f(t)}{0!} - \frac{f'(t)}{1!}(x-t) - \frac{f''(t)}{2!}(x-t)^2 - \cdots - \frac{f^{(n)}(t)}{n!}(x-t)^n,$$

we have

$$F'(t) = -f'(t) - [f''(t)(x-t) - f'(t)]$$
$$- \left[\frac{f'''(t)}{2!}(x-t)^2 - \frac{f''(t)}{2!} \cdot 2(x-t) \right]$$
$$\cdot$$
$$\cdot$$
$$\cdot$$
$$- \left[\frac{f^{(n+1)}(t)}{n!}(x-t)^n - \frac{f^{(n)}(t)}{n!} \cdot n(x-t)^{n-1} \right].$$

Here all terms cancel out, telescopically, except the first term in the last bracket; and so

$$F'(t) = - \frac{f^{(n+1)}(t)}{n!}(x-t)^n.$$

Now let

$$G(t) = \frac{(x-t)^{n+1}}{(n+1)!},$$

so that

$$G'(t) = \frac{-(x-t)^n}{n!}.$$

To the functions F and G, on the interval between a and x, we apply the parametric mean-value theorem. (This is Theorem 2 of Section 9.2.) It gives

$$\frac{F(x) - F(a)}{G(x) - G(a)} = \frac{F'(\bar{x})}{G'(\bar{x})},$$

for some $\bar{x}$ between a and x. Since $F(x) = G(x) = 0$, this means that

$$\frac{-F(a)}{-G(a)} = \frac{F'(\bar{x})}{G'(\bar{x})}.$$

And

$$\frac{F'(t)}{G'(t)} = f^{(n+1)}(t)$$

for every t. Therefore

$$\frac{F(a)}{G(a)} = f^{(n+1)}(\bar{x}).$$

By definition of $G(a)$ and $F(a)$, we have

$$R_n(x) = F(a) = f^{(n+1)}(\bar{x})G(a) = \frac{f^{(n+1)}(\bar{x})}{(n+1)!}(x-a)^{n+1},$$

which was to be proved.

In some cases we can use this theorem to prove that a formal power series converges to the expected function. For example, we may be able to find a number M such that

$$|f^{(n+1)}(\bar{x})| \leq M$$

for every n and every $\bar{x}$ between a and x. In such a case it follows that $R_n(x) \to 0$, and $f(x)$ is the sum of its formal Taylor series. Most of the time, however, estimates of $(n+1)$st derivatives are hard to come by. For example, the calculation of $f^{(n)}(x)$ is unmanageable for the function

$$f(x) = \frac{1}{1+x^3},$$

even though we can easily see what the Maclaurin series is:

$$\frac{1}{1+x^3} = \sum_{i=0}^{\infty}(-1)^i x^{3i} \qquad (|x| < 1).$$

It follows, of course that

$$f^{(3i)}(0) = (-1)^i(3i)!,$$

and that $f^{(n)}(0) = 0$ if n is not divisible by 3. But this does not give us any information about $f^{(n)}(x)$ for other values of x.

PROBLEM SET 10.10

1 through 6. In at least six of the first twelve problems in Problem Set 10.9, it is easy to get an estimate of $f^{(n)}(x)$, and then show by Taylor's Theorem that the series converges to the given function. Identify these six cases, and carry out the process.

10.11 THE COMPLEX NUMBER SYSTEM

Formally speaking, complex numbers are numbers of the type

$$z = a + bi,$$

where a and b are real numbers, and where i is some sort of number such that $i^2 = -1$. Granted that there is such a number system, and that it obeys the same manipulative rules as the real number system, the equation $i^2 = -1$ gives all that we need to perform calculations. For example,

$$(a + bi)^2 = a^2 + 2abi + b^2i^2 = (a^2 - b^2) + 2abi;$$

$$(a + bi)(c + di) = (ac - bd) + (ad + bc)i;$$

$$\left(\frac{1}{\sqrt{2}} + \frac{1}{\sqrt{2}} i\right)^2 = i;$$

$$\left(\frac{1}{\sqrt{2}} - \frac{1}{\sqrt{2}} i\right)^2 = -i;$$

$$i^4 = (i^2)^2 = 1;$$

$$i^{10,001} = i;$$

$$\left(\frac{1}{2} + \frac{\sqrt{3}}{2} i\right)^3 = \tfrac{1}{8}(1 + \sqrt{3}\, i)^3$$

$$= \tfrac{1}{8}(1 + 3\sqrt{3}\, i + 3 \cdot 3i^2 + 3\sqrt{3}\, i^3)$$

$$= \tfrac{1}{8}(1 - 9 + 3\sqrt{3}\, i - 3\sqrt{3}\, i)$$

$$= -1.$$

Obviously $0 = 0 + 0 \cdot i$ has no reciprocal. But if $a + bi \neq 0$, then $a + bi$ has a reciprocal in the complex number system. To see this, note that if $a + bi \neq 0$, then a and b are not both $= 0$. Therefore $a - bi \neq 0$, and

$$\frac{1}{a + bi} = \frac{1}{a + bi} \cdot \frac{a - bi}{a - bi} = \frac{a - bi}{a^2 - (bi)^2}$$

$$= \frac{a - bi}{a^2 + b^2} = \frac{a}{a^2 + b^2} + \frac{-b}{a^2 + b^2} i$$

$$= A + Bi.$$

This calculation begins with the assumption that $a + bi$ and $a - bi$ *have* reciprocals, but once we know the answer, it is easy to check:

$$(a + bi)(A + Bi) = (a + bi)\left(\frac{a}{a^2 + b^2} + \frac{-b}{a^2 + b^2} i\right)$$

$$= \frac{(a + bi)(a - bi)}{a^2 + b^2} = \frac{a^2 + b^2}{a^2 + b^2} = 1.$$

Therefore $A + Bi$ is the reciprocal of $a + bi$.

There are several ways to define the set of complex numbers, as a mathematical system, and check their properties. One such method is explained in Appendix J. Meanwhile we shall regard the complex numbers as known, and calculate with them, using the familiar laws of algebra and the fact that $i^2 = -1$.

The *conjugate* of the complex number

$$z = a + bi$$

is the number

$$\bar{z} = a - bi.$$

The *absolute value* of z is

$$|z| = \sqrt{a^2 + b^2}.$$

By straightforward calculations, we get the following.

Theorem 1. For all complex numbers z, z_1, z_2, we have:

$$\bar{\bar{z}} = z,$$

$$z + \bar{z} \quad \text{is a real number,}$$

$$z \cdot \bar{z} \quad \text{is a real number,}$$

$$|z|^2 = z \cdot \bar{z},$$

$$\bar{z}_1 + \bar{z}_2 = \overline{z_1 + z_2},$$

$$\bar{z}_1 \cdot \bar{z}_2 = \overline{z_1 z_2},$$

$$|z_1 z_2| = |z_1| \cdot |z_2|.$$

Proofs.

$$z = a + bi, \quad \bar{z} = a - bi, \quad \bar{\bar{z}} = a - (-b)i = a + bi = z;$$

$$z + \bar{z} = a + bi + a - bi = 2a;$$

$$z \cdot \bar{z} = (a + bi)(a - bi) = a^2 + b^2 = |z|^2;$$

$$\bar{z}_1 + \bar{z}_2 = \overline{a_1 + b_1 i} + \overline{a_2 + b_2 i} = a_1 - b_1 i + a_2 - b_2 i$$

$$= (a_1 + a_2) - (b_1 + b_2)i = \overline{z_1 + z_2};$$

$$z_1 z_2 = (a_1 + b_1 i)(a_2 + b_2 i) = a_1 a_2 - b_1 b_2 + (a_1 b_2 + a_2 b_1)i;$$

$$\bar{z}_1 \cdot \bar{z}_2 = (a_1 - b_1 i)(a_2 - b_2 i) = a_1 a_2 - b_1 b_2 - (a_1 b_2 + a_2 b_1)i = \overline{z_1 z_2};$$

$$|z_1 z_2| = \sqrt{z_1 z_2 \overline{z_1 z_2}} = \sqrt{z_1 z_2 \bar{z}_1 \bar{z}_2} = \sqrt{z_1 \bar{z}_1} \cdot \sqrt{z_2 \bar{z}_2} = |z_1| \cdot |z_2|.$$

(Note that in the last of these calculations all the radicands are real and $\geqq 0$, as they should be.)

So far, we have been treating all these ideas algebraically. We shall now interpret them geometrically, plotting each complex number $z = x + yi$ as a point (x, y) in a coordinate plane.

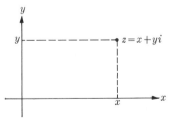

Thus real numbers $z = x$ fall on the x-axis; we shall think of this as the *real axis*. And "pure imaginary numbers," of the form $z = iy$, fall on the y-axis; we shall think of this as the imaginary axis. This explains the labels on the axes in the figure below. Evidently $\bar{z}$ is the reflection of z across the x-axis. If you reflect twice, you get back where you started; and this is the geometric meaning of the equation $\bar{\bar{z}} = z$. As the

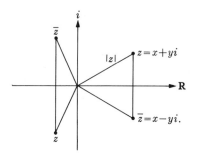

figure suggests, $|z|$ is the distance to z from the origin; the reason is that

$$|z| = \sqrt{x^2 + y^2},$$

which gives the distance. More generally, $|z_1 - z_2|$ is the distance between z_1 and z_2.

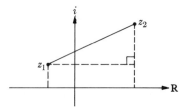

For $z_1 = x_1 + y_1 i$ and $z_2 = x_2 + y_2 i$, we have

$$z_1 - z_2 = (x_1 - x_2) + (y_1 - y_2)i,$$

so that

$$|z_1 - z_2| = \sqrt{(x_1 - x_2)^2 + (y_1 - y_2)^2},$$

which gives the distance.

If $z = x + yi$, then x is called the *real part* of z, and is denoted by Re z. The number y is called the *imaginary part* of z, and is denoted by Im z. Thus

$$z = x + yi = \text{Re } z + i \text{ Im } z.$$

It is easy to check that

$$\text{Re } z = \tfrac{1}{2}(z + \bar{z}),$$
$$\text{Im } z = (1/2i)(z - \bar{z}).$$

Since $i^2 = -1$, we have

$$i = -\frac{1}{i}, \qquad \frac{1}{i} = -i, \qquad \text{and} \qquad \text{Im } z = \frac{i}{2}(\bar{z} - z).$$

These formulas enable us to connect complex numbers with the geometry of our coordinate plane. For example, the graph of the equation $|z| = 1$ is the circle with center at the origin and radius 1; and the graph of the equation $|z - z_0| = a$ is the circle with center at z_0 and radius a. The vertical line through the point $(1, 0) = 1 + 0 \cdot i$ is the graph of the equations

$$x = 1 \iff \text{Re } z = 1$$
$$\iff \tfrac{1}{2}(z + \bar{z}) = 1$$
$$\iff z + \bar{z} = 2.$$

In the following problem set, you will be asked to carry out a variety of such processes. For short, we shall use the term C-equation to describe an equation in which complex numbers are the only variables. Thus the vertical line discussed above is the graph of the C-equation $z + \bar{z} = 2$; and a certain circle is the graph of the C-equation $z\bar{z} = 4$.

PROBLEM SET 10.11

Reduce each of the following expressions to the form $x + yi$.

1. $(1 + i)^4$

2. $(1 - i)^4$

3. $\left(\dfrac{1}{\sqrt{2}} + \dfrac{1}{\sqrt{2}}i\right)^4$

4. $\left(\dfrac{1}{\sqrt{2}} - \dfrac{1}{\sqrt{2}}i\right)^4$

5. $\left(\dfrac{1}{2} + \dfrac{\sqrt{3}}{2}i\right)^3$

6. $\left(\dfrac{1}{2} - \dfrac{\sqrt{3}}{2}i\right)^3$

7. $\left(\dfrac{\sqrt{3}}{2} + \dfrac{1}{2}i\right)^8$

8. $\left(\dfrac{\sqrt{3}}{2} - \dfrac{1}{2}i\right)^8$

9. $\left(-\dfrac{1}{\sqrt{2}} + \dfrac{1}{\sqrt{2}}i\right)^8$

10. $(\sqrt{2} - \sqrt{2}\,i)^8$

11. $(1 + i)^2$

12. $(-1 + i)^3$

13. $(2 + i)^2$

14. $(1 - 2i)^2$

15. $(\sqrt{3} + i)^4$

16. $(1 - \sqrt{3}\,i)^3$

17. $\dfrac{1}{i + 2}$

18. $\dfrac{1}{2 - i}$

19. $\dfrac{1}{2i + 1}$

20. $\dfrac{1}{2i - 1}$

21. $\dfrac{1}{1 + 3i}$

22. $\dfrac{1}{i - 3i}$

23. $\dfrac{1}{2 + \sqrt{3}\,i}$

24. $\dfrac{1}{2 - \sqrt{3}\,i}$

25. $\dfrac{1}{\sqrt{3} + 2i}$

26. $\dfrac{1}{\sqrt{3} - 2i}$

27. $\dfrac{1}{i^2 + i + 1}$

28. $\dfrac{1}{i^3 + i^2 + i - 1}$

29. $\dfrac{1}{i^2 - i + 1}$

30. $\dfrac{1}{(i + 1)^3}$

31. $\dfrac{1 + i}{1 - i}$

32. $\dfrac{2 + 3i}{1 + 2i}$

33. $\dfrac{2 - 3i}{1 - 2i}$

34. $\dfrac{1}{i^5 + i^4 + i^3 + i + 1}$

35. $\dfrac{1}{i^4 + i^3 + i^2 + i + 1}$

36. Show that $|z^n| = |z|^n$, for every z.

37. Show that $\overline{z^n} = \bar{z}^n$, for every z.   38. Show that $1/\bar{z} = \overline{(1/z)}$, for every $z \neq 0$.

Sketch the graphs of the following C-equations.

39. $\mathrm{Re}\,z + \mathrm{Im}\,z = 1$.

40. $\mathrm{Re}\,z - \mathrm{Im}\,z = 1$.

41. $\mathrm{Re}\,z = \mathrm{Im}\,z$.

42. $|z - 1| = 1$.

43. $|z - 1| < 1$.

44. $|z - 1| > 1$.

45. $|z - i| = \frac{1}{2}$.   46. $|z + i| < \frac{1}{2}$.

47. $z + \bar{z} = 0$.   48. $z + \bar{z} > 1$.

49. $|z|^2 = 4$.   50. $|\bar{z}|^2 = 4$.

51. $z^2 = 4$.   52. $z^4 = 4$.

53. $z - \bar{z} = 1$.   54. $|z + i| \leq \frac{1}{2}$.

55. $|z - i| = 1$.   56. $|i - z| + |z + i| \geq 2$.

57. $|i - z| + |z + i| = 2$.

The following two problems require the theory developed in Appendix J.

*58. Find a polynomial $q(x)$ such that $(x^5 + x^4 + x^3 + x^2 + x + 1)q(x) \equiv 1 \bmod 1 + x^2$

*59. We found that if $z = \overline{p(x)} \neq 0$, then z has a reciprocal in **C**. In the language of congruences, this says that if $p(x) \not\equiv 0 \bmod 1 + x^2$, then there is a polynomial $q(x)$ such that

$$p(x)q(x) \equiv 1 \bmod 1 + x^2.$$

Similarly, if

$$p(x) - p'(x) = r(x)(x^2 + x + 1),$$

then we write

$$p(x) \equiv p'(x) \bmod x^2 + x + 1.$$

Show that if $p(x) \not\equiv 0 \bmod x^2 + x + 1$, then there is a polynomial $q(x)$ such that $p(x)q(x) \equiv 1 \bmod x^2 + x + 1$. (In fact, the congruence classes of polynomials modulo $x^2 + x + 1$ form a field. All the conditions for a field can be checked, in the same way as for $1 + x^2$, except for the existence of reciprocals.) Is it true that every polynomial is congruent mod $x^2 + x + 1$ to some linear polynomial?

*60. In Section 10.8 we deduced the binomial formula

$$(a + b)^n = \sum_{i=0}^{n} \binom{n}{i} a^{n-i} b^i$$

from a more general result, using the methods of calculus. But the methods of Section 10.8 do not, as they stand, apply in the complex domain. Show, by induction, that

$$(1 + z)^n = \sum_{i=0}^{n} \binom{n}{j} z^j = 1 + \binom{n}{1} z + \binom{n}{2} z^2 + \cdots + \binom{n}{n-1} z^{n-1} + z^n,$$

for every positive integer n and every complex number z. Then show that for all complex numbers u and v,

$$(u + v)^n = \sum_{j=0}^{n} \binom{n}{j} u^{n-j} v^j.$$

10.12 SEQUENCES AND SERIES OF COMPLEX NUMBERS. THE COMPLEX EXPONENTIAL FUNCTION

We know that for sequences $x_1, x_2, \ldots$ of real numbers,

$$\lim_{n \to \infty} x_n = x \iff \lim_{n \to \infty} |x_n - x| = 0.$$

The second of these conditions also has a meaning if the x_n's and x are complex numbers, because the absolute values $|x_n - x|$ are real in any case. We use this idea to define limits for sequences of complex numbers.

Definition. Let $z_1, z_2, \ldots$ and z be complex numbers. If

$$\lim_{n \to \infty} |z_n - z| = 0,$$

then

$$\lim_{n \to \infty} z_n = z.$$

We can test for limits by examining the real and imaginary parts of the sequence separately.

Theorem 1. For each n, let

$$z_n = x_n + y_n i.$$

If the sequences $x_1, x_2, \ldots$ and $y_1, y_2, \ldots$ are convergent, then $z_1, z_2, \ldots$ is convergent, and

$$\lim_{n \to \infty} z_n = \lim_{n \to \infty} x_n + i \lim_{n \to \infty} y_n.$$

Proof. Let

$$x = \lim_{n \to \infty} x_n, \qquad y = \lim_{n \to \infty} y_n,$$

so that

$$\lim_{n \to \infty} (x_n - x) = \lim_{n \to \infty} (y_n - y) = 0.$$

We need to show that

$$\lim_{n \to \infty} (x_n + y_n i) = x + yi,$$

which means, by definition, that

$$\lim_{n \to \infty} |(x_n + y_n i) - (x + yi)| = 0.$$

This is trivial:

$$\lim_{n \to \infty} |x_n + y_n i - x - yi| = \lim_{n \to \infty} [(x_n - x)^2 + (y_n - y)^2]^{1/2} = (0^2 + 0^2)^{1/2} = 0.$$

The converse is also true.

Theorem 2. If $\lim_{n \to \infty} (x_n + y_n i) = x + yi$, then $\lim_{n \to \infty} x_n = x$ and $\lim_{n \to \infty} y_n = y$.

(Proof?) Once we have Theorems 1 and 2, it is a routine matter to verify that the limit of a sum, product, or quotient is the sum, product, or quotient of the limits, just as for real sequences. We shall use these rules without comment.

Just as for real numbers, we define convergence of infinite series in terms of convergence of sequences.

Definition. Given a series $\sum_{j=0}^{\infty} z_j$. For each n, let

$$S_n = \sum_{j=0}^{n} z_j.$$

If

$$\lim_{n \to \infty} S_n = S,$$

then we say that $\sum_{j=0}^{\infty} z_j$ *converges* (to S), and we write

$$\sum_{j=0}^{\infty} z_j = S.$$

For real series, we found that if $\sum |x_i|$ converges, then $\sum x_i$ also converges. The same is true in the complex domain.

Theorem 3. If $\sum_{j=0}^{\infty} |z_j|$ converges, then $\sum_{j=0}^{\infty} z_j$ converges.

Proof. For each j, let

$$z_j = x_j + iy_j.$$

Then

$$|x_j| = \sqrt{x_j^2} \leqq \sqrt{x_j^2 + y_j^2} = |z_j|,$$

and similarly

$$|y_j| \leqq |z_j|,$$

for each j. Therefore

$$\sum_{j=0}^{\infty} |x_j| < \infty, \quad \text{and} \quad \sum_{j=0}^{\infty} |y_j| < \infty.$$

Therefore $\sum x_j$ and $\sum y_j$ converge, to sums A and B. Therefore

$$\lim_{n \to \infty} \sum_{j=0}^{n} z_j = \lim_{n \to \infty} \sum_{j=0}^{n} (x_j + y_j i) = \lim_{n \to \infty} \sum_{j=0}^{n} x_j + i \lim_{n \to \infty} \sum_{j=0}^{n} y_j = A + Bi.$$

The simplicity of this theorem, and of its proof, are misleading: the theorem is powerful. It gives immediately:

Theorem 4. $\sum_{j=0}^{\infty} (z^j/j!)$ is convergent for every z.

This is so because

$$\sum_{j=0}^{\infty} \left| \frac{z^j}{j!} \right| = \sum_{j=0}^{\infty} \frac{|z|^j}{j!} < \infty$$

for every *real* number $|z|$. This enables us to extend the domain of the exponential $e^x = \exp x$ to the entire complex plane:

Definition. $e^z = \exp z = \sum_{j=0}^{\infty} (z^j/j!)$.

For the case in which z is a pure imaginary number $i\theta$, we can express e^z in another form:

$$e^{i\theta} = \sum_{j=0}^{\infty} \frac{(i\theta)^j}{j!} = \sum_{k=0}^{\infty} \frac{(i\theta)^{2k}}{(2k)!} + \sum_{k=0}^{\infty} \frac{(i\theta)^{2k+1}}{(2k+1)!}.$$

Now

$$(i\theta)^{2k} = (i^2)^k \theta^{2k} = (-1)^k \theta^{2k},$$

and

$$(i\theta)^{2k+1} = i \cdot i^{2k} \theta^{2k+1} = i(-1)^k \theta^{2k+1}.$$

Therefore

$$e^{i\theta} = \sum_{k=0}^{\infty} \frac{(-1)^k \theta^{2k}}{(2k)!} + i \sum_{k=0}^{\infty} \frac{(-1)^k \theta^{2k+1}}{(2k+1)!}.$$

This gives:

Theorem 5. For every real number θ,

$$e^{i\theta} = \cos\theta + i\sin\theta.$$

Setting $\theta = \pi$, we get a famous equation due to Leonard Euler:

$$e^{i\pi} = -1.$$

It is easy to see that if $z = e^{i\theta}$, then $|z| = 1$. The reason is that

$$|e^{i\theta}| = |\cos\theta + i\sin\theta| = \sqrt{\cos^2\theta + \sin^2\theta}.$$

Conversely, if $|z| = 1$, then $z = e^{i\theta}$ for some θ. Here $z = x + yi$, $x = \cos\theta$, and $y = \sin\theta$.

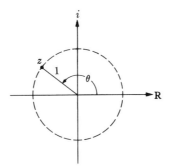

More generally, every complex number can be expressed in the form

$$z = re^{i\theta},$$

$$r = |z| \geqq 0.$$

To see this, we let

$$w = \frac{z}{|z|},$$

so that $|w| = 1$. Therefore

$$w = e^{i\theta} = \cos\theta + i\sin\theta$$

for some θ, and

$$z = |z|\, w = |z|\, e^{i\theta} = re^{i\theta} = r(\cos\theta + i\sin\theta).$$

The expression $re^{i\theta}$ (or $r(\cos\theta + i\sin\theta)$) is called the *polar form* for z, because it describes z in terms of the polar coordinates r, θ of the corresponding point.

For example, consider

$$z = 1 + 2i.$$

Here $r = |z| = \sqrt{5}$, and so

$$w = \frac{z}{|z|} = \frac{1 + 2i}{\sqrt{5}} = \frac{1}{\sqrt{5}} + \frac{2}{\sqrt{5}}\, i.$$

Let θ be any number such that

$$\cos \theta = \frac{1}{\sqrt{5}}, \qquad \sin \theta = \frac{2}{\sqrt{5}}.$$

Then

$$w = \cos \theta + i \sin \theta = e^{i\theta}$$

and

$$z = rw = r(\cos \theta + i \sin \theta)$$

$$= \sqrt{5} \left(\frac{1}{\sqrt{5}} + \frac{2}{\sqrt{5}} i \right),$$

in polar form.

Theorem 6. For every θ and α,

$$e^{i\theta} \cdot e^{i\alpha} = e^{i(\theta + \alpha)}.$$

(The same is true for all complex numbers w, z. That is, $e^w \cdot e^z = e^{w+z}$. But we are not yet in a position to prove it.)

Proof.

$$e^{i\theta} \cdot e^{i\alpha} = (\cos \theta + i \sin \theta)(\cos \alpha + i \sin \alpha)$$
$$= \cos \theta \cos \alpha - \sin \theta \sin \alpha + (\sin \theta \cos \alpha + \cos \theta \sin \alpha)i$$
$$= \cos (\theta + \alpha) + i \sin (\theta + \alpha)$$
$$= e^{i(\theta + \alpha)}.$$

In the polar form $re^{i\theta}$, r is called the *modulus* and θ is called the *amplitude*. It is a slight abuse of language to speak of θ as the amplitude of $re^{i\theta}$, because while the modulus is determined when the number z is named, the amplitude θ is not determined. In fact, when we apply the exponent $i\theta$ to e, we get a periodic function $f(\theta)$.

Theorem 7. For every integer n (positive, negative, or zero),

$$e^{i\theta} = e^{i(\theta + 2n\pi)}.$$

This is so because $\cos (\theta + 2n\pi) = \cos \theta$ and $\sin (\theta + 2n\pi) = \sin \theta$.

PROBLEM SET 10.12

Express the following complex numbers in polar form.

1. $1 + i$ 2. $3i$ 3. -7

4. $\frac{1}{2} + \frac{\sqrt{3}}{2} i$ 5. $\sqrt{3} + i$ 6. $-4 - 4i$

7. In the complex domain, the sine and cosine are defined by the series

$$\sin z = \sum_{j=0}^{\infty} (-1)^j \frac{z^{2j+1}}{(2j + 1)!},$$

$$\cos z = \sum_{j=0}^{\infty} (-1)^j \frac{z^{2j}}{(2j)!}.$$

How do we know that these series converge for every z?

8. In the text, we expressed $e^{i\theta}$ in terms of $\sin \theta$ and $\cos \theta$. More generally, express e^{iz} in terms of $\sin z$ and $\cos z$.

9. Show that for every z,
$$\sin (-z) = -\sin z,$$
$$\cos (-z) = \cos z.$$

10. Express e^{-iz} in terms of $\sin z$ and $\cos z$.

11. Express $\sin z$ and $\cos z$ in terms of the complex exponential function.

10.13 DE MOIVRE'S THEOREM

In Section 10.12 we found that every complex number could be expressed in polar form, with
$$z = re^{i\theta},$$
where
$$r = |z|.$$
And Theorem 6 said that for every θ and α,
$$e^{i\theta} \cdot e^{i\alpha} = e^{i(\theta+\alpha)}.$$
This gives us a rule for multiplying complex numbers in polar form: we multiply the moduli and add the amplitudes. For
$$z_1 = r_1 e^{i\theta_1}, \qquad z_2 = r_2 e^{i\theta_2},$$
we have
$$z_1 z_2 = r_1 r_2 e^{i(\theta_1+\theta_2)}.$$
To divide, we divide by the modulus of the divisor and subtract the amplitude.
$$z_1/z_2 = (r_1/r_2)e^{i(\theta_1-\theta_2)} \qquad (r_2 \neq 0)$$
(This is easy to check by multiplication.) By induction based on the multiplication rule, for $z = re^{i\theta}$ we have
$$z^n = r^n e^{in\theta}.$$

These ideas give us a method for extracting roots of any order. We shall now see that the number 1 has three cube roots in the complex domain. If $z^3 = 1$, then $|z|^3 = 1$, and so $|z| = 1$. Therefore
$$z = e^{i\theta},$$
for some θ. Therefore
$$z^3 = e^{i3\theta} = e^{i\cdot 0} = 1,$$
and
$$3\theta = 0 + 2n\pi$$
for some n. For three successive values of n we get:

$$n = 0, \quad \theta = 0, \quad z = z_1 = 1;$$
$$n = 1, \quad \theta = \tfrac{2}{3}\pi, \quad z = z_2 = \cos \tfrac{2}{3}\pi + i \sin \tfrac{2}{3}\pi;$$
$$n = 2, \quad \theta = \tfrac{4}{3}\pi, \quad z = z_3 = \cos \tfrac{4}{3}\pi + i \sin \tfrac{4}{3}\pi.$$

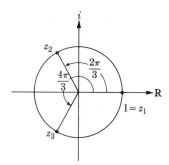

Using other values of n, we would get repetitions of the same cube roots. Thus the roots are

$$z_1 = 1,$$

$$z_2 = -\frac{1}{2} + \frac{\sqrt{3}}{2}\, i,$$

$$z_3 = -\frac{1}{2} - \frac{\sqrt{3}}{2}\, i.$$

These cube roots could have been found by elementary methods, because

$$z^3 - 1 = (z - 1)(z^2 + z + 1),$$

and the quadratic formula gives z_2 and z_3 as the roots of the equation $z^2 + z + 1 = 0$. But for roots of higher order, and for numbers less simple than 1, the elementary methods break down, and our new method still works. For example, $i = e^{i\pi/2}$. Therefore the fourth roots of i are the numbers $e^{i\theta}$ for which

$$4\theta = \frac{\pi}{2} + 2n\pi, \qquad \text{or} \qquad \theta = \frac{1 + 4n}{8}\, \pi.$$

Four successive values of n give us

$$\theta_0 = \frac{\pi}{8}, \qquad \theta_1 = \frac{5\pi}{8}, \qquad \theta_2 = \frac{9\pi}{8}, \qquad \theta_3 = \frac{13\pi}{8},$$

from which the roots $z_j = e^{i\theta_j}$ can be computed, by repeated applications of the half-angle formulas for the sine and cosine.

In general, every complex number $z \neq 0$ has exactly n nth roots in the complex domain, and the roots can be expressed in the form that we have been using.

Theorem 1 (*De Moivre's theorem*). Every complex number

$$z = re^{i\theta} \neq 0$$

has exactly n nth roots. These are the numbers

$$z_j = \sqrt[n]{r}\, e^{i\theta_j} \qquad (j = 0, 1, \ldots, n - 1),$$

where

$$\theta_j = (1/n)(\theta + 2j\pi) \qquad (j = 0, 1, \ldots, n - 1).$$

To prove this, we need to investigate two things.

1) The modulus of the roots. If $w^n = z$, then

$$|w|^n = |z| = r, \quad \text{and} \quad |w| = \sqrt[n]{r}.$$

2) The amplitudes of the roots. If $w = \sqrt[n]{r}\, e^{i\alpha}$, and $w^n = z$, then

$$n\alpha = \theta + 2j\pi \quad \text{for some } j,$$

and

$$\alpha = \frac{\theta}{n} + \frac{j}{n} \cdot 2\pi.$$

Any n successive values of j give us n different values of $e^{i\alpha}$, but thereafter the values of $e^{i\alpha}$ repeat themselves.

For example, consider

$$z = 1 + 2i, \quad n = 5.$$

Then

$$z = re^{i\theta},$$

where

$$r = \sqrt{5}, \quad \theta = \text{Sin}^{-1}\,(1/\sqrt{5}).$$

The fifth roots of z are the numbers

$$z_j = \sqrt[5]{r}\, e^{i\theta_j}$$
$$= \sqrt[5]{\sqrt{5}}\, e^{i\theta_j} = 5^{1/10} e^{i\theta_j} \quad (j = 0, 1, 2, 3, 4),$$

where

$$\theta_j = \tfrac{1}{5}(\theta + 2j\pi)$$
$$= \frac{\theta}{5} + \frac{2j\pi}{5} \quad (j = 0, 1, 2, 3, 4).$$

Thus

$$z_0 = 5^{1/10} e^{i\theta/5},$$
$$z_1 = 5^{1/10} e^{i(\theta+2\pi)/5},$$
$$z_2 = 5^{1/10} e^{i(\theta+4\pi)/5},$$

and so on. Note that it is not easy to express these numbers in the form $a + bi$.

De Moivre's theorem shows that **C** not only contains roots of all orders for all *real* numbers, but also roots of all orders for all *complex* numbers. This means, in particular, that any quadratic equation with coefficients in **C** can be solved in **C**. The method follows the derivation of the familiar quadratic formula.

$$az^2 + bz + c = 0 \quad (a \neq 0)$$
$$\Leftrightarrow z^2 + \frac{b}{a} z = -\frac{c}{a}$$
$$\Leftrightarrow z^2 + \frac{b}{a} z + \frac{b^2}{4a^2} = -\frac{c}{a} + \frac{b^2}{4a^2}$$
$$\Leftrightarrow \left(z + \frac{b}{2a}\right)^2 = \frac{b^2 - 4ac}{4a^2}.$$

If $b^2 - 4ac = 0$, then the equation takes the form

$$\left(z + \frac{b}{2a}\right)^2 = 0 \iff z + \frac{b}{2a} = 0,$$

and

$$z = -b/2a$$

is the only root. If $b^2 - 4ac \neq 0$, then the complex number $(b^2 - 4ac)/4a^2$ has two square roots z_1, z_2, and

$$az^2 + bz + c = 0 \implies \begin{cases} z + b/2a = z_1, \\ \text{or} \\ z + b/2a = z_2. \end{cases}$$

Therefore the roots are the numbers

$$-\frac{b}{2a} + z_1, \qquad -\frac{b}{2a} + z_2.$$

In fact, a much more general result holds: every polynomial equation

$$a_n z^n + a_{n-1} z^{n-1} + \cdots + a_1 z + a_0 = 0 \qquad (a_n \neq 0),$$

with coefficients in **C**, has a root in **C**. We express this by saying that **C** is *algebraically closed*. An easy proof appears eventually, in the theory of functions of a complex variable.

PROBLEM SET 10.13

Solve the following equations, algebraically in terms of radicals.

1. $z^4 + 1 = 0$.
2. $z^6 + i = 0$.
3. $z^3 + 8 = 0$.
4. $z^2 + 2z - i + 1 = 0$.
5. $z^3 + z^2 + z + 1 = 0$.
6. $z^2 + z + i + \frac{1}{4} = 0$.
7. $z^7 + z^6 + z^5 + z^4 + z^3 + z^2 + z + 1 = 0$.
8. $z^5 + z^4 + 2z^3 + 2z^2 + z + 1 = 0$.
9. We know that for each n, the number 1 has exactly n nth roots. Show that we can always find one of these, say, z_0, so that the complete set of nth roots are the powers $z_0, z_0^2, z_0^3, \ldots, z_0^n$ of z_0.
10. List, in polar form, the fifth roots of $i - 1$.
11. If z_0 is as in Problem 9, we say that z_0 is a *generator* of the nth roots of 1. Can z_0 be chosen at random? That is, if $z_0^n = 1$, and $z_0 \neq 1$, does it follow that z_0 generates the nth roots of 1? Why or why not?
12. For each n, let Z_n be the set of all nth roots of 1. Show that (a) Z_n is closed under multiplication and (b) Z_n contains the reciprocal of each of its elements.
*13. Let p be a prime, and let z_0 be any element of Z_p, other than 1. Show that the numbers $z_0, z_0^2, \ldots, z_0^{p-1}$ are all different from 1. Then show that they are all different (from each other). Explain how this result is related to one of the preceding problems.

*10.14 THE RADIUS OF CONVERGENCE. DIFFERENTIATION OF COMPLEX POWER SERIES

In dealing with real series, we observed the following phenomena.

1) The domain of convergence of $\sum a_i x^i$ was always symmetric about zero, except perhaps for the endpoints of an interval. That is, the domain of convergence always turned out to be (a) 0 alone, (b) $(-\infty, \infty)$, or (c) an interval of one of the types $(-r, r)$, $[-r, r]$, $[-r, r)$, $(-r, r]$.

2) The function $f(x) = 1/(1 + x^2)$ is defined for every x, and has derivatives of all orders. Nevertheless, its series $\sum_{i=0}^{\infty} (-1)^i x^{2i}$ converges only on the interval $(-1, 1)$. Here the series goes bad for reasons which seem unrelated to the properties of the function which it represents.

We shall now find out why these things happen. The next two theorems are modeled on theorems which are known in the real domain.

Theorem 1. If $\sum z_j$ is convergent, then $\lim_{n \to \infty} z_n = 0$.

Proof. For each n, let $S_n = \sum_{j=0}^{n} z_j$, so that $\lim_{n \to \infty} S_n = S = \sum_{j=0}^{\infty} z_j$. Then $\lim_{n \to \infty} z_n = \lim_{n \to \infty} (S_n - S_{n-1}) = S - S = 0$. (Note that this is exactly like the old proof.)

Theorem 2. Every convergent sequence of complex numbers is bounded.

Proof. Let $z_n = x_n + iy_n$. Since $z_1, z_2, \ldots$ converges, it follows that the sequences $x_1, x_2, \ldots$ and $y_1, y_2, \ldots$ also converge. Therefore they are bounded, and we have

$$|x_n| \leq a, \qquad |y_n| \leq b,$$

for every n. Therefore

$$|z_n| \leq |x_n| + |y_n| \leq a + b.$$

Throughout this section, for each $r > 0$, D_r denotes the interior of the circle with center at the origin and radius r, in the complex plane. Here D stands for *disk*: the interior of a circle is called an *open disk*. If we include the boundary circle, we get the *closed disk* $\bar{D}_r = \{z \mid |z| \leq r\}$.

Theorem 3. If a series $\sum_{j=0}^{\infty} a_j z^j$ converges for $z = z_0$, with $z_0 \neq 0$, and $0 < s < |z_0|$, then the series converges at every point of $\bar{D}_s$.

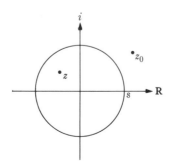

The first step in the proof is to show that $\sum |a_j| \, s^j$ is convergent. For this purpose we use a comparison test. We have

$$|a_j| \, s^j = |a_j| \cdot \left| \frac{s}{z_0} \cdot z_0 \right|^j = |a_j| \cdot \left| \frac{s}{z_0} \right|^j \cdot |z_0|^j$$

$$= |a_j z_0^j| \cdot \left| \frac{s}{z_0} \right|^j.$$

But we know that $\sum a_j z_0^j$ is convergent. Therefore $a_j z_0^j \to 0$, and so the numbers $a_j z^j$ form a bounded sequence, with

$$|a_j z_0^j| \leqq b \quad \text{for every } j.$$

And

$$\left| \frac{s}{z_0} \right| < 1,$$

because $s < |z_0|$. Therefore

$$\sum_{j=0}^{\infty} |a_j| \, s^j = \sum_{j=0}^{\infty} |a_j z_0^j| \cdot \left| \frac{s}{z_0} \right|^j \leqq \sum_{j=0}^{\infty} b \left| \frac{s}{z_0} \right|^j = \frac{b}{1 - |s/z_0|} < \infty.$$

Now for each point z of $\bar{D}_s$, we have $|z| \leqq s$. Therefore

$$\sum_{j=0}^{\infty} |a_j z^j| \leqq \sum_{j=0}^{\infty} |a_j| \, s^j < \infty \qquad (z \text{ in } \bar{D}_s).$$

Therefore $\sum a_j z^j$ converges in $\bar{D}_s$, which was to be proved.

Suppose now that we have given a series $\sum a_j z^i$ which converges for some $z \neq 0$. It follows that the series converges on some disk $\bar{D}_s$. Let S be the set of all such numbers s. That is,

$$S = \left\{ s \mid \sum_{j=0}^{\infty} a_j z^j \text{ converges on } \bar{D}_s \right\}.$$

If S is unbounded, then the series is convergent for every z. In this case, we say that the *radius of convergence* is ∞. If S is bounded, then S has a least upper bound sup S. (See page 243.) Let

$$r = \sup S.$$

In this case, we say that the *radius of convergence* is r.

Theorem 4. If a series $\sum_{j=0}^{\infty} a_j z^j$ has radius of convergence $r < \infty$, then the series converges for $|z| < r$ and diverges for $|z| > r$.

Proof.

1) Given a number z_0, with $|z_0| < r$. Since r is the *least* upper bound of the set S, the number $|z_0|$ is *not* an upper bound of S. Therefore there is an $s > |z_0|$ such that the series converges on $\bar{D}_s$. Therefore the series converges at $z = z_0$.

2) Given z_0, with $|z_0| > r$. Suppose that the series converges at $z = z_0$, and let s be such that

$$r < s < |z_0|.$$

Then the series converges on $\bar{D}_s$; and this is impossible, because r is an upper bound for such numbers s.

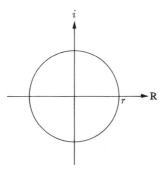

Note that while this theorem tells us what happens *inside* the circle $|z| = r$, and what happens *outside* the circle, it tells us nothing about what happens *on* the circle.

This theorem clarifies the situation for real series $\sum a_j x^j$. Suppose that the series converges for some $x \neq 0$. Then the complex series $\sum_{j=0}^{\infty} a_j z^j$ converges for some $z \neq 0$ (namely, the same x). Let the radius of convergence be r. Then $\sum a_j z^j$ converges for $|z| < r$ and diverges for $|z| > r$. Therefore, for real values of z, $\sum a_j x^j$ converges for $|x| < r$ and diverges for $|x| > r$.

The circular domain of convergence for complex power series also accounts for the behavior of the series

$$\sum_{j=0}^{\infty} (-1)^i x^{2i} = \frac{1}{1 + x^2}.$$

If this series converged for some x for which $|x| > 1$, then for the complex series

$$\sum_{j=0}^{\infty} (-1)^j z^{2j} = \frac{1}{1 + z^2},$$

we would have a radius of convergence $r > 1$. This is impossible, because the function itself blows up at a point of the unit circle: $|i| = 1$, and for $z = i$ the denominator on the right-hand side becomes 0. [Query: how do we know that the series is equal to $1/(1 + z^2)$, for complex values of z?]

The derivative of a complex-valued function is defined by an obvious analogy with the derivative of a real-valued function. That is,

$$f'(z_0) = \lim_{z \to z_0} \frac{f(z) - f(z_0)}{z - z_0}.$$

This is a complicated limiting process, because z may approach z_0 from any direction in the complex plane.

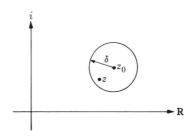

To be exact, the indicated limit means that for every $\epsilon > 0$ there is a $\delta > 0$ such that

$$0 < |z - z_0| < \delta \ \Rightarrow \ \left| \frac{f(z) - f(z_0)}{z - z_0} - f'(z_0) \right| < \epsilon.$$

Here the inequality $|z - z_0| < \delta$ allows z to lie anywhere in the interior of a circle with center at z_0 and radius δ. If $f'(z_0)$ exists, then we say that f is *differentiable at* z_0. If f is differentiable at every point of an open disk containing z_0, then we say that f is *analytic at* z_0. It is easy to see that if $f(z) = z^n$, then f is differentiable everywhere, and therefore analytic everywhere, with

$$f'(z) = nz^{n-1}.$$

The proof is exactly like the proof for $f(x) = x^n$. Similarly:

Theorem 5. If f is a polynomial, with

$$f(z) = \sum_{j=0}^{n} a_j z^j,$$

then f is analytic everywhere, and

$$f'(z) = \sum_{j=1}^{n} j a_j z^{j-1}.$$

But the proof of the following theorem is hard.

Theorem 6. If $f(z)$ has a power series $\sum_{j=0}^{\infty} a_j z^j$, converging in the open disk D_r, then f is analytic in D_r, and

$$f'(z) = \sum_{j=1}^{\infty} j a_j z^{j-1} \qquad (|z| < r).$$

Proof. Take a fixed z_0 in D_r. To compute $f'(z_0)$, we need some preliminary results. Let s be any real number such that

$$|z_0| < s < r$$

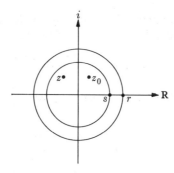

Then

$$\sum_{j=0}^{\infty} |a_j| s^j < \infty \qquad (0 < s < r).$$

(See the proof of Theorem 4, where we used the same apparatus.)

Therefore the series defines a function $g(s)$, and by Theorem A of Section 10.5 it follows that

$$g'(s) = \sum_{j=1}^{\infty} j\,|a_j|\,s^{j-1},$$

where the series on the right is convergent. Therefore

$$\sum_{j=1}^{\infty} j\,|a_j|\,s^j < \infty,$$

and so

$$\lim_{n \to \infty} \sum_{j=n+1}^{\infty} j\,|a_j|\,s^j = 0.$$

Now consider what we are trying to prove. The theorem says that

$$\lim_{z \to z_0} \left[\frac{1}{z - z_0} \cdot \left(\sum_{j=0}^{\infty} a_j z^j - \sum_{j=0}^{\infty} a_j z_0^j \right) \right] = \sum_{j=1}^{\infty} j a_j z_0^{j-1},$$

for every z_0 in D_r. Simplifying the expression in brackets, and transposing the sum on the right, we get the equivalent form

$$\lim_{z \to z_0} \sum_{j=1}^{\infty} a_j (z^{j-1} + z^{j-2} z_0 + z^{j-3} z_0^2 + \cdots + z z_0^{j-2} + z_0^{j-1} - j z_0^{j-1}) = 0.$$

This is not as bad as it looks, because the limit of the jth term is 0 for each j: each of the first j terms in the parenthesis approaches the limit z_0^{j-1}, as $z \to z_0$, and so the total expression in the parentheses approaches 0. Therefore if the sum were finite our conclusion would follow: if

$$S_n(z) = \sum_{j=0}^{n} a_j(\cdots),$$

then

$$\lim_{z \to z_0} S_n(z) = 0 \quad \text{for each } n.$$

We need to find out what happens to the remainder

$$R_n(z) = \sum_{j=n+1}^{\infty} a_j (z^{j-1} + z^{j-2} z_0 + \cdots + z_0^{j-1} - j z_0^{j-1}).$$

We know that $|z_0| < s$; and since we are discussing $\lim_{z \to z_0}$, we may assume that $|z| < s$. Under these conditions,

$$|a_j(z^{j-1} + z^{j-2} z_0 + \cdots + z_0^{j-1} - j z_0^{j-1})|$$
$$\leq |a_j| \cdot (|z^{j-1}| + |z^{j-2} z_0| + \cdots + |z_0^{j-1}| + j\,|z_0^{j-1}|) < |a_j| \cdot 2j s^{j-1}.$$

Therefore, for $|z| < s$ we have

$$|R_n(z)| \leq \sum_{j=n+1}^{\infty} |a_j| \cdot 2j s^{j-1}.$$

The sum on the right approaches 0 as $n \to \infty$.

Now let $\epsilon > 0$ be given. There is an n such that

$$|R_n(z)| < \epsilon/2 \qquad (|z| < s).$$

There is a $\delta > 0$ such that

$$0 < |z - z_0| < 0 \quad \Rightarrow \quad |S_n(z)| < \epsilon/2.$$

Since

$$\sum_{j=1}^{\infty} a_i(\cdots) = S_n(z) + R_n(z),$$

we have

$$0 < |z - z_0| < \delta \quad \Rightarrow \quad \left| \sum_{j=1}^{\infty} a_i(\cdots) \right| \leq |S_n(z)| + |R_n(z)| < \frac{\epsilon}{2} + \frac{\epsilon}{2} = \epsilon,$$

and so $\lim_{z \to z_0} \sum a_i(\cdots) = 0$, which was to be proved.

It is worthwhile to remember the device (used in this proof) of taking n large enough to make $|R_n(z)|$ small, and then taking $|z - z_0|$ small enough to make $|S_n(z)|$ small. This is not a special or isolated device, but a sample of a standard method.

Once we know how to differentiate a series, we can apply in the complex domain a variety of techniques which we have been using in the real domain. The generalizations are not hard, as you will see. In the following problem set, D is used to indicate differentiation. Thus

$$Df(z) = f'(z),$$

and

$$D \sum_{j=1}^{\infty} a_j z^j = \sum_{j=1}^{\infty} ja_j z^{j-1},$$

on any open disk D_r where the series on the left converges.

PROBLEM SET 10.14

1. Find De^z.
2. Find De^{-z}.
3. Find $D \sin z$.
4. Find $D \cos z$.
5. Find De^{z^2}.
6. Find $D \sin 2z$.

7. Show that if f is analytic in D_r, and $g = f^n$, then g is analytic in D_r, and

$$g' = Df^n = nf^{n-1}f'.$$

(The pattern of proof for the real domain works equally well here.)

8. Let f be analytic for every z, and let $g(z) = f(a + z)$. Show that $g'(z) = f'(a + z)$. (Evidently this is another simple special case of the chain rule.)

9. Given $f(z) = \sum_{j=0}^{\infty} a_j z^j$ in D_r. Show that f has not only a first derivative in D_r, but also derivatives $f^{(2)} = f''$, $f^{(3)}$, ... of all orders.

10. Given $f(z) = \sum_{j=0}^{\infty} a_j z^j = 0$ for every z in D_r. Obviously $a_0 = 0$, because $f(0) = a_0$. Show that $a_j = 0$ for every j.

11. Given $f(z) = \sum_{j=0}^{\infty} a_j z^j$ in D_r. Show that if $f'(z) = 0$ for every z, then f is a constant function and is equal to a_0 for each z.

12. Given that $f(z) = \sum a_j z^j$, $g(z) = \sum b_j z^j$ in D_r. Show that if $f' = g'$, then $f - g$ is a constant.

13. Let $\phi(z) = \cos^2 z + \sin^2 z$. Show that $\phi(z) = 1$ for every z.

14. Given that $\cos^2 z + \sin^2 z = 1$ for every z, does it follow that the complex functions $\cos z$ and $\sin z$ are bounded? Why or why not?

15. Show that $e^a \cdot e^z = e^{a+z}$, for every a and z.

16. Does the addition formula for the cosine hold in the complex domain? That is, is it true that

$$\cos (a + z) = \cos a \cos z - \sin a \sin z,$$

for every a and z?

17. Answer the same question, for the proposed identity

$$\sin (a + z) = \sin a \cos z + \cos a \sin z.$$

18. Show that $\sin 2z = 2 \sin z \cos z$.

19. Show that $\cos 2z = \cos^2 z - \sin^2 z$.

*10.15 INTEGRATION AND DIFFERENTIATION OF REAL POWER SERIES

In Section 10.5 we stated (in Theorems A and B) that a Maclaurin series could be differentiated and integrated a term at a time, on any interval $(-r, r)$ on which the series converges. We shall now prove these theorems. The ideas that are needed to do this may be easier to understand if we first show how they apply to the geometric series

$$f(x) = \frac{1}{1 - x} = \sum_{i=0}^{\infty} x^i \qquad (-1 < x < 1).$$

We want to show that

$$\int_0^k f(x)\, dx = \sum_{i=0}^{\infty} \int_0^k x^i\, dx \qquad (0 < k < 1). \tag{1}$$

Let

$$S_n(x) = \sum_{i=0}^{n} x^i,$$

$$R_n(x) = \sum_{i=n+1}^{\infty} x^i,$$

so that

$$f(x) = S_n(x) + R_n(x),$$
$$f(x) - S_n(x) = R_n(x),$$

and

$$\int_0^k f(x)\, dx - \int_0^k S_n(x)\, dx = \int_0^k R_n(x)\, dx. \tag{2}$$

Since $S_n(x)$ is a finite sum, we know that

$$\int_0^k S_n(x)\, dx = \sum_{i=0}^{n} \int_0^k x^i\, dx;$$

and

$$\lim_{n \to \infty} \sum_{i=0}^{n} \int_0^k x^i\, dx = \sum_{i=0}^{\infty} \int_0^k x^i\, dx,$$

by definition of the infinite sum. We shall show that

$$\lim_{n \to \infty} \int_0^k R_n(x)\, dx = 0. \tag{3}$$

If (3) holds, then we can take the limit in formula (2), getting

$$\lim_{n \to \infty} \left[\int_0^k f(x)\, dx - \int_0^k S_n(x)\, dx \right] = \lim_{n \to \infty} \left[\int_0^k f(x)\, dx - \sum_{i=0}^{n} \int_0^k x^i\, dx \right]$$

$$= \int_0^k f(x)\, dx - \sum_{i=0}^{\infty} \int_0^k x^i\, dx = 0.$$

This means that (1) holds.

The proof of (3) is as follows. We have

$$R_n(x) = \sum_{i=n+1}^{\infty} x^i = x^{n+1} + x^{n+2} + \cdots$$

$$= x^{n+1} \sum_{i=0}^{\infty} x^i = \frac{x^{n+1}}{1 - x}.$$

We are integrating from 0 to k. If $0 \leq x \leq k$, then

$$1 - x \geq 1 - k,$$

and

$$x^{n+1} \leq k^{n+1}.$$

Therefore

$$R_n(x) \leq \frac{k^{n+1}}{1 - k}.$$

Therefore

$$\int_0^k R_n(x)\, dx \leq \int_0^k \frac{k^{n+1}}{1 - k}\, dx = \frac{k^{n+2}}{1 - k}.$$

Therefore

$$\lim_{n \to \infty} \int_0^k R_n(x)\, dx = 0,$$

which is what we wanted.

What made this work was the fact that the functions $R_1, R_2, \ldots$ were *squeezed to 0 by a sequence of constants.* We had

$$M_n = \frac{k^{n+1}}{1 - k},$$

$$0 \leq R_n(x) \leq M_n \qquad (0 \leq x \leq k),$$

and

$$\lim_{n \to \infty} M_n = 0.$$

It followed that

$$\int_0^k R_n(x)\, dx \leq \int_0^k M_n\, dx = kM_n \to 0.$$

This method applies in general, to prove Theorem B of Section 10.5. To do this, we need some preliminaries.

Theorem 1. Every convergent sequence is bounded.

This is Theorem 7 of Section 10.1. It gives us the following result for series.

Theorem 2. If $\sum a_i x^i$ is convergent on the interval $(-r, r)$, and $0 < k < r$, then $\sum a_i k^i$ is absolutely convergent.

Proof. Let x_1 be any number between k and r.

Then $\sum a_i x_1^i$ is convergent, and so $\lim_{i \to \infty} a_i x_1^i = 0$. By Theorem 1, the sequence $a_0, a_1 x_1, a_2 x_1^2, \ldots$ is bounded. Let b be a bound for this sequence, so that

$$|a_i x_1^i| \leq b \quad \text{for every } i.$$

Now consider the series $\sum a_i k^i$. Let $s = k/|x_1|$. Then $0 < s < 1$, and so

$$\sum_{i=0}^{\infty} s^i = \frac{1}{1-s} < \infty.$$

But $k = |x_1| s$, and so

$$|a_i k^i| = |a_i x_1^i| \, s^i \leq b s^i.$$

Therefore

$$\sum_{i=0}^{\infty} |a_i k^i| = \sum_{i=0}^{\infty} |a_i x_1^i| \, s^i \leq \sum_{i=0}^{\infty} b s^i = \frac{b}{1-s} < \infty.$$

Therefore $\sum a_i k^i$ is absolutely convergent, which was to be proved.

Carrying the same ideas a little further, we get better information.

Theorem 3. If $\sum a_i x^i$ is convergent on the interval $(-r, r)$, and $0 < k < r$, then the remainders

$$R_n(x) = \sum_{i=n+1}^{\infty} a_i x^i$$

are squeezed to 0, on the interval $[-k, k]$, by a sequence of constants. That is, there is a sequence $M_1, M_2, \ldots$ of constants, such that

$$\lim_{n \to \infty} M_n = 0,$$

and

$$|R_n(x)| \leq M_n \quad (-k \leq x \leq k).$$

Proof. We know by the preceding theorem that $\sum a_i k^i$ is absolutely convergent. For each n, let

$$M_n = \sum_{i=n+1}^{\infty} |a_i k^i|.$$

Then

$$\lim_{n \to \infty} M_n = 0,$$

because

$$M_n = \sum_{i=0}^{\infty} |a_i k^i| - \sum_{i=0}^{n} |a_i k^i|.$$

On the interval $[-k, k]$, we have $|x| \leq k$. Therefore

$$|x^i| \leq k^i, \qquad |a_i x^i| \leq |a_i k^i|,$$

and

$$|R_n(x)| = \left| \sum_{i=n+1}^{\infty} a_i x^i \right| \leq \sum_{i=n+1}^{\infty} |a_i x^i| \leq \sum_{i=n+1}^{\infty} |a_i k^i| = M_n.$$

Therefore

$$|R_n(x)| \leq M_n \quad \text{for every } n,$$

and so the remainders $R_n(x)$ are squeezed to 0, on the interval $[-k, k]$, by the constants $M_1, M_2, \ldots$.

The ideas in this theorem are going to come up again, and so we need a briefer language in which to describe them.

Definition. Let $R_1, R_2, \ldots$ be a sequence of functions on the interval $[a, b]$. If there is a sequence $M_1, M_2, \ldots$ of positive constants, approaching 0, such that

$$|R_n(x)| \leq M_n,$$

for every x on $[a, b]$ and for every n, then we say that the functions $R_1, R_2, \ldots$ *approach* 0 *uniformly on* $[a, b]$, and we write

$$\text{U} \lim_{n \to \infty} R_n(x) = 0 \quad \text{on } [a, b].$$

Definition. Let $S_1, S_2, \ldots$, and S be functions on $[a, b]$. If

$$\text{U} \lim_{n \to \infty} [S(x) - S_n(x)] = 0 \quad \text{on } [a, b],$$

then we say that the functions $S_1, S_2, \ldots$ *approach* S *uniformly on* $[a, b]$, and we write

$$\text{U} \lim_{n \to \infty} S_n(x) = S(x) \quad \text{on } [a, b].$$

In the case covered by Theorem 3, we had

$$[a, b] = [-k, k],$$

$$R_n(x) = \sum_{i=n+1}^{\infty} a_i x^i,$$

$$S_n(x) = \sum_{i=0}^{n} a_i x^i,$$

$$S(x) = \sum_{i=0}^{\infty} a_i x^i,$$

$$R_n(x) = S(x) - S_n(x).$$

In our new terminology, Theorem 3 takes the following form:

Theorem 3′. If $\sum a_i x^i$ is convergent on $(-r, r)$, and $0 < k < r$, then

$$\text{U} \lim_{n \to \infty} \sum_{i=0}^{n} a_i x^i = \sum_{i=0}^{\infty} a_i x^i \quad \text{on } [-k, k].$$

At the beginning of this section, we used these ideas to integrate $\sum x^i$ a term at a time. We want to use the same idea for all power series, but we have a new problem. The series $\sum x^i$ represented a known function $f(x) = 1/(1 - x)$, which was continuous. But when $f(x)$ is given only by a series $\sum a_i x^i$, we first need to show that f is continuous, in order to conclude that the sum has an integral. Thus we need the following two theorems.

Theorem 4. If f_n is continuous for each n, and U $\lim_{n\to\infty} f_n(x) = f(x)$ on $[a, b]$, then f is continuous on $[a, b]$.

Proof. Take a fixed x_0, and let ϵ be any positive number. Let $M_1, M_2, \ldots$ be as in the definition of U lim. Then there is an n such that $M_n < \epsilon/3$. Hereafter in the proof, n is fixed. Since f_n is continuous at x_0, there is a $\delta > 0$ such that

$$|x - x_0| < \delta \quad \Rightarrow \quad |f_n(x) - f_n(x_0)| < \epsilon/3.$$

Since

$$|f(x) - f_n(x)| < M_n \quad \text{for every } x,$$

we have

$$|x - x_0| < \delta \quad \Rightarrow \quad \begin{cases} |f(x) - f_n(x)| < \epsilon/3, \\ |f_n(x) - f_n(x_0)| < \epsilon/3, \\ |f_n(x_0) - f(x_0)| < \epsilon/3. \end{cases}$$

By the triangular inequality,

$$|a + b + c| \leq |a| + |b| + |c|.$$

Therefore

$$|x - x_0| < \delta \quad \Rightarrow \quad |f(x) - f(x_0)| < \epsilon/3 + \epsilon/3 + \epsilon/3 = \epsilon,$$

which was to be proved.

To conclude that f is continuous, it is not enough to know that $\lim_{n\to\infty} f_n(x) = f(x)$ for each x of $[a, b]$; we really need to know that U $\lim_{n\to\infty} f(x) = f(x)$ on $[a, b]$. (See the following problem set, for an example showing this.) For power series, of course, we know that $S_n(x)$ is continuous for each n, because $S_n(x)$ is a polynomial; and we know that the sum of the series is not merely $\lim S_n(x)$ but also U $\lim S_n(x)$, on every closed interval lying in the interval of convergence. This gives the following theorem.

Theorem 5. If $\sum a_i x^i$ is convergent on $(-r, r)$, then $\sum a_i x^i$ is a continuous function on $(-r, r)$.

Therefore there is such a thing as the integral of $\sum a_i x^i$; and we can show that it is the sum of the integrals.

Theorem 6. If $\sum_{i=0}^{\infty} a_i x^i$ converges on $(-r, r)$, and $|x| < r$, then

$$\int_0^x \left[\sum_{i=0}^{\infty} a_i t^i \right] dt = \sum_{i=0}^{\infty} \int_0^x a_i t^i \, dt.$$

Since $\int_0^x t^i \, dt = x^{i+1}/(i+1)$, the theorem tells us that

$$\int_0^x \left[\sum_{i=0}^{\infty} a_i t^i \right] dt = \sum_{i=0}^{\infty} \frac{a_i}{i+1} x^{i+1}.$$

To prove the theorem, we let $S_n(t) = \sum_{i=0}^{n} a_i t^i$. Then for $0 < x < r$ we have

$$\text{U} \lim_{n \to \infty} S_n(t) = \sum_{i=0}^{\infty} a_i t^i \quad \text{on } [0, x].$$

(For $-r < x < 0$, the same condition holds on $[x, 0]$, and the rest of the proof is exactly the same.) Let

$$R_n(t) = \sum_{i=n+1}^{\infty} a_i t^i = \sum_{i=0}^{\infty} a_i t_i - S_n(t),$$

and let $M_1, M_2, \ldots$ be as in the definition of U lim. Then

$$|R_n(t)| = \left| \sum_{i=0}^{\infty} a_i t^i - S_n(t) \right| \leqq M_n \quad \text{for every } n.$$

Therefore

$$\left| \int_0^x \left[\sum_{i=0}^{\infty} a_i t^i \right] dt - \int_0^x S_n(t) \, dt \right| = \left| \int_0^x \left[\sum_{i=0}^{\infty} a_i t^i - S_n(t) \right] dt \right|$$

$$\leqq \int_0^x \left| \sum_{i=0}^{\infty} a_i t^i - S_n(t) \right| dt$$

$$\leqq \int_0^x M_n \, dt$$

$$= M_n \, |x|.$$

Since $\lim_{n \to \infty} M_n |x| = 0$, it follows that

$$\int_0^x \left[\sum_{i=0}^{\infty} a_i t^i \right] dt = \lim_{n \to \infty} \int_0^x S_n(t) \, dt.$$

But $S_n(t)$ is a *finite* sum, and can be integrated a term at a time. Therefore

$$\int_0^x S_n(t) \, dt = \int_0^x \left[\sum_{i=0}^{n} a_i t^i \right] dt = \sum_{i=0}^{n} \int_0^x a_i t^i \, dt,$$

and

$$\int_0^x \left[\sum_{i=0}^{\infty} a_i t^i \right] dt = \lim_{n \to \infty} \sum_{i=0}^{n} \int_0^x a_i t^i \, dt = \sum_{i=0}^{\infty} \int_0^x a_i t^i \, dt,$$

which was to be proved.

This theorem shows that if $\sum a_i x_i$ converges on $(-r, r)$, then the integral series $\sum [a_i/(i+1)] x^{i+1}$ also converges on $(-r, r)$. The same is true for the derivative series $\sum_{i=1}^{\infty} i a_i x^{i-1}$.

Theorem 7. If $\sum_{i=0}^{\infty} a_i x^i$ converges on $(-r, r)$, then $\sum_{i=1}^{\infty} i a_i x^{i-1}$ converges on $(-r, r)$.

Proof. This is going to be very similar to the proof of Theorem 2. Let x_1 be any number such that $|x| < x_1 < r$. Then $\sum a_i x_1^i$ is convergent; $\lim_{n \to \infty} a_i x_1^i = 0$; and there is a bound b such that $|a_i x_1^i| \leqq b$ for every i. Let $s = |x|/x_1$, so that

$$|x| = x_1 s,$$
$$|ia_i x^{i-1}| = |ia_i x_1^{i-1} s^{i-1}| = |a_i x_1^{i-1}| \cdot is^{i-1} \leqq bis^{i-1}.$$

Therefore

$$\sum_{i=1}^{\infty} |ia_i x^{i-1}| \leqq b \sum_{i=1}^{\infty} is^{i-1}.$$

But $s < 1$, and so the series on the right-hand side converges, by the ratio test. Therefore the series on the left-hand side converges. Therefore $\sum ia_i x^{i-1}$ converges.

It remains to show that the "derivative series" $\sum ia_i x^{i-1}$ really gives the derivative, but this is easy.

Theorem 8. If

$$\sum_{i=0}^{\infty} a_i x^i = f(x) \quad \text{on } (-r, r),$$

then f is differentiable on $(-r, r)$, and

$$f'(x) = \sum_{i=1}^{\infty} ia_i x^{i-1}.$$

Proof. Let

$$g(x) = \sum_{i=1}^{\infty} ia_i x^{i-1} \quad (-r < x < r).$$

Then

$$\int_0^x g(t)\, dt = \int_0^x \left[\sum_{i=1}^{\infty} ia_i t^{i-1}\right] dt = \sum_{i=1}^{\infty} \int_0^x ia_i t^{i-1}\, dt = \sum_{i=1}^{\infty} a_i x^i = f(x) - a_0.$$

Therefore

$$\int_0^x g(t)\, dt = f(x) - a_0.$$

But the integral on the left-hand side is a differentiable function, and

$$D \int_0^x g(t)\, dt = g(x).$$

Therefore f is differentiable, and $f' = g$, which was to be proved.

Obviously Theorem 8 can be applied again, to the derivative series, and so

$$f''(x) = \sum_{i=2}^{\infty} i(i-1)x^{i-2},$$

$$f^{(3)}(x) = \sum_{i=3}^{\infty} i(i-1)(i-2)x^{i-3},$$

and so on. Thus if $f(x)$ is represented by a power series, then f has an nth derivative for every n. In a way this is good; it means that functions given by series are in some respects manageable. But it also means that if a function f does *not* have infinitely many derivatives, then f *cannot* be represented by a power series. Later you will see that many such "irregular" functions can be represented by series of other kinds, notably by so-called Fourier series, of the form

$$a_0 + \sum_{i=1}^{\infty} [a_i \cos ix + b_i \sin ix].$$

PROBLEM SET 10.15

1. Let

$$f(x) = \sum_{i=0}^{\infty} (-1)^i \frac{x^{2i}}{(2i)!}.$$

Calculate the series for $f''(x)$, and verify that $f''(x) = -f(x)$. [This must be true, because $f(x) = \cos x$.]

2. Find a simple formula for $g''(x)$, given

$$g(x) = \sum_{i=0}^{\infty} (-1)^i \frac{x^{2i+1}}{(2i + 1)!}.$$

3. Do the same for

$$h(x) = \sum_{i=0}^{\infty} (-1)^i \frac{x^i}{i!}.$$

4. For each n, let $f_n(x) = x^n$ ($0 \leq x \leq 1$). Let $f(x) = \lim_{n \to \infty} f(x)$. Sketch the graph of f, and find out whether it is true that U $\lim_{n \to \infty} f_n(x) = f(x)$ on [0, 1]. (This throws some light on Theorem 4.)

5. Find, by any method, a power series for $f(x) = xe^x + e^x$.

6. Same question, for $g(x) = x \cos x + \sin x$.

7. Same question, for $h(x) = x \sin x - \cos x$.

8. Find a series for a positive function f such that $f'(x) = 2xf(x)$ for every x, and $f(0) = 1$. (There is a short-cut. Since $f(x) > 0$ for every x, $f'(x) = 2xf(x) \Leftrightarrow f'(x)/f(x) = 2x$. Now what?)

In each of the following cases, discuss $\lim_{n \to \infty} f_n(x)$ and U $\lim_{n \to \infty} f_n(x)$.

9. $f_n(x) = nx^n$, $-\frac{1}{2} \leq x \leq \frac{1}{2}$.

10. $f_n(x) = (1/n) \sin nx$, $-1 \leq x \leq 1$.

11. $f_n(x) = (1/n) \cos 2nx$, $-1 \leq x \leq 1$.

12. $f_n(x) = (1/n)e^{nx}$, $0 \leq x \leq 1$.

13. $f_n(x) = (1/n^2)e^{nx}$, $0 \leq x \leq 1$.

14. $f_n(x) = n \sin (x/n)^2$, $0 \leq x \leq 1$.

*15. Find out whether the following is true:

Theorem (?). If U $\lim_{n \to \infty} f_n(x) = f(x)$ on $[-k, k]$, and each of the functions f_n is differentiable, then f is differentiable, and $f'(x) = \lim_{n \to \infty} f'_n(x)$ for every x between $-k$ and k.

(If this is true, then it furnishes a straightforward proof of Theorem 8, replacing the proof using integrals.)

**16. Here we return to complex power series, as in Section 10.14. It is evident that if $\sum a_j z^j$ converges on D_r, then $\sum |a_j z^j|$ converges on D_r; in fact, every time we have proved convergence for a complex power series, we have first proved absolute convergence and then used Theorem 3 of Section 10.12. It remains, however, to consider the question of uniform convergence. Just as for sequences of real functions,

$$\text{U} \lim_{n \to \infty} f_n(z) = f(z) \quad \text{on } \bar{D}_s$$

if $|f_n(z) - f(z)|$ is squeezed to 0 by a sequence of constants. That is, the above U lim relation holds if there is a sequence $M_1, M_2, \ldots$ of positive constants such that

$$\lim_{n \to \infty} M_n = 0, \quad \text{and} \quad |f_n(z) - f(z)| \leq M_n, \quad \text{for every } z \text{ in } \bar{D}_s.$$

Prove the following:

Theorem. If $\sum a_j z^j$ has $r > 0$ as its radius of convergence, and $0 < s < r$, then

$$\text{U} \lim_{n \to \infty} \sum_{j=0}^{n} a_j z^j = \sum_{j=0}^{\infty} a_j z^j \quad \text{on } \bar{D}_s.$$

11   Vector Spaces and Inner Products

11.1   CARTESIAN COORDINATE SYSTEMS IN THREE-DIMENSIONAL SPACE

To set up a coordinate system in three-dimensional space, we use the same scheme that we used in a plane; the only difference is that we use three mutually perpendicular lines instead of two. These are the x-, y-, and z-axes. On each of the axes we take a coordinate system, in such a way that the origin O has coordinate 0. The plane containing the x- and y-axes is called the xy-plane. Similarly for the yz- and xz-planes.

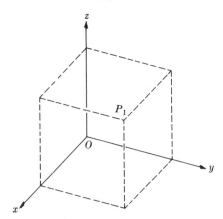

These are called the *coordinate planes*. In the figure we have indicated the position of the point P_1 by drawing the rectangular parallelepiped which has the origin O and the point P_1 as opposite corners, and sides parallel to the coordinate planes. We get the coordinates of a point P_1, as before, by dropping perpendiculars to the coordinate axes.

Here M_1, M_2, and M_3 are the feet of the perpendiculars from P_1 to the three axes. If these points have coordinates x_1, y_1, z_1, on the respective axes, then P_1 is matched with the triplet (x_1, y_1, z_1), and we write

$$P_1 \leftrightarrow (x_1, y_1, z_1).$$

As in the plane, we often fail to distinguish between a point and the number triplet with which it is matched; and so we may write

$$P_1 = (x_1, y_1, z_1).$$

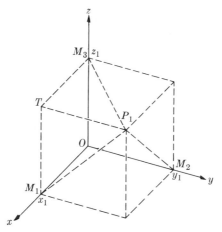

In figures, we may label a point as $P_1(x_1, y_1, z_1)$, to indicate that the given point has the given coordinates.

The coordinate planes divide space into eight parts, called *octants*. The figure above shows the *first octant*, consisting of all points of space for which all three coordinates are ≥ 0.

By two applications of the Pythagorean theorem, we see that each diagonal of a rectangular parallelepiped has length $\sqrt{a^2 + b^2 + c^2}$. This means that for each point $P_1(x_1, y_1, z_1)$ we have

$$OP_1^2 = x_1^2 + y_1^2 + z_1^2.$$

More generally, for any two points P_1, P_2, we have the distance formula given in the following theorem.

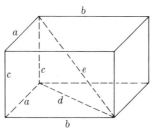

$$d^2 = a^2 + b^2; \; e^2 = d^2 + c^2 = a^2 + b^2 + c^2.$$

Theorem 1. If $P_1 \leftrightarrow (x_1, y_1, z_1)$ and $P_2 \leftrightarrow (x_2, y_2, z_2)$, then

$$P_1P_2 = \sqrt{(x_2 - x_1)^2 + (y_2 - y_1)^2 + (z_2 - z_1)^2}.$$

Proof. Suppose first that $x_1 \neq x_2, y_1 \neq y_2$, and $z_1 \neq z_2$. Then P_1 and P_2 are opposite corners of a rectangular parallelepiped. In the figure on the left below,

$$a = |x_2 - x_1|, \qquad b = |y_2 - y_1|, \qquad c = |z_2 - z_1|.$$

Therefore

$$P_1P_2^2 = a^2 + b^2 + c^2 = (x_2 - x_1)^2 + (y_2 - y_1)^2 + (z_2 - z_1)^2.$$

If some of the inequalities $x_1 \neq x_2$, $y_1 \neq y_2$, and $z_1 \neq z_2$ do not hold, then our parallelepiped reduces to a rectangle, a segment, or a point, and the same distance formula holds for simpler reasons.

We shall use this result to describe planes by equations. (See the figure on the right below.) Given a plane E, suppose first that E does not pass through the origin,

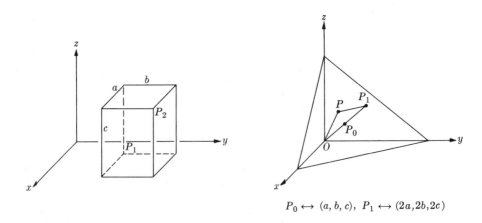

$$P_0 \longleftrightarrow (a, b, c), \quad P_1 \longleftrightarrow (2a, 2b, 2c)$$

and let $P_0 = (a, b, c)$ be the foot of the perpendicular from the origin to E. Let P_1 be the point $(2a, 2b, 2c)$. Then $OP_0 = P_0P_1$; P_0 is the midpoint of the segment from O to P_1; and E is the perpendicular bisecting plane of the segment. Therefore E is the set of all points of space that are equidistant from O and P_1. That is, E is the graph of the condition

$$OP = P_1P$$

$$\Leftrightarrow x^2 + y^2 + z^2 = (x - 2a)^2 + (y - 2b)^2 + (z - 2c)^2$$

$$\Leftrightarrow 0 = -4ax + 4a^2 - 4by + 4b^2 - 4cz + 4c^2$$

$$\Leftrightarrow ax + by + cz - (a^2 + b^2 + c^2) = 0.$$

This has the form

$$Ax + By + Cz + D = 0, \quad (A, B, C) \neq (0, 0, 0).$$

The condition on the right says that the numbers A, B, and C are not all equal to zero; and this is correct, because the point $P_0 = (a, b, c)$ is not the origin. An equation of the above type is called a *linear equation in x, y, and z*. Thus we have shown that every plane that does not pass through the origin is the graph of a linear equation in x, y, and z. For planes through the origin, the same result holds. Let $P_0(a, b, c)$ be any point of the line perpendicular to E through O, other than O itself. Let P_1 be the point $(-a, -b, -c)$. Then E is the perpendicular bisecting plane of the segment from

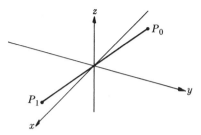

$$P_0\,(a, b, c), \quad P_1\,(-a,-b,-c)$$

P_1 to P_0; and so E is the graph of the equation

$$P_1P = P_0P$$

$$\Leftrightarrow (x + a)^2 + (y + b)^2 + (z + c)^2 = (x - a)^2 + (y - b)^2 + (z - c)^2$$

$$\Leftrightarrow 2ax + a^2 + 2by + b^2 + 2cz + c^2 = -2ax + a^2 - 2by + b^2 - 2cz + c^2$$

$$\Leftrightarrow ax + by + cz = 0.$$

This has the same form $Ax + By + Cz + D = 0$, with $D = 0$, as it must be: the origin lies in the plane E. In general:

Theorem 2. Every plane is the graph of a linear equation in x, y, and z.

This theorem can be applied in a variety of ways. For example, let E be the plane through the points $(3, 3, 4)$, $(2, 4, 4)$, and $(2, 3, 6)$. Now E has an equation of the form

$$Ax + By + Cz + D = 0,$$

and the equation must be satisfied by the coordinates of the three given points. Therefore the coefficients A, B, C, and D must satisfy the equations

$$3A + 3B + 4C + D = 0, \tag{1}$$

$$2A + 4B + 4C + D = 0, \tag{2}$$

$$2A + 3B + 6C + D = 0. \tag{3}$$

Subtracting (2) from (1) we get

$$A - B = 0. \tag{(1) $-$ (2)}$$

Setting $B = A$ in (2) and (3), we get

$$6A + 4C + D = 0, \tag{2'}$$

$$5A + 6C + D = 0. \tag{3'}$$

Subtracting (3') from (2') we get

$$A - 2C = 0, \tag{(2') $-$ (3')}$$

and so $C = A/2$. Therefore $D = -8A$. Now, to avoid fractions, we set $A = 2$. This gives

$$A = B = 2, \qquad C = 1, \qquad D = -16,$$
$$2x + 2y + z - 16 = 0.$$

This checks.

Note that any number different from 0 could have been used as A. There are some cases, however, when this is not so. For example, the graph of the equation

$$y + z = 1$$

is a plane, parallel to the x-axis. This plane is the graph of infinitely many different equations, of the form

$$ky + kz - k = 0 \qquad (k \neq 0);$$

but x does not appear (with nonzero coefficient) in any of these equations.

PROBLEM SET 11.1

1. Find the equation of the plane containing all points equidistant from the origin and the point $P_0 = (2, 6, 4)$.

2. Find the equation of the plane containing all points equidistant from $P_0 = (1, 0, 0)$ and $P_1 = (0, 2, 3)$.

3. Find the equation of the plane containing all points equidistant from the planes

$$x + y + z = 2 \quad \text{and} \quad x + y + z = -1.$$

4. Find the equation of the plane through the points $P_0 = (1, 0, 1)$, $P_1 = (1, 1, 1)$, and $P_2 = (-1, 2, 0)$.

5. Find the equation of the plane through the points $P_0 = (2, 1, 1)$, $P_1 = (-1, -1, 0)$, and $P_2 = (0, 0, 3)$.

6. Find the equation of the plane through

$$P_0 = (1, 3, 0) \qquad \text{and} \qquad P_1 = (2, 2, 3),$$

which is parallel to the z-axis.

7. The point $(1, -1, 2)$ is the foot of the perpendicular from the origin to a plane E. Find the equation for this plane.

8. Find a plane through $P_0 = (1, 0, 2)$, $P_1 = (2, 2, 1)$, and $P_2 = (3, 4, 0)$. Is there more than one such plane? Why?

9. Let $A = (1, 0, 0)$, let $B = (4, 0, 0)$, and let $K = \{P \mid 2AP = BP\}$. Find an equation whose graph is K. What sort of figure is this?

11.2 DIRECTION COSINES. THE DIRECTED NORMAL FORM

Linear equations for planes are more useful if we know the geometric significance of the coefficients in the equations. For this purpose, we need the idea of *directed distance* on a line. Given a line L with a coordinate system, and two points P and Q of L, with coordinates x_1 and x_2, we define the *directed distance* from P to Q as

$$\overline{PQ} = x_2 - x_1.$$

Since $x_1 - x_2 = -(x_2 - x_1)$, we have

$$\overline{QP} = -\overline{PQ} \qquad \text{for every} \quad P \text{ and } Q.$$

Since $(x_2 - x_1) + (x_3 - x_2) = x_3 - x_1$, we have

$$\overline{PQ} + \overline{QR} = \overline{PR} \qquad \text{for every} \quad P, Q, R.$$

And since the distance PQ is equal to $|x_2 - x_1|$, it follows that

$$\overline{PQ} = \begin{cases} PQ & \text{if } Q \text{ is in the positive} \\ & \text{direction from } P, \\ -PQ & \text{if } Q \text{ is in the negative} \\ & \text{direction from } P. \end{cases}$$

This means that directed distances are determined if the positive direction on the line L is known; we do not care where the origin is in the coordinate system.

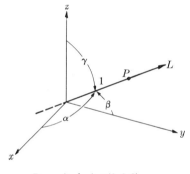

$$P \longleftrightarrow (a, b, c) \ne (0, 0, 0).$$

Consider now a directed line L, through the origin. Let $P = (a, b, c)$ be a point on the positive end of L, with $OP = 1$. Consider the angle between the positive end of the x-axis and the positive end of L. In its own plane, this angle looks like this:

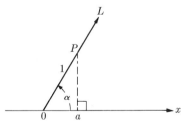

Note that the foot of the perpendicular really does have coordinate a on the x-axis; in fact, this is the definition of the x-coordinate of P. If the angle has measure α, then

$$\cos \alpha = a.$$

In the preceding three-dimensional figure, β and γ are defined similarly. They are called the *direction angles* of the directed line L. (This is an abbreviation: what we really mean is that they are the *measures* of the angles between the positive end of L and the positive ends of the axes.) The numbers

$$\cos \alpha, \qquad \cos \beta, \qquad \cos \gamma$$

are called the *direction cosines* of L. Note that they determine not merely a line through the origin but also a positive direction on the line. If the direction on the line is reversed, then

$$\alpha \to \pi - \alpha, \qquad \beta \to \pi - \beta, \qquad \gamma \to \pi - \gamma,$$

and each of the direction cosines changes in sign.

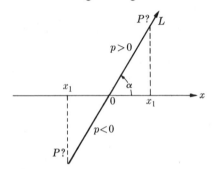

Suppose now that P is any point (x_1, y_1, z_1) of L, other than the origin. Let p be the directed distance from O to P, relative to the given positive direction on L. If $p > 0$, as above, then

$$x_1/p = \cos \alpha,$$

by definition of the cosine. If $p < 0$, as on the facing page, then

$$\cos (\pi - \alpha) = x_1/-p,$$

by definition of the cosine. Therefore

$$- \cos \alpha = - x_1/p,$$

and $x_1/p = \cos \alpha$, as before. In the same way, we get

$$x_1 = p \cos \alpha, \qquad y_1 = p \cos \beta, \qquad z_1 = p \cos \gamma.$$

We now return to our equations for planes. First we choose a direction on the normal line N, in such a way that the directed distance $\overline{OP} = p$ is ≥ 0. Then the plane is the graph of the equation

$$ax + by + cz - (a^2 + b^2 + c^2) = 0.$$

Now $p = \sqrt{a^2 + b^2 + c^2}$. The equation therefore has the form

$$\frac{a}{p} x + \frac{b}{p} y + \frac{c}{p} z - p = 0,$$

or

$$x \cos \alpha + y \cos \beta + z \cos \gamma - p = 0.$$

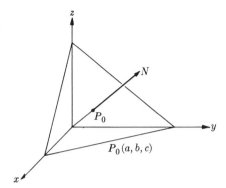

If the plane passes through the origin, then $p = 0$, but the same result holds: the equation $ax + by + cz = 0$ takes the form $x \cos \alpha + y \cos \beta + z \cos \gamma - 0 = 0$.

Here we have chosen the direction so as to make $p \geqq 0$. If we choose the opposite direction on the normal, then

$$\alpha \to \pi - \alpha, \qquad \beta \to \pi - \beta, \qquad \gamma \to \pi - \gamma,$$

$$\cos \alpha \to -\cos \alpha, \qquad \cos \beta \to -\cos \beta, \qquad \cos \gamma \to -\cos \gamma,$$

and $p \to -p$. This changes all the signs in the equation

$$x \cos \alpha + y \cos \beta + z \cos \gamma - p = 0.$$

Therefore the equation still holds; and we have the following:

Theorem 3. Let E be a plane, and let N be a directed normal to E through the origin, with direction angles α, β, and γ. Then E is the graph of the equation

$$x \cos \alpha + y \cos \beta + z \cos \gamma - p = 0,$$

where p is the directed distance from the origin to E, relative to the given direction on N.

So far, we seem to have been talking about numbers which are hard to compute. But it is easy to bring the discussion down to earth. Consider the following equation:

$$x + 2y + 3z + 4 = 0.$$

It is not reasonable to expect that an equation taken at random will be in the directed normal form; after all, if a plane E is the graph of the equation

$$Ax + By + Cz + D = 0,$$

then E is also the graph of the equation

$$kAx + kBy + kCz + kD = 0,$$

for every $k \neq 0$. In fact, Eq. (1) cannot be in the directed normal form, because of the following theorem:

Theorem 4. If α, β, γ are the direction angles of a directed line L through the origin, then

$$\cos^2 \alpha + \cos^2 \beta + \cos^2 \gamma = 1.$$

Proof. Let P be the point of L for which the directed distance $\overline{OP}$ is 1. If $P = (a, b, c)$ then

$$a = \cos \alpha, \qquad b = \cos \beta, \qquad c = \cos \gamma.$$

Since

$$\overline{OP} = OP = 1 = \sqrt{a^2 + b^2 + c^2},$$

we have $a^2 + b^2 + c^2 = 1$; and the theorem follows. We also have a converse:

Theorem 5. If $a^2 + b^2 + c^2 = 1$, then there is a directed line whose direction cosines are a, b, and c.

Proof. Let L be the line from the origin through the point $P = (a, b, c)$, directed positively from O to P. This does it.

This suggests that the equation

$$x + 2y + 3z + 4 = 0 \tag{1}$$

has the form

$$xk \cos \alpha + yk \cos \beta + zk \cos \gamma - pk = 0,$$

for some α, β, γ, p, and k. If so, we must have

$$k \cos \alpha = 1, \qquad k \cos \beta = 2, \qquad k \cos \gamma = 3,$$

$$k^2(\cos^2 \alpha + \cos^2 \beta + \cos^2 \gamma) = 1^2 + 2^2 + 3^2 = 14,$$

and

$$k = \pm\sqrt{14}.$$

Taking $k = \sqrt{14}$, we get

$$\frac{1}{\sqrt{14}} x + \frac{2}{\sqrt{14}} y + \frac{3}{\sqrt{14}} z + \frac{4}{\sqrt{14}} = 0. \tag{2}$$

Here

$$\cos \alpha = 1/\sqrt{14}, \qquad \cos \beta = 2/\sqrt{14},$$

$$\cos \gamma = 3/\sqrt{14}, \qquad p = -4/\sqrt{14}.$$

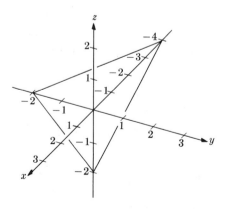

We have sketched the graph by plotting the intercepts, on the axes, and then completed a triangle just as we did in the case where the intercepts were in the first octant. The graph of (2) is the plane whose normal has the given direction cosines $1/\sqrt{14}$, $2/\sqrt{14}$, $3/\sqrt{14}$, and which lies at a directed distance $p = -4/\sqrt{14}$ from the origin.

Taking $k = -\sqrt{14}$, we get

$$-\frac{1}{\sqrt{14}}x - \frac{2}{\sqrt{14}}y - \frac{3}{\sqrt{14}}z - \frac{4}{\sqrt{14}} = 0, \tag{3}$$

which gives the opposite direction on the normal;

$$\cos\alpha' = -\frac{1}{\sqrt{14}} = -\cos\alpha, \qquad \cos\beta' = -\frac{2}{\sqrt{14}} = -\cos\beta,$$

$$\cos\gamma' = -\frac{3}{\sqrt{14}} = -\cos\gamma, \qquad p' = +\frac{4}{\sqrt{14}} = -p.$$

The same scheme works for any linear equation in x, y, z. This gives a converse of Theorem 2:

Theorem 6. The graph of every linear equation in x, y, and z is a plane.

Proof. Given

$$Ax + By + Cz + D = 0, \tag{1}$$

with A, B, and C not all $= 0$. Then $A^2 + B^2 + C^2 > 0$. Let

$$k = \sqrt{A^2 + B^2 + C^2},$$

and

$$a = \frac{A}{k}, \qquad b = \frac{B}{k}, \qquad c = \frac{C}{k}, \qquad p = -\frac{D}{k}.$$

Then Eq. (1) is equivalent to

$$ax + by + cz - p = 0. \tag{2}$$

The graph of Eq. (2) is a plane E: the direction cosines of a directed normal to E are a, b, and c; and the directed distance from the origin to E is p.

PROBLEM SET 11.2

1. Given $x + y + z - 1 = 0$. Write the two directed normal forms for the same plane, and sketch.

2. Do the same, for $x + 2z - 3 = 0$.

3. Do the same, for $x + 2y + 2z - 3 = 0$.

4. Do the same, for $x + 2y + 4z - 4 = 0$.

5. Do the same, for $x + 2y + 4z + 4 = 0$. (This is going to look awkward; the foot of the normal does not lie in the first octant.)

6. The normal to E from the origin contains the point $(2, 4, 6)$. The plane E contains the point $(1, 1, 1)$. Find the two directed normal forms of the equation of E. How far is E from the origin?

7. Let K be the set of all points which are equidistant from $A = (1, 0, 0)$, $B = (0, 1, 0)$, and $C = (0, 0, 1)$. That is,

$$K = \{P \mid AP = BP = CP\}.$$

Prove that K is a line through the origin, and find a set of direction cosines for K (more precisely, a set of direction cosines for a *direction* on K).

8. Let A, B, and C be three points which are all different, but collinear. Let

$$K = \{P \mid AP = BP = CP\}.$$

Show that K is the empty set $\{\ \}$.

9. Let A, B, and C be any points, not necessarily different. Let K be as in Problem 6. Show that K is (a) a line, (b) a plane, (c) all of space, or (d) the empty set $\{\ \}$. Give examples to show that all of these possibilities can actually arise.

10. A plane E contains the points $(a, 0, 0)$, $(0, b, 0)$, $(0, 0, c)$, $a, b, c \neq 0$. Find the two directed normal forms of the equation, and sketch, showing the case $a, b, c > 0$.

11. The normal to E from the origin lies in the xy-plane; and E contains the points $(1, 1, 1)$ and $(-1, 2, 1)$. Discuss as in Problem 10.

12. The normal to E lies in the yz-plane; and E contains the points $(2, 2, 1)$ and $(1, 1, 2)$. Discuss as in Problem 10.

13. Let E be the plane $z = -1$, and let K be the set of all points which are equidistant from E and the point $A = (1, 0, 0)$. What sort of figure is this? Sketch.

14. Let E be the plane $z = -2$ and let K be the set of all points which are equidistant from E and the x-axis. What sort of figure is this? Sketch.

11.3 THREE-DIMENSIONAL SPACE, REGARDED AS AN INNER-PRODUCT SPACE

Following the pattern of Section 9.7, we identify the point $P = (x, y, z)$ with the directed segment $\overrightarrow{OP}$ from the origin to P; we denote the resulting vector as $\overrightarrow{P}$; and for $P_1 = (x_1, y_1, z_1)$, $P_2 = (x_2, y_2, z_2)$, we define addition, scalar multiplication, and inner product by the formulas

$$\overrightarrow{P_1} + \overrightarrow{P_2} = (x_1 + x_2, y_1 + y_2, z_1 + z_2),$$

$$\alpha \overrightarrow{P_1} = (\alpha x_1, \alpha y_1, \alpha z_1),$$

$$\overrightarrow{P_1} \cdot \overrightarrow{P_2} = x_1 x_2 + y_1 y_2 + z_1 z_2.$$

To simplify the notation, however, we drop the arrows, and write P for $\overrightarrow{P}$. Thus we have an inner-product space

$$V = \{P\} = \{(x, y, z)\},$$

with addition, scalar multiplication, and inner product defined by the formulas

$$P_1 + P_2 = (x_1 + x_2, y_1 + y_2, z_1 + z_2),$$
$$\alpha P_1 = (\alpha x_1, \alpha y_1, \alpha z_1),$$
$$P_1 \cdot P_2 = x_1 x_2 + y_1 y_2 + z_1 z_2.$$

The resulting system satisfies all the vector and inner-product laws of Section 9.7. In the new notation (without the arrows) these are as follows.

A.1. $(P_1 + P_2) + P_3 = P_1 + (P_2 + P_3)$.

A.2. $\mathcal{V}$ contains a vector O such that

$$O + P = P + O = P$$

for every P.

A.3. For every P in $\mathcal{V}$ there is a vector $-P$ in $\mathcal{V}$ such that

$$P + (-P) = (-P) + P = 0.$$

A.4. $P_1 + P_2 = P_2 + P_1$.

M.1. $(\alpha\beta)P = \alpha(\beta P)$.

M.2. $(\alpha + \beta)P = \alpha P + \beta P$.

M.3. $\alpha(P_1 + P_2) = \alpha P_1 + \alpha P_2$.

M.4. $0 \cdot P = O$, for every P.

M.5. $1 \cdot P = P$, for every P.

M.6. $\alpha O = O$, for every α.

In M.4 and M.5, 0 is the real number zero, and O is the zero vector. Thus M.4 says that the scalar product of the number 0 and any vector P is the zero vector; and M.6 says that the scalar product of any number α and the zero vector is the zero vector.

It is understood that all sums and scalar products are also vectors, that is, elements of $\mathcal{V}$. But we had better make this explicit:

CA *(Closure under addition)*. For every V_1, V_2 in $\mathcal{V}$, $V_1 + V_2$ belongs to $\mathcal{V}$.

CSM *(Closure under scalar multiplication)*. For every V in $\mathcal{V}$ and every real number α, αV belongs to $\mathcal{V}$.

As in Section 9.7, any set $\mathcal{V}$, with operations satisfying the above laws, is called a *vector space*. The space that we are dealing with at the moment, in which the vectors are the triplets $P = (x, y, z)$ of real numbers, is denoted by $\mathbf{R}^3$, and is called *Cartesian three-space*. Thus

$$\mathbf{R}^3 = \{(x, y, z) \mid x, y, z \text{ in } \mathbf{R}\}.$$

The inner product that we have defined in $\mathbf{R}^3$ has the following properties:

S.1. $P_1 \cdot P_2 = P_2 \cdot P_1$.

S.2. $(\alpha P_1) \cdot P_2 = \alpha(P_1 \cdot P_2)$.

S.3. $P_1 \cdot (P_2 + P_3) = P_1 \cdot P_2 + P_1 \cdot P_3$.

S.4. $P \cdot P \geqq 0$, for every P.

S.5. If $P \cdot P = 0$, then $P = O$.

As in Section 9.7, we define an *inner-product space* to be a vector space in which an inner product is defined, satisfying S.1 through S.5.

The distance from $O = (0, 0, 0)$ to $P = (x, y, z)$ is

$$\sqrt{x^2 + y^2 + z^2} = \sqrt{P \cdot P}.$$

Hereafter this number will be called the *norm* of P, and will be denoted by $\|P\|$. (The double bars are a reminder that we are performing an operation on a vector rather than a number.) For inner-product spaces in general, the formula $\sqrt{x^2 + y^2 + z^2}$ may not apply, but the expression $\sqrt{P \cdot P}$ always has a meaning, and so we use it as our definition of the norm.

Definition. In any inner-product space,

$$\|P\| = \sqrt{P \cdot P}.$$

$\|P\|$ is called the *norm* of the vector P.

Just as in the plane, the inner product has a geometric interpretation in $\mathbf{R}^3$.

Theorem 1. In $\mathbf{R}^3$,

$$P_1 \cdot P_2 = \|P_1\| \cdot \|P_2\| \cos \theta,$$

where θ is the measure of the angle between the directed segments $\overrightarrow{OP_1}$ and $\overrightarrow{OP_2}$.

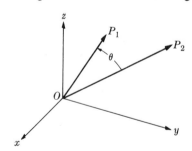

The proof is by definition of the norm, together with the law of cosines:

$$(P_1 P_2)^2 = OP_1^2 + OP_2^2 - 2 \cdot OP_1 \cdot OP_2 \cos \theta;$$

$$(x_1 - x_2)^2 + (y_1 - y_2)^2 + (z_1 - z_2)^2$$
$$= x_1^2 + y_1^2 + z_1^2 + x_2^2 + y_2^2 + z_2^2 - 2\|P_1\| \cdot \|P_2\| \cos \theta;$$

$$-2(x_1 x_2 + y_1 y_2 + z_1 z_2) = -2\|P_1\| \cdot \|P_2\| \cos \theta;$$

$$P_1 \cdot P_2 = \|P_1\| \cdot \|P_2\| \cos \theta.$$

From this we get immediately:

Theorem 2 (*The Schwarz inequality*). In $\mathbf{R}^3$,

$$(P_1 \cdot P_2)^2 \leqq \|P_1\|^2 \cdot \|P_2\|^2.$$

This is true because $\cos^2 \theta \leqq 1$.

Following the pattern of Section 9.7, we let

$$i = (1, 0, 0), \qquad j = (0, 1, 0), \qquad k = (0, 0, 1),$$

so that for each $P = (x, y, z)$ we have

$$P = xi + yj + zk.$$

In general, if $V = \alpha_1 V_1 + \alpha_2 V_2 + \cdots + \alpha_n V_n$, for some scalars $\alpha_1, \alpha_2, \ldots, \alpha_n$, then V is a *linear combination* of the vectors $V_1, V_2, \ldots, V_n$. Thus every vector in $\mathbf{R}^3$ is a linear combination of i, j, and k.

Definition. A set $\{V_1, V_2, \ldots, V_n\}$ of vectors *spans* the vector $\mathscr{V}$ if every V in $\mathscr{V}$ is a linear combination of the V_i's. (Thus $\{i, j, k\}$ spans $\mathbf{R}^3$.)

A set $\{V_1, V_2, \ldots, V_n\}$ is *linearly dependent* if there are scalars $\alpha_1, \alpha_2, \ldots, a_n$, not all equal to zero, such that

$$\alpha_1 V_1 + \alpha_2 V_2 + \cdots + \alpha_n V_n = O.$$

Thus, in $\mathbf{R}^3$, every set of vectors of the form $\{P, i, j, k\}$ is linearly dependent, because for $P = (x, y, z)$, we have $P = xi + yj + zk$, and so

$$P - xi - yj - zk = O.$$

Here $\alpha_1 = 1$, $\alpha_2 = -x$, $\alpha_3 = -y$, and $\alpha_4 = -z$; and the numbers α_i are not all $= 0$, because $\alpha_1 = 1$.

A set of vectors is *linearly independent* if it is not linearly dependent. Thus $\{V_1, V_2, \ldots, V_n\}$ is linearly independent if

$$\alpha_1 V_1 + \alpha_2 V_2 + \cdots + \alpha_n V_n = 0 \Rightarrow \alpha_1 = \alpha_2 = \cdots = \alpha_n = 0.$$

For example, $\{i, j, k\}$ is linearly independent. The reason is that in $\mathbf{R}^3$,

$$\alpha_1 i + \alpha_2 j + \alpha_3 k = (\alpha_1, \alpha_2, \alpha_3) \qquad \text{and} \qquad O = (0, 0, 0).$$

Therefore

$$\alpha_1 i + \alpha_2 j + \alpha_3 k = O \Rightarrow (\alpha_1, \alpha_2, \alpha_3) = (0, 0, 0)$$

$$\Rightarrow \alpha_1 = \alpha_2 = \alpha_3 = 0.$$

A set of vectors $\{V_1, V_2, \ldots, V_n\}$ forms a *basis* for a vector space $\mathscr{V}$ if (1) the set spans $\mathscr{V}$ and (2) the set is linearly independent. Thus we have:

Theorem 3. $\{i, j, k\}$ is a basis for $\mathbf{R}^3$.

Obviously the points of the xy-plane form a vector space in themselves; in fact, this is the vector space that was discussed in Section 9.7. In fact, all three of the coordinate planes

$$E_{xy} = \{xi + yj + 0 \cdot k\}, \qquad E_{xz} = \{xi + 0 \cdot j + zk\}, \qquad E_{yz} = \{0 \cdot i + yj + zk\},$$

form vector spaces. Such sets are called *subspaces* of $\mathbf{R}^3$. More generally:

Definition. Given a vector space $\mathcal{V}$ and a subset $\mathcal{V}'$. If $\mathcal{V}'$ also forms a vector space (under the same definitions of addition and scalar multiplication) then $\mathcal{V}'$ is called a *subspace* of $\mathcal{V}$.

Thus a subspace must satisfy all of the vector laws. But this is not as tedious to check as one might think, because of the following theorem.

Theorem 4. Let $\mathcal{V}'$ be a subset of the vector space $\mathcal{V}$. If $\mathcal{V}'$ is closed under addition and scalar multiplication, then $\mathcal{V}'$ is a subspace of $\mathcal{V}$.

Proof. Many of the laws can be checked all at once. Since A.1 and A.4 hold for all vectors in $\mathcal{V}$, they automatically hold in $\mathcal{V}'$. The same is true for M.1 through M.6 and S.1 through S.5. Therefore the only things remaining to verify are A.2 and A.3.

1) By M.4, $0 \cdot P = O$, for every P in $\mathcal{V}'$; and by CA, $0 \cdot P$ belongs to $\mathcal{V}'$. Therefore O belongs to $\mathcal{V}'$, and so A.2 holds.

2) Given P in $\mathcal{V}'$. By M.2,

$$[1 + (-1)]P = 1 \cdot P + (-1)P.$$

By M.4 and M.5, this gives

$$O = P + (-1)P.$$

Therefore $(-1)P = -P$, and $-P$ belongs to $\mathcal{V}'$.

On the basis of Theorem 4, it is easy to see that each of the three coordinate planes is a subspace of $\mathbf{R}^3$. The point is that if

$$P_1 = x_1\mathbf{i} + y_1\mathbf{j}, \qquad P_2 = x_2\mathbf{i} + y_2\mathbf{j},$$

then

$$P_1 + P_2 = (x_1 + x_2)\mathbf{i} + (y_1 + y_2)\mathbf{j},$$

so that the set E_{xy} of all linear combinations of $\mathbf{i}$ and $\mathbf{j}$ is closed under addition. Similarly, $\alpha P_1 = \alpha x_1\mathbf{i} + \alpha y_1\mathbf{j}$, and so E_{xy} is closed under scalar multiplication. Similarly for E_{yz} and E_{xz}. In fact, a more general result holds:

Theorem 5. Every plane through O forms a subspace of $\mathbf{R}^3$.

Proof. Let E be such a plane. Then E is the graph of an equation of the form

$$Ax + By + Cz = 0.$$

If $P_1 = (x_1, y_1, z_1)$ and $P_2 = (x_2, y_2, z_2)$ belong to E, then

$$Ax_1 + By_1 + Cz_1 = 0 \qquad \text{and} \qquad Ax_2 + By_2 + Cz_2 = 0.$$

By addition,

$$A(x_1 + x_2) + B(y_1 + y_2) + C(z_1 + z_2) = 0;$$

and this means that $P_1 + P_2$ is in E. Similarly, for every real number α,

$$A\alpha x_1 + B\alpha y_1 + C\alpha z_1 = \alpha \cdot 0 = 0,$$

and αP_1 is in E. Therefore E is closed under scalar multiplication. By Theorem 4, E forms a subspace.

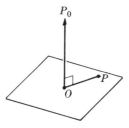

There is, however, a much better way to get this result, using vector-space methods instead of using the results of the preceding section. Given the plane E through O, let

$$P_0 = (A, B, C)$$

be any vector such that the line through O and P_0 is perpendicular to E. Then for each $P \neq O$ in E, $\overrightarrow{OP}$ and $\overrightarrow{OP_0}$ are perpendicular, and so

$$P_0 \cdot P = \|P_0\| \cdot \|P\| \cos (\pi/2) = 0.$$

For $P = O$ the same equation holds. Conversely, if $P_0 \cdot P = 0$, then P lies in E. Therefore

$$E = \{P \mid P_0 \cdot P = 0\}.$$

In terms of coordinates, this tells us that E is the graph of the equation

$$Ax + By + Cz = 0,$$

which we already knew. But when we describe E by a vector equation, using the inner product, this suggests the following theorem:

Theorem 6. Let $\mathscr{V}$ be any inner-product space; let V_0 be any vector in $\mathscr{V}$, different from O; and let

$$\mathscr{V}' = \{V \mid V_0 \cdot V = 0\}.$$

Then $\mathscr{V}'$ is a subspace of $\mathscr{V}$.

Proof. We need to show that $\mathscr{V}'$ satisfies CA and CSM. If V_1 and V_2 are in $\mathscr{V}'$, then

$$V_0 \cdot V_1 = 0 = V_0 \cdot V_2.$$

Therefore

$$V_0 \cdot (V_1 + V_2) = V_0 \cdot V_1 + V_0 \cdot V_2,$$

by S.3. Therefore $V_0 \cdot (V_1 + V_2) = 0$, and $V_1 + V_2$ is in $\mathscr{V}'$. Similarly,

$$V_0 \cdot \alpha V_1 = \alpha V_1 \cdot V_0 = \alpha(V_1 \cdot V_0) = \alpha(V_0 \cdot V_1) = \alpha \cdot 0 = 0.$$

(Reasons for these steps?)

If you rewrite these formulas, in the forms that they take when V_0, V_1, and V_2 are vectors in $\mathbf{R}^3$, with

$$V_0 = (A, B, C), \qquad V_1 = (x_1, y_1, z_1), \qquad V_2 = (x_2, y_2, z_2),$$

you will find that you are simply copying the proof of Theorem 5. (This is worth going through, to see how it works.) Thus it may seem that nothing is new in Theorem 6 except the notation. But this is not true, because Theorem 6 and its proof work in *every* vector space, including spaces of four dimensions, spaces of functions, and so on. Thus when we proved Theorem 6, we found that the method used in proving Theorem 5 had nothing to do with any special properties of $\mathbf{R}^3$; it depended only on the inner-product space laws. From now on, easy generalizations of this kind will occur often. We shall treat the vector laws (or the inner-product space laws) as basic assumptions, like postulates in geometry, and any theorems that we derive from them will be known to hold in every vector space (or any inner-product space.)

If a plane E does not contain the origin, then it never forms a subspace of $\mathbf{R}^3$, because it does not contain O. But we can still write a vector equation for E, because

$$Ax + By + Cz + D = 0 \Leftrightarrow P_0 \cdot P = -D,$$

where $P_0 = (A, B, C)$. Thus we have:

Theorem 7. Every plane in $\mathbf{R}^3$ is the graph of an equation of the form

$$P_0 \cdot P = a.$$

PROBLEM SET 11.3

Each of the following is the equation of a plane. Convert each of them to the form $P_0 \cdot P = a$, giving the value of a that you are using and the coordinates A, B, C of P_0.

1. $z = x + y$
2. $z = x - y$
3. $z = -x - y$
4. $x = 3y - 4z$
5. $y = 4z - 3x$
6. $z = 4x + 3y$
7. $z = 1$
8. $x = -4$
9. $\dfrac{x}{1} + \dfrac{y}{2} + \dfrac{z}{3} = 4$
10. $\dfrac{x}{2} - \dfrac{y}{4} + \dfrac{z}{3} = 2$

11. Let $V_1 = i + j$, $V_2 = j + k$, $V_3 = k$. Show that each of the basis vectors i, j, k is a linear combination of V_1, V_2, V_3.

12. Now show that $\{V_1, V_2, V_3\}$ spans $\mathbf{R}^3$.

13. Now show that $\{V_1, V_2, V_3\}$ is linearly independent. (By definition, this means that $\alpha_1 V_1 + \alpha_2 V_2 + \alpha_3 V_3 = 0 \Rightarrow \alpha_1 = \alpha_2 = \alpha_3 = 0$. Problems 12 and 13, in combination, tell us that $\{V_1, V_2, V_3\}$ is a basis for $\mathbf{R}^3$.)

14. Let $V_1 = i + j$, $V_2 = j + k$, $V_3 = i + k$. Proceed as in Problems 11 through 13.

15. Let $V_1 = i - j + k$, $V_2 = i + j - k$, $V_3 = -i + j + k$. Proceed as in Problems 11 through 13.

Show that the following hold, in any inner-product space. (Each of them should be derived from the inner-product space laws, with a reason given for each step.)

16. $V_1 \cdot (\alpha V_2) = \alpha(V_1 \cdot V_2)$

17. $(\alpha V_1) \cdot (\beta V_2) = \alpha\beta(V_1 \cdot V_2)$

18. $\alpha(V_1 + V_2 + V_3) = \alpha V_1 + \alpha V_2 + \alpha V_3$

19. $\|\alpha V\| = |\alpha| \cdot \|V\|$

20. $\|-\alpha V\| = |\alpha| \cdot \|V\|$

21. $(P + Q) \cdot (R + S) = P \cdot R + Q \cdot R + P \cdot S + Q \cdot S$ (Here P, Q, R, S are vectors, of course.)

22. $(P + Q)(P - R) = P \cdot P + Q \cdot P - R \cdot P - Q \cdot R$

23. $(P + Q) \cdot (P - Q) = P \cdot P - Q \cdot Q$

24. $(P + Q) \cdot (P + Q) = P \cdot P + 2(P \cdot Q) + Q \cdot Q$

25. $(P - Q) \cdot (P - Q) = P \cdot P - 2(P \cdot Q) + Q \cdot Q$

26. $(P - Q) \cdot (Q - P) = -P \cdot P + 2(P \cdot Q) - Q \cdot Q$

27. Is it true that in $\mathbf{R}^3$, $(P \cdot Q)R = (Q \cdot R)P$? (Each side of this equation has a meaning, because $P \cdot Q$ and $Q \cdot R$ are scalars.)

28. Is the following true in any inner-product space? *Theorem* (?) If $P \cdot Q = 0$ for every Q, then $P = O$. Why or why not?

29. Let the plane E be the graph of the equation $z = 2x + 3y$. Show that E contains the vectors $V_1 = i + 2k$ and $V_2 = j + 3k$.

30. Show that E contains every vector V of the form $xV_1 + yV_2$. Then show, conversely, that every vector of this form lies in E. [*Hint:* Express V in the form ( )i + ( )j + ( )k.]

31. Show that the V_1 and V_2 of Problem 29 are linearly independent.

32. Show that V_1 and V_2 span E. (Problems 30 through 32 tell us that $\{V_1, V_2\}$ is a basis for E.)

33. Let the plane E be the graph of the equation $x + y + 2z = 0$, and let $V_1 = i - j$ and $V_2 = 2j - k$. Proceed as in Problems 29 through 32.

34. Let $V_1 = i + j$, $V_2 = j + k$; and let E be the set of all vectors of the form $V = xV_1 + yV_2$. Write an equation for E, in the form $Ax + By + Cz = 0$.

35. Let $V_1 = i + 2j$, $V_2 = j + 2k$. Proceed as in Problem 34.

36. Two vectors V, V' are *orthogonal* if $V \cdot V' = 0$. A set $\{V_1, V_2, \ldots V_n\}$ of vectors is *orthogonal* if every two (different) vectors in the set are orthogonal. Verify that $\{i, j, k\}$ is an orthogonal set.

37. A set $\{V_1, V_2, \ldots, V_n\}$ is *orthonormal* if (1) the set is orthogonal, and (2) $\|V_i\| = 1$ for each i. (Thus the basis $\{i, j, k\}$ for $\mathbf{R}^3$ is orthonormal.) Given that $\{V_1, V_2, V_3\}$ is orthonormal, express the inner product

$$(a_1 V_1 + a_2 V_2 + a_3 V_3) \cdot (b_1 V_1 + b_2 V_2 + b_3 V_3)$$

in the simplest possible form.

*38. Let $\mathbf{R}^4$ be the set of all quadruplets $P = (w, x, y, z)$ of real numbers. The set $\mathbf{R}^4$ forms an inner-product space, under the obvious definitions of sum, scalar product, and inner product. As always, $\|P\| = \sqrt{P \cdot P}$. Show that $(P_1 \cdot P_2)^2 \leq \|P_1\|^2 \cdot \|P_2\|^2$.

*39. Let **P** be the set of all polynomials with real coefficients. For

$$V = \sum_{i=0}^{n} a_i x^i, \qquad W = \sum_{i=0}^{n} b_i x^i,$$

we define

$$V + W = \sum_{i=0}^{n} (a_i + b_i) x^i,$$

$$\alpha V = \sum_{i=0}^{n} \alpha a_i x^i,$$

$$V \cdot W = \sum_{i=0}^{n} a_i b_i.$$

Show that

$$(V \cdot W)^2 \leq \|V\|^2 \cdot \|W\|^2.$$

40. Does the vector space **P** of Problem 39 have a finite basis? If so, describe such a basis. If not, explain why no such basis exists.

11.4 THE DIMENSION OF A VECTOR SPACE. VARIOUS WAYS TO FORM A BASIS

For each positive integer n, let $\mathbf{R}^n$ be the set of all n-tuples of real numbers. Thus

$$\mathbf{R}^n = \{(x_1, x_2, \ldots, x_n) \mid x_i \in \mathbf{R}\},$$

and $\mathbf{R}^n$ forms an inner-product space, under the obvious definitions of sum, scalar product, and inner product. $\mathbf{R}^n$ is called *Cartesian n-space*. Let

$$\mathbf{B}^n = \{E_1, E_2, \ldots, E_n\},$$

where

$$E_1 = (1, 0, 0, \ldots, 0), \qquad E_2 = (0, 1, 0, \ldots, 0), \ldots,$$

$$E_{n-1} = (0, 0, \ldots, 1, 0), \qquad E_n = (0, 0, \ldots, 0, 1).$$

(In general, E_i has 1 in the ith position, and 0's everywhere else.) The vectors E_i span the space $\mathbf{R}^n$, with

$$(x_1, x_2, \ldots, x_n) = \sum_{i=1}^{n} x_i E_i.$$

They are linearly independent, since

$$\sum_{i=1}^{n} x_i E_i = O \Rightarrow (x_1, x_2, \ldots, x_n) = (0, 0, \ldots, 0)$$

$$\Rightarrow x_1 = x_2 = \cdots = x_n = 0.$$

Therefore $\mathbf{B}^n$ forms a basis.

Every subset of $\mathbf{B}^n$ gives a subspace of $\mathbf{R}^n$. For example,

$$\mathscr{V}_1 = \{\alpha E_1 \mid \alpha \in \mathbf{R}\}$$

is a subspace, and forms a line through the origin $(0, 0, \ldots, 0)$;

$$\mathscr{V}_2 = \{\alpha E_1 + \beta E_2 \mid \alpha, \beta \in \mathbf{R}\}$$

is a subspace, and forms a plane, and so on, for any subset of $\mathbf{B}^n$. But $\mathbf{R}^n$ has many subspaces which are not obtainable in this way. For example, we have found that in $\mathbf{R}^3$, any line or plane through the origin forms a subspace. To investigate these other subspaces, we need to use bases other than the obvious basis $\mathbf{B}^n$. Our investigation of other bases will also be useful in other connections.

The key to the theory of bases is the following theorem.

Theorem 1. Let $\mathscr{V}$ be a vector space. Let A be a set of m vectors, and let B be a set of n vectors, such that (1) A is linearly independent, and (2) B spans $\mathscr{V}$. Then $\mathscr{V}$ is spanned by a set C consisting of (a) all the elements of A, and (b) exactly $n - m$ of the elements of B.

For example, we might have

$$\mathscr{V} = \mathbf{R}^3,$$
$$A = \{E_1 + E_2, E_2 + E_3\},$$
$$B = \mathbf{B}^3 = \{E_1, E_2, E_3\}.$$

Here we can take

$$C = \{E_1 + E_2, E_2 + E_3, E_3\}.$$

This has the desired form, using all elements of A and $3 - 2 = 1$ element of B. To see that C spans $\mathbf{R}^3$, we observe that

$$E_1 = (E_1 + E_2) - (E_2 + E_3) + E_3, \qquad E_2 = (E_2 + E_3) - E_3, \qquad E_3 = E_3.$$

Therefore every basis element E_1, E_2, E_3 is a linear combination of the elements of C. Therefore every vector in $\mathbf{R}^3$ is a linear combination of the elements of C, and C spans $\mathbf{R}^3$.

We proceed to the general proof. First we list the elements of A in such a way that the vectors which also belong to B come first. Thus

$$A = \{A_1 A_2, \ldots, A_i, A_{i+1}, \ldots, A_m\},$$

where $A_1, A_2, \ldots, A_i$ belong also to B, but $A_{i+1}, \ldots, A_m$ do not. Then B can be described in the form

$$B = \{A_1, A_2, \ldots, A_i, B_1, B_2, \ldots, B_{n-i}\}.$$

One of the possibilities, of course, is that $i = 0$, as in the above example. Another possibility is that $i = m$; in this case there is nothing to prove. We shall show that given a set B, as above, with $0 \leqq i < m$, we can always delete one of the vectors B_j, and replace it by A_{i+1}, getting a new set which also spans $\mathscr{V}$. In a finite number of such steps, we get a set C of the sort that we want.

Since B spans V, it follows that A_{i+1} is a linear combination of the form

$$A_{i+1} = \alpha_1 A_1 + \alpha_2 A_2 + \cdots + \alpha_i A_i + \beta_1 B_1 + \beta_2 B_2 + \cdots + \beta_{n-i} B_{n-i}.$$

Here it cannot be true that all the numbers β_j are equal to zero, because if so it would follow that A is linearly dependent. It is a matter of notation, therefore, to suppose that $\beta_1 \neq 0$. We can therefore solve for B_1, in the above equation, getting

$$B_1 = \frac{1}{\beta_1}(A_{i+1} - \alpha_1 A_1 - \alpha_2 A_2 - \cdots - \alpha_i A_i - \beta_2 B_2 - \cdots - \beta_{n-i} B_{n-i}).$$

Now let

$$B' = \{A_1, A_2, \ldots, A_i, A_{i+1}, B_2, \ldots, B_{n-i}\}.$$

Every element of B (including B_1) is a linear combination of the elements of B'; and B spans $\mathscr{V}$. Therefore B' spans $\mathscr{V}$. In $m - i$ steps of this type, we get the desired set C.

Let us now check to see how this general scheme of proof applies to the above example. We had

$$A = \{A_1, A_2\} = \{E_1 + E_2, E_2 + E_3\},$$

$$B = \{B_1, B_2, B_3\} = \{E_1, E_2, E_3\}.$$

Here $m = 2$, $n = 3$, and at the outset, $i = 0$. Also A_1 is a linear combination of the elements of B, with

$$A_1 = E_1 + E_2.$$

This equation can be solved for E_1, giving E_1 as a linear combination of A_1 and E_2. Therefore we can replace E_1 by A_1 in B, getting

$$B' = \{A_1, B_2, B_3\} = \{E_1 + E_2, E_2, E_3\}.$$

This completes step 1. Next we express A_2 as a linear combination

$$A_2 = E_2 + E_3.$$

This equation can be solved for E_2, giving E_2 as a linear combination of A_2 and E_3. Therefore we can replace E_2 by A_2 in B', getting

$$B'' = C = \{E_1 + E_2, E_2 + E_3, E_3\};$$

and now we are done.

We shall now pursue some of the consequences of Theorem 1.

Theorem 2. Let $\mathscr{V}$ be a vector space. Let A be a set of m vectors, and let B be a set of n vectors, such that (1) A is linearly independent, and (2) B spans $\mathscr{V}$. Then $m \leq n$.

This follows from Theorem 1, because C has n elements, and contains all of A.

Theorem 3. If a vector space $\mathscr{V}$ has a basis with n elements, then every basis for $\mathscr{V}$ has exactly n elements.

Proof. Let B be a basis with n elements, and let A be any other basis, with m elements. Then A is linearly independent, and B spans $\mathscr{V}$. By Theorem 2, $m \leq n$. But we also know that B is linearly independent, and A spans $\mathscr{V}$. By Theorem 2, $n \leq m$. Therefore $n = m$, which was to be proved.

Thus the number of elements in a basis is independent of the choice of basis. This justifies the following definitions.

Definitions. A vector space $\mathscr{V}$ is *finite-dimensional* if it has a finite basis. If $\mathscr{V}$ is finite-dimensional, then the *dimension* of $\mathscr{V}$ is the number n which is the number of elements in every basis. The dimension of $\mathscr{V}$ is denoted by dim $\mathscr{V}$.

If you review the conditions for a vector space, you will see that they are all satisfied in the trivial case where $\mathscr{V}$ contains a zero vector O and nothing else. In this case we define dim $\mathscr{V} = 0$. That is,

$$\dim \{O\} = 0.$$

(Here the empty set is being regarded as a "basis" for $\{O\}$.) In a way it is a nuisance to allow this case, but to rule it out would lead to worse nuisances in the long run.

Theorem 4. In an n-dimensional vector space, every set of more than n vectors is linearly dependent.

Proof. Let B be a basis, with n elements, and let A be any set of vectors, with m elements, with $m > n$. If A were linearly independent, this would contradict Theorem 2. Therefore A is linearly dependent, which was to be proved.

Theorem 5. In an n-dimensional vector space, every linearly independent set with n elements is a basis.

Proof. Given a linearly independent set $B = \{V_1, V_2, \ldots, V_n\}$. If B spans $\mathscr{V}$, then B is a basis. If not, there is a vector V_{n+1} which is not a linear combination of elements of B. It follows that the larger set

$$B' = \{V_1, V_2, \ldots, V_n, V_{n+1}\}$$

is linearly independent; and this contradicts Theorem 4.

Theorem 6. Let $\mathscr{V}$ be a vector space, and let $B = \{V_1, V_2, \ldots, V_m\}$ be a set which spans $\mathscr{V}$. Then B contains a basis for $\mathscr{V}$.

Proof. If B is linearly independent, there is nothing to prove. If not, some V_i is a linear combination of the others. Suppose that this is V_1. Then $\{V_2, \ldots, V_m\}$ also spans $\mathscr{V}$. Repeating this process, removing superfluous vectors one at a time, we get a basis.

Theorem 7. If dim $\mathscr{V} = n$, then no set of fewer than n elements spans $\mathscr{V}$.

Proof. Any set which spans $\mathscr{V}$ contains a basis, and every basis has n elements.

Theorem 8. Let $\mathscr{V}$ be an n-dimensional vector space, and let $\mathscr{V}'$ be a subspace of $\mathscr{V}$. Then $\mathscr{V}'$ is finite-dimensional, and dim $\mathscr{V}' \leqq n$.

Proof. Let m be the largest number for which it is true that $\mathscr{V}'$ contains a linearly independent set of m vectors. (By Theorem 4, there is such a largest number m, and $m \leqq n$.) Let

$$B = \{V_1, V_2, \ldots, V_m\}$$

be a linearly independent set in $\mathscr{V}'$. We assert that B spans $\mathscr{V}'$. (*Proof:* If not, there is a vector V_{m+1} in $\mathscr{V}'$ which is not a linear combination of elements of B, and it follows, as in the proof of Theorem 5, that the larger set

$$B' = \{V_1, V_2, \ldots, V_m, V_{m+1}\}$$

is linearly independent.) Therefore B is a basis for $\mathscr{V}'$, and dim $\mathscr{V}' = m \leqq n$. Thus every subspace $\mathscr{V}$ of $\mathbf{R}^n$ has a basis.

PROBLEM SET 11.4

1. Given $V_1 = E_1 + 2E_2$, $V_2 = 2E_1 + 3E_2$, $V_3 = 3E_1 + 4E_2$, in $\mathbf{R}^2$. Theorem 4 predicts that $\{V_1, V_2, V_3\}$ is linearly dependent. Exhibit the linear dependence, by finding numbers $\alpha_1, \alpha_2, \alpha_3$, not all $= 0$, such that $\sum \alpha_i V_i = 0$.

2. Given $V_1 = E_1 + 2E_2$, $V_2 = 2E_2 + 3E_3$, $V_3 = 3E_2 + 4E_3$, $V_4 = 4E_1 - 5E_2$, in $\mathbf{R}^3$, proceed as in Problem 1.

3. Given $V_1 = E_1 + E_2$, $V_2 = E_2 + E_3$, $V_3 = E_2 + 3E_3$, $V_4 = 4E_1 + E_2 + E_3$, in $\mathbf{R}^3$, proceed as in Problem 1.

4. In $\mathbf{R}^3$, let $\mathscr{V} = \{V \mid V \cdot (E_1 + E_2) = 0\}$. Find a basis for $\mathscr{V}$.

5. Same question for $\mathscr{V} = \{V \mid V \cdot (E_1 + E_2 - 2E_3) = 0\}$.

6. In $\mathbf{R}^4$, let $\mathscr{V} = \{V \mid V \cdot (E_1 + E_2 + E_3) = 0\}$. Find a basis for $\mathscr{V}$.

7. Find a basis for $\mathbf{R}^4$, using $E_1 + E_2$ and $E_3 + E_4$ as basis elements.

8. Find a basis for $\mathbf{R}^3$ from among the vectors of

$$B = \{E_1 + E_2 + 2E_3, \quad E_1 + E_3, \quad E_3 + E_2, \quad E_1 - E_2, \quad E_1 + E_2 + E_3\}.$$

9. In $\mathbf{R}^3$, let $\mathscr{V} = \{V \mid V \cdot (E_1 + E_2) = 0, V \cdot (E_1 + 2E_3) = 0\}$. Find a basis for $\mathscr{V}$.

10. In $\mathbf{R}^4$, let $\mathscr{V} = \{V \mid V \cdot (E_1 + E_2 + E_3) = 0, V \cdot (E_2 + 2E_3 + E_4) = 0\}$. Find a basis for $\mathscr{V}$.

11.5 ORTHONORMAL BASES

In the preceding section, we found for the space $\mathbf{R}^n$ a basis

$$\mathbf{B}^n = \{E_1, E_2, \ldots, E_n\},$$

where the ith coordinate of E_i is 1 and the other coordinates of E_i are 0. Thus

$$V = (x_1, x_2, \ldots, x_n) = \sum_{i=1}^{n} x_i E_i.$$

If

$$W = (y_1, y_2, \ldots, y_n) = \sum_{i=1}^{n} y_i E_i,$$

then

$$V \cdot W = \sum_{i=1}^{n} x_i y_i.$$

Thus, for linear combinations of the E_i's, we have a simple formula for the inner product:

$$\left(\sum_{i=1}^{n} x_i E_i\right) \cdot \left(\sum_{i=1}^{n} y_i E_i\right) = \sum_{i=1}^{n} x_i y_i.$$

This formula does not hold for all bases. For example, the set

$$\{V_1, V_2\} = \{E_1, E_1 + E_2\}$$

forms a basis for $\mathbf{R}^2$, but it is *not* true that

$$(?) \quad (\alpha_1 V_1 + \alpha_2 V_2) \cdot (\beta_1 V_1 + \beta_2 V_2) = \alpha_1 \beta_1 + \alpha_2 \beta_2 \quad (?).$$

(Try taking $\alpha_1 = \beta_1 = \beta_2 = 1$.) But the above formula for the inner product does hold for a certain kind of basis, now to be defined.

Two vectors V_1 and V_2 are called *orthogonal* if $V_1 \cdot V_2 = 0$. (Thus E_1 and E_2 are orthogonal in $\mathbf{R}^n$.) More generally, a set

$$B = \{V_1, V_2, \ldots, V_n\}$$

is *orthogonal* if

$$V_i \cdot V_j = 0 \qquad \text{for } i \neq j.$$

Thus $\mathbf{B}^n = \{E_1, E_2, \ldots, E_n\}$ is an orthogonal set, but $\{E_1, E_1 + E_2\}$ is not, because

$$E_1 \cdot (E_1 + E_2) = 1 + 0 = 1 \neq 0.$$

If

$$\|V_i\| = 1 \qquad \text{for each } i,$$

then $B = \{V_1, V_2, \ldots, V_n\}$ is *normal*. If B is both orthogonal and normal, then B is *orthonormal*. Thus $\mathbf{B}^n$ is orthonormal. Since $\|V_i\|^2 = V_i \cdot V_i$, we note that B is orthonormal if and only if

$$V_i \cdot V_j = \begin{cases} 0 & \text{for } i \neq j \\ 1 & \text{for } i = j. \end{cases}$$

Theorem 1. Every finite-dimensional inner-product space has an orthonormal basis.

Proof. We shall show, by induction, that every n-dimensional inner-product space has an orthonormal basis. For $\dim \mathscr{V} = 1$, this is obvious: Given a basis $\{V_1\}$, we let $W_1 = V_1/\|V_1\|$. Then $\{W_1\}$ is an orthonormal basis. We suppose, then, that every n-dimensional inner-product space has an orthonormal basis; and we need to show that every $(n + 1)$-dimensional space has the same property.

Given $\dim \mathscr{V} = n + 1$, let

$$B = \{V_1, V_2, \ldots, V_n, V_{n+1}\}$$

be a basis for V. Let $\mathscr{V}'$ be the subspace spanned by $\{V_1, V_2, \ldots, V_n\}$. Then $\dim \mathscr{V}' = n$, and so $\mathscr{V}'$ has an orthogonal basis

$$W = \{W_1, W_2, \ldots, W_n\}.$$

Then the set $C = \{W_1, W_2, \ldots, W_n, V_{n+1}\}$ is a basis for $\mathscr{V}$. (Check that C is linearly independent, and spans $\mathscr{V}$.) We shall now find a vector V'_{n+1} such that the set

$$D = \{W_1, W_2, \ldots, W_n, V'_{n+1}\}$$

spans $\mathscr{V}$ and is orthogonal. For each i, let $a_i = W_i \cdot V_{n+1}$. Let

$$V'_{n+1} = V_{n+1} - \sum_{i=1}^{n} a_i W_i.$$

Then for each k from 1 to n we have

$$W_k \cdot V'_{n+1} = W_k \cdot V_{n+1} - \sum_{i=1}^{n} a_i W_k \cdot W_i.$$

In the sum on the right, the only nonzero term is $W_k \cdot W_k$, because W is an orthogonal set; and $W_k \cdot W_k = 1$, because W is orthonormal. Therefore

$$W_k \cdot V'_{n+1} = W_k \cdot V_{n+1} - a_k = a_k - a_k = 0.$$

Therefore D is an orthogonal set. The last step is trivial. Let

$$\alpha = \|V'_{n+1}\|,$$

and let

$$W_{n+1} = \frac{1}{\alpha} V'_{n+1}.$$

Then

$$\|W_{n+1}\| = \frac{1}{\alpha} \|V'_{n+1}\| = 1,$$

and

$$W_k \cdot W_{n+1} = \frac{1}{\alpha} (W_k \cdot V'_{n+1}) = \frac{1}{\alpha} \cdot 0 = 0.$$

Therefore the set $\{W_1, W_2, \ldots, W_n, W_{n+1}\}$ forms an orthonormal basis.

Note that the pattern of this proof supplies us with a method of actually finding an orthonormal basis, starting with a basis which is not necessarily orthonormal. The proof gives a scheme for "orthonormalizing" a given basis, a step at a time. For example, in $\mathbf{R}^3$ let V be the subspace spanned by

$$B = \{V_1, V_2\} = \{E_1 + E_2, E_2 + E_3\}.$$

Then B is a basis for V, but is neither orthogonal nor normal. We can get an orthonormal basis for V by following the pattern of the proof of Theorem 1.

1) Let

$$W_1 = V_1/\|V_1\| = \frac{1}{\sqrt{2}} (E_1 + E_2).$$

Then $\|W_1\| = 1$.

2) We shall now treat V_2 as the V_{n+1} of the above proof. Let

$$a_1 = W_1 \cdot V_2 = \frac{1}{\sqrt{2}} (E_1 + E_2) \cdot (E_2 + E_3) = \frac{1}{\sqrt{2}}.$$

Let

$$V'_2 = V_2 - a_1 W_1 = (E_2 + E_3) - \tfrac{1}{2}(E_1 + E_2)$$

$$= -\tfrac{1}{2}E_1 + \tfrac{1}{2}E_2 + E_3.$$

The theory predicts that

$$W_1 \cdot V'_2 = 0,$$

and this checks, because

$$W_1 \cdot V_2' = \frac{1}{\sqrt{2}} (E_1 + E_2) \cdot (-\tfrac{1}{2}E_1 + \tfrac{1}{2}E_2 + E_3)$$

$$= \frac{1}{\sqrt{2}} (-\tfrac{1}{2} + \tfrac{1}{2} + 0) = 0.$$

3) Now we normalize V_2', by letting

$$W_2 = V_2'/\|V_2'\|.$$

Since

$$\|V_2'\|^2 = V_2' \cdot V_2' = \tfrac{1}{4} + \tfrac{1}{4} + 1 = \tfrac{3}{2},$$

we have

$$\frac{1}{\|V_2'\|} = \sqrt{\frac{2}{3}}$$

and

$$W_2 = -\frac{1}{\sqrt{6}} E_1 + \frac{1}{\sqrt{6}} E_2 + \sqrt{\tfrac{2}{3}} E_3,$$

so that

$$\|W_2\|^2 = \tfrac{1}{6} + \tfrac{1}{6} + \tfrac{2}{3} = 1,$$

and $\|W_2\| = 1$, as it should be. Now $\{W_1, W_2\}$ is an orthonormal basis.

Orthonormal bases are what we need to get a simple formula for the inner product:

Theorem 2. If $\{V_1, V_2, \ldots, V_n\}$ is orthonormal, then

$$\left(\sum_{i=1}^{n} \alpha_i V_i\right) \cdot \left(\sum_{j=1}^{n} \beta_j V_j\right) = \sum_{i=1}^{n} \alpha_i \beta_i.$$

Proof. We know that

$$\alpha_i V_i \cdot \beta_j V_j = 0 \qquad (i \neq j),$$

and

$$\alpha_i V_i \cdot \beta_i V_i = \alpha_i \beta_i V_i \cdot V_i = \alpha_i \beta_i \|V_i\|^2 = \alpha_i \beta_i.$$

Therefore

$$\alpha_i V_i \cdot \sum_{j=1}^{n} \beta_j V_j = \alpha_i \beta_i \qquad \text{for each } i,$$

and so

$$\left(\sum_{i=1}^{n} \alpha_i V_i\right) \cdot \left(\sum_{j=1}^{n} \beta_j V_j\right) = \sum_{i=1}^{n} \alpha_i \beta_i.$$

Of course, for inner products of the form $V \cdot V = \|V\|^2$, we have

$$\left\|\sum_{i=1}^{n} \alpha_i V_i\right\|^2 = \sum_{i=1}^{n} \alpha_i^2.$$

In $\mathbf{R}^2$ and $\mathbf{R}^3$, it is easy to see by the distance formula that the distance between any two points P and Q is $\|P - Q\|$. For inner-product spaces in general, we use the latter formula as the definition of distance. Often we shall think of inner-product spaces geometrically, and so we may refer to their elements as points rather than vectors, as in the following definition.

Definition. In any inner-product space, the *distance* between two points P and Q is $\|P - Q\|$.

The distance between P and Q may be denoted by $d(P, Q)$, or simply PQ. An orthonormal basis gives us a distance formula.

Theorem 3. Let $B = \{V_1, V_2, \ldots, V_n\}$ be an orthonormal basis for the inner-product space V, and let

$$P = \sum_{i=1}^{n} \alpha_i V_i, \qquad Q = \sum_{i=1}^{n} \beta_i V_i.$$

Then

$$d(P, Q) = \sqrt{\sum_{i=1}^{n} (\alpha_i - \beta_i)^2}.$$

Proof. We have

$$d(P, Q) = \|P - Q\|, \qquad P - Q = \sum_{i=1}^{n} (\alpha_i - \beta_i) V_i,$$

and

$$\|P - Q\|^2 = (P - Q) \cdot (P - Q).$$

Since the basis is orthonormal,

$$\|P - Q\|^2 = \sum_{i=1}^{n} (\alpha_i - \beta_i)^2,$$

and the theorem follows.

PROBLEM SET 11.5

1. Let V be the subspace of $\mathbf{R}^3$ spanned by the basis $B = \{V_1, V_2\} = \{E_1 + E_2 + E_3, E_1 + E_2\}$. Find an orthonormal basis for V.

2. Same question for $B = \{V_1, V_2\} = \{E_1 + E_2 + E_3, E_1 - E_2\}$.

3. Use the orthonormalization scheme described in the proof of Theorem 1 to find an orthonormal basis for the subspace spanned by the basis

$$B = \{E_1 + E_2 + E_3, E_2 + E_3, E_3\} = \{V_1, V_2, V_3\}.$$

4. Same question for $B = \{E_2 + E_3, E_3, E_1 + E_2 + E_3\} = \{V_1, V_2, V_3\}$.

5. Same question for $B = \{E_3, E_1 + E_2 + E_3, E_2 + E_3\} = \{V_1, V_2, V_3\}$.

6. In $\mathbf{R}^3$, find an orthonormal basis for $\mathscr{V} = \{V \mid V \cdot (E_1 + E_2 + E_3) = 0\}$.

7. Same question for $\mathscr{V} = \{V \mid V \cdot (E_1 + 2E_2) = 0\}$.

8. Let $V_1 = 2E_2 + E_3$, $V_2 = 4E_1 + E_4$. Find vectors V_3, V_4 so that $\{V_1, V_2, V_3, V_4\}$ is an orthogonal basis of $\mathbf{R}^4$.

9. Given $V_1 = E_1 - E_3 + E_4$, $V_2 = E_1 + E_2 + E_3$, proceed as in Problem 8.

10. In $\mathbf{R}^4$, find an orthonormal basis of the subspace

$$\mathscr{V} = \{V \mid V \cdot (E_1 + E_2) = 0, V \cdot (E_2 + E_4) = 0\}.$$

11. Same question for $\mathscr{V} = \{V \mid V \cdot (E_1 + E_2) = 0, V \cdot E_3 = 0, V \cdot E_4 = 0\}$.

12. Show that if V_1 is a fixed nonzero vector in an n-dimensional vector space, then $\mathscr{V} = \{V \mid V \cdot V_1 = 0\}$ is an $(n - 1)$-dimensional subspace.

13. Suppose that $\{V_1, V_2, \ldots, V_n\}$ is orthogonal, but not necessarily orthonormal. Find a formula for

$$\left(\sum_{i=1}^{n} \alpha_i V_i \right) \cdot \left(\sum_{j=1}^{n} \beta_j V_j \right).$$

14. Let $\{V_1, V_2, \ldots, V_n\}$ be an orthogonal set of nonzero vectors. Let $V = \sum \alpha_i V_i$ be any linear combination of them. Show that

$$V \cdot V_i = 0 \text{ for every } i \Rightarrow V = O.$$

15. Let $\{V_1, V_2, \ldots, V_n\}$ be as in Problem 14. Show that the set is linearly independent. (This means that in an n-dimensional vector space, no orthogonal set of nonzero vectors can have more than n elements. Thus, for example, in R^3 there is no set of four concurrent lines, every two of which are perpendicular.)

16. Let $\mathscr{W}$ and $\mathscr{X}$ be subspaces of a vector space $\mathscr{V}$. If every vector in $\mathscr{W}$ is orthogonal to every vector in $\mathscr{X}$, then $\mathscr{W}$ and $\mathscr{X}$ are *orthogonal subspaces*, and we write $\mathscr{W} \perp \mathscr{X}$. Show that if $\mathscr{W} \perp \mathscr{X}$, then $\dim \mathscr{W} + \dim \mathscr{X} \leq \dim \mathscr{V}$. Give an example to show that the equality does not necessarily hold.

17. The following is a converse of Theorem 2.
    Theorem (?) Let $B = \{V_1, V_2, \ldots, V_n\}$ be a basis for $\mathscr{V}$. If for all vectors $V = \sum \alpha_i V_i$, $W = \sum \beta_j V_j$, we have
$$V \cdot W = \sum \alpha_i \beta_i,$$
    then B is orthonormal.
    Is this true? Why or why not?

18. Show that if $\{V_1, V_2, \ldots, V_n\}$ is an orthonormal basis for $\mathscr{V}$, then for every V in $\mathscr{V}$,

$$V = \sum_{i=1}^{n} (V \cdot V_i) V_i.$$

That is, for $V = \sum \alpha_i V_i$, we always have $\alpha_i = V \cdot V_i$.

11.6   THE SCHWARZ INEQUALITY.
MORE GENERAL CONCEPTS OF NORM AND DISTANCE

The "triangular inequality" for points in a plane (or in three-dimensional space) asserts that for any points P, Q, R we have

$$PR \leq PQ + QR.$$

The equality holds if the points are collinear, and Q is between P and R; and in every other case, the strict inequality holds.

We propose to show that in any inner-product space, the same inequality holds for distances. That is,

$$d(P, R) \leq d(P, Q) + d(Q, R), \tag{1}$$

for every $P, Q,$ and R. Since distance was defined by the formula

$$d(P, Q) = \|P - Q\|,$$

the proposed inequality means that

$$\|P - R\| \leq \|P - Q\| + \|Q - R\|.$$

This has the form

$$\|A + B\| \leq \|A\| + \|B\|, \tag{2}$$

where $A = P - Q$ and $B = Q - R$. Note that this is analogous to the inequality

$$|x + y| \leq |x| + |y|,$$

which is known to hold for both real and complex numbers. Obviously any general proof of (2), for all inner-product spaces, must appeal to the definition of the norm:

$$\|A\| = \sqrt{A \cdot A}, \qquad \|A\|^2 = A \cdot A. \tag{3}$$

Therefore the natural first step, in proving (2), is to restate it in terms of the definition given in (3). In these terms,

$$\|A + B\| \leq \|A\| + \|B\| \tag{2}$$

$$\Leftrightarrow \|A + B\|^2 \leq \|A\|^2 + 2\,\|A\| \cdot \|B\| + \|B\|^2$$

$$\Leftrightarrow (A + B) \cdot (A + B) \leq A \cdot A + 2\,\|A\| \cdot \|B\| + B \cdot B$$

$$\Leftrightarrow A \cdot A + 2A \cdot B + B \cdot B \leq A \cdot A + 2\,\|A\| \cdot \|B\| + B \cdot B$$

$$\Leftrightarrow A \cdot B \leq \|A\| \cdot \|B\|. \tag{4}$$

Here (4) automatically holds whenever $A \cdot B < 0$. But if (4) *always* holds, then we must have

$$|A \cdot B| \leq \|A\| \cdot \|B\|. \tag{5}$$

(If (5) were false, for some A, B, then (4) would also be false, either for A, B or $-A$, B.) And Eq. (5) is obviously equivalent to

$$(A \cdot B)^2 \leq \|A\|^2 \cdot \|B\|^2. \tag{6}$$

Formula (6) is called the *Schwarz inequality*. We shall now prove it.

Theorem 1 (*The Schwarz inequality*). In any inner-product space,

$$(A \cdot B)^2 \leq \|A\|^2 \cdot \|B\|^2, \tag{6}$$

for every A and B.

Any proof must use, at some stage, the fact that $P \cdot P \geq 0$ for every P, because this is the only inequality that is given by the inner-product space laws. But to use this law to prove the theorem, we first need to reduce the theorem to a manageable special case. This is done as follows.

If $A = O$ or $B = O$, then the inequality (6) takes the form $0 \leq 0$, which is true. We may therefore suppose that $A \neq O$ and $B \neq O$. For the same reason, we may suppose that $A \cdot B \neq 0$.

Suppose now that we replace A by αA and B by βB, with $\alpha \neq 0$ and $\beta \neq 0$.

The inequality (6) then takes the form

$$(\alpha A \cdot \beta B)^2 \leq \|\alpha A\|^2 \cdot \|\beta B\|^2 \tag{7}$$

$$\Leftrightarrow \alpha^2 \beta^2 (A \cdot B)^2 \leq \alpha^2 \|A\|^2 \cdot \beta^2 \|B\|^2$$

$$\Leftrightarrow (A \cdot B)^2 \leq \|A\|^2 \cdot \|B\|^2. \tag{6}$$

Since (7) $\Leftrightarrow$ (6), for every α, $\beta \neq 0$, we are free to choose α and β as we please. We take

$$\alpha = \frac{1}{\|A\|}, \qquad \text{so that} \qquad \|\alpha A\| = 1.$$

We then choose

$$\beta = \frac{1}{\alpha A \cdot B} \qquad \text{so that} \qquad \alpha A \cdot \beta B = 1.$$

We let $P = \alpha A$, $Q = \beta B$. Our theorem now takes the form

$$\|P\| = 1 \text{ and } P \cdot Q = 1 \Rightarrow 1 \leq \|Q\|^2.$$

This is easy to prove: for $\|P\| = 1$, $P \cdot Q = 1$, we have

$$(P - Q) \cdot (P - Q) \geq 0$$

$$\Rightarrow P \cdot P - 2P \cdot Q + Q \cdot Q \geq 0$$

$$\Rightarrow \|P\|^2 - 2P \cdot Q + \|Q\|^2 \geq 0$$

$$\Rightarrow 1 - 2 + \|Q\|^2 \geq 0$$

$$\Rightarrow 1 \leq \|Q\|^2.$$

The theorem follows. In the light of the discussion which led to the Schwarz inequality, we also have the following:

Theorem 2. In any inner-product space,

$$\|P + Q\| \leq \|P\| + \|Q\|. \tag{2'}$$

(We sometimes express this by saying that the norm is *subadditive*. In general, a real-valued function is subadditive if $f(x + y) \leq f(x) + f(y)$, for every x and y.)

Theorem 3. In any inner-product space, distance is triangular. That is,

$$d(P, R) \leq d(P, Q) + d(Q, R).$$

Equivalently,

$$\|P - R\| \leq \|P - Q\| + \|Q - R\|.$$

Let us now review what we know about norms and distance. For norms, we have

N.1. $\|A\| \geq 0$ for every A.
N.2. $\|A\| = 0 \Rightarrow A = 0$.
N.3. $\|\alpha A\| = |\alpha| \cdot \|A\|$. (Homogeneity)
N.4. $\|A + B\| \leq \|A\| + \|B\|$. (Subadditivity)

All these are easy to check, on the basis of the definition $\|A\| = \sqrt{A \cdot A}$, the vector laws, and the Schwarz inequality. For distance, we have

D.1. $d(P, Q) \geq 0$.
D.2. $d(P, Q) = 0 \Rightarrow P = Q$.
D.3. $d(P, Q) = d(Q, P)$.
D.4. $d(P, R) \leq d(P, Q) + d(Q, R)$.

On this basis, we shall define various types of mathematical systems which are more general than inner-product spaces.

Definition. A *normed vector space* is a vector space in which a norm is defined, satisfying N.1 through N.4.

Thus a normed vector space is a quadruplet

$$[\mathscr{V}, +, \text{sm}, \|\ \|],$$

where $[\mathscr{V}, +, \text{sm}]$ is a vector space, and the "norm operation" $P \mapsto \|P\|$ satisfies N.1 through N.4.

Example 1. Let $\mathscr{L}_f$ be the set of all continuous functions f, on the interval $[-1, 1]$. Addition and scalar multiplication are defined in the obvious way. We define

$$\|f\|_u = \max |f|,$$

where $\max |f|$ is the largest of the numbers $|f(x)|$ $(-1 \leq x \leq 1)$. It is easy to check that N.1, N.2, and N.3 hold. To verify N.4, we observe that for each x,

$$|f(x) + g(x)| \leq |f(x)| + |g(x)|$$
$$\leq \|f\| + \|g\|.$$

Therefore $\max \{|f(x) + g(x)|\} \leq \|f\| + \|g\|$. Thus we have a normed vector space. The norm that we have just defined is called the *uniform norm*. (Hence the notation $\|\ \|_u$.)

Example 2. Let $\mathscr{L}_f$ be as Example 1; and let $+$ and sm be as before. We can define another norm by the formula

$$\|f\|_1 = \int_{-1}^{1} |f(x)|\, dx.$$

As before, the verifications of N.1, N.2, and N.3 are easy. To verify N.4, we observe that

$$\|f + g\|_1 = \int_{-1}^{1} |f(x) + g(x)|\, dx \leq \int_{-1}^{1} [|f(x)| + |g(x)|]\, dx$$
$$= \int_{-1}^{1} |f(x)|\, dx + \int_{-1}^{1} |g(x)|\, dx$$
$$= \|f\|_1 + \|g\|_1.$$

It should be emphasized that a normed vector space is not merely a vector space in which a norm *can be* defined, but rather a linear space in which a norm *has been* defined. Thus, in Examples 1 and 2 we defined two different norms $\| \ \|_u$ and $\| \ \|_1$ in the same vector space $[\mathscr{L}_f, +, \text{sm}]$; and this gave us two different normed vector spaces

$$[\mathscr{L}_f, +, \text{sm}, \| \ \|_u], \qquad [\mathscr{L}_f, +, \text{sm}, \| \ \|_1].$$

In any normed vector space, we can define distance by means of the formula

$$d(A, B) = \|A - B\|.$$

It then follows that the distance d satisfies D.1 through D.4. (This should be checked.) More generally, we state the following.

Definition. A *metric space* is a set S which is provided with a distance d, satisfying D.1 through D.4. The distance d is called the *metric*.

Thus a metric space is a pair $[S, d]$, where d is a metric for S. It is evident that metric spaces can arise in ways that have very little to do with vector spaces or with norms. For example, S may be the surface of a sphere in $\mathbf{R}^3$; that is,

$$S = \{(x, y, z) \mid x^2 + y^2 + z^2 = 1\},$$

and for each pair of points P, Q on S, the distance $d(P, Q)$ may be the length of the shortest arc on S, joining P and Q. It is not hard to see that this system forms a metric space, that is, d satisfies D.1 through D.4. In fact, this is the metric space used in navigation on the open sea, with arc length measured in nautical miles.

PROBLEM SET 11.6

1. Show, by any method, that for any pair of pairs of real numbers (x_1, x_2) and (y_1, y_2) we have $(x_1^2 + x_2^2)(y_1^2 + y_2^2) \geq (x_1 y_1 + x_2 y_2)^2$.

2. Show, by any method, that for every pair of finite sequences $x_1, x_2, \ldots, x_n, y_1, y_2, \ldots, y_n$ of real numbers,

$$\left(\sum_{i=1}^n x_i y_i\right)^2 \leq \left(\sum_{i=1}^n x_i^2\right)\left(\sum_{i=1}^n y_i^2\right).$$

3. Let E be a coordinate plane and, for each $P = (x, y)$ and $Q = (a, b)$, let

$$d(P, Q) = (x - a)^2 + (y - b)^2.$$

Thus d is the square of the usual distance. Does $[E, d]$ form a metric space?

4. Same question, for $d(P, Q) = |x - a| + |y - b|$.

5. Same question, for $d(P, Q) = \text{maxmium of } |x - a| \text{ and } |y - b|$.

6. If $[S, d]$ is a metric space and $k > 0$, does it follow that $[S, kd]$ is a metric space?

7. The real number system $\mathbf{R}$ clearly forms a vector space. For each x in $\mathbf{R}$, let $\|x\| = \sqrt{|x|}$. Does this give a normed vector space?

8. Let S be the set of all airline passenger terminals in the world; and for each P, Q in S let $d(P, Q)$ be the minimum number of hours required to get from P to Q by a combination of regularly scheduled flights. Is $[S, d]$ a metric space?

9. Let $\mathscr{L}_f$ be the set of all continuous functions on the interval $[-1, 1]$, as in Example 1 of the text. For each f, g in $\mathscr{L}_f$, let

$$d(f,g) = \int_{-1}^{1} \frac{|f(x) - g(x)|}{1 + |f(x) + g(x)|} \, dx.$$

Does this give a metric space?

*10. For the definition of U lim, see p. 502. Let $\mathscr{L}_f$ be the normed vector space, with the "uniform norm," defined in Example 1 of the text. Show that for any sequence $f_1, f_2, \ldots$ of functions in $\mathscr{L}_f$,

$$\lim_{n \to \infty} \|f_n\|_u = 0 \quad \Leftrightarrow \quad \text{U} \lim_{n \to \infty} f_n = 0.$$

(This is why the norm defined in Example 1 is called the uniform norm.)

11. Let $\mathscr{L}_f$ be the same as in Problem 10, with the norm

$$\|f\|_1 = \int_{-1}^{1} |f(x)| \, dx,$$

as in Example 2 of the text. Is it true that

$$\lim_{n \to \infty} \|f_n\|_1 = 0 \quad \Rightarrow \quad \text{U} \lim_{n \to \infty} f_n = 0?$$

Is it true that

$$\text{U} \lim_{n \to \infty} f_n = 0 \quad \Rightarrow \quad \lim_{n \to \infty} \|f_n\|_1 = 0?$$

12. Let $C^0[-\pi, \pi]$ be the set of all continuous functions f on the interval $[-\pi, \pi]$, with $+$ and sm defined as usual. Set $f \cdot g = (\int_{-\pi}^{\pi} f(x)g(x) \, dx)$, and verify that $C^0[-\pi, \pi]$, with this inner product, forms as inner-product space.

13. Find a polynomial $f(x) = a_0 + a_1 x + x^2$ which is orthogonal to $g(x) = 1$ and $h(x) = x$, in the inner-product space defined in Problem 12.

14. Find an orthonormal basis for the subspace of $C^0[-\pi, \pi]$ spanned by the elements $\{1, x, x^2\}$.

*15. For each n, let $\mathbf{T}_n$ be the set of all functions of the form

$$f(x) = a_0 + \sum_{i=1}^{n} [a_i \cos ix + b_i \sin ix],$$

on the interval $[-\pi, \pi]$.   (Such functions are called *trigonometric polynomials*.) Evidently $\mathbf{T}_n$ forms a subspace of $C^0[-\pi, \pi]$, and the set

$$B = \{1; \cos x, \cos 2x, \ldots, \cos nx; \sin x, \sin 2x, \ldots, \sin nx\}$$

spans $\mathbf{T}_n$. Show that (a) B is orthogonal, and (b) B is a basis for $\mathbf{T}_n$. Then find an orthonormal basis for $\mathbf{T}_n$. [*Warning:* This one is long. It is easier if you note that in (a) you need not necessarily compute indefinite integrals; what you need, in each case, is the definite integral, from $-\pi$ to π. The identities

$$\cos (A + B) - \cos (A - B) = -2 \sin A \sin B,$$

$$\cos (A + B) + \cos (A - B) = 2 \cos A \cos B$$

are also useful.] Problem 15 of Problem Set 11.5 is useful at one stage.

12   Fourier Series

12.1   PROJECTIONS INTO A SUBSPACE.
TRIGONOMETRIC POLYNOMIALS AND FOURIER SERIES

The idea of a projection is taken from elementary geometry. Let E be a plane in $\mathbf{R}^3$, and let P be a point. To suit the terms of our later discussion, suppose that E passes through the origin. Then the *projection of P into E* is the point Q which is the foot of the perpendicular from P to E. (If P is in E, then the projection of P is P.) The following facts are well known:

1)   The line $\overleftrightarrow{PQ}$, through P and Q, is perpendicular to every line in E that contains Q. (In fact, this is the definition of the statement that $\overleftrightarrow{PQ} \perp E$.)

2)   If P is not in E, then there is one and only one point Q in E, satisfying (1).

3)   If R is in E, then

$$(PQ)^2 + (QR)^2 = (PR)^2.$$

It follows immediately that:

4)   The distance from P to Q is the minimum distance from P to E.

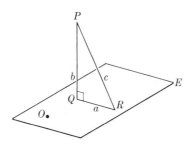

We shall now regard $\mathbf{R}^3$ as an inner-product space, regard the points $P, Q, \ldots$ as vectors, and restate these ideas in vectorial form.

In the figure below, we have completed a rectangle, by inserting the point $P - Q$. (Check that in the figure, $(P - Q) + Q = P$, as it should be.) In elementary geometric terms, O is the foot of the perpendicular from $P - Q$ to O. Therefore (1)

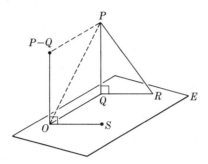

through (4) take the following forms:

1) $(P - Q) \cdot S = 0$, for every S in E.

2) There is one and only one Q in E, satisfying (1).

3) If R is in E, then

$$\|P - Q\|^2 + \|Q - R\|^2 = \|P - R\|^2.$$

Therefore

4) $\|P - Q\|$ is the minimum distance from P to E.

We shall now show that these ideas apply under far more general conditions, as follows.

Theorem 1. Let $\mathscr{V}$ be any vector space, let $\mathscr{W}$ be any finite-dimensional subspace, and let P be any vector in $\mathscr{V}$. Then there is one and only one vector Q in $\mathscr{W}$ such that

$$(P - Q) \cdot S = 0 \qquad \text{for every } S \text{ in } \mathscr{W}.$$

Proof. The easy part is to show that there is *only* one such Q. If

$$(P - Q) \cdot S = 0 \qquad \text{and} \qquad (P - Q') \cdot S = 0,$$

for every S, then

$$[(P - Q) - (P - Q')] \cdot S = 0,$$

for every S. Therefore

$$(Q' - Q) \cdot S = 0,$$

for every S. In particular, for $S = Q' - Q$, we have $(Q' - Q) \cdot (Q' - Q) = 0$, and so $Q' = Q$.

To show that there is one such Q, we need to find one; and as a guide, we look at a simple case. For

$$\mathscr{V} = \mathbf{R}^3 = \{x_1 E_1 + x_2 E_2 + x_3 E_3\},$$

$$\mathscr{W} = \mathbf{R}^2 = \{x_1 E_1 + x_2 E_2\},$$

$$P = \alpha_1 E_1 + \alpha_2 E_2 + \alpha_3 E_3,$$

we ought to have

$$Q = \alpha_1 E_1 + \alpha_2 E_2;$$

here Q is the projection of P into the xy-plane. This works:

$$P - Q = 0 \cdot E_1 + 0 \cdot E_2 + \alpha_3 E_3,$$

$$(P - Q) \cdot E_1 = (P - Q) \cdot E_2 = 0,$$

and so

$$(P - Q) \cdot S = (P - Q) \cdot (x_1 E_1 + x_2 E_2) = 0$$

for every S in $\mathcal{W}$. In our formula for Q, we have

$$Q \cdot E_1 = \alpha_1 E_1 \cdot E_1 + \alpha_2 E_2 \cdot E_1 = \alpha_1,$$

$$Q \cdot E_2 = \alpha_1 E_1 \cdot E_2 + \alpha_2 E_2 \cdot E_2 = \alpha_2;$$

and so

$$Q = (Q \cdot E_1)E_1 + (Q \cdot E_2)E_2.$$

We shall see that this pattern carries over to the general case. Let

$$B = \{W_1, W_2, \ldots, W_n\}$$

be an orthonormal basis for $\mathcal{W}$; for each i from 1 to n, let

$$\alpha_i = P \cdot W_i,$$

and let

$$Q = \sum_{i=1}^{n} \alpha_i W_i.$$

Then

$$Q \cdot W_j = \sum_{i=1}^{n} \alpha_i W_i \cdot W_j = \alpha_j \|W_j\|^2 = \alpha_j,$$

and so

$$(P - Q) \cdot W_j = P \cdot W_j - Q \cdot W_j = \alpha_j - \alpha_j = 0,$$

for each j. Therefore for every

$$S = \sum_{j=1}^{n} \beta_j W_j \in \mathcal{W},$$

we have

$$(P - Q) \cdot S = \sum_{j=1}^{n} \beta_j (P - Q) \cdot W_j = 0,$$

which was to be proved.

The point Q is called the *projection* of P into $\mathcal{W}$, and is denoted by $\operatorname{Pr} P$ (or by $\operatorname{Pr}_{\mathcal{W}} P$, if there is any doubt about the subspace into which we are projecting). To repeat:

Definition. Let $\mathcal{V}$ be any vector space, and let $\mathcal{W}$ be any finite-dimensional subspace of $\mathcal{V}$. For each P in $\mathcal{V}$, $\operatorname{Pr} P$ (or $\operatorname{Pr}_{\mathcal{W}} P$) is the point Q of $\mathcal{W}$ such that $(P - Q) \cdot S = 0$ for each S in $\mathcal{W}$.

Theorem 1 tells us that this definition defines something. And one of the ideas in the proof of Theorem 1 is worth noting for future reference:

Theorem 2. If $\operatorname{Pr} P$ is the projection of P into $\mathcal{W}$, and $\{W_1, W_2, \ldots, W_n\}$ is an orthonormal basis for $\mathcal{W}$, then

$$\operatorname{Pr} P = \sum_{i=1}^{n} (P \cdot W_i)W_i.$$

(In the proof of Theorem 1, we found that this sum satisfied the conditions for Q, and that there is only one such Q.) It remains to show that conditions (3) and (4), stated at the beginning of this section, hold on the basis of our general definition.

Theorem 3. If $Q = \text{Pr } P$, and R is in $\mathscr{W}$, then

$$\|P - Q\|^2 + \|Q - R\|^2 = \|P - R\|^2$$

Proof. Obviously $(P - Q) + (Q - R) = P - R$; and $(P - Q) \cdot (Q - R) = 0$, because $Q - R$ is in $\mathscr{W}$. Therefore Theorem 3 is a consequence of the following.

Theorem 4 (*The Pythagorean theorem*). In any inner-product space,

$$A \cdot B = 0 \;\Rightarrow\; \|A\|^2 + \|B\|^2 = \|A + B\|^2.$$

Proof

$$\|A + B\|^2 = (A + B) \cdot (A + B)$$
$$= A \cdot A + 2A \cdot B + B \cdot B$$
$$= \|A\|^2 + 0 + \|B\|^2.$$

As in the special case of $\mathbf{R}^3$, this immediately gives:

Theorem 5. The projection of a point Q into a finite-dimensional subspace $\mathscr{W}$ is the point of $\mathscr{W}$ which is closest to P.

This idea has the following unexpected application. Let

$$\mathscr{V} = \mathbf{C}^0[-\pi, \pi],$$

where $\mathbf{C}^0[-\pi, \pi]$ is the set of all continuous functions on the closed interval $[-\pi, \pi]$; and consider the inner-product space

$$[\mathscr{V}, +, \text{sm}, \cdot],$$

where $+$ and sm are defined as usual for spaces of functions, and the inner product is defined by the formula

$$f \cdot g = \int_{-\pi}^{\pi} f(x)g(x)\, dx.$$

For each positive integer n, let $\mathbf{T}_n$ be the set of all trigonometric polynomials of order n, that is, the set of all functions of the form

$$g(x) = a_0 + \sum_{i=1}^{n} [a_i \cos ix + b_i \sin ix].$$

Obviously $\mathbf{T}_n$ forms a subspace of $\mathscr{V}$. Consider the set

$$\{1; \cos x, \cos 2x, \ldots, \cos nx; \sin x, \sin 2x, \ldots, \sin nx\}.$$

This set spans $\mathbf{T}_n$. To verify that the set is orthogonal, we need to show that

$$\int_{-\pi}^{\pi} \cos ix\, dx = \int_{-\pi}^{\pi} \sin ix\, dx = 0 \qquad \text{for every } i,$$

$$\int_{-\pi}^{\pi} \cos ix \sin jx\, dx = 0 \qquad \text{for every } i, j,$$

and

$$\int_{-\pi}^{\pi} \sin ix \sin jx \, dx = \int_{-\pi}^{\pi} \cos ix \cos jx \, dx = 0 \qquad \text{for } i \neq j.$$

All of these answers can be calculated by brute force, but there are tricks that help. By more straightforward calculations, we get

$$\|1\|^2 = \int_{-\pi}^{\pi} 1^2 \, dx = 2\pi, \qquad \|1\| = \sqrt{2\pi};$$

$$\|\cos ix\|^2 = \pi = \|\sin ix\|^2.$$

To get an orthonormal basis, we divide each basis element by its norm. This gives an orthonormal basis of the form $B = \{C_0, C_1, \ldots, C_n; S_1, S_2, \ldots, S_n\}$, where

$$C_0 = \frac{1}{\sqrt{2\pi}},$$

$$C_i = \frac{1}{\sqrt{\pi}} \cos ix \qquad (i > 0),$$

$$S_i = \frac{1}{\sqrt{\pi}} \sin ix \qquad (i > 0).$$

Now let f be a function in $C^0[-\pi, \pi]$. By Theorem 2, the projection of f into the finite-dimensional subspace $\mathscr{W} = \mathbf{T}_n$ is the vector

$$\Pr_n f = \sum_{i=0}^{n} (f \cdot C_i) C_i + \sum_{i=1}^{n} (f \cdot S_i) S_i.$$

Here

$$f \cdot C_0 = \int_{-\pi}^{\pi} f(x) \cdot \frac{1}{\sqrt{2\pi}} \, dx = \frac{1}{\sqrt{2\pi}} \int_{-\pi}^{\pi} f(x) \, dx,$$

$$f \cdot C_i = \frac{1}{\sqrt{\pi}} \int_{-\pi}^{\pi} f(x) \cos ix \, dx \qquad (i > 0),$$

$$f \cdot S_i = \frac{1}{\sqrt{\pi}} \int_{-\pi}^{\pi} f(x) \sin ix \, dx.$$

Therefore

$$\Pr_n f = a_0 + \sum_{i=1}^{n} [a_i \cos ix + b_i \sin ix],$$

where

$$a_0 = \frac{1}{2\pi} \int_{-\pi}^{\pi} f(x) \, dx,$$

$$a_i = \frac{1}{\pi} \int_{-\pi}^{\pi} f(x) \cos ix \, dx \qquad (i > 0)$$

$$b_i = \frac{1}{\pi} \int_{-\pi}^{\pi} f(x) \sin ix \, dx.$$

[Check   that $(f \cdot C_0)C_0 = a_0$, $(f \cdot C_i)C_i = a_i \cos ix$,   and $(f \cdot S_i)S_i = b_i \sin ix$ $(i > 0)$.]

It now seems reasonable to hope that $\mathrm{Pr}_n f$ is in some sense an approximation of f when n is large. That is,

$$(?) \quad n \approx \infty \;\Rightarrow\; f \approx \mathrm{Pr}_n f \quad (?).$$

If we judge the approximation by observing $\|f - \mathrm{Pr}_n f\|$, it is clear, at least, that we have done our best: Theorem 5 tells us that $\mathrm{Pr}_n f$ is the element of $\mathbf{T}_n$ which minimizes $\|f - \mathrm{Pr}_n f\|$. If our best was good enough, then we should have

$$(?) \quad \lim_{n \to \infty} \|f - \mathrm{Pr}_n f\| = 0 \quad (?). \tag{1}$$

A stronger conjecture is that the approximation is good even in the *uniform* norm:

$$(?) \quad \lim_{n \to \infty} \|f - \mathrm{Pr}_n f\|_u = 0 \quad (?). \tag{2}$$

It may be disturbing, at this stage, to observe that (2) cannot be true as stated: $\mathrm{Pr}_n f(-\pi) = \mathrm{Pr}_n f(\pi)$, because all trigonometric polynomials have period 2π. Therefore (2) cannot be true unless f has the same property. We shall see, however, that this is the only way that (2) can fail to hold:

Theorem A. If f has period 2π, and f' is continuous, then $\lim_{n \to \infty} \|f - \mathrm{Pr}_n f\|_u = 0$.

This implies, of course, that

$$f(x) = \lim_{n \to \infty} \left[a_0 + \sum_{i=1}^{n} (a_i \cos ix + b_i \sin x) \right]$$

$$= a_0 + \sum_{i=1}^{\infty} (a_i \cos ix + b_i \sin ix),$$

where the limit is a pointwise limit in the elementary sense.

The last of these expressions is called the *Fourier series* for f, and the numbers $a_0, a_1, \ldots, b_1, b_2, \ldots$ are called the *Fourier coefficients* of f. Obviously every continuous function has a set of Fourier coefficients, and therefore has a Fourier series. The question is under what conditions we can conclude that the Fourier series converges to the function. This question is complicated, and the situation is not yet thoroughly understood by anybody; Theorem A, above, is the best of the simple results.

Meanwhile, the successive projections $\mathrm{Pr}_1 f$, $\mathrm{Pr}_2 f$, $\ldots$ have two encouraging properties. First, each projection $\mathrm{Pr}_{n+1} f$ is simply a continuation of the preceding one $\mathrm{Pr}_n f$; to get $\mathrm{Pr}_{n+1} f$ from $\mathrm{Pr}_n f$ we merely add a term of the form

$$a_{n+1} \cos (n + 1)x + b_{n+1} \sin (n + 1)x,$$

leaving the preceding terms unchanged. Second, the error in the approximation $\mathrm{Pr}_n f \approx f$, as measured by $\|f - \mathrm{Pr}_n f\|$, is nonincreasing.

Theorem 6. If f is continuous on $[-\pi, \pi]$, then

$$\|f - \mathrm{Pr}_{n+1} f\| \leq \|f - \mathrm{Pr}_n f\|.$$

(*Proof.* Since $\mathbf{T}_{n+1}$ contains $\mathbf{T}_n$, the minimum distance from f to $\mathbf{T}_n$ cannot be less than the minimum distance from f to $\mathbf{T}_{n+1}$.)

Note that the Fourier series for a function f depends only on the values of f on the interval $[-\pi, \pi]$. Therefore, when we set up the series for $f(x) = x$, what we are really dealing with is a discontinuous function, with period 2π, whose graph looks like this:

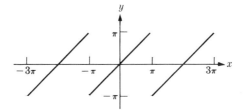

Similarly, when we set up the series for $f(x) = x^2$, the series turns out to represent a periodic function whose graph is obtained by fitting together infinitely many parabolic arcs, like this:

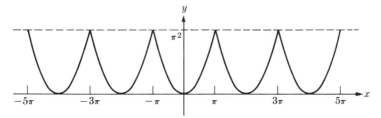

Of course the periodic function that we get from $f(x) = x$ is not continuous. But it turns out that this doesn't matter: if the graph of f is obtained by fitting together a finite number of continuous functions with continuous derivatives, then the Fourier series always converges to a function F; and $F(x) = f(x)$ at every point where f is continuous.

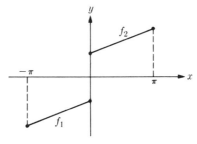

At points where the "continuous pieces" of the function fail to fit together, as at $x = 0$ in the figure above, the series makes a compromise, and converges to the *average* of the lefthand value and the righthand value. Similarly, if $f(-\pi) \neq f(\pi)$, then

$$F(-\pi) = F(\pi) = \tfrac{1}{2}[f(-\pi) + f(\pi)].$$

Thus, for the function f in the preceding figure, the graph of the function F given by the series looks like the figure below.   Here

$$F(-\pi) = F(0) = F(\pi) = 0.$$

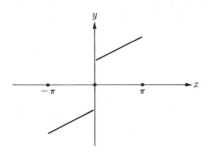

PROBLEM SET 12.1

Throughout this problem set, it should be understood that $a_0, a_1, \ldots, b_1, b_2, \ldots$ are the Fourier coefficients of the function f, and that F is the function to which the series converges.   In each case, the graph of f should be sketched; and F should be sketched also, in those cases in which F is different from f.

Compute the Fourier coefficients for each of the following functions.

1. $f(x) = x$ 2. $f(x) = x^2$ 3. $f(x) = x + x^2$

4. $f(x) = x^3$ 5. $f(x) = x - x^3/\pi^2$ 6. $f(x) = |x|$

7. $f(x) = x$ on $[0, \pi]$, $f(x) = 0$ on $[-\pi, 0]$

8. $f(x) = -x$ on $[-\pi, 0]$, $f(x) = 0$ on $[0, \pi]$

9. $f(x) = x$ on $[0, \pi]$, $f(x) = 1$ on $[-\pi, 0]$

10. $f(x) = 1$ on $[0, \pi]$, $f(x) = -1$ on $[-\pi, 0]$

11. $f(x) = 2x$ on $[0, \pi]$, $f(x) = x$ on $[-\pi, 0]$

12. For the odd functions x and x^3, you found that the series used only sines, with $a_i = 0$ for each i. Show that this happens for every odd function (f is *odd* if $f(-x) = -f(x)$ for every x).

13. Similarly, show that if f is even (with $f(-x) = f(x)$), then the series uses cosines only.

14. Show that for each f in $C^0[-\pi, \pi]$, $\|f\|^2 \leq 2\pi \|f\|_u^2$.

15. In Theorem A, f and f' are continuous, and since

$$\text{(1)} \quad \|f - \Pr_n f\|_u \to 0,$$

we have

$$\text{(2)} \quad \text{U} \lim_{n \to \infty} \Pr_n f = f \quad \text{on} \quad [-\pi, \pi].$$

(See Problem 10 of Problem Set 11.6) Can (2) hold if f is not continuous?

16. Let f be as in Theorem A. Assuming that Theorem A is true, show that

$$\lim_{n \to \infty} \| f - \mathrm{Pr}_n f \| = 0.$$

17. Working merely on the basis of the formulas for the Fourier coefficients of a continuous function f, give a geometric plausibility argument for the statement

$$\lim_{n \to \infty} a_n = \lim_{n \to \infty} b_n = 0.$$

(What does the graph of $y = \cos nx$ look like, when n is very large? How about the graph of $y = f(x) \cos nx$?)

18. In Theorem 6, under what simple conditions does the equality hold?

12.2 UNIFORM APPROXIMATIONS BY TRIGONOMETRIC POLYNOMIALS

The purpose of this section is to show that for every continuous function f, with period 2π, and every $\epsilon > 0$, there is a trigonometric polynomial

$$\phi(x) = a_0 + \sum_{i=1}^{n} (a_i \cos ix + b_i \sin ix)$$

such that

$$\| f - \phi \|_u < \epsilon.$$

That is,

$$|f(x) - \phi(x)| < \epsilon \qquad \text{for every } x.$$

Here we are not claiming that the coefficients in ϕ are Fourier coefficients. In fact, if we have a ϕ which makes $\| f - \phi \|_u < \epsilon$, and we want to improve the approximation, using an $\epsilon' < \epsilon$, we cannot always do this merely by adding new terms to the old ϕ; we may have to start afresh, with new coefficients even in the first few terms.

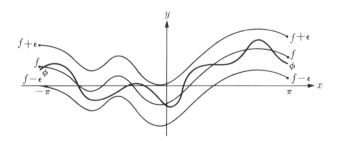

The first clue to this situation is that the trigonometric polynomials form a bigger system than one might think:

Theorem 1. The set of all trigonometric polynomials is closed under multiplication.

From this it follows immediately that $\sin^2 x$, $\cos^2 x$, ... $\sin^n x$, $\cos^n x$ are trigonometric polynomials. For example,

$$\cos^2 x = \frac{1 + \cos 2x}{2} = \tfrac{1}{2} + \tfrac{1}{2} \cos 2x.$$

The general proof depends on the trigonometric identities

$$\cos A \cos B = \tfrac{1}{2}[\cos (A + B) + \cos (A - B)],$$

$$\sin A \sin B = -\tfrac{1}{2}[\cos (A + B) - \cos (A - B)],$$

$$\sin A \cos B = \tfrac{1}{2}[\sin (A + B) + \sin (A - B)].$$

These are easy consequences of the addition formulas for the sine and cosine.
Consider now two trigonometric polynomials

$$a_0 = \sum_{i=1}^{n} (a_i \cos ix + b_i \sin ix), \qquad A_0 + \sum_{j=1}^{m} (A_j \cos jx + B_j \sin jx).$$

Every term of the product has one of the forms

$$a_0 A_0, \qquad a_0 A_j \cos jx, \qquad a_0 B_j \sin jx, \qquad a_i A_0 \cos ix,$$

$$a_i A_j \cos ix \cos jx, \qquad a_i B_j \cos ix \sin jx, \qquad b_i A_0 \sin ix,$$

$$b_i A_j \sin ix \cos jx, \qquad b_i B_j \sin ix \sin jx.$$

Each such term is a trigonometric polynomial. Therefore so also is their sum.

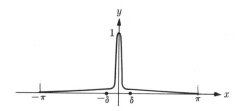

Consider now the function $g(t) = \cos^{2n} t$, which we now know to be a trigonometric polynomial. If n is large, then $\cos^{2n} t \approx 1$ only when $t \approx 0$; everywhere else, $\cos^{2n} t \approx 0$. Thus the graph looks something like the figure above, on the interval $[-\pi, \pi]$. Let δ be any number between 0 and π, and let

$$I_n = \int_{-\pi}^{\pi} \cos^{2n} t \, dt,$$

$$J_n = \int_{-\pi}^{-\delta} \cos^{2n} t \, dt = \int_{\delta}^{\pi} \cos^{2n} t \, dt,$$

$$K_n = \int_{-\delta}^{\delta} \cos^{2n} t \, dt.$$

We shall show that when n is large, $I_n \approx K_n$. That is:

Theorem 2. For each $\delta > 0$, $\lim_{n \to \infty} J_n/I_n = 0$.

This does not follow from the fact that $J_n \to 0$, because $I_n \to 0$ also. To prove the theorem, we need to get good estimates of J_n and I_n. The first of these is easy:

$$J_n = \int_{\delta}^{\pi} \cos^{2n} t \, dt < (\pi - \delta) \cos^{2n} \delta.$$

To estimate I_n, we use the reduction formula

$$\int \cos^n x \, dx = \frac{1}{n} \cos^{n-1} x \sin x + \frac{n-1}{n} \int \cos^{n-2} x \, dx.$$

(See Problem 31 of Problem Set 6.5) The formula can be derived by integration by parts. This gives a recursion formula for the definite integral:

$$\int_{-\pi}^{\pi} \cos^n x \, dx = \frac{n-1}{n} \int_{-\pi}^{\pi} \cos^{n-2} x \, dx.$$

Using $2n$ for n, we get

$$I_n = \int_{-\pi}^{\pi} \cos^{2n} x \, dx = \frac{2n-1}{2n} \int_{-\pi}^{\pi} \cos^{2(n-1)} x \, dx = \frac{2n-1}{2n} I_{n-1}.$$

Therefore

$$I_n = \frac{2n-1}{2n} \cdot \frac{2n-3}{2n-2} \cdot \frac{2n-5}{2n-4} \cdots \frac{1}{2} \int_{-\pi}^{\pi} \cos^0 x \, dx$$

$$= \frac{1 \cdot 3 \cdot 5 \cdots (2n-3)(2n-1)}{2 \cdot 4 \cdot 6 \cdots (2n-2)2n} \cdot 2\pi$$

$$= \frac{3}{2} \cdot \frac{5}{4} \cdot \frac{7}{6} \cdots \frac{2n-1}{2n-2} \cdot \frac{1}{2n} \cdot 2\pi > \frac{\pi}{n}.$$

Therefore

$$\frac{J_n}{I_n} < \frac{(\pi - \delta) \cos^{2n} \delta}{\pi/n} = \frac{\pi - \delta}{\pi} \cdot n \cos^{2n} \delta.$$

Since $0 < \cos \delta < 1$, it follows that $J_n/I_n \to 0$. [In fact, $\sum_{i=1}^{\infty} (J_i/I_i)$ converges; the easiest way to see this is to recall that $\sum ki(x^2)^i$ converges for $0 < x < 1$.]

Now let f be any continuous function with period 2π, and for each n let

$$\phi_n(x) = \frac{\int_{-\pi}^{\pi} f(x + t) \cos^{2n} t \, dt}{\int_{-\pi}^{\pi} \cos^{2n} t \, dt}.$$

It is plausible to suppose that

$$n \approx \infty \Rightarrow \phi_n(x) \approx f(x).$$

Roughly speaking, the reason is that for each $\delta > 0$,

$$n \approx \infty \Rightarrow \phi_n(x) \approx \frac{1}{I_n} \int_{-\delta}^{\delta} f(x + t) \cos^{2n} t \, dt;$$

if $\delta \approx 0$, then $f(x + t) \approx f(x)$ for $-\delta \leq t \leq \delta$, and so

$$n \approx \infty \Rightarrow \phi_n(x) \approx \frac{1}{I_n} f(x) \int_{-\delta}^{\delta} \cos^{2n} t \, dt$$

$$\approx \frac{1}{I_n} f(x) I_n = f(x).$$

As we shall see, these ideas can be built up into a proof that

$$\lim_{n \to \infty} \|f - \phi_n\|_u = 0.$$

First, however, we need the following:

Theorem 3. For each n, ϕ_n is a trigonometric polynomial.

Proof. First we observe that

$$\phi_n(x) = \frac{1}{I_n} \int_{-\pi}^{\pi} f(x + t) \cos^{2n} t \, dt$$

$$= \frac{1}{I_n} \int_{-\pi}^{\pi} f(t) \cos^{2n} (t - x) \, dt.$$

(The integrand has period 2π, and so $\int_{-\pi}^{\pi}$ is unchanged if we slide the graph back and forth horizontally.) We know that $\cos^{2n} t$ is a trigonometric polynomial; say,

$$\cos^{2n} t = a_0 + \sum_{i=1}^{m} (a_i \cos it + b_i \sin it).$$

Therefore

$$\cos^{2n} (t - x) = a_0 + \sum_{i=1}^{m} (a_i \cos it \cos ix + a_i \sin it \sin ix$$

$$+ b_i \sin it \cos ix - b_i \cos it \sin ix).$$

Therefore

$$\int_{-\pi}^{\pi} f(t) \cos^{2n} (t - x) \, dt = \int_{-\pi}^{\pi} a_0 f(t) \, dt$$

$$+ \sum_{i=1}^{m} \left[\int_{-\pi}^{\pi} (a_i \cos it + b_i \sin it) f(t) \, dt \right] \cos ix$$

$$+ \sum_{i=1}^{m} \left[\int_{-\pi}^{\pi} (a_i \sin it - b_i \cos it) f(t) \, dt \right] \sin ix.$$

The coefficients here are complicated, but they are constants, and so the indicated integral is a trigonometric polynomial.

Theorem 4. If f is continuous, and has period 2π, and

$$\phi_n(x) = \frac{\int_{-\pi}^{\pi} f(x + t) \cos^{2n} t \, dt}{\int_{-\pi}^{\pi} \cos^{2n} t \, dt},$$

then

$$\lim_{n \to \infty} \|f - \phi_n\|_u = 0.$$

Proof. Given $\epsilon > 0$, we need an n such that

$$n \geq N \Rightarrow \|f - \phi_n\|_u < \epsilon.$$

By definition of the uniform norm, this means that

$$n \geq N \Rightarrow |f(x) - \phi_n(x)| < \epsilon \text{ for } -\pi \leq x \leq \pi.$$

Step 1. Since f is continuous on $[-\pi, \pi]$, f is bounded on $[-\pi, \pi]$. Also f is periodic. Therefore there is an M such that

$$|f(t)| < M \quad \text{for every } t. \tag{1}$$

Step 2. Since f is continuous on $[-2\pi, 2\pi]$, it follows that f is uniformly continuous on $[-2\pi, 2\pi]$. And f is periodic. Therefore there is a $\delta > 0$ such that

$$|x - x'| < \delta \Rightarrow |f(x) - f(x')| < \epsilon/2.$$

This means that

$$|t| < \delta \Rightarrow |f(x) - f(x + t)| < \epsilon/2. \tag{2}$$

We are now ready to calculate. For M as in Step 1, and δ as in Step 2, we have

$$|f(x) - \phi_n(x)| = \left| f(x) - \frac{\int_{-\pi}^{\pi} f(x + t) \cos^{2n} t \, dt}{\int_{-\pi}^{\pi} \cos^{2n} t \, dt} \right|$$

$$= \frac{1}{I_n} \left| f(x) \int_{-\pi}^{\pi} \cos^{2n} t \, dt - \int_{-\pi}^{\pi} f(x + t) \cos^{2n} t \, dt \right|$$

$$= \frac{1}{I_n} \left| \int_{-\pi}^{\pi} [f(x) - f(x + t)] \cos^{2n} t \, dt \right|$$

$$\leqq \frac{1}{I_n} \int_{-\pi}^{\pi} |f(x) - f(x + t)| \cos^{2n} t \, dt$$

$$= \frac{1}{I_n} \left[\int_{-\pi}^{-\delta} |f(x) - f(x + t)| \cos^{2n} t \, dt + \int_{-\delta}^{\delta} |f(x) - f(x + t)| \cos^{2n} t \, dt \right.$$

$$\left. + \int_{\delta}^{\pi} |f(x) - f(x + t)| \cos^{2n} t \, dt \right]$$

$$\leqq \frac{1}{I_n} \left[2M \int_{-\pi}^{-\delta} \cos^{2n} t \, dt + \frac{\epsilon}{2} \int_{-\delta}^{\delta} \cos^{2n} t \, dt + 2M \int_{\delta}^{\pi} \cos^{2n} t \, dt \right]$$

$$= \frac{1}{I_n} \left[4M J_n + \frac{\epsilon}{2} \int_{-\delta}^{\delta} \cos^{2n} t \, dt \right]$$

$$< 4M \cdot \frac{J_n}{I_n} + \frac{1}{I_n} \cdot \frac{\epsilon}{2} \int_{-\pi}^{\pi} \cos^{2n} t \, dt$$

$$= 4M \frac{J_n}{I_n} + \frac{\epsilon}{2}.$$

But $J_n/I_n \to 0$. Therefore there is an N such that

$$n \geqq N \Rightarrow 4M \frac{J_n}{I_n} < \frac{\epsilon}{2}.$$

We now have

$$n \geqq N \Rightarrow \|f - \phi_n\|_u < \epsilon,$$

which was to be proved.

This theorem has an immediate consequence for Fourier series.

Theorem 5. If f is continuous, and has period 2π, and

$$\text{Pr}_n f = a_0 + \sum_{i=1}^{n} (a_i \cos ix + b_i \sin ix)$$

is the nth partial sum of the Fourier series of f, then

$$\lim_{n \to \infty} \|f - \text{Pr}_n f\| = 0.$$

Proof. Let $\epsilon > 0$ be given, and let

$$\phi(x) = A_0 + \sum_{i=1}^{N} (A_i \cos ix + B_i \sin ix)$$

be a trigonometric polynomial of order N. Then

$$\|f - \text{Pr}_N f\| \leq \|f - \phi\|,$$

because $\text{Pr}_N f$ is the element of $\mathbf{T}_N$ which is closest to f. And

$$\|f - \phi\|^2 \leq 2\pi \|f - \phi\|_u^2.$$

(See Problem 14 of Problem Set 12.1) Therefore

$$\|f - \text{Pr}_N f\|^2 \leq 2\pi \|f - \phi\|_u^2.$$

We take ϕ so that

$$2\pi \|f - \phi\|_u^2 < \epsilon^2.$$

(This will hold whenever $\|f - \phi\|_u < \epsilon/\sqrt{2\pi}$.) It follows that

$$\|f - \text{Pr}_N f\| < \epsilon.$$

Since the norm of this difference is nonincreasing (Theorem 6 of Section 12.1), it follows that $\|f - \text{Pr}_n f\| < \epsilon$ for every $n \geq N$. Therefore $\lim_{n \to \infty} \|f - \text{Pr}_n f\| = 0$.

Finally, we observe that for each n there is a point x_n at which the error in the proposed approximation $f \approx \text{Pr}_n f$ is actually equal to 0.

Theorem 6. If f is continuous, then for each n there is an x_n such that $f(x_n) = \text{Pr}_n f(x_n)$.

Proof. By definition of the projection, $f - \text{Pr}_n f$ must be orthogonal to every vector in $\mathbf{T}_n$. In particular, the inner product $(f - \text{Pr}_n f) \cdot 1$ must be 0, because 1 is in $\mathbf{T}_n$. Therefore

$$\int_{-\pi}^{\pi} [f(x) - \text{Pr}_n f(x)] \cdot 1 \, dx = 0.$$

Here the integrand must vanish at some point x_n.

At first, this theorem may seem almost like a joke, but it isn't. See the following section.

PROBLEM SET 12.2

Theorem 1 implies that each of the following functions is a trigonometric polynomial. Compute these functions in the form of trigonometric polynomials.

1. $\sin^3 x$ 2. $\sin^2 x$ 3. $\cos^3 x$

4. $\cos^2 (2x)$ 5. $\sin^2 x \cos x$ 6. $\cos^2 x \sin x$

7. $\cos x \sin 2x$ 8. $\cos^4 x$ 9. $\sin^4 x$

10. $\sin^2 x \cos^2 x$

11. Suppose that in Theorem 3 we had used

$$\phi_n(x) = \frac{\int_{-\pi}^{\pi} f(x + t) \cos^n t \, dt}{\int_{-\pi}^{\pi} \cos^n t \, dt} .$$

Would Theorem 3 still have been true? (Either prove the theorem in the more general form, with odd exponents allowed, or give an example to show that the more general theorem is false.)

12. Show, by any method, that

$$\left[\int_{-\pi}^{\pi} f(x) \, dx \right]^2 \leq 2\pi \int_{-\pi}^{\pi} [f(x)]^2 \, dx.$$

[*Hint:* There is a quick method, on the basis of what you know now.]

13. Show that if f is as in Theorem 5, then

$$\lim_{n \to \infty} \int_{-\pi}^{\pi} |f(x) - \mathrm{Pr}_n f(x)| \, dx = 0.$$

14. Show that if f is as in Theorem 5, then

$$\lim_{n \to \infty} \int_{-\pi}^{\pi} \mathrm{Pr}_n f(x) \, dx = \int_{-\pi}^{\pi} f(x) \, dx.$$

15. Now show that the same result holds on every interval $[a, b]$.

*16. In Section 10.8 we proved the binomial theorem

$$(1 + x)^n = \sum_{i=0}^{n} \binom{n}{i} x^i$$

by the methods of calculus, in the real domain. Thus the proof in Section 10.8 does not show that

$$(1 + z)^n = \sum_{j=0}^{n} \binom{n}{j} z^j$$

for every complex number z. Prove the latter theorem, by induction.

*17. Now use the result of Problem 16 to get an explicit formula for $\cos^4 x$ in the form of a trigonometric polynomial, with coefficients given numerically.

*18. Now get a general formula for $\cos^n x$ as a trigonometric polynomial.

[Note that in the text, we did not need the full force of Theorem 1; all we needed was the result of Problem 18; and the proof of the special result is neater. But Problem 18 is misleadingly special; and Theorem 1 ought to be regarded as the "real reason" why $\cos^{2n} x$ is a trigonometric polynomial.]

19. In the proof of Theorem 3, we used the function

$$\phi_n(x) = \frac{\int_{-\pi}^{\pi} f(x+t) \cos^{2n} t \, dt}{\int_{-\pi}^{\pi} \cos^{2n} t \, dt} = \frac{\int_{-\pi}^{\pi} f(t) \cos^{2n}(t-x) \, dt}{\int_{-\pi}^{\pi} \cos^{2n} t \, dt}.$$

Let $f(x) = x^2$, on the interval $[-\pi, \pi]$; and extend the graph so as to get a function of period 2π. (See the figure on p. 547.) For this function f, compute the function $\phi_2(x)$ in the form of a trigonometric polynomial, using definite integrals as coefficients. You *need not* compute the integrals numerically.

20. Given a trigonometric polynomial

$$\phi(x) = a_0 + \sum_{i=1}^{n} (a_i \cos ix + b_i \sin x).$$

Is it always true that ϕ is its own Fourier series? That is, do we have

$$\text{Pr}_m \phi = \phi \qquad \text{for } m \geq n?$$

Why or why not?

*21. Let f be a continuous function on $[0, 1]$. Show that for every $\epsilon > 0$ there is a polynomial $p(x) = \sum_{i=0}^{n} a_i x^i$ such that

$$|f(x) - p(x)| < \epsilon \qquad (0 \leq x \leq 1).$$

(This is a celebrated theorem due to Karl Weierstrass.)

*22. Given that the theorem proposed in Problem 21 is true, show that if f is continuous on $[a, b]$, and $\epsilon > 0$, then there is a polynomial $p(x)$ such that

$$|f(x) - p(x)| < \epsilon \qquad \text{for } a \leq x \leq b.$$

12.3 INTEGRATION OF FOURIER SERIES.
THE UNIFORM CONVERGENCE THEOREM

We defined the Fourier series

$$a_0 + \sum_{i=1}^{\infty} (a_i \cos ix + b_i \sin ix)$$

by using the projections

$$\text{Pr}_n f = a_0 + \sum_{i=1}^{n} (a_i \cos ix + b_i \sin ix)$$

of f into the subspace $\mathbf{T}_n$ of trigonometric polynomials of order n. We now want to show that if f has period 2π, and f' is continuous, then:

1) the Fourier series of f converges;
2) the sum of the Fourier series is the function f;
3) the series can be integrated a term at a time.

The ideas suggested by (3) are the key to the situation: to prove convergence, we first need to find out how the operations Pr_n are related to differentiation and integration.

Theorem 1. If f has period 2π, and f' is continuous, then

$$\text{Pr}_n f' = (\text{Pr}_n f)'.$$

That is, the projection of the derivative is the derivative of the projection.

Proof. Let

$$\text{Pr}_n f = a_0 + \sum_{i=1}^{n} (a_i \cos ix + b_i \sin ix),$$

$$\text{Pr}_n f' = A_0 + \sum_{i=1}^{n} (A_i \cos ix + B_i \sin ix).$$

We need to show that $A_0 = 0$, $A_i = ib_i$, $B_i = -iA_i$. The Fourier coefficients for f are given by the formulas

$$a_0 = \frac{1}{2\pi} \int_{-\pi}^{\pi} f(x)\, dx,$$

$$a_i = \frac{1}{\pi} \int_{-\pi}^{\pi} f(x) \cos ix\, dx \qquad (i > 0),$$

$$b_i = \frac{1}{\pi} \int_{-\pi}^{\pi} f(x) \sin ix\, dx.$$

Similarly,

$$A_0 = \frac{1}{2\pi} \int_{-\pi}^{\pi} f'(x)\, dx,$$

$$A_i = \frac{1}{\pi} \int_{-\pi}^{\pi} f'(x) \cos ix\, dx \qquad (i > 0),$$

$$B_i = \frac{1}{\pi} \int_{-\pi}^{\pi} f'(x) \sin ix\, dx.$$

By the fundamental theorem of integral calculus,

$$A_0 = \frac{1}{2\pi} [f(\pi) - f(-\pi)] = 0.$$

In A_i for $i > 0$, we integrate by parts, using

$$u = \cos ix, \qquad dv = f'(x)\, dx, \qquad du = -i \sin ix, \qquad v = f(x).$$

This gives

$$A_i = \frac{1}{\pi} [f(x) \cos ix]_{-\pi}^{\pi} + \frac{1}{\pi} \int_{-\pi}^{\pi} if(x) \sin ix\, dx = 0 + ib_i = ib_i.$$

Similarly, using $u = \sin ix$, $dv = f'(x)\, dx$, $du = i \cos ix\, dx$, $v = f(x)$, we get

$$B_i = \frac{1}{\pi} [f(x) \sin ix]_{-\pi}^{\pi} - \frac{1}{\pi} \int_{-\pi}^{\pi} if(x) \cos ix\, dx = 0 - ia_i = -ia_i.$$

Note that Theorem 1, as it stands, does not tell us that

$$D\left[a_0 + \sum_{i=1}^{\infty} (a_i \cos ix + b_i \sin ix)\right] = \sum_{i=1}^{\infty} (-ia_i \sin ix + ib_i \cos ix).$$

In fact, at this stage, we don't know that either of the indicated series converges to any function at all.

We now propose to find out how Pr_n is related to integration. Theorem 5 of Section 12.2 says that

$$\lim_{n \to \infty} \|f - \text{Pr}_n f\| = 0,$$

which means that

$$\|f - \text{Pr}_n f\|^2 = \int_{-\pi}^{\pi} [f(x) - \text{Pr}_n f(x)]^2 \, dx \to 0.$$

What we want is

$$\int_{-\pi}^{\pi} |f(x) - \text{Pr}_n f(x)| \, dx \to 0.$$

For this purpose we need the following:

Theorem 2. If g is continuous on $[a, b]$, then

$$\left[\int_a^b g(x) \, dx\right]^2 \leq (b - a) \int_a^b [g(x)]^2 \, dx.$$

Proof. Let $\mathbf{C}^0[a, b]$ be the set of all continuous functions on $[a, b]$. Under the usual definitions of $+$, sm, and $\cdot$, $\mathbf{C}^0[a, b]$ forms an inner-product space, and so the Schwarz inequality holds. In the inequality

$$(A \cdot B)^2 \leq \|A\|^2 \cdot \|B\|^2,$$

we take $A = g$, $B = 1$. This gives

$$\left[\int_a^b g(x) \cdot 1 \, dx\right]^2 \leq \left[\int_a^b g(x) \cdot g(x) \, dx\right]\left[\int_a^b (1 \cdot 1) \, dx\right]$$

$$= (b - a) \int_a^b [g(x)]^2 \, dx.$$

This tells us that the integral of a function is small if the norm of the function is small. Applying this principle to the function $|f - \text{Pr}_n f|$, we obtain the following theorem.

Theorem 3. Let f be a continuous function, of period 2π, and let

$$M_n = \int_{-\pi}^{\pi} |f(x) - \text{Pr}_n f(x)| \, dx.$$

Then

$$\lim_{n \to \infty} M_n = 0.$$

Proof

$$M_n^2 = \left[\int_{-\pi}^{\pi} |f(x) - \mathrm{Pr}_n f(x)| \, dx \right]^2$$

$$\leq 2\pi \int_{-\pi}^{\pi} [f(x) - \mathrm{Pr}_n f(x)]^2 \, dx$$

$$= 2\pi \, \|f - \mathrm{Pr}_n f\|^2.$$

By Theorem 5 of Section 12.2, the last of these expressions approaches 0. Therefore $M_n^2 \to 0$, and $M_n \to 0$.

We are now ready to prove a convergence theorem. If f is as in Theorem 3, and f' is continuous, then f satisfies the conditions of Theorem 3. Let

$$M_n = \int_{-\pi}^{\pi} |f'(x) - \mathrm{Pr}_n f'(x)| \, dx.$$

Then

$$\lim_{n \to \infty} M_n = 0.$$

Let x_n and x be any points of $[-\pi, \pi]$. Since $\mathrm{Pr}_n f' = (\mathrm{Pr}_n f)'$, we have

$$\int_{x_n}^{x} [f'(t) - \mathrm{Pr}_n f'(t)] \, dt = [f(t) - \mathrm{Pr}_n f(t)]_{x_n}^{x}$$

$$= f(x) - \mathrm{Pr}_n f(x) - [f(x_n) - \mathrm{Pr}_n f(x_n)].$$

By Theorem 6 of Section 12.2, we can choose each x_n so that $f(x_n) = \mathrm{Pr}_n f(x_n)$. This gives

$$|f(x) - \mathrm{Pr}_n f(x)| = \left| \int_{x_n}^{x} [f'(t) - \mathrm{Pr}_n f'(t)] \, dt \right| \leq \int_{x_n}^{x} |f'(t) - \mathrm{Pr}_n f'(t)| \, dt \leq M_n.$$

Therefore $\|f - \mathrm{Pr}_n f\|_u \leq M_n$. This gives:

Theorem 4. If f has period 2π, and f' is continuous, then

$$\lim_{n \to \infty} \|f - \mathrm{Pr}_n f\|_u = 0.$$

In the language of Section 10.15, this takes the form:

Theorem 4'. If f has period 2π, and f' is continuous, then

$$\mathrm{U} \lim_{n \to \infty} \mathrm{Pr}_n f = f \qquad \text{on } (-\infty, \infty).$$

In each case, the reason is that the differences $f(x) - \mathrm{Pr}_n f(x)$ are squeezed to 0 by a sequence of positive constants. Finally, all this can be restated in terms of the formula

$$\mathrm{Pr}_n f(x) = a_0 + \sum_{i=1}^{\infty} (a_i \cos ix + b_i \sin ix),$$

and the formulas for the Fourier coefficients a_i and b_i. This gives a third form of the theorem:

Theorem 4″. Let f be a function with period 2π and a continuous derivative. Let

$$a_0 = \frac{1}{2\pi} \int_{-\pi}^{\pi} f(x)\, dx, \qquad a_i = \frac{1}{\pi} \int_{-\pi}^{\pi} f(x) \cos ix\, dx, \qquad b_i = \frac{1}{\pi} \int_{-\pi}^{\pi} f(x) \sin ix\, dx.$$

Then

$$\mathrm{U}\lim_{n\to\infty} \left[a_0 + \sum_{i=1}^{n} (a_i \cos ix + b_i \sin ix) \right] = f(x) \quad \text{on } (-\infty, \infty).$$

Just as we found for power series, in Chapter 10, uniform convergence enables us to integrate a term at a time. In general:

Theorem 5. If the functions f_n are continuous, and $\mathrm{U}\lim_{n\to\infty} f_n = f$ on $[a, b]$, then

$$\lim_{n\to\infty} \int_a^b f_n(x)\, dx = \int_a^b f(x)\, dx.$$

This gives:

Theorem 6. If f has period 2π and f' is continuous, then the Fourier series for f can be integrated a term at a time, on any interval.

We found, at the beginning of this section, that $\mathrm{Pr}_n f' = (\mathrm{Pr}_n f)'$. If

$$f'(x) = \lim_{n\to\infty} \mathrm{Pr}_n f',$$

it follows that the series for f can be differentiated a term at a time. But we have proved Theorem 4 only for functions with continuous derivatives; and so our convergence theorem applies to f' only when f'' is continuous. Hence the heavy hypotheses in the following theorem:

Theorem 7. If f has a period 2π and f' and f'' are continuous, then the Fourier series for f can be differentiated a term at a time.

That is, for $f(x) = \lim_{n\to\infty} \mathrm{Pr}_n f$, we have

$$f'(x) = \lim_{n\to\infty} \mathrm{Pr}_n f'(x) = \lim_{n\to\infty} [\mathrm{Pr}_n f(x)]'.$$

(We get the first of these equations from Theorem 4, and the second from Theorem 1.)

In the further development of the theory, in which we allow discontinuous functions, the status of Theorem 7 is very different from that of Theorems 4 and 6, in that the latter two theorems have quite satisfactory generalizations, but Theorem 7 does not.

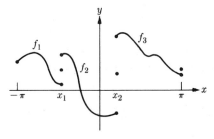

Suppose that we form a function of period 2π by fitting together the graphs of a finite set of continuously differentiable functions, end to end. We reconcile the values at common endpoints by defining

$$f(x_1) = \tfrac{1}{2}[f_1(x_1) + f_2(x_1)],$$

and so on; and we reconcile the values at $-\pi$ and π by defining

$$f(\pi) = f(-\pi) = \tfrac{1}{2}[f_1(-\pi) + f_3(\pi)].$$

A function obtained in this way will be called a function *of the Fourier type*. Evidently functions of this type are integrable, and so every such f has a Fourier series, with coefficients given by the same integral formulas as for continuous functions. The following turns out to be true:

Theorem B. Let f be a function of the Fourier type, with Fourier series

$$\sum (x) = a_0 + \sum_{i=1}^{\infty} (a_i \cos ix + b_i \sin ix).$$

Then

1) For every x, $\sum (x)$ converges to $f(x)$.

2) The convergence is uniform on every closed interval on which f is continuous.

3) The series can be integrated a term at a time on any closed interval (even if the interval contains points of discontinuity).

4) $\lim\limits_{n \to \infty} \| f - \text{Pr}_n f \| = 0$.

But it is not necessarily true that the series for a function of the Fourier type can be differentiated a term at a time, even on an interval on which f, f', and f'' are continuous. This will be brought out in the problem set below. In working on these problems, you should regard Theorem B as given.

PROBLEM SET   12.3

1. Show that if f is a function of the Fourier type, then

$$\left[\int_{-\pi}^{\pi} f(x)\, dx \right]^2 \leq 2\pi \int_{-\pi}^{\pi} [f(x)]^2\, dx.$$

(A simple proof is possible, if you use the right ideas.)

2. Let f be the function defined by the conditions

(1) $f(x) = 1$ on $(0, \pi)$,   (2) $f(0) = f(\pi) = 0$,

(3) $f(x) = -1$ on $(-\pi, 0)$,   (4) $f(x + 2n\pi) = f(x)$.

Calculate the series for f, and discuss the series obtained by termwise differentiation.

3. Let f be the function defined by the conditions

(1) $f(x) = e^x$ on $(-\pi, \pi)$,

(2) $f(-\pi) = f(\pi) = \tfrac{1}{2}(e^{-\pi} + e^{\pi})$,

(3) $f(x + 2n\pi) = f(x)$, for every x.

Evidently f has a Fourier series

$$a_0 + \sum_{i=1}^{\infty} (a_i \cos ix + b_i \sin ix).$$

Show that termwise differentiation of the series for f cannot give the Fourier series for f'.

4. Now calculate the Fourier series of f. (You did not need to do this, to solve Problem 3.)

5. Now verify, by a calculation, that for this series,

$$\int_0^x \left[a_0 + \sum_{i=1}^{\infty} a_i \cos it + b_i \sin it \right] dt = \int_0^x a_0 \, dt + \sum_{i=1}^{\infty} \int_0^x (a_i \cos it + b_i \sin it) \, dt.$$

Problems 6 through 8. Let f be the function defined by the conditions

(1) $f(x) = e^x$ on $(0, \pi)$,   (2) $f(0) = 0$, $f(\pi) = \frac{1}{2}(e^\pi - e^{-\pi})$,

(3) $f(x) = -e^x$ on $(-\pi, 0)$,   (4) $f(x + 2n\pi) = f(x)$, for every x.

Proceed as in Problems 3 through 5.

*9. Let f and g be functions of the Fourier type, and let $\sum (x)$ be the series for f, as in the text. Show that

$$\int_{-\pi}^{\pi} f(x)g(x) \, dx = \int_{-\pi}^{\pi} a_0 g(x) \, dx + \sum_{i=1}^{\infty} \int_{-\pi}^{\pi} [a_i \cos ix + b_i \sin ix]g(x) \, dx.$$

*10. Show that if f is as in Problem 1, then

$$\|f\|^2 = \sum_{i=0}^{\infty} a_i^2 + \sum_{i=1}^{\infty} b_i^2.$$

Linear Transformations, Matrices, and Determinants

13.1 LINEAR TRANSFORMATIONS

Let $\mathbf{R}^n$ and $\mathbf{R}^m$ be Cartesian spaces of any dimension. A function

$$f: \mathbf{R}^n \to \mathbf{R}^m$$

is called a *linear* function, or a *linear transformation*, if it preserves sums and scalar products. That is, f is linear if

$$f(P + Q) = f(P) + f(Q) \quad \text{for every } P, Q, \tag{1}$$

and

$$f(\alpha P) = \alpha f(P) \quad \text{for every } \alpha, P. \tag{2}$$

If f is linear, then

$$f(\alpha P + \beta Q) = f(\alpha P) + f(\beta Q) = \alpha f(P) + \beta f(Q); \tag{3}$$

and conversely, if (3) holds, then (1) and (2) both hold.

To see how linear transformations work, let us examine some special cases in low dimensions. Suppose that

$$f: \mathbf{R}^3 \to \mathbf{R}^3$$

is linear. In $\mathbf{R}^3$ we use the "standard basis"

$$\mathbf{B}^3 = \{E_1, E_2, E_3\},$$

where

$$E_1 = (1, 0, 0), \qquad E_2 = (0, 1, 0), \qquad E_3 = (0, 0, 1).$$

Now $f(E_1)$ must be a vector in $\mathbf{R}^3$; and $\mathbf{B}^3$ is a basis for $\mathbf{R}^3$. Therefore we have

$$f(E_1) = a_{11}E_1 + a_{21}E_2 + a_{31}E_3,$$

for some set of scalars a_{11}, a_{21}, a_{31}. [Here we are using double subscripts; a_{i1} is the coefficient of E_i in the expression for $f(E_1)$.] Similarly, $f(E_2)$ and $f(E_3)$ have the forms

$$f(E_2) = a_{12}E_1 + a_{22}E_2 + a_{32}E_3, \qquad f(E_3) = a_{13}E_1 + a_{23}E_2 + a_{33}E_3.$$

If $f(E_1), f(E_2)$, and $f(E_3)$ are known, then this determines $f(P)$ for every P in $\mathbf{R}^3$. The reason is that for

$$P = x_1E_1 + x_2E_2 + x_3E_3$$

we must have

$$f(P) = x_1 f(E_1) + x_2 f(E_2) + x_3 f(E_3).$$

This enables us to write a formula for $f(P)$ in terms of the numbers a_{ij}:

$$f(x_1E_1 + x_2E_2 + x_3E_3) = \quad a_{11}x_1E_1 + a_{21}x_1E_2 + a_{31}x_1E_3$$
$$+ \; a_{12}x_2E_1 + a_{22}x_2E_2 + a_{32}x_2E_3$$
$$+ \; a_{13}x_3E_1 + a_{23}x_3E_2 + a_{33}x_3E_3.$$

Collecting the coefficients of E_1, E_2, and E_3, we get:

$$f(x_1E_1 + x_2E_2 + x_3E_3) = \quad (a_{11}x_1 + a_{12}x_2 + a_{13}x_3)E_1$$
$$+ \; (a_{21}x_1 + a_{22}x_2 + a_{23}x_3)E_2$$
$$+ \; (a_{31}x_1 + a_{32}x_2 + a_{33}x_3)E_3$$
$$= \quad y_1E_1 + y_2E_2 + y_3E_3.$$

There is a special apparatus which enables us to write such formulas more compactly, and deal with them more conveniently. Evidently f is completely described by the $3 \cdot 3 = 9$ numbers a_{ij}. We write these in a rectangular array, like this:

$$M = \begin{bmatrix} a_{11} & a_{12} & a_{13} \\ a_{21} & a_{22} & a_{23} \\ a_{31} & a_{32} & a_{33} \end{bmatrix}.$$

Such an array M is called a *3 by 3 matrix*. (In general, an *m by n matrix* is a rectangular array of numbers, with m rows and n columns.)

At the beginning of Section 11.3, before we introduced a basis in $\mathbf{R}^3$, we represented the vector $P = x_1E_1 + x_2E_2 + x_3E_3$ by the ordered triplet (x_1, x_2, x_3). We now write this triplet vertically instead of horizontally, in the form of a "3 by 1 matrix"

$$P = \begin{bmatrix} x_1 \\ x_2 \\ x_3 \end{bmatrix}.$$

When P is described in this notation, we call P a *column vector*. Similarly, for

$$f(P) = y_1E_1 + y_2E_2 + y_3E_3 = (y_1, y_2, y_3),$$

we write

$$f(P) = \begin{bmatrix} y_1 \\ y_2 \\ y_3 \end{bmatrix} = \begin{bmatrix} a_{11}x_1 + a_{12}x_2 + a_{13}x_3 \\ a_{21}x_1 + a_{22}x_2 + a_{23}x_3 \\ a_{31}x_1 + a_{32}x_2 + a_{33}x_3 \end{bmatrix}.$$

Note that the array on the right is a column vector; once the indicated additions are performed, there is only one entry in each row.

Now we define the operation of *multiplication*, of the column vector P by the matrix M, in such a way that $f(P) = MP$. That is:

Definition

$$\begin{bmatrix} a_{11} & a_{12} & a_{13} \\ a_{21} & a_{22} & a_{23} \\ a_{31} & a_{32} & a_{33} \end{bmatrix} \begin{bmatrix} x_1 \\ x_2 \\ x_3 \end{bmatrix} = \begin{bmatrix} a_{11}x_1 + a_{12}x_2 + a_{13}x_3 \\ a_{21}x_1 + a_{22}x_2 + a_{23}x_3 \\ a_{31}x_1 + a_{32}x_2 + a_{33}x_3 \end{bmatrix}.$$

Under this definition of the "product" MP, we have

$$MP = \begin{bmatrix} y_1 \\ y_2 \\ y_3 \end{bmatrix} = f(P).$$

There is a usable pattern in this multiplication: to get the entry y_1 in the first row of the product, we regard the first row of M as a vector, and form its inner product with the column vector P. Similarly for the other rows.

Let us examine an example. The matrix

$$\begin{bmatrix} 1 & 2 & 1 \\ -2 & 1 & 2 \\ 1 & -1 & 2 \end{bmatrix}$$

describes a linear transformation f, with

$$f(x_1, x_2, x_3) = \begin{bmatrix} 1 & 2 & 1 \\ -2 & 1 & 2 \\ 1 & -1 & 2 \end{bmatrix} \begin{bmatrix} x_1 \\ x_2 \\ x_3 \end{bmatrix} = \begin{bmatrix} x_1 + 2x_2 + x_3 \\ -2x_1 + x_2 + 2x_3 \\ x_1 - x_2 + 2x_3 \end{bmatrix} = \begin{bmatrix} y_1 \\ y_2 \\ y_3 \end{bmatrix}.$$

Two questions arise naturally here.

Problem 1. Given a particular point (y_1, y_2, y_3), for what points P, if any, do we have $f(P) = (y_1, y_2, y_3)$? For example, for what points P is it true that

$$f(P) = (1, 2, 3)?$$

To answer this question, we need to solve the system

$$
\begin{cases}
\quad x_1 + 2x_2 + \ x_3 = 1, & \text{(1)} \\
-2x_1 + \ x_2 + 2x_3 = 2, & \text{(2)} \\
\quad x_1 - \ x_2 + 2x_3 = 3 & \text{(3)}
\end{cases}
$$

of linear equations in the unknowns x_1, x_2, x_3. Almost any method will do. We shall use a method which will be of theoretical importance later.

Step 1. Eliminate x_1 from (2) and (3), by adding twice (1) to (2) and subtracting (1) from (3). This gives a new system which is equivalent to the original system, in the sense that it has exactly the same solutions:

$$
\begin{aligned}
x_1 + 2x_2 + \ x_3 &= 1, & \text{(1)} \\
5x_2 + 4x_3 &= 4, & \text{(2')} = \text{(2)} + 2\text{(1)} \\
-3x_2 + \ x_3 &= 2. & \text{(3')} = \text{(2)} - \text{(1)}
\end{aligned}
$$

(The notations on the right indicate where the new equations come from.)

Step 2. Eliminate x_2 from (3'), by adding $\frac{3}{5}$ of (2') to (3'). This gives the equivalent system

$$
\begin{aligned}
x_1 + 2x_2 + \ x_3 &= 1, & \text{(1)} \\
5x_2 + 4x_3 &= 4, & \text{(2')} \\
\tfrac{17}{5}x_3 &= \tfrac{22}{5}. & \text{(3'')} = \text{(3')} + \tfrac{3}{5}\text{(2')}
\end{aligned}
$$

We now say that the system is in *triangular form*. In general, a system of n linear equations in n unknowns is in triangular form if the nth equation involves only x_n, the $(n-1)$st equation involves only x_{n-1} and x_n, and so on. (This means that in the matrix of coefficients, $a_{ij} = 0$ for $j < i$.)

Step 3. Solve, by successive substitutions, working from bottom to top, getting

$$x_3 = \tfrac{22}{17}, \qquad x_2 = -\tfrac{4}{17}, \qquad x_1 = \tfrac{3}{17}.$$

This means that

$$\begin{bmatrix} 1 & 2 & 1 \\ -2 & 1 & 2 \\ 1 & -1 & 2 \end{bmatrix}\begin{bmatrix} \tfrac{3}{17} \\ -\tfrac{4}{17} \\ \tfrac{22}{17} \end{bmatrix} = \begin{bmatrix} 1 \\ 2 \\ 3 \end{bmatrix}.$$

This should be checked.

Problem 2. Find, if possible, a set of formulas expressing $P = (x_1, x_2, x_3)$ in terms of (y_1, y_2, y_3).

To do this, we need to get a "general solution" of the equation

$$M\begin{bmatrix} x_1 \\ x_2 \\ x_3 \end{bmatrix} = \begin{bmatrix} 1 & 2 & 1 \\ -2 & 1 & 2 \\ 1 & -1 & 2 \end{bmatrix}\begin{bmatrix} x_1 \\ x_2 \\ x_3 \end{bmatrix} = \begin{bmatrix} y_1 \\ y_2 \\ y_3 \end{bmatrix},$$

in which the x_i's are expressed in terms of the y_i's. This is only slightly more troublesome than Problem 1; we treat (y_1, y_2, y_3) in exactly the same way as we treated $(1, 2, 3)$ in Problem 1. The solution is

$$x_1 = \tfrac{4}{17}y_1 - \tfrac{5}{17}y_2 + \tfrac{3}{17}y_3,$$
$$x_2 = \tfrac{6}{17}y_1 + \tfrac{1}{17}y_2 - \tfrac{4}{17}y_3,$$
$$x_3 = \tfrac{1}{17}y_1 + \tfrac{3}{17}y_2 + \tfrac{5}{17}y_3.$$

The coefficients in these equations give us a new matrix

$$M^{-1} = \begin{bmatrix} \tfrac{4}{17} & -\tfrac{5}{17} & \tfrac{3}{17} \\ \tfrac{6}{17} & \tfrac{1}{17} & -\tfrac{4}{17} \\ \tfrac{1}{17} & \tfrac{3}{17} & \tfrac{5}{17} \end{bmatrix}.$$

We call this matrix M^{-1} because it is the matrix of the transformation

$$f^{-1}: \mathbf{R}^3 \to \mathbf{R}^3.$$

That is, M^{-1} reverses the action of M, in the sense that

$$M^{-1}(MP) = P, \quad \text{for every } P = (x_1, x_2, x_3).$$

It may easily happen, however, that a linear transformation

$$f: \mathbf{R}^n \to \mathbf{R}^m$$

does not have an inverse

$$(?) \quad f^{-1}: \mathbf{R}^m \to \mathbf{R}^n. \quad (?)$$

There are two things that may go wrong: (1) some points Q of $\mathbf{R}^m$ may not be values of the function at all. In such cases, there is no such thing as $f^{-1}(Q)$. (2) Some points Q of $\mathbf{R}^m$ may be equal to $f(P)$ for more than one point P. In such cases, f is not invertible. An example of both these phenomena is furnished by the matrix

$$M = \begin{bmatrix} 1 & 0 & 0 \\ 0 & 1 & 0 \\ 0 & 0 & 0 \end{bmatrix}.$$

Here

$$M \begin{bmatrix} x_1 \\ x_2 \\ x_3 \end{bmatrix} = \begin{bmatrix} 1 & 0 & 0 \\ 0 & 1 & 0 \\ 0 & 0 & 0 \end{bmatrix} \begin{bmatrix} x_1 \\ x_2 \\ x_3 \end{bmatrix} = \begin{bmatrix} y_1 \\ y_2 \\ y_3 \end{bmatrix} = \begin{bmatrix} x_1 \\ x_2 \\ 0 \end{bmatrix}.$$

This function projects $\mathbf{R}^3$ into the xy-plane; no point outside the xy-plane is a value of the function, and $f(P) = f(P')$ whenever P and P' lie on the same vertical line.

To find out how a given linear transformation behaves, we try to compute its inverse, by the method used in the above example. If there is an inverse, the method gives it to us; and if there isn't, the method still tells us what is going on. Consider, for example, the transformation f described by the matrix

$$M = \begin{bmatrix} 1 & 2 & 3 \\ 2 & 5 & 8 \\ 1 & 4 & 7 \end{bmatrix}.$$

To simplify the notation, we use (x, y, z) for (x_1, x_2, x_3) and (a, b, c) for (y_1, y_2, y_3). We then have

$$MP = \begin{bmatrix} 1 & 2 & 3 \\ 2 & 5 & 8 \\ 1 & 4 & 7 \end{bmatrix} \begin{bmatrix} x \\ y \\ z \end{bmatrix} = \begin{bmatrix} x + 2y + 3z \\ 2x + 5y + 8z \\ x + 4y + 7z \end{bmatrix} = \begin{bmatrix} a \\ b \\ c \end{bmatrix}.$$

This gives the linear system

$$x + 2y + 3z = a,$$
$$2x + 5y + 8z = b,$$
$$x + 4y + 7z = c.$$

Reducing this to triangular form, we get the equivalent system

$$x + 2y + 3z = a, \tag{1}$$
$$y + 2z = b - 2a, \tag{2}$$
$$0 = 3a - 2b + c. \tag{3}$$

Thus the equation

$$f(P) = MP = (a, b, c)$$

cannot have a solution unless $3a - 2b + c = 0$. Let

$$E = \{(a, b, c) \mid 3a - 2b + c = 0\}.$$

Then E is a plane, and every point of E is $= f(P)$ for some P. The reason is that as long as (3) is satisfied, we can solve (2) and (1) in the forms

$$y = -2z + b - 2a, \tag{2'}$$

$$x = -2y - 3z + a$$

$$= z - 2b + 5a. \tag{1'}$$

Here we can choose any z; if

$$P = (z - 2b + 5a, -2z + b - 2a, z),$$

then

$$f(P) = (a, b, c).$$

As in Section 3.1, the *image* of a function is defined to be the set of all values of the function. If f is a function $A \to B$, then the image is denoted by $f(A)$. More generally, if A' is any subset of A, then

$$f(A') = \{b \mid b = f(a') \quad \text{for some } a' \text{ in } A'\}.$$

Thus, in the example above, the image is

$$f(\mathbf{R}^3) = E = \{(a, b, c) \mid 3a - 2b + c = 0\}.$$

The *kernel* of a linear transformation f is

$$\text{Ker} f = \{P \mid f(P) = O\}.$$

Obviously the kernel contains O, no matter what f may be, because

$$f(O) = O.$$

(*Proof.* $f(O) = f(P - P) = f(P) - f(P) = O$.)

If f is not one-to-one, however, then the kernel $\text{Ker} f$ contains vectors other than O

(*Proof.* If $f(P) = f(Q)$, for some $P \neq Q$, then $f(P - Q) = f(P) - f(Q) = O$, and so $P - Q$ belongs to $\text{Ker} f$.)

In the above example, the kernel is the solution set of the equation

$$\begin{bmatrix} 1 & 2 & 3 \\ 2 & 5 & 8 \\ 1 & 4 & 7 \end{bmatrix} \begin{bmatrix} x \\ y \\ z \end{bmatrix} = \begin{bmatrix} 0 \\ 0 \\ 0 \end{bmatrix}.$$

To get the solution, we set $a = b = 0$ in equations (2') and (1'). Therefore P is in the kernel if P has the form

$$P = (z, -2z, z).$$

Using α for z (to fit the usual notation of scalar multiplication), we get

$$\text{Ker} f = \{\alpha V_0\}, \qquad V_0 = (1, -2, 1).$$

Therefore $\text{Ker} f$ is a line. In other cases, the image may be an even smaller set, and the kernel even larger. For

$$M = \begin{bmatrix} 1 & 0 & 0 \\ 0 & 0 & 0 \\ 0 & 0 & 0 \end{bmatrix},$$

the image is the x-axis, because

$$\begin{bmatrix} 1 & 0 & 0 \\ 0 & 0 & 0 \\ 0 & 0 & 0 \end{bmatrix}\begin{bmatrix} x \\ y \\ z \end{bmatrix} = \begin{bmatrix} x \\ 0 \\ 0 \end{bmatrix}.$$

The kernel is the yz-plane, because the equation

$$M\begin{bmatrix} x \\ y \\ z \end{bmatrix} = \begin{bmatrix} 0 \\ 0 \\ 0 \end{bmatrix}$$

gives the system

$$x = 0, \qquad 0 = 0, \qquad 0 = 0,$$

whose solution set is obvious.

Finally, a few remarks on questions which the above discussion may suggest.

1) As far as the ideas in this section are concerned, there is nothing special about $\mathbf{R}^3$. Exactly the same methods apply, in exactly the same ways, when we deal with linear transformations $\mathbf{R}^n \to \mathbf{R}^n$. We have discussed $\mathbf{R}^3$ merely to avoid tedious notation, large matrices, and pointlessly long computations.

2) In the examples that we have discussed, the image and kernel of a linear transformation have turned out to be subspaces. This is what always happens. (The proof is easy, on the basis of Theorem 4 of Section 11.3. You will find it hardly more trouble to write it yourself than to consult the following section.)

3) If you are acquainted with determinants, and with the process of solving linear systems by Cramer's rule, then you may suspect that the method used above, by which we convert the system to a triangular system, is naive or inefficient or both. But this is not true. In computation, triangularization is about as good a method as any.

PROBLEM SET 13.1

In each of the following problems, you are given a matrix M, describing a linear transformation f. If f turns out to be invertible, compute f^{-1}. If not, find the image and the kernel. Thus the answer to each of the first ten problems below should be in one of the forms

$$\text{(a)} \quad M^{-1} = [\cdots]$$

or

$$\text{(b)} \quad f(\mathbf{R}^3) = \{(a, b, c) \mid \cdots\}, \qquad \text{and} \qquad \text{Ker } f = \{(x, y, z) \mid \cdots\}.$$

1. $\begin{bmatrix} 1 & 2 & 1 \\ 0 & 2 & 0 \\ 0 & 2 & 3 \end{bmatrix}$

2. $\begin{bmatrix} 3 & 1 & 3 \\ 1 & 1 & 1 \\ 2 & 1 & 0 \end{bmatrix}$

3. $\begin{bmatrix} 1 & 0 & 0 \\ 0 & 2 & 0 \\ 0 & 1 & 3 \end{bmatrix}$

4. $\begin{bmatrix} 2 & 1 & 1 \\ 1 & 3 & 2 \\ 2 & 8 & 6 \end{bmatrix}$

5. $\begin{bmatrix} 2 & 1 & -1 \\ 6 & 3 & -3 \\ -4 & -2 & +2 \end{bmatrix}$

6. $\begin{bmatrix} 6 & 3 & -3 \\ -4 & -3 & 2 \\ 4 & 2 & -2 \end{bmatrix}$

7. $\begin{bmatrix} 1 & 2 & 2 \\ 2 & 4 & 6 \\ 3 & 6 & 9 \end{bmatrix}$

8. $\begin{bmatrix} 5 & 0 & 1 \\ 2 & 2 & 2 \\ 0 & 0 & 0 \end{bmatrix}$

9. $\begin{bmatrix} 1 & 1 & 1 \\ 0 & 2 & 1 \\ 0 & 0 & 3 \end{bmatrix}$

10. $\begin{bmatrix} 0 & 1 & 0 \\ 1 & 0 & 0 \\ 0 & 0 & 1 \end{bmatrix}$

11. Suppose that the matrix of f has the form

$$M = \begin{bmatrix} a_1 & a_2 & a_3 \\ \alpha a_1 & \alpha a_2 & \alpha a_3 \\ \beta a_1 & \beta a_2 & \beta a_3 \end{bmatrix}.$$

What sort of subspace is $f(\mathbf{R}^3)$? How about $\operatorname{Ker} f$?

12. Same question, for matrices of the form

$$\begin{bmatrix} a_{11} & a_{12} & a_{13} \\ 0 & a_{22} & a_{23} \\ 0 & 0 & a_{33} \end{bmatrix},$$

where $a_{11}a_{22}a_{33} \neq 0$.

13. Same question, for

$$M = \begin{bmatrix} a_{11} & 0 & 0 \\ a_{21} & a_{22} & 0 \\ a_{31} & a_{32} & a_{33} \end{bmatrix},$$

with $a_{11}a_{22}a_{33} \neq 0$.

14. Same question, for

$$M = \begin{bmatrix} 0 & a_{12} & a_{13} \\ 0 & 0 & a_{23} \\ a_{31} & 0 & 0 \end{bmatrix},$$

with $a_{12}a_{23}a_{31} \neq 0$.

15. Same question, for

$$M = \begin{bmatrix} 0 & a_{12} & a_{13} \\ 0 & 0 & a_{23} \\ 0 & 0 & 0 \end{bmatrix}.$$

13.2 COMPOSITION OF LINEAR TRANSFORMATIONS AND MULTIPLICATION OF MATRICES

The preceding section was devoted almost entirely to investigation of examples of linear transformations. We shall now develop some of the theory.

Theorem 1. Let f be a linear transformation $\mathbf{R}^m \to \mathbf{R}^n$. Then the image $f(\mathbf{R}^m)$ is a subspace of $\mathbf{R}^n$, and the kernel $\operatorname{Ker} f$ is a subspace of $\mathbf{R}^m$.

Proof. By Theorem 4 of Section 11.3, we merely need to show that these sets are closed under addition and scalar multiplication. For each $f(P), f(Q)$, we have

$$f(P) + f(Q) = f(P + Q);$$

and for each $f(P)$, $\alpha f(P) = f(\alpha P)$. Therefore $f(\mathbf{R}^m)$ is a subspace. If P and Q belong to $\operatorname{Ker} f$, then

$$f(P) = f(Q) = O,$$

and so $f(P + Q) = f(P) + f(Q) = O + O = O$, and

$$f(\alpha P) = \alpha f(P) = \alpha \cdot O = O.$$

Therefore $\operatorname{Ker} f$ is a subspace.

Theorem 2. If f and g are linear transformations $\mathbf{R}^m \to \mathbf{R}^n$, then so also are $f + g$ and αf.

Here the sum and scalar product are defined by the obvious conditions

$$(f + g)(P) = f(P) + g(P),$$

$$(\alpha f)(P) = \alpha(f(P)).$$

The verification of linearity is trivial:

$$(f + g)(P + Q) = f(P + Q) + g(P + Q)$$

$$= f(P) + f(Q) + g(P) + g(Q)$$

$$= (f + g)(P) + (f + g)(Q);$$

$$(f + g)(\alpha P) = f(\alpha P) + g(\alpha P)$$

$$= \alpha f(P) + \alpha g(P) = \alpha(f(P) + g(P))$$

$$= \alpha(f + g)(P).$$

This gives:

Theorem 3. For each m and n, the linear transformations $f: \mathbf{R}^m \to \mathbf{R}^n$ form a vector space.

Theorem 4. If $g: \mathbf{R}^m \to \mathbf{R}^n$ and $f: \mathbf{R}^n \to \mathbf{R}^p$ are linear, then the composite function

$$f(g): \mathbf{R}^m \to \mathbf{R}^p$$

is also linear:

$$\mathbf{R}^m \xrightarrow{g} \mathbf{R}^n \xrightarrow{f} \mathbf{R}^p$$

$$f(g)$$

The verification is straightforward:

$$f[g(P + Q)] = f[g(P) + g(Q)] = f[g(P)] + f[g(Q)]$$

$$= f(g)(P) + f(g)(Q);$$

$$f[g(\alpha P)] = f[\alpha g(P)] = \alpha f[g(P)].$$

We found, in the preceding section, that the action of a linear transformation $\mathbf{R}^3 \to \mathbf{R}^3$ could be described by a 3 by 3 matrix. Similarly, the action of a linear transformation $g: \mathbf{R}^m \to \mathbf{R}^n$ can be described by an n by m matrix. The scheme is as follows. If

$$g(E_j) = b_{1j}E_1 + b_{2j}E_2 + \cdots + b_{nj}E_n,$$

then

$$g(x_j E_j) = b_{1j}x_j E_1 + b_{2j}x_j E_2 + \cdots + b_{nj}x_j E_n.$$

Therefore, for $P = (x_1, x_2, \ldots, x_m)$ in $\mathbf{R}^m$, we have

$$
\begin{aligned}
g(P) = \quad & b_{11}x_1E_1 + b_{21}x_1E_2 + \cdots + b_{n1}x_1E_n \\
+\ & b_{12}x_2E_1 + b_{22}x_2E_2 + \cdots + b_{n2}x_2E_n \\
& \quad\vdots \qquad\qquad\qquad\qquad\qquad\quad \vdots \\
+\ & b_{1m}x_mE_1 + b_{2m}x_mE_2 + \cdots + b_{nm}x_mE_n.
\end{aligned}
$$

Adding by columns, to get the total coefficient of each E_j on the right, we get

$$
g(P) = \left(\sum_{j=1}^{m} b_{1j}x_j\right)E_1 + \left(\sum_{j=1}^{m} b_{2j}x_j\right)E_2 + \cdots + \left(\sum_{j=1}^{m} b_{nj}x_j\right)E_n
$$
$$
= (y_1, y_2, \ldots, y_n).
$$

Describing P and $f(P)$ as column vectors, and representing g by the matrix with a_{ij} in the ith row and jth column, we get

$$
\begin{bmatrix}
b_{11} & b_{12} & \cdots & b_{1m} \\
b_{21} & b_{22} & \cdots & b_{2m} \\
\cdot & & & \cdot \\
\cdot & & & \cdot \\
\cdot & & & \cdot \\
b_{n1} & b_{n2} & \cdots & b_{nm}
\end{bmatrix}
\begin{bmatrix}
x_1 \\ x_2 \\ \cdot \\ \cdot \\ \cdot \\ x_m
\end{bmatrix}
=
\begin{bmatrix}
y_1 \\ y_2 \\ \cdot \\ \cdot \\ \cdot \\ y_n
\end{bmatrix},
$$

which has the form

$$
M_g
\begin{bmatrix}
x_1 \\ x_2 \\ \cdot \\ \cdot \\ \cdot \\ x_m
\end{bmatrix}
=
\begin{bmatrix}
y_1 \\ y_2 \\ \cdot \\ \cdot \\ \cdot \\ y_n
\end{bmatrix}.
$$

Here the pattern of the operation is the same as in the case $m = n = 3$: to get y_i in the column vector on the right, we regard the ith row of the matrix M_g as a vector, and form its inner product with the column vector.

Now if g and f are as in the preceding theorem, then each of the transformations $g, f,$ and $f(g)$ can be described by a matrix. Let these matrices be

$$
M_g = [b_{ij}] \qquad (n \text{ by } m),
$$
$$
M_f = [a_{jk}] \qquad (p \text{ by } n),
$$
$$
M_{f(g)} = [c_{ik}] \qquad (p \text{ by } m).
$$

Here $[a_{ij}]$ is a shorthand for the matrix with the number a_{ij} in the ith row and the jth column, and similarly for $[b_{jk}]$ and $[c_{ik}]$. We define the *product* of two matrices, in this case, to be the matrix of the composite function. That is,

Definition.   Given linear transformations

$$g: \mathbf{R}^m \to \mathbf{R}^n,$$

$$f: \mathbf{R}^n \to \mathbf{R}^p,$$

$$f(g): \mathbf{R}^m \to \mathbf{R}^p,$$

with associated matrices M_f, M_g, $M_{f(g)}$. By definition,

$$M_f M_g = M_{f(g)}.$$

We shall now get a formula for the product $M_f M_g$ of two matrices. The general formula looks complicated, but its pattern is easy to see by an examination of the case $m = n = p = 2$. Let the matrices of the transformations f, g, and $f(g)$ be

$$M_f = \begin{bmatrix} a_{11} & a_{12} \\ a_{21} & a_{22} \end{bmatrix}, \qquad M_g = \begin{bmatrix} b_{11} & b_{12} \\ b_{21} & b_{22} \end{bmatrix}, \qquad M_{f(g)} = \begin{bmatrix} c_{11} & c_{12} \\ c_{21} & c_{22} \end{bmatrix}.$$

Then

$$M_g \begin{bmatrix} x_1 \\ x_2 \end{bmatrix} = \begin{bmatrix} b_{11} & b_{12} \\ b_{21} & b_{22} \end{bmatrix} \begin{bmatrix} x_1 \\ x_2 \end{bmatrix} = \begin{bmatrix} b_{11}x_1 + b_{12}x_2 \\ b_{21}x_1 + b_{22}x_2 \end{bmatrix} = \begin{bmatrix} y_1 \\ y_2 \end{bmatrix},$$

and so

$$M_{f(g)} \begin{bmatrix} x_1 \\ x_2 \end{bmatrix} = M_f M_g \begin{bmatrix} x_1 \\ x_2 \end{bmatrix} = M_f \begin{bmatrix} y_1 \\ y_2 \end{bmatrix}$$

$$= \begin{bmatrix} a_{11} & a_{12} \\ a_{21} & a_{22} \end{bmatrix} \begin{bmatrix} y_1 \\ y_2 \end{bmatrix} = \begin{bmatrix} a_{11}y_1 + a_{12}y_2 \\ a_{21}y_1 + a_{22}y_2 \end{bmatrix}$$

$$= \begin{bmatrix} a_{11}(b_{11}x_1 + b_{12}x_2) + a_{12}(b_{21}x_1 + b_{22}x_2) \\ a_{21}(b_{11}x_1 + b_{12}x_2) + a_{22}(b_{21}x_1 + b_{22}x_2) \end{bmatrix}$$

$$= \begin{bmatrix} (a_{11}b_{11} + a_{12}b_{21})x_1 + (a_{11}b_{12} + a_{12}b_{22})x_2 \\ (a_{21}b_{11} + a_{22}b_{21})x_1 + (a_{21}b_{12} + a_{22}b_{22})x_2 \end{bmatrix}$$

$$= \begin{bmatrix} a_{11}b_{11} + a_{12}b_{21} & a_{11}b_{12} + a_{12}b_{22} \\ a_{21}b_{11} + a_{22}b_{21} & a_{21}b_{12} + a_{22}b_{22} \end{bmatrix} \begin{bmatrix} x_1 \\ x_2 \end{bmatrix}.$$

Therefore $M_f M_g$ is the 2 by 2 matrix in the last formula. The pattern of the operation is clear: to get the number c_{ij}, in the ith row and jth column of the product, we regard the ith row of M_f and the jth column of M_g as vectors, and compute their inner product. That is,

$$c_{ij} = a_{i1}b_{1j} + a_{i2}b_{2j}.$$

This is called the *row by column rule* of matrix multiplication. The same rule applies in the general case, and the only problem is that the formulas are complicated to write down. We have

$$M_g \begin{bmatrix} x_1 \\ x_2 \\ \cdot \\ \cdot \\ \cdot \\ x_m \end{bmatrix} = \begin{bmatrix} b_{11} & b_{12} & \cdots & b_{1m} \\ b_{21} & b_{22} & \cdots & b_{2m} \\ \cdot & & & \cdot \\ \cdot & & & \cdot \\ \cdot & & & \cdot \\ b_{n1} & b_{n2} & \cdots & b_{nm} \end{bmatrix} \begin{bmatrix} x_1 \\ x_2 \\ \cdot \\ \cdot \\ \cdot \\ x_m \end{bmatrix} = \begin{bmatrix} \sum_{j=1}^{m} b_{1j}x_j \\ \sum_{j=1}^{m} b_{2j}x_j \\ \cdot \\ \cdot \\ \cdot \\ \sum_{j=1}^{m} b_{nj}x_j \end{bmatrix} = \begin{bmatrix} y_1 \\ y_2 \\ \cdot \\ \cdot \\ \cdot \\ y_n \end{bmatrix}.$$

Here $M_g(P) = (y_1, y_2, \ldots, y_n)$, where

$$y_i = \sum_{j=1}^{m} b_{ij} x_j \qquad (i = 1, 2, \ldots, n).$$

Therefore

$$M_f M_g \begin{bmatrix} x_1 \\ x_2 \\ \cdot \\ \cdot \\ x_m \end{bmatrix} = M_f \begin{bmatrix} y_1 \\ y_2 \\ \cdot \\ \cdot \\ y_n \end{bmatrix} = \begin{bmatrix} a_{11} & a_{12} & \cdots & a_{1n} \\ a_{21} & a_{22} & \cdots & a_{2n} \\ \cdot & & & \cdot \\ \cdot & & & \cdot \\ a_{p1} & a_{p2} & \cdots & a_{pn} \end{bmatrix} \begin{bmatrix} y_1 \\ y_2 \\ \cdot \\ \cdot \\ y_n \end{bmatrix}$$

$$= \begin{bmatrix} \sum_{i=1}^{n} a_{1i} y_i \\ \sum_{i=1}^{n} a_{2i} y_i \\ \cdot \\ \cdot \\ \sum_{i=1}^{n} a_{pi} y_i \end{bmatrix} = \begin{bmatrix} \sum_{i=1}^{n} a_{1i} (\sum_{j=1}^{m} b_{ij} x_j) \\ \sum_{i=1}^{n} a_{2i} (\sum_{j=1}^{m} b_{ij} x_j) \\ \cdot \\ \cdot \\ \sum_{i=1}^{n} a_{pi} (\sum_{j=1}^{m} b_{ij} x_j) \end{bmatrix} = \begin{bmatrix} z_1 \\ z_2 \\ \cdot \\ \cdot \\ z_p \end{bmatrix}.$$

Here, for each k from 1 to p, we have

$$z_k = \sum_{i=1}^{n} a_{ki} \left(\sum_{j=1}^{m} b_{ij} x_j \right) = \sum_{j=1}^{m} \left(\sum_{i=1}^{n} a_{ki} b_{ij} \right) x_j$$

$$= \sum_{j=1}^{m} c_{kj} x_j \qquad (k = 1, 2, \ldots, p),$$

where

$$c_{kj} = \sum_{i=1}^{n} a_{ki} b_{ij}.$$

Thus

$$\begin{bmatrix} z_1 \\ z_2 \\ \cdot \\ \cdot \\ z_p \end{bmatrix} = \begin{bmatrix} c_{11} & c_{12} & \cdots & c_{1m} \\ c_{21} & c_{22} & \cdots & c_{2m} \\ \cdot & & & \cdot \\ \cdot & & & \cdot \\ c_{p1} & c_{p2} & \cdots & c_{pm} \end{bmatrix} \begin{bmatrix} x_1 \\ x_2 \\ \cdot \\ \cdot \\ x_m \end{bmatrix}.$$

But the above formula for c_{kj} says that c_{kj} is the inner product of the kth row of M_f and the jth column of M_g; these are the row vector and column vector

$$(a_{k1}, a_{k2}, \ldots, a_{kn}), \qquad \begin{bmatrix} b_{1j} \\ b_{2j} \\ \cdot \\ \cdot \\ b_{nj} \end{bmatrix},$$

and their inner product is $c_{kj} = \sum_{i=1}^{n} a_{ki}b_{ij}$. For example, in the product

$$\begin{bmatrix} 1 & 2 & 3 \\ -2 & 0 & 2 \\ 1 & 0 & 1 \\ -1 & 2 & -3 \end{bmatrix} \begin{bmatrix} -1 & 3 \\ -2 & 2 \\ 2 & 1 \end{bmatrix} = \begin{bmatrix} c_{11} & c_{12} \\ c_{21} & c_{22} \\ c_{31} & c_{32} \\ c_{41} & c_{42} \end{bmatrix},$$

c_{22} is the inner product of $(-2, 0, 2)$ and $(3, 2, 1)$. Therefore $c_{22} = -4$. The complete calculation gives the answer

$$\begin{bmatrix} 1 & 10 \\ 6 & -4 \\ 1 & 4 \\ -9 & -2 \end{bmatrix}.$$

(This should be checked.)

Similarly, we define the sum of two matrices to be the matrix of the sum:

Definition. Given the linear transformations $f: \mathbf{R}^m \to \mathbf{R}^n$ and $g: \mathbf{R}^m \to \mathbf{R}^n$, with matrices M_f and M_g. Then

$$M_f + M_g = M_{f+g}.$$

The sum is easy to calculate; the simplest possible idea works. Let

$$M_f = \begin{bmatrix} a_{11} & a_{12} & \cdots & a_{1n} \\ a_{21} & a_{22} & \cdots & a_{2n} \\ \cdot & & & \cdot \\ \cdot & & & \cdot \\ \cdot & & & \cdot \\ a_{m1} & a_{m2} & \cdots & a_{mn} \end{bmatrix}, \quad M_g = \begin{bmatrix} b_{11} & b_{12} & \cdots & b_{1n} \\ b_{21} & b_{22} & \cdots & b_{2n} \\ \cdot & & & \cdot \\ \cdot & & & \cdot \\ \cdot & & & \cdot \\ b_{m1} & b_{m2} & \cdots & b_{mn} \end{bmatrix}.$$

Then for $P = (x_1, x_2, \ldots, x_m)$, we have

$$f(P) = \begin{bmatrix} \sum_{j=1}^{n} a_{1j}x_j \\ \sum_{j=1}^{n} a_{2j}x_j \\ \cdot \\ \cdot \\ \cdot \\ \sum_{j=1}^{n} a_{mj}x_j \end{bmatrix}, \quad g(P) = \begin{bmatrix} \sum_{j=1}^{n} b_{1j}x_j \\ \sum_{j=1}^{n} b_{2j}x_j \\ \cdot \\ \cdot \\ \cdot \\ \sum_{j=1}^{n} b_{mj}x_j \end{bmatrix}.$$

Therefore

$$(f + g)(P) = f(P) + g(P) = \begin{bmatrix} \sum_{j=1}^{n} (a_{1j} + b_{1j})x_j \\ \sum_{j=1}^{n} (a_{2j} + b_{2j})x_j \\ \cdot \\ \cdot \\ \cdot \\ \sum_{n=1}^{n} (a_{mj} + b_{mj})x_j \end{bmatrix}.$$

The matrix which gives this result in one step is

$$[c_{ij}] = [a_{ij} + b_{ij}] = [a_{ij}] + [b_{ij}];$$

$$
\begin{bmatrix}
a_{11} & a_{12} & \cdots & a_{1n} \\
a_{21} & a_{22} & \cdots & a_{2n} \\
 & & & \\
 & & & \\
 & & & \\
a_{m1} & a_{m2} & \cdots & a_{mn}
\end{bmatrix}
+
\begin{bmatrix}
b_{11} & b_{12} & \cdots & b_{1n} \\
b_{21} & b_{22} & \cdots & b_{2n} \\
 & & & \\
 & & & \\
 & & & \\
b_{m1} & b_{m2} & \cdots & b_{mn}
\end{bmatrix}
$$

$$
=
\begin{bmatrix}
a_{11} + b_{11} & a_{12} + b_{12} & \cdots & a_{1n} + b_{1n} \\
a_{21} + b_{21} & a_{22} + b_{22} & \cdots & a_{2n} + b_{2n} \\
 & & & \\
 & & & \\
 & & & \\
a_{m1} + b_{m1} & a_{m2} + b_{m2} & \cdots & a_{mn} + b_{mn}
\end{bmatrix}.
$$

PROBLEM SET 13.2

Carry out the indicated operations, expressing each answer as a matrix (which may, of course, turn out to be an n by 1 matrix, that is, a column vector).

1. $\begin{bmatrix} 1 & 2 \\ 3 & 4 \\ 5 & 6 \\ 7 & 8 \end{bmatrix} \begin{bmatrix} 1 & 1 \\ 1 & 1 \end{bmatrix}$

2. $\begin{bmatrix} 1 & 2 & 3 \\ 4 & 5 & 6 \\ 7 & 8 & 9 \end{bmatrix} \begin{bmatrix} 1 & 0 & 0 \\ 0 & 1 & 0 \\ 0 & 0 & 1 \end{bmatrix}$

3. $\begin{bmatrix} 1 & 2 \\ 2 & 3 \end{bmatrix} \begin{bmatrix} 0 & 1 \\ 1 & 1 \end{bmatrix}$

4. $\begin{bmatrix} 1 & 2 & 3 \\ 4 & 5 & 6 \\ 7 & 8 & 9 \end{bmatrix} \begin{bmatrix} 0 & 0 & 1 \\ 0 & 1 & 0 \\ 1 & 0 & 0 \end{bmatrix}$

5. $\begin{bmatrix} 0 & 10 & 1 \\ 0 & 1 & 0 \\ 1 & 0 & 0 \end{bmatrix} \begin{bmatrix} 1 & 2 & 3 \\ 4 & 5 & 6 \\ 7 & 8 & 9 \end{bmatrix}$

6. $\begin{bmatrix} 0 & 0 & 1 \\ 0 & 1 & 1 \\ 1 & 1 & 0 \end{bmatrix} \begin{bmatrix} 1 & 4 \\ 2 & 5 \\ 3 & 6 \end{bmatrix}$

7. $\begin{bmatrix} 2 & 0 & 0 \\ 0 & 2 & 0 \\ 0 & 0 & 2 \end{bmatrix} \begin{bmatrix} 3 \\ 4 \\ 5 \end{bmatrix}$

8. $\begin{bmatrix} 3 & 0 & 2 \end{bmatrix} \begin{bmatrix} -2 \\ 2 \\ 3 \end{bmatrix}$

9. $\begin{bmatrix} 1 & 2 & 3 \\ 4 & 5 & 6 \end{bmatrix} \begin{bmatrix} 1 \\ 1 \\ 1 \end{bmatrix}$

10. $\begin{bmatrix} 0 & 1 & 0 \\ 0 & 0 & 1 \\ 1 & 0 & 0 \end{bmatrix} \begin{bmatrix} 1 \\ 2 \\ 3 \end{bmatrix}$

11. $\begin{bmatrix} 0 & 1 & 0 \\ 0 & 0 & 1 \\ 1 & 0 & 0 \end{bmatrix} \begin{bmatrix} 1 & 4 & 7 \\ 2 & 5 & 8 \\ 3 & 6 & 9 \end{bmatrix}$

12. $\begin{bmatrix} 0 & 1 & 0 \\ 0 & 0 & 1 \\ 1 & 0 & 0 \end{bmatrix} \begin{bmatrix} 1 & 4 \\ 2 & 5 \\ 3 & 6 \end{bmatrix}$

13. $\begin{bmatrix} 1 & 4 & 7 \\ 2 & 5 & 8 \\ 3 & 6 & 9 \end{bmatrix} \begin{bmatrix} 0 & 1 & 0 \\ 0 & 0 & 1 \\ 1 & 0 & 0 \end{bmatrix}$

14. $\begin{bmatrix} 0 & 1 & 0 \\ 0 & 0 & 1 \\ 1 & 0 & 0 \end{bmatrix}^2$

15. $\begin{bmatrix} 0 & 1 & 0 \\ 0 & 0 & 1 \\ 1 & 0 & 0 \end{bmatrix}^3$

16. $\begin{bmatrix} 0 & 1 & 0 \\ 0 & 0 & 1 \\ 1 & 0 & 0 \end{bmatrix}^4$

17. $\begin{bmatrix} 0 & 1 & 0 \\ 0 & 0 & 1 \\ 1 & 0 & 0 \end{bmatrix}^{10,000}$

18. $\begin{bmatrix} 1 & 0 & 2 \\ 3 & 4 & 0 \\ 5 & 1 & 7 \end{bmatrix} \begin{bmatrix} 2 & 1 & 1 \\ 1 & 3 & 2 \\ 1 & 0 & 1 \end{bmatrix}$

19. $\begin{bmatrix} 2 & 1 & 1 \\ 1 & 3 & 2 \\ 1 & 0 & 1 \end{bmatrix} \begin{bmatrix} 1 & 0 & 1 \\ 0 & 1 & 0 \\ 1 & 0 & 1 \end{bmatrix}$

20. $\begin{bmatrix} 1 & 0 & 1 \\ 0 & 1 & 0 \\ 1 & 0 & 1 \end{bmatrix} \begin{bmatrix} 2 & 1 & 1 \\ 1 & 3 & 2 \\ 1 & 0 & 1 \end{bmatrix}$

21. $\begin{bmatrix} 4 & 1 & 3 \\ 10 & 15 & 11 \\ 18 & 8 & 14 \end{bmatrix} \begin{bmatrix} 1 & 0 & 1 \\ 0 & 1 & 0 \\ 1 & 0 & 1 \end{bmatrix}$

22. $\begin{bmatrix} 1 & 0 & 2 \\ 3 & 4 & 0 \\ 5 & 1 & 7 \end{bmatrix}\begin{bmatrix} 3 & 1 & 3 \\ 3 & 3 & 3 \\ 2 & 0 & 2 \end{bmatrix}$

23. $\begin{bmatrix} 2 & 1 & 1 \\ 4 & 2 & 2 \\ -3 & -6 & -6 \end{bmatrix}\begin{bmatrix} 1 & -2 \\ -1 & 4 \\ -1 & 4 \end{bmatrix}$

24. $\begin{bmatrix} 0 & 0 & 1 \\ 0 & 1 & 0 \\ 1 & 0 & 0 \end{bmatrix}^2$

25. $\begin{bmatrix} 1 & 0 & 0 & 0 \\ 0 & 1 & 0 & 0 \\ 0 & 0 & 1 & 0 \\ 0 & 0 & 0 & 1 \end{bmatrix}^3$

26. $\begin{bmatrix} 0 & 0 & 0 & 1 \\ 0 & 0 & 1 & 0 \\ 0 & 1 & 0 & 0 \\ 1 & 0 & 0 & 0 \end{bmatrix}^2$

27. $\begin{bmatrix} 0 & 0 & 0 & 1 \\ 0 & 0 & 1 & 0 \\ 0 & 1 & 0 & 0 \\ 1 & 0 & 0 & 0 \end{bmatrix}^{25}$

For each of the following four matrices, find the inverse M^{-1} if there is an inverse; if not, give the simplest reason that you can for concluding that no inverse exists.

28. $\begin{bmatrix} 2 & 0 & 0 & 0 \\ 0 & 2 & 0 & 0 \\ 0 & 0 & 2 & 0 \\ 0 & 0 & 0 & 2 \end{bmatrix}$

29. $\begin{bmatrix} 1 & 2 & 3 & 0 \\ 4 & 5 & 6 & 0 \\ 7 & 8 & 9 & 0 \\ 10 & 11 & 12 & 0 \end{bmatrix}$

30. $\begin{bmatrix} 1 & 2 & 3 & 4 \\ 5 & 6 & 7 & 8 \\ 9 & 10 & 11 & 12 \\ 0 & 0 & 0 & 0 \end{bmatrix}$

31. $\begin{bmatrix} 0 & 4 \\ 4 & 0 \end{bmatrix}$

*32. Let

$$I_2 = \begin{bmatrix} 1 & 0 \\ 0 & 1 \end{bmatrix}.$$

For how many 2 by 2 matrices

$$M = \begin{bmatrix} a & b \\ c & d \end{bmatrix}$$

is it true that $MM = I_2$? (It is easy to find two such "square roots" of I_2. The question is whether there are others, and if so, what they are.)

33. Let

$$O_2 = \begin{bmatrix} 0 & 0 \\ 0 & 0 \end{bmatrix}.$$

This matrix acts like 0, in that for every 2 by 2 matrix M, we have

$$O_2 + M = M + O_2 = M.$$

Question: If A and B are 2 by 2 matrices, and $AB = O_2$, does it follow that $A = O_2$ or $B = O_2$?

34. If A and B are 2×2 matrices and $AB = O_2$, does it follow that $BA = O_2$?

13.3 FORMAL PROPERTIES OF THE ALGEBRA OF MATRICES. GROUPS AND RINGS

In the preceding problem set, you found that addition and multiplication of matrices were analogous in some ways, but not in others, to addition and multiplication of real and complex numbers. We shall now investigate the algebra of matrices systematically, and find out how far the analogy goes.

Throughout this section we shall be concerned only with square matrices. The set of all n by n matrices is denoted by $\mathcal{M}^n$. In our investigation of the formal properties of $\mathcal{M}^n$, under addition and multiplication, it will not be very useful to think about square arrays of numbers; the ideas are much easier to see if we work with the linear transformations f that the matrices represent.

Definition. $\mathscr{L}^n$ is the set of all linear transformations $f\colon \mathbf{R}^n \to \mathbf{R}^n$.

We found, in Theorem 3 of Section 13.2, that $\mathscr{L}^n$ forms a vector space, under addition and scalar multiplication. Therefore, in particular, we have:

C.1. $\mathscr{L}^n$ is closed under addition.

A.1. Addition in $\mathscr{L}^n$ is associative.

A.2 (*Existence of zero*). There is an element f_0 of $\mathscr{L}^n$ such that $f_0 + g = g + f_0 = g$ for every g.

(Obviously f_0 is the linear transformation such that
$$f_0(P) = O = (0, 0, \dots, 0) \qquad \text{for every } P \text{ in } \mathbf{R}^n.)$$

A.3 (*Existence of negatives*). For each f in $\mathscr{L}^n$ there is a $-f$ such that
$$f + (-f) = (-f) + f = f_0.$$

A.4. $f + g = g + f$, for every f and g.

A pair $[\mathscr{L}, +]$ is called a *group* if the operation $+$ satisfies C.1, A.1, A.2, and A.3. If A.4 is also satisfied, then $[\mathscr{L}, +]$ is called a *commutative group*. We can therefore sum up as follows:

Theorem 1. For each n, $[\mathscr{L}^n, +]$ is a commutative group.

We defined multiplication for matrices by composition of functions. That is,
$$M_f M_g = M_{f(g)},$$
by definition. We therefore need to investigate composition of functions in $\mathscr{L}^n$. For the sake of convenience, we shall denote the composite function $f(g)$ by the notation $f \circ g$. Theorem 4 of Section 13.2 tells us that if f and g are linear: $\mathbf{R}^n \to \mathbf{R}^n$, then so also is $f \circ g$. This gives

C.2. The set $\mathscr{L}^n$ is closed under the operation $\circ$.

For each f, g, h in $\mathscr{L}^n$,
$$f \circ (g \circ h)$$
is the composition of f and $g(h)$; and
$$(f \circ g) \circ h$$
is the composition of $f(g)$ and h. These give the same answer:
$$f \circ (g \circ h) = (f \circ g) \circ h.$$
The reason is shown by a diagram.

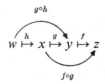

Starting at any point w, we get to the same point z, no matter how the functions f, g, h are grouped. Therefore we have:

M.1. In $\mathscr{L}^n$, the operation $\circ$ is associative.

(Obviously this has nothing to do with linearity; composition of functions is always associative, regardless of the domains, or the ranges, or the nature of the functions.)

M.2 (*Existence of unity*). There is an f_1 in $\mathscr{L}^n$ such that

$$f_1 \circ g = g \circ f_1 = g \qquad \text{for every } g.$$

This f_1 acts like the number 1, under our "multiplication." Obviously f_1 is the "identity" function, such that $f(P) = P$ for every P. We call f_1 the *unit* element.

So far, M.1 and M.2 are precisely analogous to C.1, A.1, and A.2; these conditions say the same things, about addition in one case and "multiplication" in the other. But the analogy now breaks down: not every linear function f has an inverse f^{-1}, and composition of functions is not, in general, commutative. (We have seen many examples of both of these.) But we do have:

DL (*The distributive law*). In $\mathscr{L}^n$, $f \circ (g + h) = f \circ g + f \circ h$.

The reason is that $f \circ g$ is $f(g)$; and $f(g + h) = f(g) + f(h)$ because f is linear.

A system satisfying all the conditions that we have mentioned so far is called a *ring*. More precisely:

Definition. Given a set $\mathscr{L}$, with two operations $+$ and $\circ$. The system

$$[\mathscr{L}, +, \circ]$$

is a *ring* if the following conditions are satisfied:

R.1. The pair $[\mathscr{L}, +]$ is a commutative group.

R.2. The set $\mathscr{L}$ is closed under $\circ$; $\circ$ is associative; and $\mathscr{L}$ contains a unit element.

R.3. The operation $\circ$ is distributive over $+$. (That is, $f \circ (g + h) = f \circ g + f \circ h$.)

All this discussion carries over immediately to the set $\mathscr{M}^n$ of n by n matrices, since $M_f + M_g = M_{f+g}$ and $M_f M_g = M_{f \circ g}$. This gives:

Theorem 2. For each n, the system $[\mathscr{M}^n, +, \circ]$ is a ring.

Here $\circ$ is used to denote matrix multiplication. It is easy to see that the zero-element of $\mathscr{M}^n$ is

$$M_{f_0} = O_n = \begin{bmatrix} 0 & 0 & \cdots & 0 \\ 0 & 0 & \cdots & 0 \\ \cdot & & & \cdot \\ \cdot & & & \cdot \\ \cdot & & & \cdot \\ 0 & 0 & \cdots & 0 \end{bmatrix},$$

and that the "unit element" of $\mathscr{M}^n$ is the matrix

$$I_n = \begin{bmatrix} 1 & 0 & 0 & \cdots & 0 \\ 0 & 1 & 0 & \cdots & 0 \\ \cdot & & 1 & & \cdot \\ \cdot & & & & \\ \cdot & & & & \\ 0 & & \cdots & & 1 \end{bmatrix}$$

with 1's on the main diagonal and 0's everywhere else. There is a shorthand for this: we define

$$\delta_{ij} = \begin{cases} 1 & \text{for } i = j, \\ 0 & \text{for } i \neq j. \end{cases}$$

We then have

$$I_n = [\delta_{ij}].$$

If you try proving Theorem 2 directly, carrying our calculations with n by n arrays of numbers, you will see that the use of the system $\mathscr{L}$ of linear transformations offered great advantages; most of the proofs were easier to write down than even one n by n matrix. In particular, a direct verification of the associativity of matrix multiplication, using the formula for the product of two matrices, would be extremely tedious.

A final remark, on the notation used in describing a group. In this section, the group operation is denoted by $+$. This is partly because addition was what we meant, in the case that we were discussing. Also it is customary to use the symbol $+$ when the operation is commutative. More generally, however, we can state the conditions for a group as follows:

Definition. Given a set G and an operation $*$. The system

$$[G, *]$$

is a *group* if the following conditions are satisfied:

Cl (*Closure*). G is closed under $*$.

A (*Associativity*). $a * (b * c) = (a * b) * c$, always.

EU (*Existence of Unity*). There is an element e of G such that $e * a = a * e = a$, for every a.

EI (*Existence of Inverses*). For each a in G there is an element a^{-1} of G such that $a * a^{-1} = a^{-1} * a = e$.

As before, the group is *commutative* if it satisfies:

Com (*Commutativity*). $a * b = b * a$, always.

A *field* is a system $[F, +, \cdot]$ which satisfies all the conditions which were stated for the real number system in Section 1.1. Thus $[F, +, \cdot]$ is a field if (1) $[F, +, \cdot]$ is a ring, (2) multiplication is commutative, and (3) every $x \neq 0$ has an inverse x^{-1}, such that $x \cdot x^{-1} = 1$.

Obviously the real-number system furnishes examples of all the ideas that we have been talking about in this section: $[\mathbf{R}, +, \cdot]$ is both a ring and a field, and $[\mathbf{R}, +]$ is a group. But if the real-number system were the only algebraic system that we were concerned with, there would be no advantage in using the terms *group*, *ring*, and *field*. The advantage is in other connections: already we have been dealing with vector spaces, which form groups (under addition), but do not form rings or fields; and from now on, we shall be dealing with (a) groups which are not rings, (b) rings which are not commutative, (c) commutative rings which are not fields, and so on. To find our way around in this variety of algebraic systems, we need a language in which we can explain briefly and clearly what sort of system we are dealing with at a given moment.

PROBLEM SET 13.3

1. Let G be the set of all complex numbers of the form
$$z = \cos \theta + i \sin \theta \qquad (\theta \text{ in } \mathbf{R}.)$$
Which, if any, of the following statements are true, and why, in each case?
   a) $[G, +]$ is a group.
   b) $[G, \cdot]$ is a group.
   c) $[G, +, \cdot]$ is a field.

2. Let G be the set of all "pure imaginary" numbers, of the form $z = iy$ (y in $\mathbf{R}$.) Discuss as in Problem 1.

3. Let G be the set of all complex numbers of the form $m + in$, where m and n are integers. Discuss as in Problem 1.

*4. Same problem, where m and n are rational numbers.

5. Show that if a and b are rational, and $a + b\sqrt{2} = 0$, then $a = b = 0$.

6. Let G be the set of all real numbers of the form $a + b\sqrt{2}$, where a and b are rational Discuss as in Problem 1.

7. In this problem, you may regard it as known that π is not a root of any linear or quadratic equation with rational coefficients. Let G be the set of all real numbers of the form $a + b\pi$, where a and b are rational. Is $[G, +, \cdot]$ a ring?

8. A *permutation* matrix is a square matrix with exactly one 1 in each row, exactly one 1 in each column, and 0's everywhere else. The set of all n by n permutation matrices is denoted by $\mathbf{P}^n$. Show that $[\mathbf{P}^2, \circ]$ is a group, and write a multiplication table for the group, in the form

$\circ$	I_2	A
I_2		
A		

What is the effect of a matrix in $\mathbf{P}^2$ on a vector $P = (x_1, x_2)$?

9. Now carry out the same process for $[\mathbf{P}^3, \circ]$. (You need not write a complete multiplication table, but you should find out whether the group is commutative.)

*10. An upper triangular matrix is a matrix $[a_{ij}]$ with $a_{ij} = 0$ for $j < i$. Denote the set of all $n \times n$ upper triangular matrices by V^n. Show $[V^n, +]$ is a group, but $[V^n, \circ]$ is not.

11. Show that if f and g are in $\mathscr{L}^n$, and f and g are invertible, then $f \circ g$ is also invertible. (The proof should be direct: you should produce a function which is the inverse of $f \circ g$.)

12. Let $GL(n)$ be the set of all invertible transformations in $\mathscr{L}^n$. Show that $[GL(n), \circ]$ is a group. (This is called the *general linear group*.) Then show that $[GL(n), +, \circ]$ is not a ring.

*13. Let $GLU(n)$ be the set of all upper triangular $n \times n$ matrices $[a_{ij}]$, with $a_{11}a_{22} \cdots a_{nn} \neq 0$. Show that $[GLU(n), \circ]$ is a group, but $[GLU(n), +, \circ]$ is not a ring.

13.4 THE DETERMINANT FUNCTION

The determinant function assigns, to every square matrix, a real number. The definition of this function begins as follows.

$$\begin{bmatrix} a_{11} & a_{12} & \cdots & a_{1n} \\ a_{21} & a_{22} & \cdots & a_{2n} \\ \cdot & & & \cdot \\ \cdot & & & \cdot \\ \cdot & & & \cdot \\ a_{n1} & a_{n2} & \cdots & a_{nn} \end{bmatrix}.$$

Given an n by n matrix $[a_{ij}]$, we take all possible products of the form

$$a_{1j_1}a_{2j_2} \cdots a_{ij_i} \cdots a_{nj_n},$$

using exactly one element from each row and exactly one element from each column. Thus the numbers

$$j_1, j_2, \ldots, j_i, \ldots, j_n$$

are all different. To each of these products we attach a $+$ or $-$ sign, according to a rule which will be stated presently. We then take the sum

$$\sum \pm a_{ij_1}a_{2j_2} \cdots a_{ij_i} \cdots a_{nj_n}$$

of all terms which can be formed according to the above rules. This sum is called the *determinant* of the matrix, and is denoted by det M, or det $[a_{ij}]$. Thus, when we have explained how the sign is to be chosen for each term, we shall have a function

$$\det\colon \mathscr{M}^n \to \mathbf{R}.$$

The rule for the signs takes time, to explain and justify. With each term

$$a_{1j_1}a_{2j_2} \cdots a_{nj_n}$$

there is associated a function

$$p\colon I_n \to I_n,$$

where

$$I_n = \{1, 2, 3, \ldots, n\}.$$

Here $p(i) = j_i$; that is, $p(i)$ is the column number of the element a_{ij_i} that we chose from the ith row. We can describe such a function p by a diagram in the following form:

$$p = \begin{pmatrix} 1 & 2 & 3 & \cdots & n \\ j_1 & j_2 & j_3 & \cdots & j_n \end{pmatrix},$$

with the numbers i in the top line and the numbers $p(i)$ below them. For example,

$$\begin{pmatrix} 1 & 2 & 3 & 4 \\ 3 & 1 & 4 & 2 \end{pmatrix}$$

is the function under whose action

$$1 \mapsto 3, \qquad 2 \mapsto 1, \qquad 3 \mapsto 4, \qquad 4 \mapsto 2.$$

A one-to-one function $p: I_n \to I_n$ is called a *permutation*. When permutations are described in the two-line notation, the order of the columns does not matter; all that matters is what is under what. For example,

$$\begin{pmatrix} 1 & 2 & 3 & 4 \\ 3 & 1 & 4 & 2 \end{pmatrix} = \begin{pmatrix} 4 & 3 & 2 & 1 \\ 2 & 4 & 1 & 3 \end{pmatrix} = \begin{pmatrix} 1 & 3 & 2 & 4 \\ 3 & 4 & 1 & 2 \end{pmatrix},$$

and so on.

If p interchanges two integers a and b, and leaves every other integer in I_n fixed, then p is called a *transposition*. For example,

$$p = \begin{pmatrix} 1 & 2 & 3 & 4 \\ 3 & 2 & 1 & 4 \end{pmatrix}$$

is a transposition; it interchanges 1 and 3. We denote this permutation by the shorthand (13). In general, (ab) is the permutation which interchanges a and b.

Theorem 1. Every permutation can be expressed as a product of transpositions.

Here the word *product* is used in the sense of composition of functions. For example, for

$$p = \begin{pmatrix} 1 & 2 & 3 & 4 \\ 3 & 1 & 4 & 2 \end{pmatrix}, \qquad \qquad \bullet$$

we can use the following transpositions:

$$\begin{pmatrix} 1 & 2 & 3 & 4 \\ 3 & 2 & 1 & 4 \end{pmatrix} = (13) \qquad \begin{pmatrix} 3 & 2 & 1 & 4 \\ 3 & 1 & 2 & 4 \end{pmatrix} = (21) \qquad \begin{pmatrix} 3 & 1 & 2 & 4 \\ 3 & 1 & 4 & 2 \end{pmatrix} = (24).$$

We now have

$$p = (24)(21)(13).$$

As always, for composition of functions, the operations are performed in the order from right to left. Thus

$$1 \xrightarrow{\;(13)\;} 3 \xrightarrow{\;(21)\;} 3 \xrightarrow{\;(24)\;} 3,$$

$$2 \xrightarrow{\;(13)\;} 2 \xrightarrow{\;(21)\;} 1 \xrightarrow{\;(24)\;} 1,$$

$$3 \xrightarrow{\;(13)\;} 1 \xrightarrow{\;(21)\;} 2 \xrightarrow{\;(24)\;} 4,$$

$$4 \xrightarrow{\;(13)\;} 4 \xrightarrow{\;(21)\;} 4 \xrightarrow{\;(24)\;} 2.$$

This checks, with

$$1 \mapsto 3, \qquad 2 \mapsto 1, \qquad 3 \mapsto 4, \qquad 4 \mapsto 2.$$

The scheme that we used on this example always works. Given

$$p = \begin{pmatrix} 1 & 2 & 3 & \cdots & n \\ j_1 & j_2 & j_3 & \cdots & j_n \end{pmatrix},$$

we first take the transposition $(1\ j_1)$; this puts j_1 under 1, where we want it to be. To the resulting sequence, we apply a transposition which puts j_2 in the second position, and so on. A further example:

$$p = \begin{pmatrix} 1 & 2 & 3 & 4 & 5 & 6 & 7 \\ 2 & 5 & 4 & 1 & 7 & 3 & 6 \end{pmatrix}.$$

Here we could use the following stages:

$$
\begin{array}{ccccccc}
1 & 2 & 3 & 4 & 5 & 6 & 7 \\
\end{array}
$$
$$(12)$$
$$
\begin{array}{ccccccc}
2 & 1 & 3 & 4 & 5 & 6 & 7 \\
\end{array}
$$
$$(15)$$
$$
\begin{array}{ccccccc}
2 & 5 & 3 & 4 & 1 & 6 & 7 \\
\end{array}
$$
$$(34)$$
$$
\begin{array}{ccccccc}
2 & 5 & 4 & 3 & 1 & 6 & 7 \\
\end{array}
$$
$$(31)$$
$$
\begin{array}{ccccccc}
2 & 5 & 4 & 1 & 3 & 6 & 7 \\
\end{array}
$$
$$(37)$$
$$
\begin{array}{ccccccc}
2 & 5 & 4 & 1 & 7 & 6 & 3 \\
\end{array}
$$
$$(36)$$
$$
\begin{array}{ccccccc}
2 & 5 & 4 & 1 & 7 & 3 & 6 \\
\end{array}
$$

It is not claimed, in Theorem 1, that every p can be expressed in only one way as a product of transpositions; and in fact this is not true. For example, the above diagram gives

$$p = \begin{pmatrix} 1 & 2 & 3 & 4 & 5 & 6 & 7 \\ 2 & 5 & 4 & 1 & 7 & 3 & 6 \end{pmatrix} = (36)(37)(31)(34)(15)(12).$$

But it is also true that

$$p = (36)(73)(16)(43)(57)(24)(16)(27)(14)(57),$$

which looks different, and uses ten transpositions instead of six. Nevertheless all such expressions for a given p have a common property, now to be described.

A permutation p is called *even* if it can be expressed as the product of an even number of transpositions; p is *odd* if p is the product of an odd number of transpositions.

Theorem 2. No permutation is both odd and even.

Proof. The *alternating function on n variables* is defined by the formula

$$f(x_1, x_2, \ldots, x_n) = \prod_{i<j} (x_i - x_j),$$

where the expression on the right is the product of all differences $x_i - x_j$ for which $i < j$. (Analogously, we might use

$$\sum_{1 \leq i \leq n} x_i$$

to denote $\sum_{i=1}^{n} x_i$). For example,

$$f(x_1, x_2, x_3) = (x_1 - x_2)(x_1 - x_3)(x_2 - x_3),$$

and

$$f(x_1, x_2, x_3, x_4) = (x_1 - x_2)(x_1 - x_3)(x_1 - x_4)(x_2 - x_3)(x_2 - x_4)(x_3 - x_4).$$

Now consider what happens to f when we apply the transposition $(i\,j)$, thus interchanging x_i and x_j. The factors of f are of the following types:

1) $(x_r - x_i), \quad (x_r - x_j) \quad (r < i),$

2) $(x_i - x_s), \quad (x_s - x_j) \quad (i < s < j),$

3) $(x_i - x_t), \quad (x_j - x_t) \quad (j < t),$

4) $(x_i - x_j).$

When we apply the transposition $(i\,j)$, interchanging x_i and x_j, the effect is

1) $\begin{cases} (x_r - x_i) \mapsto (x_r - x_j), \\ (x_r - x_j) \mapsto (x_r - x_i); \end{cases}$

2) $\begin{cases} (x_i - x_s) \mapsto (x_j - x_s) = -(x_s - x_j), \\ (x_s - x_j) \mapsto (x_s - x_i) = -(x_i - x_s); \end{cases}$

3) $\begin{cases} (x_i - x_t) \mapsto (x_j - x_t), \\ (x_j - x_t) \mapsto (x_i - x_t); \end{cases}$

4) $(x_i - x_j) \mapsto (x_j - x_i) = -(x_i - x_j).$

Thus the factors of the first three types fit together in pairs, and in each case, the *products* of the pairs are left unchanged. But the sign of $x_i - x_j$ is changed, and so the sign of f is changed. Briefly,

$$(i\,j)f(x_1, x_2, \ldots, x_n) = -f(x_1, x_2, \ldots, x_n)$$

or, more briefly still,

$$(i\,j)f = -f.$$

It follows that $pf = f$ if p is an even permutation, and $pf = -f$ if p is an odd permutation. No permutation can have both these effects, and so the theorem follows.

On this basis, we can finally define the determinant function:

$$\det [a_{ij}] = \sum \pm a_{1j_1} \; a_{2j_2} \; \cdots \; a_{nj_n},$$

where we use $+$ if the permutation

$$p = \begin{pmatrix} 1 & 2 & 3 & \cdots & n \\ j_1 & j_2 & j_3 & \cdots & j_n \end{pmatrix}$$

is even, and $-$ if p is odd.

In practice, when we have developed some of the theory, we shall never have to make direct use of the above definition; and this is fortunate, because the definition is even more tedious to handle than one might think. In order to form a term of det $[a_{ij}]$, for an n by n matrix, we have to choose an element from each row, in such a way as never to use the same column twice. Thus we have n possibilities to choose from in the first row; there are then $n - 1$ possibilities in the second row; and so on. Therefore the total number of terms is

$$n(n - 1)(n - 2) \cdots 3 \cdot 2 \cdot 1 = n!$$

Therefore the determinant of an n by n matrix is the sum of $n!$ terms. In particular, for $n = 20$, the number of terms of det M is

$$20! = 2{,}432{,}902{,}008{,}176{,}640{,}000.$$

The number of seconds in a year is only 31,526,000. This is why nobody asks even an electronic computer to calculate the determinants of large matrices by brute force.

Nevertheless, the definition of the function det is usable conceptually, as the basis of a theory which leads quickly to efficient techniques. In the rest of this section, we shall begin to develop the portion of the theory which makes direct use of the idea of odd and even permutations.

Theorem 3. Every permutation is invertible.

Obviously, since every permutation is one-to-one.

Theorem 4. For each n, let S_n be the set of all permutations $p: I_n \to I_n$. Then $[S_n, \circ]$ is a group.

Proof. (1) S_n is closed under $\circ$, because the product $p \circ q$ of two one-to-one functions $I_n \to I_n$ is another such function.

2)   The operation $\circ$ is associative; composition of functions always is.

3)   There is an identity

$$e = \begin{pmatrix} 1, 2, 3, \ldots, n \\ 1, 2, 3, \ldots, n \end{pmatrix}.$$

4)   By Theorem 3, every p in S_n has an inverse.

Theorem 5. In any group, the inverse of a product is the product of the inverses, in reverse order.

Thus

$$(p \circ q)^{-1} = q^{-1} \circ p^{-1},$$

because

$$(p \circ q) \circ (q^{-1} \circ p^{-1}) = p \circ (q \circ q^{-1}) \circ p^{-1} = p \circ e \circ p^{-1} = p \circ p^{-1} = e.$$

Hereafter, we shall omit the operation sign $\circ$. For products of n factors, the theorem says that

$$(p_1p_2\cdots p_{k-1}p_k)^{-1} = p_k^{-1}p_{k-1}^{-1}\cdots p_2^{-1}p_1^{-1},$$

and this is true, because in the product

$$p_1p_2\cdots p_{k-1}p_kp_k^{-1}p_{k-1}^{-1}\cdots p_2^{-1}p_1^{-1},$$

all the factors cancel each other in pairs, starting in the middle.

Note that if the group is commutative, then the order doesn't matter, and the theorem takes a simpler form. But the group that we are working with at the moment is not commutative.

Theorem 6. For each p in S_n, p and p^{-1} are either both even or both odd.

Proof. We express p as a product of transpositions:

$$p = p_1p_2\cdots p_{k-1}p_k,$$

where each p_i is a transposition. Every transposition is its own inverse. Therefore

$$p^{-1} = p_kp_{k-1}\cdots p_2p_1.$$

If k is even, then p and p^{-1} are both even. If not, p and p^{-1} are both odd.

The *transpose* of a matrix $\mathscr{M}$ is the matrix obtained by reflecting $\mathscr{M}$ across its main diagonal. The transpose is denoted by $\mathscr{M}^t$. Thus

$$\begin{bmatrix} a_{11} & a_{12} & a_{13} \\ a_{21} & a_{22} & a_{23} \\ a_{31} & a_{32} & a_{33} \end{bmatrix}^t = \begin{bmatrix} a_{11} & a_{21} & a_{31} \\ a_{12} & a_{22} & a_{32} \\ a_{13} & a_{23} & a_{33} \end{bmatrix},$$

and in general

$$[a_{ij}]^t = [a_{ji}].$$

Theorem 7. For each M in $\mathscr{M}^n$,

$$\det M = \det M^t.$$

Proof. The terms of $\det M$ are of the form

$$\pm a_{1j_1}a_{2j_2}\cdots a_{nj_n},$$

with $+$ or $-$ according as the permutation

$$p = \begin{pmatrix} 1 & 2 & 3 & \cdots & n \\ j_1 & j_2 & j_3 & \cdots & j_n \end{pmatrix}$$

is even or odd. The corresponding term of $\det M^t$ is

$$\pm a_{j_11}a_{j_22}\cdots a_{j_nn},$$

with $+$ or $-$ according as the permutation

$$q = \begin{pmatrix} j_1 & j_2 & j_3 & \cdots & j_n \\ 1 & 2 & 3 & \cdots & n \end{pmatrix}$$

is even or odd. Obviously $q = p^{-1}$, and so p and q are both even or both odd. There-fore the terms of det M and det M^t have the same signs, and det M = det M^t.

Theorem 8. If two rows of M are interchanged, then the determinant of the resulting matrix is $-$det M.

For example,

$$\begin{bmatrix} a_{31} & a_{32} & a_{33} \\ a_{21} & a_{22} & a_{23} \\ a_{11} & a_{12} & a_{13} \end{bmatrix} = -\det \begin{bmatrix} a_{11} & a_{12} & a_{13} \\ a_{21} & a_{22} & a_{23} \\ a_{31} & a_{32} & a_{33} \end{bmatrix}.$$

The reason is that when two rows are interchanged, this contributes exactly one transposition to the permutation

$$p = \begin{pmatrix} 1 & 2 & 3 & \cdots & n \\ j_1 & j_2 & j_3 & \cdots & j_n \end{pmatrix}.$$

For example, if the first and third rows are interchanged, then the sign of the term $a_{1j_1}a_{2j_2} \cdots a_{nj_n}$ in the new determinant is determined by the permutation

$$q = \begin{pmatrix} 1 & 2 & 3 & \cdots & n \\ j_3 & j_2 & j_1 & \cdots & n \end{pmatrix} = (j_1 \ j_3) \circ p.$$

PROBLEM SET 13.4

1. Working directly from the definition of det, get an explicit formula for

$$\det \begin{bmatrix} a_{11} & a_{12} \\ a_{21} & a_{22} \end{bmatrix}.$$

2. Similarly, get a formula for

$$\det \begin{bmatrix} a_{11} & a_{12} & a_{13} \\ a_{21} & a_{22} & a_{23} \\ a_{31} & a_{32} & a_{33} \end{bmatrix}.$$

3. Similarly, get a formula for

$$\det \begin{bmatrix} 1 & 0 & a_{13} & a_{14} \\ 0 & 1 & a_{23} & a_{24} \\ 0 & 0 & a_{33} & a_{34} \\ 0 & 0 & 0 & a_{44} \end{bmatrix}.$$

Calculate the following.

4. $\det \begin{bmatrix} 1 & 0 & 0 & 0 \\ 0 & 1 & 0 & 0 \\ 0 & 0 & 1 & 0 \\ 0 & 0 & 0 & 1 \end{bmatrix}$

5. $\det \begin{bmatrix} 0 & 0 & 0 & 1 \\ 0 & 0 & 1 & 0 \\ 0 & 1 & 0 & 0 \\ 1 & 0 & 0 & 0 \end{bmatrix}$

6. $\det \begin{bmatrix} 0 & 2 & 0 & 0 & 0 \\ 1 & 0 & 0 & 0 & 0 \\ 0 & 0 & 0 & 4 & 0 \\ 0 & 0 & 0 & 0 & 1 \\ 0 & 0 & 7 & 0 & 0 \end{bmatrix}$

7. $\det \begin{bmatrix} 1 & 0 & 0 & 0 \\ 0 & 1 & 0 & 0 \\ 0 & 0 & 2 & 3 \\ 0 & 0 & 4 & 5 \end{bmatrix}$

8. $\det \begin{bmatrix} 0 & 2 & 0 & 1 & 0 \\ 1 & 0 & 0 & 0 & 0 \\ 0 & 0 & 0 & 4 & 0 \\ 0 & 0 & 0 & 0 & 1 \\ 0 & 0 & 7 & 0 & 0 \end{bmatrix}$

9. $\det \begin{bmatrix} 0 & 2 & 0 & 1 & 0 \\ 0 & 0 & 0 & 0 & 1 \\ 0 & 0 & 0 & 4 & 0 \\ 1 & 0 & 0 & 0 & 0 \\ 0 & 0 & 7 & 0 & 0 \end{bmatrix}$

10. $\det \begin{bmatrix} 0 & 2 & 3 & 1 & 0 \\ 0 & 0 & 0 & 0 & 1 \\ 0 & 0 & 0 & 4 & 0 \\ 0 & 0 & 7 & 0 & 0 \\ 1 & 0 & 0 & 0 & 0 \end{bmatrix}$

11. $\det \begin{bmatrix} 1 & 0 & 0 & 0 & 0 \\ 0 & 1 & 0 & 0 & 0 \\ 0 & 0 & 2 & 0 & 3 \\ 0 & 0 & 0 & 1 & 0 \\ 0 & 0 & 4 & 0 & 5 \end{bmatrix}$

12. $\det \begin{bmatrix} 0 & 4 & 6 & 2 & 0 \\ 0 & 0 & 0 & 0 & 1 \\ 0 & 0 & 0 & 4 & 0 \\ 0 & 0 & 7 & 0 & 0 \\ 1 & 0 & 0 & 0 & 0 \end{bmatrix}$

13. $\det \begin{bmatrix} 0 & 2 & -6 & 1 & 0 \\ 0 & 0 & 0 & 0 & 1 \\ 0 & 0 & 0 & 4 & 0 \\ 0 & 0 & -14 & 0 & 0 \\ 1 & 0 & 0 & 0 & 0 \end{bmatrix}$

14. $\det \begin{bmatrix} 0 & 4 & 0 & 6 & 1 \\ 2 & 0 & 0 & 4 & 0 \\ 0 & 0 & 0 & 0 & 0 \\ 6 & 0 & 1 & 1 & 0 \\ 2 & 0 & 0 & 1 & 7 \end{bmatrix}$

15. $\det \begin{bmatrix} 0 & 0 & 0 & 1 & 0 & 0 \\ 0 & -2 & 0 & 0 & 0 & 0 \\ 3 & 0 & 0 & 0 & 0 & 0 \\ 0 & 0 & 0 & 0 & -4 & 0 \\ 0 & 0 & 0 & 0 & 0 & 5 \\ 0 & 0 & -6 & 0 & 0 & 0 \end{bmatrix}$

16. $\det \begin{bmatrix} 1 & 2 & 3 & 4 & 5 & 6 \\ 6 & 5 & 4 & 3 & 2 & 1 \\ 1 & 2 & 3 & 4 & 5 & 6 \\ 2 & 3 & 4 & 5 & 6 & 7 \\ 3 & 4 & 5 & 6 & 7 & 8 \\ 4 & 5 & 6 & 7 & 8 & 9 \end{bmatrix}$

17. $\det \begin{bmatrix} 0 & 1 & 0 & 0 & 0 & 0 \\ 1 & 0 & 0 & 0 & 0 & 0 \\ 0 & 0 & 0 & 1 & 0 & 0 \\ 0 & 0 & 1 & 0 & 0 & 0 \\ 0 & 0 & 0 & 0 & 2 & 3 \\ 0 & 0 & 0 & 0 & 4 & 5 \end{bmatrix}$

18. Let A_n be the set of all even permutations in S_n. Is $[A_n, \circ]$ a group?

19. Let B_n be the set of all odd permutations in S_n. Is $[B_n, \circ]$ a group?

20. Let C_n denote the set of permutations in S_n, for which $j_1 = 1$. Is $[C_n, \circ]$ a group?

21. Let D_n denote the set of permutations in S_n such that either $j_1 = 1$ and $j_2 = 2$ or $j_1 = 2$ and $j_2 = 1$. Is $[D_n, \circ]$ a group?

22. Can any general statement be made about the evenness or oddness of the following permutation?

$$p = \begin{pmatrix} 1 & 2 & 3 & \cdots & n-2 & n-1 & n \\ n & n-1 & n-2 & \cdots & 3 & 2 & 1 \end{pmatrix}$$

23. What can you say about the sign of the following permutation?

$$q = \begin{pmatrix} 1 & 2 & 3 & \cdots & n-2 & n-1 & n \\ n & 1 & 2 & \cdots & n-3 & n-2 & n-1 \end{pmatrix}$$

24. Suppose that $[G, \circ]$ satisfies all the conditions for a group, except that some elements of G may not have inverses. Let H be the set of all elements of G that have inverses. Does it follow that $[H, \circ]$ is a group?

25. Find the roots of the equation

$$\det \begin{bmatrix} x_1^2 & x_2^2 & x^2 \\ x_1 & x_2 & x \\ 1 & 1 & 1 \end{bmatrix} = 0.$$

Express the left-hand member as a quadratic expression in factored form.

26. Find the roots of the equation

$$\det \begin{bmatrix} x_1^3 & x_2^3 & x_3^3 & x^3 \\ x_1^2 & x_2^2 & x_3^2 & x^2 \\ x_1 & x_2 & x_3 & x \\ 1 & 1 & 1 & 1 \end{bmatrix} = 0.$$

Then express the lefthand member as a cubic equation in factored form.

*27. Find a matrix M whose determinant is the alternating function $f(x_1, x_2, \ldots, x_n)$.

13.5 EXPANSIONS BY MINORS.
CRAMER'S RULE AND INVERSION OF MATRICES

In the preceding section we showed that

$$\det M^t = \det M$$

for every square matrix M; and we showed that if two rows of M are interchanged, the effect is to change the sign of $\det M$. These statements in combination give us the following:

Theorem 1. If two rows of M are interchanged, or two columns are interchanged, then the determinant of the resulting matrix is $-\det M$.

To get the second half of this theorem, we take the transpose, perform the appropriate interchange of two rows, and take the transpose of the resulting matrix. The first and third of these operations leave the determinant unchanged, and the second one reverses the sign.

In fact, since $\det M^t = \det M$, every theorem about rows automatically gives us a theorem about columns.

Theorem 2. If M has two identical rows, then $\det M = 0$. Similarly for columns.

The reason is that when the two identical rows are interchanged, nothing happens to the determinant (or even to the matrix). Therefore $\det M = -\det M$, and $\det M = 0$.

The *minor* of an element a_{ij}, in a square matrix M, is the matrix that we get by deleting the ith row and the jth column of M. The minor is denoted by M_{ij}, and its determinant $\det M_{ij}$ is denoted by D_{ij}.

It is easy to see that the sum of all terms of $\det M$ that include a_{11} is $a_{11} D_{11}$:

$$M = \begin{bmatrix} a_{11} & a_{12} & a_{13} & \cdots & a_{1n} \\ a_{21} & a_{22} & a_{23} & \cdots & a_{2n} \\ a_{31} & a_{32} & a_{33} & \cdots & a_{3n} \\ \vdots & & & & \vdots \\ a_{n1} & a_{n2} & a_{n3} & \cdots & a_{nn} \end{bmatrix}.$$

Every term that involves a_{11} has the form

$$\pm a_{11}a_{2j_2}a_{3j_3}\cdots a_{nj_n};$$

here $a_{2j_2}a_{3j_3}\cdots a_{nj_n}$ is (except possibly for sign) a term of $D_{11} = \det M_{11}$. And these two corresponding terms of $\det M$ and $\det M_{11}$ have the same sign, because the permutations

$$\begin{pmatrix} 1 & 2 & 3 & \cdots & n \\ 1 & j_2 & j_3 & \cdots & j_n \end{pmatrix}, \qquad \begin{pmatrix} 2 & 3 & \cdots & n \\ j_2 & j_3 & \cdots & j_n \end{pmatrix}$$

are either both even or both odd.

This leads to a more general result:

Theorem 3. The sum of all terms of $\det M$ that involve a_{ij} is $(-1)^{i+j}a_{ij}D_{ij}$.

Proof. This is known for the case $i = j = 1$. We shall reduce the theorem to this case.

By a *simple row transposition* we mean an operation which interchanges two consecutive rows of a matrix. Similarly for *simple column transpositions*. We assert that a_{ij} can be moved into the first column by $j - 1$ simple column transpositions:

$$[c_1, c_2, c_3, c_4, c_5, \ldots, c_n].$$

Here c_j denotes the jth column. For $j = 5$, the transpositions are $(c_4 c_5)$, $(c_3 c_5)$, $(c_2 c_5)$, $(c_1 c_5)$. The new order of columns is

$$[c_5, c_1, c_2, c_3, c_4, c_6, \ldots, c_n].$$

Thus the fifth column becomes the first, and the other columns are in the same order, among themselves, as they were before.

Similarly, we can then move a_{ij} into the first row, by $i - 1$ simple row transpositions. Let the new matrix be M'. Then

$$\det M' = (-1)^{(j-1)+(i-1)} \det M$$

$$= (-1)^{i+j} \det M,$$

and so

$$\det M = (-1)^{-i-j} \det M' = (-1)^{i+j} \det M'.$$

But the sum of all terms of $\det M'$ that involve a_{ij} is $a_{ij} \det M'_{ij}$, where M'_{ij} is the minor of a_{ij} in M'; and M'_{ij} is M_{ij}, because our total operations on the rows and columns of M did not disturb the order of the rows and columns of M_{ij}. Therefore the sum of all terms of $\det M$ that involve a_{ij} is

$$(-1)^{i+j}a_{ij} \det M_{ij} = (-1)^{i+j}a_{ij}D_{ij},$$

which was to be proved.

Theorem 4 (*Expansion about the minors of a row*). For each i,

$$\det M = \sum_{j=1}^{n} (-1)^{i+j}a_{ij}D_{ij}.$$

This is true because every term of $\det M$ involves exactly one element in the ith row. Thus the above formula separates the $n!$ terms of $\det M$ into n classes, with

$(n - 1)!$ terms in each class. Similarly:

Theorem 5 (*Expansion about the minors of a column*). For each j,

$$\det M = \sum_{i=1}^{n} (-1)^{i+j} a_{ij} D_{ij}.$$

(At this stage you should check to see how these formulas apply to a 3 by 3 matrix, using, say, the second row and the second column.)

If we multiply the elements of one column by the determinants of the minors of some *other* column, with the appropriate signs, and add, we get 0:

Theorem 6. If $k \neq j$, then

$$\sum_{i=1}^{n} (-1)^{i+j} a_{ik} D_{ij} = 0.$$

The reason is this. Let M' be the matrix obtained by changing the jth column so as to make it identical with the kth column of the given matrix M. Then the above sum is the expansion of M' about the minors of its jth column. Therefore the sum is $\det M'$. But $\det M'$ is 0, because M' has two identical columns. Similarly for rows:

Theorem 7. If $k \neq i$, then

$$\sum_{j=1}^{n} (-1)^{i+j} a_{kj} D_{ij} = 0.$$

Theorems 6 and 7 may seem, at first glance, to be merely descriptions of what happens if somebody makes a mistake, but in fact they are very pointed statements. They enable us to write explicit formulas for the solution of a set of n linear equations in n unknowns, in the case in which the solution exists and is unique. To avoid tedious notation, we show how the method applies in the case $n = 3$. (The general case is exactly the same in principle.) Given the system

$$a_{11}x_1 + a_{12}x_2 + a_{13}x_3 = b_1,$$

$$a_{21}x_1 + a_{22}x_2 + a_{23}x_3 = b_2,$$

$$a_{31}x_1 + a_{32}x_2 + a_{33}x_3 = b_3.$$

Let M be the matrix $[a_{ij}]$ of the system; let $D = \det M$, and suppose that $D \neq 0$. In the first equation, we multiply by D_{11}, in the second by $-D_{21}$, and in the third by D_{31}. Then we add:

$$
\begin{array}{rcl}
a_{11}D_{11}x_1 + a_{12}D_{11}x_2 + a_{13}D_{11}x_3 = & & b_1 D_{11} \\
-a_{21}D_{21}x_1 - a_{22}D_{21}x_2 - a_{23}D_{21}x_3 = & & -b_2 D_{21} \\
a_{31}D_{31}x_1 + a_{32}D_{31}x_2 + a_{33}D_{31}x_3 = & & b_3 D_{31} \\
\hline
Dx_1 \quad + \quad 0 \cdot x_2 + \quad 0 \cdot x_3 = & & \sum_{i=1}^{3} (-1)^{1+i} b_i D_{i1}.
\end{array}
$$

On the lefthand side, the coefficient of x_1 is D, by Theorem 5, and the coefficients of x_2 and x_3 are 0, by Theorem 6. Thus the use of the minors as multipliers has given us an equation in which x_1 is the only unknown. The sum on the right-hand side, in the last equation, is easy to describe: it is the determinant D_1 of the matrix

$$M_1 = \begin{bmatrix} b_1 & a_{12} & a_{13} \\ b_2 & a_{22} & a_{23} \\ b_3 & a_{32} & a_{33} \end{bmatrix},$$

obtained by replacing the first column of M by the b_i's.

To solve for x_2, we multiply in the three equations by $-D_{12}$, D_{22}, and $-D_{32}$ respectively, and add. This gives

$$0 \cdot x_1 + D x_2 + 0 \cdot x_3 = \sum_{i=1}^{3} (-1)^{2+i} b_i D_{i2}.$$

Here the sum on the righthand side is the determinant D_2 of the matrix obtained by replacing the second column of M by the b_i's. The same scheme works for x_3. Therefore, if $D \neq 0$, the system has one and only one solution, namely,

$$x_1 = D_1/D, \qquad x_2 = D_2/D, \qquad x_3 = D_3/D.$$

Obviously none of the above discussion depended on the condition $n = 3$. In general, we have:

Theorem 8 (*Cramer's rule*). Given a linear system of the form

$$M \begin{bmatrix} x_1 \\ x_2 \\ \cdot \\ \cdot \\ \cdot \\ x_n \end{bmatrix} = \begin{bmatrix} b_1 \\ b_2 \\ \cdot \\ \cdot \\ \cdot \\ b_n \end{bmatrix}.$$

Let $D = \det M$. If $D \neq 0$, then the system has one and only one solution; and the solution is given by the formula

$$x_j = D_j/D,$$

where D_j is the determinant of the matrix M_j obtained by replacing the jth column of M by the vector $(b_1, b_2, \ldots, b_n)$.

Cramer's rule has the following consequence. A square matrix M is called *non-singular* if M has an inverse.

Theorem 9. If M is a square matrix, and $\det M \neq 0$, then M is nonsingular, and its inverse is given by the formula

$$M^{-1} = [c_{ij}],$$

where for each i and j,

$$c_{ij} = \frac{1}{D}(-1)^{i+j} D_{ji}.$$

That is, M^{-1} is the *transpose* of the matrix

$$\left[\frac{1}{D}(-1)^{i+j} D_{ij} \right].$$

To see why this is true, consider the matrix equation

$$M \begin{bmatrix} x_1 \\ x_2 \\ \cdot \\ \cdot \\ \cdot \\ x_n \end{bmatrix} = \begin{bmatrix} y_1 \\ y_2 \\ \cdot \\ \cdot \\ \cdot \\ y_n \end{bmatrix}.$$

Here

$$Dx_j = \sum_{i=1}^{n} (-1)^{i+j} y_i D_{ij},$$

by Cramer's rule. Therefore

$$x_j = \sum_{i=1}^{n} (-1)^{i+j} (1/D) D_{ij} y_i.$$

What we want is a matrix $M' = M^{-1}$, such that

$$M' \begin{bmatrix} y_1 \\ y_2 \\ \cdot \\ \cdot \\ \cdot \\ y_n \end{bmatrix} = \begin{bmatrix} x_1 \\ x_2 \\ \cdot \\ \cdot \\ \cdot \\ x_n \end{bmatrix}.$$

Here the y's form a *column* vector, and if $M' = [c_{ij}]$, then

$$x_j = \sum_{j=1}^{n} c_{ij} y_j.$$

To convert our previous formula for x_j to this form, we interchange i and j on the right, getting

$$x_j = \sum_{j=1}^{n} (-1)^{i+j} \left(\frac{1}{D}\right) D_{ji} y_j.$$

The value of the sum on the right is unchanged when we use j as an index of summation. Therefore

$$M^{-1} = M' = [c_{ij}] = \left[\frac{1}{D} (-1)^{i+j} D_{ji}\right],$$

which was to be proved.

PROBLEM SET 13.5

1. Find multipliers which eliminate y and z from the system

$$x + 2y + 3z = 4,$$
$$2x + 3y + 4z = 5,$$
$$3x + 4y + 5z = 6.$$

Carry out the multiplications, add, and solve for x. (Here you are not supposed to use Cramer's rule; you should use the scheme used in deriving Cramer's rule.)

2. Similarly, solve for y in the system

$$x - 2y + 3z = -4,$$
$$2x + 3y + 4z = -5,$$
$$3x - 4y + 5z = -6.$$

3. Similarly, solve for z in the system

$$x - 2y - 3z = 4,$$
$$2x + 3y - 4z = -5,$$
$$-3x + 4y - 5z = 6.$$

Find the inverses of the following matrices, by a direct application of Theorem 9, and check your answers by matrix multiplication.

4. $\begin{bmatrix} 1 & 0 \\ 0 & 2 \end{bmatrix}$.

5. $\begin{bmatrix} 0 & 2 \\ 3 & 0 \end{bmatrix}$.

6. $\begin{bmatrix} 1 & -2 \\ 3 & 4 \end{bmatrix}$.

7. $\begin{bmatrix} 1 & 1 \\ 1 & 0 \end{bmatrix}$.

8. $\begin{bmatrix} 1 & 1 & 0 \\ 0 & 1 & 1 \\ 1 & 0 & 1 \end{bmatrix}$.

9. $\begin{bmatrix} 0 & 1 & 0 \\ 1 & 0 & 0 \\ 0 & 0 & 1 \end{bmatrix}$.

10. $\begin{bmatrix} 1 & 1 & 1 & 1 \\ 0 & 1 & 0 & 1 \\ 0 & 0 & 1 & 0 \\ 0 & 0 & 0 & 1 \end{bmatrix}$.

11. $\begin{bmatrix} 1 & 0 & 0 & 0 \\ 0 & 2 & 0 & 3 \\ 0 & 0 & 1 & 0 \\ 0 & 4 & 0 & 5 \end{bmatrix}$.

Find, by any method, the inverses of the following matrices. (You need not calculate the determinants unless you need to, as a step in finding the inverse.)

12. $\begin{bmatrix} 0 & 0 & 0 & 1 \\ 0 & 0 & 1 & 0 \\ 0 & 1 & 0 & 0 \\ 1 & 0 & 0 & 0 \end{bmatrix}$.

13. $\begin{bmatrix} 0 & 1 & 0 & 0 \\ 1 & 0 & 0 & 0 \\ 0 & 0 & 0 & 1 \\ 0 & 0 & 1 & 0 \end{bmatrix}$.

14. $\begin{bmatrix} 1 & 0 & 0 & 0 \\ 0 & 1 & 0 & 0 \\ 0 & 0 & 0 & 1 \\ 0 & 0 & 1 & 0 \end{bmatrix}$.

15. $\begin{bmatrix} 0 & 0 & 0 & -1 & 0 \\ 0 & -2 & 0 & 0 & 0 \\ 3 & 0 & 0 & 0 & 0 \\ 0 & 0 & 0 & 0 & 4 \\ 0 & 0 & -5 & 0 & 0 \end{bmatrix}$.

16. $\begin{bmatrix} 0 & 0 & 0 & 0 & 1 \\ 0 & 0 & 0 & 1 & 0 \\ 0 & 0 & 1 & 0 & 0 \\ 0 & 1 & 0 & 0 & 0 \\ 1 & 0 & 0 & 0 & 0 \end{bmatrix}$.

17. Suppose we form a 4 by 4 matrix by fitting together four 2 by 2 matrices, like this:

$$M = \begin{bmatrix} M_{11} & M_{12} \\ M_{21} & M_{22} \end{bmatrix}.$$

Let $D = \det M$, and let $D_{ij} = \det M_{ij}$ for each i, j. Is it true that

$$D = \det \begin{bmatrix} D_{11} & D_{12} \\ D_{21} & D_{22} \end{bmatrix}?$$

If so, verify it. If not, give an example of a case in which it fails.

18. Similarly, discuss the case in which a $2n$ by $2n$ matrix is formed by fitting together four n by n matrices.

19. Similarly, discuss the case in which a 6 by 6 matrix is formed by fitting together nine 2 by 2 matrices.

13.6   ROW AND COLUMN OPERATIONS.
LINEAR INDEPENDENCE OF SETS OF FUNCTIONS

We shall now show that when we apply to a square matrix the "triangularization" process that we applied to systems of linear equations in Section 13.1, the determinant of the matrix is unchanged.

Theorem 1. If one row of a square matrix is multiplied by a scalar, and the resulting vector added to another row, the determinant of the matrix is unchanged. Similarly for columns.

Proof. Suppose that the kth row of the matrix $M = [a_{ij}]$ is multiplied by α and added to the ith row, giving a matrix M'. Expanding M' about the minors of the ith row, we get

$$\det M' = \sum_{j=1}^{n} (-1)^{i+j}(a_{ij} + \alpha a_{kj})D_{ij}$$

$$= \sum_{j=1}^{n} (-1)^{i+j}a_{ij}D_{ij} + \alpha \sum_{j=1}^{n}(-1)^{i+j}a_{kj}D_{ij} = \det M + \alpha \cdot 0,$$

by Theorems 4 and 7 of Section 13.5. We get the other half of the theorem by taking transposes, as in the proof of Theorem 1 of Section 13.5.

Iterations of this procedure constitute the most efficient scheme for computing determinants; by appropriate row (or column) operations, we can introduce 0's into a particular row (or column), so that when we use an expansion by minors, only one of the minors needs to be computed. Note that without these preliminaries, an expansion by minors is not a short cut in computation, but merely a device for systematizing our work; in an expansion by minors, the same number of terms appear as under the original definition of the determinant; they have merely been sorted into n sets of $(n-1)!$ terms each.

Theorem 2. If the rows of a matrix M form a linearly dependent set, then $\det M = 0$. Similarly for the columns.

Proof. Let $M = [a_{ij}]$; let the rows be $r_i = (a_{i1}, a_{i2}, \ldots, a_{in})$; suppose that

$$\sum_{i=1}^{n} \alpha_i r_i = 0,$$

for some set of numbers α_i, not all equal to 0. Then some r_i, say, r_1, is a linear combination of the others:

$$r_1 = \sum_{i=2}^{n} \beta_i r_i \qquad (\beta_i = \alpha_i/\alpha_1).$$

By $i - 1$ row operations as in Theorem 1, we get a new matrix M' in which r_1 is replaced by a row of 0's. Therefore det $M' = 0$, and so det $M =$ det $M' = 0$.

The converse is a little harder.

Theorem 3. If det $M = 0$, then the rows of M form a linearly dependent set (and so also do the columns).

Proof. The proof is by induction. Obviously the theorem holds for 1 by 1 matrices. We need to show that if it holds for $n - 1$ by $n - 1$ matrices, then it also holds for n by n matrices.

Given an n by n matrix $M = [a_{ij}]$, with rows $r_1, r_2, \ldots, r_n$. If any row r_i is the zero vector, then the linear dependence of the set

$$R = \{r_1, r_2, \ldots, r_n\}$$

is obvious. Therefore we may assume that $r_1 \neq 0$. We may also assume that $a_{11} \neq 0$, since the linear dependence or independence of the rows is unaffected by permutations of the columns.

Now consider the matrix M' whose rows form the set

$$R' = \{r_1', r_2', \ldots, r_n'\} = \{r_1, r_2 - \alpha_2 r_1, \ldots, r_n - \alpha_n r_1\}.$$

The r_i's are linear combinations of the r_i''s, and vice versa. Therefore R and R' span the same subspace of $\mathbf{R}^n$. This gives:

1) If R' is linearly dependent, then R is linearly dependent, and conversely.

But the α_i's can be chosen so that a_{11} is the only nonzero element in the first column of M'. Therefore det $M' = a_{11}D_{11}'$, where D_{11}' is the determinant of the minor M_{11}' of a_{11} in M'. Since

$$\det M' = \det M = 0 \qquad \text{and} \qquad a_{11} \neq 0,$$

we have

2) det $M_{11}' = 0$.

But M_{11}' is an $n - 1$ by $n - 1$ matrix. Therefore the rows of M_{11}' form a linearly dependent set. Therefore R' is linearly dependent. Therefore R is linearly dependent, which was to be proved.

This theorem easily gives:

Theorem 4. If the rows of a matrix M are linearly dependent, then so also are the columns, and conversely.

Proof. By the preceding two theorems, each of the following conditions is equivalent to the next:

1) The rows of M are linearly dependent.
2) det $M = 0$.
3) The columns of M are linearly dependent.

Here, and throughout this chapter so far, we have been talking about matrices of *numbers*. We shall now discuss linear independence of sets of functions; for this purpose, we shall use matrices of *functions;* and the first thing that we need to understand is that *for matrices of functions, the analogue of Theorem 4 is false.* This can be shown by a very simple example, as follows:

$$M(x) = \begin{bmatrix} 1 & 2 & 3 \\ x & 2x & 3x \\ x^2 & 2x^2 & 3x^2 \end{bmatrix}.$$

Here the columns are linearly dependent, obviously; but the rows are not: if

$$\alpha_1 + \alpha_2 x + \alpha_3 x^2 = 0 \qquad \text{for every } x,$$

for some real numbers α_1, α_2, α_3, then $\alpha_1 = \alpha_2 = \alpha_3 = 0$, because the only polynomial that vanishes for every x is the zero polynomial.

The simplest general test for linear independence of functions uses the determinant of a matrix of functions. Let $f_1, f_2, \ldots, f_n$ be functions on an interval I. The *Wronskian* of the sequence $f_1, f_2, \ldots, f_n$ is the function

$$W(x) = W(f_1, f_2, \ldots, f_n) = \det \begin{bmatrix} f_1 & f_1' & f_1'' & \cdots & f_1^{(n-1)} \\ f_2 & f_2' & f_2'' & \cdots & f_2^{(n-1)} \\ \cdot & & & & \cdot \\ \cdot & & & & \cdot \\ \cdot & & & & \cdot \\ f_n & f_n' & f_n'' & \cdots & f_n^{(n-1)} \end{bmatrix}.$$

Thus

$$W(x) = \det [f_i^{(j-1)}];$$

the Wronskian matrix has the $(j-1)$-derivative of f_i in the ith row and jth column. Note that the Wronskian really depends on a *sequence* of functions, and not merely on a *set* of functions; if the same functions are taken in a different order, the sign of W may change. The notation $W(x)$ is meant to emphasize that the Wronskian is a function and not a number. The following is easy:

Theorem 5. If $\{f_1, f_2, \ldots, f_n\}$ is linearly dependent, then $W(f_1, f_2, \ldots, f_n) = 0$. That is, $W(x) = 0$ for every x.

The reason is that for each x, the rows of the Wronskian matrix are linearly dependent: if

$$\sum_{i=1}^{n} \alpha_i f_i(x) = 0 \qquad \text{for every } x,$$

for some numbers α_i which are not all $= 0$, then automatically

$$D^j \sum_{i=1}^{n} \alpha_i f_i(x) = 0 \qquad (j = 1, 2, \ldots, n-1).$$

(Here $D^j[\cdot\cdot\cdot]$ denotes the jth derivative.) Therefore

$$\sum_{i=1}^{n} \alpha_i f_i^{(j)}(x) = 0 \qquad (j = 1, 2, \ldots, n-1),$$

for every x, and the rows are linearly dependent.

Ordinarily, we apply this theorem backwards, using the following equivalent form:

Theorem 5'. If $W(x_0) \neq 0$ for some x_0, then the set $\{f_1, f_2, \ldots, f_n\}$ is linearly independent.

For example, consider

$$f_1(x) = \sin x, \qquad f_2(x) = \cos x.$$

Here

$$W(x) = \det \begin{bmatrix} \sin x & \cos x \\ \cos x & -\sin x \end{bmatrix}$$

$$= -\sin^2 x - \cos^2 x = -1 \qquad \text{for every } x.$$

Here $W(x) \neq 0$ for every x, and it follows that sin and cos are linearly independent. But sometimes we have to choose a particular x_0. For example, consider

$$f_1(x) = \sin x, \qquad f_2(x) = \sin 2x.$$

Here

$$W(x) = \det \begin{bmatrix} \sin x & \cos x \\ \sin 2x & 2 \cos 2x \end{bmatrix}.$$

Here $W(0) = 0$, but

$$W\left(\frac{\pi}{2}\right) = \det \begin{bmatrix} 1 & 0 \\ 0 & -2 \end{bmatrix} = -2.$$

By Theorem 5', $\{\sin x, \sin 2x\}$ is linearly independent. Similarly for

$$f_1(x) = xe^x, \qquad f_2(x) = x^2 e^x.$$

Here

$$W(x) = \det \begin{bmatrix} xe^x & xe^x + e^x \\ x^2 e^x & x^2 e^x + 2xe^x \end{bmatrix} = \det \begin{bmatrix} xe^x & e^x \\ x^2 e^x & 2xe^x \end{bmatrix}$$

$$= xe^{2x} \det \begin{bmatrix} 1 & 1 \\ x & 2x \end{bmatrix} = x^2 e^{2x},$$

and so $W(x) \neq 0$ for every $x \neq 0$.

Elaborations of these techniques will build up gradually, in the following problem set. Meanwhile, it is natural to inquire about the converse of Theorem 5.

(?) Theorem (?). If $W(x) = 0$ for every x, then $\{f_1, f_2, \ldots, f_n\}$ is linearly dependent.

As it stands, this is *false;* and strong hypotheses need to be added to make it a true theorem. For example, let

$$f_1(x) = x^2, \qquad f_2(x) = \begin{cases} 2x^2 & \text{for } x \leq 0, \\ 3x^2 & \text{for } x \geq 0. \end{cases}$$

Then $W(x) = 0$ on the interval $[0, \infty)$, because f_1 and f_2 are linearly dependent on $[0, \infty)$; and for the same reason, $W(x) = 0$ on $(-\infty, 0]$. Therefore $W(x) = 0$ for every x. Nevertheless, $\{f_1, f_2\}$ is linearly independent, because there is no *one* pair of nonzero constants α_1, α_2 for which

$$\alpha_1 f_1(x) + \alpha_2 f_2(x) = 0 \qquad \text{for every } x.$$

If the above equation holds for every x, then setting $x = -1$ and $x = 1$ we get

$$\begin{cases} \alpha_1 + 2\alpha_2 = 0, \\ \alpha_1 + 3\alpha_2 = 0. \end{cases}$$

By subtraction, $\alpha_2 = 0$. It follows that $\alpha_1 = 0$.

This indicates that the Wronskian can give proofs of linear dependence only for special types of functions. Fortunately, these functions are of special interest and importance, as we shall see.

In the following problem set, you will be writing long strings of equations between determinants, and it will be convenient to use $|\cdots|$ as an abbreviation for $\det [\cdots]$. For example,

$$\begin{vmatrix} a & b \\ c & d \end{vmatrix} = \det \begin{bmatrix} a & b \\ c & d \end{bmatrix} = ad - bc.$$

PROBLEM SET 13.6

Investigate the following sets of functions for linear dependence.

1. $\{e^x, e^{2x}\}$ 2. $\{e^x, e^{2x}, e^{3x}\}$ 3. $\{e^{-x}, e^x\}$

4. $\{e^x, e^{2x}, xe^x\}$ 5. $\{\sin x, \sin^2 x\}$ 6. $\{\cos 2x, \sin^2 x\}$

7. $\{\sin x, \cos x, \sin^2 x\}$ 8. $\{x, \sin x\}$ 9. $\{e^{ax}, e^{bx}\} \quad (a \neq b)$

10. $\{e^{ax}, e^{bx}, xe^{ax}\} \quad (a \neq b)$ 11. $\{e^{ax}, e^{bx}, xe^{ax}, xe^{bx}\} \quad (a \neq b)$

12. $\{e^x, xe^x, x^2 e^x\}$ 13. $\{\sin x, \sin 2x, \sin 3x\}$

14. $\{\cos x, \cos 2x, \cos 3x\}$ 15. $\{\cos x, x \cos x\}$

16. $\{\cos x, \cos 2x, \cos^2 x\}$ 17. $\{\cos^2 x, \sin^2 x\}$

18. $\{\cos^4 x, \sin^4 x, \sin^2 (2x)\}$ 19. $\{e^x, \sin x, \cos x, e^x \sin x\}$

20. $\{e^x, e^x \sin x, e^x \cos x\}$ 21. $\{x, e^x, xe^x\}$

22. Since a polynomial equation of degree n has at most n roots, it follows that for each n, $\{1, x, x^2, \ldots, x^n\}$ is linearly independent. Get an alternative proof of this statement by calculating the Wronskian.

23. Find the roots of the equation

$$D_4 = \begin{vmatrix} 1 & x_1 & x_1^2 & x_1^3 \\ 1 & x_2 & x_2^2 & x_2^3 \\ 1 & x_3 & x_3^2 & x_3^3 \\ 1 & x & x^2 & x^3 \end{vmatrix} = 0.$$

Here the numbers x_1, x_2, and x_3 are all different. How do you know that D_4 is a polynomial of degree three? Express D_4 as a product of linear factors, not involving determinants.

*24. Express the following determinant as a product of linear factors:

$$\begin{vmatrix} 1 & x_1 & x_1^2 & \cdots & x_1^{n-1} \\ 1 & x_2 & x_2^2 & \cdots & x_2^{n-1} \\ \cdot & & & & \cdot \\ \cdot & & & & \cdot \\ \cdot & & & & \cdot \\ 1 & x_{n-1} & x_{n-1}^2 & \cdots & x_{n-1}^{n-1} \\ 1 & x_n & x^2 & \cdots & x_n^{n-1} \end{vmatrix}.$$

*25. Investigate for linear dependence: $\{e^{a_1 x}, e^{a_2 x}, e^{a_3 x}, \ldots, e^{a_n x}\}$, where the a_i's are all different.

13.7 LINEAR DIFFERENTIAL EQUATIONS

Consider a differential equation of the form

$$f'' + bf' + cf = 0, \tag{1}$$

where b and c are constants. As always, when we study a differential equation, we want to answer the following three questions:

i) Does the equation have any solutions?
ii) If so, what are they?
iii) How do we know that the solutions that we found are the only ones?

For Eq. (1), and for many others like it, we can give complete answers to all these questions. (In fact, the only hard one is the third.) As a guide to what to try, we look first at cases in which some of the solutions are obvious. The equation

$$f'' - f = 0$$

has the solutions

$$f_1(x) = e^x, \qquad f_2(x) = e^{-x},$$

because $D^2 e^x = e^x$ and $D^2 e^{-x} = e^{-x}$. And it is easy to see that if f_1 and f_2 are solutions, then so also is $\alpha_1 f_1 + \alpha_2 f_2$, for every pair of scalars α_1 and α_2. We had better make a note of this, more generally:

Theorem 1. Given a differential equation

$$f^{(n)} + a_{n-1} f^{(n-1)} + \cdots + a_1 f' + a_0 f = 0,$$

where a_i's are constants. Let $\mathscr{V}$ be the set of all solutions of the equation. Then $\mathscr{V}$ forms a vector space.

That is, $\mathscr{V}$ is closed under addition and scalar multiplication. This is trivial to check. Note, however, that if the zero on the right is replaced by a nonzero function, or even a nonzero constant k, the solutions of the resulting equation never form a vector space. (If the sum of two solutions is a solution, then $2k = k$, and $k = 0$.)

The above example suggests that we try solutions of the form

$$f(x) = e^{mx}, \qquad f'(x) = me^{mx}, \qquad f''(x) = m^2 e^{mx}.$$

If such a function is a solution, then

$$m^2 e^{mx} + bm e^{mx} + c e^{mx} = 0,$$

and since $e^{mx} \neq 0$ for every x, this is equivalent to the equation

$$m^2 + bm + c = 0. \tag{2}$$

Equation (2) is called the *auxiliary* equation. There are three possibilities for its solutions.

I. If $b^2 - 4c > 0$, then there are two roots

$$m_1 = \frac{-b + \sqrt{b^2 - 4c}}{2}, \qquad m_2 = \frac{-b - \sqrt{b^2 - 4c}}{2},$$

and both these roots are real.

II. If $b^2 - 4c = 0$, then there is only one root

$$m_1 = -b/2,$$

and this is a real root of multiplicity 2, with

$$m^2 + bm + c = (m - m_1)^2.$$

III. If $b^2 - 4c < 0$, then the roots are two conjugate complex numbers

$$m_1 = \alpha + \beta i, \qquad m_2 = \alpha - \beta i,$$

where $\alpha = -b/2$ and $\beta = \sqrt{4c - b^2}$.

In case I, the functions

$$f_1(x) = e^{m_1 x}, \qquad f_2 = e^{m_2 x}$$

are solutions, and so also is every linear combination

$$f(x) = \alpha_1 e^{m_1 x} + \alpha_2 e^{m_2 x}.$$

Also $\{f_1, f_2\}$ is linearly independent, because

$$W(f_1, f_2) = \det \begin{bmatrix} e^{m_1 x} & m_1 e^{m_1 x} \\ e^{m_2 x} & m_2 e^{m_2 x} \end{bmatrix} = e^{m_1 x} e^{m_2 x} \det \begin{bmatrix} 1 & m_1 \\ 1 & m_2 \end{bmatrix}$$

$$= e^{m_1 x} e^{m_2 x} (m_2 - m_1) \neq 0.$$

Therefore the solutions that we have found for case I include all the solutions, if the following theorem is true:

Theorem 2. Let $\mathscr{V}$ be the solution space of the equation

$$f'' + bf' + cf = 0.$$

Then dim $\mathscr{V} = 2$.

This is true, and will be proved in the following section. Meanwhile we shall use it. In case II, we seem to have only one solution

$$f_1(x) = e^{m_1 x};$$

and on the basis of Theorem 2, we need to find another one, f_2, such that $\{f_1, f_2\}$ is linearly independent. We do not know how somebody first thought of trying

$$f_2(x) = xe^{m_1 x},$$

but at any rate, it works:

$$f_2'(x) = m_1 xe^{m_1 x} + e^{m_1 x},$$

$$f_2''(x) = m_1^2 xe^{m_1 x} + m_1 e^{m_1 x} + m_1 e^{m_1 x}$$

$$= m_1^2 xe^{m_1 x} + 2m_1 e^{m_1 x};$$

$$f_2''(x) + bf_2'(x) + cf_2(x) = xe^{m_1 x}[m_1^2 + bm_1 + c] + e^{m_1 x}[2m_1 + b] = 0,$$

because both the expressions in the brackets are equal to 0.

Also, $\{f_1, f_2\}$ is linearly independent, because

$$W(f_1, f_2) = \det \begin{bmatrix} e^{m_1 x} & m_1 e^{m_1 x} \\ xe^{m_1 x} & (m_1 x + 1)e^{m_1 x} \end{bmatrix}$$

$$= e^{2m_1 x} \det \begin{bmatrix} 1 & m_1 \\ x & m_1 x + 1 \end{bmatrix} = e^{2m_1 x} \neq 0.$$

Case III looks peculiar. Taken at face value, the roots of the auxiliary equation give us

$$f_1(x) = e^{(\alpha + \beta i)x} = e^{\alpha x}(\cos \beta x + i \sin \beta x),$$

$$f_2(x) = e^{(\alpha - \beta i)x} = e^{\alpha x}(\cos \beta x - i \sin \beta x).$$

At the outset, we did not intend to get into the complex domain; but in the complex domain, our formulas still make sense: if m is complex, then the function $f(x) = e^{mx}$ is well defined; in fact, we have a function

$$\phi(z) = e^{mz},$$

with

$$\phi'(z) = me^{mz}, \qquad \phi''(z) = m^2 e^{mz}.$$

In particular, when z is real, $= x$, we have

$$\phi(z) = f(x) = e^{mx}, \qquad \phi'(z) = f'(x) = me^{mx}, \qquad \phi''(z) = f''(x) = m^2 e^{mx}.$$

Therefore f_1 and f_2 really are solutions. But at the moment we are interested only in *real* solutions (in another sense), and so we take the real and imaginary parts separately, getting

$$g_1(x) = e^{\alpha x} \cos \beta x, \qquad g_2(x) = e^{\alpha x} \sin \beta x.$$

(Check that if a complex-valued function f is a solution, then its real and imaginary parts are also solutions. This is easier than checking g_1 and g_2 by a brute-force calculation.) It remains to verify that g_1 and g_2 are linearly independent. We have

$$g_1'(x) = \alpha e^{\alpha x} \cos \beta x - \beta e^{\alpha x} \sin \beta x, \qquad g_2'(x) = \alpha e^{\alpha x} \sin \beta x + \beta e^{\alpha x} \cos \beta x.$$

Therefore

$$W(g_1, g_2) = \det \begin{bmatrix} e^{\alpha x} \cos \beta x & \alpha e^{\alpha x} \cos \beta x - \beta e^{\alpha x} \sin \beta x \\ e^{\alpha x} \sin \beta x & \alpha e^{\alpha x} \sin \beta x + \beta e^{\alpha x} \cos \beta x \end{bmatrix}$$

$$= e^{2\alpha x} \det \begin{bmatrix} \cos \beta x & \alpha \cos \beta x - \beta \sin \beta x \\ \sin \beta x & \alpha \sin \beta x + \beta \cos \beta x \end{bmatrix}$$

$$= e^{2\alpha x} \det \begin{bmatrix} \cos \beta x & -\beta \sin \beta x \\ \sin \beta x & \beta \cos \beta x \end{bmatrix} = \beta e^{2\alpha x} \neq 0,$$

because $\beta \neq 0$.

In case III, the linear combinations

$$f(x) = k_1 g_1(x) + k_2 g_2(x) = k_1 e^{\alpha x} \cos \beta x + k_2 e^{\alpha x} \sin \beta x$$

can be described in a better form. We have

$$f(x) = e^{\alpha x}(k_1 \cos \beta x + k_2 \sin \beta x)$$

$$= \sqrt{k_1^2 + k_2^2}\, e^{\alpha x} \left(\frac{k_1}{\sqrt{k_1^2 + k_2^2}} \cos \beta x + \frac{k_2}{\sqrt{k_1^2 + k_2^2}} \sin \beta x \right)$$

$$= k e^{\alpha x} \cos \beta(x - x_0),$$

where

$$k = \sqrt{k_1^2 + k_2^2},$$

and x_0 is any number such that $\cos \beta x_0 = k_1/k$ and $\sin \beta x_0 = k_2/k$. Using t for x, we get

$$f(t) = k e^{\alpha t} \cos \beta(t - t_0),$$

which describes the motion of a particle along a line, with the position given as a function of the time. This kind of motion is called *damped oscillation*. To get the graph of f, we start with the "simple oscillating function" $\cos \beta t$; we move the graph βt_0 units to the right, so that t_0 acts like 0; and then we damp the function by multiplying each value by $k e^{\alpha t}$. (For $\alpha < 0$, this damps the oscillations as $t \to \infty$; for $\alpha > 0$, the oscillations are damped as $t \to -\infty$.)

Note that in our formula for f, the constants α and β play a very different part from k and t_0: α and β are determined by the coefficients b and c in the differential equation, while k and t_0 range arbitrarily.

It is a fact that a solution f of the equation

$$f'' + bf' + cf = 0$$

is completely determined if $f(x_0)$ and $f'(x_0)$ are known, for some x_0. This can be verified by a calculation, for the three types of solutions that we have found, but the theorem is best postponed until the next section, where we can give the "right proof." Meanwhile, in the following problem set, you will find that such initial conditions always determine an answer.

Case III, in which $b^2 - 4c < 0$, and we get real solutions by making a detour into complex variables, may seem peculiar, but it is case III that has the most elementary application in physics: it describes the behavior of a vibrating spring. This problem is as follows. Suppose that you hang a coiled steel spring from a rigid support, like this:

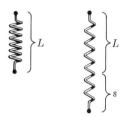

The spring has a certain natural length L. If you hang an object of weight w to the bottom end, the spring will be stretched by a distance s. It turns out experimentally that if the weight w is not too great, then the ratio w/s is a constant k; that is, $s = w/k$; the stretch is proportional to the weight. This statement is called Hooke's law. The proportionality constant k depends on the physical properties of the spring; the thicker and stiffer the spring, the larger k will be. This law, of course, applies only within certain limits: if you hang a brick on the hairspring of a watch, the result will not be an illustration of the law. Note, however, that the validity of the law for a given spring and a given range of weights is capable of being tested by static experiments; and this is important, because we are about to deduce from Hooke's law first a differential equation and then a law of motion.

If the spring is in equilibrium, when stretched to a length $L + s$, with a weight w at the bottom, then the spring must be exerting a force of magnitude $w = ks$, upward, to balance the force w exerted downward by gravity. Let us now set up a coordinate system on the line which is the axis of the spring, in such a way that the origin is at the equilibrium point for the given weight. In the figure below, we omit the spring itself, to clarify the labeling.

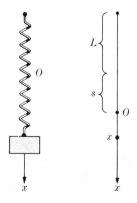

Suppose that the spring has been stretched to a point with coordinate x. Then two forces are acting:

1) The force F_1 exerted by the spring. This is

$$F_1 = -k(x + s),$$

because $x + s$ is the total stretch. We use the minus sign because the x-axis is directed downward.

2) The weight w. This counts positively, because weight acts downward, in the positive direction on the x-axis. Therefore the total force is

$$F = -k(x + s) + w = -kx.$$

Now suppose that the weight is pulled down to a certain point x_0 and then released. Then the weight will bob up and down, with its position given as a function of the time. For $x = f(t)$, the velocity and acceleration are

$$v(t) = f'(t),$$
$$a(t) = v'(t) = f''(t).$$

Newton's second law says that

$$F(t) = ma(t),$$

where m is the mass. The force represented by the weight is equal to the mass m times the acceleration g of gravity. Thus $w = mg$, and $m = w/g$. This gives

$$F(t) = \frac{w}{g} a(t).$$

But we know that

$$F(t) = -kx.$$

Therefore the function f which describes the motion must satisfy the differential equation

$$\frac{w}{g} a(t) = -kf(t),$$

which can be written in the form

$$f''(t) + \frac{kg}{w} f(t) = 0.$$

Since k, g, and w are all positive, this has the form

$$f'' + 0f' + cf = 0 \qquad (c > 0),$$

where $b^2 - 4c = -4c < 0$.

PROBLEM SET 13.7

In each of the following eight problems, find the solution space of the given differential equation, and then find the scalars which give the solution satisfying the initial conditions on the right. In each of these cases, you should use the methods but not the results of this section of the text. That is, set up the auxiliary equation, solve it, and then use the root(s) to get two solutions which form a linearly independent set.

1. $f'' - 5f' + 6f = 0$; $f(0) = 1$, $f'(0) = 2$.
2. $f'' - f' = 0$; $f(0) = 2$, $f'(0) = 3$.
3. $f'' - 6f' + 9f = 0$; $f(0) = 1$, $f'(0) = 1$.
4. $f'' + 4f = 0$; $f(1) = 1$, $f'(1) = 2$.
5. $f'' + f' + f = 0$; $f(0) = f'(0) = 0$.
6. $f'' - 7f' + 6f = 0$; $f(0) = 0$; $f'(0) = 1$.
7. $f'' - 7f = 0$; $f(1) = 0$, $f'(1) = 2$.
8. $4f'' - 4f' + f = 0$; $f(0) = 1$, $f'(0) = 0$.

Solve by any method:

9. A spring is such that an 8-lb weight stretches it 6 in. A 4-lb weight is attached, allowed to reach equilibrium, then pulled 2 in. below the equilibrium point and released. What happens? What is the period?

10. A spring is such that a 10-lb weight stretches it 18 in. A 1-lb weight is attached, allowed to reach equilibrium, pushed 6 in. above the equilibrium point, and released. What happens? What is the period?

11. You found, in Problem 27 of Problem Set 4.3, that the sine and cosine are the only functions f and g for which it is true that

$$f' = g, \quad g' = -f, \quad f(0) = 0, \quad \text{and} \quad g(0) = 1.$$

Show that there is only one function f for which it is true that

$$f'' = -f, \quad f(0) = 0, \quad \text{and} \quad f'(0) = 1.$$

*12. We know that the set of all infinite sequences of real numbers forms a vector space. Let f_1, f_2, f_3 be solutions of the differential equation

$$f'' + bf' + cf = 0;$$

and for $i = 1, 2, 3$, and every j, let

$$y_{ij} = f_i^{(j)}(0).$$

This gives three "vectors"

$$Y_1 = (y_{11}, y_{12}, \ldots), \quad Y_2 = (y_{21}, y_{22}, \ldots), \quad Y_3 = (y_{31}, y_{32}, \ldots),$$

which form a "3 by infinity matrix." Show that the rows of this matrix form a linearly dependent set.

13.8 THE DIMENSION THEOREM FOR THE SPACE
OF SOLUTIONS. THE NONHOMOGENEOUS CASE

In the preceding section, we found that for every equation of the form

$$f'' + bf' + cf = 0, \tag{1}$$

the solutions formed a linear space $\mathscr{V}$. In each case, we found solutions f_1, f_2 such that $\{f_1, f_2\}$ is linearly independent. We shall now show that the linear combinations

$$f = \alpha_1 f_1 + \alpha_2 f_2$$

are the only solutions.

Lemma. Every solution of (1) has derivatives of all orders.

Proof. Obviously every solution of (1) has a first and a second derivative, and

$$f'' = -bf' - cf.$$

Here the righthand side is differentiable, and so also is the lefthand side. Therefore

$$f^{(3)} = -bf^{(2)} - cf^{(1)}.$$

Similarly,

$$f^{(4)} = -bf^{(3)} - cf^{(2)};$$

and by induction,

$$f^{(n)} = -bf^{(n-1)} - cf^{(n-2)},$$

for every n, which proves the lemma.

It follows that for every solution of (1) we can write a "formal Taylor series"

$$\sum_{i=0}^{\infty} a_i(x-a)^i = \sum_{i=0}^{\infty} \frac{f^{(i)}(a)}{i!}(x-a)^i.$$

We shall now show that f is real-analytic; that is, the Taylor series converges, for every x, and its sum is the function f that we started with.

Theorem 1. If f is a solution of (1), defined in a neighborhood of a point a, then

$$f(x) = \sum_{i=0}^{\infty} \frac{f^{(i)}(a)}{i!}(x-a)^i \qquad \text{for every } x.$$

In the proof, we shall use Taylor's theorem. (This is Theorem 1 of Section 10.10. Note that we are now using it for the first time.) The theorem says that for each x, the remainder

$$R_n(x) = f(x) - \sum_{i=0}^{n} \frac{f^{(i)}(a)}{i!}(x-a)^i$$

is given by the formula

$$R_n(x) = \frac{f^{(n+1)}(\bar{x})}{(n+1)!}(x-a)^{n+1},$$

for some $\bar{x}$ between a and x. We want to conclude that $R_n(x) \to 0$; and to do this, we need to show that the numbers $f^{(n+1)}(\bar{x})$ cannot increase fast enough to overcome the effect of the $(n+1)!$ in the denominator. It should be understood that x is fixed, throughout the following discussion.

Let k be any number such that

$$k \geq |b|, \quad k \geq |c|, \quad \text{and} \quad k \geq 1;$$

and let M be a number which is an upper bound for both $|f(t)|$ and $|f'(t)|$, on the interval

$$I = \{t \mid |t - a| \leq |x - a|\}.$$

For each t on I, we then have

$$|f''(t)| = |-bf'(t) - cf(t)|$$
$$\leq |b| \cdot |f'(t)| + |c| \cdot |f(t)|$$
$$\leq kM + kM = 2kM.$$

Similarly,

$$|f^{(3)}(t)| = |-bf''(t) - cf'(t)|$$
$$\leq k \cdot 2kM + kM$$
$$< (2k)^2 M.$$

We now claim that

$$|f^{(n+1)}(t)| \leq (2k)^n M \qquad \text{for every } n.$$

This is known for $n = 1$ and $n = 2$. And if it holds for two successive integers $n - 2$ and $n - 1$, then it holds for the next integer n:

$$\begin{cases} |f^{(n-1)}(t)| \leq (2k)^{n-2}M \\ |f^{(n)}(t)| \leq (2k)^{n-1}M \end{cases} \Rightarrow \begin{aligned} |f^{(n+1)}(t)| &= |-bf^{(n)}(t) - cf^{(n-1)}(t)| \\ &\leq k\,|f^{(n)}(t)| + k\,|f^{(n-1)}(t)| \\ &\leq k(2k)^{n-1}M + k(2k)^{n-2}M \\ &= (2k)^{n-2}(2k^2 + k)M < (2k)^n M, \end{aligned}$$

because $k \geq 1$. From this the theorem follows, because it gives

$$|R_n(x)| \leq \frac{(2k)^n M}{(n + 1)!} |x - a|^{n+1},$$

which obviously approaches 0; in fact, the expression on the right is the $(n + 1)$st term of the series for

$$\frac{M}{2k} e^{2k|x-a|}.$$

It is not an accident that the series for f converges with the rapidity of an exponential series: the solutions that we have found, so far, for our differential equation have been combinations of exponentials, sines, and cosines; and we are about to find that these are the only solutions.

Theorem 2. If $f'' + bf' + cf = 0$, and $f(a) = f'(a) = 0$ for some a, then $f(x) = 0$ for every x.

Proof. Since $f^{(n)} = -bf^{(n-1)} - cf^{(n-2)}$, it follows by induction that $f^{(n)}(a) = 0$ for every n, and so all the coefficients in the Taylor series are equal to 0. This gives the result which was used without proof in the last problem set:

Theorem 3. If f_1 and f_2 are solutions of the equation $f'' + bf' + cf = 0$, and

$$f_1(a) = f_2(a) \quad \text{and} \quad f_1'(a) = f_2'(a) \quad \text{for some } a,$$

then $f_1(x) = f_2(x)$ for every x.

This is because the difference function $f = f_1 - f_2$ is also a solution, and

$$\dot{f}(a) = f'(a) = 0.$$

The dimension theorem is now easy:

Theorem 4 (*The dimension theorem*). Let $\mathscr{V}$ be the space of solutions of the equation $f'' + bf' + cf = 0$. Then $\dim \mathscr{V} = 2$.

Proof. We found, in the preceding section, that every equation of this form has two linearly independent solutions. Therefore $\dim \mathscr{V} \geq 2$. It remains to show that every three solutions f_1, f_2, f_3 form a linearly dependent set. Consider the matrix

$$M = \begin{bmatrix} f_1(0) & f_1'(0) & f_1''(0) \\ f_2(0) & f_2'(0) & f_2''(0) \\ f_3(0) & f_3'(0) & f_3''(0) \end{bmatrix}.$$

This is *not* the Wronskian; it is not a matrix of functions but a matrix of numbers. Therefore Theorem 4 of Section 13.6 can be applied to it: since the columns of M are linearly dependent, so also are the rows. Thus

$$\alpha_1 f_1(0) + \alpha_2 f_2(0) + \alpha_3 f_3(0) = 0, \quad \alpha_1 f_1'(0) + \alpha_2 f_2'(0) + \alpha_3 f_3'(0) = 0,$$

for some scalars α_1, α_2, α_3, not all equal to 0. Let

$$f = \alpha_1 f_1 + \alpha_2 f_2 + \alpha_3 f_3.$$

Then $f(0) = f'(0) = 0$, and so $f(x) = 0$ for every x. Therefore $\{f_1, f_2, f_3\}$ is linearly dependent, which was to be proved.

All the results that we have been getting for equations of order 2 can be generalized, in a straightforward way, to nth-order equations, of the form

$$f^{(n)} + \sum_{i=0}^{n-1} b_i f^{(i)} = 0.$$

As before, we try $f(x) = e^{mx}$, and form the auxiliary equation

$$m^n + \sum_{i=0}^{n-1} b_i m^i = 0.$$

If m_1 is a root of the equation, with multiplicity k_1 (so that $(m - m_1)^{k_1}$ divides the lefthand member), then the functions

$$e^{m_1 x}, xe^{m_1 x}, \ldots, x^{k_1-1}e^{m_1 x}$$

are solutions, and form a linearly independent set. If $\alpha + \beta i$, $\alpha - \beta i$ are a pair of conjugate complex roots, with multiplicity k_2, then the functions

$$e^{\alpha x} \cos \beta x, xe^{\alpha x} \cos \beta x, \ldots, x^{k_2-1}e^{\alpha x} \cos \beta x$$

and

$$e^{\alpha x} \sin \beta x, xe^{\alpha x} \sin \beta x, \ldots, x^{k_2-1}e^{\alpha x} \sin \beta x$$

are solutions, and are linearly independent. Moreover, the total set of functions obtained in this way forms a linearly independent set, and the number of elements in the set is n, because n is the sum of the multiplicities of the roots of the auxiliary equation. Therefore the dimension of the solution space is at least equal to n. As for equations of order 2, it can be shown that all solutions of the equation are real-analytic; and matrix theory then furnishes a proof that every set of $n + 1$ solutions forms a linearly dependent set. It follows that the dimension of the solution space is exactly n.

Thus the results follow the pattern that we found for equations of order 2. However, to derive them, in a reasonably efficient and natural way, requires new theoretical ideas, and, in particular, a new kind of algebraic formalism. This theory is best postponed to a systematic course in differential equations.

Meanwhile we consider what happens, in a linear differential equation with constant coefficients, when the 0 on the right is replaced by a function. Consider, for example,

$$f''(x) + 5f'(x) + 6f(x) = e^x. \tag{1}$$

We know how to find all solutions of the equation

$$f'' + 5f' + 6f = 0; \tag{2}$$

the solution space is

$$\mathscr{V} = \{\alpha_1 e^{-2x} + \alpha_2 e^{-3x}\}.$$

Equation (2) is called the *reduced* equation. In general, a linear differential equation with 0 as its righthand member is called *homogeneous*.

If f and f_0 are solutions of (1), then

$$f''(x) + 5f'(x) + 6f(x) = e^x \quad \text{and} \quad f_0''(x) + 5f_0'(x) + 6f_0(x) = e^x,$$

and so by subtraction we get

$$(f - f_0)'' + (f - f_0)' + 6(f - f_0) = 0.$$

Therefore the function $f - f_0$ is a solution of the reduced equation (2). This means that if we can find *one* solution of (1), then we can express *all* solutions of (1) in the form

$$f(x) = \alpha_1 e^{-2x} + \alpha_2 e^{-3x} + f_0(x);$$

every solution of (1) has this form, because $f - f_0$ is of the form $\alpha_1 e^{-2x} + \alpha_2 e^{-3x}$.

If the function on the right is real-analytic, then there is a systematic scheme for looking for solutions of a nonhomogeneous equation: we assume that

$$f(x) = \sum_{i=0}^{\infty} a_i x^i,$$

and solve for the coefficients a_i one at a time. But if the function on the right is simple, then the method of trial and error may work faster and lead to a simpler formula. In the example above, we try

$$f(x) = Ae^x, \quad f'(x) = Ae^x, \quad f''(x) = Ae^x,$$

and try to find A so as to make f a solution:

$$Ae^x + 5Ae^x + 6Ae^x = e^x,$$

$$12A = 1, \quad A = \tfrac{1}{12}, \quad f(x) = \tfrac{1}{12}e^x.$$

Therefore the solutions of (1) form a set

$$H = \{\alpha_1 e^{-2x} + \alpha_2 e^{-3x} + \tfrac{1}{12}e^x\}.$$

The set H does not form a subspace. A set of this kind is called a *hyperplane*. In general, if $\mathscr{W}$ is a subspace of a linear space $\mathscr{V}$, and f_0 is any point of $\mathscr{V}$, then the set

$$H = \{f + f_0 \mid f \text{ in } \mathscr{W}\}$$

is called a *hyperplane*. Note that every subspace is automatically a hyperplane, because f_0 may be zero. The term *hyperplane* is suggested by the language of geometry in Cartesian 3-space. If E is a plane through the origin, and P_0 is any point of $\mathbf{R}^3$, then the set

$$H = \{P + P_0 \mid P \text{ in } E\}$$

is a plane. (We use the prefix *hyper* because in vector spaces of higher dimension, the dimension of a hyperplane may easily be greater than 2. The set H may, of course, be of dimension 1 or 0; every line in $\mathbf{R}^3$ forms a hyperplane, under the above definition, because every line through the origin forms a subspace. The same applies for a point, although we rarely have any occasion to say so.)

Similar devices work for various other functions on the right in a nonhomogeneous equation. For example, consider

$$f''(x) + 5f'(x) + 6f(x) = \sin x.$$

First we try

$$(?) \quad f(x) = A \sin x \quad (?).$$

This gives

$$(?) \quad -A \sin x + 5A \cos x + 6A \sin x = \sin x \quad (?),$$

which is impossible, because $\{\sin, \cos\}$ is linearly independent. Next we try

$$(?) \quad f(x) = A \sin x + B \cos x \quad (?),$$

which gives

$$-A \sin x - B \cos x + 5(A \cos x - B \sin x)$$

$$+ 6A \sin x + 6B \cos x = \sin x$$

$$\Leftrightarrow (-A - 5B + 6A) \sin x$$

$$+ (-B + 5A + 6B) \cos x = \sin x$$

$$\Leftrightarrow \begin{cases} 5A - 5B = 1 \\ 5A + 5B = 0 \end{cases}$$

$$\Leftrightarrow A = \tfrac{1}{10}, \qquad B = -\tfrac{1}{10}.$$

This gives

$$f_0(x) = \tfrac{1}{10}(\sin x - \cos x),$$

so that the hyperplane of solutions is

$$H = \{\alpha_1 e^{-2x} + \alpha_2 e^{-3x} + \tfrac{1}{10}(\sin x - \cos x)\}.$$

PROBLEM SET 13.8

For each of the following equations, find the space $\mathscr{W}$ of solutions. Answers should be in the form

$$\mathscr{W} = \{\alpha_1 f_1 + \alpha_2 f_2 + \cdots + \alpha_n f_n\}.$$

1. $f^{(4)} + 4f^{(3)} + 6f'' + 4f' + f = 0.$ 2. $f^{(4)} + f^{(3)} - 3f'' - 5f' - 2f = 0.$
3. $f^{(4)} + 2f^{(3)} - 3f'' - 4f' + 4f = 0.$ 4. $f^{(4)} + 2f'' + f = 0.$
5. $f^{(3)} - f'' + f' - f = 0.$ 6. $f^{(4)} + f^{(3)} - f' - f = 0.$
7. $f^{(5)} - 2f^{(3)} + f' = 0.$

For each of the following, find (a) the space $\mathscr{W}$ of solutions of the reduced equation, in the same form as in the preceding problems, and (b) the hyperplane H of solutions of the given equation, in the form

$$H = \{\alpha_1 f_1 + \alpha_2 f_2 + \cdots + \alpha_n f_n + f_0\}.$$

8. $f'' + f = e^x$ 9. $f'' + f = \sin x$ 10. $f'' + f = 1$
11. $f'' + f = \cos x$ 12. $f'' + f = x + 1$ 13. $f'' + f = x$
14. $f'' + f = x^3$ 15. $f'' + f = e^x \sin x$ 16. $f'' + f = x^2 + 1$
17. $f'' + f = e^x \cos x$ 18. $f'' + f = xe^x$ 19. $f'' + f = xe^x + e^x$
20. $f'' + f = x \sin x$ 21. $f'' + f = x \cos x$
22. $f'' + f = \sin x + 2x \cos x$

14   Functions of Several Variables

14.1 SURFACES AND SOLIDS IN R³

We recall, from elementary geometry, the definition of a right cylinder. Given a set B in a plane E, the *right cylinder with base B* is the union of all lines that intersect B and are perpendicular to E.

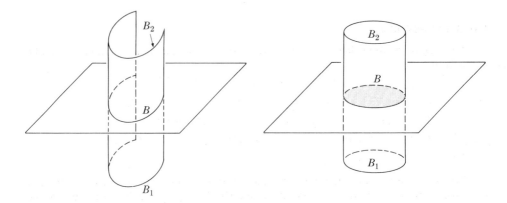

Hereafter in this chapter, when we speak of a cylinder we shall always mean a right cylinder. Some other remarks are in order here.

1) The figures above suggest that a cylinder is a bounded figure, with a lower base B_1 and an upper base B_2. But this is merely a device for clarifying the meaning of the pictures; according to our definition, cylinders are of infinite extent, in each of two directions. In the same way, planes are unbounded, although we indicate them in pictures by drawing parallelograms.

2) The base may be any set of points in a plane. If the base is a curve, as on the left above, then the cylinder is a surface. If the base is a region, as on the right above, then the cylinder is a solid. The definition applies to each of these cases in exactly the same way. In each case, if the plane of the base is regarded as horizontal, then the cylinder is the union of all vertical lines that intersect the base.

To avoid possible confusion, we may distinguish these cases by speaking of *cylindrical surfaces* and *cylindrical solids*.

If the base is in the xy-plane, and is described by an equation in x and y, then the same equation can be regarded as a description of the cylinder.

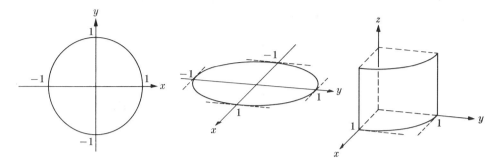

On the left above we show the unit circle in the xy-plane; this is the graph of the equation $x^2 + y^2 = 1$. In the center above, we see the same figure in perspective, as it appears when we are about to draw in a z-axis. In the three-dimensional figure on the right above we see the portion of the cylinder that lies in the first octant. The cylinder is the graph of the same equation $x^2 + y^2 = 1$; since the equation imposes no restriction on z, the graph includes the vertical line through each of its points. To be more precise, the circle is

$$\{(x, y) \mid x^2 + y^2 = 1\},$$

and the cylinder is

$$\{(x, y, z) \mid x^2 + y^2 = 1\}.$$

The relations among these figures deserve careful examination. At the left above, the tangents to the circle at the y-intercepts are horizontal, that is, parallel to the x-axis. This should be true also in the perspective drawings at the center and right. Hence the dotted guide lines. Similarly, the tangents to the circle at the x-intercepts are vertical, that is, parallel to the y-axis. This should also be true in the perspective drawings; it is indicated by the dotted guide lines. Often a correctly drawn figure looks peculiar, unless you analyze it in this way. For example:

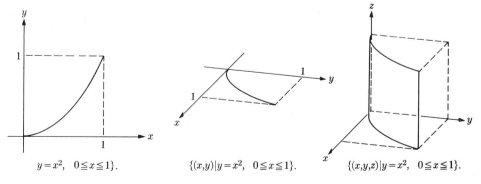

$y = x^2, \ 0 \leq x \leq 1\}.$ $\{(x,y) | y = x^2, \ 0 \leq x \leq 1\}.$ $\{(x,y,z) | y = x^2, \ 0 \leq x \leq 1\}.$

To get a circular disk, instead of a circle, we use the inequality $x^2 + y^2 \leq 1$ instead of the corresponding equation. Using this as base we get the solid cylinder,

which is

$$\{(x, y, z) \mid x^2 + y^2 \leqq 1\}.$$

If we use the xz-plane or the yz-plane as the plane of the base, then the same scheme works, in a similar way. For example:

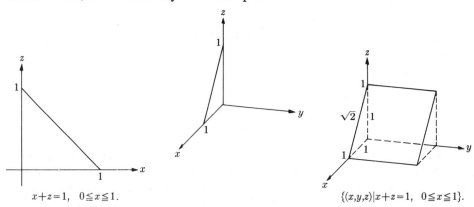

$x+z=1, \quad 0\leqq x\leqq 1.$ $\{(x,y,z)|x+z=1, \quad 0\leqq x\leqq 1\}.$

As usual, the figure is cut off at the ends, to clarify it in a pictorial sense. In its own plane, the cylinder is an infinite strip, of width $\sqrt{2}$.

If we had used the entire line $x + z = 1$, $y = 0$ as base, then the cylinder would have been the entire plane

$$\{(x, y, z) \mid x + z = 1\}.$$

This plane is parallel to the y-axis. Thus any plane parallel to one of the coordinate axes can be described as a cylinder. In fact, for appropriate choice of the base plane, any plane whatever can be regarded as a cylinder.

We have seen that cylindrical surfaces with their bases in the coordinate planes are easy to describe by equations (if their bases are so describable). The next simplest surfaces are the *surfaces of revolution*, whose areas we learned to compute in Section 7.5. Given a curve in, say, the yz-plane, we may rotate the curve about the y-axis. This generates a surface.

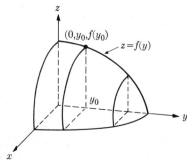

The cross sections of the surface, in planes parallel to the xz-plane, are all circles, with their centers on the y-axis. If the generating curve is described by a function, say,

$$z = f(y) \geqq 0,$$

then for each y_0, the cross section in the plane $y = y_0$ is the circle with center at $(0, y_0, 0)$ and radius $f(y_0)$. Thus the cross section is the graph of the condition

$$x^2 + z^2 = [f(y_0)]^2, \qquad y = y_0,$$

and the surface of revolution is the graph of the equation

$$x^2 + z^2 = [f(y)]^2.$$

Two important special cases are as follows:

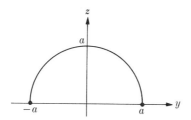

1)   Consider the generating curve

$$z = \sqrt{a^2 - y^2}, \qquad x = 0.$$

This is a semicircle. We rotate about the y-axis. The surface of revolution is the graph of the equation

$$x^2 + z^2 = [\sqrt{a^2 - y^2}]^2 = a^2 - y^2$$
$$\Leftrightarrow \ x^2 + y^2 + z^2 = a^2.$$

This is as it should be, because the surface of revolution is the sphere with center at the origin and radius a; it is easy to see by the distance formula that the sphere must be the graph of the equation

$$\sqrt{(x - 0)^2 + (y - 0)^2 + (z - 0)^2} = a$$
$$\Leftrightarrow \ x^2 + y^2 + z^2 = a^2.$$

2)   Consider the line

$$z = my, \qquad x = 0,$$

in the yz-plane. When we rotate about the y-axis, we get a *cone* (that is, a conical surface).

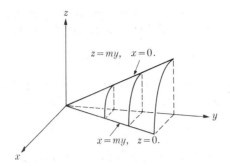

As usual, the figure shows only the first octant. The conical surface is the graph of
the equation

$$x^2 + z^2 = (my)^2$$

$$\Leftrightarrow \quad x^2 - m^2 y^2 + z^2 = 0.$$

If we had taken a line through the origin and rotated it about one of the other
coordinate axes, we would have gotten an equation of one of the forms

$$\text{(a)} \quad -m^2 x^2 + y^2 + z^2 = 0,$$

$$\text{(b)} \quad x^2 + y^2 - m^2 z^2 = 0.$$

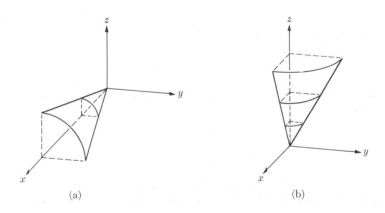

(a)   (b)

Each of the surfaces that we have investigated so far has been the graph of an
equation of the second degree in x, y, and z, that is, an equation of the form

$$Ax^2 + By^2 + Cz^2 + Dxy + Exz + Fyz + Gx + Hy + Iz + J = 0,$$

where the first six coefficients are not all equal to 0. Using the method of rotation of
axes in a plane, as in Section 8.4, we can find out what the plane cross sections of such
surfaces are like. Let E_0 be any plane, and let N be the normal line to E_0 through the
origin. Let F_0 be the plane which contains N and the z-axis, and let L be the line in
which F_0 intersects the xy-plane. By a rotation of axes in the xy-plane, we can make

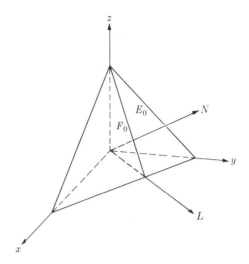

L the x'-axis. The equations of the rotation are of the form

$$x = x' \cos \theta - y' \sin \theta \quad \text{and} \quad y = x' \sin \theta + y' \cos \theta.$$

In the new coordinate system, the equation of the surface that we started with has the form

$$A'x'^2 + B'y'^2 + C'z^2 + D'x'y' + E'x'z + F'y'z + Gx' + Hy' + I'z + J = 0.$$

(*Query:* How do we know that the constant term is unchanged? And how do we know that the first six coefficients are not all equal to 0?) In the $x'z$-plane, we now perform another rotation of axes, in such a way that N becomes the new x-axis. The equations for this rotation are of the form

$$x' = x'' \cos \phi - z' \sin \phi, \quad z = x'' \sin \phi + z' \cos \phi;$$

and in the x''-, y'-, z'-coordinate system, the equation of our surface is still of the second degree, for the same reason as before. The plane F_0 is the graph of an equation of the form

$$x'' = k,$$

where k is the distance between the origin and F_0. To get the equation of the inter-section of the surface with F_0, we should set $x'' = k$ in the equation of the surface. This gives an equation of the second degree in y' and z'. By Theorem 2 of Section 8.4, this means that every plane cross section of a second-degree surface is (a) a circle, (b) a parabola, (c) an ellipse, (d) a hyperbola, (e) a point, (f) the empty set, (g) a line, or (h) the union of two lines (either parallel or intersecting).

In particular, every plane cross section of a cone is a "conic section" of the sort that we investigated in Chapter 8.

PROBLEM SET 14.1

Sketch the graphs of the following, in the first octant only. All the equations are to be regarded as equations in (x, y, z). For example, $x + y = 1$ is the equation of a plane, $x^2 + y^2 - 1 = 0$ is the equation of a cylindrical surface, and so on.

1. $x + y = 1$ 2. $x + z = 1$ 3. $y - z = 1$

4. $x^2 + z^2 = 1$ 5. $x^2 + y^2 = 1$ 6. $y^2 + z^2 = 4$

7. $z = y^2, \quad 0 \leq y \leq 1$ 8. $x = z^2, \quad 0 \leq z \leq 1$

9. $x = 4y^2, \quad 0 \leq y \leq 1$ 10. $(x^2/4) + y^2 = 1, \quad x, y \geq 0$

11. $(y^2/4) + z^2 = 1, \quad y, z \geq 0$ 12. $(x^2/4) + (z^2/9) = 1, \quad x, z \geq 0$

13. $x^2 + (z^2/4) = 1, \quad x, z \geq 0$ 14. $x = |y - 1|$

15. $|x| = y + 1$

Find equations for the surfaces described as follows, and sketch in the first octant.

16. The graph of $z = \sin y, 0 \leq y \leq \pi$ is rotated about the y-axis.

17. The graph of $y + z = 1$ is rotated about the y-axis.

18. The same graph is rotated about the z-axis.

19. The line which passes through the origin and the point $(1, 3, 1)$ is rotated about the y-axis.

20. The same line is rotated about the z-axis.

21. The same line is rotated about the x-axis.

22. The graph of $y = e^x, 0 \leq x \leq 1$ is rotated about the y-axis.

23. The same graph is rotated about the x-axis.

24. The graph of $y = |x| + 1$ is rotated about the x-axis.

25. The graph of $y = \cos z, 0 \leq z \leq \pi/2$, is rotated about the z-axis.

26. The same graph is rotated about the y-axis.

14.2 THE QUADRIC SURFACES

A *quadric surface* is a surface which is the graph of an equation of the second degree in x, y, and z. Thus all the surfaces discussed in the preceding section are quadric surfaces. We now consider less simple cases.

1) *Spheres and ellipsoids.* Obviously the graph of the equation

$$x^2 + y^2 + z^2 = a^2$$

is a sphere of radius a. More generally, the graph of the equation

$$\frac{x^2}{a^2} + \frac{y^2}{b^2} + \frac{z^2}{c^2} = 1$$

is called an ellipsoid.

The sketch on the left shows the entire surface. On the right we show only the part of the surface that lies in the first octant. Such partial sketches are much easier to draw, and sometimes they are actually easier to interpret and to use.

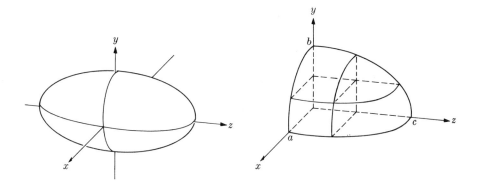

It is easy to see, algebraically, that every cross section of an ellipsoid, parallel to one of the coordinate axes, is an ellipse, a circle, a point, or the empty set. For example, setting $x = x_0$ we get

$$\frac{y^2}{b^2} + \frac{z^2}{c^2} = 1 - \frac{x_0^2}{a^2} = k.$$

For $k > 0$ the cross section is an ellipse or a circle; for $k = 0$ the graph is the point $(x_0, 0, 0)$; and for $k < 0$ the graph is empty. Similarly for $y = y_0$ or $z = z_0$.

2) *Elliptic and circular cones.* We found in the last section that the graph of the equation

$$x^2 - m^2 y^2 + z^2 = 0$$

is a circular cone. More generally, the graph of

$$\frac{x^2}{a^2} - y^2 + \frac{z^2}{b^2} = 0$$

is a cone, either elliptic or circular.

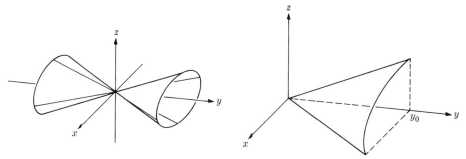

The figure on the left shows the entire cone, and the one on the right shows only the portion that lies in the first octant. The cross section in the yz-plane is obviously a pair of lines, because it is the graph of

$$x = 0, \qquad y^2 = \frac{z^2}{b^2}.$$

Similarly, the cross section in the xy-plane is a pair of lines, because it is the graph of

$$z = 0, \qquad y^2 = \frac{x^2}{a^2}.$$

The cross section in the plane $y = y_0$ is the graph of

$$y = y_0, \qquad \frac{x^2}{a^2} + \frac{z^2}{b^2} = y_0^2.$$

This is a point for $y_0 = 0$, and is a circle or an ellipse for $y_0 \neq 0$. Note that the cross section in the plane $x = x_0$ ($x_0 \neq 0$) is the graph of

$$\frac{z^2}{b^2} - y^2 = \frac{x_0^2}{a^2},$$

which is a hyperbola.

3) *The hyperboloid of one sheet.* Consider the equation

$$\frac{x^2}{a^2} + \frac{y^2}{b^2} - \frac{z^2}{c^2} = 1.$$

Setting $z = z_0$ and transposing, we get

$$\frac{x^2}{a^2} + \frac{y^2}{b^2} = 1 + \frac{z_0^2}{c^2} = k \geqq 1.$$

Therefore all horizontal cross sections of the graph are ellipses. Rewriting in the form

$$\frac{x^2}{a^2 k} + \frac{y^2}{b^2 k} = 1,$$

we see that as $|z_0|$ increases the ellipses get bigger, but their shape does not change.

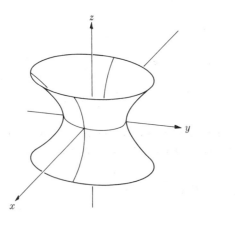

The cross sections in the other coordinate planes are hyperbolas; they are the graphs of the conditions

$$x = 0, \qquad \frac{y^2}{b^2} - \frac{z^2}{c^2} = 1,$$

$$y = 0, \qquad \frac{x^2}{a^2} - \frac{z^2}{c^2} = 1.$$

4) *The hyperboloid of two sheets.* Consider the equation

$$\frac{x^2}{a^2} - \frac{y^2}{b^2} - \frac{z^2}{c^2} = 1 \quad \Leftrightarrow \quad \frac{y^2}{b^2} + \frac{z^2}{c^2} = \frac{x^2}{a^2} - 1.$$

Again we investigate cross sections. For $|x_0| < a$, the cross section in the plane $x = x_0$ is empty; for $x = \pm a$, the graph is a point; and for $x_0^2 > a$, the graph is an ellipse (or a circle), being the graph of

$$x = x_0, \qquad \frac{y^2}{b^2} + \frac{z^2}{c^2} = \frac{x_0^2}{a^2} - 1 > 0.$$

The cross sections in the xz-plane and the xy-plane are obviously hyperbolas.

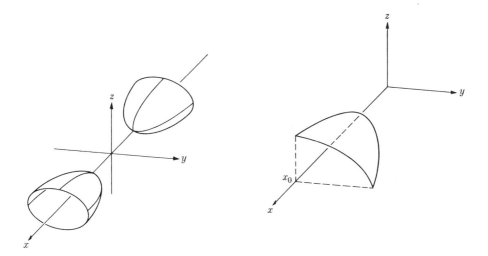

5) *The hyperbolic paraboloid.* This one is hard to visualize and hard to sketch. It is the graph of the equation

$$cz = \frac{y^2}{b^2} - \frac{x^2}{a^2} \qquad (c \neq 0).$$

We give the sketch for the case $a = b = c = 1$. Thus the equation of the surface becomes

$$z = y^2 - x^2.$$

To sketch, we use the cross sections in the planes $z = -1$, $z = 1$, $x = 0$, and $y = 0$. (Rather oddly, it is a bad idea to draw the cross section in the plane $z = 0$; this is a pair of intersecting lines, and when we indicate it correctly, the result is very hard to interpret pictorially.)

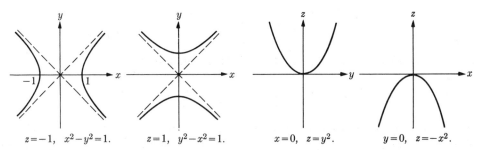

$z = -1$, $x^2 - y^2 = 1$. $\qquad$ $z = 1$, $y^2 - x^2 = 1$. $\qquad$ $x = 0$, $z = y^2$. $\qquad$ $y = 0$, $z = -x^2$.

Using these cross sections in a perspective drawing, we get the result shown below.

For other values of a and b, we get hyperbolas of different shapes in the horizontal cross sections. And when the sign of c is changed, the effect is to reflect the surface across the xy-plane.

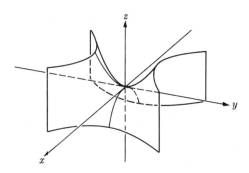

PROBLEM SET 14.2

Sketch the graphs of the following equations, and identify the surfaces.

1. $x^2 + \dfrac{y^2}{4} + \dfrac{z^2}{9} = 1$

2. $x^2 - \dfrac{y^2}{4} + \dfrac{z^2}{9} = 1$

3. $x^2 + \dfrac{y^2}{4} - \dfrac{z^2}{9} = 1$

4. $x^2 - \dfrac{y^2}{4} - \dfrac{z^2}{9} = 1$

5. $x^2 - \dfrac{y^2}{4} - \dfrac{z^2}{9} = 0$

6. $x^2 + \dfrac{y^2}{4} - \dfrac{z^2}{9} = 0$

7. $-x^2 + \dfrac{y^2}{4} - \dfrac{z^2}{9} = 0$

8. $x^2 - \dfrac{y^2}{4} - \dfrac{z^2}{9} = -1$

9. $z = x^2 - \dfrac{y^2}{4}$

10. $z^2 = x^2 + \dfrac{y^2}{4}$

11. $z = \dfrac{y^2}{4} - x^2$ 12. $z^2 = \dfrac{y^2}{4} - x$

13. $z = xy$ [*Hint:* This is easier to sketch if you first identify it.]

14. $z^2 = x^2 + 2xy + y^2$

15. Consider the hyperbolic paraboloid which is the graph of the equation

$$cz = \frac{y^2}{b^2} - \frac{x^2}{a^2} \qquad (c \neq 0).$$

Let (x_0, y_0) be any point of the xy-plane. For each α, consider the path whose coordinate functions are

$$x = x_0 + t \cos \alpha, \qquad y = y_0 + t \sin \alpha.$$

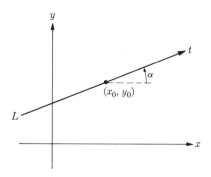

(In effect, we have set up a coordinate system on a line L through (x_0, y_0), in such a way that L becomes the t-axis.) Now consider the path in space defined by the coordinate functions

$$x = x_0 + t \cos \alpha,$$

$$y = y_0 + t \sin \alpha,$$

$$z = \frac{1}{c}\left(\frac{y^2}{b^2} - \frac{x^2}{a^2}\right) = \frac{1}{b^2 c}(y_0 + t \sin \alpha)^2 - \frac{1}{a^2 c}(x_0 + t \cos \alpha)^2.$$

What kinds of curves can be the loci of such paths?

16. A surface S is said to be *ruled* if for each point P of S there is a line L which contains P and lies entirely in S. (Thus every cone is automatically a ruled surface, and so also is every cylinder.) Show that every hyperbolic paraboloid is a ruled surface.

*17. Show that any hyperboloid of one sheet is a ruled surface (see Problem 16 for definition).

18. Find the volume of the region inside the graph of the equation

$$x^2/a^2 + y^2/b^2 + z^2/c^2 = 1.$$

19. Find the volume of the solid defined by the inequalities

$$0 \leqq y \leqq 1, \qquad -y \leqq x \leqq y, \qquad 0 \leqq z \leqq y^2 - x^2.$$

[This is the solid which lies (a) above the xy-plane, (b) below the hyperbolic paraboloid $z = y^2 - x^2$, and (c) between the planes $y = 0$ and $y = 1$.]

20. Find the volume of the solid which lies between the planes $z = 0$ and $z = 1$, and inside the one-sheeted hyperboloid

$$x^2 + \frac{y^2}{4} = z^2 + 1.$$

21. Same question for

$$\frac{x^2}{4} + \frac{y^2}{9} = z^2 + 1.$$

22. Find the volume of the solid which lies between the planes $z = 1$ and $z = 2$ and the two-sheeted hyperboloid

$$z^2 = x^2 + y^2 + 1.$$

14.3 FUNCTIONS OF TWO VARIABLES.
SLICE FUNCTIONS AND PARTIAL DERIVATIVES

So far, most of the functions that we have been studying have been of the following types.

1) Functions whose domains are sets of real numbers. In these cases, the domain D was usually an interval.

2) Functions of one vector space into another. These were always linear, and were referred to as linear transformations.

3) Sequences. (Every sequence $a_1, a_2, \ldots$ can be regarded as a function, with $f(i) = a_i$, and sometimes it is useful to think of sequences in this way. You may be able to recall such an occasion.)

We shall now investigate functions of the type

$$f\colon D \to \mathbf{R},$$

where D is a region in an inner-product space, and f is not necessarily linear.

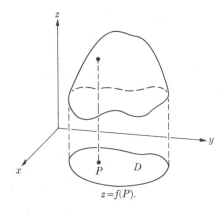

$z = f(P).$

Suppose that a rule is given under which to each point P of D there corresponds a real number z. We then say that we have a function

$$f\colon D \to \mathbf{R}.$$

As usual, the number corresponding to the point P of D is denoted by $f(P)$. If D is a region in $\mathbf{R}^2$, and $P = (x, y)$, then we may write $f(x, y)$ for $f(P)$; in this case, f is called a *function of two variables*, by which we mean two real variables. In this case, the graph of f is

$$\{(x, y, z) \mid (x, y) \text{ in } D \quad \text{and} \quad z = f(x, y)\},$$

or, equivalently,

$$\{(P, z) \mid P \text{ in } D \quad \text{and} \quad z = f(P)\},$$

as in the figure above.

Roughly speaking, f is continuous at a point P_0 of D if

$$P \approx P_0 \quad \Rightarrow \quad f(P) \approx f(P_0).$$

As for a function defined on an interval, the limit of f is defined, more generally, for the case in which P_0 does not necessarily lie in D. In this case,

$$\lim_{P \to P_0} f(P) = L$$

means that

$$P \approx P_0 \quad \text{and} \quad P \neq P_0 \quad \Rightarrow \quad f(P) \approx L.$$

To make these ideas precise, we interpret $P \approx P_0$ to mean that $\|P - P_0\|$ is small, and we interpret $f(P) \approx L$ to mean that $|f(P) - L|$ is small. This gives the following definition:

Definition. Let D be a region in an inner-product space $\mathscr{V}$, let f be a function $D \to \mathbf{R}$, let P_0 be a point of $\mathscr{V}$ (not necessarily lying in D), and let L be a real number. Suppose that for every $\epsilon > 0$ there is a $\delta > 0$ such that

$$0 < \|P - P_0\| < \delta \quad \Rightarrow \quad |f(P) - L| < \epsilon.$$

Then

$$\lim_{P \to P_0} f(P) = L.$$

Note that this is a quite straightforward generalization of the definition of the statement $\lim_{x \to x_0} f(x) = L$, with $|x - x_0|$ replaced by $\|P - P_0\|$. Following, in the same way, the pattern of our earlier work, we state:

Definition. Let f be a function $D \to \mathbf{R}$, and let P_0 be a point of D. If

$$\lim_{P \to P_0} f(P) = f(P_0),$$

then f is continuous at P_0.

The elementary theory of limits can now be generalized very easily.

Theorem 1. Let D be a region in a vector space $\mathscr{V}$, and let f and g be functions $D \to \mathbf{R}$. If

$$\lim_{P \to P_0} f(P) = L \qquad \text{and} \qquad \lim_{P \to P_0} g(P) = L',$$

then

$$\lim_{P \to P_0} [f(P) + g(P)] = L + L' \qquad \text{and} \qquad \lim_{P \to P_0} [f(P)g(P)] = LL'.$$

If $L' \neq 0$, then

$$\lim_{P \to P_0} \frac{f(P)}{g(P)} = \frac{L}{L'}.$$

The proofs of these results are exactly the same as in Appendix B; we merely need to translate the old proofs into the new language in the same way that we have just translated the definitions. In this process, there is no advantage in using coordinates, writing $f(x, y)$ for $f(P)$, $g(x, y)$ for $g(P)$, and so on, even if domains in $\mathbf{R}^2$ are the only ones that we are concerned with. In fact, the use of coordinates in these particular proofs merely complicates the notation and obscures the ideas. But for other purposes, we need to know how continuity is related to coordinates, as in the following:

Theorem 2. Let D be a region in $\mathbf{R}^2$, let f be a function $D \to \mathbf{R}$, and let $P_0 = (x_0, y_0)$ be a point of $\mathbf{R}^2$. Suppose that for every $\epsilon > 0$ there is a $\delta > 0$ such that

$$|x - x_0| < \delta \qquad \text{and} \qquad |y - y_0| < \delta \ \Rightarrow \ |f(x, y) - L| < \epsilon.$$

Then

$$\lim_{P \to P_0} f(P) = L.$$

To see this, we merely need to examine the geometric meaning of the inequalities on the left (preceding the $\Rightarrow$).

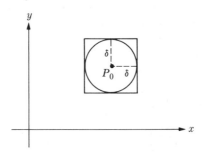

The inequalities $|x - x_0| < \delta$, $|y - y_0| < \delta$ hold inside the square, and the inequality $\|P - P_0\| < \delta$ holds inside the inscribed circle. Therefore

$$\|P - P_0\| < \delta \ \Rightarrow \ |x - x_0| < \delta \qquad \text{and} \qquad |y - y_0| < \delta$$
$$\Rightarrow \ |f(x, y) - L| < \epsilon;$$

and so

$$\lim_{P \to P_0} f(P) = L.$$

From this it follows, as for functions of one variable, that for functions of two variables, continuity is preserved under composition of functions:

Theorem 3. Let f, g, and h be continuous functions of two variables. For each $P = (x, y)$, let

$$\phi(x, y) = f[g(x, y), h(x, y)].$$

(Equivalently, $\phi(P) = f[g(P), h(P)]$.) Then ϕ is continuous.

We give the proof in outline:

$$P \approx P_0 \quad \Leftrightarrow \quad (x, y) \approx (x_0, y_0) \tag{1}$$

$$\Rightarrow \quad x \approx x_0 \quad \text{and} \quad y \approx y_0 \tag{2}$$

$$\Rightarrow \quad g(x, y) \approx g(x_0, y_0) \quad \text{and} \quad h(x, y) \approx h(x_0, y_0) \tag{3}$$

$$\Rightarrow \quad f[g(x, y), h(x, y)] \approx f[g(x_0, y_0), h(x_0, y_0)]. \tag{4}$$

The reasons for these implications are as follows:

1) $P = (x, y)$ and $P_0 = (x_0, y_0)$.
2) Any circle $\|P - P_0\| < \delta$ lies in its circumscribed square, as in the figure above.
3) g and h are continuous.
4) f is continuous, and Theorem 2 holds.

To verify that various functions defined by algebraic formulas are continuous, we need the following:

Theorem 4. Let ϕ be a function of one variable; and for each x and y (in a domain D) let

$$f(x, y) = \phi(x).$$

If ϕ is continuous, then f is continuous.

Note that the graph of such an f is always a portion of a cylinder, like this:

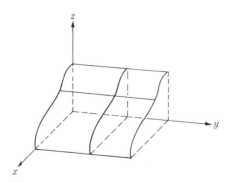

Geometrically it is obvious that the cylinder must be a continuous surface, and an algebraic proof is almost equally easy. Given (x_0, y_0) in D, and $\epsilon > 0$. We know that there is a $\delta > 0$ such that

$$|x - x_0| < \delta \quad \Rightarrow \quad |\phi(x) - \phi(x_0)| < \epsilon.$$

Therefore

$$|x - x_0| < \delta \quad \text{and} \quad |y - y_0| < \delta \;\Rightarrow\; |\phi(x) - \phi(x_0)| < \epsilon$$
$$\Leftrightarrow |f(x, y) - f(x_0, y_0)| < \epsilon.$$

Therefore, by Theorem 2, f is continuous.

These theorems enable us to infer that various simple functions, which obviously ought to be continuous, really are. Consider, for example,

$$f(x, y) = \cos \frac{x^2 y^2 - x^3 y}{1 + x^2 + y^2}.$$

By Theorem 4 the functions

$$x^2, \quad y^2, \quad x^3, \quad y, \quad 1,$$

are all continuous, considered as functions of two variables. By Theorem 1 it follows that x^2, y^2, $x^2 y^2$, and $x^3 y$ are continuous. Therefore $x^2 y^2 - x^3 y$ and $1 + x^2 + y^2$ are continuous (Theorem 1 again). By Theorem 1, $(x^2 y^2 - x^3 y)/(1 + x^2 + y^2)$ is continuous. By Theorem 3, so also is f.

We investigated the quadric surfaces by examining cross sections. We do the same for the graphs of functions of two variables. Given a region D in $\mathbf{R}^2$, and a function

$$f: D \to \mathbf{R},$$

take a fixed y_0, and consider the intersection of the graph with the plane $y = y_0$.

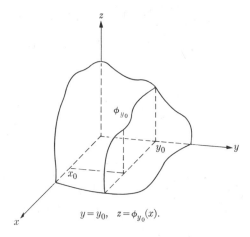

$$y = y_0, \quad z = \phi_{y_0}(x).$$

The cross section is a curve, and is the graph of a function ϕ_{y_0}: for each x for which (x, y_0) is in the domain of f, we have

$$\phi_{y_0}(x) = f(x, y_0).$$

Such a function is called a *slice function* of f. If the slice function is differentiable at x_0, then its derivative is denoted by

$$f_x(x_0, y_0).$$

Geometrically, $f_x(x_0, y_0)$ is the slope of the surface in the x-direction at the point (x_0, y_0).

Naturally, we can restate this in purely analytic terms, without any reference to the geometry, and without mentioning a slice function. We can state:

Definition

$$f_x(x_0, y_0) = \lim_{x \to x_0} \frac{f(x, y_0) - f(x_0, y_0)}{x - x_0},$$

if such a limit exists.

The function f_x is called the *partial derivative of f with respect to x*.

Standard differentiation formulas give us partial derivatives very easily. For example, given

$$f(x, y) = x^3 + xy^2 + x^2y + y^4,$$

we set $y = y_0$, to get

$$\phi_{y_0}(x) = x^3 + xy_0^2 + x^2y_0 + y_0^4.$$

This gives

$$\phi'_{y_0}(x) = 3x^2 + y_0^2 + 2xy_0 = f_x(x, y_0).$$

Dropping the subscript 0, we get

$$f_x(x, y) = 3x^2 + y^2 + 2xy.$$

The formal rule is simple: regard y as a constant, regard x as the dummy letter defining the function, and differentiate.

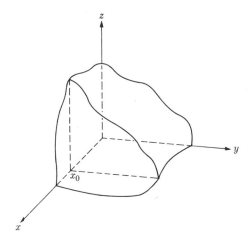

We can equally well consider slice functions in the y-direction, setting $x = x_0$. The derivative of the slice function is now the partial derivative of f with respect to y. More precisely:

Definition

$$f_y(x_0, y_0) = \lim_{y \to y_0} \frac{f(x_0, y) - f(x_0, y_0)}{y - y_0},$$

if such a limit exists.

The partial derivatives f_x and f_y, once we get them, are also functions; and *their* partial derivatives are defined in the same way. For example, consider

$$f(x, y) = x^3 + xy^2 + x^2y + y^4.$$

Here

$$f_x(x, y) = 3x^2 + y^2 + 2xy,$$
$$f_y(x, y) = 2xy + x^2 + 4y^3.$$

The partial derivative of f_x with respect to x is denoted by f_{xx}. Thus

$$f_{xx}(x, y) = 6x + 2y.$$

Similarly

$$f_{yy}(x, y) = 2x + 12y^2.$$

Now f_{xy} is the partial derivative of f_x with respect to y. We have

$$f_{xy} = D_y(3x^2 + y^2 + 2xy) = 2y + 2x.$$

And similarly

$$f_{yx} = D_x(2xy + x^2 + 4y^3) = 2y + 2x.$$

Note that while f_{xy} and f_{yx} turned out to be the same function, they were not defined in the same way, and they were not arrived at by the same process. Therefore the fact that $f_{xy} = f_{yx}$, for this particular function f, must be due either to an accident or to a nontrivial theorem.

Finally, some remarks on notation. Often, people write

$$\frac{\partial f}{\partial x} \text{ for } f_x, \qquad \frac{\partial f}{\partial y} \text{ for } f_y,$$

$$\frac{\partial^2 f}{\partial x^2} \text{ for } f_{xx}, \qquad \frac{\partial^2 f}{\partial y^2} \text{ for } f_{yy},$$

$$\frac{\partial^2 f}{\partial y\, \partial x} \text{ for } f_{xy}, \qquad \frac{\partial^2 f}{\partial x\, \partial y} \text{ for } f_{yx},$$

and so on. Note, in the last line, that in the symbols f_{xy} and f_{yx}, the letters indicating partial differentiation accumulate on the *right*; while in the symbols

$$\frac{\partial^2 f}{\partial y\, \partial x} \qquad \text{and} \qquad \frac{\partial^2 f}{\partial x\, \partial y}$$

the letters indicating partial differentiation accumulate on the *left*. Thus

$$\frac{\partial^3 f}{\partial x\, \partial y\, \partial y} = f_{yyx} \qquad \text{and} \qquad \frac{\partial^3 f}{\partial y\, \partial y\, \partial x} = f_{xyy}.$$

Note that in the ∂-notation, the symbols for higher derivatives look like "products" of "factors" of the types $\partial/\partial x$, $\partial/\partial y$. Thus

$$f_{xyy} = \frac{\partial}{\partial y}\frac{\partial}{\partial y}\frac{\partial}{\partial x} f(x, y) = \frac{\partial^3 f}{\partial y\, \partial y\, \partial x}.$$

This is why the symbols accumulate on the left instead of the right.

PROBLEM SET 14.3

Citing the theorems of this section, at the points where you need them, show that each of the following functions is continuous.

1. $f(x, y) = \sqrt{x^2 + y^2}$ 2. $f(x, y) = \sqrt{x^4 + y^4 + 1}$ 3. $f(x, y) = \sqrt[3]{x^2 + y^2}$

4. $f(x, y) = \dfrac{xy}{x^2 + y^2}$ $[(x, y) \neq (0, 0)]$ 5. $f(x, y) = \sin(x^2y + y^2x)$

6. $f(x, y) = \dfrac{x^2 - y^2}{x^2 + y^2}$ $[(x, y) \neq (0, 0)]$ 7. $f(x, y) = \dfrac{xy}{x^2 + y^2 + 1}$

8. $f(x, y) = \dfrac{\sin x}{y}$ $(y \neq 0)$ 9. $f(x, y) = \dfrac{\cos y - 1}{x}$ $(x \neq 0)$

10. $f(x, y) = \dfrac{\sin xy}{x^2 + y^2}$ $[(x, y) \neq (0, 0)]$ 11. $f(x, y) = \dfrac{\cos y}{x^2 + y^2}$ $[(x, y) \neq (0, 0)]$

Problems 12 through 22. For each of the functions f given in Problems 1 through 11, find f_x, f_y, f_{xy}, and f_{yx}.

23. Obviously the definition of f, in Problem 4, is valid only for $(x, y) \neq (0, 0)$. Is it possible to give a separate definition of $f(0, 0)$, in such a way that the resulting function is continuous? That is, is there any such thing as

$$(?) \quad \lim_{(x,y)\to(0,0)} \frac{xy}{x^2 + y^2} \quad ?$$

Why or why not?

24. Same question, for the function defined in Problem 10.

25. Same question, for the function defined in Problem 11.

26. By a *polynomial* in x and y we mean a function f which is the sum of a finite number of terms of the form $a_{ij}x^iy^j$, where the a_{ij}'s are constants. Thus

$$f(x, y) = \sum_{i=0}^{n} \sum_{j=0}^{n} a_{ij}x^iy^j.$$

Show that if f is a polynomial, then $f_{xy} = f_{yx}$.

27. Let us say, for short, that a function f is *regular* if $f_{xy} = f_{yx}$. Show that if f is regular then so also is f^2.

28. Show that if f is regular and positive, then $\sqrt{f}$ is regular.

29. Show that if f is regular, then so also is f^3.

30. Show that if f is regular and never zero, then $(1/f)$ is regular.

31. Show that if f and g are regular, then so also is f/g, at every point (x, y) where $g(x, y) \neq 0$.

32. Given a function f, and a point (x_0, y_0). Let $\Delta f = f(x, y) - f(x_0, y_0)$. For the function

$$f(x, y) = x^2y + y^2x,$$

show that Δf can be expressed in the form

$$\Delta f = f_x(x_0, y_0)\, \Delta x + f_y(x_0, y_0)\, \Delta y + E(\Delta x, \Delta y)\, \Delta x + F(\Delta x, \Delta y)\, \Delta y,$$

where E and F are functions such that

$$\lim_{(\Delta x, \Delta y) \to (0,0)} E(\Delta x, \Delta y) = \lim_{(\Delta x, \Delta y) \to (0,0)} F(\Delta x, \Delta y) = 0.$$

*33. Write a complete proof of Theorem 3, showing that for every $\epsilon > 0$ there is a $\delta > 0$ such that . . .

14.4 DIRECTIONAL DERIVATIVES AND DIFFERENTIABLE FUNCTIONS

The partial derivatives of a function $f: D \to \mathbf{R}$, where D is a region in $\mathbf{R}^2$, were defined as the derivatives of the slice functions $z = f(x, y_0)$ and $z = f(x_0, y)$; and we got the slice functions by taking cross sections of the graph of f, parallel to the yz-plane and the xz-plane. We proceed to consider more general slice functions, obtained by taking cross sections in any vertical plane whatever.

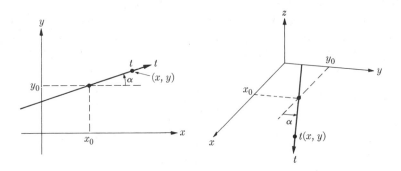

Given a function

$$f: D \to \mathbf{R}$$

with continuous partial derivatives f_x and f_y. Let (x_0, y_0) be a point of D, let L be a directed line through (x_0, y_0), and let α be the angle between the positive direction on L and the positive direction on the x-axis. Then L is the locus of the path

$$x = x_0 + t \cos \alpha,$$

$$y = y_0 + t \sin \alpha.$$

We now form the composite function

$$\phi_\alpha(t) = f(x_0 + t \cos \alpha, y_0 + t \sin \alpha).$$

We call ϕ_α the *slice function* in the direction α. If ϕ_α has a derivative at $t = 0$, then $\phi_\alpha'(0)$ is called the derivative of f in the direction α, and is denoted by $f_\alpha(x_0, y_0)$. That is,

$$f_\alpha(x_0, y_0) = \lim_{t \to 0} \frac{f(x_0 + t \cos \alpha, y_0 + t \sin \alpha) - f(x_0, y_0)}{t}.$$

For $\alpha = 0$, the slice is parallel to the x-axis, and so it ought to be true that $f_0 = f_x$. And this is true: for $\alpha = 0$ we have $\cos \alpha = 1$, $\sin \alpha = 0$, and

$$f_0(x_0, y_0) = \lim_{t \to 0} \frac{f(x_0 + t, y_0) - f(x_0, y_0)}{t}$$

$$= \lim_{\Delta x \to 0} \frac{f(x_0 + \Delta x, y_0) - f(x_0, y_0)}{\Delta x} = f_x(x_0, y_0).$$

Similarly, for $\alpha = \pi/2$ we have

$$\cos \alpha = 0, \qquad \sin \alpha = 1,$$

and

$$f_{\pi/2}(x_0, y_0) = f_y(x_0, y_0).$$

We now want a general formula for f_α. As a guide to what we should be aiming at, we consider first the simplest case, in which f is linear, with

$$f(x, y) = Ax + By + C.$$

For each point (x_0, y_0) and each α, we have

$$\phi_\alpha(t) = f(x_0 + t \cos \alpha, y_0 + t \sin \alpha)$$
$$= A(x_0 + t \cos \alpha) + B(y_0 + t \sin \alpha) + C,$$
$$\phi'_\alpha(t) = A \cos \alpha + B \sin \alpha.$$

Since

$$f_x(x, y) = A, \qquad f_y(x, y) = B,$$

for each x and y, it follows that these equations hold at the particular point (x_0, y_0); and so we have the following:

Theorem A. If f is a linear function of two variables, then

$$f_\alpha(x_0, y_0) = f_x(x_0, y_0) \cos \alpha + f_y(x_0, y_0) \sin \alpha,$$

for each α.

We shall now see that this formula holds under much more general conditions, when f is not necessarily linear, but is "approximately linear near (x_0, y_0)," in a sense which we shall define presently.

We recall that if f is a differentiable function of one variable, the difference

$$\Delta f = f(x_0 + \Delta x) - f(x_0)$$

can be expressed in the form

$$\Delta f = f'(x_0) \Delta x + E(\Delta x) \Delta x,$$

where

$$\lim_{\Delta x \to 0} E(\Delta x) = 0.$$

We want to get an analogous expression for the difference

$$\Delta f = f(x_0 + \Delta x, y_0 + \Delta y) - f(x_0, y_0).$$

It is fairly easy to find out what form the formula *has* to take if it exists at all. If f is linear, with

$$f(x, y) = Ax + By + C,$$

then

$$\Delta f = Ax + By + C - (Ax_0 + By_0 + C)$$
$$= A(x - x_0) + B(y - y_0)$$
$$= A\,\Delta x + B\,\Delta y.$$

Here, as before,

$$A = f_x(x, y), \qquad B = f_y(x, y);$$

for a linear function, the partial derivatives are simply the coefficients of x and y. This suggests that our expression for Δf ought to take the form

$$\Delta f = f_x(x_0, y_0)\,\Delta x + f_y(x_0, y_0)\,\Delta y + [\text{- - -}],$$

where [- - -] is, we hope, a function which approaches 0 very rapidly as $(\Delta x, \Delta y) \rightarrow (0, 0)$. Consider, for example,

$$f(x, y) = x^2 + xy + y^2.$$

Here

$$\Delta f = (x_0 + \Delta x)^2 + (x_0 + \Delta x)(y_0 + \Delta y) + (y_0 + \Delta y)^2 - x_0^2 - x_0 y_0 - y_0^2$$
$$= 2x_0\,\Delta x + \Delta x^2 + x_0\,\Delta y + y_0\,\Delta x + \Delta x\,\Delta y + 2y_0\,\Delta y + \Delta y^2$$
$$= (2x_0 + y_0)\,\Delta x + (x_0 + 2y_0)\,\Delta y + [\Delta x^2 + \Delta x\,\Delta y + \Delta y^2]$$
$$= f_x(x_0, y_0)\,\Delta x + f_y(x_0, y_0)\,\Delta y + [\text{- - -}],$$

where [- - -] $\rightarrow 0$ rapidly, as $\Delta x, \Delta y \rightarrow 0$.

We now attack the general problem, for functions with continuous partial derivatives f_x, f_y. We can get from (x_0, y_0) to $(x_0 + \Delta x, y_0 + \Delta y)$ by moving first vertically and then horizontally. Algebraically,

$$\Delta f = f(x_0 + \Delta x, y_0 + \Delta y) - f(x_0, y_0 + \Delta y) + f(x_0, y_0 + \Delta y) - f(x_0, y_0).$$

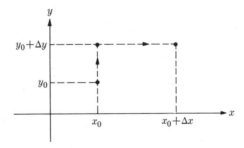

We apply the mean-value theorem MVT to the function

$$\phi(x) = f(x, y_0 + \Delta y),$$

on the interval from x_0 to $x_0 + \Delta x$.

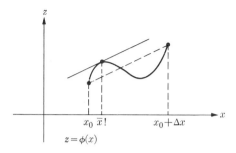

$$z = \phi(x)$$

The mean-value theorem says that there is an $\bar{x}$, between x_0 and $x_0 + \Delta x$, such that

$$\phi(x_0 + \Delta x) - \phi(x_0) = \phi'(\bar{x}) \, \Delta x.$$

This means that

$$f(x_0 + \Delta x, y_0 + \Delta y) - f(x_0, y_0 + \Delta y) = f_x(\bar{x}, y_0 + \Delta y) \, \Delta x.$$

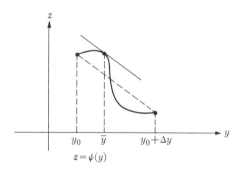

$$z = \psi(y)$$

Similarly, we apply MVT to the function

$$\psi(y) = f(x_0, y),$$

on the interval from y_0 to $y_0 + \Delta y$. Thus

$$\psi(y_0 + \Delta y) - \psi(y_0) = \psi'(\bar{y}) \, \Delta y,$$

for some $\bar{y}$ between y_0 and $y_0 + \Delta y$; and so

$$f(x_0, y_0 + \Delta y) - f(x_0, y_0) = f_y(x_0, \bar{y}) \, \Delta y.$$

Fitting these two results together, we get

$$\Delta f = f_x(\bar{x}, y_0 + \Delta y) \, \Delta x + f_y(x_0, \bar{y}) \, \Delta y,$$

where $\bar{x}$ is between x_0 and $x_0 + \Delta x$, and $\bar{y}$ is between y_0 and $y_0 + \Delta y$.

We are now almost finished. For each $\Delta x, \Delta y$, let

$$E_1(\Delta x, \Delta y) = f_x(\bar{x}, y_0 + \Delta y) - f_x(x_0, y_0),$$
$$E_2(\Delta x, \Delta y) = f_y(x_0, \bar{y}) - f_y(x_0, y_0).$$

Then
$$f_x(\bar{x}, y_0 + \Delta y) = f_x(x_0, y_0) + E_1(\Delta x, \Delta y),$$
and
$$f_y(x_0, \bar{y}) = f_y(x_0, y_0) + E_2(\Delta x, \Delta y).$$
Therefore
$$\Delta f = f_x(x_0, y_0)\, \Delta x + f_y(x_0, y_0)\, \Delta y + E_1(\Delta x, \Delta y)\, \Delta x + E_2(\Delta x, \Delta y)\, \Delta y.$$
Note that
$$\lim_{(\Delta x, \Delta y) \to (0,0)} E_1(\Delta x, \Delta y) = 0,$$
$$\lim_{(\Delta x, \Delta y) \to (0,0)} E_2(\Delta x, \Delta y) = 0,$$

because f_x and f_y are continuous. Thus, if the partial derivatives of f are continuous, then Δf is well approximated by the linear function $f_x(x_0, y_0)\, \Delta x + f_y(x_0, y_0)\, \Delta y$. For functions of two or more variables, the idea of approximation by linear functions is used as the *definition of differentiability*. To be exact:

Definition. Let D be a domain in $\mathbf{R}^2$, let f be a function $D \to \mathbf{R}$, and let $P_0 = (x_0, y_0)$ be a point of D. Suppose that there is a linear function
$$L(\Delta x, \Delta y) = A\, \Delta x + B\, \Delta y,$$
and functions E_1 and E_2, defined in a neighborhood of $(0, 0)$, such that
$$\Delta f = A\, \Delta x + B\, \Delta y + E_1(\Delta x, \Delta y)\Delta x + E_2(\Delta x, \Delta y), \tag{1}$$
and
$$\lim_{(\Delta x, \Delta y) \to (0,0)} E_1(\Delta x, \Delta y) = \lim_{(\Delta x, \Delta y) \to (0,0)} E_2(\Delta x, \Delta y) = 0. \tag{2}$$
Then f is said to be *differentiable* at (x_0, y_0).

This definition was modeled on the preceding discussion, and so we have already proved the following theorem.

Theorem 1. If f_x and f_y are continuous at (x_0, y_0), then f is differentiable at (x_0, y_0), with
$$\Delta f \approx A\, \Delta x + B\, \Delta y = f_x(x_0, y_0)\, \Delta x + f_y(x_0, y_0)\, \Delta y.$$

For functions of one variable, we defined the differential to be the linear function
$$df = df(\Delta x) = f'(x_0)\, \Delta x,$$
which gives good approximations of $\Delta f = f(x_0 + \Delta x) - f(x_0)$. For functions of two variables, the differential is defined analogously:

Definition. If f is differentiable at (x_0, y_0), then
$$df = df(\Delta x, \Delta y) = f_x(x_0, y_0)\, \Delta x + f_y(x_0, y_0)\, \Delta y.$$

The definition of differentiability is complicated to state, but it is easy to use. It gives us the formula that we wanted, for directional derivatives:

Theorem 2. If f is differentiable at (x_0, y_0), then for every direction α,
$$f_\alpha(x_0, y_0) = f_x(x_0, y_0) \cos \alpha + f_y(x_0, y_0) \sin \alpha.$$

Proof. By definition,

$$f_\alpha(x_0, y_0) = \lim_{t \to 0} \frac{f(x_0 + t \cos \alpha, y_0 + t \sin \alpha) - f(x_0, y_0)}{t}.$$

This has the form

$$\lim_{t \to 0} \frac{f(x_0 + \Delta x, y_0 + \Delta y) - f(x_0, y_0)}{t} = \lim_{t \to 0} \frac{\Delta f}{t},$$

where

$$\Delta x = t \cos \alpha, \qquad \Delta y = t \sin \alpha.$$

Since f is differentiable,

$$\begin{aligned}
\Delta f &= A \, \Delta x + B \, \Delta y + E_1 \, \Delta x + E_2 \, \Delta y \\
&= f_x(x_0, y_0) \, \Delta x + f_y(x_0, y_0) \, \Delta y + E_1 \, \Delta x + E_2 \, \Delta y \\
&= f_x(x_0, y_0) t \cos \alpha + f_y(x_0, y_0) t \sin \alpha + E_1 t \cos \alpha + E_2 t \sin \alpha.
\end{aligned}$$

Therefore

$$\begin{aligned}
\frac{\Delta f}{t} &= f_x(x_0, y_0) \cos \alpha + f_y(x_0, y_0) \sin \alpha \\
&\quad + E_1(t \cos \alpha, t \sin \alpha) \cos \alpha + E_2(t \cos \alpha, t \sin \alpha) \sin \alpha,
\end{aligned}$$

and so

$$\begin{aligned}
f_\alpha(x_0, y_0) &= \lim_{t \to 0} \frac{\Delta f}{t} \\
&= f_x(x_0, y_0) \cos \alpha + f_y(x_0, y_0) \sin \alpha + 0 \cdot \cos \alpha + 0 \cdot \sin \alpha,
\end{aligned}$$

which was to be proved.

In the first five problems in the following problem set, you are asked to "verify directly" that certain functions are differentiable at certain points. In each of these cases, you should go through an elementary calculation to express Δf in the form

$$\Delta f = A \, \Delta x + B \, \Delta y + E_1(\Delta x, \Delta y) \, \Delta x + E_2(\Delta x, \Delta y) \, \Delta y.$$

For example, given

$$f(x, y) = x^2 y \qquad (x_0, y_0) = (1, 1),$$

you would proceed as follows:

$$\begin{aligned}
\Delta f &= f(x_0 + \Delta x, y_0 + \Delta y) - f(x_0, y_0) \\
&= (1 + \Delta x)^2 (1 + \Delta y) - 1 \\
&= 1 + 2 \, \Delta x + \Delta x^2 + \Delta y + 2 \, \Delta x \, \Delta y + \Delta x^2 \, \Delta y - 1 \\
&= 2 \, \Delta x + \Delta y + (\Delta x + 2 \, \Delta y) \, \Delta x + (\Delta x^2) \, \Delta y.
\end{aligned}$$

The answer can now be written in the form

$$A = 2, \qquad B = 1, \qquad E_1 = \Delta x + 2 \, \Delta y, \qquad E_2 = \Delta x^2.$$

Note that other choices of E_1 and E_2 would have worked just as well. For example,

$$\Delta f = 2 \, \Delta x + \Delta y + (\Delta x) \, \Delta x + (\Delta x^2 + 2 \, \Delta x) \, \Delta y.$$

Therefore each of the first nine problems below has more than one right answer.

PROBLEM SET 14.4

Verify directly that each of the following nine functions is differentiable at the indicated point.

1. $f(x, y) = xy,$ $(2, 1)$
2. $f(x, y) = x^2 y^2,$ $(-1, -1)$
3. $f(x, y) = xy^2,$ $(1, 1)$
4. $f(x, y) = x^3,$ $(0, 0)$
5. $f(x, y) = x^2 - y^2,$ $(1, 1)$
6. $f(x, y) = y^4,$ $(-1, 1)$
7. $f(x, y) = x^2 + y^2,$ $(-1, 1)$
8. $f(x, y) = 4x^2 + y^2,$ $(1, 1)$
9. $f(x, y) = x^2 - 4y^2,$ $(1, -1)$

10. Given $f(x, y) = \sqrt{x^2 + y^2}$, $(x_0, y_0) = (1, 1)$, get a general formula for $f_\alpha(x_0, y_0)$. For which α does $f_\alpha(x_0, y_0)$ take on its maximum value? For which α do we get the minimum value?

11. Same question, for $f(x, y) = x^2 - y^2$, $(x_0, y_0) = (1, 1)$.

12. Same question for $f(x, y) = xy$, $(x_0, y_0) = (1, 1)$.

13. Suppose that f has a directional derivative f_α in every direction α, at a point (x_0, y_0). Is it possible that $f_\alpha(x_0, y_0) > 0$ for every α? Why or why not? (Try to answer this one merely on the basis of the definition of f_α, without appealing to Theorem 2.)

14. Show that if $f_\alpha(x_0, y_0) = 0$ for every α, then $f_x(x_0, y_0) = f_y(x_0, y_0) = 0$.

15. Give an example to show that the following "Theorem" is false:
Theorem (?) "Given $f: D \to \mathbf{R}$. If $f_x(x, y) = f_y(x, y) = 0$, for every (x, y) in D, then f is a constant."

16. Show that the following theorem is true:

Theorem A. Given

$$z = f(x, y)   (a < x < b, c < y < d).$$

If $f_x(x, y) = f_y(x, y) = 0$, for every (x, y) in the given domain, then f is a constant. Here we are requiring that the domain be a rectangular region with sides parallel to the x- and y-axes.

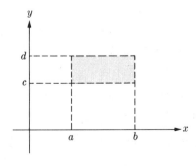

17. Theorem A, stated in Problem 16, is artificially special; it does not apply, as it stands, to domains like the following:

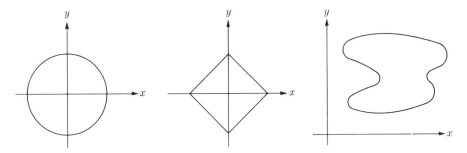

Find a way of describing the property of D that is really needed in the proof of Theorem A, and prove a theorem which uses your more general hypothesis.

14.5 THE CHAIN RULE FOR PATHS

In the preceding section, we defined the directional derivative f_α as the derivative of f along a linear path

$$x = g(t) = x_0 + t \cos \alpha,$$

$$y = h(t) = y_0 + t \sin \alpha,$$

and we found that if f is differentiable, then

$$f_\alpha(x_0, y_0) = f_x(x_0, y_0) \cos \alpha + f_y(x_0, y_0) \sin \alpha.$$

This result can be generalized, so as to apply to derivatives along paths which are not necessarily linear. Suppose that a path P is defined by a pair of coordinate functions. Strictly speaking, we should write

$$x = g(t), \qquad y = h(t), \qquad a < t < b.$$

But it is easier to keep track if we use the letters x and y as the names of the coordinate functions. Thus we write

$$x = x(t), \qquad y = y(t), \qquad a < t < b.$$

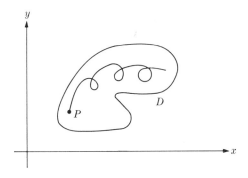

Let $F: D \to \mathbf{R}$ be a differentiable function, and suppose that the locus of the path P lies in D. We can then form the composite function

$$\phi(t) = F(x(t), y(t)),$$

and we have the following theorem:

Theorem 1 (*The chain rule for paths*). Let $\phi(t) = F(x(t), y(t))$. If F, $x(t)$, and $y(t)$ are differentiable, then ϕ is differentiable, and

$$\phi(t) = F_x(x(t), y(t))x'(t) + F_y(x(t), y(t))y'(t).$$

For example, consider

$$F(x, y) = x^2 y + y^2 x,$$

$$x(t) = \cos t,$$

$$y(t) = \sin t.$$

Here the locus of the path P is a circle. And

$$\phi(t) = \cos^2 t \sin t + \sin^2 t \cos t,$$

$$\phi'(t) = -2 \cos t \sin^2 t + \cos^3 t + 2 \sin t \cos^2 t - \sin^3 t.$$

Since

$$F_x(x, y) = 2xy + y^2 \quad \text{and} \quad F_y(x, y) = x^2 + 2xy,$$

Theorem 3 gives us

$$\phi'(t) = (2 \cos t \sin t + \sin^2 t)(-\sin t) + (\cos^2 t + 2 \sin t \cos t) \cos t,$$

which is the right answer.

We proceed to the proof. Take a fixed t_0, and let

$$x_0 = x(t_0), \qquad y_0 = y(t_0),$$

$$\Delta x = x(t_0 + \Delta t) - x(t_0),$$

$$\Delta y = y(t_0 + \Delta t) - y(t_0).$$

In this notation,

$$\phi'(t_0) = \lim_{\Delta t \to 0} \frac{F(x_0 + \Delta x, y_0 + \Delta y) - F(x_0, y_0)}{\Delta t}.$$

(Re-examine the definition of ϕ.) Now

$$F(x_0 + \Delta x) - F(x_0, y_0) = \Delta F$$

$$= F_x(x_0, y_0) \Delta x + F_y(x_0, y_0) \Delta y + E_1 \Delta x + E_2 \Delta y,$$

where

$$E_1, E_2 \to 0 \quad \text{as} \quad (\Delta x, \Delta y) \to (0, 0).$$

Therefore

$$\frac{\Delta F}{\Delta t} = F_x(x_0, y_0) \frac{\Delta x}{\Delta t} + F_y(x_0, y_0) \frac{\Delta y}{\Delta t} + E_1 \frac{\Delta x}{\Delta t} + E_2 \frac{\Delta y}{\Delta t}.$$

Therefore

$$\phi'(t_0) = \lim_{\Delta t \to 0} \frac{\Delta F}{\Delta t}$$

$$= F_x(x_0, y_0)x'(t_0) + F_y(x_0, y_0)y'(t_0) + 0 \cdot x'(t_0) + 0 \cdot y'(t_0)$$

$$= F_x(x_0, y_0)x'(t_0) + F_y(x_0, y_0)y'(t_0).$$

This holds for every t_0. Therefore

$$\phi'(t) = F_x\big(x(t), y(t)\big)x'(t) + F_y\big(x(t), y(t)\big),$$

which was to be proved.

Briefly,

$$\phi' = F_x x' + F_y y'.$$

In the "fractional" notation,

$$\frac{d\phi}{dt} = \frac{\partial F}{\partial x}\frac{dx}{dt} + \frac{\partial F}{\partial y}\frac{dy}{dt}.$$

Both of these short formulas should be regarded as abbreviations of the longer formula preceding them. The last of these three formulas is the easiest to remember, especially if we give it an intuitive interpretation, as follows. The derivative $d\phi/dt$ measures the rate of change of ϕ, when t changes slightly. The change in ϕ is due to (1) the change in x and (2) the change in y. These two effects are measured by the quantities

$$\frac{\partial F}{\partial x}\frac{\partial x}{\partial t}, \qquad \frac{\partial F}{\partial y}\frac{dy}{dt}.$$

The formula says that to combine these two effects, we simply add them.

PROBLEM SET 14.5

1. Given $F(x, y) = \cos xy$, $x(t) = t^2 + 1$, $y(t) = t^3$, and $\phi(t) = F(x(t), y(t))$, find ϕ'. Did you need to use Theorem 3?

2. Same question, for $F(x, y) = \sin xy$, $x(t) = t^2 + 1$, $y(t) = t^2 - 1$, $\phi(t) = F(x(t), y(t))$.

3. Same question, for $F(x, y) = 2xy$, $x(t) = \cos t$, $y(t) = \sin t$, $\phi(t) = F(x(t), y(t))$.

4. Same question, for $F(x, y) = x^2 + y^2$, $x(t) = \cos t$, $y(t) = \sin t$, $\phi(t) = F(x(t), y(t))$.

5. Same question, for $F(x, y) = xy$, $x(t) = t^2$, $y(t) = t^3$, $\phi(t) = F(x(t), y(t))$.

6. Same question, for $F(x, y) = x^2 + y^2$, $x(t) = \cos t$, $y(t) = \sin t$, $\phi(t) = F(x(t), y(t))$.

7. Given $F(x, y) = xy$, $x = x(t)$, $y = y(t)$, and $\phi(t) = F(x(t), y(t))$, find ϕ', using Theorem 3. (This will give you a circuitous derivation of the formula for the derivative of a product.)

8. Same question, for $F(x, y) = x/y$. (This will give you an equally circuitous derivation of the formula for the derivative of a quotient.)

9. Same question, for $F(x, y) = x^y$, $x > 0$. This will give you the formula

$$D[g^h] = hg^{h-1}g' + (g^h \ln g)h'.$$

This formula is easy to remember: first we differentiate as though the exponent were a constant, then we differentiate as though the base were a constant, and then we add the results.

10. Now derive the same differentiation formula, without using the theory developed in this chapter, appealing only to the basic definition

$$a^b = e^{b \ln a} \qquad (a > 0).$$

14.6 DIFFERENTIABLE FUNCTIONS OF MANY VARIABLES. THE CHAIN RULE

All the ideas which we developed in the last section, for functions of two variables, can be generalized immediately for functions of any number of variables. Limits and continuity have already been defined in the general case, in Section 14.3. Following the pattern of Section 14.4, we say that a function of n variables is differentiable at a point if the difference function is well approximated by a linear function, in small neighborhoods of the point. The definition is as follows:

Definition. Let D be a region in $\mathbf{R}^n$, let f be a function $D \to \mathbf{R}$, and let

$$P_0 = (a_1, a_2, \ldots, a_n)$$

be a point of D. For each point $P = (x_1, x_2, \ldots, x_n)$ of D, let

$$\Delta x_i = x_i - a_i,$$

let

$$\Delta P = P - P_0 = (\Delta x_1, \Delta x_2, \ldots, \Delta x_n),$$

and let

$$\Delta f = f(P) - f(P_0).$$

Suppose that there is a linear function

$$L(\Delta P) = L(\Delta x_1, \Delta x_2, \ldots, \Delta x_n) = A_1 \, \Delta x_1 + A_2 \, \Delta x_2 + \cdots + A_n \, \Delta x_n,$$

and a set of n functions $E_1, E_2, \ldots, E_n$, defined in a neighborhood of O, such that

$$\Delta f = L(\Delta x_1, \Delta x_2, \ldots, \Delta x_n) + E_1(\Delta P) \, \Delta x_1 + E_2(\Delta P) \, \Delta x_2 + \cdots + E_n(\Delta P) \, \Delta x_n,$$

$$(1)$$

and

$$\lim_{\Delta P \to O} E_i(\Delta P) = 0 \qquad (2)$$

for each i. Then f is *differentiable* at P_0.

Partial derivatives are defined in exactly the same way as for functions of two variables. For example,

$$f_{x_1}(P_0) = \lim_{\Delta x_1 \to 0} \frac{f(x_1, a_2, a_3, \ldots, a_n) - f(a_1, a_2, \ldots, a_n)}{\Delta x_1},$$

$$f_{x_2}(P_0) = \lim_{\Delta x_2 \to 0} \frac{f(a_1, x_2, a_3, \ldots, a_n) - f(a_1, a_2, \ldots, a_n)}{\Delta x_2},$$

and so on. Sometimes it is convenient to write $f_1(P_0)$ for $f_{x_1}(P_0)$, and in general $f_i(P_0)$ for $f_{x_i}(P_0)$; that is, f_i is the derivative of f with respect to the ith coordinate in $\mathbf{R}^n$. Thus, for $\mathbf{R}^n = \mathbf{R}^2$, $P = (x, y)$, we may write f_1 for f_x and f_2 for f_y.

Just as in the preceding section, if a function is differentiable, then it has all its first partial derivatives, and these are the coefficients in the linear approximation $L(\Delta P)$:

Theorem 1. If f is differentiable at P_0, with $\Delta f \approx L(\Delta P)$, then $f_i(P_0)$ is defined for each i, and

$$L(\Delta P) = f_1(P_0) \, \Delta x_1 + f_2(P_0) \, \Delta x_2 + \cdots + f_n(P_0) \, \Delta x_n.$$

The proof is just the same as for two variables: we take a fixed integer k, and set $\Delta x_i = 0$ for $i \neq k$, so that $\Delta P = \Delta x_k$. Then

$$\Delta f = L(\Delta P) + E_k(\Delta P) \, \Delta x_k,$$

because all other error terms on the right get multiplied by 0. Since $\Delta P = \Delta x_k$, we have $L(\Delta P) = A_k \, \Delta x_k$. Therefore

$$\Delta f = A_k \, \Delta x_k + E_k(\Delta P) \, \Delta x_k \qquad \text{and} \qquad f_k(P_0) = \lim_{\Delta x_k \to 0} \frac{\Delta f}{\Delta x_k} = A_k + 0.$$

As before, the differential of f at P_0 is defined to be the linear function which gives good approximations of Δf. Thus

$$df = f_1(P_0) \, \Delta x_1 + f_2(P_0) \, \Delta x_2 + \cdots + f_n(P_0) \, \Delta x_n.$$

By now it should be clear that we are in much the same situation as we were when dealing with n by n matrices, in Chapter 13: the ideas are adequately conveyed by the case $n = 3$, but the notation of the general case is tedious. For this reason, we shall often deal hereafter with only three variables, using

$$\mathbf{R}^n = \mathbf{R}^3, \qquad P = (w, x, y), \qquad P_0 = (w_0, x_0, y_0), \qquad \Delta P = (\Delta w, \Delta x, \Delta y).$$

You will probably find it easier to generalize the ideas for yourself, in your head, than to read a generalized version.

Theorem 2. Let D be a domain in $\mathbf{R}^3$, and let f be a function $D \to \mathbf{R}$. If f has all three of its first partial derivatives in D, and these are continuous at the point P_0, then f is differentiable at P_0.

Proof. Let

$$\Delta f = f(P) - f(P_0) = f(w, x, y) - f(w_0, x_0, y_0).$$

By a slight extension of the device that we used in the proof of the same theorem for two variables, we write

$$\Delta f = [f(w, x, y) - f(w_0, x, y)] + [f(w_0, x, y) - f(w_0, x_0, y)]$$
$$+ [f(w_0, x_0, y) - f(w_0, x_0, y_0)].$$

In the first bracket, we regard x and y as constants, and apply the mean-value theorem

to the function
$$\phi(w) = f(w, x, y).$$
Then
$$\phi(w) - \phi(w_0) = \phi'(\bar{w}) \, \Delta w,$$
where $\bar{w}$ is between w_0 and w. Since $\phi'(\bar{w}) = f_w(\bar{w}, x, y)$, we have
$$f(w, x, y) - f(w_0, x, y) = f_w(\bar{w}, x, y) \, \Delta w.$$
By two more such applications of the mean-value theorem, we get
$$\Delta f = f_w(\bar{w}, x, y) \, \Delta w + f_x(w_0, \bar{x}, y) \, \Delta x + f_y(w_0, x_0, \bar{y}) \, \Delta y. \tag{1}$$
Let
$$E_1(\Delta P) = f_w(\bar{w}, x, y) - f_w(w_0, x_0, y_0),$$
$$E_2(\Delta P) = f_x(w_0, \bar{x}, y) - f_x(w_0, x_0, y_0),$$
$$E_3(\Delta P) = f_y(w_0, x_0, \bar{y}) - f_y(w_0, x_0, y_0).$$
Then
$$\Delta f = f_w(P_0) \, \Delta w + f_x(P_0) \, \Delta x + f_y(P_0) \, \Delta y$$
$$+ E_1(\Delta P) \, \Delta w + E_2(\Delta P) \, \Delta x + E_3(\Delta P) \, \Delta y,$$
as in the definition of differentiability.

The chain rule for paths takes the same form as for two variables, and has the same proof.

Theorem 3 (*The chain rule for paths*). Let P be a path, with coordinate functions $w(t)$, $x(t)$, $y(t)$, and with locus lying in the domain D in $\mathbf{R}^3$. Let f be a function $D \to \mathbf{R}$, and for each t, let
$$\phi(t) = f(w(t), x(t), y(t)).$$
If f and the three coordinate functions are differentiable, then ϕ is differentiable, and
$$\phi'(t) = f_w(w, x, y)w' + f_x(w, x, y)x' + f_y(w, x, y)y'.$$
That is,
$$\phi(t) = \quad f_w(w(t), x(t), y(t))w'(t)$$
$$+ f_x(w(t), x(t), y(t))x'(t)$$
$$+ f_y(w(t), x(t), y(t))y'(t)$$
for every t.

This automatically gives us a chain rule for composite functions in which w, x, and y are functions of several variables. As in Theorem 3, let D be a domain in $\mathbf{R}^3$, and let f be a function $D \to \mathbf{R}$. But now let w, x, and y be functions of three variables, defined in a domain D'. Suppose that for each point (t, u, v) of D', the point $(w(t, u, v), x(t, u, v), y(t, u, v))$ lies in D. We then have a composite function
$$\phi : D' \to \mathbf{R},$$
defined by the formula
$$\phi(t, u, v) = f(w(t, u, v), x(t, u, v), y(t, u, v)),$$

and we want to find the partial derivatives ϕ_t, ϕ_u, and ϕ_v. But this is not a new problem, really: in calculating ϕ_t, we regard u and v as constants, and this means that we can calculate ϕ_t by means of the chain rule for paths. The only difference is that the derivatives $w'(t)$, $x'(t)$, $y'(t)$ in Theorem 3 are now the partial derivatives $w_t(t, u, v)$, $x_t(t, u, v)$, $y_t(t, u, v)$, and the final answer $\phi'(t)$ in Theorem 3 now becomes $\phi_t(t, u, v)$. This gives the formula

$$\phi_t(t, u, v) = f_w(w, x, y)w_t(t, u, v)$$
$$+ f_x(w, x, y)x_y(t, u, v)$$
$$+ f_y(w, x, y)y_t(t, u, v).$$

This may be easier to remember in the ∂-notation. In this notation,

$$\frac{\partial \phi}{\partial t} = \frac{\partial f}{\partial w}\frac{\partial w}{\partial t} + \frac{\partial f}{\partial x}\frac{\partial x}{\partial t} + \frac{\partial f}{\partial y}\frac{\partial y}{\partial t}.$$

Similarly for ϕ_u and ϕ_v. Thus we have

Theorem 4 (*The chain rule*). Let f be a differentiable function of w, x, and y; and let w, x, and y be differentiable functions of t, u, and v. Let

$$\phi(t, u, v) = f\big(w(t, u, v), x(t, u, v), y(t, u, v)\big).$$

Then the partial derivatives of ϕ are given by the formulas

$$\frac{\partial \phi}{\partial t} = \frac{\partial f}{\partial w}\frac{\partial w}{\partial t} + \frac{\partial f}{\partial x}\frac{\partial x}{\partial t} + \frac{\partial f}{\partial y}\frac{\partial y}{\partial t},$$

$$\frac{\partial \phi}{\partial u} = \frac{\partial f}{\partial w}\frac{\partial w}{\partial u} + \frac{\partial f}{\partial x}\frac{\partial x}{\partial u} + \frac{\partial f}{\partial y}\frac{\partial y}{\partial u},$$

$$\frac{\partial \phi}{\partial v} = \frac{\partial f}{\partial w}\frac{\partial w}{\partial v} + \frac{\partial f}{\partial x}\frac{\partial x}{\partial v} + \frac{\partial f}{\partial y}\frac{\partial y}{\partial v}.$$

For example, we might have

$$f(w, x, y) = w^2 + x^2 + y^2,$$
$$w = t + 2u + 3v, \qquad x = 2t + 3u + 4v, \qquad y = 3t + 4u + 5v.$$

Here

$$\frac{\partial f}{\partial w} = 2w = 2(t + 2u + 3v),$$

$$\frac{\partial f}{\partial x} = 2x = 2(2t + 3u + 4v),$$

$$\frac{\partial f}{\partial y} = 2y = 2(3t + 4u + 5v),$$

$$\frac{\partial w}{\partial u} = 2, \qquad \frac{\partial x}{\partial u} = 3, \qquad \frac{\partial y}{\partial u} = 4.$$

Therefore, by the chain rule,

$$\frac{\partial \phi}{\partial u} = 2(t + 2u + 3v)2 + 2(2t + 3u + 4v)3 + 2(3t + 4u + 5v)4$$

$$= 40t + 58u + 76v.$$

This is the right answer; by a direct calculation, we get

$$\phi(t, u, v) = 14t^2 + 29u^2 + 50v^2 + 40tu + 76uv + 52tv,$$

so that

$$\frac{\partial \phi}{\partial u} = \phi_u(t, u, v) = 58u + 40t + 76v,$$

as before.

PROBLEM SET 14.6

1. Given $f(t, u) = t^2 u$, $g(t, u) = t + u^2$, and $\phi(t, u) = f^2 + g^2$, find ϕ_t and ϕ_u.
2. Given $f(t, u) = t^3 - u$, $g(t, u) = t - u^2$, and $\phi(t, u) = f^2 g^2$, find ϕ_t and ϕ_u.
3. Given $f(s, t, u, v) = s^2 + t^3 + u^4 + v^5$, $g(s, t, u, v) = s + t + u + v$, and $\phi(s, t, u, v) = \sin f \cos g$, find ϕ_t and ϕ_u.
4. Under the conditions of Problem 3, find ϕ_s and ϕ_v.
5. Given $f(t, u) = t \cos u$, $g(t, u) = t \sin u$, and $\phi(t, u) = f^2 + g^2$, find ϕ_t and ϕ_u.
6. Given $f(t, u) = \cos u \cos t$, $g(t, u) = \cos u \sin t$, $h(t, u) = \sin u$, and $\phi(t, u) = f + g + h$, find ϕ_t and ϕ_u.
7. Given $f(t, u) = \cos u \cos t$, $g(t, u) = \cos u \sin t$, $h(t, u) = \sin u$, and $\phi(t, u) = f^2 + g^2 + h^2$, find ϕ_t and ϕ_u.
8. Given $f(t, u, v) = t \cos u \cos v$, $g(t, u, v) = t \sin u \cos v$, $h(t, u, v) = t \sin v$, and $\phi(t, u, v) = f^2 + g^2 - h$, find ϕ_t, ϕ_u, and ϕ_v.
9. Given $f(s, t, u, v, w) = stuvw$, verify by a direct calculation that f is differentiable at $(1, 0, 0, 1, 0)$. That is, find error functions $E_1(\Delta P), \ldots, E_5(\Delta P)$ as in the definition of differentiability of a function at a point.
10. Same question, for $f(s, t, u, v) = s^2 + tu + v^3$, at $(0, 1, 1, -1)$.

14.7 DIRECTIONAL DERIVATIVES AND GRADIENTS

We shall now generalize the idea of the directional derivative f_α, in such a way that it applies to functions of any number of variables. Given a region D in $\mathbf{R}^n$, and a differentiable function $f: D \to \mathbf{R}$, let P_0 be any point of D, and let V be any vector in $\mathbf{R}^n$, with $\|V\| = 1$. (A vector of norm 1 will be called a *direction* in $\mathbf{R}^n$.) The *derivative of f in the direction V*, at the point P_0, is defined to be

$$f_V(P_0) = \lim_{t \to 0} \frac{f(P_0 + tV) - f(P_0)}{t}.$$

To compute $f_V(P_0)$, we let

$$P_0 = (a_1, a_2, \ldots, a_n),$$
$$V = (c_1, c_2, \ldots, c_n).$$

We now have a path, defined by the coordinate functions

$$x_i(t) = a_i + c_i t \qquad (i = 1, 2, \ldots, n),$$

and a composite function

$$\phi(t) = f\big(x_1(t), x_2(t), \ldots, x_n(t)\big);$$

and

$$f_V(P_0) = \phi'(0).$$

We can now calculate $\phi'(0)$ by the chain rule for paths:

$$\phi'(0) = f_{x_1}(P_0)c_1 + f_{x_2}(P_0)c_2 + \cdots + f_{x_n}(P_0)c_n.$$

Thus we have

Theorem 1. If f is differentiable in D, then f has a directional derivative at every point of D, in every direction. If V is a unit vector $(c_1, c_2, \ldots, c_n)$, then

$$f_V(P_0) = f_{x_1}(P_0)c_1 + f_{x_2}(P_0)c_2 + \cdots + f_{x_n}(P_0)c_n$$
$$= f_1(P_0)c_1 + f_2(P_0)c_2 + \cdots + f_n(P_0)c_n.$$

(You should check that for the case $n = 2$, our definition of the directional derivative, and the formula given in Theorem 1, agree with the definition and formula given in Section 14.4.)

The *gradient* of a differentiable function, at a point P_0, is the vector whose components are the partial derivatives of f at P_0. The gradient vector is denoted by grad f. Thus if f is a differentiable function $D \to \mathbf{R}$, with real numbers as its values, then grad f is a *vector-valued* function $D \to \mathbf{R}^n$, with

$$\operatorname{grad} f = (f_1, f_2, \ldots, f_n),$$

where the f_i's are the first partial derivatives of f. That is,

$$\operatorname{grad} f(P) = \big(f_{x_1}(P), f_{x_2}(P), \ldots, f_{x_n}(P)\big),$$

for each P in D. For example, if

$$f(P) = f(x, y) = x^2 + xy + y^3,$$

then

$$f_1(x, y) = f_x(x, y) = 2x + y, \qquad f_2(x, y) = f_y(x, y) = x + 3y^2,$$

and

$$\operatorname{grad} f(x, y) = (2x + y, x + 3y^2),$$

which is a vector in $\mathbf{R}^2$, as it should be, for each point $P = (x, y)$.

The definition of the gradient may seem arbitrary, but it is not; the gradient has a geometric meaning, now to be explained. First we observe that for each unit vector V, with

$$V = (c_1, c_2, \ldots, c_n),$$

$$\|V\|^2 = c_1^2 + c_2^2 + \cdots + c_n^2 = 1,$$

the directional derivative f_V can be expressed as an inner product:

$$f_V = f_1 c_1 + f_2 c_2 + \cdots + f_n c_n,$$

where the f_i's are the first partial derivatives of f, and so

$$f_V = (\text{grad } f) \cdot V.$$

We shall prove the following:

Theorem 2. If f is differentiable at P, then (1) the direction of grad $f(P)$ is the direction which gives the maximum value of the directional derivative $f_V(P)$, and (2) the norm of grad $f(P)$ is the maximum value of $f_V(P)$.

Proof. Let $G = \text{grad } f(P)$, so that

$$G = (f_1(P), f_2(P), \ldots, f_n(P)),$$

and let V be the unit vector with the same direction as G; that is,

$$V = \frac{1}{\|G\|} G,$$

so that

$$V = (c_1, c_2, \ldots, c_n),$$

where

$$c_i = \frac{f_i(P)}{\|G\|}.$$

Then

$$f_V(P) = \big(\text{grad } f(P)\big) \cdot V = G \cdot V$$

$$= G \cdot \left(\frac{G}{\|G\|}\right) = \frac{1}{\|G\|}(G \cdot G) = \|G\| = \|\text{grad } f(P)\|.$$

Thus *the directional derivative, in the direction of the gradient, is the norm of the gradient.* And this is the direction which maximizes the directional derivative: if W is any unit vector, then we know by the Schwarz inequality (Theorem 1 of Section 11.6) that

$$(G \cdot W)^2 \leqq \|G\|^2 \|W\|^2 = \|G\|^2,$$

and so

$$f_W(P) = G \cdot W \leqq \|G\| = f_V(P).$$

This completes the proof.

A continuous function $D \to \mathscr{V}$, where D is a region in a Cartesian space $\mathbf{R}^n$ and $\mathscr{V}$ is a vector space, is called a *vector field*. We ordinarily draw the graphs of vector fields by using free vectors. For example, for

$$f(x, y) = x^2 + y^2,$$

$$\operatorname{grad} f(x, y) = (2x, 2y),$$

we can indicate the vector field grad f by drawing sample vectors in the xy-plane, like this:

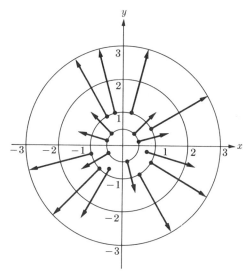

At each point $P = (x, y)$, the direction of grad $f(P)$ is the direction of the ray from the origin through P, and the length $\|\operatorname{grad} f(P)\|$ is twice the distance from the origin to P. At the origin, the gradient vector vanishes. Such a point is called a *singularity* of a vector field.

PROBLEM SET 14.7

For each of the following functions f, sketch the graph of grad f.

1. $f(x, y) = x + y$ 2. $f(x, y) = \sqrt{x^2 + y^2}$ 3. $f(x, y) = x - y$
4. $f(x, y) = y^2$ 5. $f(x, y) = xy$ 6. $f(x, y) = x^3$
7. $f(x, y) = x^2 - y^2$ 8. $f(x, y) = \sqrt{1 - x^2 - y^2}$
9. $f(x, y) = \sqrt{1 - \dfrac{x^2}{4} - \dfrac{y^2}{4}}$ 10. $f(x, y) = x^2 + y$
11. $f(x, y) = (x^2 + y^2)^2$ 12. $f(x, y) = \dfrac{1}{x^2 + y^2 + 1}$ 13. $f(x, y) = 4y^2 - x^2$

**14.8 INTERIOR LOCAL MAXIMA AND MINIMA,
FOR FUNCTIONS OF TWO VARIABLES. LEVEL CURVES**

For functions of one variable, defined on a closed interval, we had two kinds of maxima. In the figure on the left below, the maximum occurs at the endpoint b; at

x_1 the function has a *local* maximum, but not a maximum, because $f(x_1) < f(b)$. In the figure on the right, the function has a maximum at x_1; this is an interior maximum, and so $f'(x_1)$ must be 0.

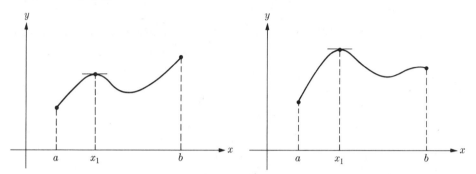

One of the simplest theorems for functions of one variable was the following:

Theorem 1. Suppose that f, f', and f'' are continuous, in a neighborhood of x_0. If $f'(x_0) = 0$ and $f''(x_0) < 0$, then f has a local maximum at x_1.

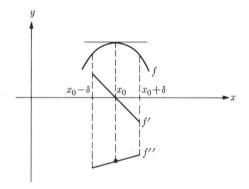

$f'(x_0) = 0$, $f''(x_0) < 0$, f has a LMax at x_0.

The proof is simple. Since $f''(x_0) < 0$, and f'' is continuous, it follows that there is a neighborhood $(x_0 - \delta, x_0 + \delta)$ of x_0 such that

$$f''(x) < 0 \qquad \text{for} \quad x_0 - \delta < x < x_0 + \delta,$$

as indicated in the figure. Therefore f' is decreasing on the interval

$$(x_0 - \delta, x_0 + \delta).$$

Therefore

$$f'(x) > 0 \qquad \text{for} \quad x_0 - \delta < x < x_0$$

and

$$f'(x) < 0 \qquad \text{for} \quad x_0 < x < x_0 + \delta.$$

Therefore f is increasing, from $x_0 - \delta$ to x_0, and f is decreasing, from x_0 to $x_0 + \delta$. Therefore $f(x_0)$ is the maximum value of f on the interval from $x_0 - \delta$ to $x_0 + \delta$.

The same proof proves the following theorem, which is going to be more useful:

Theorem 2. Given f, f', and f'', on an interval $(x_0 - \delta, x_0 + \delta)$. If $f'(x_0) = 0$, and

$$f''(x) < 0 \qquad \text{for} \quad x_0 - \delta < x < x_0 + \delta,$$

then $f(x_0)$ is the maximum value of f on the interval $(x_0 - \delta, x_0 + \delta)$.

We return now to functions of two variables. Let $P_0 = (x_0, y_0)$ be a point of the xy-plane. For each $\delta > 0$, the δ-*neighborhood* of P_0 is the interior of the circle with center at P_0 and radius δ. This is denoted by $N(P_0, \delta)$. Thus

$$N(P_0, \delta) = \{P \mid P_0P < \delta\} = \{(x, y) \mid (x - x_0)^2 + (y - y_0)^2 < \delta^2\}.$$

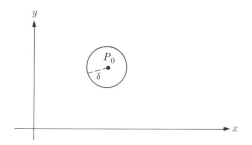

Let D be a set of points in the xy-plane. If P_0 is a point of D, and D contains a neighborhood of P_0 (for some δ), then P_0 is called an *interior point* of D.

Thus P is an interior point of D if P_0 lies in D, with at least a little room to spare. Consider, for example,

$$D = \{(x, y) \mid x^2 + y^2 \leqq 1\}.$$

Here D consists of the unit circle, plus its interior. If $OP_0 < 1$, as in the figure, then P_0 is an interior point; if we let

$$\delta = 1 - OP_0,$$

then $N(P_0, \delta)$ lies in D.

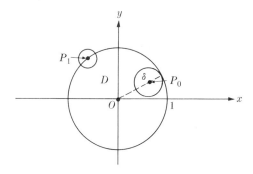

This works, no matter how close P_0 may be to the circle, as long as P_0 isn't actually *on* the circle; no matter how small the positive number $1 - OP_0$ may be, we can use it as our positive δ. On the other hand, if $OP_1 = 1$, so that P_1 is *on* the circle, then P_1 is not an interior point of D; no matter how small we take δ, the neighborhood $N(P_1, \delta)$ contains points outside of D.

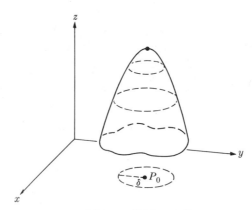

f has an ILMax at P_0.

Consider now a function of two variables

$$f: D \to \mathbf{R},$$

and let P_0 be a point of D. If $f(P) \leq f(P_0)$ for every P in D, then we say that f has a *maximum* at P_0. If P_0 is an interior point of D, and P_0 has a neighborhood $N(P_0, \delta)$ such that

$$f(P) \leq f(P_0) \qquad \text{for every } P \text{ in } N(P_0, \delta),$$

then we say that f has an *interior local maximum* (ILMax) at P_0.

This discussion has been rather lengthy, but if you review the figures which have been given in this section so far, you will find that they convey, by themselves, most of the ideas that we have been talking about.

Our purpose at this stage is to find conditions under which we can conclude that a function of two variables has an ILMax at a given point. At an ILMax, we must have

$$f_x(x_0, y_0) = 0 = f_y(x_0, y_0),$$

because an ILMax of f must be an ILMax of both the slice functions

$$\phi_0(t) = f(x_0 + t, y_0),$$
$$\phi_{\pi/2}(t) = f(x_0, y_0 + t).$$

(At this point you may want to review the definition of slice functions, at the beginning of Section 14.3.) Obviously, however, the vanishing of the partial derivatives f_x and f_y is not enough to guarantee an ILMax; we might have a minimum or a saddle point. For example, for

$$f(x, y) = x^2 - y^2,$$

we have

$$f_x(0, 0) = f_y(0, 0) = 0,$$

but the point $(0, 0)$ is a saddle point; the slice function

$$\phi_0(t) = f(t, 0) = t^2$$

has a minimum at 0, and the slice function

$$\phi_{\pi/2}(t) = f(0, t) = -t^2$$

has a maximum at 0. One way to see the difference between the behavior of a function at an ILMax or ILMin and its behavior at a saddle point is to consider the so-called *level curves*, in the xy-plane, on which the function takes on various constant values.

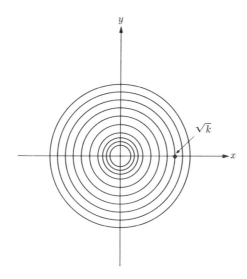

For the function

$$f(x, y) = x^2 + y^2,$$

the level curves are circles with center at the origin, as shown above. For each $k > 0$, the level curve on which $f(x, y) = k$ is the circle with center at the origin and radius $\sqrt{k}$. The origin is a singular point of this family of curves; and this is the point at which the function takes on its obvious minimum value 0.

For the function

$$f(x, y) = x^2 - y^2,$$

there is no maximum and no minimum. Since f is defined for every x and y, any Max or Min would have to be an ILMax or ILMin; at any such point, both the partial derivatives f_x and f_y would have to vanish; f_x and f_y vanish simultaneously only at

(0, 0), and at (0, 0) the function has a saddle point. The level curves for this function look like this:

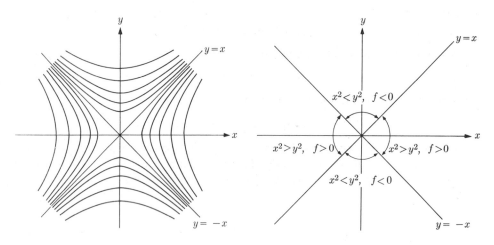

For $k > 0$, the level curve on which $f(x, y) = k$ is a hyperbola in standard position, and the level curve on which $f(x, y) = -k < 0$ is conjugate to it. The level curve on which $f(x, y) = 0$ is the union of two lines, which intersect each other at the origin, where f has a saddle point. These examples are typical of the way level curves behave in simple cases.

Even if *each* of the slice functions in the x- and y-directions has an ILMax at a point (x_0, y_0), we may still have a saddle point, on which a man in the saddle would be facing in some third direction. Consider

$$f(x, y) = -xy - \tfrac{1}{4}x^2 - \tfrac{1}{4}y^2.$$

Here

$$\phi_0(t) = -\tfrac{1}{4}t^2,$$

so that ϕ_0 has an ILMax at 0; and similarly for

$$\phi_{\pi/2}(t) = -\tfrac{1}{4}t^2.$$

But for $\alpha = 3\pi/4$ we have

$$\cos \alpha = -\frac{1}{\sqrt{2}}, \qquad \sin \alpha = \frac{1}{\sqrt{2}},$$

$$\phi_\alpha(t) = f(t \cos \alpha, t \sin \alpha)$$

$$= f\left(\frac{-t}{\sqrt{2}}, \frac{t}{\sqrt{2}}\right)$$

$$= -\left(\frac{-t}{\sqrt{2}}\right)\left(\frac{t}{\sqrt{2}}\right) - \frac{1}{4}\frac{t^2}{2} - \frac{1}{4}\frac{t^2}{2}$$

$$= \frac{t^2}{2} - \frac{t^2}{4} = \frac{1}{4}t^2,$$

which has a *minimum* at 0.

Thus, if we want to infer that f has an ILMax at (x_0, y_0), we need to consider every direction α, and examine all the slice functions

$$\phi_\alpha(t) = f(x_0 + t \cos \alpha, y_0 + t \sin \alpha).$$

This is the basis on which we shall attack the problem. Now

$$\phi_\alpha'(t) = f_x(x_0 + t \cos \alpha, y_0 + t \sin \alpha) \cos \alpha + f_y(x_0 + t \cos \alpha, y_0 + t \sin \alpha) \sin \alpha;$$

here we are using the chain rule. Applying the chain rule again, to each term, we get

$$
\begin{aligned}
\phi_\alpha''(t) = \;& f_{xx}(x_0 + t \cos \alpha, y_0 + t \sin \alpha) \cos^2 \alpha \\
&+ f_{xy}(x + t \cos \alpha, y_0 + t \sin \alpha) \cos \alpha \sin \alpha \\
&+ f_{yx}(x_0 + t \cos \alpha, y_0 + t \sin \alpha) \sin \alpha \cos \alpha \\
&+ f_{yy}(x_0 + t \cos \alpha, y_0 + t \sin \alpha) \sin^2 \alpha \\
= \;& c^2 f_{xx} + 2cs f_{xy} + s^2 f_{yy}.
\end{aligned}
$$

Here we are using the abbreviations

$$c = \cos \alpha, \qquad s = \sin \alpha,$$

$$f_{xx} = f_{xx}(x_0 + t \cos \alpha, y_0 + t \sin \alpha),$$

and so on. We are also assuming that all our derivatives are continuous; in this case it is a fact that $f_{yx} = f_{xy}$. (See Appendix K, where this is proved.)

The following is easy to see:

Theorem A. Suppose that

$$\phi_\alpha'(0) = 0$$

for every α. Suppose also that there is a number $\delta > 0$ such that for $|t| < \delta$ we have

$$\phi_\alpha''(t) < 0 \qquad \text{for every } \alpha.$$

Then f has an ILMax at (x_0, y_0).

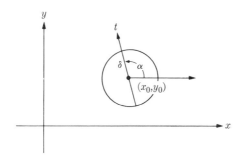

The reason is that for every α, $\phi_\alpha(0)$ is the maximum value of ϕ_α on the interval $(-\delta, \delta)$. It follows that $f(x_0, y_0)$ is the maximum value of f in the δ-neighborhood of (x_0, y_0).

We therefore need to find conditions under which such a δ exists. We have

$$\phi_\alpha''(t) = c^2 f_{xx} + 2cs f_{xy} + s^2 f_{yy}.$$

We now use an "ingenious device." We write

$$\phi_\alpha''(t) = f_{xx}\left[c^2 + 2\cdot\frac{cs f_{xy}}{f_{xx}} + \frac{s^2 f_{yy}}{f_{xx}}\right]$$

$$= f_{xx}\left[\left(c + s\cdot\frac{f_{xy}}{f_{xx}}\right)^2 - \frac{s^2 f_{xy}^2}{f_{xx}^2} + \frac{s^2 f_{yy}}{f_{xx}}\right]$$

$$= f_{xx}\left[\left(c + s\cdot\frac{f_{xy}}{f_{xx}}\right)^2 + \frac{f_{xx}f_{yy} - f_{xy}^2}{f_{xx}^2}s^2\right].$$

Suppose now that *at the point* (x_0, y_0) we have

$$f_{xx} < 0, \qquad f_{xx}f_{yy} - f_{xy}^2 > 0.$$

We are assuming that all the partial derivatives that we are dealing with are continuous. It follows that the same inequalities hold in the δ-neighborhood of (x_0, y_0), for some $\delta > 0$. Thus for $|t| < \delta$ we have

$$\phi_\alpha''(t) < 0 \qquad \text{for every } \alpha.$$

If we also know that

$$f_x(x_0, y_0) = f_y(x_0, y_0) = 0,$$

then we have

$$\phi_\alpha'(0) = 0 \qquad \text{for every } \alpha.$$

Therefore, by Theorem A, f has an ILMax at (x_0, y_0). We sum all this up in the following theorem:

Theorem 3. Suppose that f has continuous second partial derivatives in a neighborhood of (x_0, y_0). If

$$f_x(x_0, y_0) = f_y(x_0, y_0) = 0, \tag{1}$$

$$f_{xx}(x_0, y_0) < 0, \tag{2}$$

and

$$f_{xy}^2(x_0, y_0) - f_{xx}(x_0, y_0)f_{yy}(x_0, y_0) < 0, \tag{3}$$

then f has an ILMax at (x_0, y_0).

Not only the proof of this theorem, but also the theorem itself, are hard to read and hard to remember. This is typical of what you can expect from now on: when we pass from one variable to two or more, the calculus takes on a higher order of difficulty.

The following is a corollary of Theorem 3:

Theorem 4. Suppose that f has continuous second partial derivatives in a neighborhood of (x_0, y_0). If

$$f_x(x_0, y_0) = f_y(x_0, y_0) = 0, \tag{1}$$

$$f_{xx}(x_0, y_0) > 0, \tag{2}$$

and
$$f_{xy}^2(x_0, y_0) - f_{xx}(x_0, y_0)f_{yy}(x_0, y_0) < 0, \tag{3}$$
then f has an ILMin at (x_0, y_0).

Proof. If f satisfies the hypothesis of Theorem 4, then $-f$ satisfies the hypothesis of Theorem 3. Therefore $-f$ has an ILMax at (x_0, y_0). Therefore f has an ILMin at (x_0, y_0).

When we were studying functions of one variable, we found simple cases in which a function had an ILMax or an ILMin, but the second derivative test failed to reveal the fact. For example, $f(x) = x^4$ has an ILMin at $x = 0$, but $f''(0) = 0$. Here the trouble seems to be that the function approaches its ILMin value "very flatly." The same sort of thing can happen for functions of two variables. For example, the function $f(x, y) = x^4 + y^4$ has an ILMin at $(0, 0)$, but Theorem 3 does not apply, because at the point $(0, 0)$, all the second partial derivatives f_{xx}, f_{yy}, and f_{xy} are equal to 0.

Sometimes, however, we can get negative information by examining the quantity
$$f_{xy}^2(x_0, y_0) - f_{xx}(x_0, y_0)f_{yy}(x_0, y_0).$$
If this quantity is positive, then we can infer that f has neither an ILMax nor an ILMin at (x_0, y_0). The proof is as follows. We found that
$$\phi_\alpha''(t) = c^2 f_{xx} + 2cs f_{xy} + s^2 f_{yy}.$$
We set $t = 0$. Let
$$A = f_{xx}(x_0, y_0), \qquad B = f_{xy}(x_0, y_0), \qquad C = f_{yy}(x_0, y_0).$$
Then
$$\phi_\alpha''(0) = A \cos^2 \alpha + 2B \sin \alpha \cos \alpha + C \sin^2 \alpha,$$
and we are assuming that
$$B^2 - AC > 0.$$
Consider the function
$$\psi(u) = A + 2Bu + Cu^2.$$
The graph is a parabola; the discriminant
$$(2B)^2 - 4AC = 4(B^2 - AC) > 0,$$
and so $\psi(u_1) > 0$ for some u_1 and $\psi(u_2) < 0$ for some u_2. But for $\cos \alpha \neq 0$ we have
$$\phi_\alpha''(0) = \cos^2 \alpha [A + 2B \tan \alpha + C \tan^2 \alpha];$$
for $\tan \alpha = u_1$, we have $\phi_\alpha''(0) > 0$; and for $\tan \alpha = u_2$, we have $\phi_\alpha''(0) < 0$. Therefore the direction of concavity of the slice functions ϕ_α is different for different values of α, and we cannot have an ILMax or an ILMin. To sum up:

Theorem 5. If $f_{xy}^2 - f_{xx}f_{yy} > 0$ at P_0, then f has neither an ILMax nor an ILMin at P_0.

PROBLEM SET 14.8

Investigate the following functions for interior local maxima and minima. Not all of these problems can be worked by straightforward applications of the theorems in Section 14.8; you may need to examine slice functions, or use other elementary methods.

1. $f(x, y) = xy$ 2. $f(x, y) = x^2 + xy$

3. $f(x, y) = x^2 - y^2$ 4. $f(x, y) = (x + y + 1)^2 + (x - y + 1)^2$

5. $f(x, y) = x^2 + y^2 + x + y + 1$ 6. $f(x, y) = x^2 + y^2 + 2x + 1$

7. $f(x, y) = x^2 + xy + y^2 + x + y + 1$ 8. $f(x, y) = x^2 + 2xy + y^2$

9. $f(x, y) = 1 - x^4 - y^2$ 10. $f(x, y) = x^4 - 2x^2 - y^2$

11. At what point does the function $f(x, y) = x^2(1 - x^2 - y^2)$ take on its maximum value? What is the maximum value of the function?

12. At what point or points does the function

$$f(x, y) = \sqrt{(x - 1)^2 + (y - 2)^2} + \sqrt{x^2 + y^2}$$

take on its minimum value? What is the minimum value of the function?

13. Consider the ellipsoid

$$x^2 + y^2/4 + z^2/9 = 1.$$

In this surface we are to inscribe a rectangular parallelepiped, with sides parallel to the coordinate planes. What is the maximum possible volume of such a parallelepiped? Give the coordinates of its corner in the first octant.

14. Same question, for the ellipsoid

$$x^2 + \frac{y^2}{4} + \frac{z^2}{\pi} = 1.$$

15. Same question, for the ellipsoid

$$x^2 + \frac{y^2}{4} + \frac{z^2}{4} = 1.$$

16. Let $A_1 = (0, 0)$, $A_2 = (1, 2)$, and $A_3 = (2, 1)$. For each $P = (x, y)$, let

$$f(P) = (A_1P)^2 + (A_2P)^2 + (A_3P)^2.$$

At what point P_0 does $f(P)$ take on its minimum value?

17. Same question, for $A_i = (a_i, b_i)$ $(i = 1, 2, \ldots, n)$, and $f(P) = \sum_{i=1}^{n} (A_iP)^2$.

14.9 DOUBLE INTEGRALS, INTUITIVELY CONSIDERED

You recall that in Section 3.7 we gave a preliminary intuitive definition of the definite integral of a continuous function over a closed interval. Here the A_i's are *areas*, in the elementary geometric sense, so that $A_i \geq 0$ for every i. To get the integral, we count areas above the x-axis positively, and areas below the x-axis negatively.

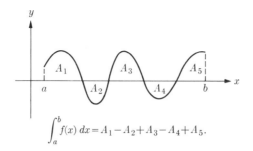

$$\int_a^b f(x)\,dx = A_1 - A_2 + A_3 - A_4 + A_5.$$

Later, in Section 7.2, we gave a new definition of the integral, as the limit of the sample sums of the function as the mesh of the net approaches 0:

$$\int_a^b f(x)\,dx = \lim_{|N| \to 0} \sum_{i=1}^n f(\bar{x}_i)\,\Delta x_i,$$

provided, of course, that such a limit exists. The new definition was necessary for two reasons. First, we needed it to clarify the underlying theory. Second, we wanted to use the definite integral to solve problems which did not, at the outset, look like area problems at all. For example, to calculate arc lengths, surface areas, volumes, and moments, we regarded them as limits of sample sums, as the mesh approaches zero. Thus our second definition of the definite integral was not only more exact but also more widely applicable.

We shall follow the same scheme with multiple integrals, first giving an intuitive definition, and then reformulating it when the need arises (which will be soon). Suppose that we have given a nonnegative continuous function

$$f\colon D \to \mathbf{R},$$

defined in a domain D in the xy-plane. (See figure on the left below.)

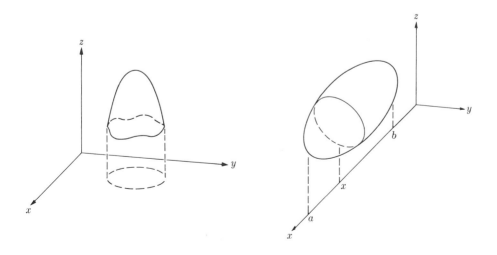

The expression

$$\iint_D f(P)\, dA$$

denotes the volume of the region lying above the xy-plane and below the graph of f. This is called the *integral of f over D*. Thus the integral is the volume of the solid

$$S = \{(x, y, z) \mid (x, y) \in D \quad \text{and} \quad 0 \leq z \leq f(x, y)\}.$$

In "reasonable" cases, double integrals can be calculated by the method of cross sections, developed in Section 7.4. The scheme is shown on the right above.

Given a solid S, in space, lying between the planes $x = a$ and $x = b$. Suppose that for each x_0 from a to b we can compute, somehow, the area of the cross section in the plane $x = x_0$. If for each such x_0 we let $A(x_0)$ be the area of this cross section, then the volume of our solid S is

$$vS = \int_a^b A(x)\, dx.$$

This method works for many solids whose volumes are not given by standard formulas. Consider the following.

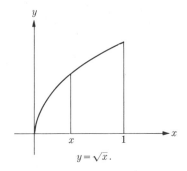

$y = \sqrt{x}$.

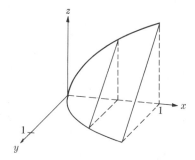

We start with the region

$$R = \{(x, y) \mid 0 \leq x \leq 1, \quad 0 \leq y \leq \sqrt{x}\},$$

in the xy-plane. For each x, we join the point $(x, 0)$ to the point (x, y^2), by a segment. On each such segment we set up an isosceles right triangle, as shown above on the right. Let S be the union of all these triangles (including, of course, their interiors). For each x, the area of the triangle at x is

$$A(x) = \tfrac{1}{2}\sqrt{x}\,\sqrt{x} = \tfrac{1}{2}x.$$

Therefore the volume is

$$\int_0^1 A(x)\, dx = \tfrac{1}{2}\int_0^1 x\, dx = \tfrac{1}{2}[\tfrac{1}{2}x^2]_0^1 = \tfrac{1}{4}.$$

Here $A(x)$ was computable by an elementary formula, because the cross sections for constant x were triangular. But no matter what method you use to compute $A(x)$, you can still find the volume by integrating $A(x)$ between the appropriate limits. In particular, *you can use the method when $A(x)$ is itself computed as a definite integral.*

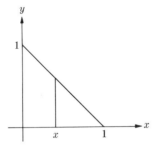

Consider the following. Let

$$D = \{(x, y) \mid 0 \leq x \leq 1, 0 \leq y \leq 1 - x\}.$$

For each (x, y) in D, let

$$f(x, y) = x^2 + y^3.$$

We want to find the volume of the solid lying above D and below the graph of f. Now for each x_0, the cross section in the plane $x = x_0$ looks like the drawing on the left below. Therefore the area of the cross section at x_0 is

$$
\begin{aligned}
A(x_0) &= \int_0^{1-x_0} (x_0^2 + y^3) \, dy = [x_0^2 y + \tfrac{1}{4} y^4]_0^{1-x_0} \\
&= x_0^2(1 - x_0) + \tfrac{1}{4}(1 - x_0)^4 \\
&= x_0^2 - x_0^3 + \tfrac{1}{4}(1 - x_0)^4.
\end{aligned}
$$

Dropping the subscript we get

$$A(x) = x^2 - x^3 + \tfrac{1}{4}(1 - x)^4.$$

Therefore the volume is

$$
\iint_D f(P) \, dA = \int_0^1 A(x) \, dx = [\tfrac{1}{3} x^3 - \tfrac{1}{4} x^4 - \tfrac{1}{4} \cdot \tfrac{1}{5}(1 - x)^5]_0^1
$$

$$
= [\tfrac{1}{3} - \tfrac{1}{4}] - [-\tfrac{1}{20}] = \tfrac{2}{15}.
$$

The method works more generally. Suppose that we have a region D in the xy-plane, lying between the graphs of two functions, as on the right below.

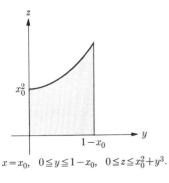

$x = x_0, \quad 0 \leq y \leq 1 - x_0, \quad 0 \leq z \leq x_0^2 + y^3.$

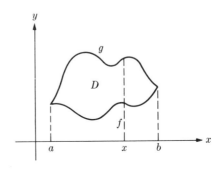

We have given $F(x, y) \geqq 0$ on D, and we want to find

$$\iint_D F(x, y) \, dA.$$

This is the volume of the solid lying above D and below the graph of F. For each x, the cross-sectional area is

$$A(x) = \int_{f(x)}^{g(x)} F(x, y) \, dy.$$

Here x is being held constant, and we are integrating from $f(x)$ to $g(x)$. But $A(x)$, once you get it, is a function. Therefore the total volume is

$$\iint_D F(x, y) \, dA = \int_a^b A(x) \, dx = \int_a^b \left[\int_{f(x)}^{g(x)} F(x, y) \, dy \right] dx.$$

This takes a very simple form when D is a rectangular region defined by inequalities of the form

$$a \leqq x \leqq b, \qquad c \leqq y \leqq d.$$

Here

$$\iint_D F(x, y) \, dA = \int_a^b \int_c^d F(x, y) \, dy \, dx.$$

Here it is to be understood that the "inside integration" is to be performed first, giving the cross-sectional area

$$A(x) = \int_c^d F(x, y) \, dy,$$

and the resulting function is to be integrated from a to b.

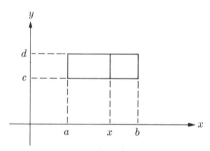

Of course, we could equally well have used cross sections for constant y. This would give a different cross-sectional area function

$$B(y) = \int_a^b F(x, y) \, dx;$$

and we would have

$$\iint_D F(x, y) \, dA = \int_c^d B(y) \, dy = \int_c^d \int_a^b F(x, y) \, dx \, dy.$$

The expressions

$$\int_a^b \int_c^d F(x, y)\, dy\, dx, \qquad \int_c^d \int_a^b F(x, y)\, dx\, dy$$

are called *iterated integrals*. If the general assumptions that we are making in this section are correct, and in fact they are, then it follows that the two iterated integrals are equal; that is, the order of integration does not matter. The reason is that each of the two iterated integrals is equal to the double integral.

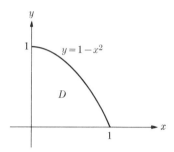

The same phenomenon occurs, in a less simple form, if the domain is not rectangular. In the figure above, the domain D can equally well be described by the inequalities

$$0 \leqq x \leqq 1, \qquad 0 \leqq y \leqq 1 - x^2, \tag{1}$$

or

$$0 \leqq y \leqq 1, \qquad 0 \leqq x \leqq \sqrt{1 - y}. \tag{2}$$

Thus, for any continuous nonnegative function $f\colon D \to \mathbf{R}$ we have

$$\iint_D F(x, y)\, dA = \int_0^1 \int_0^{1-x^2} F(x, y)\, dy\, dx,$$

and we also have

$$\iint_D F(x, y)\, dA = \int_0^1 \int_0^{\sqrt{1-y}} F(x, y)\, dx\, dy.$$

Therefore the two iterated integrals must have the same value.

PROBLEM SET 14.9

In each problem below, we have a domain D described by a pair of inequalities, and a function defined by a formula. In each case, express the double integral

$$\iint_D F(x, y)\, dA$$

as an iterated integral in two different ways, evaluate both of your iterated integrals, and check by observing that they ought to have the same value.

1. $D: 0 \leq x \leq 2, \quad 0 \leq y \leq x^3; \quad F(x, y) = x + y$

2. $D: 0 \leq x \leq 2, \quad 0 \leq y \leq \dfrac{x^2}{4}; \quad F(x, y) = x - y$

3. $D: 0 \leq x \leq 2, \quad x^3 \leq y \leq 8; \quad F(x, y) = x^2 + y$

4. $D: 0 \leq x \leq 1, \quad x^2 \leq y \leq x; \quad F(x, y) = x + y$

5. $D: 0 \leq x \leq 1, \quad x \leq \bar{y} \leq 1; \quad F(x, y) = x^3 y^3$

6. $D: 0 \leq y \leq 1, \quad y \leq x \leq 1; \quad F(x, y) = x^2 + y^2$

7. $D: -1 \leq x \leq 1, \quad 0 \leq y \leq 1 - x^2; \quad F(x, y) = xy$

8. $D: 0 \leq x \leq 1, \quad -\sqrt{1 - x^2} \leq y \leq \sqrt{1 - x^2}; \quad F(x, y) = (x^2 + y^2)^2$

9. $D: 0 \leq x \leq 1, \quad 0 \leq y \leq x^2; \quad F(x, y) = \sqrt{xy}$

10. Let

$$f(\alpha) = \int_0^1 \int_0^\alpha (x^2 + y^2)^2 \, dx \, dy.$$

Find $f'(\alpha)$.

11. Let ϕ be positive and continuous, and let

$$f(\alpha) = \int_a^b \int_c^\alpha \phi(x, y) \, dx \, dy.$$

Find the simplest formula that you can for $f'(\alpha)$.

12. Same question for

$$f(\alpha) = \int_\alpha^b \int_c^d \phi(x, y) \, dx \, dy.$$

13. Let ϕ be a positive function, with continuous first and second partial derivatives. Get the simplest formula that you can for

$$\int_a^b \int_c^d \phi_{xy}(x, y) \, dx \, dy.$$

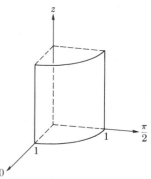

14.10 CYLINDRICAL COORDINATES IN SPACE.
THE DEFINITION OF THE INTEGRAL

To set up a system of cylindrical coordinates in space, we use polar coordinates in the xy-plane, and leave the z-coordinate unchanged.

For some solids and surfaces, this leads to a considerable simplification. For example, the cylindrical surface of radius 1, with the z-axis as its axis of symmetry, is the graph of the equation $r = 1$. (See the figure above.)

The unit sphere with center at the origin is the graph of the equation $r^2 + z^2 = 1$. (See the figure on the left below.)

Recalling the familiar formulas giving x and y in terms of r and θ, we see that rectangular and cylindrical coordinates are related by the formulas

$$x = r \cos \theta,$$

$$y = r \sin \theta,$$

$$z = z,$$

$$x^2 + y^2 = r^2.$$

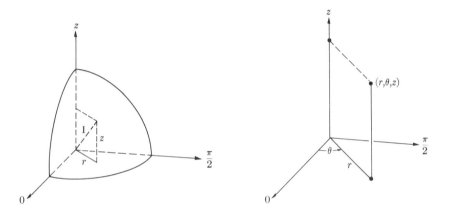

As for polar coordinates in the plane, these formulas work in only one direction: when r and θ are named, x and y are determined, but when x and y are named, there are two possibilities for r and infinitely many possibilities for θ. (See figure on the right above.)

Suppose now that we have given a domain D in the plane $z = 0$. The plane $z = 0$ may be regarded as the xy-plane or the $r\theta$-plane; sometimes we shall refer to it simply as the *base plane*. Suppose that we have given a continuous function $f \colon D \to \mathbf{R}$. If we describe a point P of D by its polar coordinates (r, θ), then we have

$$z = f(P) = f(r, \theta).$$

We now want to compute

$$\iint_D f(P)\, dA,$$

and we want to do this without transforming to rectangular coordinates. In some cases we might not be able to transform; and in other cases we wouldn't want to, because the rectangular form would turn out to be unmanageable. Therefore we need to know how to deal with cylindrical coordinates in their own terms. This can be done as follows.

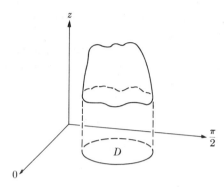

Given a domain D in the base plane. By a *net* over D we mean a finite collection

$$N: D_1, D_2, \ldots, D_n$$

of regions such that (1) D is the union of the D_i's, (2) each D_i has an area (i.e., is *measurable*, in the sense defined in Appendix G), and (3) if D_i intersects D_j, then the area of the intersection is 0. The sets D_i are called the *cells* of the net. The figure indicates, at long last, why we use the word *net* in integration theory.

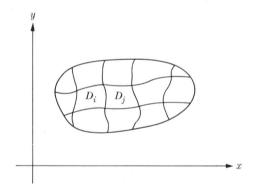

By the *diameter* of a set D_i we mean the supremum of the distances between its points. The diameter is denoted by δD_i. Thus

$$\delta D_i = \sup \{PQ \mid P, Q \text{ in } D_i\}.$$

Note that if D_i is a circular region, then δD_i is the diameter of D_i in the elementary sense. The *mesh* of the net N is the greatest of the diameters of the cells of the net. The mesh is denoted by $|N|$. Thus $|N| = \text{Max } \{\delta D_i\}$.

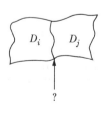

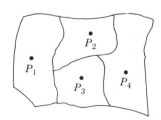

A *sample* of the net

$$N: D_1, D_2, \ldots, D_n$$

is a sequence

$$X: P_1, P_2, \ldots, P_n$$

of points, where P_i belongs to D_i for each i (see figure at the right above).

For each i, let ΔA_i be the area of D_i. A *sample sum* of f over the net N is a sum of the form

$$\sum_{i=1}^{n} f(P_i)\,\Delta A_i.$$

We are now finally ready to give our definition of the double integral. By definition,

$$\iint\limits_{D} f(P)\,dA = \lim_{|N|\to 0}\sum_{i=1}^{n} f(P_i)\,\Delta A_i,$$

if such a limit exists. If the limit exists, then f is said to be *integrable on D*, or simply *integrable*. In this definition, $\lim_{|N|\to 0}$ means the same thing that it meant in the definition of the integral for functions of one variable; when we write

$$\lim_{|N|\to 0}\sum_{i=1}^{n} f(P_i)\,\Delta A_i = L,$$

this means that for every $\epsilon > 0$ there is a $\delta > 0$ such that

$$|N| < \delta \Rightarrow \left|\sum_{i=1}^{n} f(P_i)\,\Delta A_i - L\right| < \epsilon.$$

We recall that if f is continuous on the closed interval $[a, b]$, then f is integrable on $[a, b]$. We want to state an analogous theorem for functions of two variables. It would hardly do to restrict ourselves to "two-dimensional closed intervals" $a \leq x \leq b,\ c \leq y \leq d$. On the other hand, we cannot allow all sets D in the xy-plane as domains, because continuous functions on some domains may not even be bounded. (Examples?) What is needed here is the following:

Definition. A point P is a *limit point* of a set D if every neighborhood $U(P, \delta)$ of P contains a point of D other than P.

Definition. A set D is *closed* if it contains all its limit points.

Thus a closed interval is closed, but an open interval is not; the region

$$D = \{(x, y) \mid x^2 + y^2 \leq 1\}$$

is closed, but the region

$$D' = \{(x, y) \mid x^2 + y^2 < 1\}$$

is not.

We recall that a set D in a plane is *bounded* if it lies in the interior of some circle (or, equivalently, if it lies in the interior of some rectangle). We can now finally state our theorem:

Theorem 1. Let D be a closed, bounded, measurable set in the xy-plane, and let f be a function which is continuous on D. Then f is integrable on D.

You may be able to convince yourself of this, for positive functions, by thinking of the integral as a volume, and thinking of the sample sums as approximations of the volume; the idea is that we can approximate the volume as closely as we please, by cutting up the base domain into sufficiently small pieces. If the function is negative somewhere, then we need to use volumes with signs attached, but the idea is much the same. But a mathematical proof that all this works is far beyond the scope of this book, and we make no attempt to present one.

Meanwhile we assume that the theorem is true, and return to the problem of integration in cylindrical coordinates. For the sake of simplicity, we consider first a domain of the type

$$D = \{(r, \theta) \,|\, a \leqq r \leqq b, \alpha \leqq \theta \leqq \beta\}.$$

This is the polar equivalent of a rectangular region.

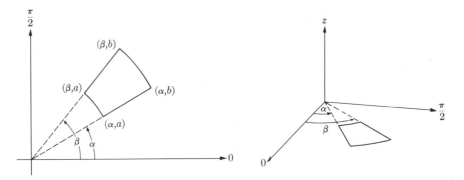

The first step in the calculation of the integral is to set up a net N_r on the interval $[a, b]$ and a net N_θ on the interval $[\alpha, \beta]$. Thus we have

$$N_r: r_0, r_1, \ldots, r_n,$$
$$N_\theta: \theta_0, \theta_1, \ldots, \theta_m.$$

The circles $r = r_i$ and the rays $\theta = \theta_j$ now cut up the domain D into nm little pieces, like this:

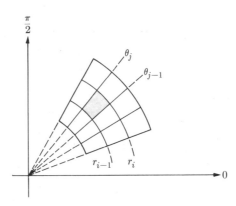

Let D_{ij} be the shaded region in the figure. Thus

$$D_{ij} = \{(r, \theta) \mid r_{i-1} \leqq r \leqq r_i, \theta_{j-1} \leqq \theta \leqq \theta_j\}.$$

Evidently the D_{ij}'s form a net N over D. For each i, j, the area of D_{ij} is

$$\Delta A_{ij} = \tfrac{1}{2}r_i^2 \Delta\theta_j - \tfrac{1}{2}r_{i-1}^2 \Delta\theta_j,$$

where $\Delta\theta_j = \theta_j - \theta_{j-1}$. We let

$$\Delta r_i = r_i - r_{i-1}.$$

Then

$$r_{i-1} = r_i - \Delta r_i,$$

and

$$r_i^2 - r_{i-1}^2 = r_i^2 - (r_i^2 - 2r_i \Delta r_i + \Delta r_i^2)$$
$$= 2r_i \Delta r_i - \Delta r_i^2.$$

Therefore

$$\Delta A_{ij} = \tfrac{1}{2}\Delta\theta_j[r_i^2 - r_{i-1}^2] = \tfrac{1}{2}\Delta\theta_j[2r_i \Delta r_i - \Delta r_i^2]$$
$$= r_i \Delta r_i \Delta\theta_j - \tfrac{1}{2}\Delta r_i^2 \Delta\theta_j.$$

In each cell D_{ij} of the net we pick the sample point $P_{ij} = (r_i, \theta_i)$. We now form the sample sum

$$\Sigma = \sum_{i=1}^{n} \sum_{j=1}^{m} f(r_i, \theta_j) \Delta A_{ij}$$
$$= \sum_{i=1}^{n} \sum_{j=1}^{m} f(r_i, \theta_j)[r_i \Delta r_i \Delta\theta_j - \tfrac{1}{2}\Delta r_i^2 \theta_j].$$

Evidently Σ is a sample sum of f over D; and our problem is to find

$$\lim_{|N| \to 0} \Sigma = \iint_D f(P) \, dA.$$

But this is much easier than it looks:

$$\Sigma = \sum_{i=1}^{n} \sum_{j=1}^{m} f(r_i, \theta_j) r_i \Delta r_i \Delta\theta_j - \tfrac{1}{2}\Delta r_i \sum_{i=1}^{n} \sum_{j=1}^{m} f(r_i, \theta_j) \Delta r_i \Delta\theta_j.$$

We interpret each of these double sums as a sample sum in rectangular coordinates.

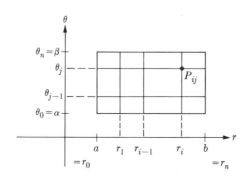

Let D' be the rectangular region shown in the figure. That is,

$$D' = \{(r, \theta) \mid a \leq r \leq b, \alpha \leq \theta \leq \beta\}.$$

The limits of our sums are now known:

$$\lim_{|N| \to 0} \sum_{i=1}^{n} \sum_{j=1}^{m} f(r_i, \theta_j) r_i \Delta r_i \Delta \theta_j = \iint_{D'} f(r, \theta) r \, dr \, d\theta,$$

and

$$\lim_{|N| \to 0} \sum_{i=1}^{n} \sum_{j=1}^{m} f(r_i, \theta_j) \Delta r_i \Delta \theta_j = \iint_{D'} f(r, \theta) \, dr \, d\theta.$$

Since $\lim_{|N| \to 0} \Delta r_i = 0$, we have

$$\iint_{D} f(r, \theta) \, dA = \iint_{D'} f(r, \theta) r \, dr \, d\theta + 0 \cdot \iint_{D'} f(r, \theta) \, dr \, d\theta = \iint_{D'} f(r, \theta) r \, dr \, d\theta.$$

Thus the second integral has dropped out.

We usually evaluate double integrals by converting them into iterated integrals. For the special type of domain that we have been discussing, we can sum up our results in the following theorem:

Theorem 2.   Let

$$D = \{(r, \theta) \mid a \leq r \leq b, \alpha \leq \theta \leq \beta\},$$

in polar coordinates, and let f be a function which is continuous on D. Then

$$\iint_{D} f(r, \theta) \, dA = \int_{\alpha}^{\beta} \int_{a}^{b} f(r, \theta) r \, dr \, d\theta = \int_{a}^{b} \int_{\alpha}^{\beta} f(r, \theta) r \, d\theta \, dr.$$

Let us try this out in a simple case in which we know the answer. Consider the hemisphere under the graph of

$$z = f(x, y) = \sqrt{1 - x^2 - y^2}.$$

In cylindrical coordinates,

$$z = f(r, \theta) = \sqrt{1 - r^2}.$$

Let

$$D = \{(x, y) \mid x^2 + y^2 \leq 1\} = \{(r, \theta) \mid r^2 \leq 1\}.$$

Then the volume of the hemisphere is

$$\iint_{D} \sqrt{1 - r^2} \, dA = \int_{0}^{2\pi} \int_{0}^{1} \sqrt{1 - r^2} \, r \, dr \, d\theta.$$

Now

$$\int \sqrt{1 - r^2} \, r \, dr = \{-\tfrac{1}{2} \cdot \tfrac{2}{3}(1 - r^2)^{3/2} + C\}.$$

Therefore

$$\int_0^1 \sqrt{1 - r^2}\, r\, dr = [-\tfrac{1}{3}(1 - r^2)^{3/2}]_0^1 = \tfrac{1}{3}.$$

Therefore

$$\iint_D \sqrt{1 - r^2}\, dA = \int_0^{2\pi} \tfrac{1}{3}\, d\theta = \frac{2\pi}{3}.$$

This is right, because the volume of the whole sphere is $4\pi/3 \cdot 1^3 = 4\pi/3$.

PROBLEM SET 14.10

1. Let $D = \{(x, y) \mid x^2 + y^2 \leq 1\}$. Find

$$\iint_D (x^2 + y^2)^{7/2}\, dy\, dx.$$

2. Let $D = \{(x, y) \mid x^2 + y^2 \leq 1\}$. Find

$$\iint_D \sqrt{1 + x^2 + y^2}\, dy\, dx.$$

3. Find the volume of the solid which lies under the paraboloid $z = x^2 + y^2$ and over the interior of the cardioid $r = 1 - \sin \theta$.

4. Find the volume of the solid lying inside the cylinder $x^2 + y^2 \leq 1$ and inside the sphere $x^2 + y^2 + z^2 = 4$.

5. Find the volume of the solid lying inside the cylinder $x^2 + y^2 \leq 1$ and inside the ellipsoid

$$\frac{x^2}{4} + \frac{y^2}{4} + z^2 = 1.$$

6. Find

$$\int_{-1}^1 \int_{-\sqrt{1-x^2}}^{\sqrt{1-x^2}} (x^2 + y^2)^{10}\, dy\, dx.$$

7. Let D be the circular region with center at $(0, 1)$ and radius 1, in the xy-plane. Find

$$\iint_D \frac{x}{\sqrt{x^2 + y^2}}\, dy\, dx.$$

8. Let D be as in Problem 7. Find

$$\iint_D \frac{y}{\sqrt{x^2 + y^2}}\, dy\, dx.$$

9. Let S be the part of the disk D lying in the half-plane $\{(x, y) \mid x \geq 0\}$. Find

$$\iint_S \frac{xy}{x^2 + y^2}\, dy\, dx.$$

10. Find

$$\int_{-1}^{1} \int_{-\sqrt{1-x^2}}^{\sqrt{1-x^2}} \frac{\sin^7 x \cos^7 y}{(x^2 + y^2)^{3/2}} \, dy \, dx.$$

11. Find

$$\int_{0}^{2} \int_{0}^{\sqrt{4-x^2}} \frac{x^2 - y^2}{x^2 + y^2} \, dy \, dx.$$

14.11 MOMENTS AND CENTROIDS OF NONHOMOGENEOUS BODIES

We recall, from Section 7.6, the definitions of moments and centroids for finite systems of point masses in a coordinate plane. Suppose that we have given a set of particles $P_1, P_2, \ldots, P_n$, with masses $m_1, m_2, \ldots, m_n$, at the points (x_1, y_1), $(x_2, y_2), \ldots, (x_n, y_n)$. The moment of the system about the y-axis is defined to be

$$M_y = \sum_{i=1}^{n} m_i x_i,$$

and the moment about the x-axis is

$$M_x = \sum_{i=1}^{n} m_i y_i.$$

More generally, the moment about the line $x = x_0$ is

$$M_{x=x_0} = \sum_{i=1}^{n} m_i (x_i - x_0),$$

and the moment about the line $y = y_0$ is

$$M_{y=y_0} = \sum_{i=1}^{n} m_i (y_i - y_0) .$$

If

$$M_{x=\bar{x}} = M_{y=\bar{y}} = 0,$$

then the point $(\bar{x}, \bar{y})$ is called the *centroid* of the system. By easy calculations we get

$$\bar{x} = \frac{1}{m} \sum_{i=1}^{n} m_i x_i, \qquad \bar{y} = \frac{1}{m} \sum_{i=1}^{n} m_i y_i \qquad \left(m = \sum_{i=1}^{n} m_i \right).$$

It is easy to see that if the axes are translated, the centroid is unchanged: for

$$x = x' + h, \qquad y = y' + k, \qquad x' = x - h, \qquad y' = y - k,$$

the coordinates of the centroid in the new coordinate system are given by the formulas

$$\bar{x}' = \frac{1}{m} \sum_{i=1}^{n} m_i x_i' = \frac{1}{m} \sum_{i=1}^{n} m_i (x_i - h)$$

$$= \bar{x} - \frac{1}{m} h \sum_{i=1}^{n} m_i = \bar{x} - h,$$

$$\bar{y}' = \frac{1}{m} \sum_{i=1}^{n} m_i y_i' = \bar{y} - k,$$

so that in the new coordinate system we get the same centroid as before. Similarly, if we reverse the direction of the x-axis, or the y-axis, or both, we get the same centroid as before. Finally, we observe that the centroid is unchanged if we rotate the axes through an angle of measure θ. We have

$$x = x' \cos \theta - y' \sin \theta,$$

$$y = x' \sin \theta + y' \cos \theta.$$

Therefore

$$\bar{x} = \frac{1}{m} \sum_{i=1}^{n} m_i(x_i' \cos \theta - y_i' \sin \theta)$$

$$= \frac{1}{m} \left(\sum_{i=1}^{n} m_i x_i' \right) \cos \theta - \frac{1}{m} \left(\sum_{i=1}^{n} m_i y_i' \right) \sin \theta$$

$$= \bar{x}' \cos \theta - \bar{y}' \sin \theta,$$

where $\bar{x}'$ and $\bar{y}'$ are the new coordinates of the centroid. A similar calculation gives

$$\bar{y} = \bar{x}' \sin \theta + \bar{y}' \cos \theta.$$

Thus the old coordinate system and the new one give us the same point as centroid.

Suppose now that we have a thin rod, lying on an interval $[a, b]$ on the x-axis. We do not suppose that its mass per unit length is constant. But in any case there is a function f which gives, for each x, the mass of the part of the rod that lies on the interval $[a, x]$. If f has a continuous derivative f', then

$$\int_a^x f'(t)\, dt = f(x) - f(a) = f(x),$$

because $f(a) = 0$. More generally,

$$f(x_2) - f(x_1) = \int_{x_1}^{x_2} f'(x)\, dx.$$

A function ρ which behaves in the way that we have just observed for f' is called a *density function* for the rod. That is:

Definition. Given a rod on $[a, b]$. For $a \leq x_1 < x_2 \leq b$, let $m(x_1, x_2)$ be the mass of the part of the rod that lies on $[x_1, x_2]$. A *density function* for the rod is a function ρ such that

$$m(x_1, x_2) = \int_{x_1}^{x_2} \rho(x)\, dx.$$

It follows, of course, that the total mass m of the rod is $m(a, b) = \int_a^b \rho(x)\, dx$. And the definition agrees with our intuitive notion of what density at a point ought to mean. If $\rho(x_0)$ is the density at x_0, then it ought to be true that

$$\frac{m(x_0, x_0 + \delta)}{\delta} \approx \rho(x_0) \qquad \text{when } \delta \approx 0.$$

Here the lefthand side is the average mass per unit length on the interval $[x_0, x_0 + \delta]$, and this ought to be approximately $\rho(x_0)$ when $\delta \approx 0$. And this is true, at any point where ρ is continuous:

$$\lim_{\delta \to 0} \frac{1}{\delta} m(x_0, x_0 + \delta) = \lim_{\delta \to 0} \frac{1}{\delta} \int_{x_0}^{x_0 + \delta} \rho(x)\, dx = \rho(x_0),$$

by the general formula for the derivative of the integral. We assume hereafter that ρ is a continuous function. Let us take a net N: $x_0, x_1, \ldots, x_n$ over $[a, b]$, and form the sum

$$\sum_{i=1}^{n} x_i \rho(x_i)\, \Delta x_i.$$

This sum is the moment, about the origin, of a finite system of particles of mass

$$\rho(x_1)\, \Delta x_1, \ldots, \rho(x_n)\, \Delta x_n,$$

at the points $x_1, x_2, \ldots, x_n$. The limit of the sum, as the mesh of the net approaches 0, is $\int_a^b x\rho(x)\, dx$. This integral is defined to be the moment of the rod about the origin. Thus

$$M_0 = \int_a^b x\rho(x)\, dx.$$

More generally, the moment about the point $x = k$ is

$$M_k = \int_a^b (x - k)\rho(x)\, dx.$$

We now define the centroid as the point $\bar{x}$ such that

$$M_{\bar{x}} = 0.$$

It is easy to calculate that

$$\bar{x} = \frac{\int_a^b x\rho(x)\, dx}{\int_a^b \rho(x)\, dx}.$$

By the definition of the density function, the integral in the denominator is the total mass m of the system. Thus, briefly,

$$\bar{x} = \frac{1}{m} \int_a^b x\rho(x)\, dx.$$

Suppose, for example, that the rod lies on the interval $[0, 2]$, and that the density is proportional to the distance from the origin. Here we have

$$\rho(x) = kx,$$

$$m = \int_0^2 kx\, dx = [\tfrac{1}{2}kx^2]_0^2 = 2k,$$

$$\bar{x} = \frac{1}{m} \int_0^2 x \cdot kx\, dx = \frac{1}{2k} \cdot k[\tfrac{1}{3}x^3]_0^2 = \tfrac{1}{2} \cdot \tfrac{1}{3} \cdot 8 = \tfrac{4}{3}.$$

This is greater than one, as it should be.

Consider next a thin plate, occupying a region D in the xy-plane. Again we do not suppose that the density per unit area is constant.

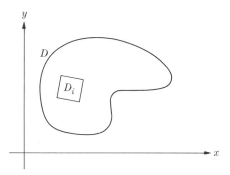

For each subregion D_i of D let $m(D_i)$ be the mass of the portion of the plate that lies in D_i. Following the analogy of the rod, we define a *density function* for the plate to be a function ρ such that

$$m(D_i) = \iint\limits_{D_i} \rho(P)\, dP$$

for every D_i lying in D. In particular, D_i may be all of D; and in this case the total mass of the plate is

$$m = m(D) = \iint\limits_{D} \rho(P)\, dA.$$

Hereafter, we assume that ρ is continuous. We take a net

$$N = D_1, D_2, \ldots, D_n$$

over D; we take a sample $P_1, P_2, \ldots, P_n$ of N, with $P_i = (x_i, y_i)$; and we form the sum

$$\sum_{i=1}^{n} x_i \rho(x_i, y_i)\, \Delta A_i,$$

where ΔA_i is the area of D_i. This sum is the moment about the y-axis of a system of particles of mass $\rho(x_i, y_i)\, \Delta A_i$, with x-coordinates x_i. As the mesh of the net approaches zero, these sums approach the limit

$$\iint\limits_{D} x\rho(x, y)\, dA.$$

By definition, the moment of the plate about the y-axis is this integral. More generally, the moment about the line $x = k$ is

$$M_{x=k} = \iint\limits_{D} (x - k)\rho(x, y)\, dA;$$

and similarly,

$$M_{y=k} = \iint_D (y - k)\rho(x, y)\, dA.$$

The *centroid* is the point $(\bar{x}, \bar{y})$ such that

$$M_{x=\bar{x}} = 0 = M_{y=\bar{y}}.$$

Since the total mass is

$$m = \iint_D \rho(x, y)\, dA,$$

an easy calculation gives

$$\bar{x} = \frac{1}{m} \iint_D x\rho(x, y)\, dA, \qquad \bar{y} = \frac{1}{m} \iint_D y\rho(x, y)\, dA.$$

In the preceding discussion, we have assumed for the sake of simplicity that the density is continuous, so that we don't need to worry about whether our integrals exist. In some very simple cases, however, the density is not continuous. Suppose, for example, that we take a rod of unit length, with constant density 1, and another rod of unit length, with constant density 2, and lay them end to end.

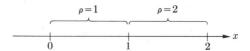

Thus $\rho(x) = 1$ for $0 \le x < 1$, and $\rho(x) = 2$ for $1 < x \le 2$. At the midpoint 1, we split the difference, and take $\rho(1) = \frac{3}{2}$. We now have a discontinuous density function whose graph looks like this:

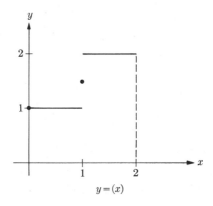

$$y = (x)$$

Note, however, that ρ is integrable, and that

$$\int_0^2 \rho(x)\, dx = 1 + 2 = 3,$$

which is equal to the mass, as it should be. The function $x\rho(x)$ looks like this:

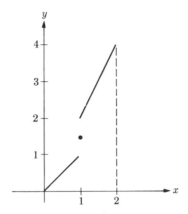

and

$$M_0 = \int_0^2 x\rho(x)\,dx = \tfrac{1}{2} + \tfrac{1}{2}(2 + 4) = \tfrac{7}{2}.$$

Therefore

$$\bar{x} = \frac{1}{m} M_0 = \tfrac{1}{3} \cdot \tfrac{7}{2} = \tfrac{7}{6}.$$

This is the right answer. If we assume that the masses of the two halves of the rod are concentrated at their centroids, then we get two particles, of masses 1 and 2, at the points $\tfrac{1}{2}$ and $\tfrac{3}{2}$.

Here

$$M_0 = \tfrac{1}{2} \cdot 1 + \tfrac{3}{2} \cdot 2 = \tfrac{7}{2}, \qquad m = 3,$$

and

$$\bar{x} = \tfrac{1}{3} \cdot \tfrac{7}{2} = \tfrac{7}{6},$$

as before.

This illustrates the way in which our formulas work, for discontinuous density functions. The general theory, however, is hard, and we make no attempt to discuss it here. Meanwhile the above example shows that some very simple physical situations lead naturally to discontinuous functions.

PROBLEM SET 14.11

1. A thin rod occupies the interval $[2, 4]$. Its density is proportional to the distance from the origin. Find the centroid.

2. A thin rod occupies the interval $[1, 2]$. Its density is proportional to the square root of the distance from the origin. Find the centroid.

3. A thin plate occupies the unit disk with center at the origin. Its density is proportional to $e^{-(x^2+y^2)^4}$. Find the centroid.

4. A thin plate occupies the unit disk with center at the origin. Its density is proportional to $\sqrt{1 + x^2 + y^2}$. Find the centroid.

5. A thin plate occupies the righthand half $(x \geq 0)$ of the unit disk with center at the origin. Its density is proportional to the distance from the origin. Find the centroid.

6. Same question, where the density is proportional to the square of the distance from the origin.

7. A thin plate occupies the interior of the cardioid $r = 1 - \sin \theta$. Its density is constant, say, $= 1$. Find the centroid. (The computation is long, even if the appropriate short-cuts are used.)

8. A function f is defined by the conditions

$$f(x) = \begin{cases} x^2 & \text{for } 0 \leq x \leq 1, \\ (x - 2)^2 & \text{for } 1 \leq x \leq 2. \end{cases}$$

Find $\int_0^2 f(x)\, dx$.

9. A function f is defined by the condition

$$f(x) = \begin{cases} x - 1 & \text{for } 0 \leq x \leq 1, \\ 1 - x^2 & \text{for } 1 \leq x \leq 2. \end{cases}$$

Find $\int_0^2 f(x)\, dx$.

10. A thin plate occupies the square region whose corners are $(0, 0)$, $(1, 1)$, $(2, 0)$, and $(1, -1)$. Its density is proportional to the distance from the y-axis. Find the centroid.

11. A thin plate occupies a triangular region with vertices $(0, 0)$, $(1, 1)$, and $(1, -1)$. Its density is proportional to the distance from the x-axis. Find the centroid.

12. Given a thin plate, occupying a region D, with density function ρ. The *moment of inertia* of the plate about the point $P_0 = (x_0, y_0)$ is defined to be

$$I_{P_0} = \iint_D (P_0 P)^2 \rho(P)\, dA = \iint_D [(x - x_0)^2 + (y - y_0)^2]\rho(x, y)\, dA.$$

Suppose that the plate occupies the unit circle with center at the origin, and that the density is constant. Find the moment of inertia about the origin.

13. Under the conditions of Problem 12, find out which point P_0 gives the minimum value of the moment of inertia.

14. The moment of inertia of a thin plate about the line $x = x_0$ is

$$I_{x=x_0} = \iint_D (x - x_0)^2 \rho(x, y)\, dA.$$

The moment of inertia $I_{y=y_0}$ about the line $y = y_0$ is defined similarly. A thin plate occupies the righthand half of the unit disk with center at the origin, and its density is proportional to the distance from the origin. Find $I_{x=0}$ and $I_{y=0}$.

15. Given a thin plate, with density function ρ, on a domain D. For what point P_0 does the moment of inertia I_P take on its minimum value?

14.12 LINE INTEGRALS

Suppose that we have given a path $P: I \to D$, where I is a closed interval $[a, b]$ and D is a region in a coordinate plane.

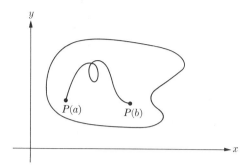

Let f and g be the coordinate functions of the path P, so that

$$P(t) = \big(f(t), g(t)\big)$$

for each t, and suppose that f and g have continuous derivatives. Let F and G be continuous functions defined on D. The *line integral* $\int_P F\,dx + G\,dy$, of F and G over the path P, is defined as follows.

Let

$$N: a = t_0, t_1, \ldots, t_n = b$$

be any net over $I = [a, b]$. For each i, let

$$x_i = f(t_i), \qquad\qquad y_i = g(t_i),$$
$$\Delta x_i = x_i - x_{i-1}, \qquad \Delta y_i = y_i - y_{i-1}.$$

Then

$$\int_P F\,dx + G\,dy = \lim_{|N| \to 0} \sum_{i=1}^{n} [F(x_i, y_i)\,\Delta x_i + G(x_i, y_i)\,\Delta y_i].$$

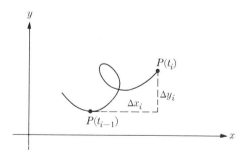

We need, of course, to show that the limit exists. We shall do this by deriving a formula for the limit, as follows:

Theorem 1. If F and G are continuous, and the coordinate functions f and g of the path P have continuous derivatives, then

$$\int_P F\,dx + G\,dy = \int_a^b F\big(f(t), g(t)\big)f'(t)\,dt + \int_a^b G\big(f(t), g(t)\big)g'(t)\,dt.$$

Proof. On each little interval $[t_{i-1}, t_i]$ of the net, we take a $\bar{t}_i$ such that

$$\Delta x_i = f'(\bar{t}_i) \, \Delta t_i$$

(where $\Delta t_i = t_i - t_{i-1}$). Then

$$\sum_{i=1}^{n} F(x_i, y_i) \, \Delta x_i = \sum_{i=1}^{n} F\big(f(t_i), g(t_i)\big) f'(\bar{t}_i) \, \Delta t_i.$$

This is almost, but not quite, a sample sum of the function

$$\phi(t) = F\big(f(t), g(t)\big) f'(t)$$

over the net N; the only trouble is that we have substituted two different sample points in two different places in the formula for ϕ. But in the limit, this does not matter. (See Appendix I, where a very similar case is discussed in detail.) Therefore

$$\lim_{|N| \to 0} \sum_{i=1}^{n} F(x_i, y_i) \, \Delta x_i = \int_a^b \phi(t) \, dt = \int_a^b F\big(f(t), g(t)\big) f'(t) \, dt.$$

In exactly the same way, we get

$$\lim_{|N| \to 0} \sum_{i=1}^{n} G(x_i, y_i) \, \Delta y_i = \int_a^b G\big(f(t), g(t)\big) g'(t) \, dt;$$

and from this the theorem follows.

For example, we might have

$$P(t) = \big(f(t), g(t)\big) = (t + 1, t^2) \qquad (0 \le t \le 1),$$

$$F(x, y) = x + y, \qquad G(x, y) = x + y^2.$$

Then

$$\int_P F \, dx + G \, dy = \int_0^1 [(t + 1 + t^2) + (t + 1 + t^4)2t] \, dt$$

$$= \int_0^1 (2t^5 + 3t^2 + 3t + 1) \, dt = \tfrac{23}{6}.$$

Line integrals have the following quite natural physical interpretation. We regard the path $P: I \to D$ as a description of the motion of a particle in the plane, during the time interval $a \le t \le b$. Suppose that at each point (x, y) of D there is a resisting force $R(x, y) = F(x, y)\mathbf{i} + G(x, y)\mathbf{j}$.

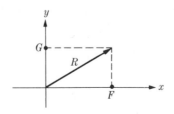

Here R is a vector, with components F and G in the x- and y-directions, and the indicated addition is vector addition. As the particle moves from $P(t_{i-1})$ to $P(t_i)$, the *work* should be approximately

$$W_i = F(x_i, y_i)\,\Delta x_i + G(x_i, y_i)\,\Delta y_i,$$

where the first term is the "work in the x-direction" and the second term is the "work in the y-direction." Therefore the total work W should be

$$W \approx \sum_{i=1}^{n} [F(x_i, y_i)\,\Delta x_i + G(x_i, y_i)\,\Delta y_i].$$

Passing to the limit, as $|N| \to 0$, we get

$$W = \int_P F\,dx + G\,dy,$$

which is the formula ordinarily used as the definition of work.

In the above discussion we have described the path P and the resistance R in terms of the coordinate functions f, g (for P) and F, G (for R). Thus, in effect, we have been using a particular basis $\{i, j\}$ for the base plane R^2, with

$$P(t) = f(t)i + g(t)j,$$
$$R(Q) = F(Q)i + G(Q)j.$$

Note, however, that

$$(\Delta x_i, \Delta y_i) = \Delta P_i = \Delta x_i i + \Delta y_i j,$$

so that the sum

$$\sum_{i=1}^{n} [F(x_i, y_i)\,\Delta x_i + G(x_i, y_i)\,\Delta y_i],$$

whose limit is the line integral, is really a sum of inner products:

$$\sum_{i=1}^{n} R(Q_i) \cdot \Delta P_i,$$

where $Q_i = P(t_i) = (x_i, y_i)$. Therefore the line integral depends merely on the vector-valued functions P and R; it is independent of the coordinate system in the base plane. Of course this must be true, for any mathematical concept which has a physical meaning. In vector notation, the line integral is denoted by

$$\int_P R \cdot dP.$$

A very important special case is the one in which the function

$$R\colon D \to \mathbf{R},$$
$$\colon (x, y) \to F(x, y)i + G(x, y)j$$

is an *exact differential*. This is the case in which there is a function ϕ such that

$$\phi_x = F, \qquad \phi_y = G,$$

so that the line integral takes the form

$$\int_P R \cdot dP = \int_P F\,dx + G\,dy = \int_P \phi_x\,dx + \phi_y\,dy$$

$$= \int_a^b [\phi_x(f(t), g(t))f'(t) + \phi_y(f(t), g(t))g'(t)]\,dt.$$

Here, by the chain rule for paths, the integrand is the derivative of the function

$$\Phi(t) = \phi(f(t), g(t)).$$

It follows that

$$\int_P R \cdot dP = \phi(f(b), g(b)) - \phi(f(a), g(a)).$$

The answer here depends only on the *endpoints*

$$P(b) = (f(b), g(b)), \qquad P(a) = (f(a), g(a))$$

of the path; it is independent of the way in which the path proceeds from the initial point $P(a)$ to the terminal point $P(b)$.

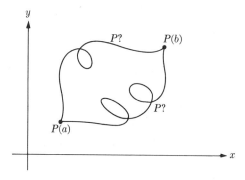

In such a case, we may describe the line integral merely by using the endpoints of the path as limits of integration, writing

$$\int_{(c,d)}^{(c',d')} \phi_x\,dx + \phi_y\,dy$$

for

$$\int_P \phi_x\,dx + \phi_y\,dy.$$

In these formulas, we use ϕ_x, ϕ_y for F and G to emphasize that the notation $\int_{(a,b)}^{(a'b')}$ can be used *only* in the case where the integrand is an exact differential.

PROBLEM SET 14.12

Calculate the line integral $\int_P R \cdot dP$, by any method, for the following functions R and P.

1. $R(x, y) = x^2i + y^2j;$ $P(t) = i \sin t + j \cos t$ $(0 \leqq t \leqq 2\pi)$
2. $R(x, y) = x^2i + y^2j;$ $P(t) = it + jt^2$ $(0 \leqq t \leqq 1)$
3. $R(x, y) = yi + xj;$ $P(t) = i \sin t + j \cos t$ $(0 \leqq t \leqq \pi)$
4. $R(x, y) = yi + xj;$ $P(t) = i \sin^{11} t + j \cos^{11} t$ $(0 \leqq t \leqq \pi)$
5. $R(x, y) = xyj;$ $P(t) = it + jt^3$ $(0 \leqq t \leqq 1)$
6. $R(x, y) = iye^{xy} + jxe^{xy};$ $P(t) = it^3 + j(1 - t^6)$ $(-1 \leqq t \leqq 1)$
7. $R(x, y) = iye^{xy} + jxe^{xy};$ $P(t) = it$ $(-1 \leqq t \leqq 1)$
8. $R(x, y) = xi - yj,$ $P(t) = i \sin^2 t + j \cos^2 t$ $(0 \leqq t \leqq 2\pi)$
9. $R(x, y) = xi - yj,$ $P(t) = it^2 + jt^3$ $(0 \leqq t \leqq 1)$
10. $R(x, y) = xi - yj,$ $P(t) = i \sin t + j \cos t$ $(0 \leqq t \leqq \pi/2)$
11. $R(x, y) = yi - xi,$ $P(t) = i \sin t + j \cos t$ $(4 \leqq t \leqq 7)$
12. $R(x, y) = (x^2 - y^2)i + 2xyj,$ $P(t) = i \cos t + j \sin t$ $(0 \leqq t \leqq \pi)$

Appendix A

The Shorthand of Logic and Set Theory

In this book, the use of logical symbols is held to a minimum, on the ground that words are usually easier to read. But the symbolism explained below will at some points be convenient in the text, and is even more useful in notebooks and on blackboards.

We explained in Chapter 1 that $\Leftrightarrow$ means "is equivalent to." Thus

$$x < 1 \iff 1 > x.$$

And $\Rightarrow$ means "implies." Thus

$$x > 2 \implies x^2 > 4.$$

Occasionally we write this symbol backwards:

$$x^2 > 4 \impliedby x > 2,$$

which means the same thing. We also recall from Chapter 1 that

$$\{x \mid P(x)\}$$

denotes the set of all objects x such that $P(x)$ is true. That is, $\{x \mid P(x)\}$ is the solution set of the open sentence $P(x)$. Thus the closed interval from 0 to 4 is

$$[0, 4] = \{x \mid 0 \leq x \leq 4\},$$

and the open interval from 1 to 2 is

$$(1, 2) = \{x \mid 1 < x < 2\}.$$

If A and B are sets, of any kind whatever, then

$$A \subset B$$

means that A is a subset of B; that is, every element of A is also an element of B. For example, if

$$A = (1, 2) \quad \text{and} \quad B = [0, 4],$$

then

$$A \subset B.$$

We allow the possibility that $A = B$, so that $A \subset A$ for every set A. Thus $\subset$ is like $\leq$, not like $<$. If $A \subset B$, we can also write

$$B \supset A.$$

If x is an element of the set A, then we write

$$x \in A.$$

This is read "x belongs to A." The denial of this statement is indicated by a diagonal stroke. That is,

$$x \notin A$$

means that x does not belong to A. The union of A and B is denoted by

$$A \cup B.$$

Thus

$$A \cup B = \{x \mid x \in A \quad \text{or} \quad x \in B\}.$$

The formula $A \cup B$ is read "A cup B." The intersection of A and B is denoted by

$$A \cap B.$$

Thus

$$A \cap B = \{x \mid x \in A \text{ and } x \in B\}.$$

The formula $A \cap B$ is read "A cap B." The difference

$$A - B$$

of A and B is the set of all elements of A that are not elements of B. To write $A - B$, we need not suppose that $B \subset A$. For example, if $A = [0, 2]$ and $B = [1, 4]$, then

$$A - B = [0, 1).$$

These sets are intervals, of course, described in the notation of Chapter 1. The empty set is denoted by $\{\ \}$. Thus

$$A \cap B = \{\ \}$$

means that A and B have no element in common. And

$$A - B = \{\ \}$$

means that $A \subset B$. When we write

$$(x - 1)^2 = x^2 - 2x + 1 \qquad \forall_x,$$

we mean that the equation on the left holds true for every x. Similarly,

$$x^2 - y^2 = (x - y)(x + y) \qquad \forall_{x,y}$$

means that the equation holds true for every x and y. The symbol "$\forall$" is read "for every." More informally, we may write

$$x^3 - y^3 = (x - y)(x^2 + xy + y^2) \qquad \forall,$$

where the symbol "$\forall$" means that the preceding equation holds true for all values of all the variables that appear in it. When "$\forall$" stands alone, it may be pronounced *always*.

The symbol $\exists$ stands for "there exists," and the symbol $\ni$ stands for "such that."

Used in moderation, this symbolism is a convenience. But its use can be over-done; and it takes practice to read formulas like

$$(x \in \mathbf{R} \text{ and } y \in \mathbf{R} \text{ and } x < y) \implies \exists z \ni z \in \mathbf{R} \text{ and } x < z < y.$$

This says that between any two real numbers there is a third.

This symbolism is introduced merely as a scheme of abbreviations of English words and phrases. This book makes no attempt to deal with symbolic logic; and its use of the "theory of sets" is entirely intuitive. But the shorthand of logic is useful simply as a shorthand; the point is that we are more likely to say what we mean if we have a quick and easy way to do so.

Algebraic Operations with Limits of Functions

In Section 3.4 a number of theorems on limits were stated without proofs. Here we give the proofs. First we recall some of the results of Section 3.4.

Theorem 1. If $\lim_{x \to x_0} f(x) = L$, then $\lim_{x \to x_0} [f(x) - L] = 0$.

Theorem 2. If $\lim_{x \to x_0} [f(x) - L] = 0$, then $\lim_{x \to x_0} f(x) = L$.

(These were Theorems 2 and 3 of Section 3.4.)

Theorem 3. If $\lim_{x \to x_0} f(x) = 0$ and $\lim_{x \to x_0} g(x) = 0$, then

$$\lim_{x \to x_0} [f(x) + g(x)] = 0.$$

Proof. Let $\epsilon > 0$ be given. There is a $\delta_1 > 0$ such that

$$0 < |x - x_0| < \delta_1 \Rightarrow |f(x)| < \epsilon/2.$$

(Here we are using $\epsilon/2$ for ϵ in the definition of the statement $\lim_{x \to x_0} f(x) = 0$.) Similarly, there is a $\delta_2 > 0$ such that

$$0 < |x - x_0| < \delta_2 \Rightarrow |g(x)| < \epsilon/2.$$

Let δ be the smaller of the numbers δ_1 and δ_2. Briefly, $\delta = \min(\delta_1, \delta_2)$. Then

$$0 < |x - x_0| < \delta \Rightarrow |f(x)| < \epsilon/2 \quad \text{and} \quad |g(x)| < \epsilon/2.$$

Since

$$|f(x) + g(x)| \leq |f(x)| + |g(x)|,$$

we have

$$0 < |x - x_0| < \delta \Rightarrow |f(x) + g(x)| \leq |f(x)| + |g(x)|$$
$$< \epsilon/2 + \epsilon/2 = \epsilon,$$

which is what we wanted.

Theorem 4. If $\lim_{x \to x_0} f(x) = L$ and $\lim_{x \to x_0} g(x) = L'$, then

$$\lim_{x \to x_0} [f(x) + g(x)] = L + L'.$$

Proof. By Theorem 1 we have

$$\lim_{x \to x_0} [f(x) - L] = 0, \qquad \lim_{x \to x_0} [g(x) - L'] = 0.$$

By Theorem 3,

$$\lim_{x \to x_0} [f(x) + g(x) - (L + L')] = 0.$$

By Theorem 2,

$$\lim_{x \to x_0} [f(x) + g(x)] = L + L'.$$

A function f is *locally bounded* at a point x_0 if there are positive numbers M and δ such that

$$0 < |x - x_0| < \delta \implies |f(x)| < M.$$

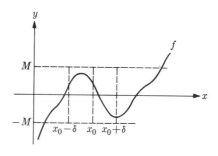

Theorem 5. If f approaches a limit, as $x \to x_0$, then f is locally bounded at x_0.

Proof. Let

$$\lim_{x \to x_0} f(x) = L.$$

Let ϵ be any positive number, and let δ be as in the definition of a limit.

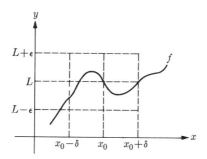

Thus

$$0 < |x - x_0| < \delta \implies L - \epsilon < f(x) < L + \epsilon.$$

The δ that we now have is the δ that we wanted. We let M be the larger of the numbers $|L + \epsilon|$, $|L - \epsilon|$.

Theorem 6. If $\lim_{x \to x_0} f(x) = 0$, and g is locally bounded at x_0, then

$$\lim_{x \to x_0} [f(x)g(x)] = 0.$$

Proof. Let a positive number ϵ be given. Take $\delta_1 > 0$ and $M > 0$ such that

$$0 < |x - x_0| < \delta_1 \tag{1}$$
$$\Rightarrow \quad |g(x)| < M. \tag{2}$$

Next, using ϵ/M in place of ϵ in the definiton of a limit, take $\delta_2 > 0$ such that

$$0 < |x - x_0| < \delta_2 \tag{3}$$

$$\Rightarrow \quad |f(x)| < \frac{\epsilon}{M}. \tag{4}$$

Now let $\delta = \min(\delta_1, \delta_2)$. If

$$0 < |x - x_0| < \delta, \tag{5}$$

then (1) and (3) *both* hold, and so (2) and (4) *both* hold. Therefore

$$0 < |x - x_0| < \delta \quad \Rightarrow \quad \begin{cases} g(x) < M, \\ f(x) < \dfrac{\epsilon}{M}. \end{cases}$$

Since

$$|f(x)g(x)| = |f(x)| \cdot |g(x)|,$$

we have

$$0 < |x - x_0| < \delta \quad \Rightarrow \quad |f(x)g(x)| < \frac{\epsilon}{M} \cdot M = \epsilon.$$

Thus, given an $\epsilon > 0$ we have found a $\delta > 0$ such that

$$0 < |x - x_0| < \delta \quad \Rightarrow \quad |f(x)g(x)| < \epsilon.$$

This means that

$$\lim_{x \to x_0} [f(x)g(x)] = 0,$$

which was to be proved.

Theorem 7. If $\lim_{x \to x_0} f(x) = L$ and $\lim_{x \to x_0} g(x) = L'$, then

$$\lim_{x \to x_0} [f(x)g(x)] = LL'.$$

Proof. By Theorem 2, we need to show that

$$\lim_{x \to x_0} [f(x)g(x) - LL'] = 0;$$

Theorem 7 will then follow. Now

$$\begin{aligned} f(x)g(x) - LL' &= f(x)g(x) - Lg(x) + Lg(x) - LL' \\ &= [(f(x) - L)g(x)] + [(g(x) - L')L]. \end{aligned}$$

In each of these brackets, the first factor approaches 0 as $x \to x_0$, and the second factor is locally bounded at x_0. Therefore

$$\lim_{x \to x_0} [(f(x) - L)g(x)] = 0,$$

and

$$\lim_{x \to x_0} \left[(g(x) - L')L\right] = 0.$$

Therefore the sum of the two bracketed expressions approaches 0, which was to be proved.

Roughly speaking, a function f is *locally bounded away from* 0 at x_0 if $f(x)$ is not very close to 0 when x is close to x_0 and different from x_0.

Definition. Suppose that there are numbers $\epsilon > 0$ and $\delta > 0$ such that

$$0 < |x - x_0| < \delta \;\Rightarrow\; |f(x)| > \epsilon.$$

Then f is *locally bounded away from* 0 at $x = x_0$.

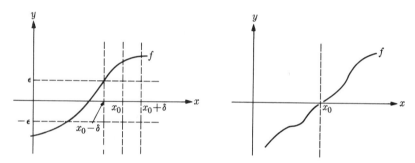

Note that, if $f(x)$ is never 0 when $x \neq x_0$ and x is close to x_0, it does not follow that f is locally bounded away from 0 at x_0. The situation shown in the figure on the right above can easily occur. Here f is undefined at x_0, $f(x) \neq 0$ for $x \neq x_0$, and $\lim_{x \to x_0} f(x) = 0$. In this case f is not locally bounded away from 0 at x_0.

Theorem 8. If $\lim_{x \to x_0} f(x) = L$, and $L \neq 0$, then f is locally bounded away from 0 at x_0.

Proof. Suppose first that $L > 0$. Let $\epsilon = L/2$, and let δ be as in the definition of a limit.

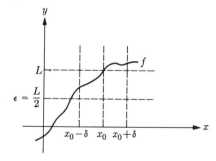

Then

$$0 < |x - x_0| < \delta \;\Rightarrow\; L - \frac{L}{2} < f(x) < L + \frac{L}{2}.$$

Thus

$$0 < |x - x_0| < \delta \;\Rightarrow\; f(x) > \epsilon$$
$$\Rightarrow\; |f(x)| > \epsilon.$$

Suppose now that $L < 0$. Then $-L > 0$, and $\lim_{x \to x_0} [-f(x)] = -L$. Therefore $-f$ is locally bounded away from 0 at x_0. Therefore so also is f. (Look again at the definition: it is a statement about $|f|$.)

Theorem 9. If f is locally bounded away from 0 at x_0, then $1/f$ is locally bounded at x_0.

Proof. We know that there are positive numbers ϵ and δ such that

$$0 < |x - x_0| < \delta \;\Rightarrow\; |f(x)| > \epsilon.$$

Therefore

$$0 < |x - x_0| < \delta \;\Rightarrow\; \frac{1}{|f(x)|} < \frac{1}{\epsilon} \;\Rightarrow\; \left|\frac{1}{f(x)}\right| < \frac{1}{\epsilon}.$$

Therefore $1/f$ is locally bounded at x_0; we use the δ that was given for f, and we use the bound $M = 1/\epsilon$.

Theorem 10. If $\lim_{x \to x_0} f(x) = L$, and $L \neq 0$, then

$$\lim_{x \to x_0} \frac{1}{f(x)} = \frac{1}{L}.$$

Proof. We need to show that

$$\lim_{x \to x_0} \left[\frac{1}{f(x)} - \frac{1}{L}\right] = 0.$$

Now

$$\frac{1}{f(x)} - \frac{1}{L} = \frac{L - f(x)}{Lf(x)} = [L - f(x)] \cdot \frac{1}{Lf(x)}.$$

The bracket on the left approaches 0. The fraction on the right is locally bounded at x_0. Therefore the product approaches 0.

Theorem 11. If $\lim_{x \to x_0} f(x) = L$, $\lim_{x \to x_0} g(x) = L'$, and $L' \neq 0$, then

$$\lim_{x \to x_0} \frac{f(x)}{g(x)} = \frac{L}{L'}.$$

Proof by Theorems 7 and 10.

Appendix C

Algebraic Operations with Limits of Sequences

Nearly everything in this appendix is analogous to something in the preceding one. The analogy begins with the definition of the limit of a sequence.

Definition. Given a sequence $a_1, a_2, \ldots$ and a number L. Suppose that for every $\epsilon > 0$ there is an integer N such that

$$n > N \quad \Rightarrow \quad |a_n - L| < \epsilon.$$

Then

$$\lim_{n \to \infty} a_n = L.$$

The following theorems are modeled on the theorems of Appendix B. We therefore give only a few of the proofs; you ought to be able to supply the rest of the proofs yourself.

Theorem 1. If $\lim_{n \to \infty} a_n = L$, then $\lim_{n \to \infty} (a_n - L) = 0$.

Proof. Let $\epsilon > 0$ be given. There is an N such that

$$n > N \quad \Rightarrow \quad |a_n - L| < \epsilon.$$

Therefore

$$n > N \quad \Rightarrow \quad |(a_n - L) - 0| < \epsilon.$$

Since for every $\epsilon > 0$ there is such an N, it follows that $\lim_{n \to \infty} (a_n - L) = 0$.

Note that statements of the form $n > N$ are playing exactly the same part as statements of the form $0 < |x - x_0| < \delta$.

Theorem 2. If $\lim_{n \to \infty} (a_n - L) = 0$, then $\lim_{n \to \infty} a_n = L$.

Theorem 3. If $\lim_{n \to \infty} a_n = 0$ and $\lim_{n \to \infty} b_n = 0$, then

$$\lim_{n \to \infty} (a_n + b_n) = 0.$$

Theorem 4. If $\lim_{n \to \infty} a_n = L$ and $\lim_{n \to \infty} b_n = L'$ then

$$\lim_{n \to \infty} (a_n + b_n) = L + L'.$$

Definition. A sequence $a_1, a_2, \ldots$ is *bounded* if there is a number M such that $|a_n| \leq M$ for every n.

(For sequences, the question of local boundedness does not arise.)

Theorem 5. If $\lim_{n\to\infty} a_n = L$, then $a_1, a_2, \ldots$ is bounded.

Proof. Take any $\epsilon > 0$. There is an N such that

$$n > N \quad \Rightarrow \quad L - \epsilon < a_n < L + \epsilon.$$

Therefore the sequence $a_{N+1}, a_{N+2}, \ldots$ is bounded. (The larger of the numbers $|L - \epsilon|, |L + \epsilon|$ is a bound.)

And the finite sequence $a_1, a_2, \ldots, a_N$ is bounded. We now get a bound M for the entire sequence $a_1, a_2, \ldots$: let M be the largest of the numbers $|L - \epsilon|, |L + \epsilon|$, $|a_1|, |a_2|, \ldots, |a_N|$.

Theorem 6. If $\lim_{n\to\infty} a_n = 0$, and $b_1, b_2, \ldots$ is bounded, then

$$\lim_{n\to\infty} a_n b_n = 0.$$

Theorem 7. If $\lim_{n\to\infty} a_n = L$ and $\lim_{n\to\infty} b_n = L'$, then

$$\lim_{n\to\infty} a_n b_n = LL'.$$

Definition. Suppose that there are numbers $\epsilon > 0$ and N such that

$$n > N \quad \Rightarrow \quad |a_n| > \epsilon.$$

Then $a_1, a_2, \ldots$ is *bounded away from* 0.

Theorem 8. If $\lim_{n\to\infty} a_n = L \neq 0$. then $a_1, a_2, \ldots$ is bounded away from 0.

Theorem 9. If $a_1, a_2, \ldots$ is bounded away from 0, and $a_n \neq 0$ for each n, then the sequence $1/a_1, 1/a_2, \ldots$ is bounded.

Theorem 10. If $\lim_{n\to\infty} a_n = L$, and $L \neq 0$, and $a_n \neq 0$ for each n, then

$$\lim_{n\to\infty} \frac{1}{a_n} = \frac{1}{L}.$$

Note that we must require that $a_n \neq 0$ for each n; otherwise the sequence of reciprocals is not defined.

Theorem 11. If $\lim_{n\to\infty} a_n = L$, $\lim_{n\to\infty} b_n = L' \neq 0$, and $b_n \neq 0$ for every n, then

$$\lim_{n\to\infty} \frac{a_n}{b_n} = \frac{L}{L'}.$$

Theorem 12 (*The squeeze principle*). If $\lim_{n\to\infty} a_n = L$, $\lim_{n\to\infty} c_n = L$, and $a_n \leqq b_n \leqq c_n$ for every n, then $\lim_{n\to\infty} b_n = L$.

The Error in the
Approximation $\Delta f \approx df$

At the beginning of Section 4.4, we gave numerical examples of the use of the approximation $\Delta f \approx df$, and we found that when we checked our approximate answers against the exact answers, the approximations looked good. But numerical approximation methods are important precisely in those cases where their accuracy *cannot* be checked in this way: if you can find the exact answer, then you use it; you don't get an inexact answer to compare with it. This brings up the problem of setting a limit on the error that results when you use df in place of Δf. The solution of this problem is as follows. We have

$$\Delta f = f(x_0 + \Delta x) - f(x_0), \qquad df = f'(x_0)\,\Delta x.$$

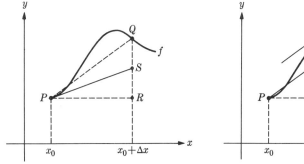
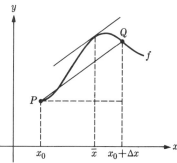

Applying the mean-value theorem (MVT) to the function f, on the interval from x_0 to $x_0 + \Delta x$, we conclude that

$$\frac{f(x_0 + \Delta x) - f(x_0)}{\Delta x} = f'(\bar{x}),$$

for some $\bar{x}$ between x_0 and $x_0 + \Delta x$. (For $x_0 < x_0 + \Delta x$, as in the figure, this result follows. You should check that it also follows when $x_0 + \Delta x < x_0$. Hereafter we shall assume that $\Delta x > 0$. The case $\Delta x < 0$ needs to be checked separately.) Therefore

$$\Delta f = f'(\bar{x})\,\Delta x, \qquad x_0 < \bar{x} < x_0 + \Delta x,$$

and

$$\Delta f - df = f'(\bar{x})\,\Delta x - f'(x_0)\,\Delta x = [f'(\bar{x}) - f'(x_0)]\,\Delta x.$$

697

We now apply MVT to the function f', on the interval $[x_0, x]$. MVT tells us that

$$\frac{f'(\bar{x}) - f'(x_0)}{\bar{x} - x_0} = f''(\bar{x}'),$$

for some $\bar{x}'$ between x_0 and $\bar{x}$. Therefore

$$f'(\bar{x}) - f'(x_0) = f''(\bar{x}')(\bar{x} - x_0),$$

and

$$\Delta f - df = f''(\bar{x}')(\bar{x} - x_0)\, \Delta x.$$

Now $|\bar{x} - x_0| \leq |\Delta x|$. Therefore $|\Delta f - df| \leq |f''(\bar{x}')|\, \Delta x^2$, where $\bar{x}'$ is between x_0 and $x_0 + \Delta x$.

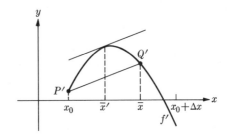

It often happens that we can find a bound M for the numbers $|f''(x)|$, for x between x_0 and $x_0 + \Delta x$. If so, we can conclude that

$$|\Delta f - df| \leq M\, \Delta x^2.$$

For example, if $f(x) = \sin x$, then $f''(x) = -\sin x$, and $|f''(x)| \leq 1$, no matter what x may be. Before giving further examples, let us write down the theorem that we have proved:

Theorem 1. Suppose that f has a second derivative f'', and that $|f''(x)| \leq M$, for every x between x_0 and $x_0 + \Delta x$. Then

$$|\Delta f - df| \leq M\, \Delta x^2.$$

Let us see how this applied to Example 1 of Section 4.4. Here we had

$$f(x) = \sqrt{x}, \qquad x_0 = 25, \qquad \Delta x = 0.4.$$

Now

$$f(x) = x^{1/2}, \qquad f'(x) = \tfrac{1}{2}x^{-1/2},$$

$$f''(x) = -\tfrac{1}{4}x^{-3/2} = \frac{-1}{4\sqrt{x^3}}.$$

We want to find a bound for this function, on the interval $[25, 25.4]$. As x increases, so also does x^3, and therefore so also does $4\sqrt{x^3}$. Hence the maximum value of

$|f''(x)|$ on the interval $[25, \infty)$, is $|f''(25)|$. Now

$$|f''(25)| = \frac{1}{4\sqrt{25^3}} = \frac{1}{500} \, .$$

We can therefore take $M = \frac{1}{500}$. This gives

$$|\Delta f - df| \leqq \tfrac{1}{500} \Delta x^2 = \tfrac{1}{500}(0.4)^2 = 0.00032.$$

In fact, we had

$$\Delta f = 0.039841, \qquad df = 0.04, \qquad |\Delta f - df| = 0.000159.$$

Thus the error was considerably smaller than Theorem 1 predicted that it had to be.

Appendix E

The Continuity of Composite Functions

In Section 4.5 we stated the following theorem, with only rough indications of proof.

Theorem. The composition of two continuous functions is continuous. That is, if

$$\lim_{x \to x_0} g(x) = g(x_0) = u_0, \tag{1}$$

and

$$\lim_{u \to u_0} f(u) = f(u_0), \tag{2}$$

then

$$\lim_{x \to x_0} f\big(g(x)\big) = f\big(g(x_0)\big). \tag{3}$$

Proof. Let ϵ be any positive number. By (2), there is a number δ_1 such that

$$|u - u_0| < \delta_1 \ \Rightarrow \ |f(u) - f(u_0)| < \epsilon.$$

Now take δ_1 as ϵ, in the definition of hypothesis (1). By (1), there is a $\delta > 0$ such that

$$|x - x_0| < \delta \ \Rightarrow \ |g(x) - g(x_0)| < \delta_1.$$

This is the δ that we wanted: we have

$$|x - x_0| < \delta \ \Rightarrow \ |g(x) - g(x_0)| < \delta_1 \ \Rightarrow \ |f\big(g(x)\big) - f\big(g(x_0)\big)| < \epsilon,$$

and so (3) holds.

Limits of composite functions are trickier than one might think. For example, the following "theorem" is false.

Theorem (?). If $\lim_{x \to x_0} g(x) = u_0$, and $\lim_{u \to u_0} f(u) = L$, then $\lim_{x \to x_0} f\big(g(x)\big) = L$.

To see that this is false, consider the following example. Let

$$f(x) = \begin{cases} 1 & \text{for } x \neq 1, \\ 2 & \text{for } x = 1. \end{cases}$$

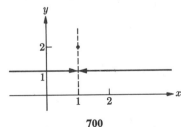

Let g be the same as f, and let $x_0 = u_0 = 1$. Then

$$\lim_{x \to x_0} g(x) = \lim_{x \to 1} g(x) = 1 = u_0,$$

and

$$\lim_{u \to u_0} f(u) = \lim_{u \to 1} f(u) = 1,$$

but it does not follow that

$$(?) \quad \lim_{x \to x_0} f(g(x)) = 1.$$

In fact,

$$x \neq 1 \implies g(x) = 1 \implies f(g(x)) = 2,$$

and so

$$\lim_{x \to 1} f(g(x)) = 2.$$

Appendix F   The Error in Simpson's Rule

The results of our calculations in Section 4.8 and in later sections suggest two questions:

1) Why does Simpson's rule give such good approximations?

2) In a particular computation, how can we tell how good the approximation is? That is, how can we determine a bound for the error?

These questions have the following answer. In the theorem below, $f^{(4)}$ denotes the fourth derivative of f. Thus $f^{(1)}$ is f', $f^{(2)} = f''$, $f^{(3)} = Df'' = f'''$, and $f^{(4)} = Df^{(3)}$. As usual,

$$y_0 = f(-k), \qquad y_1 = f(0), \qquad y_2 = f(k).$$

Theorem 1. If f has a fourth derivative, on the interval $[-k, k]$, then the error in Simpson's rule is equal to

$$E(k) = \frac{k^5}{90} f^{(4)}(\bar{x}),$$

where $\bar{x}$ is some number between $-k$ and k. That is

$$\int_{-k}^{k} f(x)\, dx - \frac{k}{3}(y_0 + 4y_1 + y_2) = \frac{k^5}{90} f^{(4)}(\bar{x}) \qquad (-k < \bar{x} < k).$$

It follows, of course, that if $|f^{(4)}(x)| \le M\ (-k < x < k)$, then

$$|E(k)| \le \tfrac{1}{90}k^5 M.$$

The latter is the statement which is most convenient to apply. Before proceeding to the proof of the theorem, let us look at an application of it.

Example. Suppose that we want to compute

$$\int_1^2 \frac{dx}{x},$$

to five decimal places. Here

$$f(x) = \frac{1}{x} = x^{-1}, \qquad f'(x) = -x^{-2}, \qquad f''(x) = 2x^{-3},$$

$$f'''(x) = -6x^{-4}, \qquad f^{(4)}(x) = 24x^{-5} = \frac{24}{x^5}.$$

Since $f^{(4)}$ decreases as x increases, its maximum value on the interval $[1, 2]$ is $f^{(4)}(1) = 24$. Therefore $|f^{(4)}(x)| \leq 24$ $(1 \leq x \leq 2)$. Therefore, if we cut up the interval $[1, 2]$ into $2n$ parts, each of length k, we have

$$|E(k)| \leq \tfrac{1}{90}k^5 \cdot 24.$$

We want

$$|E(k)| < 5 \cdot 10^{-6},$$

for the fifth decimal place in our approximation to be correct. Thus we want to take k such that

$$\tfrac{1}{90}k^5 \cdot 24 < 5 \cdot 10^{-6},$$

or

$$k^5 < \tfrac{90}{24} \cdot 5 \cdot 10^{-6}.$$

Arithmetically, this reduces to

$$k^5 < \tfrac{15}{8} \cdot 10^{-5},$$

which surely holds if $k = 0.05$. Therefore $E(0.05) < 5 \cdot 10^{-6}$.

This example was selected for its simplicity. For most functions, the calculation of fourth derivatives is tedious.

We proceed to the proof of Theorem 1. Let F be any function such that

$$F' = f.$$

(How do we know that there is such a function?) Then

$$\int_{-k}^{k} f(x)\, dx = F(k) - F(-k),$$

so that

$$E(k) = \int_{-k}^{k} f(x)\, dx - \frac{k}{3}(y_0 + 4y_1 + y_2)$$

$$= F(k) - F(-k) - \frac{k}{3}[f(-k) + 4f(0) + f(k)].$$

Therefore

$$E'(k) = f(k) + f(-k) - \frac{k}{3}[-f'(-k) + f'(k)] - \tfrac{1}{3}[f(-k) + 4f(0) + f(k)]$$

$$= \tfrac{2}{3}f(k) + \tfrac{2}{3}f(-k) - \tfrac{4}{3}f(0) + \frac{k}{3}f'(-k) - \frac{k}{3}f'(k),$$

$$E''(k) = \tfrac{2}{3}f'(k) - \tfrac{2}{3}f'(-k) - \frac{k}{3}f''(-k) + \tfrac{1}{3}f'(-k) - \tfrac{1}{3}f'(k) - \frac{k}{3}f''(k)$$

$$= \tfrac{1}{3}f'(k) - \tfrac{1}{3}f'(-k) - \frac{k}{3}f''(k) - \frac{k}{3}f''(-k),$$

$$E'''(k) = \tfrac{1}{3}f''(k) + \tfrac{1}{3}f''(-k) - \frac{k}{3}f'''(k) - \tfrac{1}{3}f''(k) + \frac{k}{3}f'''(-k) - \tfrac{1}{3}f''(-k)$$

$$= -\frac{k}{3}[f'''(k) - f'''(-k)].$$

We need all of these formulas, not just the last. It is easy to check that

$$E(0) = E'(0) = E''(0) = E'''(0) = 0.$$

From now on, k is going to be regarded as a *constant*. For each t on $[-k, k]$, let

$$G(t) = E(t) - \frac{E(k)}{k^5} t^5.$$

It is easy to check that

$$G(0) = G(k) = 0,$$

and

$$G'(0) = G''(0) = G'''(0) = 0.$$

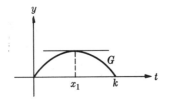 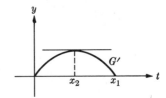

On the interval $[0, k]$ we apply the mean-value theorem (MVT) to the function G. This gives

$$G'(x_1) = 0, \qquad 0 < x_1 < k.$$

We next apply MVT to the function G', on the interval $[0, x_1]$. This gives

$$G''(x_2) = 0, \qquad 0 < x_2 < x_1.$$

Applying MVT to G'', on $[0, x_2]$, we get

$$G'''(x_3) = 0, \qquad 0 < x_3 < x_2.$$

By a straightforward calculation,

$$G'''(t) = E'''(t) - \frac{E(k)}{k^5}(60t^2)$$

$$= -\frac{t}{3}[f'''(t) - f'''(-t)] - 60\frac{E(k)}{k^5}t^2.$$

Setting $t = x_3$, and solving for $E(k)$, we get

$$E(k) = -\frac{k^5}{180} \cdot \frac{f'''(x_3) - f'''(-x_3)}{x_3} = \frac{k^5}{90} \cdot \frac{f'''(x_3) - f'''(-x_3)}{2x_3}.$$

We now apply MVT for the last time. By MVT there is an $\bar{x}$, between $-x_3$ and x_3, such that the second fraction on the right is equal to $f^{(4)}(\bar{x})$. This gives

$$E(k) = \frac{k^5}{90}f^{(4)}(\bar{x}) \qquad (-k < \bar{x} < k),$$

which was to be proved.

Appendix G

The Idea of a Measurable Set

If you reexamine Section 2.10, you will see that at the end of the section we were in a peculiar position: we had gotten an answer for the area under the graph of $y = kx^2$, from $x = a$ to $x = b$, but we were not in a position to prove it, because we had no definition of area. The trouble, however, is easy to remedy. For the sake of simplicity, consider first the case in which R is the region under the graph of $y = x^2$, from $x = 0$ to $x = h$. In Section 2.10, we proved the following two things:

1) There is a sequence $R_1, R_2, \ldots$ of *polygonal* regions containing R, with areas $A_1, A_2, \ldots$, such that

$$\lim_{n \to \infty} A_n = L.$$

(Here R_n was the union of the outer rectangles, A_n was $(h^3/3)(1 + 1/n)(1 + 2/n)$, and L was $h^3/3$.)

2) There is a sequence $R_1', R_2', \ldots$ of polygonal regions lying in R, with areas $A_1', A_2', \ldots$, such that $\lim_{n \to \infty} A_n'$ is the same number L.
(Here R_n' was the union of the inner rectangles, A_n' was $(h^3/3) \sum_{i=1}^{n} (i - 1)^2$, and L was $h^3/3$, as in condition 1.)

These ideas can be used to give a definition of area, in the following way.

Definition. Let R be a region in the plane. If R satisfies conditions (1) and (2), then R is said to be *measurable*, and the number L is called its *area*.

Under this definition, the plane regions discussed in Chapter 2 are measurable, and their areas are the numbers that we computed. The same conclusion follows whenever we compute an area by means of a definite integral. In Section 7.8 we showed that every continuous function is integrable. This gives the following:

Theorem. Let f be continuous and nonnegative on $[a, b]$, and let R be the region under the graph of f. Then R is measurable, and the area of R is

$$A = \int_a^b f(x) \, dx.$$

Proof. Take a sequence of nets

$$N_1, N_2, \ldots$$

over $[a, b]$, with $|N_i| \to 0$. For each i, let A_i be the upper sum $S(N_i)$ and let A_i' be the

705

lower sum $s(N_i)$. Then A_i is the area of a polygonal region containing R, and A_i' is the area of a polygonal region lying in R, as in the definition of a measurable set. And

$$A_i' \leqq \int_a^b f(x)\, dx \leqq A_i.$$

The sequences $A_1, A_2, \ldots$ and $A_1', A_2', \ldots$ have the same limit, namely, the integral. Therefore R is measurable, and its area is the integral.

This theorem can be extended so as to apply to the region between the graphs of two continuous functions.

It might seem that we could simplify the preceding discussion by *defining* the area to be the integral, in the first place. But this will not work. The point is that some regions can be represented in many different ways as the regions between the graphs of two continuous functions. Different directions for the axes give different limits of integration, and also different integrands, even for so simple a figure as an ellipse. In the theory that we have just developed, we know that all the resulting integrals give the same answer, because they all give the right answer for the area of the region. But if we *defined* the area to be the integral, we would have the problem of showing, by the methods of calculus, that all the integrals have the same value, and this would be hard.

Proof of the
Northeast Theorem

The Northeast theorem asserts that if

$$\lim_{t \to \infty} f(t) = \lim_{t \to \infty} g(t) = \infty, \tag{1}$$

and

$$\lim_{t \to \infty} \frac{g'(t)}{f'(t)} = L, \tag{2}$$

then

$$\lim_{t \to \infty} \frac{g(t)}{f(t)} = L.$$

To start the proof, we first observe that since $g'(t)/f'(t) \to L$ as $t \to \infty$, we must have $f'(t) \neq 0$ when t is sufficiently large, say, for $t \geqq t_0$. Since we are taking the limit as $t \to \infty$, we may regard t_0 as the initial point of the path. Since $g'(t)/f'(t) \to L$, the function g'/f' must be bounded on some interval $[t_1, \infty)$ $(t_1 \geqq t_0)$. The reason is that for every $\epsilon > 0$, we have

$$L - \epsilon < \frac{g'(t)}{f'(t)} < L + \epsilon,$$

for $t \geqq$ a certain t_1. Therefore g'/f' is bounded on the interval $[t_1, \infty)$. We now take t_1 as the initial point of the path.

As a further simplification, we translate the point $(f(t_1), g(t_1))$ to the origin, replacing $f(t)$ and $g(t)$ by $F(t) = f(t) - f(t_0)$, $G(t) = g(t) - g(t_0)$, where $t \geqq t_0$. Obviously G'/F' is bounded, and

$$\lim_{t \to \infty} \frac{G'(t)}{F'(t)} = L,$$

because $F' = f'$ and $G' = g'$. And if we can prove that

$$\lim_{t \to \infty} \frac{G(t)}{F(t)} = L,$$

then it will follow immediately that

$$\lim_{t \to \infty} \frac{g(t)}{f(t)} = L.$$

The reason is that

$$\frac{g(t)}{f(t)} = \frac{G(t) + g(t_1)}{F(t) + f(t_1)} = \frac{G(t)/F(t) + g(t_1)/F(t)}{1 + f(t_1)/F(t)}.$$

Therefore it will be sufficient to prove the theorem in the following special form.

Theorem A. Let F and G be differentiable functions on the interval $[t_1, \infty)$, such that

$$\lim_{t \to \infty} F(t) = \lim_{t \to \infty} G(t) = \infty, \tag{3}$$

$$\lim_{t \to \infty} \frac{G'(t)}{F'(t)} = L, \tag{4}$$

$$F'(t) \neq 0 \quad \text{for } t \geqq t_1, \tag{5}$$

$$G'/F' \text{ is bounded,} \tag{6}$$

and

$$F(t_1) = G(t_1) = 0. \tag{7}$$

Then

$$\lim_{t \to \infty} \frac{G(t)}{F(t)} = L. \tag{8}$$

We now make a final simplification: under the conditions of Theorem A, the locus of the path must be the graph of a function ϕ, defined on the interval $[0, \infty)$, with

$$\phi(0) = 0, \quad \text{and} \quad \phi'(x) = \frac{G'(t)}{F'(t)} \quad (x = F(t)).$$

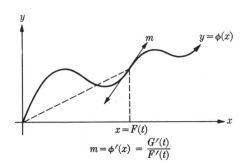

Evidently

$$\frac{G(t)}{F(t)} = \frac{\phi(x)}{x} \quad (x = F(t)).$$

Therefore, rewriting Theorem A in terms of ϕ, we get the following:

Theorem B (*The Northeast theorem, rectangular form*). Let ϕ be a function on the interval $[0, \infty)$, such that

$$\phi(0) = 0, \tag{9}$$

$$\phi' \text{ is bounded}, \tag{10}$$

$$\lim_{x \to \infty} \phi(x) = \infty, \tag{11}$$

and

$$\lim_{x \to \infty} \phi'(x) = L. \tag{12}$$

Then

$$\lim_{x \to \infty} \frac{\phi(x)}{x} = L. \tag{13}$$

The proof is in several steps.

Step 1.

$$\lim_{x \to \infty} \frac{\phi(x^2) - \phi(x)}{x^2 - x} = L.$$

Proof. By the mean-value theorem (MVT), for each x there is an $\bar{x}$, between x and x^2, such that

$$\frac{\phi(x^2) - \phi(x)}{x^2 - x} = \phi'(\bar{x}).$$

As $x \to \infty$, $\bar{x} \to \infty$, and so $\phi'(\bar{x}) \to L$. Therefore the fraction on the left also $\to L$

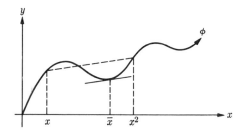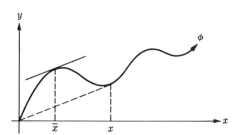

Step 2.

$$\lim_{x \to \infty} \left[\frac{\phi(x^2) - \phi(x)}{x^2 - x} - \frac{\phi(x^2)}{x^2} \right] = 0.$$

Proof.

$$\frac{\phi(x^2) - \phi(\bar{x})}{x^2 - x} - \frac{\phi(x^2)}{x^2} = \frac{x\phi(x^2) - x\phi(x) - x\phi(x^2) + \phi(x^2)}{x^2(x - 1)}$$

$$= -\frac{1}{x - 1}\left[\frac{\phi(x)}{x} + \frac{\phi(x^2)}{x^2} \right].$$

By MVT, there is an $\bar{x}$ between 0 and x such that

$$\frac{\phi(x)}{x} = \phi'(\bar{x}).$$

Since ϕ' is bounded, it follows that $\phi(x)/x$ is bounded. Therefore $\phi(x^2)/x^2$ is bounded. Since $-1/(x - 1) \to 0$, it follows that

$$-\frac{1}{x-1}\left[\frac{\phi(x)}{x} + \frac{\phi(x^2)}{x^2}\right] \to 0.$$

Step 3.

$$\lim_{x \to \infty} \frac{\phi(x^2)}{x^2} = L,$$

by Steps 1 and 2. From Step 3 it follows immediately that $\phi(x)/x \to L$, which was to be proved.

Proof of the Formula for Path Length

Here we complete the proof of Theorem 1 of Section 9.6, which asserts that the length of a path is given by the formula

$$s = \int_a^b \sqrt{f'(t)^2 + g'(t)^2}\, dt,$$

where f and g are the coordinate functions and f' and g' are continuous. The notation is that of Section 9.6. By definition,

$$s = \lim_{|N| \to 0} \sum_{i=1}^n P_{i-1} P_i;$$

we know that

$$\sum_{i=1}^n P_{i-1} P_i = \sum_{i=1}^n \sqrt{f'(\bar{t}_i)^2 + g'(\bar{t}_i')^2}\, \Delta t_i,$$

where $\bar{t}_i$ and $\bar{t}_i'$ are between t_{i-1} and t_i; and

$$\lim_{|N| \to 0} \sqrt{f'(\bar{t}_i)^2 + g'(\bar{t}_i)^2}\, \Delta t_i = \int_a^b \sqrt{f'(t)^2 + g'(t)^2}\, dt,$$

by definition of the integral. Therefore what remains to be proved is that

$$\lim_{|N| \to 0} \left[\sum_{i=1}^n \sqrt{f'(\bar{t}_i)^2 + g'(\bar{t}_i')^2}\, \Delta t_i - \sum_{i=1}^n \sqrt{f'(\bar{t}_i)^2 + g'(\bar{t}_i)^2}\, \Delta t_i \right] = 0.$$

In terms of the definition of $\lim_{|N| \to 0}$, this can be restated as follows:

Lemma. For every $\epsilon > 0$ there is a $\delta > 0$ such that

$$|N| < \delta \quad \Rightarrow \quad \left| \sum_{i=1}^n [\sqrt{f'(\bar{t}_i)^2 + g'(\bar{t}_i')^2} - \sqrt{f'(\bar{t}_i)^2 + g'(\bar{t}_i)^2}] \Delta t_i \right| < \epsilon.$$

Proof. 1) Since f' and g' are continuous on $[a, b]$, so also is the function $f'^2 + g'^2$. Let M be such that

$$f'(t)^2 + g'(t)^2 \leqq M \quad \text{for } a \leqq t \leqq b.$$

2) The function $z \mapsto \sqrt{z}$ is continuous on the interval $[0, M]$. Therefore this function is uniformly continuous on $[0, M]$ (Section 7.8, Theorem 4). Therefore,

given any $\epsilon > 0$, there is a $\theta > 0$ such that

$$|z - z'| < \theta \;\Rightarrow\; |\sqrt{\bar{z}} - \sqrt{\bar{z}'}| < \frac{\epsilon}{b - a}.$$

3) The function g'^2 is continuous, and therefore uniformly continuous, on $[a, b]$. Therefore, given $\theta > 0$ there is a $\delta > 0$ such that

$$|t_i' - t_i| < \delta \;\Rightarrow\; |g'(t_i')^2 - g'(t_i)^2| < \theta.$$

4) The δ given by (3) is the δ that we need:

$$
|N| < \;\; \delta \Rightarrow \;|t_i' - t_i| < \delta \quad\;\; \text{(for every } i)
$$

$$
\Rightarrow \; |g'(t_i')^2 - g'(t_i)^2| < \theta \quad \text{(for every } i)
$$

$$
\Rightarrow \; \left| \sqrt{f'(t_i)^2 + g'(t_i')^2} - \sqrt{f'(t_i)^2 + g'(t_i)^2} \right| < \frac{\epsilon}{b - a}
$$

$$
\Rightarrow \; \left| \left[\sqrt{f'(t_i)^2 + g'(t_i')^2} - \sqrt{f'(t_i)^2 + g'(t_i)^2} \right] \Delta t_i \right| < \frac{\epsilon}{b - a} \Delta t_i
$$

$$
\Rightarrow \; \sum_{i=1}^{n} \left| \left[\sqrt{f'(t_i)^2 + g'(t_i')^2} - \sqrt{f'(t_i)^2 + g'(t_i)^2} \right] \Delta t_i \right|
$$

$$
< \frac{\epsilon}{b - a} \sum_{i=1}^{n} \Delta t_i = \frac{\epsilon}{b - a} (b - a) = \epsilon.
$$

Since the absolute value of the sum is less than or equal to the sum of the absolute values, it follows that δ satisfies the conditions of the lemma.

A Method for Constructing the Complex Numbers

In Section 10.11 the complex numbers were presented as a formal system of symbols $a + bi$, with $i^2 = -1$. We shall now define a mathematical system of this kind, and show that it has the properties that we want. There are various ways to do this. The following method has the advantage of copying the pattern of the manipulative processes that we would be using anyway. It has the further advantage of introducing ideas that will be useful later, in modern algebra.

Let $\mathbf{P}(x)$ be the set of all polynomials $p(x) = \sum_{i=0}^{n} a_i x^i$. In $\mathbf{P}(x)$ we can add and multiply. We know that in $\mathbf{P}(x)$, these operations obey the CAD laws, that is, they are commutative, associative, and distributive:

$$pq = qp, \qquad p + q = q + p,$$

$$p(qr) = (pq)r, \qquad p + (q + r) = (p + q) + r,$$

$$p(q + r) = pq + pr.$$

These follow immediately from the corresponding laws for the real numbers $p(x)$ which are the values of our polynomial functions.

Two polynomials p, q will be called *congruent modulo* $1 + x^2$ if their difference is a multiple of $1 + x^2$. We then write

$$p(x) \equiv q(x) \bmod 1 + x^2,$$

or briefly $p \equiv q$. Thus

$$p(x) \equiv q(x) \qquad \text{if} \qquad p(x) - q(x) = r(x)(1 + x^2),$$

for some polynomial $r(x)$. For example, $x^3 \equiv -x$, because $x^3 - (-x) = x^3 + x = x(1 + x^2)$; and $x^2 \equiv -1$, because $x^2 - (-1) = x^2 + 1 = 1 \cdot (1 + x^2)$.

In the following theorem, the primes do not indicate differentiation; $p, q, p',$ and q' are supposed to be any polynomials.

Theorem 1. If $p \equiv p'$ and $q \equiv q'$, then $p + q \equiv p' + q'$, and $pq \equiv p'q'$.

This is a straightforward calculation. Given

$$p' = p + r \cdot (x^2 + 1), \qquad q' = q + s \cdot (x^2 + 1),$$

we get

$$p' + q' = p + q + (r + s) \cdot (x^2 + 1),$$

so that $p' + q' \equiv p + q$. And

$$p'q' = pq + (ps + qr)(x^2 + 1) + rs(x^2 + 1)^2$$
$$= pq + (ps + qr + rsx^2 + rs)(x^2 + 1),$$

so that $p'q' \equiv pq$.

For each $p = p(x)$ in $\mathbf{P}(x)$, let $\bar{p}$ be the set of all polynomials that are congruent to p. That is,

$$\bar{p} = \{p' \mid p' \equiv p\}.$$

These sets are called *congruence classes*. Let $\mathbf{C}$ be the set of all such congruence classes $\bar{p}$. In $\mathbf{C}$ we define addition and multiplication by the conditions

$$\bar{p} + \bar{q} = \overline{p + q}, \quad \text{and} \quad \bar{p} \cdot \bar{q} = \overline{pq}.$$

These definitions make sense, because the congruence class that contains $p + q$ does not depend on the choice of p and q; it depends only on the *congruence classes* $\bar{p}$ and $\bar{q}$. The same applies for the product.

Theorem 2. The CAD laws hold in $\mathbf{C}$.

Proof. We know that these laws hold in $\mathbf{P}(x)$. Therefore, under our definitions of addition and multiplication in $\mathbf{C}$, we have

$$\bar{p} \cdot \bar{q} = \overline{pq} = \overline{qp} = \bar{q}\bar{p};$$
$$\bar{p} + \bar{q} = \overline{p + q} = \overline{q + p} = \bar{q} + \bar{p};$$
$$\bar{p}(\bar{q} + \bar{r}) = \bar{p} \cdot \overline{q + r} = \overline{p(q + r)} = \overline{pq + pr} = \overline{pq} + \overline{pr} = \bar{p}\bar{q} + \bar{p}\bar{r};$$

and similarly for the other laws.

We now observe that

$$x^{2n} = (x^2)^n \equiv (-1)^n, \quad \text{and} \quad x^{2n+1} = (x^2)^n x \equiv (-1)^n x.$$

Therefore every power of x is congruent to a linear polynomial. Therefore:

Theorem 3. Every polynomial $p(x)$ is congruent to a linear polynomial $a + bx$.

For example,

$$p(x) = 7x^7 - 5x^3 + 6x^2 - 3$$
$$= 7(x^2)^3 \cdot x - 5x^2 \cdot x + 6x^2 - 3$$
$$= -7x + 5x - 6 - 3$$
$$= -2x - 9.$$

In fact, the system $\mathbf{C}$ that we have just defined has all the properties of the number system that we wanted. To describe it in the familiar notation, we denote each congruence class $\overline{p(x)}$ by the formal expression $p(i)$, in which x is replaced by i. Thus, if $p(x)$ is as in the preceding example, we have

$$\overline{p(x)} = p(i) = 7i^7 - 5i^3 + 6i^2 - 3$$
$$= 7(i^2)^3 \cdot i - 5i^2 \cdot i + 6i^2 - 3$$
$$= -7i + 5i - 6 - 3$$
$$= -9 - 2i.$$

Here we have simplified by substituting -1 for i^2, and this is right; since

$$x^2 \equiv -1,$$

we have

$$\overline{x^2} = i^2 = -1;$$

any *congruence* between two polynomials $p(x)$ and $q(x)$ gives an *equation* between their congruence classes $p(i)$ and $q(i)$. And our number system satisfies the conditions for a field, given in Chapter 1: the CAD laws hold; there are numbers 0 and 1, such that if

$$z = a + bi,$$

then

$$0 \cdot z = 0, \quad \text{and} \quad 1 \cdot z = z;$$

for each z there is a number

$$-z = -a - bi,$$

such that

$$z + (-z) = 0.$$

Finally, every $z \neq 0$ has a reciprocal. To prove this, we first observe that

$$a + bi = 0 \;\Rightarrow\; a = b = 0.$$

The reason is that

$$a + bi = 0 \;\Leftrightarrow\; a + bx \equiv 0 \bmod 1 + x^2$$
$$\Leftrightarrow\; a + bx = r(x)(1 + x^2).$$

Here $r(x)$ must be 0, because otherwise $r(x)(1 + x^2)$ would be of degree $\geqq 2$. Since $r = 0$, $a + bx$ is the zero polynomial, and so $a = b = 0$. Similarly,

$$a - bi = 0 \;\Rightarrow\; a = b = 0,$$

and so

$$a + bi \neq 0 \;\Rightarrow\; a - bi \neq 0, \quad \text{and} \quad a^2 + b^2 > 0.$$

We shall now find a reciprocal for any $a + bi \neq 0$, by assuming, experimentally, fractions with complex denominators make sense, and then checking that our answer works:

$$\frac{1}{a + bi} = \frac{1}{a + bi} \cdot \frac{a - bi}{a - bi} = \frac{a}{a^2 + b^2} - \frac{bi}{a^2 + b^2}.$$

The last expression really is the reciprocal of $a + bi$, because

$$(a + bi)\left[\frac{a}{a^2 + b^2} - \frac{bi}{a^2 + b^2}\right] = \frac{1}{a^2 + b^2}(a + bi)(a - bi) = \frac{a^2 + b^2}{a^2 + b^2} = 1.$$

To sum up:

Theorem 4. **C** is a field.

Note that when we passed from $\mathbf{P}(x)$ to **C**, by forming congruence classes modulo $1 + x^2$, the algebraic character of the system changed: in $\mathbf{P}(x)$, only the constant polynomials $p(x) = a \neq 0$ have reciprocals; but every congruence class $p(i) = \overline{p(x)} \neq 0$ has a reciprocal.

To set up this number system, and check its properties, we need to use equivalence classes of polynomials (or some equivalent device). But now that we have such a system, we need not remember where we got it. In the future, we shall never have occasion to refer to the fact that $a + bi$ was defined to be

$$\{p(x) \mid p(x) \equiv a + bx\}.$$

APPENDIX K

Iterated Limits.
Mixed Partial Derivatives

In discussing double limits of the type

$$\lim_{(x,y)\to(x_0,y_0)} f(x, y),$$

for functions of two variables, we assume that f is defined in a neighborhood of (x_0, y_0), except, perhaps, at the point (x_0, y_0) itself. Under the same conditions, we can discuss limits of the type

$$\lim_{x\to x_0}\lim_{y\to y_0} f(x, y) \qquad \text{and} \qquad \lim_{y\to y_0}\lim_{x\to x_0} f(x, y).$$

These are called *iterated limits*.

There are simple examples in which the iterated limits both exist but have different values. That is, the order in which we take the limits may make a difference. Consider

$$f(x, y) = \frac{x^2 - y^2}{x^2 + y^2}.$$

Here

$$\lim_{x\to 0}\lim_{y\to 0} \frac{x^2 - y^2}{x^2 + y^2} = \lim_{x\to 0} \frac{x^2}{x^2} = 1,$$

and

$$\lim_{y\to 0}\lim_{x\to 0} \frac{x^2 - y^2}{x^2 + y^2} = \lim_{y\to 0} \frac{-y^2}{y^2} = -1.$$

This sort of thing cannot happen, however, if f is continuous in D and the double limit exists. That is, we have the following theorem:

Theorem 1. If f is continuous in D, and

$$\lim_{(x,y)\to(x_0,y_0)} f(x, y) = L,$$

then

$$\lim_{x\to x_0}\lim_{y\to y_0} f(x, y) = \lim_{y\to y_0}\lim_{x\to x_0} f(x, y) = L.$$

Proof. If $f(x_0, y_0)$ is defined at all, then $f(x_0, y_0)$ must be L. If $f(x_0, y_0)$ is *not* defined, we define it to be L. Thus we may assume that f is continuous in a neighborhood of (x_0, y_0), including (x_0, y_0).

717

The proof is now trivial: For each x, we have

$$\lim_{y \to y_0} f(x, y) = f(x, y_0), \tag{1}$$

because f is continuous. Therefore, for the same reason we have

$$\lim_{x \to x_0} \lim_{y \to y_0} f(x, y) = \lim_{x \to x_0} f(x, y_0) = f(x_0, y_0). \tag{2}$$

Similarly,

$$\lim_{y \to y_0} \lim_{x \to x_0} f(x, y) = \lim_{y \to y_0} f(x_0, y) = f(x_0, y_0). \tag{3}$$

In (1), all we are saying is that if f is continuous (as a function of two variables) then the slice functions, for each fixed x, are also continuous. This may be easier to keep straight if we rewrite (1) in the form

$$\lim_{y \to y_0} f(a, y) = f(a, y_0), \tag{1'}$$

which reminds us that x is fixed as $y \to y_0$. Equations (2) and (3) follow by repeated applications of the same principle.

Since in some cases the two iterated limits are different, we always have to investigate, in the cases where we need to know that they are the same. One such case comes up when we consider the "mixed partial derivatives" f_{xy} and f_{yx}. If we write in full the definitions of $f_{xy}(x_0, y_0)$ and $f_{yx}(x_0, y_0)$, we see that they are iterated limits of the same function:

$$f_{xy}(x_0, y_0) = \lim_{\Delta y \to 0} \frac{f_x(x_0, y_0 + \Delta y) - f_x(x_0, y_0)}{\Delta y}$$

$$= \lim_{\Delta y \to 0} \frac{1}{\Delta y} \left[\lim_{\Delta x \to 0} \frac{f(x_0 + \Delta x, y_0 + \Delta y) - f(x_0, y_0 + \Delta y)}{\Delta x} \right.$$

$$\left. - \lim_{\Delta x \to 0} \frac{f(x_0 + \Delta x, y_0) - f(x_0, y_0)}{\Delta x} \right]$$

$$= \lim_{\Delta y \to 0} \lim_{\Delta x \to 0} \frac{1}{\Delta y \, \Delta x} \{ [f(x_0 + \Delta x, y_0 + \Delta y) - f(x_0 + \Delta x, y_0)]$$

$$- [f(x_0, y_0 + \Delta y) - f(x_0, y_0)] \}.$$

Note that in the last step we have changed the order of two of the terms. The reason for this will soon be clear. Thus we have

$$f_{xy}(x_0, y_0) = \lim_{\Delta y \to 0} \lim_{\Delta x \to 0} \frac{1}{\Delta y \, \Delta x} F(\Delta x, \Delta y),$$

where $F(\Delta x, \Delta y)$ is the function defined by the expression in the braces. An entirely analogous calculation tells us that $f_{yx}(x_0, y_0)$ is the iterated limit, in reverse order, of exactly the same function:

$$f_{yx}(x_0, y_0) = \lim_{\Delta x \to 0} \lim_{\Delta y \to 0} \frac{1}{\Delta y \, \Delta x} F(\Delta x, \Delta y).$$

Let us now investigate the function F. This function can be regarded as the difference of two values of the function

$$\phi(x) = f(x, y_0 + \Delta y) - f(x, y_0).$$

That is,

$$F(\Delta x, \Delta y) = \phi(x_0 + \Delta x) - \phi(x_0).$$

Applying the mean-value theorem MVT to the function ϕ, on the interval from x_0 to $x_0 + \Delta x$, we get

$$F(\Delta x, \Delta y) = \phi'(\bar{x})\, \Delta x = [f_x(\bar{x}, y_0 + \Delta y) - f_x(\bar{x}, y_0)]\, \Delta x,$$

where $\bar{x}$ is between x_0 and $x_0 + \Delta x$. Now the quantity in brackets can be regarded as the difference of two values of the function

$$\psi(y) = f_x(\bar{x}, y);$$

we have

$$F(\Delta x, \Delta y) = [\psi(y_0 + \Delta y) - \psi(y_0)]\, \Delta x.$$

By MVT,

$$\psi(y_0 + \Delta y) - \psi(y_0) = \psi'(\bar{y})\, \Delta y,$$

where $\bar{y}$ is between y_0 and $y_0 + \Delta y$. This gives

$$F(\Delta x, \Delta y) = f_{xy}(\bar{x}, \bar{y})\, \Delta y\, \Delta x.$$

Suppose now that f_{xy} is continuous. We then have

$$\lim_{(\Delta x, \Delta y) \to (0,0)} \frac{1}{\Delta y\, \Delta x} F(\Delta x, \Delta y) = \lim_{(\Delta x, \Delta y) \to 0} f_{xy}(\bar{x}, \bar{y}) = f_{xy}(x_0, y_0).$$

[Since $|\bar{x} - x_0| < \Delta x$ and $|\bar{y} - y_0| < \Delta y$, we must have $(\bar{x}, \bar{y}) \to (x_0, y_0)$ as $(\Delta x, \Delta y) \to (0, 0)$.]

Let us now take stock. We had

1) $f_{xy}(x_0, y_0) = \lim_{\Delta y \to 0} \lim_{\Delta x \to 0} \dfrac{1}{\Delta x\, \Delta y} F(\Delta x, \Delta y)$ (by definition).

2) If f_{xy} is continuous at (x_0, y_0), then the double limit

$$\lim_{(\Delta x, \Delta y) \to (0,0)} \frac{1}{\Delta x\, \Delta y} F(\Delta x, \Delta y)$$

exists, and is equal to $f_{xy}(x_0, y_0)$. This is what we have just proved. Suppose now that f_{yx} is also defined in a neighborhood of (x_0, y_0), and is continuous. Then

3) $$f_{yx}(x_0, y_0) = \lim_{\Delta x \to 0} \lim_{\Delta y \to 0} \frac{1}{\Delta x\, \Delta y} F(\Delta x, \Delta y).$$

Since the double limit in (2) exists, the iterated limit in (3) must be equal to it. Therefore $f_{xy}(x_0, y_0) = f_{yx}(x_0, y_0)$. Thus we have proved the following theorem:

Theorem 2. If f_{xy} and f_{yx} exist and are continuous, in a domain D, then $f_{xy} = f_{yx}$.

By repeated applications of this theorem, we can draw a more general conclusion about partial derivatives of higher order: if these exist and are continuous, then all that matters is the number of times we differentiate with respect to each variable. Thus, for example,

$$f_{xyxy} = f_{xxyy}.$$

Proof

$$f_{xyx} = (f_x)_{yx} = (f_x)_{xy} = f_{xxy}.$$

Therefore

$$(f_{xyx})_y = (f_{xxy})_y,$$

and $f_{xyxy} = f_{xxyy}$, which was to be proved.

Warning: It is not true that if f_{xy} exists and is continuous, then f_{yx} also exists and is the same. To see this, let

$$f(x, y) = \phi(y),$$

where ϕ is any function which is not differentiable. Then

$$f_x(x, y) = 0 \qquad \text{for every } x, y,$$

because the slice functions for constant y are constant. Therefore, trivially, f_{xy} exists, and $f_{xy}(x, y) = 0$ for every x, y, so that f_{xy} is continuous. But f_y is not defined, because $\phi'(y)$ is not defined. Therefore f_{yx} is not defined either.

APPENDIX L

Possible Peculiarities of Functions of Two Variables

Some of the definitions that we have used in Chapter 14, and the hypotheses of some of our theorems, may seem needlessly strong. In fact, they are not. The theory of functions of two variables includes some rather odd and unexpected phenomena; and if we want to draw simple conclusions, we need to use hypotheses sufficiently strong to rule out the oddities. Some of these are as follows.

Example 1. There is a function f such that

1) all the slice functions $f(x, y_0)$ (with y held constant) are continuous, and
2) all the slice functions $f(x_0, y)$ (with x held constant) are continuous, but
3) f is not continuous.

Proof. In the first quadrant of the xy-plane we take an infinite sequence $D_1, D_2, \ldots$ of circular disks, not intersecting each other, with radii approaching 0, and approaching the origin as a limit.

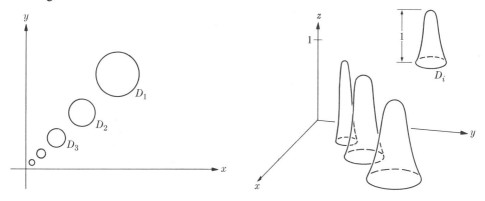

As indicated in the figure on the left, we take these disks with their centers on the line $y = x$, in such a way that no horizontal or vertical line intersects more than one of them. This is easy to arrange, because we can make the disks as small as we want.

If (x, y) lies in *none* of the disks D_i, then we define $f(x, y)$ to be 0. Over each disk D_i, the graph of f is a "blister" of height 1, shown on the right.

Obviously f is not continuous at $(0, 0)$. But all the slice functions $\phi(x) = f(x, y_0)$ are continuous. Since no horizontal line intersects more than one of the disks D_i, it follows that the graph of ϕ looks, at worst, like the graph shown below.

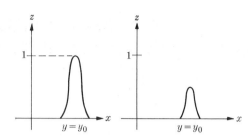

On the left we see what happens if the line $y = y_0$ passes through the center of a disk. If this doesn't happen, then the maximum of ϕ is smaller; and of course $\phi(x)$ may be 0 for every x. Similarly for the slice functions for constant x.

Here, of course, the slice function

$$\phi_{\pi/4}(t) = f\left(\frac{t}{\sqrt{2}}, \frac{t}{\sqrt{2}}\right)$$

is *not* continuous. Its graph is shown below. But even if a function f has slice functions

$$\phi_\alpha(t) = f(x_0 + t \cos \alpha, y_0 + t \sin \alpha)$$

which are continuous, for every (x_0, y_0) and every α, we still cannot conclude that f is continuous.

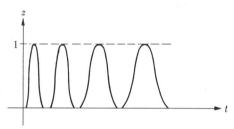

Example 2. There is a function f such that

1) Every slice function ϕ_α is continuous, but
2) f is not continuous.

Proof. Consider the parabola $y = x^2$, in the xy-plane. Between the parabola and the x-axis we take a sequence $D_1, D_2, \ldots$ of circular disks, with radii approaching 0, approaching the origin as a limit. On each disk, the graph of f is a blister of height 1, as in Example 1; everywhere else, $f(x, y) = 0$.

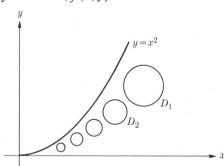

As before, f is not continuous. But all the slice functions are. The reason is that no line L intersects more than a finite number of the disks D_i:

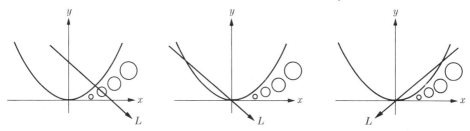

If L does not pass through the origin (as on the left above) or if L passes through the origin and has negative slope (as in the center), this conclusion is trivial. The interesting case is shown on the right. Here L passes through $(0, 0)$ and has positive slope. Near the origin, in the first quadrant, the line lies above the parabola and the disks lie below it. Therefore L cannot intersect infinitely many disks. Therefore the slice functions defined along any line L are continuous.

Our next peculiar function is going to be continuous. We recall that in Section 14.8 we proved the following theorem:

Theorem A. Given a function f, defined in a neighborhood of (x_0, y_0). For each α, let

$$\phi_\alpha = f(x_0 + t \cos \alpha, y_0 + t \sin \alpha).$$

Suppose that $\phi_\alpha'(0) = 0$ for every α; and suppose that there is a number $\delta > 0$ such that for $|t| < \delta$ we have $\phi_\alpha''(t) < 0$ for every α. Then f has an ILMax at (x_0, y_0).

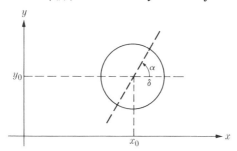

The reason is that for each α, $\phi_\alpha(0)$ is the maximum value of ϕ on the interval $(-\delta, \delta)$. See the figure below.

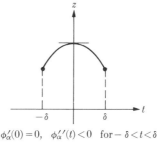

$$\phi_\alpha'(0) = 0, \quad \phi_\alpha''(t) < 0 \quad \text{for} -\delta < t < \delta$$

Here it is essential that there be a single number $\delta > 0$ which works for every α. The following plausible-looking variation on Theorem A is false.

? Theorem B?   Given a function f, defined in a neighborhood of (x_0, y_0). For each α, let

$$\phi_\alpha(t) = f(x_0 + t \cos \alpha, y_0 + t \sin \alpha).$$

If (1) each function ϕ_α has an ILMax at 0, then (2) f has an ILMax at (x_0, y_0).

The falsity of this "theorem" is demonstrated by:

Example 3.   There is a continuous function f such that (1) every slice function through the origin has an ILMax at the origin, but (2) f does not have an ILMax at the origin.

This is similar to Example 2.   As before, we take a sequence of disks lying under a parabola.   We define $f(x, y)$ to be 0 everywhere except on the disks.   But this time, we take the blister over the ith disk D_i in such a way that its height is $1/i$.   Now our function f is continuous.

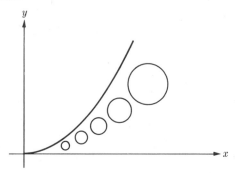

As before, no line L in the xy-plane intersects more than a finite number of the disks. Therefore every slice function

$$\phi_\alpha(t) = f(t \cos \alpha, t \sin \alpha)$$

is equal to 0 in a *neighborhood* of 0.   That is, for each α there is a $\delta_\alpha > 0$ such that $\phi_\alpha(t) = 0$ for $|t| < \delta_\alpha$.   Therefore every ϕ_α has an ILMax at 0.   But obviously f does not have an ILMax at $(0, 0)$; $f(0, 0) = 0$, but every neighborhood of $(0, 0)$ contains a disk D_i, on which $f(x, y) > 0$.

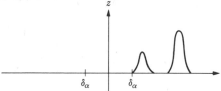

The trouble here is that while for every α there is a δ_α with the desired property, there is no *one* δ which works for every α.   If $\alpha > 0$ and $\alpha \approx 0$, then $\delta_\alpha \approx 0$; and so inf $\{\delta_\alpha\} = 0$.   If you reread the proof of Theorem 3, Section 14.8, you will see how this trouble was avoided: using the continuity of f_{xy}, f_{xx}, and f_{yy}, we found a single $\delta > 0$ which worked for every α.   Thus the proof of Theorem 3 was not merely complicated in a technical way but was also subtle, in a way which is not likely to be understood unless we re-examine the proof in the light of Example 3.

APPENDIX M

Maxima and Minima, for Functions of Two Variables

Here we give a brief sample of the way the theory of continuous functions of one variable can be extended so as to apply to functions $f\colon D \to \mathbf{R}$, where D is a domain in a Cartesian space $\mathbf{R}^n$ $(n > 1)$.

Theorem 1. Let D be a closed rectangular region in $\mathbf{R}^2$, defined by the inequalities $a \leqq x \leqq b$ and $c \leqq y \leqq d$. Let f be a continuous function $D \to \mathbf{R}$. Then f is bounded above.

Proof. Suppose that f is not bounded above. We shall show that this leads to a contradiction.

The region D is the union of four closed rectangular regions, shown in the figure on the left below. These will be called *quarters* of D. These are like the "halves" of an interval $[a, b]$, as defined in Section 5.6; and they are going to be used in exactly the same way.

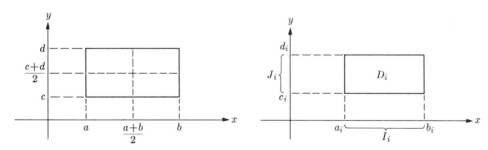

Following the pattern of Section 5.6, we say that a closed rectangular region D' is *good* if f is bounded on D'; and D' is *bad* if f is unbounded on D'. We are assuming that the given D is bad. It follows that one of the quarters of D must be bad. (Why? See Lemma 1 on page 240.) Let D_1 be a bad quarter of D. Similarly, let D_2 be a bad quarter of D_1. Proceeding in this way, we get a sequence

$$D_1, D_2, \ldots$$

of closed rectangular regions, each of which is bad, such that for each i, D_{i+1} is a quarter of D_i. As indicated in the figure on the right above, let I_i and J_i be the closed intervals which are the projections of I_i and J_i onto the x- and y-axes. Then $I_1, I_2, \ldots$ is a nested sequence. By the Nested Interval Postulate (NIP) there is an $\bar{x}$ which lies on each interval I_i. Similarly, there is a $\bar{y}$ which lies on every interval J_i. It follows that the point $\bar{P} = (\bar{x}, \bar{y})$ lies in every region D_i.

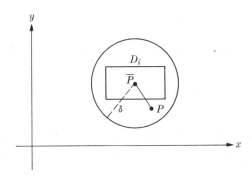

But f is continuous at $\bar{P}$. Thus for every $\epsilon > 0$ there is a $\delta > 0$ such that

$$P\bar{P} < \delta \;\Rightarrow\; |f(P) - f(\bar{P})| < \epsilon,$$

and this means that

$$P\bar{P} < \delta \;\Rightarrow\; f(P) < f(\bar{P}) + \epsilon;$$

that is, f is bounded on the circular disk with center at $\bar{P}$ and radius δ. But this circular disk contains some D_i, because $\bar{P}$ lies in all the D_i's, and the height and width of D_i both $\to 0$ as $i \to \infty$. Therefore D_i must be *good* for some i, which contradicts our hypothesis.

Theorem 2. If f is continuous on a closed rectangular region D, then f has a maximum value on D.

The proof is exactly like the proof of Theorem 3 of Section 5.6. Let $k = \sup f$. If $k = f(P)$ for some P, then f has its maximum value at P. If $f(P) < k$ for every P in D, let

$$g(P) = \frac{1}{k - f(P)}\,.$$

Then g is continuous on D, but is not bounded above; and this contradicts Theorem 1.

As before, the existence of maxima gives, as a corollary, the existence of minima:

Theorem 3. If f is continuous on a closed rectangular region D, then f has a minimum value on D.

(*Proof.* Any maximum value of $-f$ is a minimum value of f.)

The same scheme works for continuous functions defined on an "n-dimension interval"

$$D = \{(x_1, x_2, \ldots, x_n) \mid a_i \leqq x_i \leqq b_i \;\; \text{for } i = 1, 2, \ldots, n\}.$$

We use a subdivision process just as in $\mathbf{R}^1$ and $\mathbf{R}^2$, dividing our "interval," at each stage, into 2^n parts.

APPENDIX N

An Exact Definition of the Idea of a Function

In Chapter 3 we explained that a function $f: A \to B$ is defined if a rule is given under which for each element of the set A, there exists one and only one corresponding element of the set B.

This formulation of the idea of a function is adequate for the purposes of elementary calculus, and so, in the text, we have let it stand. But eventually we need a more exact definition, now to be explained. To see what we are driving at, in the new definition, consider the function

$$f: \mathbf{R} \to \mathbf{R}$$
$$: x \mapsto x^2 \quad \text{for every } x.$$

The graph is a parabola.

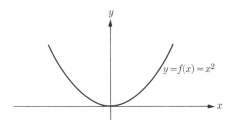

We shall now approach our new definition in the following two steps.

Step 1. We regard the function as being indistinguishable from its graph, so that the function f becomes a set of points P, in a coordinate plane. (In this case, the function is a parabola.)

Step 2. We regard a point P, in a coordinate plane, as indistinguishable from the ordered pair (x, y) which gives its coordinates.

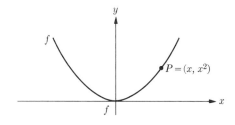

The graph now becomes a collection of ordered pairs of real numbers, namely,

$$f = \{(x, x^2)\}.$$

This collection of ordered pairs has the property that each real number x is the first term of exactly one ordered pair (x, y) in the set. (This is because the graph intersects every vertical line in exactly one point.)

This final description of f, as a collection of ordered pairs $\{(x, x^2)\}$, can be generalized to apply to functions of any kind, on any domain. The final definition is as follows.

Definition. Let A and B be sets. Let f be a collection of ordered pairs (a, b). Suppose that

1) if (a, b) belongs to f, then a belongs to A and b belongs to B, and
2) every element a of A is the first term of exactly one pair belonging to f.

Then f is *a function of A into B*, and we write

$$f: A \rightarrow B.$$

For each a in $A, f(a)$ denotes the second term of the ordered pair whose first term is a.

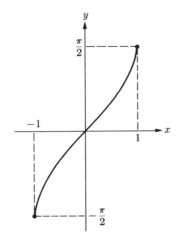

For the function whose graph was the parabola, we had

$$A = B = \mathbf{R}, \qquad f = \{(x, x^2)\}.$$

Similarly, for the function Sin^{-1}, we have

$$A = [-1, 1], \qquad B = \mathbf{R} \qquad (\text{or } B = [-\pi/2, \pi/2]),$$

$$f = \text{Sin}^{-1} = \left\{(x, y) \,\middle|\, -1 \leq x \leq 1, -\frac{\pi}{2} \leq y \leq \frac{\pi}{2}, x = \sin y\right\}.$$

Here again the idea is that the function is defined to *be* its graph, and the graph is regarded as a set of ordered pairs of real numbers. Note, however, that our general definition of a function applies in a variety of other contexts, in which A and B may be sets of quite different kinds. For example, A may be a vector space, or a region in a vector space, or the set of all positive integers. In this book, B has usually, but not always, been a set of numbers.

Eventually, the exact definition of a function becomes useful, as a matter of technique. A little reflection will convince us, however, that it cannot be *quite* right, at any stage, to define a function to be a rule. The point is that rules are formed with words (or with a combination of words and symbols). Therefore there is such a thing as a fifteen-word rule. But surely there is no such thing as a fifteen-word function. Analogously, there is such a thing as a three-syllable name, but there is no such thing as a three-syllable man. As a matter of common sense, a man is different from his name; and in the same way, a function is different from the phrases and formulas that we use to describe it. Therefore functions ought to be defined in such a way as to be mathematical objects.

Table 1
Natural Trigonometric Functions

Angle					Angle				
De-gree	Ra-dian	Sine	Co-sine	Tan-gent	De-gree	Ra-dian	Sine	Co-sine	Tan-gent
0°	0.000	0.000	1.000	0.000					
1°	0.017	0.017	1.000	0.017	46°	0.803	0.719	0.695	1.036
2°	0.035	0.035	0.999	0.035	47°	0.820	0.731	0.682	1.072
3°	0.052	0.052	0.999	0.052	48°	0.838	0.743	0.669	1.111
4°	0.070	0.070	0.998	0.070	49°	0.855	0.755	0.656	1.150
5°	0.087	0.087	0.996	0.087	50°	0.873	0.766	0.643	1.192
6°	0.105	0.105	0.995	0.105	51°	0.890	0.777	0.629	1.235
7°	0.122	0.122	0.993	0.123	52°	0.908	0.788	0.616	1.280
8°	0.140	0.139	0.990	0.141	53°	0.925	0.799	0.602	1.327
9°	0.157	0.156	0.988	0.158	54°	0.942	0.809	0.588	1.376
10°	0.175	0.174	0.985	0.176	55°	0.960	0.819	0.574	1.428
11°	0.192	0.191	0.982	0.194	56°	0.977	0.829	0.559	1.483
12°	0.209	0.208	0.978	0.213	57°	0.995	0.839	0.545	1.540
13°	0.227	0.225	0.974	0.231	58°	1.012	0.848	0.530	1.600
14°	0.244	0.242	0.970	0.249	59°	1.030	0.857	0.515	1.664
15°	0.262	0.259	0.966	0.268	60°	1.047	0.866	0.500	1.732
16°	0.279	0.276	0.961	0.287	61°	1.065	0.875	0.485	1.804
17°	0.297	0.292	0.956	0.306	62°	1.082	0.883	0.469	1.881
18°	0.314	0.309	0.951	0.325	63°	1.100	0.891	0.454	1.963
19°	0.332	0.326	0.946	0.344	64°	1.117	0.899	0.438	2.050
20°	0.349	0.342	0.940	0.364	65°	1.134	0.906	0.423	2.145
21°	0.367	0.358	0.934	0.384	66°	1.152	0.914	0.407	2.246
22°	0.384	0.375	0.927	0.404	67°	1.169	0.921	0.391	2.356
23°	0.401	0.391	0.921	0.424	68°	1.187	0.927	0.375	2.475
24°	0.419	0.407	0.914	0.445	69°	1.204	0.934	0.358	2.605
25°	0.436	0.423	0.906	0.466	70°	1.222	0.940	0.342	2.748
26°	0.454	0.438	0.899	0.488	71°	1.239	0.946	0.326	2.904
27°	0.471	0.454	0.891	0.510	72°	1.257	0.951	0.309	3.078
28°	0.489	0.469	0.883	0.532	73°	1.274	0.956	0.292	3.271
29°	0.506	0.485	0.875	0.554	74°	1.292	0.961	0.276	3.487
30°	0.524	0.500	0.866	0.577	75°	1.309	0.966	0.259	3.732
31°	0.541	0.515	0.857	0.601	76°	1.326	0.970	0.242	4.011
32°	0.559	0.530	0.848	0.625	77°	1.344	0.974	0.225	4.332
33°	0.576	0.545	0.839	0.649	78°	1.361	0.978	0.208	4.705
34°	0.593	0.559	0.829	0.675	79°	1.379	0.982	0.191	5.145
35°	0.611	0.574	0.819	0.700	80°	1.396	0.985	0.174	5.671
36°	0.628	0.588	0.809	0.727	81°	1.414	0.988	0.156	6.314
37°	0.646	0.602	0.799	0.754	82°	1.431	0.990	0.139	7.115
38°	0.663	0.616	0.788	0.781	83°	1.449	0.993	0.122	8.144
39°	0.681	0.629	0.777	0.810	84°	1.466	0.995	0.105	9.514
40°	0.698	0.643	0.766	0.839	85°	1.484	0.996	0.087	11.43
41°	0.716	0.656	0.755	0.869	86°	1.501	0.998	0.070	14.30
42°	0.733	0.669	0.743	0.900	87°	1.518	0.999	0.052	19.08
43°	0.750	0.682	0.731	0.933	88°	1.536	0.999	0.035	28.64
44°	0.768	0.695	0.719	0.966	89°	1.553	1.000	0.017	57.29
45°	0.785	0.707	0.707	1.000	90°	1.571	1.000	0.000	

Table 2
Exponential Functions

x	e^x	e^{-x}	x	e^x	e^{-x}
0.00	1.0000	1.0000	2.5	12.182	0.0821
0.05	1.0513	0.9512	2.6	13.464	0.0743
0.10	1.1052	0.9048	2.7	14.880	0.0672
0.15	1.1618	0.8607	2.8	16.445	0.0608
0.20	1.2214	0.8187	2.9	18.174	0.0550
0.25	1.2840	0.7788	3.0	20.086	0.0498
0.30	1.3499	0.7408	3.1	22.198	0.0450
0.35	1.4191	0.7047	3.2	24.533	0.0408
0.40	1.4918	0.6703	3.3	27.113	0.0369
0.45	1.5683	0.6376	3.4	29.964	0.0334
0.50	1.6487	0.6065	3.5	33.115	0.0302
0.55	1.7333	0.5769	3.6	36.598	0.0273
0.60	1.8221	0.5488	3.7	40.447	0.0247
0.65	1.9155	0.5220	3.8	44.701	0.0224
0.70	2.0138	0.4966	3.9	49.402	0.0202
0.75	2.1170	0.4724	4.0	54.598	0.0183
0.80	2.2255	0.4493	4.1	60.340	0.0166
0.85	2.3396	0.4274	4.2	66.686	0.0150
0.90	2.4596	0.4066	4.3	73.700	0.0136
0.95	2.5857	0.3867	4.4	81.451	0.0123
1.0	2.7183	0.3679	4.5	90.017	0.0111
1.1	3.0042	0.3329	4.6	99.484	0.0101
1.2	3.3201	0.3012	4.7	109.95	0.0091
1.3	3.6693	0.2725	4.8	121.51	0.0082
1.4	4.0552	0.2466	4.9	134.29	0.0074
1.5	4.4817	0.2231	5	148.41	0.0067
1.6	4.9530	0.2019	6	403.43	0.0025
1.7	5.4739	0.1827	7	1096.6	0.0009
1.8	6.0496	0.1653	8	2981.0	0.0003
1.9	6.6859	0.1496	9	8103.1	0.0001
2.0	7.3891	0.1353	10	22026	0.00005
2.1	8.1662	0.1225			
2.2	9.0250	0.1108			
2.3	9.9742	0.1003			
2.4	11.023	0.0907			

Table 3
Natural Logarithms of Numbers

n	$\log_e n$	n	$\log_e n$	n	$\log_e n$
0.0	*	4.5	1.5041	9.0	2.1972
0.1	7.6974	4.6	1.5261	9.1	2.2083
0.2	8.3906	4.7	1.5476	9.2	2.2192
0.3	8.7960	4.8	1.5686	9.3	2.2300
0.4	9.0837	4.9	1.5892	9.4	2.2407
0.5	9.3069	5.0	1.6094	9.5	2.2513
0.6	9.4892	5.1	1.6292	9.6	2.2618
0.7	9.6433	5.2	1.6487	9.7	2.2721
0.8	9.7769	5.3	1.6677	9.8	2.2824
0.9	9.8946	5.4	1.6864	9.9	2.2925
1.0	0.0000	5.5	1.7047	10	2.3026
1.1	0.0953	5.6	1.7228	11	2.3979
1.2	0.1823	5.7	1.7405	12	2.4849
1.3	0.2624	5.8	1.7579	13	2.5649
1.4	0.3365	5.9	1.7750	14	2.6391
1.5	0.4055	6.0	1.7918	15	2.7081
1.6	0.4700	6.1	1.8083	16	2.7726
1.7	0.5306	6.2	1.8245	17	2.8332
1.8	0.5878	6.3	1.8405	18	2.8904
1.9	0.6419	6.4	1.8563	19	2.9444
2.0	0.6931	6.5	1.8718	20	2.9957
2.1	0.7419	6.6	1.8871	25	3.2189
2.2	0.7885	6.7	1.9021	30	3.4012
2.3	0.8329	6.8	1.9169	35	3.5553
2.4	0.8755	6.9	1.9315	40	3.6889
2.5	0.9163	7.0	1.9459	45	3.8067
2.6	0.9555	7.1	1.9601	50	3.9120
2.7	0.9933	7.2	1.9741	55	4.0073
2.8	1.0296	7.3	1.9879	60	4.0943
2.9	1.0647	7.4	2.0015	65	4.1744
3.0	1.0986	7.5	2.0149	70	4.2485
3.1	1.1314	7.6	2.0281	75	4.3175
3.2	1.1632	7.7	2.0412	80	4.3820
3.3	1.1939	7.8	2.0541	85	4.4427
3.4	1.2238	7.9	2.0669	90	4.4998
3.5	1.2528	8.0	2.0794	95	4.5539
3.6	1.2809	8.1	2.0919	100	4.6052
3.7	1.3083	8.2	2.1041		
3.8	1.3350	8.3	2.1163		
3.9	1.3610	8.4	2.1282		
4.0	1.3863	8.5	2.1401		
4.1	1.4110	8.6	2.1518		
4.2	1.4351	8.7	2.1633		
4.3	1.4586	8.8	2.1748		
4.4	1.4816	8.9	2.1861		

Selected Answers

PROBLEM SET 1.3

1. $x < -3$
3. $x < \frac{1}{6}$
5. $x > -3$
7. $x > 13$
9. $x < 2$
11. $x < -3$

PROBLEM SET 1.4

1. $(-\infty, -1)$
3. $(-1, 1)$
5. $(0, 10)$
7. $(-\infty, -1]$ and $[3, \infty)$
9. a) $[0, \infty)$ b) $[-1, \infty)$
11. $(-\infty, \infty)$
13. $\{-2, \frac{8}{3}\}$
15. $\{ \}$
17. $(-\infty, \infty)$
19. $\{1\}$
21. $(-\infty, -2)$ and $(-\frac{2}{3}, \infty)$
23. $(-2, 2)$
25. $[\frac{5}{4}, \frac{7}{4}]$
27. $[\frac{5}{4}, \frac{7}{4}]$
29. $[-1, 3]$
31. $[\frac{1}{2}, 3)$

PROBLEM SET 2.2

1. a) $2\sqrt{2}$ b) $6\sqrt{2}$ c) $2\sqrt{2}$ d) $\sqrt{2}$
5. $\left(1 + \sqrt{3}, 1 - \sqrt{3}\right)$ and $\left(1 - \sqrt{3}, 1 + \sqrt{3}\right)$
7. $(\frac{25}{14}, \frac{5}{14}), \frac{5}{14}\sqrt{26}$
9. a) $|y|$ b) $|x|$
11. $(3, 0)$

PROBLEM SET 2.3

1. $x = 3$
3. $y = 4, x \geq 0$
5. $x^2 + y^2 + 4x - 4y + 4 = 0$
7. $y = 2x, x \geq -1$
9. $x^2 - 2x + y^2 - 3 = 0$
21. b) $y = x$
23. $x = \frac{3}{2}$

PROBLEM SET 2.4

1. $y - 1 = -\frac{1}{5}(x - 2), y = -\dfrac{x}{5} + \dfrac{7}{5}$
3. $y = \frac{3}{2}(x - 1), y = \frac{3}{2}x - \frac{3}{2}$
5. $4y + 3x = 25$
7. $(1, 1)$ and $(1, -1)$

733

PROBLEM SET 2.6

5. $F = (0, \frac{1}{12})$, $D:y = -\frac{1}{12}$

7. $F = (0, \frac{5}{4})$, $D:y = \frac{3}{4}$

9. $F = (2, \frac{5}{4})$, $D:y = \frac{3}{4}$

11. $F = (-1, \frac{1}{4})$, $D:y = -\frac{1}{4}$

13. $F = (-\frac{1}{2}, \frac{1}{16})$, $D:y = -\frac{1}{16}$

PROBLEM SET 2.7

9. crosses x-axis at $(0, 0)$, $(2, 0)$, $(-2, 0)$; tangent horizontal where $x = \pm\dfrac{2}{\sqrt{3}}$; slope at $(0, 0) = -4$; $y > 0$ if $-2 < x < 0$ or $2 < x$; $y < 0$ if $x < -2$ or $0 < x < 2$

PROBLEM SET 2.8

1. 14

3. 50

5. 33

7. $b_3 + 2c_3 + b_4 + 2c_4 + b_5 + 2c_5$

9. $m^7 + (m + 1)^7 + \cdots + n^7$

11. $\displaystyle\sum_{i=3}^{k} i^2$

13. $\displaystyle\sum_{i=2}^{k} \frac{1}{i}$

PROBLEM SET 2.9

5. $2n^2$

7. $\frac{1}{3}n(n + 1)(n - 1)$

PROBLEM SET 2.10

1. $\frac{320}{3}$

3. $\frac{40}{3}$

5. $\frac{80}{3}$

7. $\frac{5}{3}$

9. $\frac{9}{8}$

11. $\frac{8}{3}$

15. a) $n \geqq 4$ b) $n \geqq 2 \cdot 10^{10}$ c) $n > -\frac{1}{2} + \dfrac{1}{2\epsilon}\sqrt{\epsilon(8 - 3\epsilon)}$

$\left(n > \dfrac{2}{\epsilon} \text{ also works. Note that you were not asked to find the smallest possible } n.\right)$

17. a) $n > 98$ b) $n > \dfrac{1}{\epsilon} - 2$

PROBLEM SET 3.4

7. bounded with $M = 1$

9. bounded with $M = 1$

11. bounded with $M = 8$

13. bounded with $M = \frac{1}{2}$

15. bounded with $M = \frac{1}{2}$

17. unbounded

19. bounded with $M = 1$

PROBLEM SET 3.5

1. $70x^9 - 8x^7$

3. $\dfrac{1}{(x + 1)^2}$

5. $\dfrac{-2y^3 - 3}{(y^3 - 3)^2}$

7. $\dfrac{-1}{(x - 1)^2}$

9. $3(1 + x)^2$

11. a) $3ay + 2x$ b) $3xy + 3a^2$

13. $\dfrac{2(x^2 - 1)}{(x^2 + x + 1)^2}$

15. $4x^3 + 6x^2 + 2x$

17. $\dfrac{1}{2\sqrt{x + 1}}$

19. $\dfrac{-x}{\sqrt{1 - x^2}}$

21. $\dfrac{-1}{2\sqrt{x^3}}$

23. $\dfrac{1}{\sqrt{(1 - x^2)^3}}$

PROBLEM SET 3.6

1. $\dfrac{2x + 3}{2\sqrt{(x + 1)(x + 2)}}$

3. $\dfrac{-3x + 1}{(x + 1)^5}$

5. $712(x^3 + x^2 - x + 7)^{711}(3x^2 + 2x - 1)$ 7. $\dfrac{x - 1}{\sqrt{x(x - 2)}}$

9. $\dfrac{-3x^2 + 4x + 1}{2\sqrt{x - 1}(x^2 + 1)^2}$

11. $\dfrac{-1}{\sqrt{1 - x}\,\sqrt{(1 + x)^3}}$

13. 1 if $x > 0$, -1 if $x < 0$

15. $3(2x^3y - 3x^2y^2)(x^3y^2 - x^2y^3)^2$

19. $\frac{3}{2}(3x^2 + 1)(x^3 + x)^{1/2}$

21. $\frac{2}{3}f^{1/2}f'$

23. $\frac{5}{2}(x^2 + 3x + 1)^{3/2}(2x + 3)$

25. b) $\frac{1}{3}x^{-2/3}$

27. $\dfrac{p}{q}\,x^{(p/q) - 1}$

PROBLEM SET 3.7

1. a) $\dfrac{x^2}{2}$ b) $-\dfrac{x^2}{2}$ c) $\frac{1}{2}x\,|x|$ d) x e) $-x$ f) $|x|$

3. a) x b) $-x$ c) $|x|$ d) 1 e) -1 f) sig x

PROBLEM SET 3.8

1. $\frac{1}{4}$

3. a) $\dfrac{b^{11}}{11}$ b) $\dfrac{1}{11}(b^{11} - a^{11})$ c) $\dfrac{1}{101}(b^{101} - a^{101})$ d) $\dfrac{1}{n + 1}(x^{n+1} - a^{n+1})$

5. a) $-\frac{1}{4}$ b) $\dfrac{x^4}{4} + \dfrac{x^2}{2} - x$

7. a) $\sqrt{5} - \sqrt{2}$ b) 0

9. $(1 + x^{10})^{100}$ 11. $2(\sqrt{2} - 1)$

13. $\frac{15}{16}$ 15. $\frac{7}{54}$

PROBLEM SET 3.9

1. $\frac{3}{2}t^2 + 4t + 4$ 3. $-\frac{t^2}{2} + 2t + 3$ 5. $\frac{t^5}{20} + t - \frac{21}{20}$

7. $-\sqrt{1 - x^2}$ 9. $2\sqrt{t} - 2\sqrt{2} + 5$ 11. $\frac{t^3}{3} + t + \frac{2}{3}$

13. $\frac{-1}{3(1 + t^3)} + \frac{7}{6}$ 15. $\frac{20}{g}$, $a(t) = -g$ in the time interval $\left[0, \frac{20}{g}\right]$

17. $v_0 = -100 + g$ 19. $g_L = 6$ ft/sec

PROBLEM SET 3.10

1. $x^2, 2x$ 3. $x^4 - x, 4x^3 - 1$

5. $\sqrt{1 + x^8}, \quad \frac{4x^7}{\sqrt{1 + x^8}}$ 7. $\sqrt{x}, \quad \frac{1}{2\sqrt{x}}$

9. $\frac{1}{1 + x^2}, \quad \frac{-2x}{(1 + x^2)^2}$ 11. $-\sqrt{x}(x^2 + 1), -\frac{5}{2}x\sqrt{x} - \frac{1}{2\sqrt{x}}$

13. $-\frac{1}{x}, \frac{1}{x^2}$ 15. $\frac{1}{x^4 + 1}, \frac{-4x^3}{(x^4 + 1)^2}$

17. $\sqrt{\frac{1 + x}{1 - x}}, \frac{1}{\sqrt{1 + x}\sqrt{(1 - x)^3}}$ 19. $2\sqrt{1 + (2x)^8}$

PROBLEM SET 4.1

1. $\csc \theta$ 3. $\cot x$ 5. $\cos y$
7. $\tan \theta$ 9. $\tan \theta$ 11. $\sec^2 \theta$
13. $\tan x$ 15. $-\tan \theta$ 17. $\sec \theta$
19. $\tan \theta$ 21. $-\sec \theta$ 23. $\sin \theta$
25. $-\tan \theta$ 27. $-\sec \theta$

PROBLEM SET 4.2

9. a) -1 b) $-\cos \theta$

11. a) ∞ b) $-\cot \theta$ 13. $\cos \theta$ 15. $\left|\cos \frac{\theta}{2}\right|$

17. $\left|\sin \frac{\theta}{2}\right|$ 27. $4 \cos^3 \theta - 3 \cos \theta$

PROBLEM SET 4.3

1. $\sec^2 x$

3. $\sec x \tan x$

5. $-\sin x \dfrac{\cos x}{|\cos x|}$

7. $-2 \sin x \cos x$

9. $2 \cos 2x$

11. 0

13. $\dfrac{-1}{1 + \sin x}$

PROBLEM SET 4.4

7. 1

PROBLEM SET 4.5

1. $g(x) = \sin x, f(u) = u^2, f'(u) = 2u, g'(x) = \cos x, f'(g) = 2 \sin x, \varphi'(x) = 2 \sin x \cos x$

3. $g(x) = \sin x + \cos x, f(u) = u^2, f'(u) = 2u, g'(x) = \cos x - \sin x, f'(g) = 2(\sin x + \cos x), \varphi'(x) = 2 \cos 2x$

5. $g(x) = 2x, f(u) = \tan u, f'(u) = \sec^2 u, g'(x) = 2, f'(g) = \sec^2 2x, \varphi'(x) = 2 \sec^2 2x$

7. $g(x) = 1 - x^2, f(u) = \sqrt{u}, f'(u) = \dfrac{1}{2\sqrt{u}}, g'(x) = -2x, f'(g) = \dfrac{1}{2\sqrt{1 - x^2}},$

$\varphi'(x) = \dfrac{-x}{\sqrt{1 - x^2}}$

9. $g(x) = 1 + x, \ f(u) = u^{1/3}, \ f'(u) = \frac{1}{3}u^{-2/3}, \ g'(x) = 1, \ f'(g) = \frac{1}{3}(1 + x)^{-2/3}, \ \varphi'(x) = \frac{1}{3}(1 + x)^{-2/3}$

11. $g(x) = \cos x, f(u) = \int_0^u (t^2 + 1) \, dt, f'(u) = u^2 + 1, g'(x) = -\sin x, f'(g) = \cos^2 x + 1,$

$\varphi'(x) = -\sin x(\cos^2 x + 1)$

13. $-3x_0^2 \cos (x_0^3)$

15. $g(x) = x^3, \ f(u) = \sin u, \ f'(u) = \cos u, \ g'(x) = 3x^2, \ f'(g) = \cos x^3, \ \varphi'(x) = 3x^2 \cos x^3$

17. $\varphi'(x) = \cos x \sqrt{1 + \sin^2 x}$

19. 0 21. 0

PROBLEM SET 4.6

1. $2x \cos x^2$

3. $-3x^2 \sin x^3$

5. $2t \sec^2 (t^2 + 1)$

7. $(3x^2 + 1) \cos (x^3 + x)$

9. $\dfrac{-1}{2\sqrt{x}} \sin \sqrt{x}$

11. $\frac{1}{2} \sec^2 \left(\dfrac{x - 1}{2} \right)$

13. 1

15. a) $2 \sec^2 x \tan x$

b) $2x \sec x^2 \tan x^2$

17. $-2 \sin 2x$

19. $-2 \sin 2x$

21. $\cos x$

23. $\dfrac{1}{1 + \cos x}$

25. $\dfrac{-x}{\sqrt{x^2 + 1}} \sin \sqrt{x^2 + 1}$

27. a) $\frac{1}{3}x^{-2/3} \cos x^{1/3}$

b) $\frac{1}{3}(\sin x)^{-2/3} \cos x$

29. $\cos x$

31. $-\sin x \cos \cos x$

33. $\cos x \sec^2 \sin x$

35. $-\sin x \cos^3 x$

37. 0

39. 0

41. 0

43. 0

PROBLEM SET 4.7

1. $\dfrac{1}{\sqrt{2x - x^2}}$

3. $\dfrac{1}{2 + 2x + x^2}$

5. 1

7. $2x$

9. $\dfrac{-x}{|x|\sqrt{1 - x^2}}$

11. $\dfrac{2x}{2 + 2x^2 + x^4}$

13. $\dfrac{2}{x\sqrt{x^4 - 1}}$

15. $\dfrac{-1}{\sqrt{1 - x^2}}$

17. $\dfrac{1}{x\sqrt{x^2 - 1}}$

19. $\dfrac{x}{\sqrt{1 + x^2}}$

21. $\dfrac{1}{(1 + x^2)^{3/2}}$

23. $\dfrac{1}{(1 - x^2)^{3/2}}$

25. $\dfrac{1}{\sqrt{-x - x^2}}$

27. $\dfrac{2x}{\sqrt{1 - x^4}}$

1. $\dfrac{\pi}{3}$

33. $1 - \dfrac{1}{\sqrt{2}}$

35. $f^{-1}(x) = \sqrt{1 - x^2},\ 0 \leqq x \leqq 1$

37. $\dfrac{\pi}{3}$

PROBLEM SET 4.9

1. $\log_e x + 1,\quad \dfrac{1}{x}$

3. $(1 + 2x)e^{2x},\ (4 + 4x)e^{2x}$

5. $(\sin x + \cos x)e^x,\quad 2e^x \cos x$

7. $2/x,\ -2/x^2$

9. $\dfrac{500}{x},\ \dfrac{-500}{x^2}$

11. $(\log_e 10)\, 10^x,\ (\log_e 10)^2\, 10^x$

13. $0, 0$

15. $(2x + x^2)e^x,\ (2 + 4x + x^2)e^x$

17. $(2x^2 + 1)e^{x^2},\ (4x^3 + 6x)e^{x^2}$

19. $\dfrac{-1}{1 - x},\ \dfrac{-1}{(1 - x)^2}$

21. $e^{x-1},\ e^{x-1}$

23. $\tan x,\ \sec^2 x$

25. $\cot x,\ -\csc^2 x$

27. $\sec x,\ \sec x \tan x$

29. $-\csc x,\ \csc x \cot x$

PROBLEM SET 4.10

1. $\dfrac{2}{x} \ln x$

3. $\dfrac{2x}{x^2 + 1}$

5. $2x \exp x^2$

7. $2x$

9. $\cos x \exp \sin x$

11. $(\ln x + 1) \exp (x \ln x)$

13. $\cot x$

15. $(\ln x + 1)x^x$

19. 2

21. $\sec^3 x$

23. 1

25. 2

27. 2

29. 2

PROBLEM SET 4.11

23. $4 \cosh^3 x - 3 \cosh x$ 31. $\sqrt{1 + x^2}$ 33. $\dfrac{2}{\sqrt{1 + 4x^2}}$

39. $\dfrac{2x}{\sqrt{x^4 - 1}}$ 43. $x = \ln 2$ 45. $x = \ln (2y)$

47. $\text{Cosh}^{-1}(x) = \ln \left(x + \sqrt{x^2 - 1} \right)$

PROBLEM SET 5.1

1. increasing on $[-\pi/2, \pi/2]$, decreasing on $[-\pi, -\pi/2]$ and $[\pi/2, \pi]$

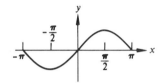

3. increasing on $[-2, 0]$, decreasing on $[0, 2]$

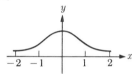

5. increasing on $[-2, -1]$ and $[1, 2]$, decreasing on $[-1, 1]$

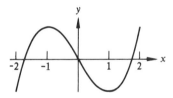

7. increasing on $[-\pi, -3\pi/4]$, $[-\pi/4, \pi/4]$, and $[3\pi/4, 5\pi/4]$, decreasing on $[-3\pi/4, -\pi/4]$ and $[\pi/4, 3\pi/4]$

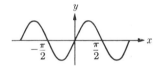

9. increasing on $[0, 1]$

11. increasing on $[\pi, 2\pi]$, decreasing on $[0, \pi]$

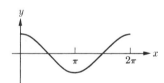

13. increasing on $[-2, -1]$ and $[0, 1]$, decreasing on $[-1, 0]$ and $[1, 2]$

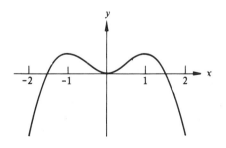

15. increasing on $[-1, 0]$, decreasing on $[0, 1]$

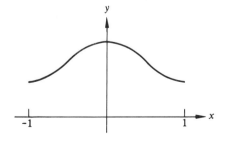

17. increasing on $[-\pi, -\pi/2]$ and $[\pi/2, \pi]$, decreasing on $[-\pi/2, \pi/2]$

19. increasing on $[\ln 2, 2]$, decreasing on $[0, \ln 2]$

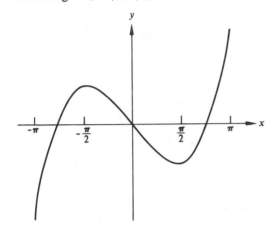

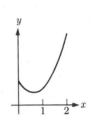

PROBLEM SET 5.2

1. local maxima $-\pi$, $\pi/2$; local minima $-\pi/2$, π; maximum $\pi/2$; minimum $-\pi/2$; inflection point 0; image $[-1, 1]$; concave upward $[-\pi, 0]$; concave downward $[0, \pi]$

3. local maximum 1; local minima -2, 2; maximum 1; minima -2, 2; inflection points $-1/\sqrt{3}$, $1/\sqrt{3}$; image $[\frac{1}{5}, 1]$; concave upward $\left[-2, -1/\sqrt{3}\right]$ and $\left[1/\sqrt{3}, 2\right]$; concave downward $\left[-1/\sqrt{3}, 1/\sqrt{3}\right]$

5. local maxima -1, 2; local minima -2, 1; maxima -1, 2; minima -2, 1; inflection point 0; image $[-2, 2]$; concave upward $[0, 2]$; concave downward $[-2, 0]$

7. local maxima $-3\pi/4$, $\pi/4$, π; local minima $-\pi$, $-\pi/4$, $3\pi/4$; maxima $-3\pi/4$, $\pi/4$; minima $-\pi/4$, $3\pi/4$; inflection points $-\pi/2$, 0, $\pi/2$; image $[-1, 1]$; concave upward $[-\pi/2, 0]$ and $[\pi/2, \pi]$; concave downward $[-\pi, -\pi/2]$ and $[0, \pi/2]$

9. local maximum 1; local minimum 0; maximum 1; minimum 0; inflection points none; image $[1, e-2]$; concave upward $[0, 1]$; concave downward $\{\}$

11. local maxima 0, 2π; local minimum π; maxima 0, 2π; minimum π; inflection points $\pi/2$, $3\pi/2$; image $[-1, 1]$; concave upward $[\pi/2, 3\pi/2]$; concave downward $[0, \pi/2]$ and $[3\pi/2, 2\pi]$

13. local maxima -1, 1; local minima -2, 0, 2; maxima -1, 1; minima -2, 2; inflection points $-\sqrt{\frac{1}{3}}$, $+\sqrt{\frac{1}{3}}$; image $[-8, 1]$; concave upward $\left[-\sqrt{\frac{1}{3}}, \sqrt{\frac{1}{3}}\right]$; concave downward $\left[-2, -\sqrt{\frac{1}{3}}\right]$ and $\left[\sqrt{\frac{1}{3}}, 2\right]$

15. local maximum 0; local minima -1, 1; maximum 0; minima -1, 1; inflection points $-\sqrt[4]{\frac{3}{5}}$, $\sqrt[4]{\frac{3}{5}}$; image $[\frac{1}{2}, 1]$; concave upward $\left[-1, -\sqrt[4]{\frac{3}{5}}\right]$ and $\left[\sqrt[4]{\frac{3}{5}}, 1\right]$; concave downward $\left[-\sqrt[4]{\frac{3}{5}}, \sqrt[4]{\frac{3}{5}}\right]$

17. local maxima $-\pi/2$, π; local minima $-\pi$, $\pi/2$; maximum π; minimum $-\pi$; inflection point 0; image $[-\pi, \pi]$; convex upward $[0, \pi]$; concave downward $[-\pi, 0]$

19. local maxima 0, 2; local minimum $\ln 2$; maximum 2; minimum $\ln 2$; inflection points none; image $[2 - 2\ln 2, e^2 - 4]$; concave upward $[1, 2]$; concave downward $\{\}$

21. No

PROBLEM SET 5.3

1. maxima none; minima none; local maximum 1; concave upward $(-\infty, 0)$ and $(2, \infty)$; concave downward $(0, 2)$; inflection points none;
$$\lim_{x\to\infty} f(x) = 0, \quad \lim_{x\to-\infty} f(x) = 0, \quad \lim_{x\to0+} f(x) = -\infty, \quad \lim_{x\to0-} f(x) = \infty, \quad \lim_{x\to2+} f(x) = \infty,$$
$$\lim_{x\to2-} f(x) = -\infty$$

3. maxima none; minima none; local maximum $\frac{1}{2}$; concave upward $(-\infty, -2)$ and $(3, \infty)$; concave downward $(-2, 3)$; inflection points none;
$$\lim_{x\to\infty} f(x) = 0, \quad \lim_{x\to-\infty} f(x) = 0, \quad \lim_{x\to-2+} f(x) = -\infty, \quad \lim_{x\to-2-} f(x) = \infty,$$
$$\lim_{x\to3+} f(x) = \infty, \quad \lim_{x\to3-} f(x) = -\infty$$

5. maxima none; minima none; local maxima none; concave upward $(-\infty, 0)$ and $(0, \infty)$; concave downward $\{\}$; inflection points none;
$$\lim_{x\to\infty} f(x) = 0, \quad \lim_{x\to-\infty} f(x) = 0, \quad \lim_{x\to0+} f(x) = \infty, \quad \lim_{x\to0-} f(x) = \infty$$

7. maxima none; minimum 0; local maxima none; concave upward $[-1/\sqrt{3}, 1/\sqrt{3}]$; concave downward $(-\infty, -1/\sqrt{3}]$ and $[1/\sqrt{3}, \infty)$; inflection points $-1/\sqrt{3}, 1/\sqrt{3}$;
$$\lim_{x\to\infty} f(x) = 1, \lim_{x\to-\infty} f(x) = 1$$

9. maxima $\frac{1}{2}$; minima none; local maxima $\frac{1}{2}$; no local minima; concave upward $(-\infty, 0]$ and $[1, \infty)$; concave downward $[0, 1]$; inflection points 0, 1;
$$\lim_{x\to-\infty} f(x) = 0, \quad \lim_{x\to\infty} f(x) = 0, \quad \lim_{x\to-1-} f(x) = \frac{1}{3}; \quad \lim_{x\to-1+} f(x) = \frac{1}{3}$$

11. no maxima; no minima; no local maxima; no local minima; concave upward $(-\infty, -1)$ and $[0, \sqrt[3]{\frac{1}{2}}]$; concave downward $(-1, 0]$ and $[\sqrt[3]{\frac{1}{2}}, \infty)$; inflection points $0, \sqrt[3]{\frac{1}{2}}$;
$$\lim_{x\to-\infty} f(x) = 1, \quad \lim_{x\to\infty} f(x) = 1, \quad \lim_{x\to-1-} f(x) = \infty, \quad \lim_{x\to-1+} f(x) = -\infty$$

13. e 15. e 17. e

19. $e^{1/2}$ 21. e 23. $e^{2/3}$

25. 0   27. 0   29. $\$e$

PROBLEM SET 5.4

1. a^2 3. $2a^2$ 5. $\dfrac{k^2}{8}$

7. $x = \dfrac{2}{\sqrt{3}} a, \ y = \dfrac{2\sqrt{2}}{\sqrt{3}} a$ 9. 128 in.³   11. $r = 3\sqrt{\dfrac{10}{\pi}}, \ h = 3\sqrt{\dfrac{10}{\pi}}$

13. $\dfrac{w^2}{8}$ 15. $\dfrac{k^3}{1728}$ 17. $\dfrac{8}{5\sqrt[4]{5}}$

19. 2   21. $\dfrac{1}{\sqrt[4]{3}}$ 23. $\dfrac{\pi}{2} + 2\pi k, \ k$ an integer

25. $-\dfrac{1}{\sqrt[4]{3}}$ 27. $-\dfrac{\pi}{2} + 2\pi k, \ k$ an integer   33. $\dfrac{\pi d^2 h}{27}$

35. $\dfrac{4\pi r^3}{3\sqrt{3}}$ 39. $(3 + 2\sqrt{2})r^2$

PROBLEM SET 5.5

1. $\frac{1}{2}$ 3. $\frac{1}{2}$ 5. 1   7. $1/\sqrt{3}$

9. h/d 11. $\sqrt{2}$ 13. $4x^3 + 4ff' = 0$

PROBLEM SET 5.6

1. $\sqrt{2}$ 3. 2

PROBLEM SET 5.8

1. $e^g (= e^{x^2})$

3. $\exp \mathrm{Tan}^{-1} g \; \dfrac{1}{1 + g^2} \; (= e^x \cos^2 x)$

5. $\dfrac{1}{3x^2} (\sin x + x \cos x)e^{x \sin x}$

7. $\dfrac{3}{4x} e^{x^4}$

9. $3x^2 \cos^2 x$

11. $\varphi(u) = e^{u^2/4}, \; \varphi'(u) = (u/2)e^{u^2/4}, \; \varphi'(g) = xe^{x^2}$

13. $\varphi(u) = e^{u \sin u}, \; \varphi'(u) = (\sin u + u \cos u)e^{u \sin u}, \; \varphi'(g) = (\sin x + x \cos x)e^{x \sin x}$

15. $\varphi(u) = e^{u^3/2}, \; \varphi'(u) = \frac{3}{2}u^{1/2}e^{u^3/2}, \; \varphi'(g) = \frac{3}{2} \times e^{x^3}$

17. $\varphi(u) = \sqrt{1 - u^2}, \; \varphi'(u) = - \dfrac{u}{\sqrt{1 - u^2}}, \; \varphi'(g) = - \dfrac{\cos x}{\sin x}$

19. $\cos (t)/e^t$ 21. $2t^3$ 23. $\dfrac{1}{xe^x}$

25. $\varphi(u) = \dfrac{1}{\sqrt{1 + u^2}}, \; \varphi'(u) = \dfrac{-u}{(1 + u^2)^{3/2}}, \; \varphi'(g) = -\tan t \cos^3 t = -\sin t \cos^2 t$

27. $\varphi(u) = (\mathrm{Tan}^{-1} u)^6, \; \varphi'(u) = \dfrac{6}{1 + u^2} (\mathrm{Tan}^{-1} u)^5, \; \varphi'(g) = \dfrac{6t^5}{1 + (\tan t)^5}$

33. $df/dt = -t/f$ for any function $f(t)$ such that $f^2 + t^2 + 1 = 0$

35. $df/dt = -t^3/f^3, f_1(t) = (1 - t^4)^{1/4}, f_1'(t) = \frac{1}{4}(-4t^3)(1 - t^4)^{-3/4} = -t^3/f_1^3,$
$f_2(t) = -(1 - t^4)^{1/4}, f_2'(t) = \frac{1}{4}(4t^4)(1 - t^4)^{-3/4} = -t^3/f_2^3$

PROBLEM SET 6.2

1. $\{\frac{1}{8}(1 + x^2)^4 + C\}$

3. $\{\frac{3}{4}(2 + u^2)^2 + C\}$

5. $\{x^7/7 + \frac{3}{5}x^5t^2 + x^3t^4 + xt^6 + C\}$

7. $\{(x/8)(x^2 + t^2)^4 + C\}$

9. $\{\frac{2}{33}(t^{3/2} + 5)^{11} + C\}$

11. $\{\frac{1}{3}(1 + \sin x)^3 + C\}$

13. $\{-\frac{2}{3}(\cos x)^{3/2} + C\}$

15. $\{\frac{1}{2}e^{2x} - 6e^x + 12x + 8e^{-x} + C\}$

17. $\{\frac{1}{3}e^{3x} + 3e^x - 3e^{-x} - \frac{1}{3}e^{-3x} + C\}$

19. $\{-\frac{1}{6}(1 + x^3)^{-2} + C_1\}, x > -1; \{-\frac{1}{6}(1 + x^3)^{-2} + C_2\}, x < -1$

21. $\{\frac{1}{3} \ln (1 + x^3) + C_1\}, x > -1; \{\frac{1}{3} \ln (-1 - x^3) + C_2\}, x < -1$

23. $\{\ln^2(x^2) + C_1\}, x > 0; \{\ln^2(x^2) + C_2\}, x < 0$

25. $\{\frac{1}{3} \sin^3 x + C\}$ 27. $\{\frac{1}{102} \sin^{102} x + C\}$ 29. $\{-\frac{1}{4} \cos^4 x + C\}$

31. $\{\tan \theta + C\}$ 33. $\{-\cot \theta + C\}$ 35. $\{\sec \theta + C\}$

37. $\{-\frac{1}{2} \cos 2\theta + C\}$ 39. $\{\frac{1}{2} \sin 2\theta + C\}$ 41. $\{\frac{1}{2} \sin 2\theta + C\}$

43. $\{\frac{1}{2}\sin 2\theta + C\}$ 45. $\left\{\frac{\theta}{2} - \frac{1}{4}\sin 2\theta + C\right\}$ 47. $\left\{\frac{\theta}{8} - \frac{1}{32}\sin 4\theta + C\right\}$

49. $\{-\frac{1}{3}\cos^3\theta + C\}$ 51. $\{2\sin\theta/2 + C\}$ 53. $\{\frac{2}{3}(1 - \cos\theta)^{3/2} + C\}$

55. $\{\frac{1}{3}e^{t^3} + C\}$ 57. $\{\frac{1}{2}e^{2x} + C\}$ 59. $\{\frac{1}{2}e^{t^2} + C\}$

61. $\{\tan x + C\}$ 63. $\{-e^{\cos t} + C\}$ 65. $\left\{\frac{1}{2\ln 10}10^{x^2} + C\right\}$

67. $\{-2(2 + t)^{-1/2} + C\}$ 69. $\{2t - 2t^{-1/2} + C\}$

71. $\{\mathrm{Sin}^{-1}(t) + C\},\ |t| < 1$ 73. $\{\frac{1}{2}(1 + t^3)^{2/3} + C\}$

75. $\{\frac{1}{4}\ln(1 + t^4) + C\}$ 77. $\{-2\sqrt{1 - e^x} + C\}$

79. $\{\ln|\sin x| + C\}$ on any interval where $\sin x \neq 0$

81. $\{\ln|\sec x| + C\}$ on any interval where $\cos x \neq 0$

83. $\{-\ln|\csc x + \cot x| + C\}$ on any interval where $\csc x + \cot x \neq 0$

PROBLEM SET 6.3

1. $\{\frac{1}{2}\mathrm{Sin}^{-1}(x^2) + C\}$ 3. $\{\frac{1}{4}\mathrm{Sin}^{-1}(2y^2) + C\}$

5. $\{\frac{1}{6}\mathrm{Tan}^{-1}(3x^2) + C\}$ 7. $\{\frac{1}{4}\ln(9 + x^4) + C\}$

9. $\{\frac{1}{3}\mathrm{Tan}^{-1}(z^3) + C\}$ 11. $\{(1/3\sqrt{2})\mathrm{Tan}^{-1}(\sqrt{2}\,z^3) + C\}$

13. $\{\frac{1}{6}\ln(5 + z^6) + C\}$ 15. $\{-\frac{1}{4}\sqrt{1 - x^8} + C\}$

17. $\{\frac{1}{8}\ln(1 + x^8) + C\}$ 19. $\{\ln(1 + e^z) + C\}$

21. $\{\mathrm{Tan}^{-1}e^x + C\}$ 23. $\{\ln|e^x + e^{-x}| + C\}$ on $(-\infty, 0)$ or $(0, \infty)$

25. $\{(e^x - e^{-x})^3 - e^x + e^{-x} + C\}$ 27. $\{(1/3)\mathrm{Sin}^{-1}(x^3/\sqrt{2}) + C\}$

29. $\{(1/3\sqrt{2})\mathrm{Sin}^{-1}(\sqrt{2}\,x^3) + C\}$ 31. $\{\frac{1}{2}(\mathrm{Sin}^{-1}x)^2 + C\}$

33. $\{xe^x + C\}$ 35. $\left\{\frac{x^3}{3} + C\right\}$

37. $\{x\sin x + C\}$ 39. $\{-x\cos x + C\}$

41. $\left\{\frac{x^2}{2}\ln x + C\right\}$ 43. $\{x^3\ln x + c\}$

45. $\{\frac{1}{4}\ln^4 x + C\}$

47. $\{\frac{1}{4}\ln^2|x^2 + 2x| + C\}$ on any interval where $x^2 + 2x \neq 0$

49. $\{\frac{1}{2}\mathrm{Sec}^{-1}(x^2) + C\}$ 51. $\{x\mathrm{Sin}^{-1}x + C\}$

53. $\{x\mathrm{Sin}^{-1}x + \sqrt{1 - x^2} + C\}$ 55. $\{x\mathrm{Cos}^{-1}x - \sqrt{1 - x^2} + C\}$

57. $\{\frac{1}{2}\ln(1 + e^{2u}) + C\}$ 59. $\{\frac{1}{2}(e^{2u} - \ln(1 + e^{2u})) + C\}$

61. $\{\frac{1}{2}\ln(1 + x^2) + C\}$ 63. $\{\frac{1}{2}(\mathrm{Sin}^{-1}(z) + z\sqrt{1 - z^2}) + C\}$

PROBLEM SET 6.4

1. $\{x\ln^2 x - 2x\ln x + 2x + C\}$ 3. $\{a(x - 1)e^x + C\}$

5. $\{-(x/a)\cos ax + (1/a^2)\sin ax + C\}$ 7. $\{[1/(1 + a^2)]e^{ax}(-\cos x + a\sin x) + C\}$

9. $\{[1/(a^2 + b^2)]e^{ax}(a\sin bx - b\cos bx) + C\}$

11. $\{-x^2 \cos x + 2x \sin x + 2 \cos x + C\}$

13. $\{x^3 e^x - 3x^2 e^x + 6x e^x - 6e^x + C\}$ 15. $\{(x^3/3)(\ln^2 x - \frac{2}{3} \ln x + \frac{2}{9}) + C\}$

17. $\{x \operatorname{Sin}^{-1} x + \sqrt{1 - x^2} + C\}$ 19. $\{x \operatorname{Tan}^{-1} x - \frac{1}{2} \ln (1 + x^2) + C\}$

21. $\{\frac{1}{2}(x \sin x e^x + e^x \cos x - x e^x \cos x) + C\}$

23. $\{x \ln^3 x - 3x \ln^2 x + 6x \ln x - 6x + C\}$

25. $\int x^n e^x \, dx = x^n e^x - n \int x^{n-1} e^x \, dx$

27. $\{(x/2)(\sin \ln x - \cos \ln x) + C\}$

PROBLEM SET 6.5

1. $\{\frac{1}{3} \sin^3 x - \frac{1}{5} \sin^5 x + C\}$ 3. $\{\frac{1}{2}x - \frac{1}{4} \sin 2x + C\}$

5. $\{\frac{1}{8}x - \frac{1}{32} \sin 4x + C\}$ 7. $\{x + \cot x - \frac{1}{3} \cot^3 x + C\}$

9. $\{-\frac{1}{4} \cot^4 x + \frac{1}{2} \cot^2 x - \ln |\cos x| + C\}$, $\cos x \neq 0$

11. $\{-\frac{1}{3} \cot^3 x - \cot x + C\}$

13. $\{-1/2(\csc x \cot x + \ln |\csc x + \cot x|) + C\}$, $\cot x + \csc x \neq 0$

15. $\{-\frac{1}{4} \csc^3 x \cot x + \frac{3}{8} \csc x \cot x + \frac{3}{8} \ln |\csc x + \cot x| + C\}$, $\cot x + \csc x \neq 0$

17. $\{\ln |\sin x| + C\}$ 19. $\{\frac{1}{2} \sin^4 x + C\}$

21. $\left\{\dfrac{x}{2} + \frac{1}{8} \sin (4x) + C\right\}$ 23. $\{\ln|\sec x + \tan x| + C\}$

25. $\{- \csc x + C\}$ 27. $\{- \cot x - \frac{1}{3} \cot^3 x + C\}$

29. $\{+ \ln |\sec x + \tan x| - \sin x + C\}$ 31. $A = \dfrac{1}{n}, B = \dfrac{n-1}{n}$

PROBLEM SET 6.6

1. $\{x/\sqrt{1 - x^2} + C\}$ 3. $\{\ln |x + \sqrt{x^2 - 1}| + C\}$

5. $\{\operatorname{Sec}^{-1} x + C\}$ 7. $\{-1/x - \operatorname{Tan}^{-1} x + C\}$

9. $\{\frac{1}{2}(\operatorname{Sin}^{-1} x + x\sqrt{1 - x^2}) + C\}$ 11. $\{\sqrt{x^2 - 1} + C\}$

13. $\{\frac{1}{2}(\operatorname{Sin}^{-1} x - x\sqrt{1 - x^2}) + C\}$ 15. $\{x - \operatorname{Tan}^{-1} x + C\}$

17. $\{-\frac{1}{3}(1 - x^2)^{3/2} + C\}$ 19. $\{\frac{1}{3}(1 + x^2)^{3/2} + C\}$

21. $\{\frac{1}{2}x^2 - \frac{1}{2} \ln (1 + x^2) + C\}$

PROBLEM SET 6.7

1. $\left\{\left(\dfrac{-1 - 2\sqrt{x}}{(1 + \sqrt{x})^2}\right) + C\right\}$ 3. $\left\{\dfrac{x}{a^2\sqrt{a^2 - x^2}} + C\right\}$

5. $\{x - 2 \ln (\sqrt{e^x + 1} + 1) + C\}$

7. $\{\frac{3}{2}(1 + \sqrt[3]{x})^2 - 6(1 + \sqrt[3]{x}) + 3 \ln |1 + \sqrt[3]{x}| + C\}$

9. $\{\frac{4}{3}(\sqrt{\sqrt{x} + 1})^3 - 4\sqrt{\sqrt{x} + 1} + C\}$ 11. $\{\sqrt{z^2 - 1} + \frac{1}{3}(\sqrt{z^2 - 1})^3 + C\}$

13. $\left\{\frac{6}{5}\left(\sqrt[3]{1 + \sqrt{x}}\right)^5 - 3\left(\sqrt[3]{1 + \sqrt{x}}\right)^2 + C\right\}$ 15. $\left\{x - \frac{4}{3}x^3 + \frac{6}{5}x^5 - \frac{4}{7}x^7 + \frac{1}{9}x^9 + C\right\}$

17. $\left\{x - \ln\left(\sqrt{1 + e^{2x}} + 1\right) + C\right\}$ 19. $\left\{2x - \ln\left(\sqrt{1 + e^{4x}} + 1\right) + C\right\}$

21. $\left\{\frac{1}{2}x^2 \ln|x| - \frac{1}{4}x^2 + C\right\}$ 23. $\left\{x \operatorname{Tan}^{-1} x - \frac{1}{2}\ln\left(1 + x^2\right) + C\right\}$

25. $\left\{2\ln\left(1 + \sqrt{x}\right) + 2\left(1 + \sqrt{x}\right)^{-1} + C\right\}$

27. $\left\{4\left(\frac{1}{3}x^{3/4} - \frac{1}{2}x^{1/2} + x^{1/4} - \ln\left(1 + x^{1/4}\right)\right) + C\right\}$

29. $\left\{x - \ln\left(1 + e^x\right) + \dfrac{1}{1 + e^x} + C\right\}$

31. $\left\{\ln|x| - \dfrac{1}{x} - \dfrac{1}{2x^2} - \ln|1 - x| + C\right\}$

33. $\left\{\frac{1}{3}x^3 \operatorname{Tan}^{-1} x - \dfrac{x^2}{6} + \frac{1}{6}\ln\left(1 + x^2\right) + C\right\}$

PROBLEM SET 6.8

1. $\left\{\dfrac{1}{2}\operatorname{Tan}^{-1}\left(\dfrac{x + 1}{2}\right) + C\right\}$ 3. $\left\{\dfrac{1}{4}\ln\left|\dfrac{x - 2}{x + 2}\right| + C\right\}$

5. $\left\{\ln\left|\dfrac{2\sqrt{x^2 + x - 4}}{\sqrt{17}} + \dfrac{2\left(x + \frac{1}{2}\right)}{\sqrt{17}}\right| + C\right\}$ 7. $\left\{\operatorname{Sin}^{-1}\left(\dfrac{x + 1}{\sqrt{3}}\right) + C\right\}$

9. $\left\{\operatorname{Tan}^{-1}(x + 3) + C\right\}$ 11. $\left\{\operatorname{Tan}^{-1}(x - 3) + C\right\}$

13. $\left\{2\ln\left|\dfrac{x - 2}{x - 1}\right| + C\right\}$ 15. $\left\{\ln\left|\dfrac{x - 2}{x - 1}\right| + C\right\}$

17. $\left\{\dfrac{1}{x - 1} - \ln|x - 1| + \ln|x - 2| + C\right\}$

19. $\left\{-\dfrac{1}{x} + 2\ln x - \dfrac{1}{x - 1} - 2\ln|x - 1| + C\right\}$

21. $\left\{27\ln|x - 2| - 27\ln|x + 1| - \dfrac{1}{6(x - 2)^2} + \dfrac{1}{9(x - 2)} + C\right\}$

23. $\left\{\ln|x| - \frac{1}{2}\ln(x^2 + 1) + C\right\}$

25. $\left\{\ln|x| + \dfrac{1}{2(x^2 + 1)} - \frac{1}{2}\ln(x^2 + 1) + C\right\}$

27. $\sin\theta = \dfrac{2x}{1 + x^2}, \cos\theta = \dfrac{1 - x^2}{1 + x^2}$

29. $\{-\sec\theta + \tan\theta + C\}$

31. $\left\{-\dfrac{1}{x + 3} + C\right\}$

33. $\{\ln(1 + \sin x) + C\}, \quad \sin x \neq -1$

35. $\{\tan x + \sec x + C\}$

37. $\left\{\dfrac{\sqrt{2}}{2}\ln|\sec(\theta + \pi/4) + \tan(\theta + \pi/4)| + C\right\}$

PROBLEM SET 7.1

1. $\frac{8}{27}((\frac{11}{2})^{3/2} - 1)$ 3. $\ln\left(\sqrt{2} + 1\right)$ 5. $\frac{2}{3}(5^{3/2} - 1)$

7. a) $\frac{1}{2}(e - 1/e)$ b) $\frac{1}{2a}(e^a - e^{-a})$

PROBLEM SET 7.3

1. $f(x) = (r/h)x$,   volume $= \pi r^2 h/3$ 3. $\pi/2$ 5. $\pi/4$

7. $\pi/2$ 9. $\pi^3/4 - 2\pi$ 11. $\frac{2\pi}{3}\left(1 - \frac{1}{\sqrt{2}}\right)$

13. (b) $x = 1$

PROBLEM SET 7.4

1. $40\pi^2$ 3. 32π 5. a) $\pi(e - 2)$

7. a) $\frac{\pi}{2}(e^2 + 1)$ 9. $2\sqrt{3}\pi$ 11. $\frac{4\pi}{3}(\sqrt{2} - 1)$

PROBLEM SET 7.5

1. $2\pi a^2$ 3. $4\pi^2 ab$ 5. $4\pi(a + k)k^2$

7. $4\pi k^3$ 9. $16\pi ak$ 13. $10\sqrt{2}\,\pi^2$

PROBLEM SET 7.6

1. $(\bar{x}, \bar{y}) = \left(\frac{a + b}{3}, \frac{c}{3}\right)$ 5. $\frac{\pi}{3}ac^2$

7. $\pi c(\sqrt{b^2 + c^2} + \sqrt{(a - b)^2 + c^2})$ 9. $(\bar{x}, \bar{y}) = \left(\frac{b}{2}, \frac{c}{6}\frac{(3b - 4a)}{(b - a)}\right)$

11. $\pi cb(b - a)$ 13. $4\pi^2 ab$

15. $8\sqrt{2}\,\pi k(a + b)$ 17. $8\sqrt{2}\,\pi k^2$

19. $a = \frac{1}{4\pi}(-\pi + \sqrt{\pi^4 - 8\pi^2 + 120\pi})$

PROBLEM SET 7.7

1. $\pi/4$ 3. $\frac{1}{2}$ 5. 2
7. 10,000   9. ∞ 11. ∞
13. $\frac{1}{2}$ 15. 1   17. 2
19. $\pi/8$ 23. not finite   25. finite
27. finite   29. not finite   31. not finite
33. not finite   35. finite

PROBLEM SET 8.1

1. $x' = x - 5, y' = y - 6$
3. $x' = x + \frac{1}{2}, y' = y + \frac{1}{2}$
5. $x' = x + 2, y' = y + 1$
7. $x' = x + \frac{1}{3}, y' = y + \frac{1}{3}$

PROBLEM SET 8.2

1. $x^2/4 + y^2/3 = 1$
3. $(x - 1)^2/3 + (y - 3)^2/4 = 1$
5. $3x^2 + 2xy + 3y^2 = 8$
7. $x^2/5 + y^2/9 = 1$
9. foci $(\pm\sqrt{3}, 0)$; focal sum 4
11. foci $\left(-2, 1 + \sqrt{8}\right), \left(-2, 1 - \sqrt{8}\right)$; focal sum 6
13. foci $(0, \pm\sqrt{3}/2)$; focal sum 2

PROBLEM SET 8.3

11. $4x^2/9 - 4y^2/7 = 1$
13. $-4x^2/7 + 4(y - 2)^2/9 = 1$
15. $xy = \frac{1}{2}$
17. $x^2 - y^2/3 = 1$
19. $-x^2/4 + y^2/5 = 1$
23. $x + y = 1, x - y = 1$
25. $x = 0, y = 0$

PROBLEM SET 8.4

1. hyperbola
3. hyperbola
5. straight line
7. $A' = 2 \pm \sqrt{2}, C' = 2 \mp \sqrt{2}$
9. $A' = \frac{5}{2} \pm \sqrt{3}, C' = \frac{5}{2} \mp \sqrt{3}$

PROBLEM SET 9.1

11. $x = b \cot\theta + a \cos\theta, y = b + a \sin\theta$
13. $x = 2a \cos\theta, y = 0$

PROBLEM SET 9.2

1. 1
3. 1
5. 1
7. 1
9. $-8/(\pi^3 - 8)$
11. 0
13. 0
15. 1
17. 1
19. 2
21. $\frac{1}{e}$
23. e
25. 0
27. 0
29. $x = a\theta + a \sin\theta, y = a - a \cos\theta$
31. $x = (a + b) \cos\theta + b \cos\left(\dfrac{a + b}{b}\right)\theta, y = (a + b) \sin\theta + b \sin\left(\dfrac{a + b}{b}\right)\theta$

PROBLEM SET 9.3

1. 0
3. 0
5. 0
9. 1
11. 0
13. e^{-2}
15. 0
17. e
19. 0
21. 0
23. $-\infty$
27. 1
29. e^k
31. 1
33. e^{-3}

PROBLEM SET 9.4

1. $y = 2$

3. $x + y = \sqrt{2}$

7. $y = x^3$

9. $(x^2 + y^2)^2 = 2xy$

13. $2x + y^2 - 1 = 0$

19. $x - y = 1$

29. $(x^2 + y^2)^2 = a^2(x^2 - y^2)$

31. $r = 1/(1 + \sin \theta)$

33. $3r^2 - 16r \cos \theta - 16r \sin \theta + 32 = 0$

PROBLEM SET 9.5

1. $3\pi/2$

3. $\pi/4$

5. $\frac{1}{2}$

7. $\frac{1}{4}$

9. 2

11. 2

13. $\frac{1}{8}(e^{8\pi} - 1)$

PROBLEM SET 9.6

1. π

3. 3

5. 2π

7. $\frac{2}{3}\left((1 + \frac{9}{4})^{3/2} - 1\right)$

9. $\frac{1}{2}\sqrt{5} + \frac{1}{4} \ln\left(2 + 2\sqrt{5}\right)$

11. $\frac{3}{4}$

PROBLEM SET 9.7

23. $j = \frac{1}{2}c - \frac{1}{2}d$

25. a) $i = \frac{1}{5}e + \frac{2}{5}f$ b) $j = \frac{2}{5}e - \frac{1}{5}f$

PROBLEM SET 9.8

1. a) $\overrightarrow{OS} = \frac{1}{2}\overrightarrow{OR} + \frac{1}{2}\overrightarrow{OP}$ b) $\overrightarrow{OT} = \overrightarrow{OP} + \frac{1}{2}(\overrightarrow{OR} - \overrightarrow{OP}) = \frac{1}{2}\overrightarrow{OP} + \frac{1}{2}\overrightarrow{OR}$

3. a) $\overrightarrow{OS} = \frac{1}{3}\overrightarrow{OR} + \frac{1}{3}\overrightarrow{OP}$ b) $\overrightarrow{OT} = \frac{1}{2}\overrightarrow{OP} + \frac{1}{3}(\overrightarrow{OR} - \frac{1}{2}\overrightarrow{OP}) = \frac{1}{3}\overrightarrow{OP} + \frac{1}{3}\overrightarrow{OR}$

11. c) $a = \frac{7}{2}$

PROBLEM SET 9.9

1. maximum at $x = 0$, $\kappa = 2$

3. maximum at $x = (45)^{-1/4}$, $\kappa = 5^{3/2} (45)^{-1/4} 6^{-1/2}$; minimum at $x = -(45)^{-1/4}$, $\kappa = -5^{3/2} (45)^{-1/4} 6^{-1/2}$

5. $\dfrac{1}{a}, \dfrac{1}{b}$

PROBLEM SET 10.1

1. 0

3. 0

5. 0

7. 0

9. $\dfrac{1}{e}$

11. ∞

13. $-\infty$

15. 1

17. 2

19. ∞

21. converges

23. 2

25. $\ln 2$

27. $e - 1$

29. 1

31. 0

33. ∞

35. converges

PROBLEM SET 10.2

1. not convergent

3. not convergent

5. $\frac{3}{8}$

7. $\dfrac{\pi}{1 + \pi}$

9. convergent

11. convergent

13. convergent

15. not convergent

17. convergent

19. not convergent

21. convergent

23. convergent

25. convergent

27. not convergent

29. not convergent

33. $\sum\limits_{i=0}^{\infty} (-1)^i x^{4i}, \ -1 < x \leq 1$

35. convergent

PROBLEM SET 10.3

1. alternating, absolutely convergent

3. alternating, absolutely convergent

5. alternating, absolutely convergent

7. alternating, absolutely convergent

9. not alternating, not absolutely convergent

PROBLEM SET 10.4

1. $|R_n| \leq \frac{1}{8}(\frac{1}{3})^{2n-1}$

3. $|R_n| \leq \pi^{-(n+1)}$

5. $|R_n| \leq \dfrac{1}{n}$

7. $|R_n| \leq \dfrac{1}{(n + 1)^4}$

9. $|R_n| \leq \dfrac{1}{(n + 1)^{.9}}$

11. $|R_n| \leq \dfrac{1}{n}$ (This is the estimate which is easiest to derive. Much better estimates are possible.)

13. $|R_n| \leq \dfrac{1}{n}$

15. $|R_n| \leq \dfrac{1}{2n^2}$

PROBLEM SET 10.5

1. 0.019997

5. a) $f(x) = \sum\limits_{i=0}^{\infty} (-1)^i (\sqrt{x})^i$ b) $|R_n(x)| \leq (0.49)^{(n+1)/2}$ c) $\sum\limits_{i=1}^{\infty} (-1)^i \dfrac{2(0.49)^{(i+2)/2}}{i + 2}$

7. a) $f(x) = \sum\limits_{i=0}^{\infty} (-1)^i x^{5i/2}$ b) $|R_n(x)| \leq (0.25)^{5(n+1)/2}$ c) $\sum\limits_{i=1}^{\infty} (-1)^i \dfrac{2(0.25)^{(5i+2)/2}}{5i + 2}$

PROBLEM SET 10.6

1. $(-1, 1)$

3. $(-1, 1)$

5. $[-1, 1)$

7. $[-1, 1)$

9. $(-1, 1)$

11. $(-1, 1)$

13. $(-\infty, \infty)$

15. $(-\infty, \infty)$

17. $(-\infty, \infty)$

19. $[-1, 1]$

21. $[4 - 1/e, 4 + 1/e]$

PROBLEM SET 10.7

1. $\displaystyle\sum_{i=0}^{\infty} (-1)^i \frac{x^{i+2}}{i+1}$, $(-1, 1]$

3. $\displaystyle\sum_{i=0}^{\infty} (-1)^i \frac{x^{i+3}}{i+1}$, $(-1, 1]$

5. $\displaystyle\sum_{i=0}^{\infty} (-1)^i \frac{2^{2i+1}x^{2i+1}}{(2i+1)!}$, $(-\infty, \infty)$

7. $\displaystyle\sum_{i=0}^{\infty} (-1)^i \frac{x^{2i+1}}{2^{2i+1}(2i+1)!}$, $(-\infty, \infty)$

9. $\displaystyle\sum_{i=0}^{\infty} (-1)^i \frac{x^{2i}}{(3)^{2i}(2i)!}$, $(-\infty, \infty)$

11. $\displaystyle\sum_{i=0}^{\infty} \frac{x^{3i+3}}{i!}$, $(-\infty, \infty)$

13. $\displaystyle\sum_{i=0}^{\infty} \frac{x^{3i+4}}{i!(3i+4)}$, $(-\infty, \infty)$

15. $\displaystyle\sum_{i=0}^{\infty} (-1)^i \frac{x^{2i+1}}{(2i+1)^2}$, $[-1, 1]$

17. $\displaystyle\sum_{i=1}^{\infty} \frac{x^i}{i(i!)}$, $(-\infty, \infty)$

19. $\displaystyle\sum_{i=0}^{\infty} (-1)^i \frac{x^{2i+1}}{(2i+1)(2i+1)!}$, $(-\infty, \infty)$

21. $\frac{1}{2}\displaystyle\sum_{i=0}^{\infty} (-1)^i \frac{2^{2i}x^{2i+3}}{(2i)!} + \frac{x^3}{2}$, $(-\infty, \infty)$

23. $\displaystyle\sum_{i=0}^{\infty} (-1)^i \frac{2^{2i}x^{2i}}{(2i)!}$, $(-\infty, \infty)$

25. $f(x) = e^{x/2}$

29. $j(j-1)\cdots(j-i+1)x^{j-1}$

PROBLEM SET 10.8

1. $\displaystyle\sum_{i=0}^{\infty} \binom{\frac{1}{2}}{i} x^i$

3. $\displaystyle\sum_{i=0}^{\infty} \binom{\frac{1}{4}}{i} x^i$

5. $\dbinom{n+1}{i} = \dfrac{(n+1)!}{(n+1-i)!\, i!}$, $\dbinom{n}{i-1} = \dfrac{n!}{(n-i+1)!\, (i-1)!}$

9. $\displaystyle\sum_{i=0}^{\infty} \binom{-\frac{1}{2}}{i} x^{i+2}$ on $(-1, 1)$

11. $\displaystyle\sum_{i=0}^{\infty} \binom{-\frac{1}{3}}{i} x^{2i+1}$ on $(-1, 1)$

13. $\displaystyle\sum_{i=0}^{\infty} \binom{\frac{3}{2}}{i} x^i$ on $(-1, 1)$

15. $2^{1/3}\displaystyle\sum_{i=0}^{\infty} \binom{\frac{1}{3}}{i} \left(\frac{x}{2}\right)^i$ on $(-2, 2)$

17. $2^{1/3}\displaystyle\sum_{i=0}^{\infty} \binom{\frac{1}{3}}{i} \frac{x^{2i+1}}{2^i(2i+1)}$ on $(-\sqrt{2}, \sqrt{2})$

19. $2^k\displaystyle\sum_{i=0}^{\infty} \binom{k}{i} \frac{x^{2i+1}}{2^{2i}(2i+1)}$ on $(-\sqrt{2}, \sqrt{2})$

21. $\displaystyle\sum_{i=0}^{\infty} \binom{-\frac{1}{2}}{i} \frac{x^{i+2}}{i+2}$ on $(-1, 1)$

23. $f(x) = e^{\sin(x)}$

PROBLEM SET 10.9

1. $a_n = 0$ if n is even, $a_n = \dfrac{(-1)^m}{(2m+1)!}$ if $n = 2m+1$

3. $a_0 = 0$, $a_1 = 1$, $a_2 = 0$

5. $a_n = 0$ if n is even, $a_n = \dfrac{(-1)^m}{(2m+1)}$ if $n = 2m+1$

7. $a_n = \dfrac{1}{n!}$

9. $a_0 = \ln 2$, $a_1 = \frac{1}{2}$, $a_2 = -\frac{1}{8}$

11. $a_n = 0$ if n is odd, $a_n = \dfrac{(-1)^{m-1}}{2m}$ if $n = 2m$, $m > 0$ $(a_0 = 0)$

PROBLEM SET 10.10

1. $|R_n(x)| \leqq \dfrac{|x|^{n+1}}{(n+1)!}$

7. $|R_n(x)| \leqq e^x \dfrac{|x|^{n+1}}{(n+1)!}$

9. $|R_n(x)| \leqq \dfrac{e^{x-a}|x-a|^{n+1}}{(n+1)!}$

PROBLEM SET 10.11

1. -4

3. -1

5. -1

7. $-\frac{1}{2} - \dfrac{\sqrt{3}}{2}i$

9. 1

11. $2i$

13. $3 + 4i$

15. $-9 + 8\sqrt{3}\,i$

17. $\dfrac{2}{5} - \dfrac{i}{5}$

19. $\dfrac{1}{5} - \dfrac{2i}{5}$

21. $\dfrac{1}{10} - \dfrac{3i}{10}$

23. $\dfrac{2}{7} - \dfrac{\sqrt{3}\,i}{7}$

25. $\dfrac{\sqrt{3}}{7} - \dfrac{2i}{7}$

27. $-i$

29. i

31. i

33. $\dfrac{8}{5} + \dfrac{i}{5}$

35. 1

PROBLEM SET 10.12

1. $\sqrt{2}\,e^{i\pi/4}$

3. $7e^{i\pi}$

5. $2e^{i\pi/6}$

11. $\sin z = \dfrac{1}{2i}(e^{iz} - e^{-iz})$, $\cos z = \frac{1}{2}(e^{iz} + e^{-iz})$

PROBLEM SET 10.13

1. $z = \dfrac{1}{\sqrt{2}} + \dfrac{i}{\sqrt{2}}, \dfrac{1}{\sqrt{2}} - \dfrac{i}{\sqrt{2}}, -\dfrac{1}{\sqrt{2}} + \dfrac{i}{\sqrt{2}}, -\dfrac{1}{\sqrt{2}} - \dfrac{i}{\sqrt{2}}$

3. $z = -2, 1 + \sqrt{3}\,i, 1 - \sqrt{3}\,i$

5. $z = -1, i, -i$

7. $z = -1, i, -i, \dfrac{1}{\sqrt{2}} + \dfrac{i}{\sqrt{2}}, \dfrac{1}{\sqrt{2}} - \dfrac{i}{\sqrt{2}}, -\dfrac{1}{\sqrt{2}} + \dfrac{1}{\sqrt{2}}, -\dfrac{1}{\sqrt{2}} - \dfrac{i}{\sqrt{2}}$

PROBLEM SET 10.14

1. e^z 3. $\cos z$ 5. $2ze^{z^2}$

PROBLEM SET 10.15

5. $\displaystyle\sum_{i=0}^{\infty} \frac{(i+1)x^i}{i!}$ 7. $\displaystyle\sum_{i=0}^{\infty} (-1)^{i+1} \frac{(2i+1)x^{2i}}{(2i)!}$

9. $\lim\limits_{n\to\infty} f_n(x) = 0 = \text{U}\lim\limits_{n\to\infty} f_n(x)$ 11. $\lim\limits_{n\to\infty} f_n(x) = 0$, U lim does not exist

13. $\lim\limits_{n\to\infty} f_n(0) = 1$, $\lim\limits_{n\to\infty} f_n(x) = \infty$ if $0 < x \leq 1$, U lim does not exist

PROBLEM SET 11.1

1. $x + 3y + 2z = 14$ 3. $x + y + z = \frac{1}{2}$ 5. $5x - 8y + z = 3$

7. $x - y + 2z = 6$ 9. The figure is a sphere with equation $x^2 + y^2 + z^2 = 4$.

PROBLEM SET 11.2

1. $\dfrac{x}{\sqrt{3}} + \dfrac{y}{\sqrt{3}} + \dfrac{z}{\sqrt{3}} - \dfrac{1}{\sqrt{3}} = 0$, $-\dfrac{x}{\sqrt{3}} - \dfrac{y}{\sqrt{3}} - \dfrac{z}{\sqrt{3}} + \dfrac{1}{\sqrt{3}} = 0$

3. $\dfrac{x}{3} + \dfrac{2y}{3} + \dfrac{2z}{3} - 1 = 0$, $-\dfrac{x}{3} - \dfrac{2y}{3} - \dfrac{2z}{3} + 1 = 0$

5. $\dfrac{x}{\sqrt{21}} + \dfrac{2y}{\sqrt{21}} + \dfrac{4z}{\sqrt{21}} + \dfrac{4}{\sqrt{21}} = 0$, $-\dfrac{x}{\sqrt{21}} - \dfrac{2y}{\sqrt{21}} - \dfrac{4z}{\sqrt{21}} - \dfrac{4}{\sqrt{21}} = 0$

7. $\left(\dfrac{1}{\sqrt{3}}, \dfrac{1}{\sqrt{3}}, \dfrac{1}{\sqrt{3}}\right)$

11. $\dfrac{x}{\sqrt{5}} + \dfrac{2y}{\sqrt{5}} - \dfrac{3}{\sqrt{5}} = 0$, $\dfrac{-x}{\sqrt{5}} - \dfrac{2y}{\sqrt{5}} + \dfrac{3}{\sqrt{5}} = 0$

PROBLEM SET 11.3

1. $a = 0$, $P_0 = (1, 1, -1)$ 3. $a = 0$, $P_0 = (1, 1, 1)$

5. $a = 0$, $P_0 = (-3, -1, 4)$ 7. $a = 1$, $P_0 = (0, 0, 1)$

9. $a = 4$, $P_0 = (1, \frac{1}{2}, \frac{1}{3})$ 11. $i = V_1 - V_2 + V_3$, $j = V_2 - V_3$, $k = V_3$

13. $\quad \alpha_1 V_1 + \alpha_2 V_2 + \alpha_3 V_3 = 0$
$\Rightarrow \alpha_1(i + j) + \alpha_2(j + k) + \alpha_3(k) = 0$
$\Rightarrow \alpha_1 i + (\alpha_1 + \alpha_2)j + (\alpha_1 + \alpha_2 + \alpha_3)k = 0$
$\Rightarrow \alpha_1 = 0$, $\alpha_1 + \alpha_2 = 0$ so $\alpha_2 = 0$, and
$\quad \alpha_1 + \alpha_2 + \alpha_3 = 0$ so $\alpha_3 = 0$

15. $i = \frac{1}{2}(V_1 + V_2)$, $j = \frac{1}{2}(V_2 + V_3)$, $k = \frac{1}{2}(V_1 + V_3)$

$$xi + yj + zk = \frac{x}{2}(V_1 + V_2) + \frac{y}{2}(V_2 + V_3) + \frac{z}{2}(V_1 + V_3)$$

$$= \tfrac{1}{2}(x + z)V_1 + \tfrac{1}{2}(x + y)V_2 + \tfrac{1}{2}(y + z)V_3,$$

$$\alpha_1 V_1 + \alpha_2 V_2 + \alpha_3 V_3 = 0$$

$$\Rightarrow \alpha_1(i - j + k) + \alpha_2(i + j - k) + \alpha_3(-i + j + k) = 0$$

$$\Rightarrow \alpha_1 + \alpha_2 - \alpha_3 = 0, \; -\alpha_1 + \alpha_2 + \alpha_3 = 0, \; \alpha_1 - \alpha_2 + \alpha_3 = 0$$

$$\Rightarrow \alpha_1 = 0, \; \alpha_2 = 0, \; \alpha_3 = 0 \text{ by elimination}$$

33. $V_1 = i - j = (1, -1, 0)$ satisfies $x + y + 2z = 0$ as does $V_2 = 2j - k = (0, 2, -1)$. Any vector of the form $aV_1 + bV_2$ can be written $a(i - j) + b(2j - k) = ai + (2b - a)j - bk$ and this satisfies the equation $x + y + 2z = 0$. If $\alpha_1 V_1 + \alpha_2 V_2 = 0$, then $\alpha_1(i - j) + \alpha_2(2j - k) = 0$; therefore $\alpha_1 i + (2\alpha_2 - \alpha_1)j - \alpha_2 k = 0$; therefore $\alpha_1 = 0$, and $\alpha_2 = 0$ since i, j, and k are linearly independent. Any vector $(-y - 2z, y, z)$ in E can be written as $(-y - 2z)V_1 - zV_2$, and so V_1 and V_2 span E.

35. $4x - 2y + z = 0$

PROBLEM SET 11.4

1. $\alpha_1 = 1, \alpha_2 = -2, \alpha_3 = 1$

3. $\alpha_1 = 4, \alpha_2 = -5, \alpha_3 = 2, \alpha_4 = -1$

5. $\{E_1 - E_2, 2E_2 + E_3\}$

7. $\{E_1 + E_2, E_2, E_3 + E_4, E_4\}$

9. $\{2E_1 - 2E_2 - E_3\}$

PROBLEM SET 11.5

1. $\left\{\dfrac{1}{\sqrt{3}}(E_1 + E_2 + E_3), \dfrac{1}{\sqrt{6}}(-E_1 - E_2 + 2E_3)\right\}$

3. $\left\{\dfrac{1}{\sqrt{3}}(E_1 + E_2 + E_3), \dfrac{1}{\sqrt{6}}(2E_1 - E_2 - E_3), \dfrac{1}{\sqrt{2}}(E_2 - E_3)\right\}$

5. $\left\{E_3, \dfrac{1}{\sqrt{2}}(E_1 + E_2), \dfrac{1}{\sqrt{2}}(E_1 - E_2)\right\}$

7. $\left\{\dfrac{1}{\sqrt{5}}(2E_1 - E_2), E_3\right\}$

9. $V_3 = E_1 - E_2 - E_4, \; V_4 = -E_2 + E_3 + E_4$

11. $\left\{\dfrac{1}{\sqrt{2}}(E_1 - E_2)\right\}$

PROBLEM SET 11.6

1. [*Hint:* Expand $(x_1 y_2 - x_2 y_1)^2$.]

3. No. **D.4** fails for an obtuse triangle.

5. **D.1** $d(P, Q) = \max\{|x - a|, |y - b|\} \geq 0$ since both are ≥ 0.

   D.2 $d(P, Q) = 0 \Rightarrow \max\{|x - a|, |y - b|\} = 0 \Rightarrow |x - a| = 0, |y - b| = 0, \Rightarrow P = Q$

   D.3 $d(P, Q) = \max\{|x - a|, |y - b|\} = \max\{|y - b|, |x - a|\} = d(Q, P)$

   D.4 $d(P, R) = \max\{|x - u|, |y - v|\} = \max\{|x - a + a - u|, |y - b + b - v|\}$

   $\leq \max\{|x - a| + |a - u|, |y - b| + |b - v|\}$

   $\leq \max\{|x - a|, |y - b|\} + \max\{|a - u|, |b - v|\}$

   $= d(P, Q) + d(Q, R)$

13. $f(x) = x^2 - \dfrac{\pi^2}{3}$

PROBLEM SET 12.1

1. $a_0 = 0$, $a_i = 0$, $b_i = \dfrac{2}{i}(-1)^{i+1}$

3. $a_0 = \dfrac{\pi^2}{3}$, $a_i = \dfrac{(-1)^i 4}{i^2}$, $b_i = \dfrac{2}{i}(-1)^{i+1}$

5. $a_i = 0$, $b_i = \left(\dfrac{12}{i^3\pi^2}\right)(-1)^{i+1}$

7. $a_0 = \dfrac{\pi}{4}$, $a_i = \dfrac{1}{i^2\pi}((-1)^i - 1)$, $b_i = \dfrac{1}{i}(-1)^{i+1}$

9. $a_0 = \dfrac{\pi}{4} + \dfrac{1}{2}$, $a_i = \left(\dfrac{1}{i^2\pi}\right)((-1)^i - 1)$, $b_i = \dfrac{1}{i}(-1)^{i+1} + \dfrac{1}{i\pi}((-1)^i - 1)$

11. $a_0 = \dfrac{\pi}{4}$, $a_i = \dfrac{1}{i^2\pi}((-1)^i - 1)$, $b_i = \dfrac{3}{i}(-1)^{i+1}$

PROBLEM SET 12.2

1. $-\frac{1}{4}\sin 3x + \frac{3}{4}\sin x$
5. $-\frac{1}{4}\cos 3x + \frac{1}{4}\cos x$
9. $\frac{3}{8} - \frac{1}{2}\cos 2x + \frac{1}{8}\cos 4x$

3. $\frac{1}{4}\cos 3x + \frac{3}{4}\cos x$
7. $-\frac{1}{2}\cos 3x + \frac{1}{2}\cos x$

PROBLEM SET 12.3

7. $a_0 = \dfrac{1}{2\pi}(-2 + e^\pi + e^{-\pi})$

$a_i = \dfrac{1}{\pi(1 + i^2)}(-2 + (e^\pi + e^{-\pi})(-1)^i)$

$b_i = \dfrac{i}{\pi(1 + i^2)}(+2 + (e^\pi + e^{-\pi})(-1)^{i+1})$

PROBLEM SET 13.1

1. $M^{-1} = \frac{1}{6}\begin{bmatrix} 6 & -4 & -2 \\ 0 & 3 & 0 \\ 0 & -2 & 2 \end{bmatrix}$

3. $M^{-1} = \frac{1}{6}\begin{bmatrix} 6 & 0 & 0 \\ 0 & 3 & 0 \\ 0 & -1 & 2 \end{bmatrix}$

5. $f(\mathbf{R}^3) = \{(a, b, c) \mid b = 3a, c = -2a\}$

$\operatorname{Ker} f = \{(x, y, z) \mid 2x + y - z = 0\}$

7. $f(\mathbf{R}^3) = \{(a, b, c) \mid 3b = 2c\}$

$\operatorname{Ker} f = \{(x, y, z) \mid x + 2y = 0, z = 0\}$

9. $M^{-1} = \frac{1}{6}\begin{bmatrix} 6 & -3 & -1 \\ 0 & 3 & -1 \\ 0 & 0 & 2 \end{bmatrix}$

PROBLEM SET 13.2

1.
$$\begin{bmatrix} 3 & 3 \\ 7 & 7 \\ 11 & 11 \\ 15 & 15 \end{bmatrix}$$

3.
$$\begin{bmatrix} 2 & 3 \\ 3 & 5 \end{bmatrix}$$

5.
$$\begin{bmatrix} 47 & 58 & 69 \\ 4 & 5 & 6 \\ 1 & 2 & 3 \end{bmatrix}$$

7.
$$\begin{bmatrix} 6 \\ 8 \\ 10 \end{bmatrix}$$

9.
$$\begin{bmatrix} 6 \\ 15 \end{bmatrix}$$

11.
$$\begin{bmatrix} 2 & 5 & 8 \\ 3 & 6 & 9 \\ 1 & 4 & 7 \end{bmatrix}$$

13.
$$\begin{bmatrix} 7 & 1 & 4 \\ 8 & 2 & 5 \\ 9 & 3 & 6 \end{bmatrix}$$

15.
$$\begin{bmatrix} 1 & 0 & 0 \\ 0 & 1 & 0 \\ 0 & 0 & 1 \end{bmatrix}$$

17.
$$\begin{bmatrix} 0 & 1 & 0 \\ 0 & 0 & 1 \\ 1 & 0 & 0 \end{bmatrix}$$

19.
$$\begin{bmatrix} 3 & 1 & 3 \\ 3 & 3 & 3 \\ 2 & 0 & 2 \end{bmatrix}$$

21.
$$\begin{bmatrix} 7 & 1 & 7 \\ 21 & 15 & 21 \\ 32 & 8 & 32 \end{bmatrix}$$

23.
$$\begin{bmatrix} 0 & 4 \\ 0 & 8 \\ 9 & -42 \end{bmatrix}$$

25.
$$\begin{bmatrix} 1 & 0 & 0 & 0 \\ 0 & 1 & 0 & 0 \\ 0 & 0 & 1 & 0 \\ 0 & 0 & 0 & 1 \end{bmatrix}$$

27.
$$\begin{bmatrix} 0 & 0 & 0 & 1 \\ 0 & 0 & 1 & 0 \\ 0 & 1 & 0 & 0 \\ 1 & 0 & 0 & 0 \end{bmatrix}$$

29. no inverse

31. $M^{-1} = \begin{bmatrix} 0 & \dfrac{1}{4} \\ \dfrac{1}{4} & 0 \end{bmatrix}$

PROBLEM SET 13.4

1. $a_{11}a_{22} - a_{12}a_{21}$
3. $a_{33}a_{44}$
5. 1
7. -2
9. 56
11. -2
13. 112
15. -720
17. -2

PROBLEM SET 13.5

1. $D_{11} = -1,\ -D_{21} = 2,\ D_{31} = -1$
3. $z = -\dfrac{50}{47}$

5. $\begin{bmatrix} 0 & \dfrac{1}{3} \\ \dfrac{1}{2} & 0 \end{bmatrix}$

7. $\begin{bmatrix} 0 & 1 \\ 1 & -1 \end{bmatrix}$

9. $\begin{bmatrix} 0 & 1 & 0 \\ 1 & 0 & 0 \\ 0 & 0 & 1 \end{bmatrix}$

11. $\begin{bmatrix} 1 & 0 & 0 & 0 \\ 0 & -\dfrac{5}{2} & 0 & +\dfrac{3}{2} \\ 0 & 0 & 1 & 0 \\ 0 & +2 & 0 & -1 \end{bmatrix}$

13. $\begin{bmatrix} 0 & 1 & 0 & 0 \\ 1 & 0 & 0 & 0 \\ 0 & 0 & 0 & 1 \\ 0 & 0 & 1 & 0 \end{bmatrix}$

15. $\begin{bmatrix} 0 & 0 & \dfrac{1}{3} & 0 & 0 \\ 0 & -\dfrac{1}{2} & 0 & 0 & 0 \\ 0 & 0 & 0 & 0 & -\dfrac{1}{5} \\ -1 & 0 & 0 & 0 & 0 \\ 0 & 0 & 0 & \dfrac{1}{4} & 0 \end{bmatrix}$

PROBLEM SET 13.6

1. independent
3. independent
5. independent
7. independent
9. independent
11. independent
13. independent
15. independent
17. dependent
19. independent
21. independent

PROBLEM SET 13.7

1. $\{e^{2x}, e^{3x}\}, f(x) = e^{2x}$

3. $\{e^{3x}, xe^{3x}\}, f(x) = (1 - 2x)e^{3x}$

5. $\left\{e^{-x/2} \cos\left(\dfrac{\sqrt{3}}{2} x\right), e^{-x/2} \sin\left(\dfrac{\sqrt{3}}{2} x\right)\right\}, f(x) = 0$ for all x

7. $\{e^x, e^{-x}\}, f(x) = \dfrac{1}{e}(e^x) - e(e^{-x})$

PROBLEM SET 13.8

1. $\mathscr{W} = \{\alpha_1 e^{-x} + \alpha_2 xe^{-x} + \alpha_3 x^2 e^{-x} + \alpha_4 x^3 e^{-x}\}$ 3. $\mathscr{W} = \{\alpha_1 e^x + \alpha_2 xe^x + \alpha_3 e^{-2x} + \alpha_4 xe^{-2x}\}$

5. $\mathscr{W} = \{\alpha_1 e^x + \alpha_2 \cos x + \alpha_3 \sin x\}$ 7. $\mathscr{W} = \{\alpha_1 + \alpha_2 e^x + \alpha_3 xe^x + \alpha_4 e^{-x} + \alpha_5 xe^{-x}\}$

9. $H = \{\alpha_1 \sin x + \alpha_2 \cos x - \frac{1}{2}x \cos x\}$ 11. $H = \{\alpha_1 \sin x + \alpha_2 \cos x + \frac{1}{2}x \sin x\}$

13. $H = \{\alpha_1 \sin x + \alpha_2 \cos x + x\}$

15. $H = \{\alpha_1 \sin x + \alpha_2 \cos x + \frac{1}{5}e^x \sin x - \frac{2}{5}e^x \cos x\}$

17. $H = \{\alpha_1 \sin x + \alpha_2 \cos x + \frac{2}{5}e^x \sin x + \frac{1}{5}e^x \cos x\}$

19. $H = \{\alpha_1 \sin x + \alpha_2 \cos x + \frac{1}{2}xe^x\}$

21. $H = \{\alpha_1 \sin x + \alpha_2 \cos x + \frac{1}{4}x \cos x + \frac{1}{4}x^2 \sin x\}$

PROBLEM SET 14.1

17. $x^2 + z^2 = (1 - y)^2$

19. $9(x^2 + z^2) = 2y^2$

21. $z^2 + y^2 = 10x^2$

23. $y^2 + z^2 = e^{2x}$

25. $x^2 + y^2 = \cos^2 z,\ 0 \leqq z \leqq \pi/2$

PROBLEM SET 14.2

1. ellipsoid

3. one-sheeted hyperboloid

5. elliptic cone

7. elliptic cone

9. hyperbolic paraboloid

11. hyperbolic paraboloid

13. hyperbolic paraboloid

19. $\dfrac{1}{3}$

21. 8π

PROBLEM SET 14.3

13. $f_x = \dfrac{2x^3}{\sqrt{x^4 + y^4 + 1}},\ f_y = \dfrac{2y^3}{\sqrt{x^4 + y^4 + 1}},\ f_{xy} = \dfrac{-4x^3y^3}{(x^4 + y^4 + 1)^{3/2}} = f_{yx}$

15. $f_x = \dfrac{y(y^2 - x^2)}{(x^2 + y^2)^2},\ f_y = \dfrac{x(x^2 - y^2)}{(x^2 + y^2)^2},\ f_{xy} = \dfrac{-y^4 + 6x^2y^2 - x^4}{(x^2 + y^2)^3} = f_{yx}$

17. $f_x = \dfrac{4xy^2}{(x^2 + y^2)^2},\ f_y = \dfrac{-4yx^2}{(x^2 + y^2)^2},\ f_{xy} = \dfrac{8xy(x^2 - y^2)}{(x^2 + y^2)^3} = f_{yx}$

19. $f_x = \dfrac{\cos x}{y},\ f_y = \dfrac{-\sin x}{y^2},\ f_{xy} = \dfrac{-\cos x}{y^2} = f_{yx}$

21. $f_x = \dfrac{(x^2 + y^2)y \cos xy - 2x \sin xy}{(x^2 + y^2)^2},\ f_y = \dfrac{(x^2 + y^2)x \cos xy - 2y \sin xy}{(x^2 + y^2)^2}$

$f_{xy} = \dfrac{(-\cos xy - xy \sin xy)(x^2 + y^2)^2 + 8xy \sin xy}{(x^2 + y^2)^3} = f_{yx}$

PROBLEM SET 14.4

1. $A = 1,\ B = 2,\ E_1 = \Delta y,\ E_2 = 0$

3. $A = 1,\ B = 2,\ E_1 = \Delta y^2,\ E_2 = \Delta y + 2\Delta x$

5. $A = 2,\ B = -2,\ E_1 = \Delta x,\ E_2 = -\Delta y$

7. $A = 2,\ B = 2,\ E_1 = \Delta x,\ E_2 = \Delta y$

9. $A = 2,\ B = -8,\ E_1 = \Delta x,\ E_2 = 4\Delta y$

11. $f_\alpha(1, 1) = 2 \cos \alpha - 2 \sin \alpha$, maximum at $\alpha = -\pi/4$, minimum at $3\pi/4$

PROBLEM SET 14.5

1. $\phi'(t) = -(5t^4 + 3t^2) \sin (t^5 + t^3)$

3. $\phi'(t) = 2 \cos 2t$

5. $\phi'(t) = 5t^4$

PROBLEM SET 14.6

1. $\phi_t = 4t^3u^2 + 2t + 2u^2,\ \phi_u = 2t^4u + 4ut + 4u^3$

3. $\phi_t = 3t^2 \cos f \cos g - \sin f \sin g,\ \phi_u = 4u^3 \cos f \cos g - \sin f \sin g$

5. $\phi_t = 2t,\ \phi_u = 0$ 7. $\phi_t = 0,\ \phi_u = 0$

9. $E_1 = \Delta t\,\Delta u\,\Delta w(1 + \Delta v),\ E_2 = \Delta u\,\Delta w(1 + \Delta v),\ E_3 = 0,\ E_4 = 0,\ E_5 = 0$

PROBLEM SET 14.8

1. no ILMax, no ILMin   3. no ILMax, no ILMin

5. no ILMax, ILMin at $(-\frac{1}{2}, -\frac{1}{2})$ 7. no ILMax, ILMin at $(-\frac{1}{3}, -\frac{1}{3})$

9. ILMax at $(0, 0)$, no ILMin

11. maximum value at $x = (1/\sqrt{2}, 0)$ and $x = (-1/\sqrt{2}, 0)$, maximum value $= \frac{1}{4}$

13. maximum volume $\dfrac{16}{\sqrt{3}}$, coordinates $\left(\dfrac{1}{\sqrt{3}}, \dfrac{2}{\sqrt{3}}, \dfrac{3}{\sqrt{3}}\right)$

15. maximum volume $= \dfrac{32\sqrt{3}}{9}$, coordinates $= \left(\dfrac{1}{\sqrt{3}}, \dfrac{2}{\sqrt{3}}, \dfrac{2}{\sqrt{3}}\right)$

PROBLEM SET 14.9

1. $\dfrac{544}{35}$ 3. $\dfrac{1376}{21}$ 5. $\dfrac{1}{32}$

7. 0   9. $\dfrac{4}{27}$

PROBLEM SET 14.10

1. $2\pi/11$ 3. $15\pi/16$ 5. $(\frac{8}{3} - \sqrt{3})2\pi$

7. 0   9. $\frac{1}{4}$ 11. 0

PROBLEM SET 14.11

1. $\dfrac{28}{9}$ 3. $\bar{x} = 0,\ \bar{y} = 0$

5. $\bar{x} = 3/2\pi,\ \bar{y} = 0$ 7. $\bar{x} = 0,\ \bar{y} = -\frac{5}{6}$

9. $-11/6$ 11. $\bar{x} = \dfrac{3}{4},\ \bar{y} = 0$ 13. $(0, 0)$

PROBLEM SET 14.12

1. 0   3. 0   5. $3/7$

7. 0   9. 0   11. 3

Index

absolute convergence, 447
absolute value, of a real number, 9 ff
  of a complex number, 481
acceleration, 119, 422
  of gravity, 120
addition formula, for the cosine, 136
  for the sine, 138
algebraic substitutions, 291 ff
ALO, 7
alternating function, 585
alternating series, 446
alternating series test, 447
amplitude, of a complex number, 488
annihilation theorem, 435
antidifferentiation, 255
AO, 4
apocryphal anecdote, 120
approximations by trigonometric polynomials, 549
arc length, 303 ff
  formula for, 306
Archimedes, 57
area, of a parabolic sector, 57 ff
  of a surface, 327 ff
areas in polar coordinates, 402 ff
arithmetic series, 49
associativity, 1
asymptote, 368 ff
axis, of a parabola, 38

Banchoff, Thomas F., 56
base plane, 667
basis, for a vector space, 521
betweenness theorem for integrals, 312
binomial series, 468 ff
binomial theorem, 468
  induction proof of, 472, 484
bound, of a function, 86
bounded above, 193
bounded away from 0, 696
bounded below, 194
bounded function, 86
  sequence, 517
bounded sequence, 433, 695
  set, 669
box, for a function, 77

cardioid, 400
Cartesian n-space, 526
center, of an ellipse, 362
  of any point-symmetric set, 365
centroid, 335 ff
  of nonhomogeneous bodies, 674
chain rule, 159 ff
  proof of, 162
  for paths, 642, 646
  for functions of many variables, 647
chord, of a function, 72
circle, 23
circular cone, 621
closed interval, 12
  disk, 493
  set, 669
commutative group, 578
commutative law, for real numbers, 1
  for positive series, 476
comparison theorem, for positive series, 441
completeness, of the real number system, 238 ff
completing the square, in integration, 297 ff
complex numbers, 479 ff
  construction of, 535, 713
components of a vector, 418
composition, of functions, 154 ff
  for functions of two variables, 629
concave, 215
cone, 41
conical surface, 618
conjugate, of a complex number, 480
conjugate hyperbolas, 370
constant function, 89
continuity, of a function, 75 ff
  of a function of two variables, 627
  of composite functions, 520
continuous, definition of, 77
continuous function, 75
  elementary theorems on, 85
continuous interest, 223
convergent, 431
coordinate functions, 382
coordinate system, 16 ff
Cos, Cos^{-1}, 170
cosh, 203
cosine, series for, 466

coth, 203
Cramer's rule, 593
cross section, 321
csch, 203
curvature, 425 ff
cycloid, 391
cylinder, 316, 614
cylindrical coordinates, 666
cylindrical shell, 316
    solid, 614
    surface, 614

D, 95
D_x, 96
damped oscillation, 604
DCP, 239
dd'-box, 79
decreasing function, 206
    sequence, 433
Dedekind cut postulate, 239
definite integral, definition of, 308 ff, 312
De Moivre's theorem, 489 ff
density function, 675, 677
derivative, of a function, 69 ff
    of an arbitrary power of x, 202
    of one function with respect to another, 250 ff
    of the integral, 109 ff
    of the integral, proof of formula for, 124 ff
determinant function, 582
df/dg, 250
diameter, of a plane set, 668
differences, Δx and Δf, 140
differentiability, for functions of two variables, 638
differentiable function, 89
differentials, 148 ff
differentiation, 89 ff
    of complex power series, 496
    of series, 453, 505
    for functions of many variables, 644
    dimension, of a vector space, 529
dimension theorem, 610
directed angles, 128
    distance, 512
    normal form, 512
    segments, 415
direction angles, 513
    cosines, 514
directional derivatives, 634, 648
directrix, of a parabola, 38
disk, 493
distance formula, 19
    from a point to a line, 38
    in an inner-product space, 534
    in polar coordinates, 401
distributive law, 1
diverges to infinity, 435
divisor, 55
domain, of a function, 64
    of convergence, 462

dot product, 411
double integrals, 660, 666

e, definition of, 198
    existence of, 198
    as a limit as $x \to \infty$, 221
    series for, 464
eccentricity of a conic section, 380
ellipse, 360 ff
    area of, 62
ellipsoid, 620
elliptic cone, 621
epicycloid, 392
epsilon-delta box, 77
equivalence, 5
    of directed segments, 415
equivalence class, of directed segments, 416
equivalent to, 509
error in Simpson's rule, 702
estimates of remainders, 448 ff
even function, 339
    permutation, 584
exact differential, 683
existence, of maxima, 225, 244
    of minima, 227, 244
existence theorems, use in finding maxima and minima, 223 ff
exp, 194
    laws for, 196
    series for, 464
exponential function, in the complex domain, 484 ff
exponentials and logarithms, 185 ff

factor, 55
falling body, 120
field postulates, 1, 580
finite covering theorem, 350
finite dimensional vector space, 529
first octant, 509
focal difference, 366
    sum, 360
focus, of a hyperbola, 366
    of a parabola, 38
    of an ellipse, 360
force, 119
Fourier coefficients, 546
    type, 561
fractional exponents, 101
free vectors, 415 ff
frustum, 329
function, 63
    of g, 252
    of x, 255
    of two variables, 627
function-graph, 67
functional equations, 232 ff
    differentiation of, 235
fundamental theorem of integral calculus, 255

Galileo, 120
general equation of the second degree, 372 ff
geometric series, 49
gradient, 649
graph, of a condition, 21 ff
    of a function, 66
    of an inequality, 33 ff
group, 578

half, of an interval, 241
half-open interval, 13
half-plane, 34
harmonic series, 439
Heaviside function, 76
homogeneous differential equation, 611
Hooke's law, 605
hyperbola, 366 ff
hyperbolic functions, 203 ff
hyperboloid of one sheet, 622
    of two sheets, 623
hypocycloid, 385, 391

ILMax, 213
ILMin, 212
Im z, 482
image, of a function, 67
implies, 6, 507
improper integrals, 344 ff
increasing function, 206
    sequence, 433
indefinite integral, 256
    integration, 257
independent variable, 255
induction principle, 51 ff
inequalities, 4 ff
infinite interval, 13
infinity, 13
inflection point, 215
initial side, 128
inner product, 411
    space, 412, 518
inscribed broken line, 303
inside function, 155
integers modulo 2, 3
    modulo 4, 3
integrability of continuous functions, 353
integrable function, 312
integral, of a nonnegative function, 102 ff
integral test for infinite series, 445
integrand, 106
integration by parts, 273 ff
integration, of power series, 454, 503
    of Fourier series, 556
interior local maximum, 213
    local minimum, 212
interior of a circle, 22
interior point, of an interval, 176, 207
    of a set in a plane, 653
interval, 12
inverse, of an invertible function, 165

invertible function, 165
    derivative of, 168
iterated integral, 665
    limits, 717

kernel, 568

law of cosines, 135
Least Upper Bound Postulate, 243
left-handed coordinate system, 129
length of a path, 405 ff
    formula for, 407
level curves, 655
l'Hôptal's rule, 388, 389, 393 ff
$\lim_{x \to \infty}$, $\lim_{x \to -\infty}$, $\lim_{x \to a^+}$, $\lim_{x \to a^-}$,
    217 ff
limit, 45
    definition of, 80, 81
    of a function of two variables, 627
    of a sequence, definition, 431
    of integration, 105
limit point of a set, 669
limits, theorems on, 82 ff
line integrals, 680
linear combination, 413, 521
    differential equation, 601
    equation in x, y, and z, 510
    function, 563
    transformation, 563
linearly dependent, 421, 521
    independent, 421, 521, 596
Lipschitzian function, 82, 315, 355
LMax, 212
LMin, 212
ln, 191 ff
    laws for, 196
    series for, 455
local maximum, 212
    minimum, 212
locally bounded function, 88, 511, 691
locally bounded away from 0, 693
locus of a path, 381
logarithms, 186 ff
    laws of, 189
logic and set theory, 687
lower bound, 194
lower limit of integration, 105
lower sum, 309
LUBP, 243

Maclaurin series, 473
matrix, 564
maxima, 211 ff
    for functions of several variables, 651
mean-value theorem, 72 ff, 246
measurable set, in a plane, 527, 668, 705
measurable solid, 318
"measure," of a directed angle, 133
mesh of a net over an interval, 304
    over a plane region, 668

metric space, 539
minima, 211 ff
minor, of an element of a matrix, 590
mixed partial derivatives, 717
MLO, 7
MO, 4
modulus, of a complex number, 488
moments, 335 ff
    of nonhomogeneous bodies, 674
multiplication of matrices, 573
MVT, 73, 246

natural logarithm, 191
neighborhood of a point, on a line, 152
    in a plane, 653
Nested Interval Postulate, 240
net, over an interval, 303
    over a plane region, 668
Newton's second law, 120
NIP, 240
no-jump theorem, 191, 249
    for derivatives, 314
nonoverlapping solids, 318
nonsingular matrix, 593
norm, 520
normal component, 423
normal set of vectors, 531
normal vector space, 538
Northeast theorem, 393
    proof of, 527, 707

octant, 509
odd function, 339
odd permutation, 584
open disk, 493
    interval, 12
    sentence, 5
order, 3
orthogonal vectors, 531
orthonormal basis, 531
outside function, 155

Pappus' theorem, for surfaces, 342
    for volumes, 340
parabola, 38 ff
paraboloid of revolution, 41
parallel, 30
parallelepiped, 316
parallelogram law, 412
parameter, 382
parametric mean-value theorem, 387
parametric slope formula, 386
partial derivatives, 631
partial fractions, 298 ff
partial sum, 431
path, 381
path length, formulas for, 407, 408
    proof of formula for, 531, 711
path of motion, 41

permutation, 583
perpendicular, 30
plane path, 381
p(N), 303
point of contact, 43
point-slope form, 30
polar coordinates, 397 ff
    distance formula in, 401
polar form, of a complex number, 487
polygonal inequality, 55
polynomial, 86
    in x and y, 633
power series, 453
powers of functions, 99
prime number, 55
projection into a sub space, 541
Pythagorean theorem, 19, 22

quadrant, 20
quadratic function, 179
quadric surface, 620

radius of convergence, 494
range, of a function, 64
ratio test, 457 ff
ray, 25, 128
real-analytic, 468
rectangular equation of a locus, 382
rectangular hyperbola, 370
rectifiable, 304
reduction formula, 277, 284
reflecting property, of a parabola, 48
regular path, 392
Re z, 482
right cylinder, 614
right-handed coordinate system, 129
ring, 579
RO, 7
Rolle's theorem, 246
roots of functions, 97
rotation of axes, 374 ff
ruler postulate, 17
saddle point, 655
sample, of a net, 310
    of a net over a plane region, 669
sample sum, 310
scalar, 411
Schwarz inequality, 521, 536
Sec, Sec^{-1}, 173
secant line, 45
sech, 203
second derivative, 119
segment, 25
signum, 108
Simpson's rule, 176 ff
    error in, 524
Sin, Sin^{-1}, 169
sine, series for, 466
sinh, 203
slice function, 630, 634

slope, of a segment, 28
   of a line, 29
slope-intercept form, 29
S(N), 309
s(N), 309
solid torus, 327
solution set, 12
span, 414, 521
sphere, 620
square root, 10
squeeze principle, for functions, 145, 222
   for sequences, 435, 518, 696
   for volumes of solids, 318
standard position, for an angle, 133
   for an ellipse, 362
   for a parabola, 40
subadditive function, 537
subspace, 522
substitution, integration by, 284 ff, 291 ff
   justification of, 290
summation notation, 49
sup, 243
supremum, 243
surface of revolution, 327, 616
symmetric, 338

Tan, Tan^{-1}, 171–172
Tan^{-1}, series for, 455
tangent, 43 ff
   to a parabola, 45
tangent vector, 423
tangential component, 423
tanh, 203
Taylor series, 473 ff
Taylor's theorem, 477 ff
terminal side, 128

term-wise integration of series, 453
transitivity, 4
translation, of axes, 356 ff
   of a coordinate plane, 415
transpose, of a matrix, 587
transposition, 583
triangular inequality, 11
trichotomy, 4
trigonometric functions, of angles, 129
   of numbers, 131

U lim, 502
unbounded, 193–194
uniform convergence, 502
   norm, 538
uniformly accelerated motion, 119
uniformly continuous, 315, 353
uniqueness theorem, 113
unit element, 579
unit tangent vector, 423
upper bound, of a function, 193
   of a set, 243
upper limit of integration, 105
upper sum, 309

value, of a function, 67
vector space, 411
vectors, in a plane, 409 ff
   operations on, 410 ff
velocity, 119, 422
vertex, of a parabola, 39

weight, 120
well-ordering principle, 54
winding function, 130